it's your body

AN EXPLANATORY TEXT
IN BASIC REGIONAL ANATOMY
WITH FUNCTIONAL
AND CLINICAL CONSIDERATIONS

it's your body

lawrence m. elson, ph.d.

San Francisco City College
University of California
School of Medicine, San Francisco

McGraw-Hill
Book Company

New York
St. Louis
San Francisco
Düsseldorf
Johannesburg
Kuala Lumpur
London
Mexico
Montreal
New Delhi
Panama
Paris
São Paulo
Singapore
Sydney
Tokyo
Toronto

it's your body

2 3 4 5 6 7 8 9 0 D O D O 7 9 8 7

Library of Congress Cataloging in Publication Data

Elson, Lawrence M date
It's your body.

1. Anatomy, Surgical and Topographical.
2. Anatomy, Human. I. Title. [DNLM: 1. Anatomy, Regional—Popular works. 2. Physiology—Popular works.
WO101 E49i 1974]
QM531.E43 611'.9 74-4113
ISBN 0-07-019299-5

This book was set in Helvetica by Progressive Typographers.
The editors were William J. Willey and Stuart D. Boynton;
the designer was J. E. O'Connor;
the production supervisor was Leroy A. Young.
The illustrations were done by Russell Peterson;
photography was done by Robert Beaubein.
The printer and binder was R. R. Donnelley & Sons Company.

contents

part four: appendix

for the instructor

This text is designed for students taking a general human anatomy course in community colleges, allied health profession schools, and universities. I have assumed that such students have done little academic work in the life sciences (an assumption that rarely offends even prospective physicians).

Offering instructors a different approach to teaching a first course in human structure, this book: (1) is regionally oriented but considers the structures within each region systemically; (2) is primarily concerned with concepts and explanations rather than descriptions; (3) directs the students investigation of his or her own body through palpation and observation.

Instructors sometimes shy away from a fully regional approach because of lack of appropriate teaching materials or a breakdown in functional continuity. I have found that models and charts available to anatomy instructors can be easily used in elementary regional courses. Now that such teaching materials as cadavers are often more easily procured, regional anatomy seems all the more appropriate. The second problem is overcome, I believe, by treating the tissues and organs in each region systemically. In the region of the upper limb, for example, bones are considered first, then muscles, nerves, and vessels. In going from one region to another, a good deal of reinforcement is provided with each major system. As a result, the student comes through it all with an intellectual appreciation of the body as an organization of integrated tissues and organs rather than as a bag of isolated functional systems. Further, I have consistently found that anatomy taught according to region has more everyday relevance than the systemic approach. Relevance and fascination with one's body often provides the motivation necessary for understanding structure and function.

Sure and rapid communication is the principal objective in this text, and I employ a variety of means to achieve it: examination of one's own body, line drawings, photographs, and roentgenograms, self-testing material, consideration of relevant anomalies, diseases or injuries, and perhaps most important of all, digestible concepts and explanations of the arrangement of body structure and the tasks these structures accomplish. I use *concept* to mean a verbal description of a structure or process. As teachers, we deal in concepts all the time. It is the students concern to create understanding from our concepts even though they have probably not perceived the structure or process as we have. This is not always easy to do. It is unfortunate that many students, out of frustration, merely memorize the words without creating a perception from the concept. To help them over the hurdle, I have used in a number of instances familiar analogies or situations which they have experienced at one time or another. I have found that such explanations enhance and accelerate understanding. I have tried to eliminate unnecessary new words which serve only to muddle the explanation.

We anatomists have a lot going for our chosen discipline, because everybody is interested in anatomy, i.e., themselves. What we want to do is transform their superficial curiosity into an intellectual one. We must step away from the classical approach ("Here's 50,000 facts . . . memorize them!"). We must bring the topic of anatomy to life: more self-inspection, fewer dried bones. We must strive to present the human body as it really is: a beautiful, fascinating, dynamic structure to be appreciated at any level of study. This book is an attempt to do these things.

Lawrence M. Elson

Sample order of presentation of units

	Unit topic	*Number of lectures or seminars*	*Number of laboratory periods*	*Organization by week*	
				11-week quarter	*17-week semester*
	Introduction	1–3	0–1	1	1
1	The Cell	1–2	½–1	2	2
2	The Tissues	2–4	½–1	2	3
3	Introduction to Regional Anatomy	2–5	1–2	3	4, 5
4	The Upper Limb	2–4	1–2	4	6, 7
5	The Lower Limb	2–4	1–2	5	8, 9
6	The Head and Neck	4–6	2–3	6, 7	10–12
7	The Body Wall	2–3	1	8	13
8	Thoracic Viscera	2–3	1	9	14
9	Abdominal Viscera	2–4	1–2	10	15, 16
10	The Pelvis and Perineum	2–4	1	11	17
		22–42	11–17		

acknowledgments

Although this page is rarely of interest to anyone except those whose name is imprinted hereon, it is my favorite page. I am very happy to recognize certain people who cared enough about the development of this book to contribute their special talent to it. With sincere thanks to:

Dr. Marian Diamond—
who started the ball rolling for national publication.

Russell Peterson—
for the fine illustrations. We shared a common philosophy as to the goals of this book.

Robert Beaubein—
for the magnificent photographs of the human body.

Mr. James Runner—
for the excellent light photomicrographs.

Nathan Cotlar, M.D.—
for his assistance in preparing the clinical considerations.

Drs. Robert Liebelt, Jackson Wagner, Vaughn Critchlow, Russell Deter, and David Caley—
my colleagues who proofread the original manuscript and kept me honest.

Ms. Claudine Lewis, Edwina Stancil, Linda Taylor, and Sheila Ottino—
my typists who persevered and probably thought this book would never end.

Dr. Dale Spence, Ann Hodson, Ken Yoder, Janet Robbins, and Miss Hazelwood—
my models who may never recover from the grueling sessions.

James Young, David Beckwith, and Stuart Boynton—
my editors: sources of encouragement and partners in controversy.

Capt. and Mrs. John Chesley—
for a bit of technical assistance that I could not do without.

Conley V. Baker—
for a beautiful writing sanctuary nestled in the hills of Sausalito, California.

Eugene H. Mattingly—
for helping me define the term *concept*.

Penelope Vaillancourt—
for a great editorial assist

Finally, this book is dedicated to:

Herbert H. Srebnik, Ph.D.—an anatomist of anatomists—a teacher, advisor and friend.

My students in Anatomy at San Francisco City College—
It was for them that this all had a beginning. They taught me how to teach, and for this I shall always be grateful.

how to use this book

Dear Student,

Welcome to a truly fascinating topic—yourself! That's what this book and the course you are taking is all about—to give you new insight into your physical being. I hope that the concepts and related facts you pick up here will seem significant and relevant for the rest of your life. The orientation of this book is, in part, a product of a variety of solicited comments from over a hundred students of mine in anatomy at San Francisco City College. They seemed to find the topic exciting and offered me ideas as to what would make an interesting and motivating text. Your instructor and I are interested in your educational development and we are undoubtedly in agreement that a *meaningful understanding* of the structure and function of your body is our principal goal in the course you are about to take.

A quick look at the Contents will tell you how this book is organized. There are three parts:

1. The introduction, where you will learn about the body's organization, basic terminology, and how the "big words" are really *combinations* of "little words."
2. Basic histology, a brief survey of the materials which make up our body as seen under the microscope.
3. Regional anatomy, a presentation of concepts and related facts about the various regions of your body, for example, head, neck, arm, leg. A Synopsis of Systems is presented in the Appendix for cross reference.

The units are written informally and each topic is approached quite fundamentally. Self-testing devices are employed so you can periodically check your progress in achieving understanding. You will be asked to feel (palpate) various parts of your body as you study each region. It's *your* body which is the subject of our study, so you should use it as a source of information without inhibition or embarrassment.

A good understanding of the topics presented in each unit will make the laboratory experience much more meaningful. You may or may not have a laboratory period in your course. If you do not, then you should create your own "laboratory experience" through self-inspection—a worthwhile and fascinating endeavor. The laboratory investigation is the culmination of the learning experience, for it is here that you will see things not otherwise encountered.

For those who are inclined, a number of references are listed at the end of each unit—references which will lead you to a more refined study of the topic. A subject of intense interest to you may be pursued in this manner.

Thumb through the book and note the many illustrations, beautiful photography, and Atlas of X-Rays (see Appendix). The illustrations are the principal source of information on distribution and arrangement of blood vessels and nerves. The microscopic photographs have been carefully selected to give you a visualization of body structure not seen with the unaided eye. Some of these extraordinary photographs show structure at magnifications just above the molecular dimensions. The photographs of the body have been provided to give you insight on how some of the structures underlying the skin can be demonstrated on the body surface. There are also various surface landmarks which aid one in locating deeper, invisible structures. Finally, and just as importantly, the photographs have been designed to illustrate the simple grace and harmony of body structure—no memorization required here—just a very normal, subjective appreciation of the magnificent human body.

May this book and your course provide you with a clear and everlasting insight as to how *beautiful* you really are.

Sincerely,

Lawrence M. Elson

it's your body

part one
the study of your body

introduction

the organization of your body

There is a definite hierarchy of structure to the human body, just as there is a hierarchy of people in a large organization. To continue the analogy, the rank (in terms of service to the organization) of these people ranges from president to janitor, let us say. Now you would not deny that the president has a very important (and prestigious) role. But how about the janitor, would you play down his role? Could the organization function without janitors? Of course not, no more than they could function without a president. So, in general, *all* of the people are important, even essential, in order that the whole organization may function. Back to the body. The human body is made up of a large number of structures which may be ranked from very complex to relatively simple structures. *The* complex structure in this arrangement might be the brain. A *relatively* simple structure might be a mitochondrion. It is true that you cannot live without a brain, but you cannot live without mitochondria either, since the latter supplies the energy so the former can function. Which is *more* important? Neither. They are interdependent. Each is as important as the other. Thus, the point: Although there is an ordered system of structure, from large to small, from complex to simple, it certainly does not follow, in general, that one part of the system is *more* important than another; the individual parts are functionally and frequently structurally interrelated in order that the whole body may function as a living organism.

The classification of structure may have a *functional* or a *regional* basis (Tables I-1 and I-2). In courses of human anatomy at the introductory or elementary level, the functional organization of structure is usually presented. This is **systemic human anatomy.** In courses of human anatomy at the advanced or graduate level, the regional organization of structure is usually presented. This is **regional human anatomy.** In general, the human body is described systemically, and dissected (dis-sected) regionally.

Table I-1 *General systemic organization of the body.**

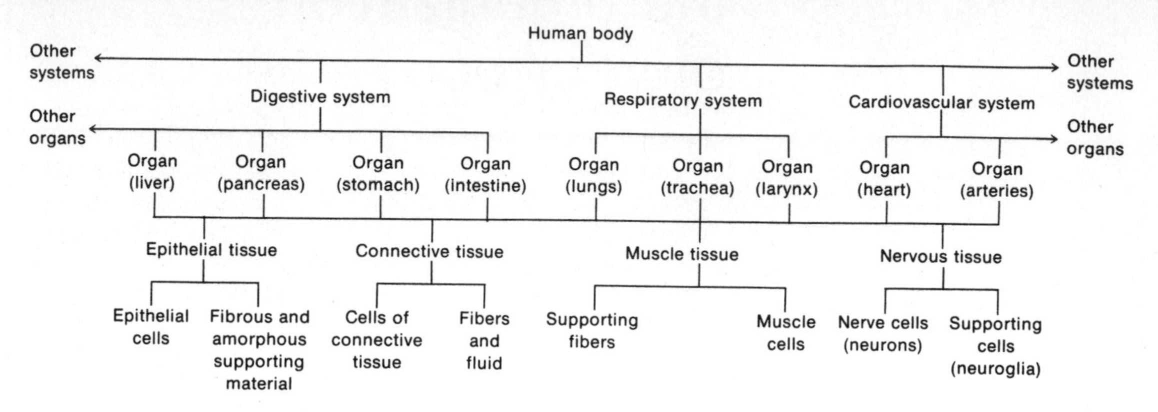

* Partial listing of systems and organs.

Table I-2 *General regional organization of the body**

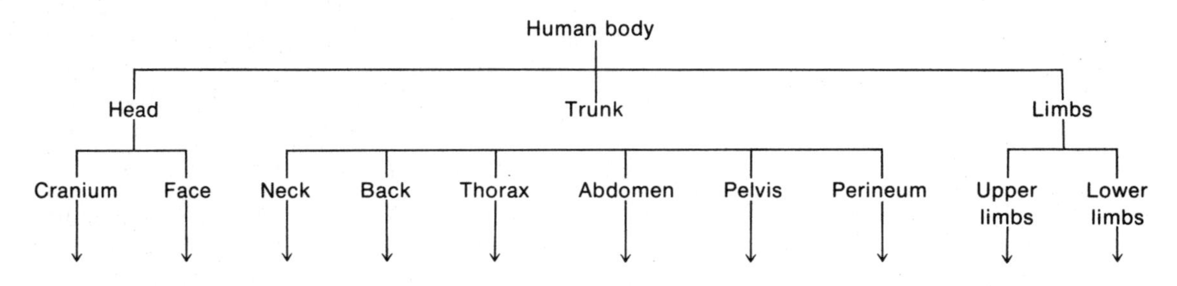

* Below the arrows the student should list the smaller regions within the ones identified. Refer to the text and Figs. I-1 to I-3.

SYSTEMIC ORGANIZATION OF THE BODY

Cells

Cells (L., *cella,* a small room) are the simplest *living* units of the body. Smaller and less complex structures are either components of the cell or products of the cells. Larger and more complex structures are merely aggregations of cells and their products (supporting fibers) infused in a matrix or ground substance (a product of cells) (Table I-1). The ground substance may be watery or it may be solid depending on the structure concerned. For example, note the constitution of the brain:

nerve cells and processes (neurons)	+	supporting cells and processes (neuroglia)	+	connective tissue cells and fibers plus ground substance

Tissues

Most cells of the body are neither isolated nor randomly placed. Instead, they are congregated into units having a common function. These aggregations of cells, often attached by intercellular "cement" or by fibers, are termed **tissues** [Fr., *tissu,* woven (fabric)].

The whole body is structured from four basic tissues:

1. **Epithelial tissue:** aggregations of similar cells which line all the cavities of the body as well as cover the external surfaces; may occur in single or multiple layers.
2. **Connective tissue:** a variety of cellular aggregations and their fibrous products; the "ground substance" ranges in texture from fluid to granitelike density; probably occupies more space in the body than any other tissue; and functions to bind, support, and link other structures in the body.
3. **Muscle tissue:** groups of similar cells specialized for contraction; cells are often called fibers because of their shape.
4. **Nervous tissue:** collections of cells with threadlike extensions which are specialized for generating and conducting excitations in response to physical and chemical stimuli; also includes a kind of nonconducting supporting cell.

Now, having the above firmly in mind, consider how these tissues are organized (or might be organized) in various regions or parts of your body. For example, look at your hand. *Feel* the texture of the surface of your hand. *Push down* on the palm of your hand. Try to imagine what is *in* the hand. What is the name of the tissue which makes up the surface of the hand? *Epithelium.* Notice that this surface material "fits like a glove"—bound securely to deeper structures. What is the tissue that does the binding? . . . *Connective tissue.* What is the tissue that forms the skeleton of the hand? . . . *Connective tissue.* What is the tissue that brings nutrition to all of the cells of the hand? . . . *Connective tissue.* Note the two "pads" on either side of the palm of the hand. Touch that pad nearest the base of the thumb and then move the thumb. Now, what tissue do you think comprises the bulk of that "pad"? . . . *Muscle.* And finally, what tissue stimulates the "pad" to operate? . . . *Nervous tissue.* What tissues may be found in the following structures:

1. Arm: ______________________________
2. Head: ______________________________
3. Liver: ______________________________
4. Urinary bladder: ______________________
5. Artery: ______________________________

That's right, each structure incorporates *all four tissues* in its make-up (Tables I-1 and I-2).

Organs

In most cases the four fundamental tissues are structurally and functionally interrelated to such a degree that they form a complex or structure in themselves. Such a structure is called an **organ** (L., *organum,* instrument), and it has one or more functions *as an organ.* Organs should not be confused with **regions,** the latter being an *area* consisting of various organs (Table I-2). Examples of organs are: arteries, veins, skin, glands, nerves, liver, kidney, gallbladder, etc. How is it that a nerve such as the large sciatic nerve of the lower limb is considered an organ?

Systems

In most cases, certain organs are associated together in performing an overall function, just as the carburetor, pistons, cylinders, drive shaft, etc. all work together as an engine which makes a vehicle move. Those organs which are associated with a common function constitute a **system** (L., *systema,* to place together). The ten or eleven systems of the body include all of the cells, tissues, and organs. How many systems can you think of right "off the top of your head"?:*

1. respiratory
2. reproductive
3. excretory
4. cardio-vascular
5. nervous
6. digestive
7. lymphatic
8. skeletal
9. muscular
10. integumentary
11. Endocrine

REGIONAL ORGANIZATION OF THE BODY

This classification of the body is based on the structural arrangement of the body without regard to function or composition. In laboratory dissections of the body, the dissector works in a specific region and considers the structures within. In this way, he will consider several systems, not to mention a multitude of organs, but only *four* tissues.

The primary divisions of the body (Table I-2) are:

1. Head
2. Trunk
3. Limbs

The head includes the Face and Cranium. The trunk includes Back, Thorax, Abdomen, neck, pelvis, and perineum. The limbs include upper limb and lower limb components.

The regions listed below are subdivisions of the primary regions cited above; they are not for memorization but are presented to provide an overall regional breakdown of the body, available for later reference. As you read, refer to Figs. I-1 to I-3.

Regions of the *cranium* include:

- occipital
- parietal
- temporal
- mastoid
- auricular
- frontal

* For a full listing of systems, see Synopsis of Systems in the appendix.

Regions of the *face* include:
- orbital
- nasal
- oral
- cheeks
- mental (lower jaw)

Regions of the *neck* include:
- anterior cervical
- lateral cervical
- posterior cervical
- deeper contents of neck

Regions of the *thorax* include:
- sternal
- thoracic cavity and contents
- epigastric (in part)
- anterior thoracic (pectoral)
- axillary (in part)

Regions of the *abdomen* include:
- upper right and left quadrant
- lower right and left quadrant
- abdominal cavity and contents

Regions of the *pelvis* include:
- pelvic contents
- perineum

Regions of the *back* include:
- scapular
- lumbar
- sacral

Regions of the *upper limb* include:
- axillary (in part)
- shoulder (acromial and deltoid)
- arm (brachium)
- forearm (antebrachium)
- hand (carpus or wrist; palm, posterior hand)

Regions of the *lower limb* include:
- gluteal (including hips)
- femoral (thigh)
- knee (patellar and popliteal)
- leg (crural)
- foot (tarsus or ankle; dorsum of foot; sole or plantus of foot)

CLOSED CAVITIES AND SEROUS MEMBRANES

The body is so organized that the various viscera of the body occupy closed spaces called **cavities.** Those cavities behind the midline frontal plane of the body (separating body into equal front and back halves) are the posterior cavities and those in front are the anterior cavities.

The anterior cavities, from top to bottom (Fig. I-1), are as follows:

1. **Thoracic cavity:** the cavity of the chest bordered by the muscular diaphragm below, the vertebral column behind, the sternum in front, and the ribs connecting vertebral column and sternum. The thoracic cavity houses lungs, heart, esophagus, trachea, ducts, vessels, and nerves. The lungs themselves are embraced by a double layer of tissue (pleurae) between which there is a "potential" cavity; it would become an "actual" cavity only if the layers were forced apart. This potential cavity around the lungs is the **pleural cavity.** The same situation exists around the heart, where the potential cavity is termed the **pericardial cavity.**
2. **Abdominopelvic cavity:** the cavity of the abdomen and pelvis bordered by the vertebral column behind and muscular walls elsewhere, with the exception of parts of the hip bone at the sides (laterally). The roof of the abdominopelvic cavity is the **thoracic diaphragm.** The floor is a muscle, called, with its coverings of tissue, the **pelvic diaphragm.** The abdominopelvic cavity houses the alimentary tract, liver, gallbladder, pancreas, spleen, urinary organs, and the internal organs of reproduction, as well as innumerable ducts, vessels, and nerves. Many of these organs are more or less enclosed by a double layer of tissue (peritoneum) between which is another "potential" cavity, the **peritoneal cavity.** The two layers enclosing the three potential cavities already named—peritoneal, pericardial, and pleural—are called **serous membranes.** Each serous membrane consists of a single layer of cells and supporting tissue. These cells secrete a watery (serous) fluid which prevents friction when two such membranes rub against one another, as they are prone to do with organ movement.

The posterior cavities, from the head down (Fig. I-2), are:

1. **Cranial cavity,** housing the brain.
2. **Vertebral cavity,** housing the spinal cord.

Strictly speaking, these two are really parts of one cavity. The cranial cavity is surrounded by the flat bones of the skull (vault of the cranium). The spinal cavity is a narrow passageway surrounded by the arches of the vertebral bones laterally and behind and by the bodies of the vertebra in front.

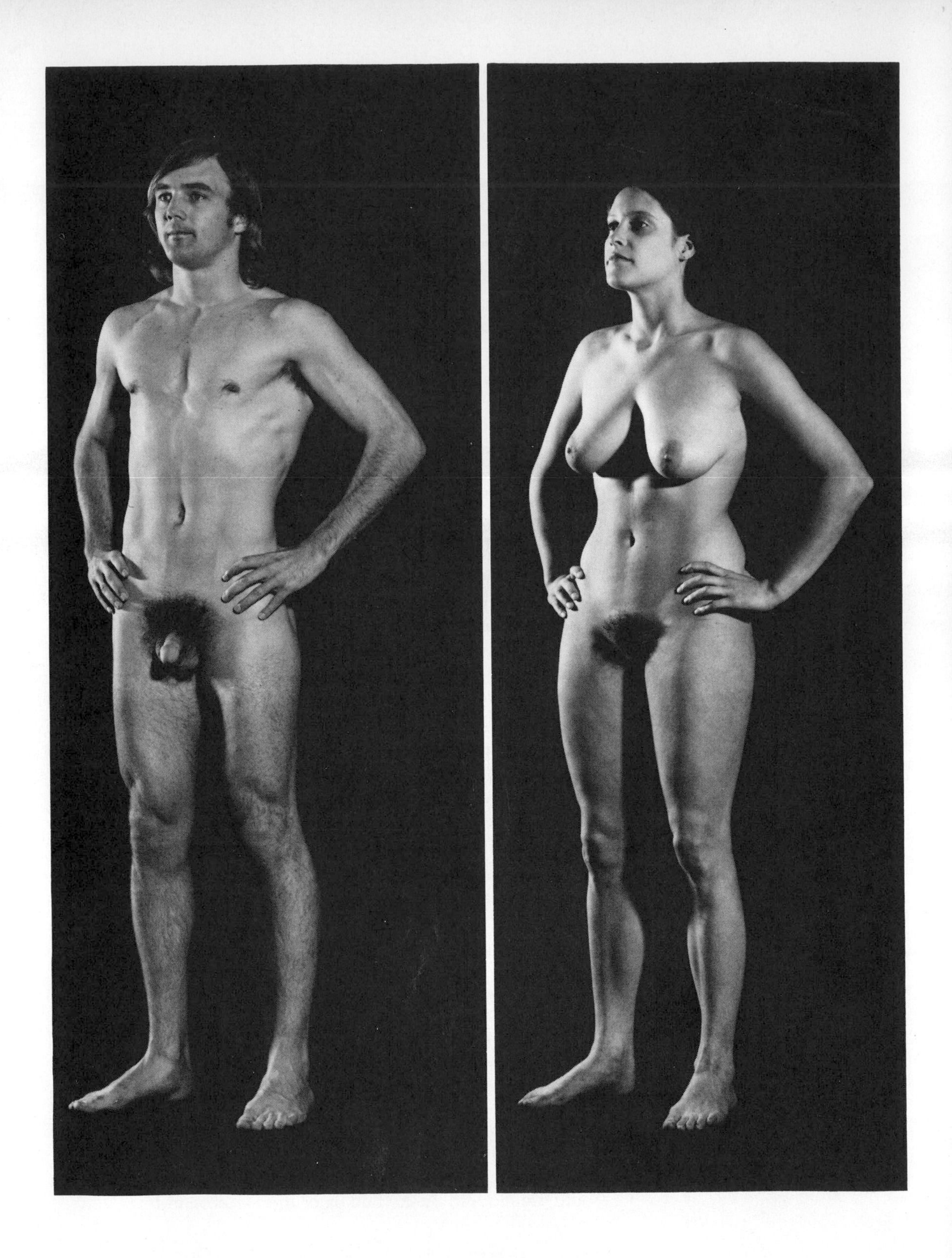

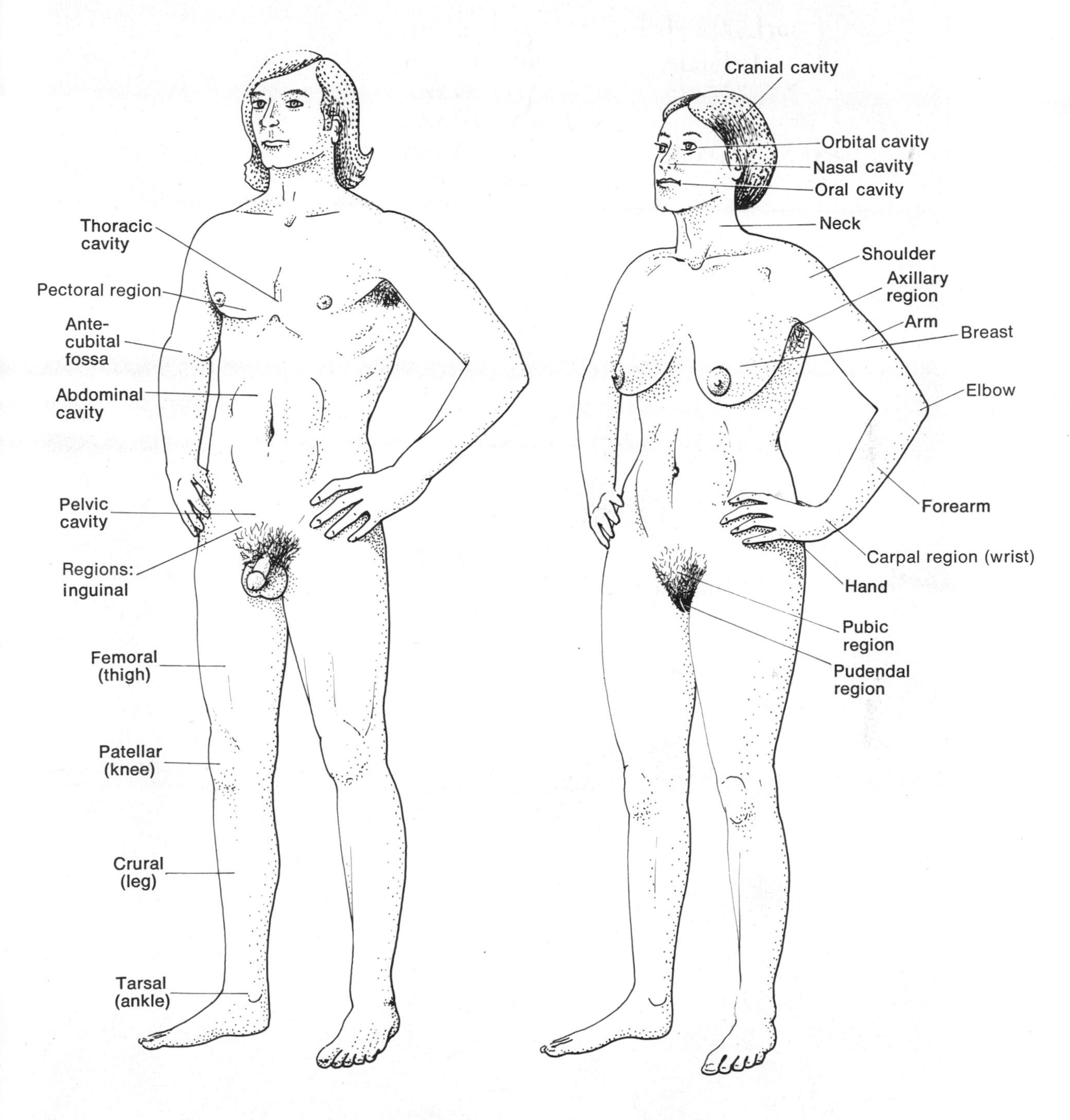

Fig. I-1 General regions and cavities of the human body.

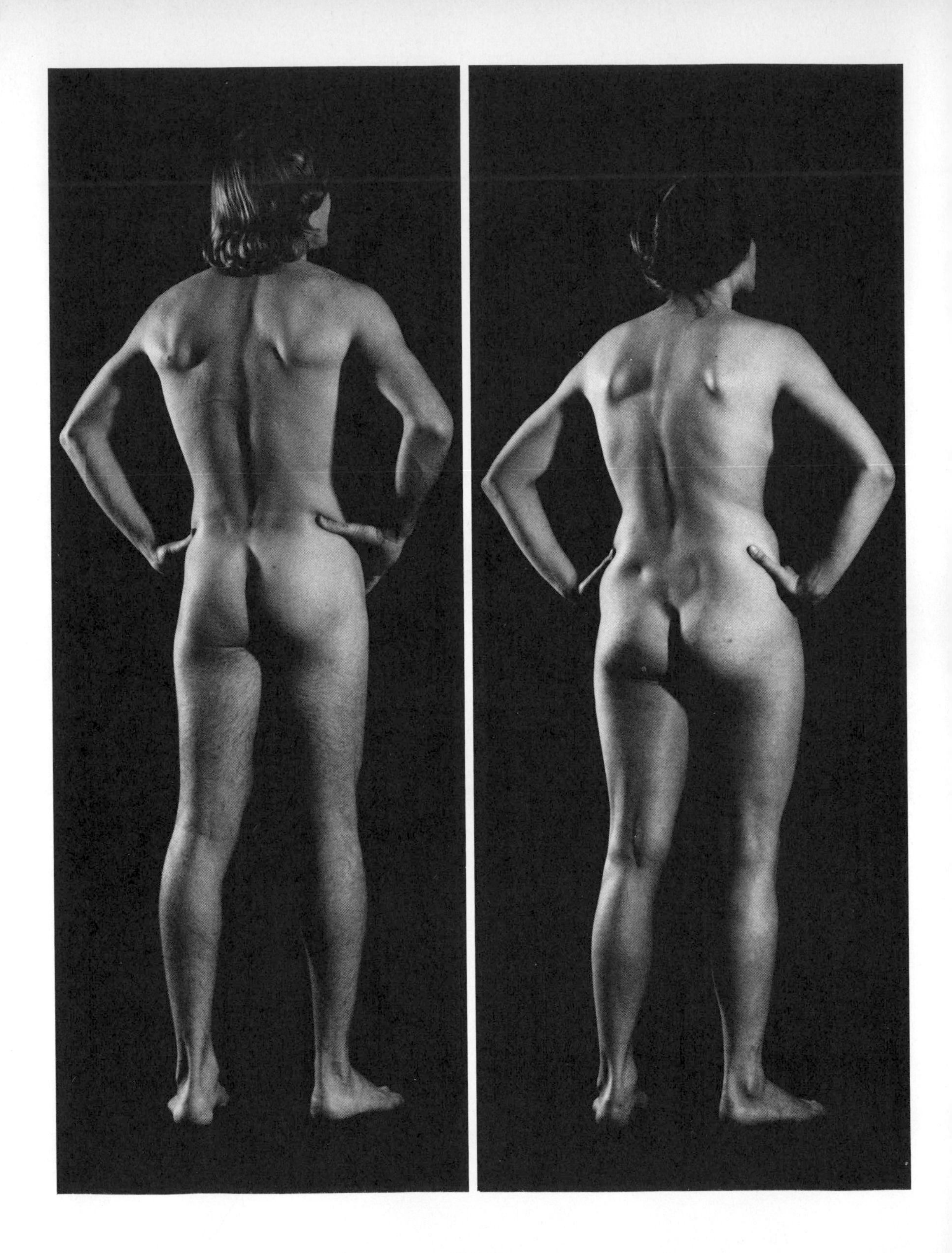

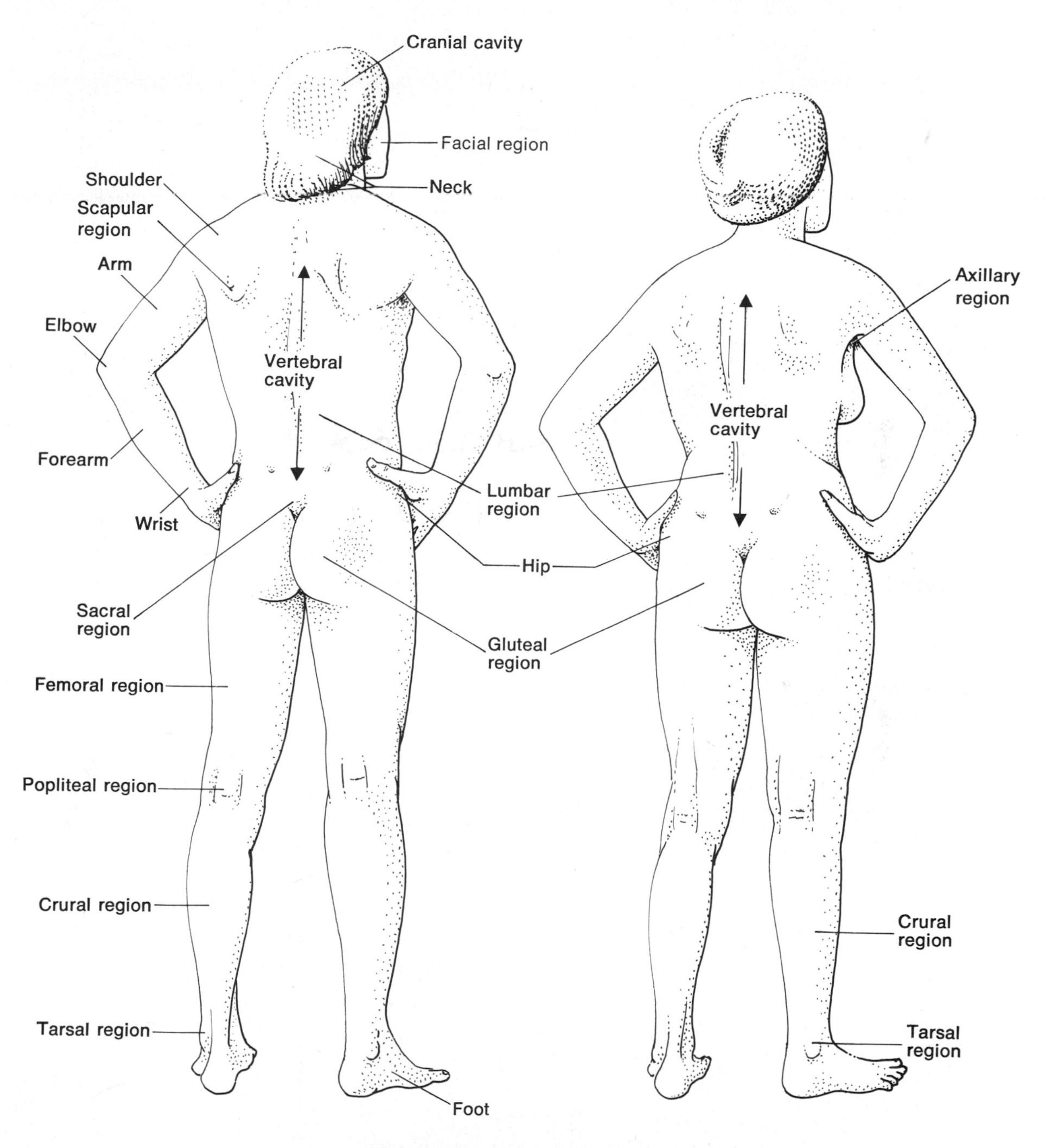

Fig. I-2 General regions and cavities of the human body.

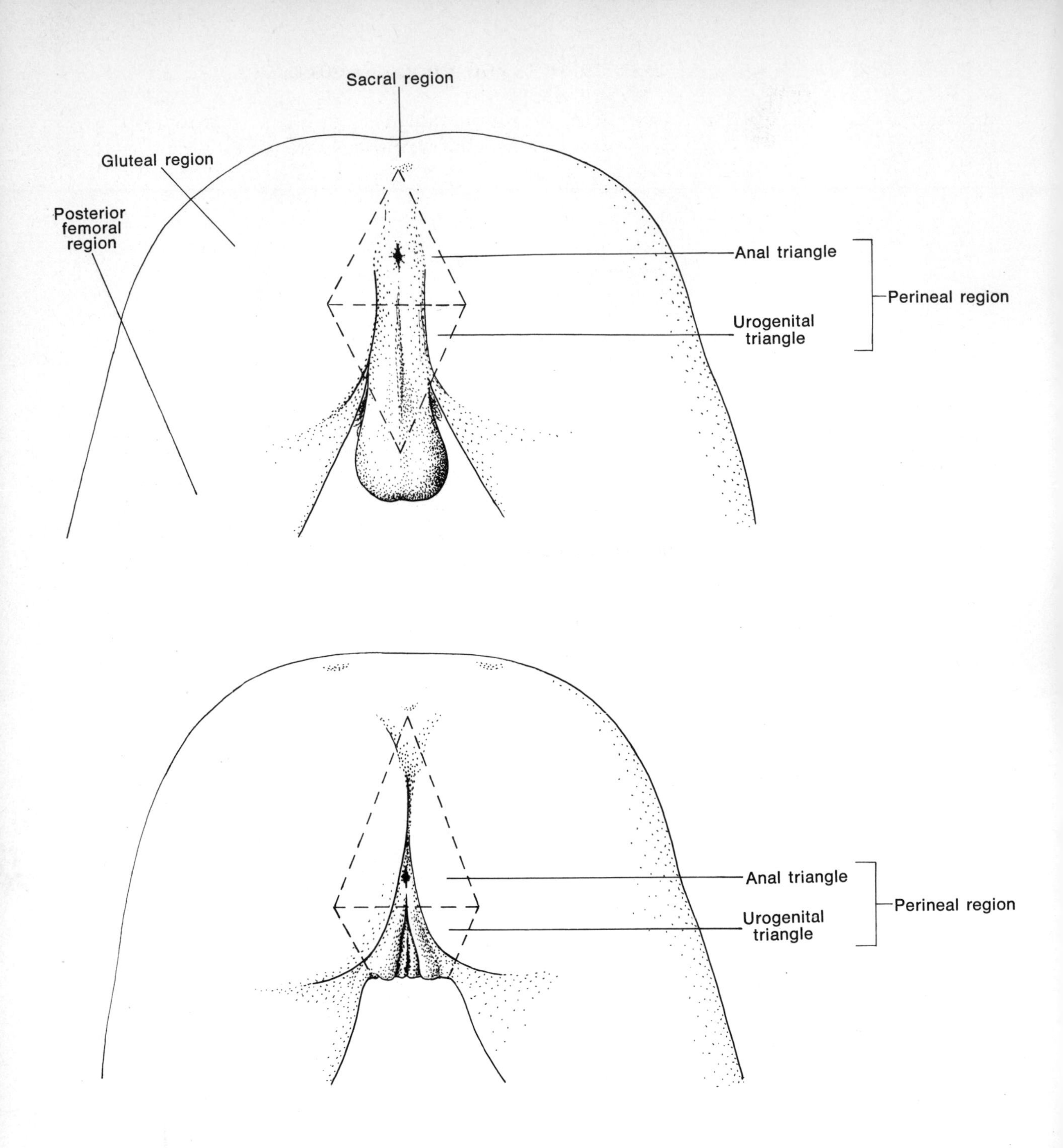

Fig. I-3 General regions and cavities of the human body, viewed from below.

OPEN CAVITIES AND MUCOUS MEMBRANES

Many of the organs of the body have their own cavities in which materials passed to or through it may be conducted, processed, or stored (Fig. I-4). Organs are not ends in themselves. They function to process, store, or transport some material, either ingested or inhaled from outside (as in the case of the alimentary or respiratory tracts) or secreted from inside (as in the case of the reproductive and urinary tracts). These visceral cavities are, at one end or at both ends, continuous with the outside of your body (Fig. I-5). They are all lined with a multilayered membrane called a **mucous membrane.**

Mucous membranes (mucosa) contain organized strata of lining cells (epithelia), connective tissue, and frequently muscle. Mucous membranes supply the labor force for the tasks of the organ. Epithelial cells do the processing (secretion of enzymes, absorption, etc.), and muscle provides for the mixing of materials in the cavity, or **lumen.** The mucosa also serves to keep the lumen moist. The mucosa is supported by an underlying connective tissue layer (submucosa) and a thick layer, or layers, of muscle.

UPON REFLECTION

It has been said that in order for one to find a pin in a haystack, one must find the haystack first. So it is in the study of anatomy—before getting into the "nitty-gritty" of detail, you should concentrate on the overall systemic and regional picture . . . see "the forest" before you study "the trees," so to speak. Periodic reference to Tables I-1 and I-2 and Figs. I-1 to I-3 throughout the course should help you keep an appreciation of "the big picture."

on terminology

One of the aspects of studying anatomy which provides some difficulty and not a little displeasure for the beginning student is terminology. I wish that some sort of "gimmick" could be introduced here as a quick aid to learning new words. Unfortunately there is none. Some practice and self-discipline are required if one is to learn some 3,000 new words or terms in a relatively short period of time.

Happily, most words which have biological applications are Greek or Latin in origin. This often means that the word to be learned has a *prefix*, a *root-word*, and/or a *suffix*. Once you have learned a few basic prefixes, root-words, and suffixes, you can reason out the meaning of many, many medical and biological terms.

Fig. I-4 The body trunk and its cavities. The thoracic and abdominopelvic cavities are closed to the outside. Viscera located within these closed cavities often have cavities (lumina) open to the outside, e.g., digestive tract, uterus, bladder.

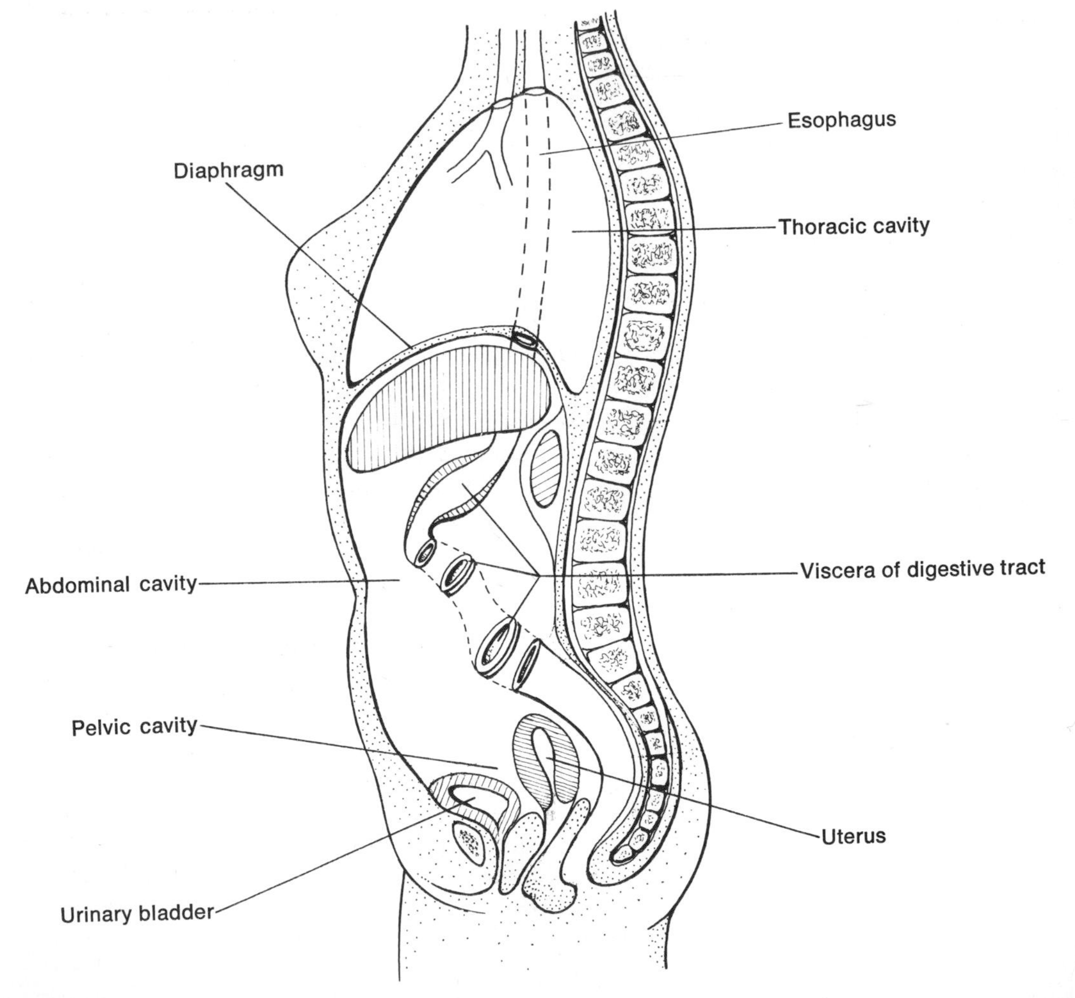

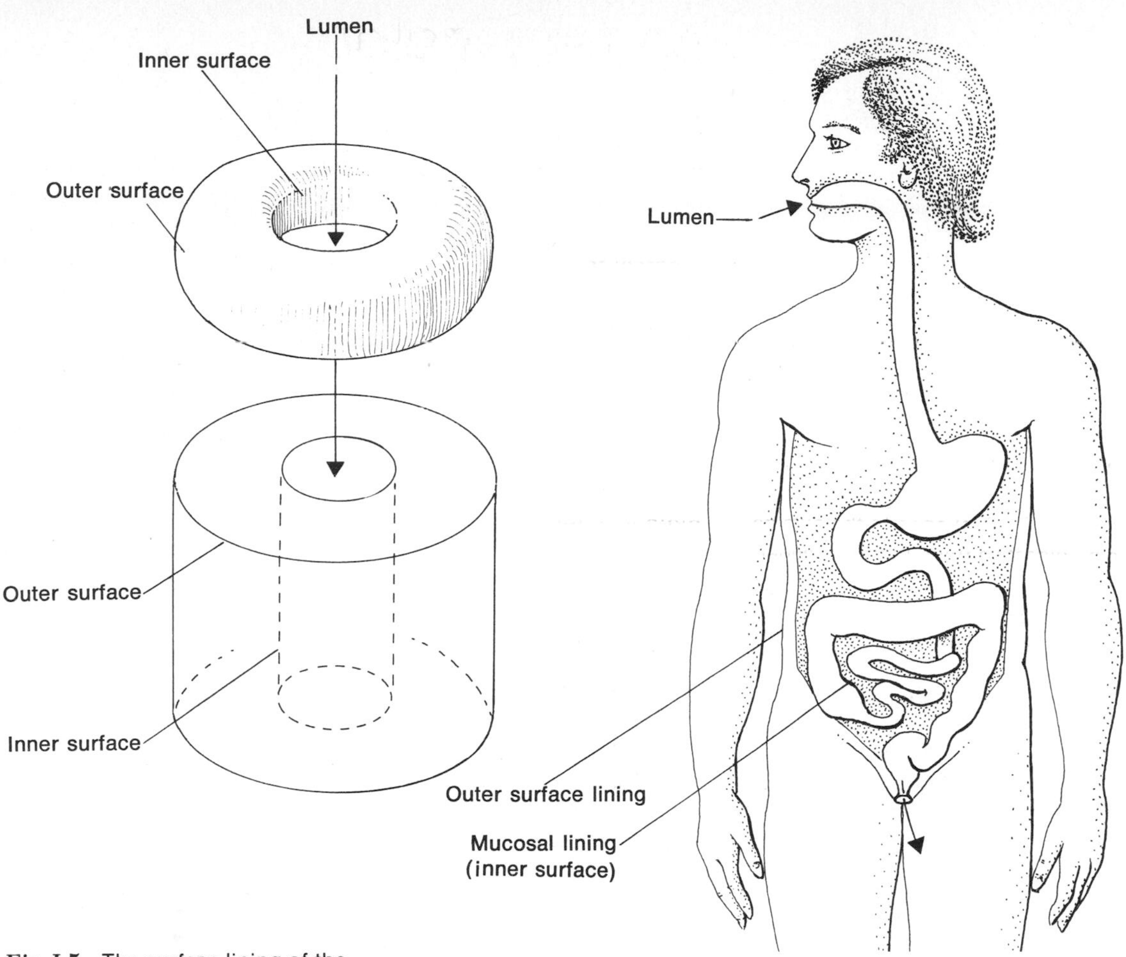

Fig. I-5 The surface lining of the body is continuous with the mucosa lining the visceral cavities, just as the inner surface of a doughnut or pipe is continuous with the external surface.

For example, consider the word *spondylos.* This is a Greek word meaning vertebra. A vertebra is one of the bones of the spinal column. The word *spondylos* is not commonly known, but how about the suffix *-itis?* Many associate this term with pain, and correctly so, since this term means *inflammation of,* which, of course, implies pain. Put the two terms together and you have *spondylositis,* or more properly, *spondylitis,* meaning inflammation of the vertebra.

In this section of the book, you will learn about some new and frequently used terms. You will be using them throughout the course again and again. Use a dictionary frequently—it is indispensable in a course such as this. Whenever you come upon a word or term of which the definition is obscure, consult a dictionary immediately. Become addicted to this habit and your vocabulary will increase exponentially throughout the years.

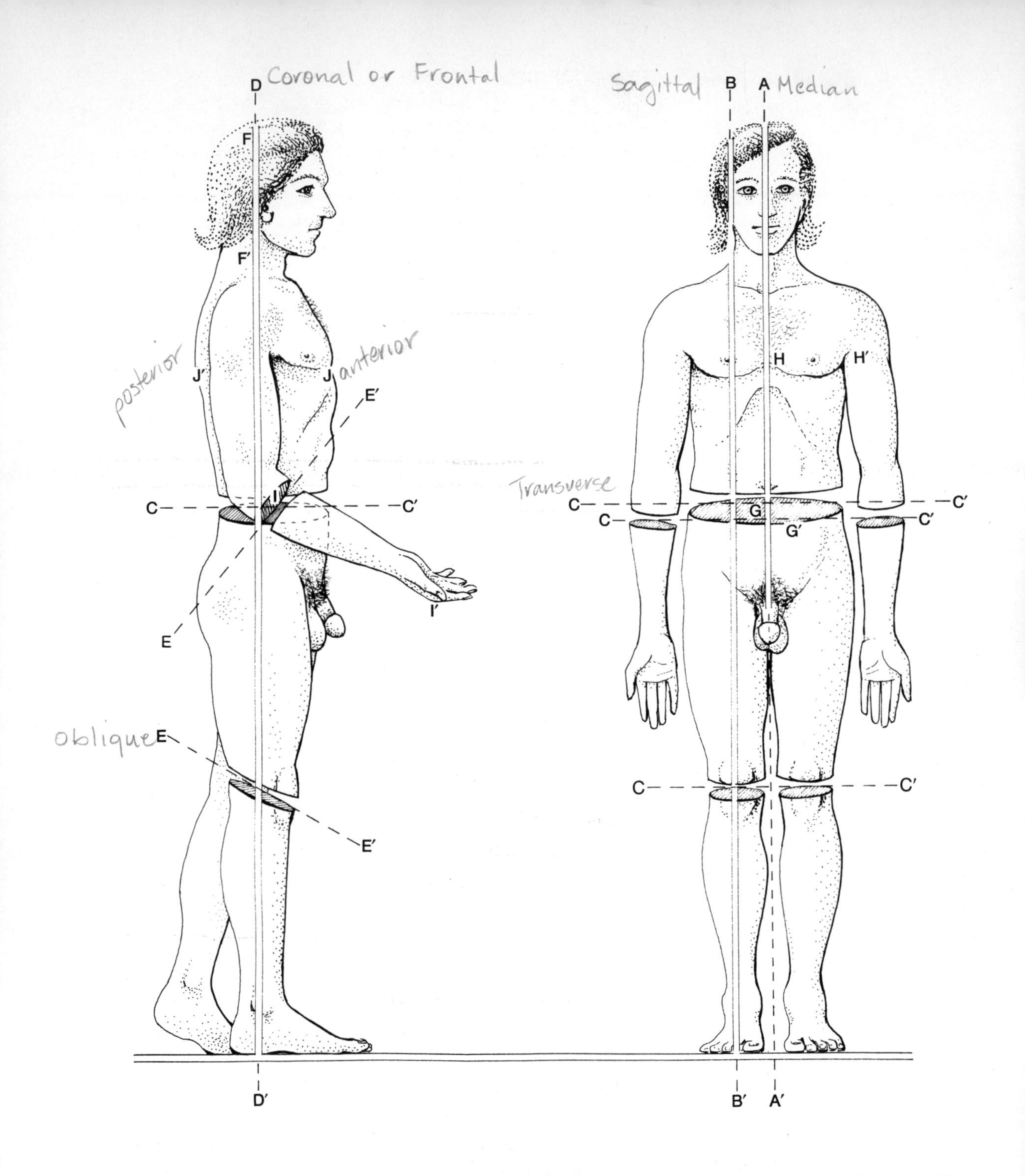

Fig. I-6 Planes of reference and terms of position and relationship. See text for key to letters.

TERMS OF RELATIONSHIP

Refer to Fig. I-6. Note the letters on the figure. Each letter and its complementary letter with a prime (′) refer to the relationship between two parts, that is, area F is *above* F′ and, conversely, F′ is *below* F. In anatomy, we have more specific terms for above and below; for you see, although F *is* above F′ when the body is upright, F is certainly not above F′ when the body lies in the horizontal plane. In Fig. I-6, the following relationships can be appreciated:

1. Point F is **superior** (cephalic) to point F′, while point F′ is **inferior** (caudal) to point F.
2. Point G is **deep** (internal) to point G′, while point G′ is **superficial** (external) to point G.
3. Point H is **medial** to H′, while point H′ is **lateral** to point H.
4. Point I is **proximal** to point I′, while point I′ is **distal** to point I. (These terms are generally applicable to the limbs only.)
5. Point J is **anterior** (ventral) to point J′, while point J′ is **posterior** (dorsal) to point J.

QUIZ

Now let's put your new-found knowledge to the test.*

Assume that the body is upright and in the *anatomical position* (see the right-hand body in Fig. I-6). The nose is (1) Anterior______ to the back of the head. The shoulder is (2) Lateral______ to and inferior to the neck. The little finger is (3) Medial______ to the thumb of the same hand (palm facing forward) while both the thumb and little finger are (4) Distal______ to the wrist. (In the limbs, the terms *distal* and *proximal,* meaning *away from* or *closer to* the body trunk, respectively, are more appropriate than *inferior* and *superior*.) The stomach is (5) Deep______ to the skin overlying the stomach. The head is (6) Superior______ to the neck, while the navel is (7) Inferior______ to the neck. The ears are both (8) Lateral______ and (9) Posterior______ to the eyes. The knee joint is (10) proximal______ to the ankle joint. The layer of fat just deep to the skin is said to be (11) Superficial______ to a deeper layer of connective tissue.

* *Answers:*
(1) Anterior
(2) Lateral
(3) Medial
(4) Distal
(5) Deep
(6) Superior
(7) Inferior
(8) Lateral
(9) Posterior
(10) Proximal
(11) Superficial

Note in most cases, one structure always stands *in relation* to another when using the above terms. When someone says, "This organ is lateral," you should retort, "Lateral to *what?*"

BODY PLANES

Refer again to Fig. I-6. Note the dotted lines which serve to divide parts of the body. The ends of certain of these lines are labeled with a capital letter or its prime ('). These lines indicate the various **axes** (sing. = axis) of the body. The axes or planes are:

A to A′: **Median** (midsagittal) plane. Relative to the whole body, this is also the **longitudinal** axis. It divides the body into equal left and right halves.

B to B′: **Sagittal** plane, a longitudinal axis on either side of the median plane, dividing body into unequal left and right parts.

C to C′: **Transverse** plane or **cross section.** Relative to the whole body, this is also the **horizontal** plane.

D to D′: **Coronal** or **frontal** plane. It divides the body into anterior and posterior parts.

E to E′: **Oblique** plane.

QUIZ

After studying these body planes in Fig. I-6, try the following:*
That plane which divides the body into equal or unequal anterior and posterior parts is the (1) Frontal or (2) Coronal plane. This plane is at right angles to the (3) Transverse plane as well as the (4) Sagittal plane. That sagittal plane, occurring along the (5) Longitudinal axis of the body, which divides the body into *equal* left and right halves is the (6) Median plane. The plane which may occur at any angle other than the sagittal, transverse, or coronal plane is called the (7) oblique plane. A section of a structure made at right angles to the longitudinal axis of that structure is a (8) cross section.

* *Answers:*
(1) Frontal
(2) Coronal
(3) Transverse
(4) Sagittal
(5) Longitudinal
(6) Median
(7) Oblique
(8) Cross section

TERMS OF POSITION AND MOVEMENT

Before defining any terms of body movement, it is essential we define the starting point of such movements. This is the

1. Joints of neck
2. R. Shoulder joint
3. L. Shoulder joint
4. L. Elbow joint
5. R. Elbow joint
6. R. Hip joint
7. L. Hip joint
8. R. Knee joint
9. L. Knee joint
10. L. Ankle joint

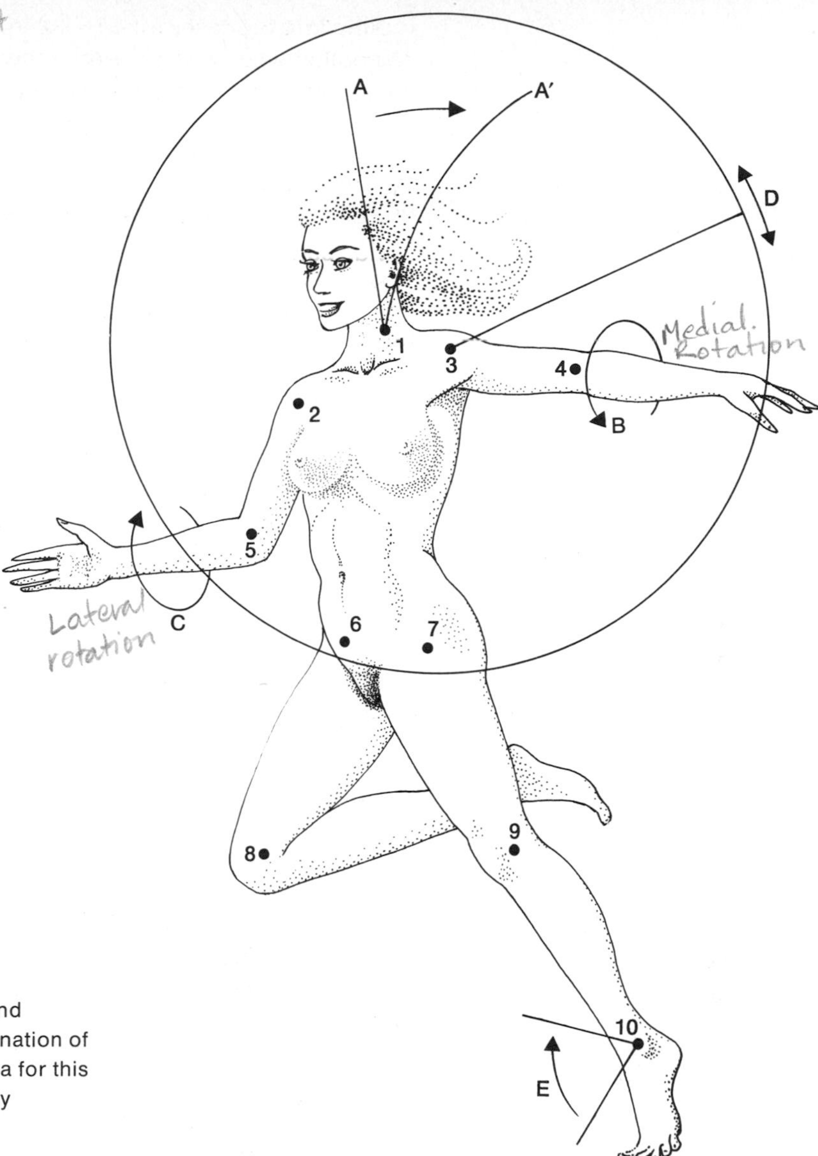

Fig. I-7 Terms of position and movement. See text for explanation of numbers and letters. (The idea for this illustration was contributed by Virginia Junker and is greatly appreciated.)

anatomical position, as shown on the right in Fig. I-6. The body is erect, the arms are at the sides with the palms facing anteriorly. The lower limbs are parallel with the toes facing anteriorly. *All terms of movement are relative to the anatomical position.* When practicing movements and attempting to define those movements with the proper term, always start at the anatomical position . . . or confusion will reign. Having the above firmly in mind, see Fig. I-7.

The dots (·) on the body refer to some of the *joints,* which occur between two (or more) bones. They are numbered 1, 2, 3, etc. It is at joints that movement can occur. Nowhere else. The joints represented in Fig. I-7 are: joints of the neck (1), right shoulder joint (2), left shoulder joint (3), left elbow joint (4), right elbow joint (5), right hip joint (6), left hip joint (7), right knee joint (8), left knee joint (9), and left ankle joint (10).

Flexion: to bend, or make an angle. Flexion may be observed to have taken place at the following joints: 2, 5, 6, 8.

Extension: to straighten out. The body is in a state of "relaxed" extension when in the anatomical position. Extension may be observed to have taken place at the following joints: 1, 4, 7, 9. The exaggerated state or motion of extension, indicated at the neck joint by letters A–A′ is referred to as hyperextension.

Abduction: to take away laterally, from the median plane. You should easily be able to associate this word with a term seen frequently in newspapers when referring to one being kidnapped: *abducted.* Abduction may be observed to have taken place at joint 3.

Adduction: to bring toward the median plane; the opposite action of *ab*duction. Adduction will have taken place if the arm is returned to the side of the body at joint 3.

Circumduction: the act of flexion, abduction, extension, and adduction in sequence and without hesitation. This movement describes a cone with the joint effecting the movement acting as the apex. In Fig. I-7, the motion of circumduction is indicated by letter D. Circumduction has taken place at joint 3.

Rotation: the act of turning or rotating about an axis, as a propeller rotates about its shaft or a wheel about its axle. In Fig. I-7, the act of rotation is indicated by the letters B and C. *Lateral* rotation is indicated by letter C at joint 5. *Medial* rotation is indicated by the letter B at joint 4. In distinguishing lateral from medial rotation, follow the thumb. If it swings outward from the body, it is external or lateral rotation; if it swings inward toward the body, it is medial rotation. Rotation can occur at several joints. At the elbow, there are two separate but neighboring joints—one allows the hinge movement between forearm and arm, the other allows rotation of the forearm. The former movement has taken place at joint 5 in Fig. I-7. Remember, these are two different joints at the elbow.

Supination: refers to being face up. The body lying face up is said to be supine. Rotation of the forearm so that the palm of the hand is facing anteriorly is called supination. In Fig. I-7, the state of supination or the motion of supinating is indicated by the letter C. Supination has taken place at joint 5.

Pronation: refers to being face down. The body lying face down is said to be prone. Rotation of the forearm so that the palm of the hand is facing posteriorly is called pronation. Pronation has taken place at joint 4 in Fig. I-7.

Protraction: to move forward (not shown in Fig. I-7). The lower jaw, when jutted out and forward, is said to be protracted. Movement of the scapula laterally away from the midline of the body is said to be protraction of the scapula (see Fig. 4-10,B, page 195).

Retraction: to move backward (not shown in Fig. I-7). Returning the lower jaw to normal position after having protracted it is termed retraction. Movement of the scapula medially toward the midline of the body is said to be retraction of the scapula (see Fig. 4-10,A, page 195).

Plantar flexion: refers to extension of the ankle joint such that the toes of the foot are pushed down and away from the leg, as in certain ballet movements. The term is derived from the fact that the sole or underside of the foot is the plantar surface; so bending the foot toward the plantar surface at the ankle joint is plantar flexion. But relative to the *anatomical position,* the motion is one of extension. (Confused?) Plantar flexion may be seen at joint 10 in Fig. I-7.

Dorsiflexion: the opposite of plantar flexion. This refers to flexion of the ankle joint such that the toes of the foot are brought up and toward the leg. The upper surface of the foot is the dorsal (posterior) surface. (The term dorsal here seems erroneous, but actually the limbs rotated in fetal life and the dorsal surface *was* dorsal.) Bending the foot toward that surface is dorsiflexion. Dorsiflexion is represented at joint 10 by the letter E.

Inversion: a position of the foot in which the sole faces inward (medially) such as when one stands on the lateral borders of the feet, thus elevating the medial borders (Fig. 5-23).

Eversion: a position of the foot in which the sole faces outward (laterally), as when one stands on the medial borders of the feet, elevating the lateral borders (Fig. 5-23).

It would be advisable at this point in your study to review these terms of position and movement by practicing them and naming them as you do them. After doing this, try the following quiz on for size.

QUIZ

Match the definition at right with the proper term*:

6 Flexion
12 Inversion
7 Supination
1 Abduction
11 Retraction
4 Circumduction
5 Adduction
10 Pronation
13 Eversion

1. To take away from the midline laterally
2. To turn about an axis
3. To straighten out
4. Flexion, abduction, extension, and adduction in sequence
5. To bring toward the midline
6. To bend
7. Face up
8. Exaggerated straightening
9. To move forward; to draw forward
10. *Medial* rotation of the forearm
11. To move backward; to draw back
12. To stand on lateral borders of feet
13. To stand on medial borders of feet

GENERAL TERMS

The terms presented below are critical ones—used frequently throughout the course and in everyday life as well. Others may be learned as they become appropriate in the course and notations made on the following pages. Consultation of any standard dictionary would be appropriate, and a medical or biological dictionary may be necessary in some cases. At the end of this section is space for additional words you may wish to add to the list.

Prefixes	*Definition*	*Example*
1. a-, an-	a. without	*a*graphia
2. ab-	b. motion or direction *away from*	*ab*duction
3. ad-	c. motion or direction *toward*	*ad*duction
4. ana-	d. again, up, excessive	*ana*plasia
5. ante-, pre-	e. before or in front of, forward	*ante*flexion
6. arthr-	f. referring to *joints*	*arthr*itis
7. auto-	g. referring to *self*	*auto*graph
8. bi-	h. two	*bi*polar
9. cauda-	i. referring to the *tail*	*cauda*l
10. ceph-	j. referring to the *head*	*ceph*alic

* *Answers:* 6, 12, 7, 1, 11, 4, 5, 10, 13

Suffixes	*Definition*	*Example*
1. -ac	a. of or pertaining to	hypochondri*ac*
2. -cyte	b. referring to a cell	fibro*cyte*
3. -genesis	c. development of	tumoro*genesis*
4. -ia	d. noun suffix in names of diseases	hemophil*ia*
5. -ic	e. of or pertaining to	hydrophob*ic*
6. -itis	f. inflammation of	appendic*itis*
7. -oid	g. like; in the form of	human*oid*
8. -ology	h. refers to a science of knowledge	ec*ology*
9. -oma	i. denoting a tumor	lip*oma*
10. -osis	j. condition of	neur*osis*

Vocabulary drill

(Based on above prefixes and suffixes, match the terms with the proper definition.)

____ agenesis	a. toward the head end
____ arthritis	b. inflammation of joints
____ arthrology	c. tail-shaped
____ cephalic	d. without growth or development
____ caudal	e. study of joints
____ caudate	f. toward the tail

How about another set?

Prefixes	*Definition*	*Example*
1. di-	a. two	*di*gastric
2. dis-	b. apart	*dis*sect
3. en-, endo-	c. within, inside, in	*endo*plasm
4. epi-	d. on, above	*epi*thelium
5. ex-, exo-	e. outside of	*ex*crete
6. hetero-	f. unlike	*hetero*sexual
7. histo-	g. refers to tissues	*histo*logy
8. homo-	h. like, similar	*homo*geneous
9. hyper-	i. over, excess	*hyper*activity
10. hypo-	j. under, less than normal	*hypo*baric
11. inter-	k. between	*inter*cellular
12. intra-	l. within, inside, in	*intra*venous
13. macro-	m. large, big	*macro*scopic
14. mega-	n. great, mighty, huge	*mega*lomaniac
15. meta-	o. between, with, after (implying a change)	*meta*plasia

Suffixes	*Definition*	*Example*
1. -sect	a. to cut	dis*sect*
2. -troph	b. refers to state of nutrition or health	a*trophic*
3. -stasis	c. stoppage, standing still	bacterio*stasis*
4. -some, somatic	d. body	micro*some*
5. -plastic, -plasty	e. growth, development	hyper*plastic*
6. -phobic	f. fear of, dislike	hydro*phobic*
7. -philic	g. love of, like, agreement	hydro*philic*
8. -pathy	h. suffering, sickness, or a feeling for	sym*pathy*, *path*ology
9. -ectomy	i. removal of	append*ectomy*
10. -blast	j. early, undeveloped form	neuro*blast*

Just a few more:

Prefixes	*Definition*	*Example*
1. micro-	a. small	*micro*scope
2. milli-	b. thousand	*milli*liter
3. morpho-	c. refers to body or structure	a*morphous*
4. myo-	d. refers to muscle	*myo*logy
5. neo-	e. new	*neo*plasm
6. neuro-	f. refers to nerves or nervous tissue	*neur*itis
7. osteo-	g. refers to bone	*osteo*pathy
8. peri-	h. around	*peri*osteum
9. post-	i. after, afterward	*post*operative
10. supra-	j. above	*supra*clavicular
11. syn-	k. with, along with, together	*syn*ergy

A final test. Can you figure out the meaning of:

1. pathologic: science of disease
2. histology: science of tissues
3. aplastic: without growth
4. microsome: small body
5. atrophy: decline of healthy state
6. amorphous: die
7. dissect: cut in two
8. endogenous: internal development
9. intravenous: within the vein

10. periosteum: area around the bone
11. myology: science of muscles
12. osteology: science of bones
13. neurosis: nervous condition
14. macrocyte: large cell
15. hypertrophy: increase health
16. neoplasia: new growth (disease)
17. somatic: body

Additional vocabulary list for you to fill in

1.
2.
3.
4.
5.
6.
7.
8.
9.
10.
11.
12.
13.
14.
15.
16.
17.
18.
19.
20.
21.
22.
23.
24.
25.
26.
27.
28.
29.
30.
31.
32.
33.
34.
35.
36.
37.
38.
39.
40.
41.
42.
43.
44.
45.
46.

part two basic histology

unit 1 the cell

A STRUCTURAL AND FUNCTIONAL CONCEPT

In this unit you will study and investigate the fundamental unit of all living things, the cell (Fig. 1-1). First, what does one mean by the term "fundamental unit?" To answer that, let's start with some principles you are all aware of but which many of you have not thought about. We all know that the universe is composed of things which we refer to as *structures* or *forms.* Some of these structures are so small we cannot accurately fathom their smallness without very sophisticated and complex instruments. Such structures would be subatomic particles (protons, neutrons, electrons) which make up the structure of an atom. It is probable that these structures are composed of even smaller structures, and so on to an infinity of size in a direction we call small. Some structures are so big we cannot accurately fathom their bigness. Such a structure would be the universe, and so on to an infinity of size in a direction we call big. But, if we could agree that the universe is *the* structure consisting of an order of progressively small structures, one could say further that every structure is composed of an organization of *smaller* structures or units. Such a structure would be the human body.

In the human body, as in the structure of most organisms, there are structures which *in themselves and only in them-*

Fig. 1-1 Some isolated cells of the human body. Note the variety of shapes (which can change rapidly in certain cases) and nuclear/cytoplasmic size relationships.

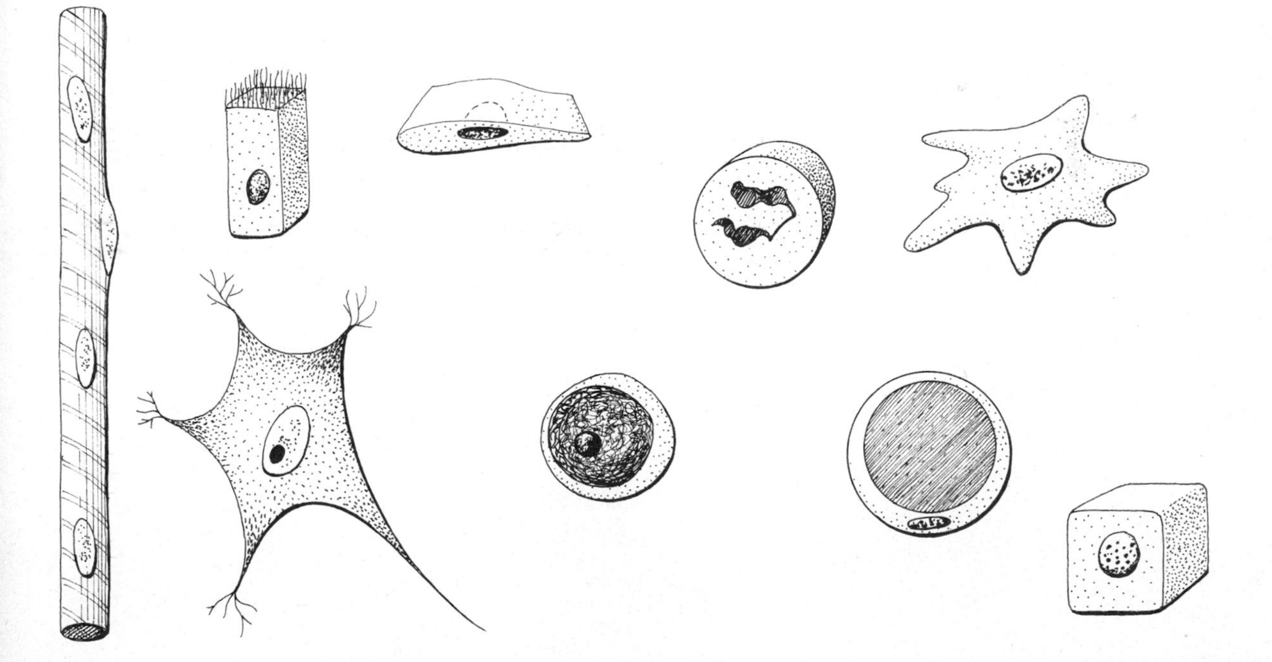

selves exhibit the phenomenon, property, or quality called life*. These simple units of life are the common denominator, then, of all living things. They are, in fact, the *fundamental unit of biological life* and are called **cells.** Any living structure in the body more complex (e.g., bigger) than a cell is simply a collection of cells. There is, of course, a continuum of decreasing complexity from cells to lesser forms. Structures which have a less complex organization than cells are *not* considered capable of independent life by standards based on our present level of knowledge.

QUIZ

Before going on, let's test your comprehension of some of the important elements of what has been stated:

1. Every structure is composed of an organization of ______________ structures.
2. In the body, the smallest, simplest structure exhibiting independent life is a ______________.
3. Structures exhibiting life which are more complex than cells must be collections of ______________.
4. What about forms which are simpler than cells? Are they living things (i.e., capable of independent life)? __________.

A score of less than 4/4 suggests that you might wish to review the previous paragraphs.

If one is to understand that the cell is the fundamental unit of biological life, the term *biological life* should be defined. It may be said that biological life is a quality generated by certain events having taken or capable of taking place. These events show up in certain activities characteristically observed in living things. By extrapolating backward from known activities, one can deduce those "certain events" which collectively are responsible for the property of life. (This is an exercise in reasoning . . . much more reliable than memorizing.) Here we go:

What is one of the most obvious activities in living organisms? *Movement.* Now realizing that organisms are basically collections of cells, what basic "event" on the cellular level, must take place for movement to occur? *Production of energy.* Obviously movement consumes energy, and cells pro-

* The term *life* might be better received by cell biologists if it were prefixed by the word *independent,* that is independent life. Many structures, such as viruses, appear to live and can even reproduce yet are not capable of independent life as are cells.

vide this energy. What is a well-known biological term meaning the exchange of energy, or production and consumption of energy? *Metabolism.* This is one of three of those "certain events" which together produce the quality we call "life." What basic activities of cells and multicellular organisms are initiated by metabolic activity?

1. Ingestion of food.
2. Assimilation.
3. Respiration.
4. Synthesis and degradation.
5. Elimination.
6. Excitability.
7. Movement.

Cells capable of metabolic activity didn't achieve this ability like a "bolt from the blue."* They inherited it from parent cells. To pass on "blueprints" for metabolic machinery and synthesis, most cells divide into daughter cells in *reproduction. Reproduction,* then, is the second event.

The third "event" is *adaptation.* It is essential that cells be able to adapt to their environment. Over a period of many years, it is conceivable that a given environment may change; for example, there may be an increase in the average temperature. Cells are able to adapt to these changes over many generations in the slow process of evolution.

In conclusion, it may be said that independent biological life is a quality produced by the combined and interrelated events of *metabolism, reproduction,* and *adaptation.*

Before going on, it is advisable to stop and contemplate this conceptual analysis of the cell and its activities. If possible, discuss the analysis with your fellow students. You may find reasons for disagreement and argument. It is hoped that this will lead you to sources for information in depth. (See suggested references at end of this unit.)

COMPOSITION OF THE CELL

A cell is the anatomical sum of its parts. Those parts which seem to be involved in some basic task necessary for normal cell operation are called **organelles.** Those parts which are

* This is not entirely true. Laboratory synthesis of certain amino acids (constituents of protein) requires an electrical charge to initiate the reaction. Proteins preceded cells. Thus it is surmised that the electrical activity in the earth's atmosphere may have provided the energy required for protein synthesis and thus creation of the First Cell.

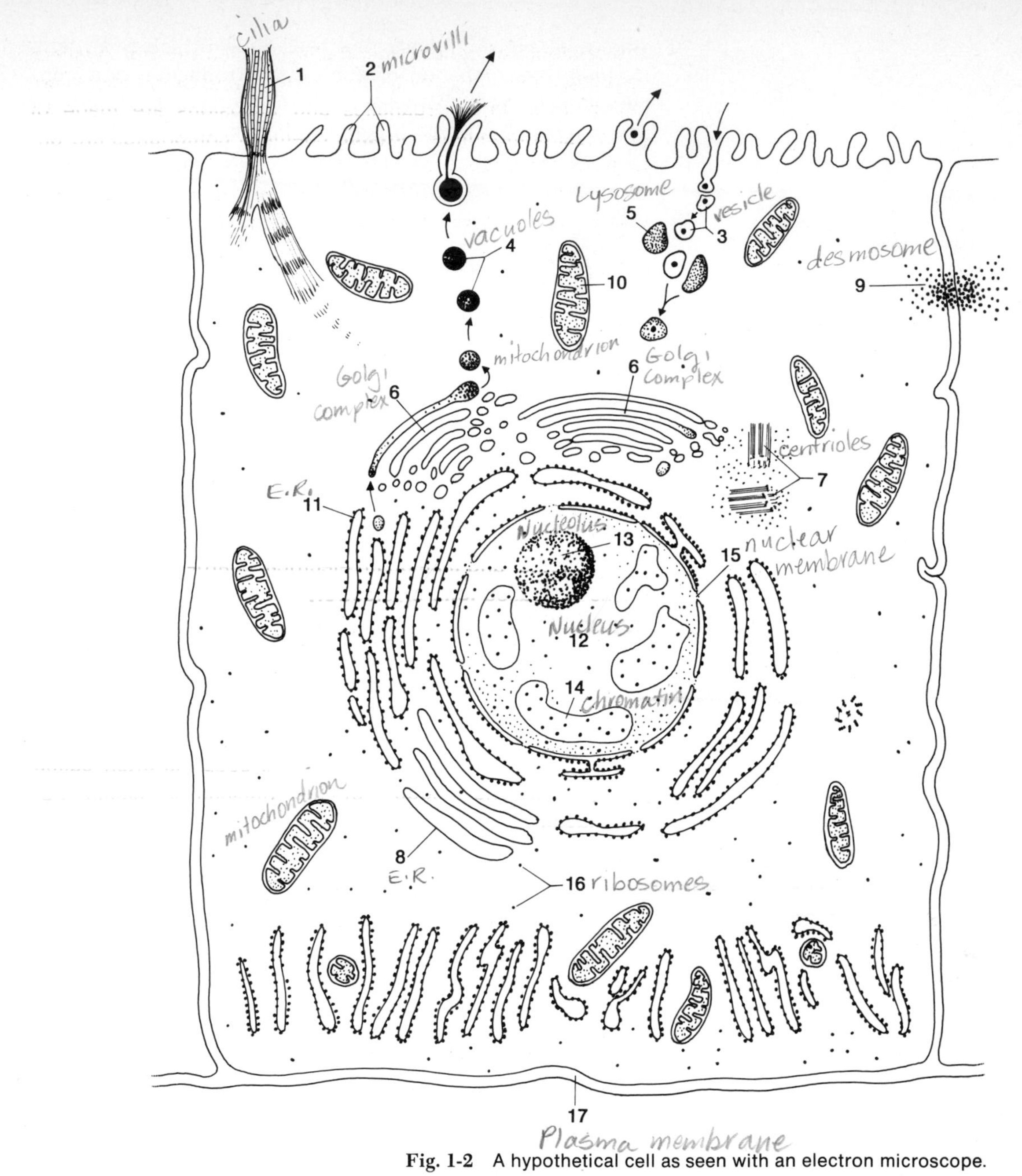

Fig. 1-2 A hypothetical cell as seen with an electron microscope. Refer to text for explanation of numbers. (Adapted from De Robertis and Pellegrino de Iraldi.)

the products of cellular metabolism, which are not required for proper cellular activity, and which are often transient, are called **inclusions.** Organelles and inclusions are made of chemical compounds. **Organic** chemical compounds are defined as compounds which incorporate carbon in their structure. All other compounds are called **inorganic.** The following classes of organic compounds are found in all cells:

1. Nucleic acids.
2. Proteins.
3. Lipids (that is, fat).
4. Carbohydrates.
5. Conjugated compounds.

In cells, inorganic compounds exist primarily in the form of **salts,** examples of which are sodium chloride (NaCl), potassium chloride (KCl), and calcium nitrate [$Ca(NO_3)_2$]. In the cell interior, these salts are dissolved in water and, as a consequence, their atoms become free of the bonds that held them together and take on an electrical charge. Such charged atoms are called **ions** or **electrolytes.** These ions are vital to life and are basically responsible for the fact that the body is a conductor of electricity.

Proteins and lipids are the basic building materials for cells. The general composition of a typical cell is 15 percent protein, 3 percent lipid, 1 percent carbohydrate, 1 percent salts, and 80 percent water. This "chemical soup" is often called **protoplasm.** The "soup" of the nucleus is called **nucleoplasm.**

Fig. 1-3 Nucleus of a cell demonstrating a double membrane with a less dense middle layer (trilaminar). The gaps ("pores") actually have a very thin membrane. These presumably permit communication between nucleus and cytoplasm. Drawn from an electron micrograph.

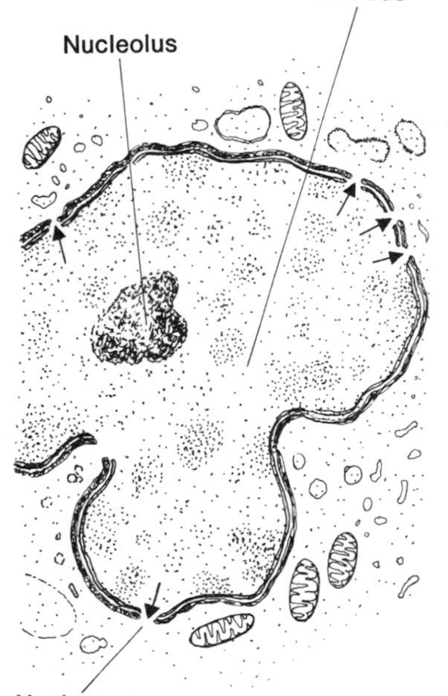

ORGANELLES OF THE CELL AND THEIR ROLE IN PROTEIN SYNTHESIS

Refer now to Fig. 1-2. This illustration of a hypothetical cell shows most of its working components and how some of them work, as well. Each organelle is identified by number. In the commentary below, each organelle discussed is referred to by that number, enclosed in brackets.

The cell is "ruled" by the nucleus [12], surrounded by a porous membrane (Fig. 1-3). It is the administrative center in which the metabolic and synthetic activities of the cell are organized, initiated, and directed. The nucleus contains the hereditary material received from the parent cell. This hereditary material, the "blueprint" directing cellular activity, is **DNA** (*d*eoxyribo*n*ucleic *a*cid). It also determines cell shape and size. DNA has a protein framework. When the cell is not undergoing division, the DNA-protein complex can be seen in

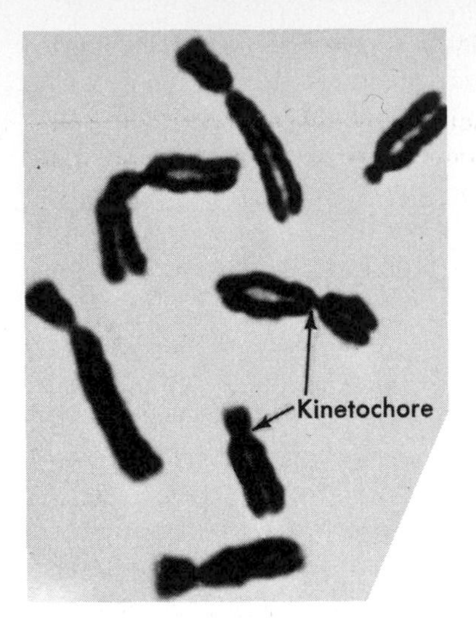

Fig. 1-4 A few of the chromosomes from a human white blood cell. Each chromosome has duplicated itself and the two parts (chromatids) are joined at a kinetochore. Shown here in metaphase of mitosis, the chromatids separate in the next stage of cell division.

the nucleus as **chromatin** [14] (Gk., *chroma*, color). As the derivation suggests, chromatin is quite colorful when properly stained. During cell division, chromatin condenses into **chromosomes** (Fig. 1-4). In protein synthesis the DNA first makes **RNA** (*r*ibo*n*ucleic *a*cid) and the RNA distributes DNA's "instructions" to the organelles outside the nucleus, getting out through the perforations of the nuclear membrane [15]. Since it acts as a messenger, this RNA is called "messenger RNA." A highly stainable granular structure can be seen particularly well within the nucleus of cells that make protein. This structure is the **nucleolus** [13] (see also Fig. 1-3), which may be the source of RNA. So much for the nucleus now.

The protoplasm of the cell less the nucleoplasm is called the **cytoplasm.** Cytoplasm is surrounded by an actively metabolic membrane composed of an orderly arrangement of lipid and protein. This **plasma membrane** [17] (see also Fig. 1-5) allows a variety of materials to pass through it into or out of the cell interior, but it is *selectively permeable;* that is, some things pass through and others do not.

Within the cytoplasm there is a network of tubules known as the **endoplasmic reticulum** [8, 11] (see also Fig. 1-6). Endoplasmic reticulum is believed to serve as a "highway" for the movement of materials used in construction (synthesis). It may also play a specific role in that synthetic activity.

Not infrequently, endoplasmic reticulum is studded with little round bodies of RNA in a framework of protein different from that of messenger RNA. These round bodies, known as **ribosomes** [16] (see also Fig. 1-6), are derived from RNA of the nucleolus and are distributed through the cytoplasm. Ribosomes are believed to be the site where protein components are brought together and connected. The procedure is thought to be about as follows: A length of messenger RNA leaves the nucleus and drapes itself over one or more ribosomes, like a strip of carpet lying on a row of bowling balls. A special type of RNA which transfers molecules from one place to another, called "transfer RNA," collects amino acids in the cytoplasm. (Amino acids are the basic building blocks of protein. They get into the body as protein in the food we eat, which is broken down in digestion into amino acids and distributed to the cells.) Once collected, the amino acids are led by the transfer RNA to the messenger RNA–lined ribosomes.

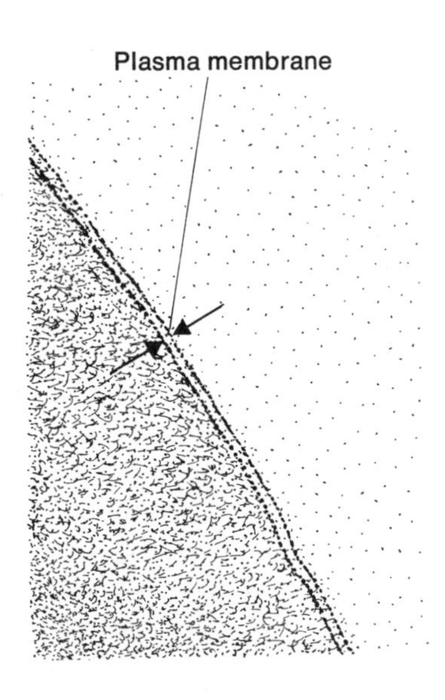

Fig. 1-5 Plasma membrane of a cell. Note the trilaminar character. Drawn from an electron micrograph.

Each combined transfer RNA and amino acid is fitted precisely onto the messenger RNA and linked to the other ones of its kind in the sequence prescribed by the DNA. A chain forms and grows until it is a specific kind of protein, when it is released into the endoplasmic reticulum, if it is to be discharged from the cell as a secretion. If it is to be retained for use in rebuilding cell structure, the protein is released directly into the cytoplasm.

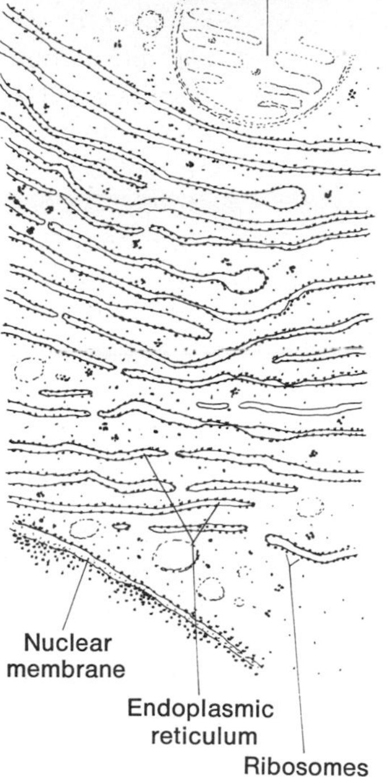

Fig. 1-6 Liver cell endoplasmic reticulum seen as flattened, elongated vesicles arranged in parallel. Ribosomes often lie in close association. Drawn from an electron micrograph.

In a certain part of the cell the endoplasmic reticulum is near a complex of tubules and vesicles named Golgi, after the man who described it. The **Golgi complex** [6] (see also Fig. 1-7) is believed to receive the newly synthesized protein from the endoplasmic reticulum and concentrate, package, store, and release it for final discharge in membrane-lined containers called **vesicles** or vacuoles [4] (see also Fig. 1-7).

The secretory material in vesicles is moved through the cytoplasm to the free (unattached) border of the cell where, between fingerlike extensions of the plasma membrane, it is released. These tiny, undulating extensions are called **microvilli** [2] (see also Fig. 1-8) and they characterize certain active epithelial (lining) cells. They serve to increase the absorptive surface of the cell, yet they exercise selective permeability. Note the intercellular junction, a fibrous connection called a **desmosome** [9].

Foreign particles that manage to cross the barrier of the cell membrane carry along a bit of the membrane which is then pinched off to form a **vesicle** [3]. The "visitor" may be greeted by a **lysosome** [5] (see also Fig. 1-9), which arrives in its own membranous sac. The two membranous sacs fuse, the lysosome enzymes mix with and digest the foreign body, and the debris may be taken to the plasma membrane and ejected from the cell. Lysosomes are capable also of digesting "injured" intracellular structures.

The energy for these intracellular activities comes from structures called **mitochondria** [10], which are found close to the site of any energy-consuming activity. Mitochondria are complex in organization (see Fig. 1-10) and the site of great enzymatic activity.

Centrioles are shaped like small barrels. They are found in the cytoplasm in pairs [7] (see also Fig. 1-11). They act in cell division (see ahead). A centriole is composed of radially ar-

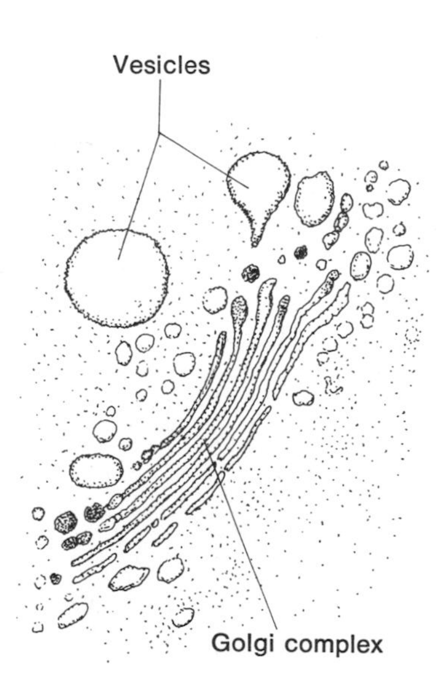

Fig. 1-7 A Golgi complex. The nearby vesicles contain secretory material probably packaged by the Golgi. Note how the small vesicles seem to bud off from the swollen ends of the Golgi tubules. Drawn from an electron micrograph.

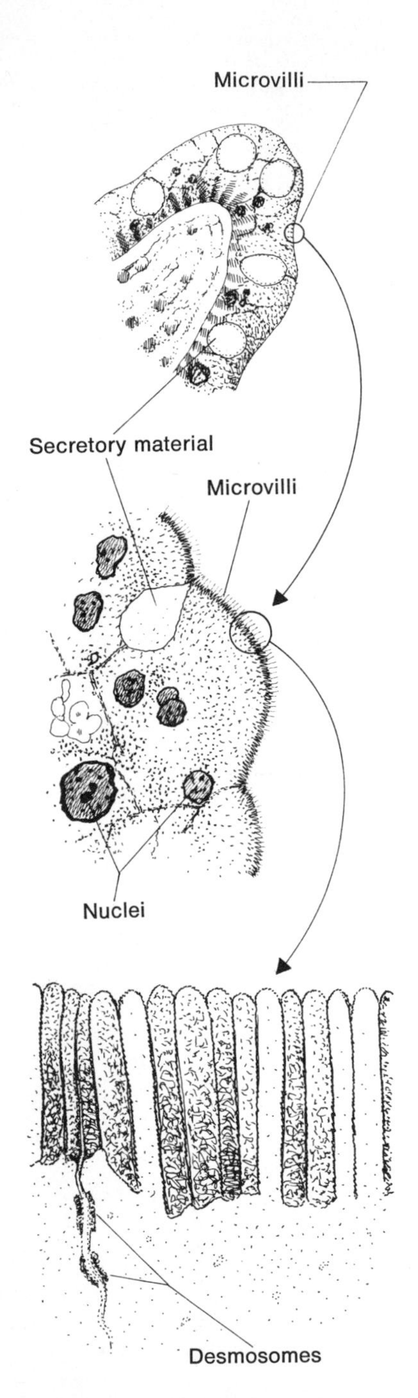

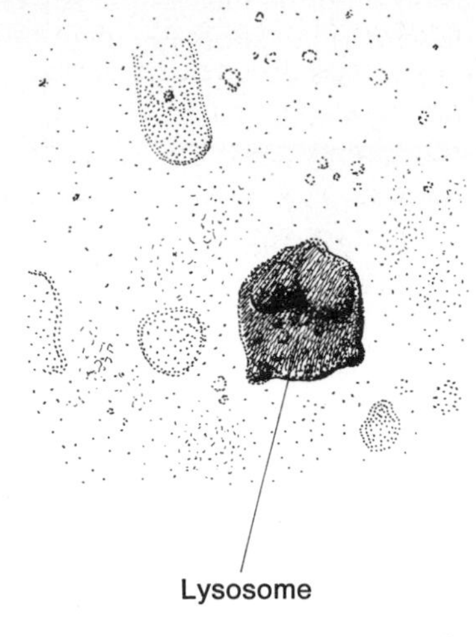

Fig. 1-9 (above) Lysosome. Note its multivesicular appearance. Drawn from an electron micrograph.

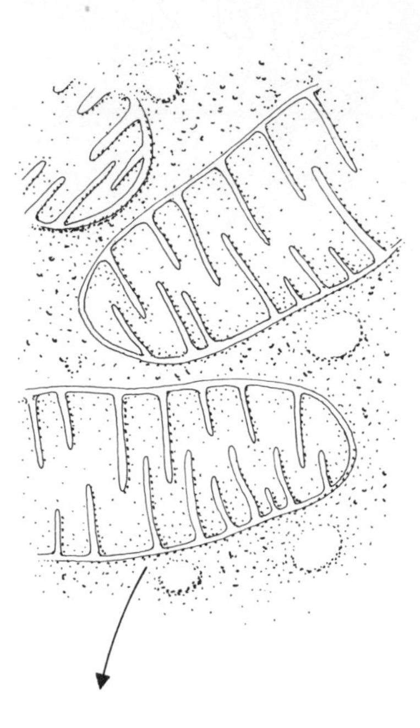

Fig. 1-8 (left) Microvilli, drawn in stages of increasing magnification from light and electron micrographs. Note the desmosome (intercellular junction) in the lowest illustration.

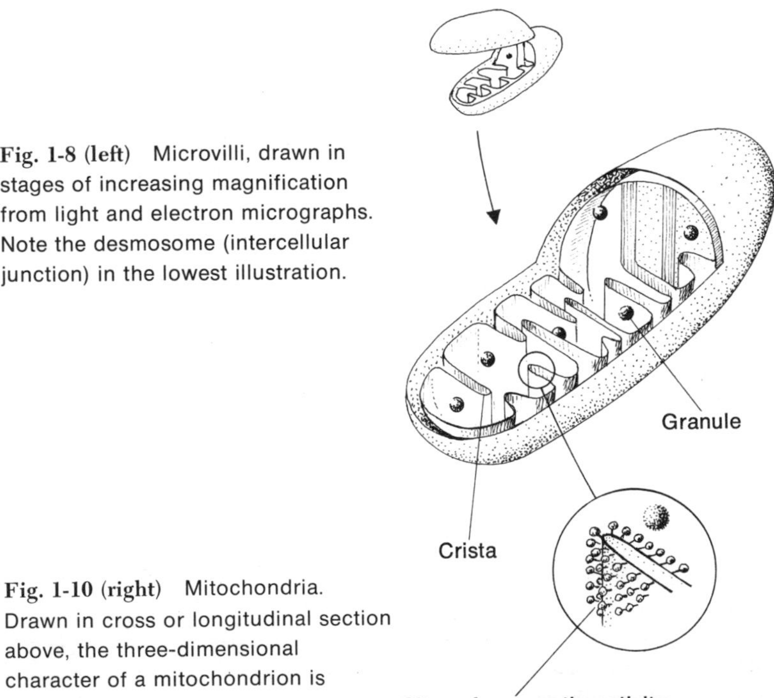

Fig. 1-10 (right) Mitochondria. Drawn in cross or longitudinal section above, the three-dimensional character of a mitochondrion is shown below.

ranged tubules and is also believed to produce **cilia** [1], which are short, filamentous organelles that project from the free surface of certain epithelial cells. Each cilium (Fig. 1-12) is composed of nine sets of radially arranged fibers and two that are centrally located. Cilia characterize the lining of the respiratory and reproductive tracts. Their sweeping strokes move material along the cell surface; for instance, the many cells lining the airway move mucus from the trachea to the pharynx.

INCLUSIONS OF THE CELL

Inclusions are products of cell metabolism, but they are not to be regarded as pure litter. At least two kinds of inclusions—starch (glycogen) and fat—may be stored sources of energy.

FUNCTION OF THE CELL

This book will deal with the subject of cell function in only a general way. The basic function of any cell is to live. To do this, it must perform certain basic work: ingestion, assimi-

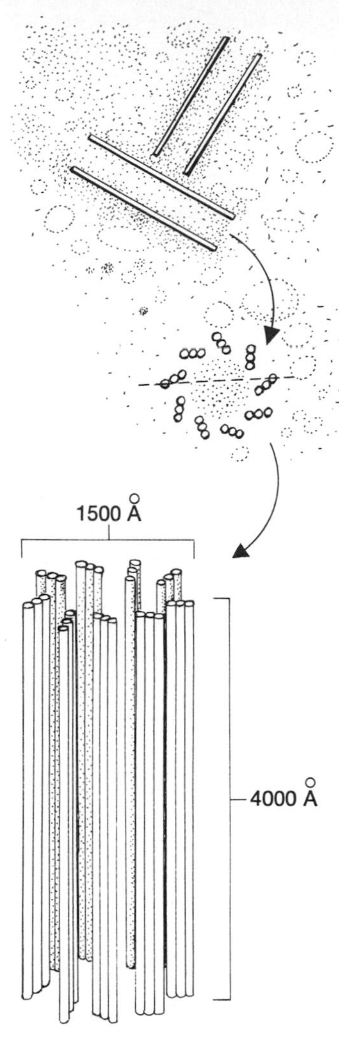

Fig. 1-11 (above) Two centrioles, one drawn at right angle to the other. The tubular arrangement is shown below. Drawn from an electron micrograph (×143,000).

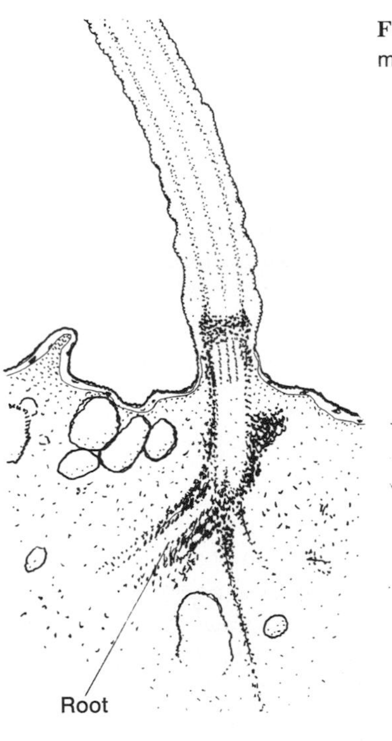

Fig. 1-12 (right) A cilium in longitudinal section demonstrating a striated root (which may be related to a centriole). Drawn from an electron micrograph (×100,000).

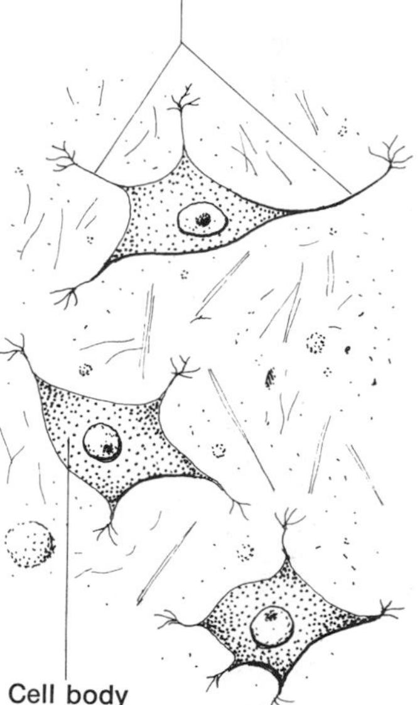

Fig. 1-13 Nerve cells of the multipolar (many-process) variety.

lation, synthesis, degradation, and elimination. These are aspects of cell metabolism. In addition, to achieve a state of complex activity, a cell must experience certain growth. To be sensitive to and responsive to stimuli in its environment, it must be irritable. To survive changes in its internal and external environment, it must be able to adapt. To perpetuate its quality of life, it must reproduce. All cells, except those which are highly specialized and do not normally reproduce, perform these tasks. The tasks may be considered as the general background against which each cell carries out its specific function.

There are many different kinds of cells with as many types of function. The functions are different because the DNA-initiated instructions for protein synthesis are different. And it is the synthesis or lack of synthesis of certain proteins which is responsible for a cell accomplishing a certain function. Thus there are cells which can generate and transmit nervous impulses (something like electrical impulses) along their processes over considerable distances. These are, obviously, nerve cells (Fig. 1-13). There are cells which have the capacity of shortening (contraction). These are muscle cells (Fig. 1-14). There are cells which secrete substances, cells which absorb substances, and cells which filter substances. There are cells which, in mass, protect other cells and tissues from a variety of noxious stimuli. There are cells which collectively support, bind, and connect other tissues and organs. Such cells—approximately 30 trillion of them—and their products, make up the human body.

Fig. 1-14 Muscle cells of the heart. Some of these are shown in a contracted state. Note the folded nuclei of such cells.

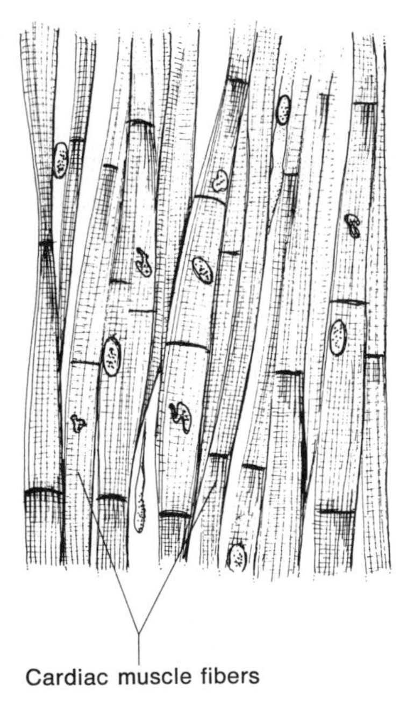

CELL DIVISION

It will be remembered that one of the fundamental activities of all living organisms is reproduction. Reproduction is necessary in the event of a requirement for cellular replacement or tissue repair. Reproduction of cells is necessary for growth of the organism. Reproduction is essential for the continuity of life. In the cells of man, as well as in many other organisms, the reproductive process is started by the duplication of DNA. This action is followed by a partial duplication and then exact division of cellular contents. This kind of cell division is called **mitosis** (Fig. 1-15).

In observing the stages of mitosis in the laboratory, one will be primarily concerned with the changes taking place in the nucleus. The changes in the cytoplasm are apparent only in the final phase, upon cleavage and formation of the two daughter cells.

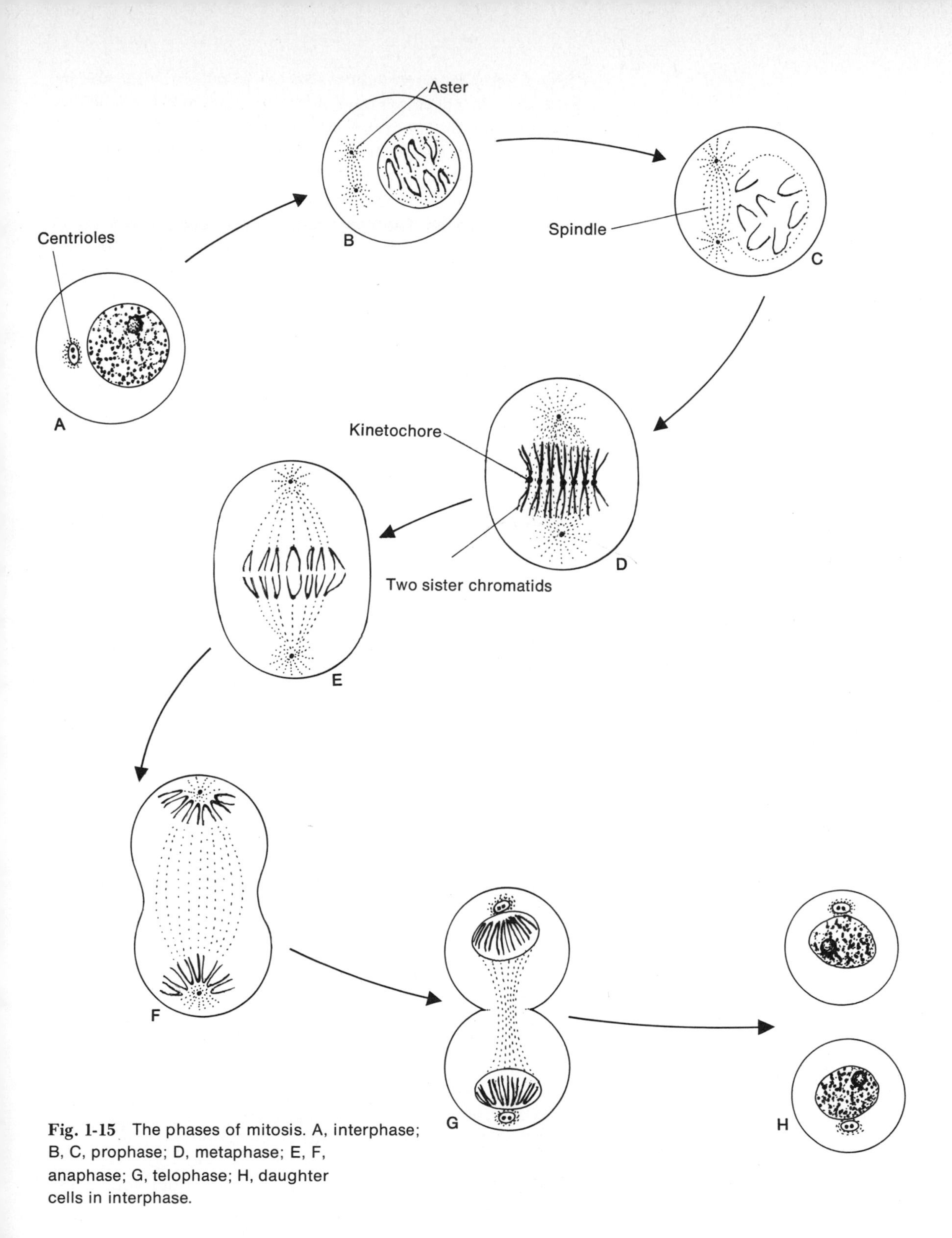

Fig. 1-15 The phases of mitosis. A, interphase; B, C, prophase; D, metaphase; E, F, anaphase; G, telophase; H, daughter cells in interphase.

The period between successive cell divisions is termed **interphase,** (A) during which the chromosomes are not visible but are dispersed as chromatin granules.

Refer now to Fig. 1-15 and note the changes taking place within the nucleus during mitosis:

Prophase (B,C): the chromosomes (46) begin to condense and appear. The centrioles appear, each with a set of starlike rays, the *aster.* Connecting the two centrioles are strands of fibers, the *spindle.* The centrioles move to opposite poles of the cell, stretching the spindle between them. The nucleolus disappears and the nuclear membrane disintegrates.

Metaphase (D): the 46 chromosomes line up along the middle of the cell in close relation to the spindle. Each chromosome is seen to split longitudinally, and each resultant half is called a *chromatid,* of which there are 92. The two halves of each chromosome are attached to one another at the *kinetochore.* The kinetochore is attached to the spindle.

Anaphase (E,F): each chromatid splits off from its sister chromatid and they head for opposite poles of the cell, guided by the spindle as a train is guided by a track. These 92 freed chromatids now constitute *new chromosomes,* and there are 46 at each of the two poles of the cell. The cytoplasm begins to cleave in the midline.

Telophase (G): a new nuclear membrane begins to appear about the chromosomes in each daughter cell. The chromosomes begin to disperse into chromatin. As the two daughter cells mature into final form, the spindle fibers disappear. The two cells enter the interphase condition. (H)

UPON REFLECTION

You have now seen that the cell is the common denominator of life. If you are going to understand the structure of your body, it is necessary to know the basic organization and structure of the cell so that the tasks of the cell can be appreciated. Protein synthesis and mitosis are characteristic of these tasks. To understand these phenomena is to understand how skin wounds, deep lacerations, and organic diseases can be naturally repaired; how a child can grow to adulthood; and how that natural development can be impaired by insufficient protein in the diet. Do you need more reasons than these for finding out a little about your cells?

SUGGESTED REFERENCES

1. Weisz, Paul B.: *The Science of Biology,* 4th ed., McGraw-Hill Book Company, New York, 1971. (Excellent introduction to the science of biology; includes sections on basic chemistry and physics as they apply to living things.)
2. Bloom, William, and Fawcett, Don W.: *A Textbook of Histology,* 9th ed., W. B. Saunders Co., Philadelphia, 1968. (An excellent text for a study of cells and cell division in a depth just short of physicochemical applications.)
3. Giese, Arthur C.: *Cell Physiology,* 3d ed., W. B. Saunders Company, Philadelphia, 1968. (Cellular activity in depth—an exceptional source of information for those with the proper background in mathematics, chemistry, and physics.)
4. Pfeiffer, John (ed.): *The Cell,* from the Life Science Library, Time Inc., New York, 1964.
5. Scientific American offprints:
   - 36 Powerhouse of the Cell (Siekevitz, 1957).
   - 47 The Origin of Life (Wald, 1954).
   - 61 Life and Light (Wald, 1959).
   - 76 Pores in the Cell Membrane (Solomon, 1960).
   - 77 Polyoma Virus (Stewart, 1960).
   - 79 Cilia (Satir, 1961).
   - 90 The Living Cell (Brachet, 1961).
   - 91 How Cells Transform Energy (Lehninger, 1961).
   - 92 How Cells Make Molecules (Allfrey and Mirsky, 1961).
   - 93 How Cells Divide (Mazia, 1961).
   - 94 How Cells Specialize (Fischberg and Blackler, 1961).
   - 95 How Cells Associate (Moscona, 1961).
   - 96 How Things Get Into Cells (Holter, 1961).
   - 97 How Cells Move (Hayashi, 1961).
   - 98 How Cells Communicate (Katz, 1961).
   - 99 How Cells Receive Stimuli (Miller, et al; 1961).
   - 119 Messenger RNA (Hurwitz and Furth, 1962).
   - 151 The Membrane of the Living Cell (Robertson, 1962).
   - 156 The Lysosome (DeDuve, 1963).
   - 161 Sex Differences in Cells (Mittwock, 1963).
   - 165 Autobiographies of Cells (Leski and Leski, 1963).
   - 1024 Reversal of Tumor Growth (Braun, 1965).
   - 1061 Repair of DNA (Hanawalt and Haynes, 1967).
   - 1069 Induction of Cancer by Viruses (Dulbecco, 1967).
   - 1085 Lysosomes and Disease (Allison, 1967).
   - 1103 Human Cells and Aging (Hayflick, 1968).

clinical considerations

As you can now appreciate, after having studied cells and their characteristics, cells carefully regulate their own rate of mitosis. Cell size during growth and development and during repair after injury is also very carefully regulated. Regulation of growth and rate of mitosis are dependent upon proper functioning of nuclear structures (for example, DNA), and upon the presence of necessary enzymes, which is further dependent upon appropriate protein synthesis. The above is true for all cells of all living things, plant and animal.

Occasionally, certain physical trauma, chemical irritants, or viral infections can cause cell growth- and mitosis-regulating systems to go awry, and the result is that the stressed cells commence dividing at a rapid and uncontrollable rate, resulting in the appearance of a palpable swelling in the affected area. Such a swelling is called a **tumor**—a true tumor, as opposed to a swelling caused by parasites, inflammation, or physical conditioning (such as hypertrophied muscle).

True tumors may occur anywhere on the body. Tumors may be **benign** or **malignant.** Most tumors, by far, are benign, and are composed of cells quite normal in appearance. Such tumors are usually encapsulated, and are frequently multiple in the affected area. Such tumors are usually functionless and can be surgically removed without fear that they will recur. Benign tumors are classified according to the tissue in which they are found, and the word is terminated with the suffix *-oma,* for example, lipoma (fat tumor), fibroma (connective tissue tumor), adenoma (glandular epithelial tumor), papilloma (skin tumor).

Malignant tumors (Fig. 1-16) are characterized by cells which are generally different in shape, size, and appearance than the cell population from which the first tumor cell arose, hence the term **neoplasm** (new tissue) for such tumors. Often the tumor cells regress to a primitive, functionless but prodigiously dividing type (anaplasia). Such cells are further characterized by tending to break off from the initial mass of malignant cells and passing into the lymphatic and vascular

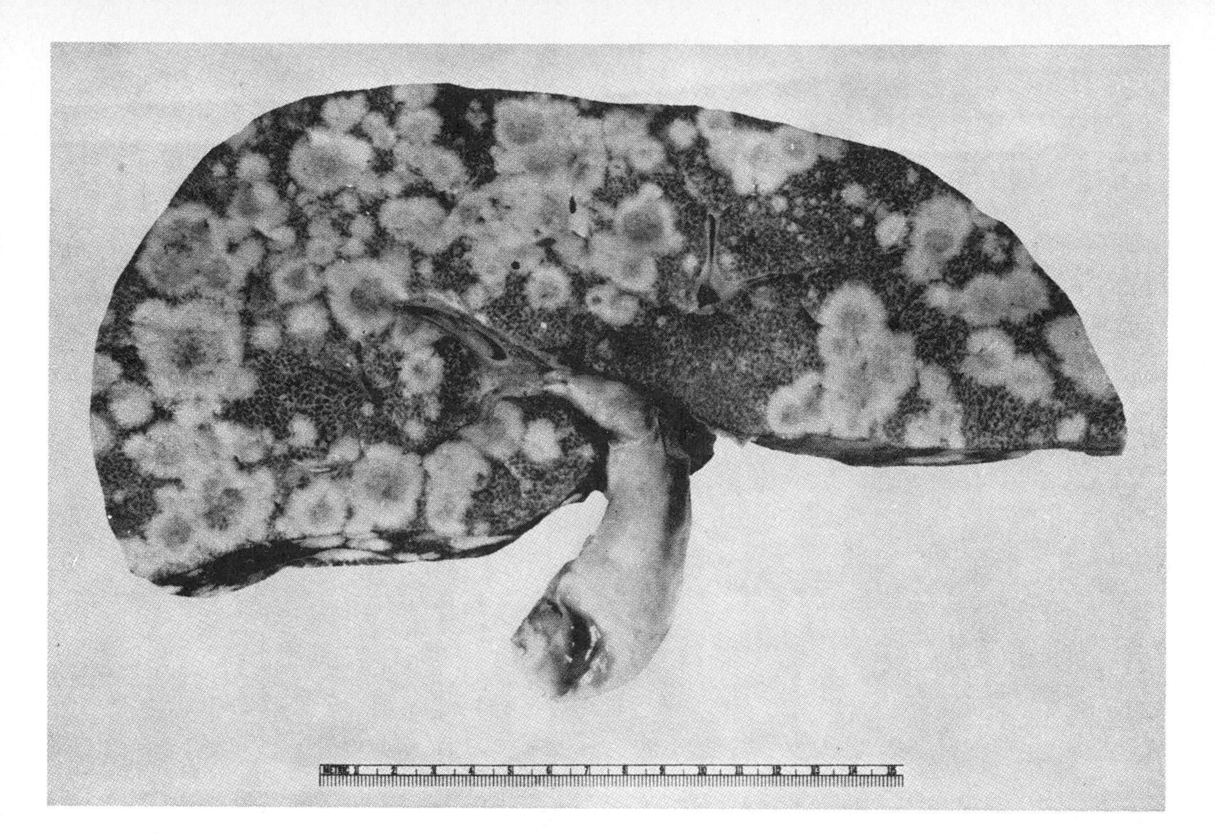

Fig. 1-16 Metastatic carcinoma of the liver (white spots). Malignant cells crowd out normal liver cells, suffocating them or otherwise inhibiting their functioning. (From R. P. Morehead, *Human Pathology*, McGraw-Hill, New York, 1965.)

circulation to ultimately stop and grow on any number of organs, for example, lymph nodes, muscle, lung, liver. This spreading is called **metastasis.** Ultimately the metastasized cells will overpower the normal cells, strangling the latter, with loss of organ function. Meanwhile, the organ gets larger in size. Obviously, early diagnosis of cancer (that is, neoplasm) is essential in order to arrest the development of the malignant process which, in many cases, will lead to death.

The substance or agent which causes cancer is called a **carcinogen.** Malignant tumors are classified as **carcinomas** (originating among epithelial cells) and **sarcomas** (originating in any of the connective tissues). Prefixes added to the basic term indicate specific origin, for example, osteogenic sarcoma (originating in the bone cells) and fibrosarcoma (originating in connective tissue proper). These terms will become clearer in meaning once you have studied Unit 2, on tissues.

unit 2 the tissues

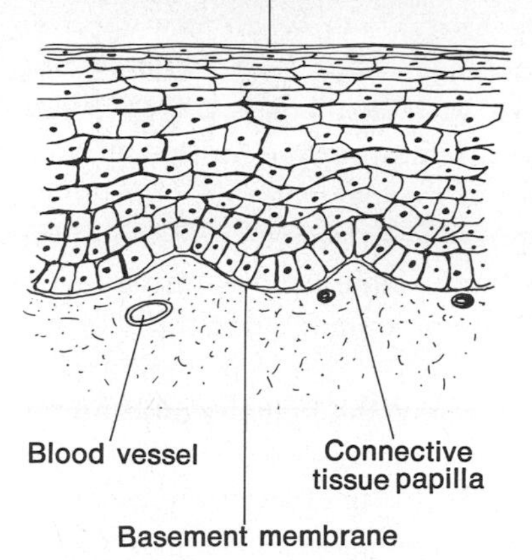

Fig. 2-1 A section of epithelial tissue. Cells are supported by an underlying basement membrane. The cells are attached to one another by "cement" and desmosomes. Note the free surface.

The entire human body—consisting of trillions of cells, hundreds of organs, and many systems—demonstrates only four basic *tissues*. The tissues are the fabric of which the body is woven, and the cells are its thread. The classification of tissues is based on both morphology (shape) and function. Within each classification, there are several distinguishable types. It is helpful to understand these tissue types in order that you can visualize m*a*croscopic structures m*i*croscopically and so that you can appreciate how the structure of some organ enhances or makes clear its function.

Generally, tissues are defined as collections of *like* cells, often attached to one another by fibers or intercellular "cement," and usually having a *common function* (Fig. 2-1). The four tissues are usually combined such that an organ is formed. Consider the stomach: It is a cavity with walls. These walls consist of the four basic tissues, all working together to perform the common function of protein digestion. The four tissues are as follows:

1. The lining of the cavity, consisting of a single layer of cells which secrete hydrochloric acid, mucus, and an enzyme—all of which aid in the chemical degradation of protein foodstuffs. Cells which *line* a cavity or any surface constitute **epithelial tissue (epithelium).** There are several varieties of epithelium.
2. The material which *supports* the epithelium and gives it a degree of elasticity. Because it also supports local blood vessels and nerves, as well as glands and certain sense receptors, this tissue is appropriately termed **connective tissue.** Connective tissue comes in a variety of textures, densities, and compositions.
3. The *contractile* matter which gives the stomach motility and provides for mechanical digestion. This is **muscle tissue** and, throughout the body, it comes in only three assortments.
4. The *excitable* tissue which provides the spark of stimulus for muscular contraction of the stomach. This is **nervous tissue,** and, working in networks complexly interconnected, it provides us with the ability to be not only sensitive to our environment but responsive to it as well.

As we consider organ after organ, system after system, region after region, keep in mind the reality that their complex activities are based on the interrelated functioning of four humble tissues.

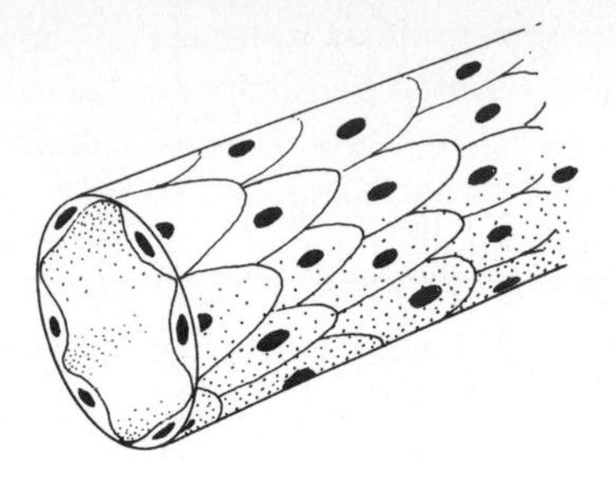

Fig. 2-2 Simple squamous epithelial lining (endothelium) of a blood vessel.

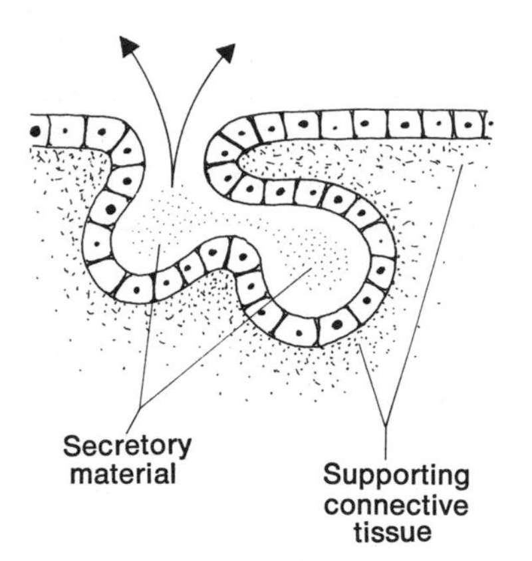

Fig. 2-3 Cross section of a gland whose duct opens into an epithelial surface.

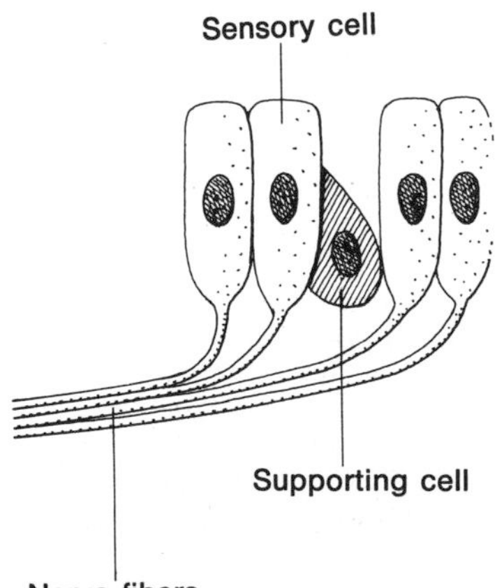

EPITHELIAL TISSUE

Epithelium is one or more layers of cells which form the free surface of the closed and open cavities of the body, including all tubes, ducts, and vessels. In fact, there is an unbroken epithelial layer passing through your body. The skin of your body is lined with epithelium. It extends without interruption in this sequence: skin → mouth → throat → esophagus → stomach → intestines → rectum → anal canal → anus → external skin. In a way, you're like a "human" doughnut! (Turn back to Fig. I-5.) There is continuity of epithelium within and to all parts of your body by virtue of the millions of interconnecting blood vessels. The kinds of epithelium found in all these areas may differ according to functional requirements.

First, what does epithelium mean—where does the term come from? It is derived from two Greek words: *epi* (upon) and *thelium* (nipple). The term was based on the appearance of the cellular surface of skin areas which exhibit nipplelike projections (papillae) of connective tissue directed into the overlying cellular layers (Fig. 2-1). Epithelium which lines the interior of blood vessels and the heart is specially titled—**endothelium** (Fig. 2-2), the epithelium making up the cellular lining of serous membranes (see page 9) is called **mesothelium.**

Epithelial tissue is generally characterized by the following:

1. Cells situated next to one another.
2. Little intercellular material.
3. A *free* (unattached) *surface.*
4. Cells supported by and bound to the underlying connective tissue by a *basement membrane.*
5. Absence of blood vessels (avascularity), nutrition being received by diffusion.

In the embryo, epithelial tissue may differentiate into one of three modified forms:

1. Lining (surface) epithelium (Fig. 2-2).
2. Glandular epithelium (Fig. 2-3).
3. Sense receptor epithelium (Fig. 2-4).

Classification of epithelium

Epithelial tissue is categorized according to the shape of its cells and the number of cellular layers.

Fig. 2-4 Specialized epithelial cells which are sensitive to mechanical stimulation and respond by initiating an electrical-like impulse.

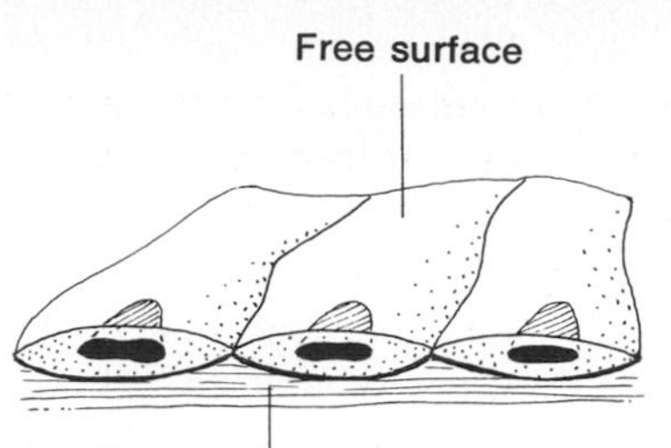

Fig. 2-5 Simple squamous epithelium.

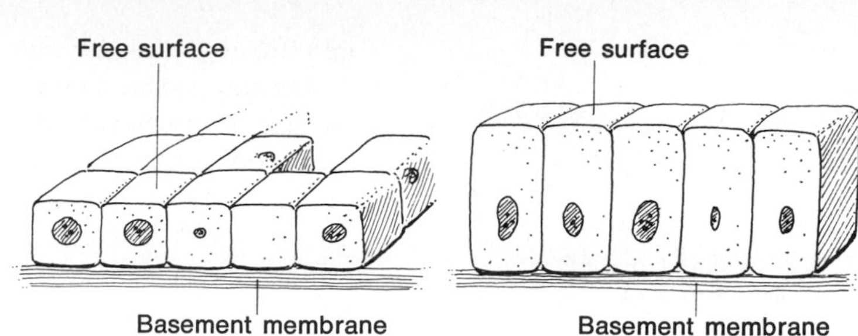

Fig. 2-6 Simple cuboidal epithelium.

Fig. 2-7 Simple columnar epithelium.

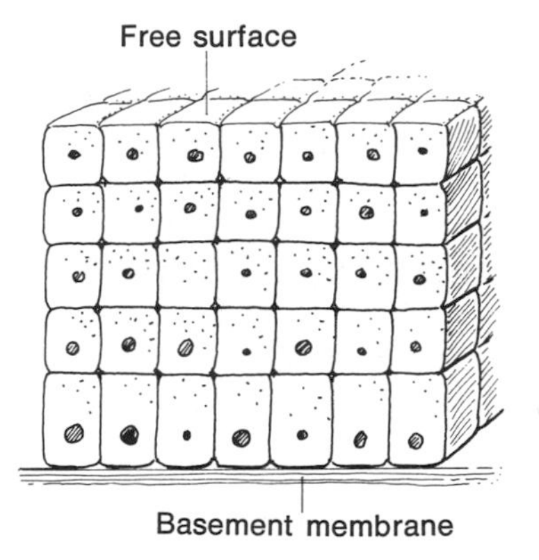

Fig. 2-8 Stratified cuboidal epithelium. Note basal layer of cells: they are columnar. The tissue type is based on appearance of cells at the free surface.

Refer now to Fig. 2-5: cells so shaped are called **squamous** (platelike). If the cells are in a single layer, the tissue is **simple squamous epithelium.** If the cells are in multiple layers, the tissue is **stratified squamous epithelium.** Is it not apparent that simple squamous epithelium has a completely different function from stratified squamous epithelium? Endothelium is simple squamous epithelium found in the cardiovascular system, as is mesothelium in serous membranes.

Now see Fig. 2-6: cells so shaped are called **cuboidal.** In single layers, a tissue like this is **simple cuboidal epithelium.** In multiple layers it is **stratified cuboidal epithelium.**

Now look at Fig. 2-7: cells of this shape are called **columnar.** In single layers the tissue is **simple,** in multiple layers, **stratified columnar epithelium.**

To determine a type of stratified epithelium, it is necessary to note the cells on the free surface layer, because generally the lowest (basal) layer of cells is columnar in all stratified epithelium but the surface layer is not.*

Now see Fig. 2-8. Since the cells on the free surface are cube-shaped, this tissue is classified **stratified cuboidal epithelium.** But look at Fig. 2-9. How would you classify this?†

Now, there are two epithelial tissues which do not fit into the descriptive categories defined above. See Fig. 2-10: Looking at the nuclei of the cells, you may think this is stratified

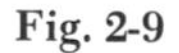
Fig. 2-9

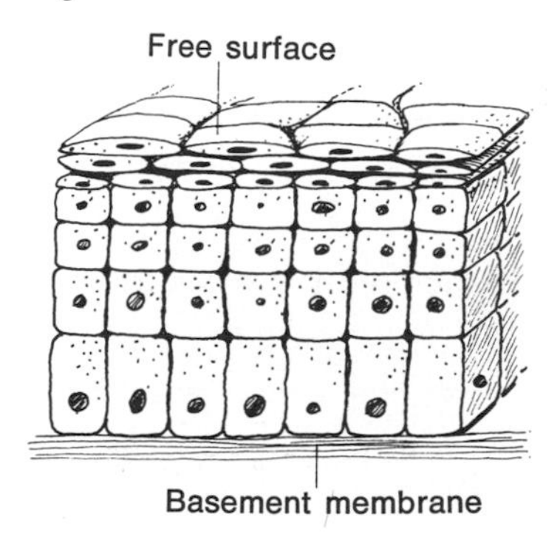

*Sometimes it is very difficult to determine an epithelial classification. Keep in mind: Within the body there is a spectrum of epithelial shapes from squamous to columnar, and in some cases it is not easy to differentiate squamous from cuboidal, cuboidal from columnar, and so on. Remember that the classifications were developed for general descriptive purposes. Don't be rigid in your eagerness to categorize. Some epithelial tissues just do not fit neatly into a specific category but lie between two classifications.

† Stratified *squamous* epithelium.

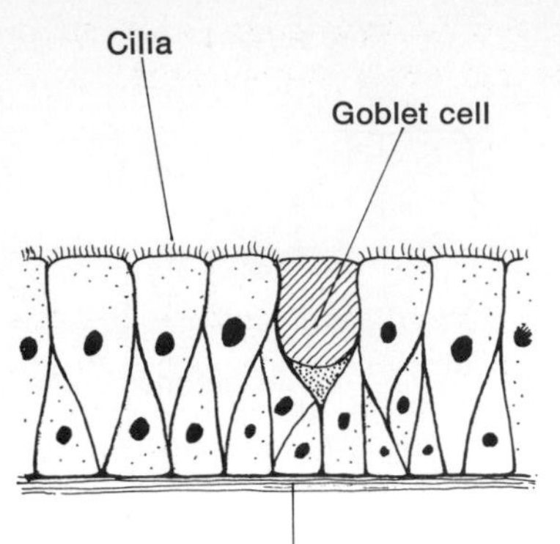

Fig. 2-10 Pseudostratified columnar epithelium with cilia and goblet cells.

epithelium. However, the bases of *all* these cells contact the basement membrane, while not all the cells reach the free surface of the tissue. Thus, this tissue is properly **pseudostratified** columnar epithelium. In the illustration, the cells are ciliated and one mucus-secreting (goblet) cell is shown.

See Fig. 2-11: The stratified epithelium shown here lines the urinary bladder and is said to be contracted, such as in a partially filled or empty bladder. It appears to be a stratified cuboidal type. However, look at Fig. 2-12: Here the same epithelium is stretched (dilated) and looks like a stratified squamous type, seen in a distended, full bladder. To avoid confusion, this epithelium of the urinary tract is called **transitional** or **uro-epithelium.**

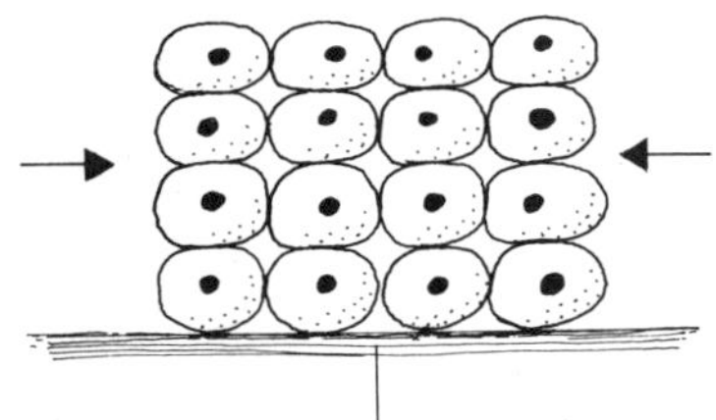

Fig. 2-11 Transitional (uro-) epithelium of the urinary bladder and ureters. Shown here in a contracted state (empty bladder).

QUIZ

Let's check your comprehension of the significant facts brought to light in this unit so far:

1. Tissues are usually defined as being collections of ___like___ cells and having a ___common___ ___function___.
2. That tissue which ___lines___ a cavity constitutes ___epithelial tissue___
3. The tissue which has a ___supportive___ function is called connective tissue.
4. Muscle tissue is characterized by being ___contractile___.
5. Nervous tissue is characterized as being ___excitable___.
6. In general all epithelial tissue is characterized by having a ___free___ ___surface___, supported by a ___basement membrane___ and being ___unattached___.
7. Epithelial cells are categorized according to ___shape___ and number of ___cellular layers___.
8. Platelike epithelial cells are termed ___squamous___ and prefixed with the term ___simple___ (one layer) or ___stratified___ (multiple layers).
9. The other shapes of cells found in either single or multiple layers are ___cuboidal___ and ___columnar___.
10. Two unorthodox epithelial tissues are the tissue that appears to be stratified but isn't (___pseudostratified___ epithelium) and the tissue that changes character depending on volume variations (___Uro-epithelium___ epithelium).

Fig. 2-12 Transitional epithelium shown in a dilated or stretched state. Compare with Fig. 2-11.

Grade yourself on this quiz and be your own judge as to whether you should review the material just presented.

Functional activity

Aside from normal intracellular tasks, epithelial cells are capable of certain specialized activities which contribute to the function of the particular organ of which they are a part. In many cases, the function of the epithelial cell reflects the shape of the cell. For instance, you would not expect that multiple layers of epithelial cells would be found lining a vessel whose walls are constantly traversed by molecules of CO_2 and oxygen, as in the air sacs of the lungs, where quick gaseous exchange is essential to life. And, on the other hand, you would not expect the body skin to be but one cell thick, and of course it is not. The thinner the lining, the more improbable it is that protection of a wall or organ is one of its functions. Some of the functions carried out by epithelial tissue include:

1. Protection against abrasion, ultraviolet light, etc.
2. Filtration or diffusion.
3. Absorption of material into a cell.
4. Secretion of material to the outside of the cell.
5. Sweeping material away by ciliary action.
6. Flexibility, ability to withstand variable pressures and volumes.
7. Reception of sensory stimuli, such as touch and pressure.

Consider the function of protection. What surfaces of the body require protection against abrasion, "weathering," wear-and-tear forces, and the like? Certainly the skin, for starters. And what is the epithelium lining the skin? Stratified squamous epithelium (Fig. 2-13A). Other surfaces requiring the services of this kind of epithelium include the mouth, anal canal, and vagina.

Simple squamous epithelium, on the other hand, cannot handle any trauma at all, and lines surfaces through which substances of molecular dimensions shuttle, as in filtration. It is found in such places as blood vessels and lymph vessels (endothelium), the filtering capsules and certain tubules of the kidney, and lining of closed cavities (mesothelium) (Fig. 2-13B).

Cuboidal and columnar epithelial tissue is generally single-layered and is involved in the secretion and absorption of

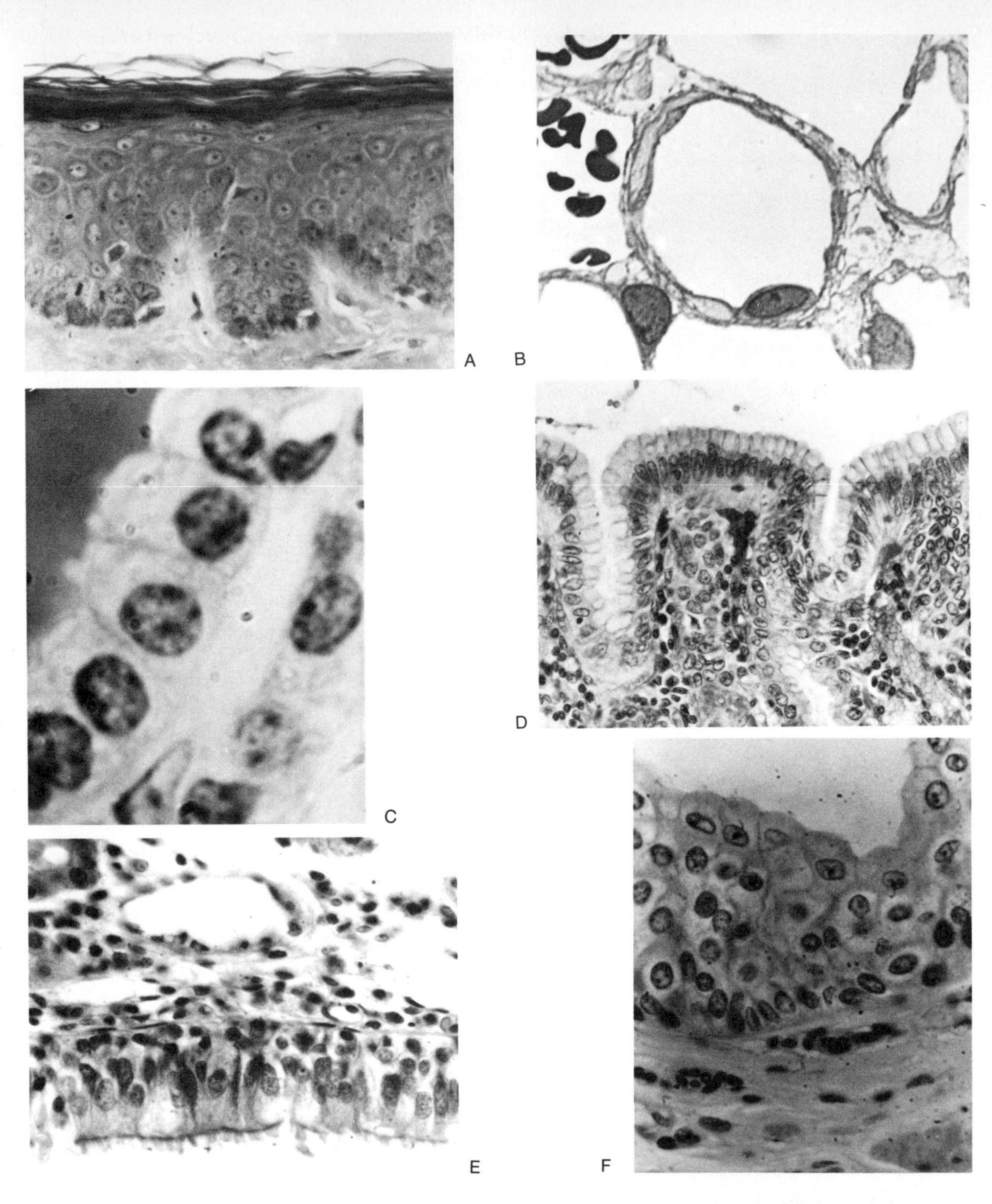

Fig. 2-13 Some types of epithelia. A. Stratified squamous epithelium, esophagus. B. Simple squamous epithelium, capillary. C. Simple cuboidal epithelium, thyroid gland. D. Simple columnar epithelium, intestine. E. Pseudostratified columnar epithelium with cilia and goblet cells, trachea. F. Transitional epithelium (contracted), urinary bladder. (A, C–F, courtesy of James Runner. B, courtesy of Dr. C. Bordelon.)

substances, as seen in the lining of the gastrointestinal tract, glands, and other places (Fig. 2-13C and D).

The respiratory structures (trachea and related tubes, nasal cavity, etc.) require a multifunctional epithelium—one which will secrete mucus to trap foreign particles and has motile "hairs" on its free surface to sweep the particles away to an orifice. The obvious candidate for such a lining is pseudostratified columnar epithelium with cilia and (secretory) goblet cells (Fig. 2-13E).

Certain parts of the kidney, the ureters (kidney-to-bladder tubes), and the urinary bladder are subject to stretching due to variations in volume of urine. The epithelium best adapted for this kind of stress is transitional epithelium (Fig. 2-13F).

Now although not every surface of every cavity or tube in the body was discussed, you should have a generalized understanding of how surface lining function can be related to one of various types of epithelium.

QUIZ

Test yourself now, by matching epithelium type with function below (more than one answer may apply*):

4,7 Simple squamous
1 Simple cuboidal
1 Simple columnar
2,5 Stratified squamous
3 Pseudostratified columnar with cilia and goblet cells
6 Transitional

1. Secretion and/or absorption
2. Lines external surface of body, mouth, etc.
3. Lines respiratory tract
4. Lines blood vessels
5. Protection against wear and tear
6. Subject to pressure and volume variations
7. Lines surfaces involved in filtration, diffusion, etc.

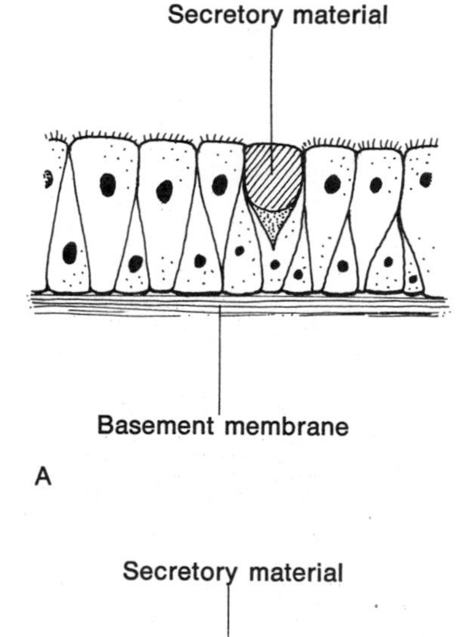

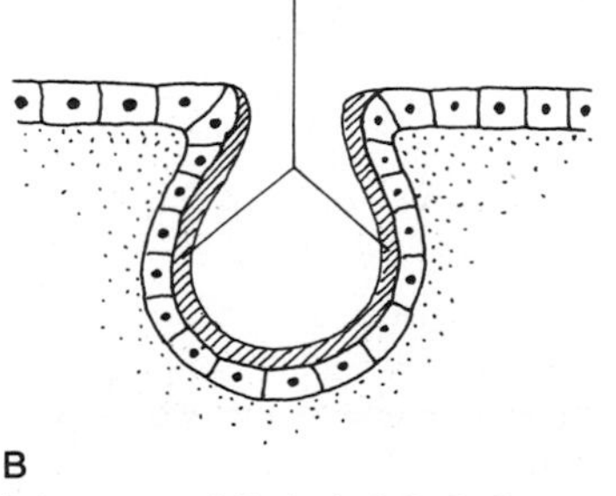

* Answers: 4,7; 1; 1; 2,5; 3; 6.

Glands

Epithelial cells and tissue having a major function of secretion are called **glands.** Glands may be composed of one cell (unicellular) or many cells (multicellular) (Fig. 2-14). They may secrete their product via a duct or ducts onto a surface (exocrine glands) or directly into capillaries (endocrine glands)

Fig. 2-14 A. Unicellular gland (goblet) cell in pseudostratified columnar epithelium. B. A multicellular gland opening onto an epithelial layer.

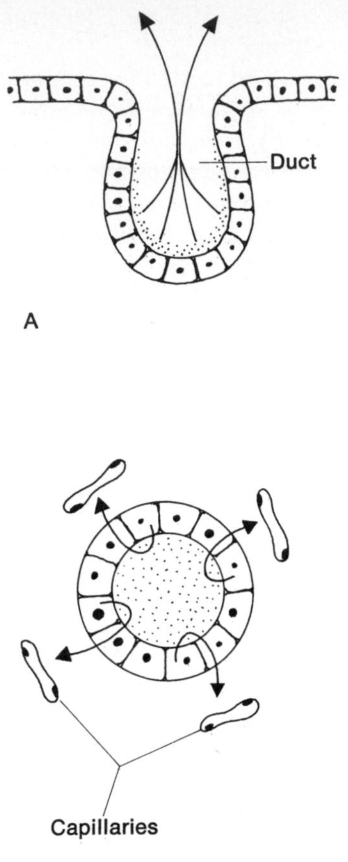

Fig. 2-15 A. Section of an exocrine gland. B. Cross section of an endocrine gland.

(Fig. 2-15). Multicellular exocrine and endocrine glands develop from in-dippings (invaginations) of surface epithelium. However, while the exocrine glands retain their connection with the surface epithelium, the endocrine glands lose theirs (Fig. 2-16).

Multicellular exocrine glands make up the majority of glands found throughout the skin and mucous membranes. They are classified according to the number of ducts (simple or compound) and the shape of the secretory units (tubular, coiled, or alveolar) (Fig. 2-17). Some will be discussed in Unit 3.

Endocrine glands (pituitary, thyroid, adrenal, etc.) will be discussed in the context of the region where they are found. A synopsis of the glands of the endocrine system may be studied in the Synopsis of Systems in the appendix.

CONNECTIVE TISSUE

In the overall sense, connective tissue may be defined as a variety of cellular aggregations and their secretory products which generally function to *bind, support,* or *link* other structures in the body (Fig. 2-18). Put more simply, connective tissue connects. All organs are bound in connective tissue to one degree or another. The body is supported by this tissue. Furthermore, it occupies more space in the body than any other tissue.

The classification of connective tissue is based on the relative density and character of the extracellular material in the tissue. You will remember that the body is composed of three basic elements: cells, fibers, and fluid. That connective tissue which consists of cells and fibers in variable ratios in a fluid or gel substance is called **connective tissue proper.** Connective tissue which consists of cells and fibers embedded in a solid substance is called **supporting tissue** (cartilage and bone). **Blood** and **lymph** are connective tissues in which the extracellular material is liquid and which contain a variety of cells and compounds.

Fig. 2-16 (opposite, top) A and B. Development of an exocrine gland. A to D. Development of an endocrine gland.

Fig. 2-17 (opposite, bottom) Some types of glands. A. Simple tubular. B. Branched tubular. C. Simple coiled. D. Branched alveolar. E. Compound tubuloalveolar.

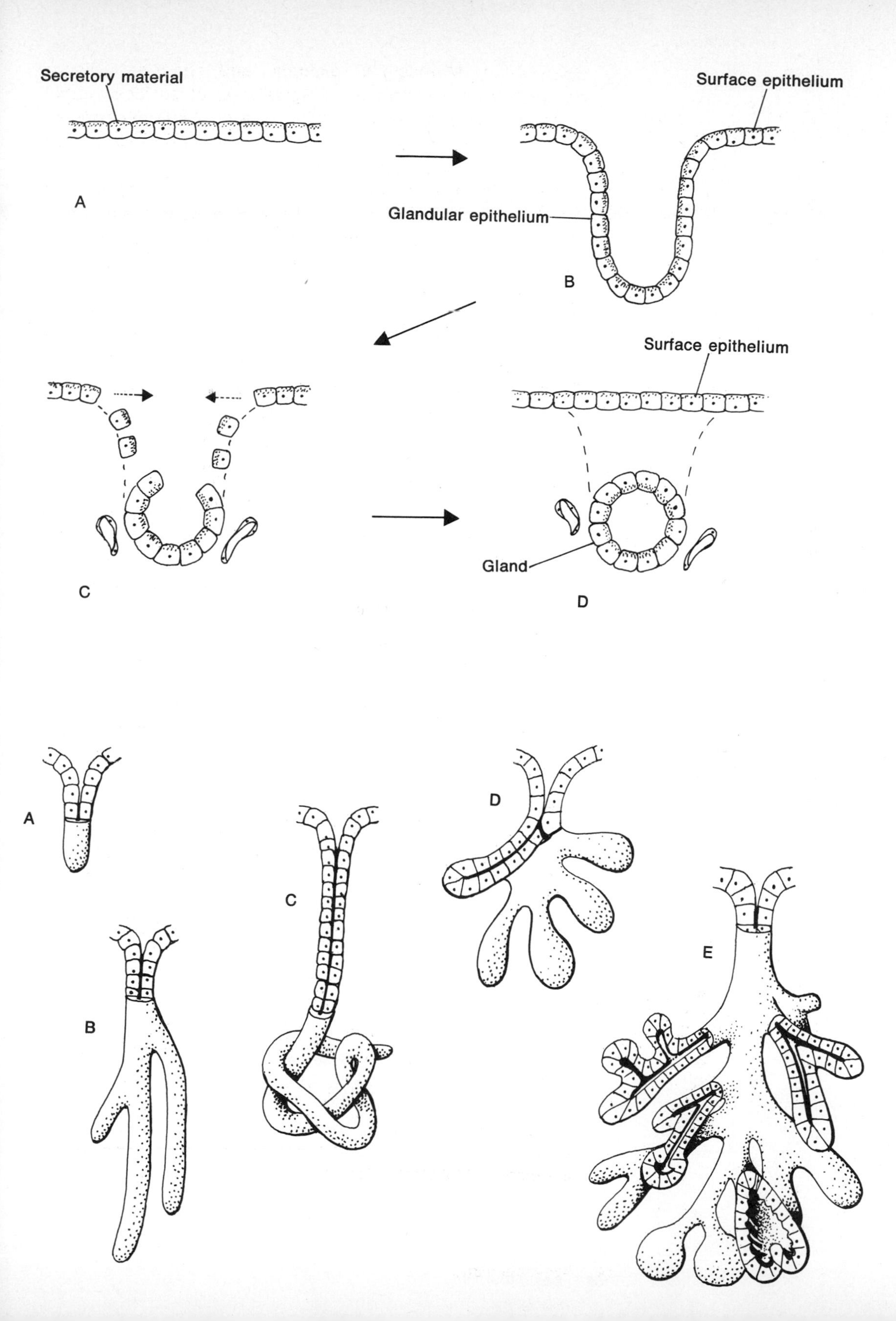

Secretory material
Surface epithelium
A
Glandular epithelium
B
Surface epithelium
C
Gland
D
A
B
C
D
E

Fig. 2-18 Some of the varieties of connective and supporting tissue as seen in the knee joint.

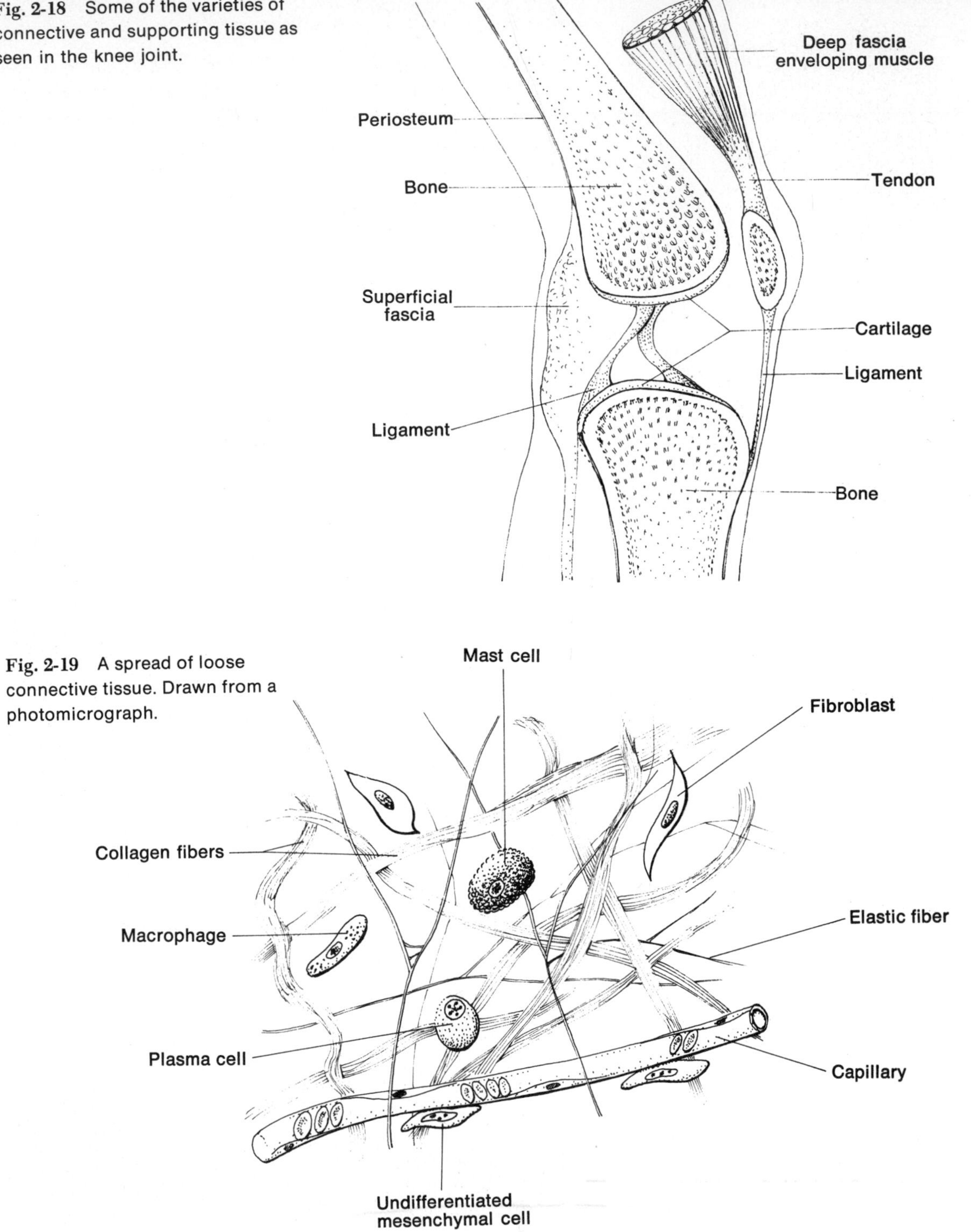

Fig. 2-19 A spread of loose connective tissue. Drawn from a photomicrograph.

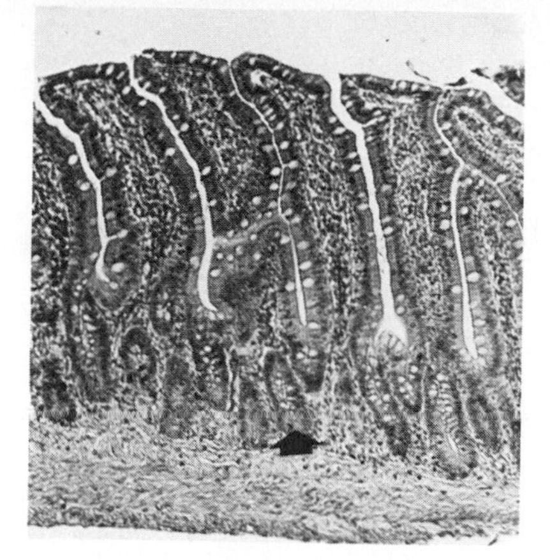

F. 2-20 The mucous membrane of the stomach. The cell layer at the free surface is epithelium, with lamina propria (loose connective tissue) beneath it. The head of the arrow touches a gland; the tail lies in muscularis mucosa. Below that is submucosa (dense irregular connective tissue). (Courtesy of Dr. C. Bordelon.)

Connective tissue proper

Connective tissue (c.t.) proper comes in several varieties:

1. **Loose** c.t. (Fig. 2-19) is characterized by having many cells and fewer fibers. This most common of all connective tissues fills any and all unoccupied spaces in the body, binds all skeletal muscle, and makes up the connective tissue layers in mucous membranes, as well as the subcutaneous tissue "holding down" the skin.
2. **Dense** c.t. characteristically has many fibers. The chief functional feature here is strength. The fibers may be arranged in coarse, interwoven bundles, as in dense **irregular** c.t. (Fig. 2-20) or arranged neatly in parallel, as in dense **regular** c.t. (Fig. 2-21). Dense irregular c.t. is significantly stronger than loose c.t. and is found encapsulating many organs of the body. As periosteum it covers bone; as perichondrium it covers cartilage (Fig. 2-18). Dense regular c.t. is particularly suited for tissues which experience high tensile (pulling) forces, such as

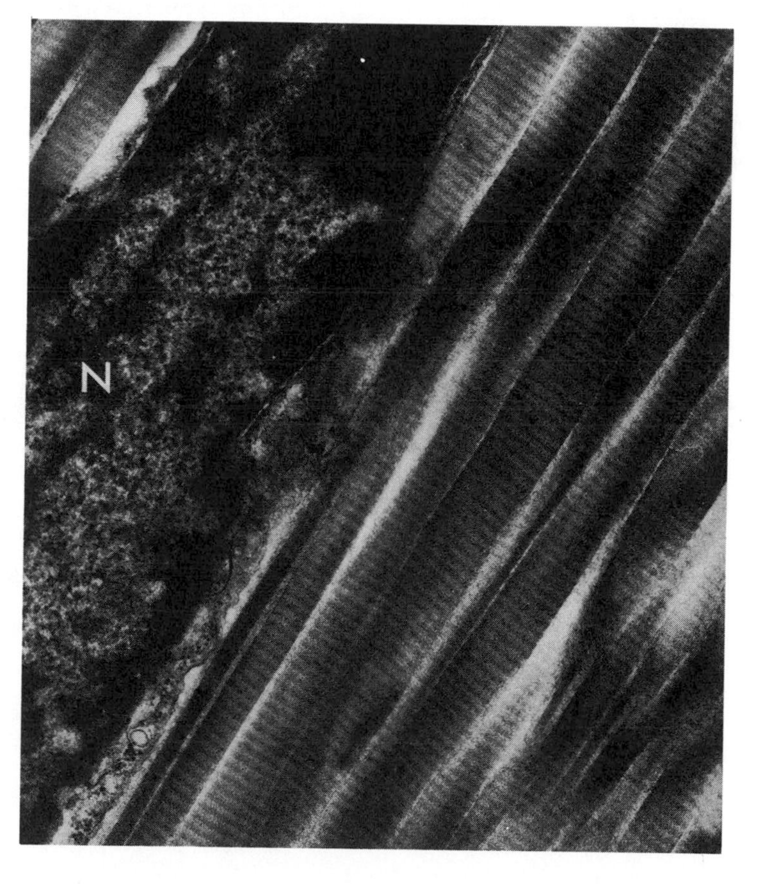

Fig. 2-21 Ultrastructure of a tendon. Note the parallel arrangement of collagen fibrils passing by the nucleus (N). Electron micrograph. ×17,600. (From R. O. Greep, *Histology*, 2d ed., McGraw-Hill, New York, 1966.)

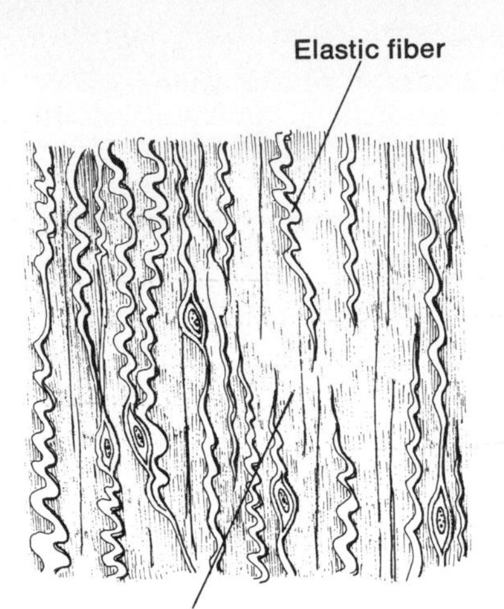

Fig. 2-22 Elastic tissue. This section shows broad, straight collagen fibers and unstretched elastic fibers. Drawn from a photomicrograph.

ligaments (which connect bone to bone) and tendons (which connect muscle to bone or muscle to muscle).

3. **Elastic** c.t. (Fig. 2-22) is found in the walls of large arteries and in certain ligaments. Elastic c.t. offers a degree of resiliency not found in other tissues
4. **Adipose** c.t. (Fig. 2-23) is composed of large spherical cells whose cytoplasm is filled with fat. Thus it is popularly called fatty tissue. The cells are often pressed together and the nuclei of the cells are pushed off to one side. The cytoplasmic organelles occupy a thin peripheral rim of cytoplasm. The cells are supported by reticular fibers (see later). Fatty tissue is active metabolically; its ability to synthesize fat from carbohydrate is influenced by hormones. Fatty tissue is fairly rich in its supply of nerves and blood vessels.
5. **Reticular** c.t. consists of delicate fibers and cells which form a supportive framework for free and fixed cells composing such structures as lymphatic organs, bone marrow, and liver.
6. **Embryonic** c.t.: the connective tissue found in the embryo. One kind of this tissue, mesenchyme, includes cells which have the capacity to change into one of several more specialized connective tissue cell types, i.e., chondroblasts, osteoblasts, or fibroblasts.

A variety of cells may be found in the connective tissues, the two most common of which are fibroblasts and macrophages. **Fibroblasts** (Fig. 2-24) are the source of all connec-

Fig. 2-23 Adipose tissue. Fat droplets occupied cell interior but were washed out in processing. Drawn from a photomicrograph.

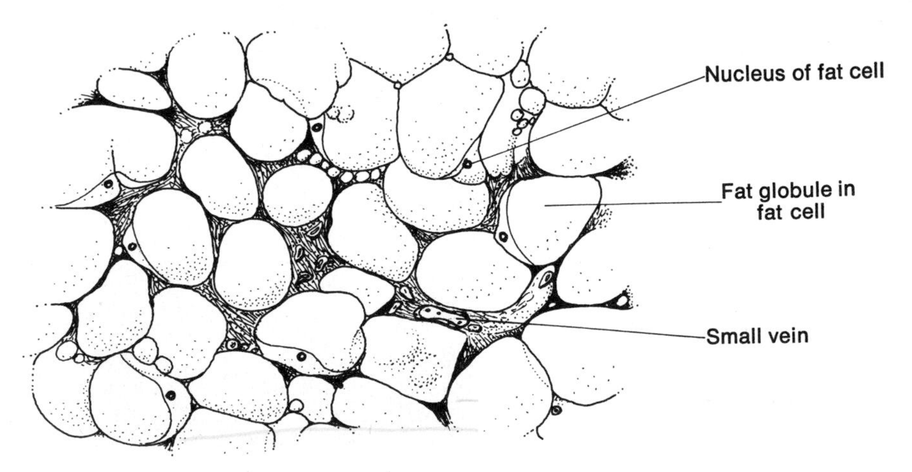

Fig. 2-24 Isolated fibroblasts. Drawn from a section of superficial fascia.

tive tissue fibers and are found in all the fibrous connective tissues from loose areolar to dense. They are especially important in secreting collagenous fibers at an inflammation site, the result being to wall off the inflammation.

Macrophages (Fig. 2-25) are fixed or wandering cells of the connective tissues which "eat up" injured cells or foreign material in a process called **phagocytosis.** They are often difficult to identify unless filled with ingested material.

Other cells of the connective tissues include certain white blood cells, fat cells, antibody-producing cells (plasma cells), and heparin- and histamine-secreting cells (mast cells) (Fig. 2-19).

The fibers of connective tissues are of three assortments:

1. **Collagenous** (white) (Fig. 2-26): present in all connective tissue including bone, these fibers, made of the protein collagen, have the tensile strength of steel.
2. **Elastic** (Fig. 2-27): present in most connective tissues to one degree or another; imparts a yellow color if present in large concentrations. Elastic fibers make the principally collagenous tissues elastic and flexible.
3. **Reticular** (Fig. 2-28): of small, delicate proportions, forming networks for support of cells.

These varieties of connective tissues are all immersed in a "soup" called **ground substance** or **matrix.** The organic compounds which make up this substance are primarily protein-

Fig. 2-25 Isolated macrophages with ingested particles. Drawn from a section of superficial fascia.

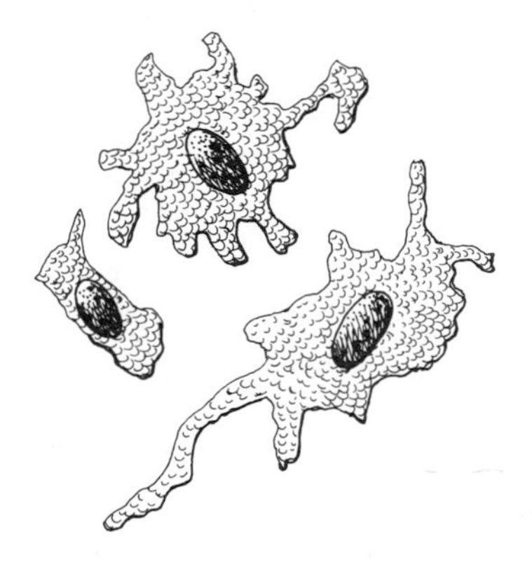

Fig. 2-26 Collagen fibers. Photomicrograph. ×450. (From R. O. Greep, *Histology*, 2d ed., McGraw-Hill, New York, 1966.)

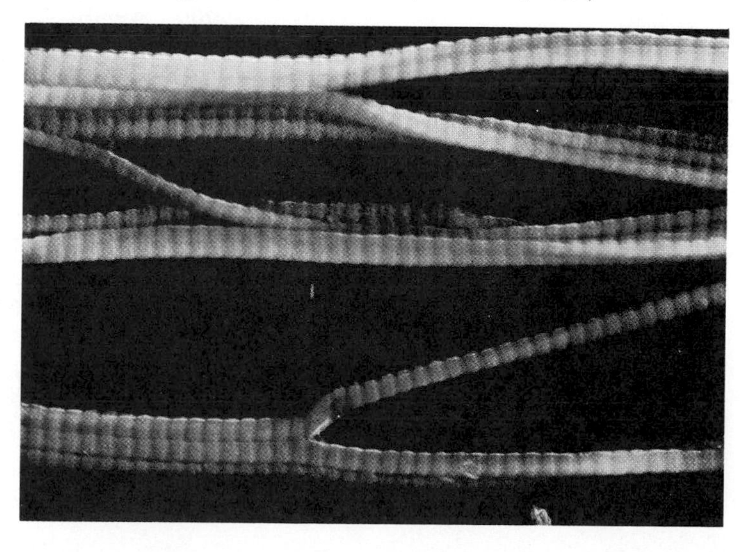

Fig. 2-27 Elastic fibers in a spread of mesentery. Photomicrograph. ×450. (From R. O. Greep, *Histology*, 2d ed., McGraw-Hill, New York, 1966.)

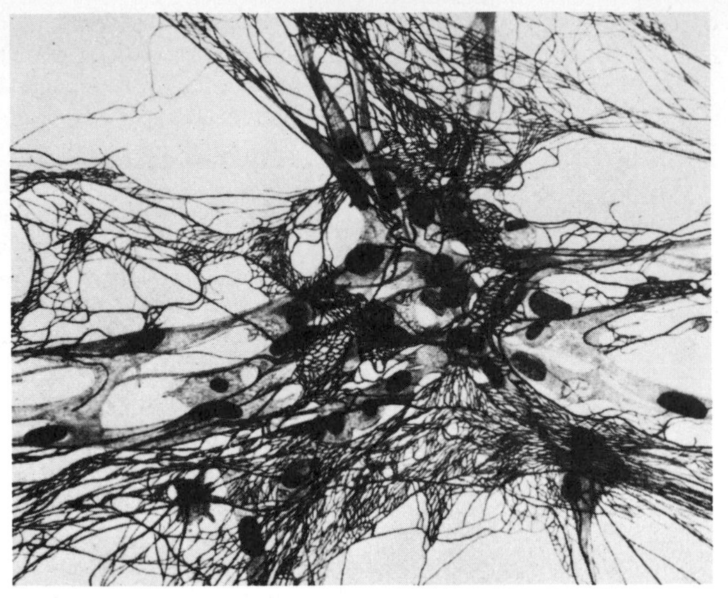

Fig. 2-28 Reticular fibers in a tissue culture of thymus gland. About ×900. (From W. F. Windle, *Textbook of Histology*, 4th ed., McGraw-Hill, New York, 1969.)

sugar complexes (mucopolysaccharides), and the consistency of the material varies from a viscous to a jellylike state.

Supporting tissue

This connective tissue occurs as **bone** and **cartilage** and is characterized by a dense, hard matrix, which gives it the ability to support weight. Cartilage, however, is less hard than bone because it lacks the minerals (calcium crystals) of bone. Cartilage is a little like a cement mix without the cement. Cartilage consists of cells (chondrocytes) in small cavities in an organic matrix that is supported by collagenous fibers, much as the concrete of road pavements is supported by steel rods.

Cartilage grows faster than bone but slower than connective tissue proper. Cartilage, while hard, is somewhat flexible and is responsible for supporting the embryo during its development. Cartilage makes up the skeleton of the embryo and is progressively replaced by bone beginning in the fetus and ending in the young adult, in whom the remaining skeletal cartilage occupies only the ends of long bones (articular surfaces) in the skeleton. There are three types of cartilage: (1) hyaline, (2) elastic, and (3) fibrous.

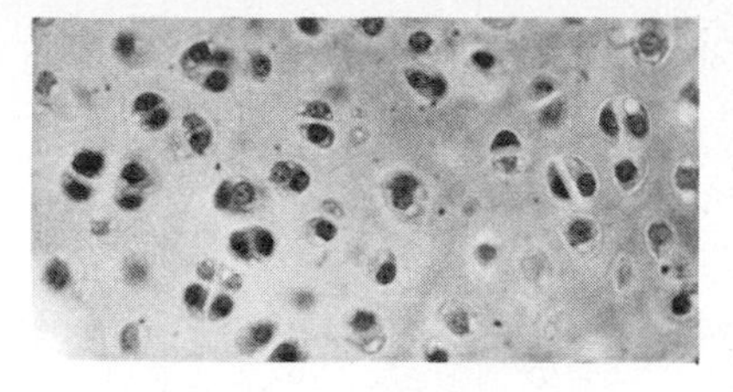

Fig. 2-29 Hyaline cartilage. Note the clear matrix and the nests of chondrocytes. Photomicrograph. (Courtesy of James Runner.)

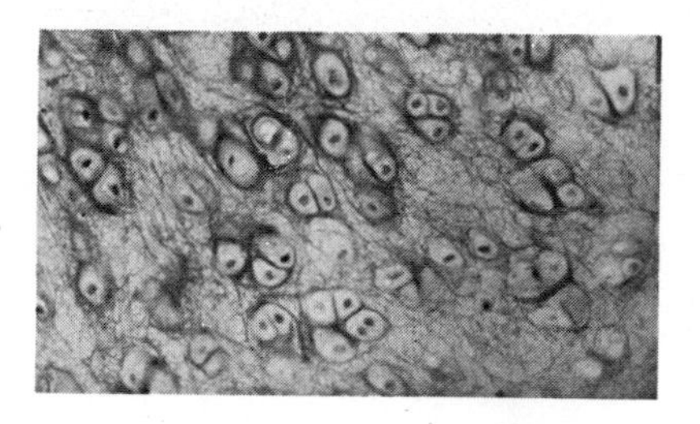

Fig. 2-30 Elastic cartilage. Note the elastic fibers intertwined between nests of cells. Photomicrograph. (Courtesy of James Runner.)

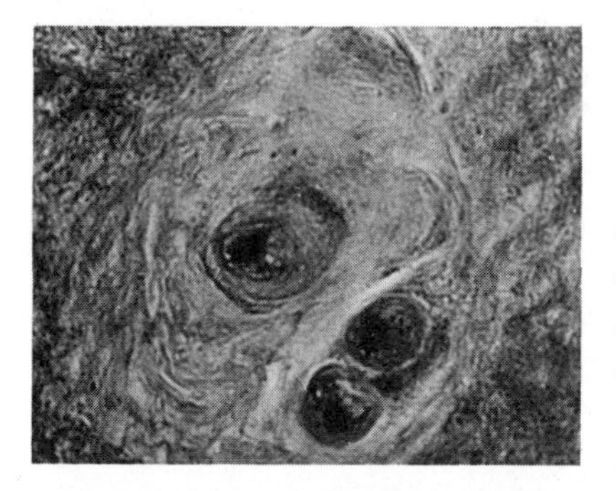

Fig. 2-31 Fibrocartilage. Note dense masses of collagen fibers and sparse nests of chondrocytes. Photomicrograph. (Courtesy of James Runner.)

Hyaline cartilage (Fig. 2-29) has a glassy appearance with a blue tint. When cut, it chips like ice. It is found at the end of bone (articular surfaces); in the nose, the trachea, and parts of the larynx; and it composes the cartilaginous portions of the ribs. You can feel hyaline cartilage on yourself: bend your nose back and forth—the supporting framework of this proboscis is hyaline cartilage. Now run your fingers over the cartilages of the "adam's apple" and other parts of the larynx and the trachea below that. Compare the density of these cartilages with the bones of your hand.

The transition zone between hyaline cartilage and its loose c.t. binding is a dense fibrous tissue called **perichondrium.**

Elastic cartilage (Fig. 2-30) is a hyaline cartilage infiltrated with elastic fibers, thus rendering the tissue flexible and resilient. Elastic cartilage can be explored on yourself: twist and bend your external ear. Note its supporting framework. This is elastic cartilage. Compare it with the hyaline cartilage of the larynx or nose and with the bones of your hand. Elastic cartilage can also be found supporting the external auditory canal, the auditory tube, and some of the smaller cartilages of the larynx.

Fibrous cartilage or **fibrocartilage** (Fig. 2-31) is really dense regular c.t. with cartilage cells (chondrocytes) interspersed. Frequently, it seems that fibrocartilage is a transition tissue between hyaline cartilage and the surrounding dense connective tissue (regular or irregular). Fibrocartilage is found between adjacent vertebral bodies of the spinal column, between the pubic bones of the hip, and in most transitional tissues between joint capsules, ligaments, tendons, and articular cartilage.

Cartilage lacks blood vessels and nerve fibers. The chondroblasts and chondrocytes synthesize and secrete the fibers and matrix of cartilage.

Cartilage develops from chondroblasts which arise from mesenchyme. When cartilage is damaged, the injured portions are "cleared out" by macrophages and replaced with faster growing loose connective tissue. The "fibroblasts" later transform into chondroblasts which secrete cartilaginous matrix and fibers.

Cartilage is not inert—it is actively metabolic and comes under hormonal influences. Nutritional deficiencies can cause alterations in cartilage which have serious life-impairing consequences.

Bone is the densest and hardest of all the connective tissues, in fact, of all tissues of the body except teeth. Bone is composed of:

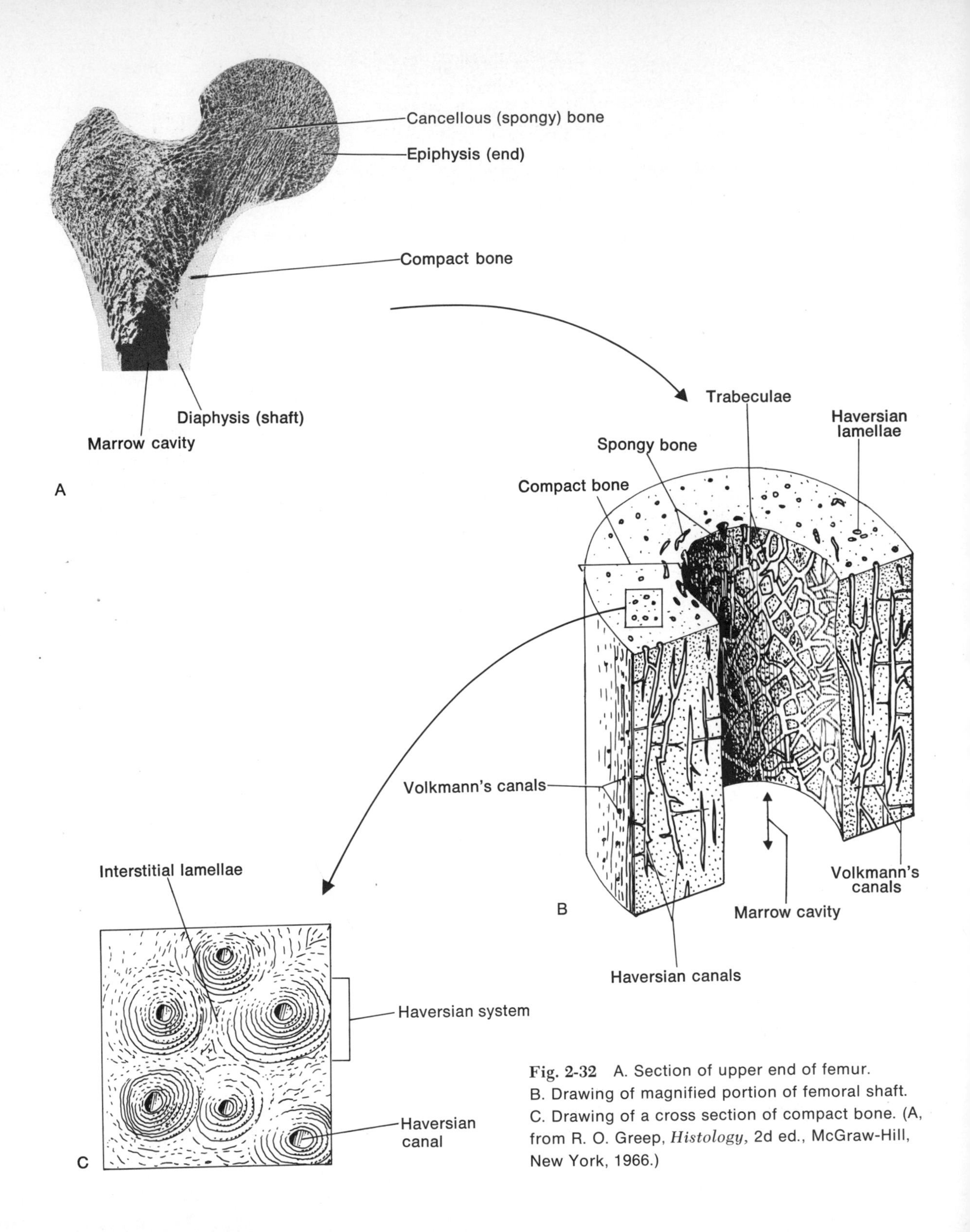

Fig. 2-32 A. Section of upper end of femur. B. Drawing of magnified portion of femoral shaft. C. Drawing of a cross section of compact bone. (A, from R. O. Greep, *Histology*, 2d ed., McGraw-Hill, New York, 1966.)

1. **Osteocytes** (bone cells).
2. **Osteoblasts** (developing bone cells).
3. **Osteoclasts** (bone-destroying cells).
4. Organic ground substance (collagenous fibers, mucopolysaccharides).

} 35 percent

5. Inorganic (mineral) substance (calcium and phosphate complexes).

} 65 percent

Based on structural differences, there are two types of bone (Fig. 2-32):

1. **Compact** (dense).
2. **Cancellous** (spongy).

These two types are essentially the same in constitution; the difference lies in the organization of the tissue. In compact bone, the tissue is densely packed into circular or cylindrical structures called **Haversian systems** (Fig. 2-32C). In cancellous bone, the tissue is arranged into thin spicules (trabeculae) with spaces (cavities between and among them) (Fig. 2-32 A and B). Haversian systems are absent in cancellous bone. Bone carries out a number of important structural and physiological activities, to wit:

1. It serves as an internal skeleton for support of the body.
2. It is the site of attachment for muscles, tendons, and ligaments.
3. Bone offers protection of cranial, thoracic, and pelvic viscera.
4. It is a center of blood-forming (hemopoietic) activity.
5. Bone is a source of calcium needed by the body.

Bone development

The functional relationships of the cells and other structures associated with bone can best be appreciated in a consideration of the development of bone—a process taking 20 to 25 years in a human being. Bone development (osteogenesis) begins during embryonic growth. Bone may develop directly from mesenchyme (a method called **intramembranous ossification**) or it may develop by replacing the cartilage models described earlier (termed **endochondral ossification**). All the bones of the body develop by the latter process with the exception of the flat bones of the skull and the clavicle.

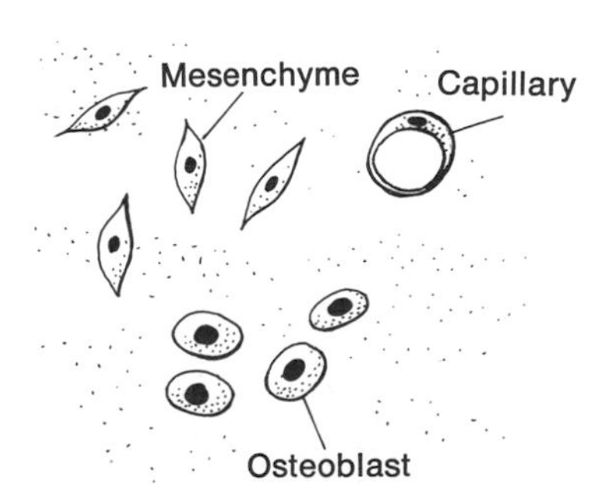

Fig. 2-33

Process of intramembranous ossification

Figure 2-33: Osteoblasts differentiate from mesenchyme. Note the capillary cut in cross section.

Figures 2-34 and 2-35: Osteoblasts form clusters and start secreting unmineralized bone (osteoid tissue). Some osteoblasts have become trapped in their own secretions. These cells are now osteocytes and reside in spaces called lacunae. The surrounding mesenchyme will differentiate into more osteoblasts or periosteum.

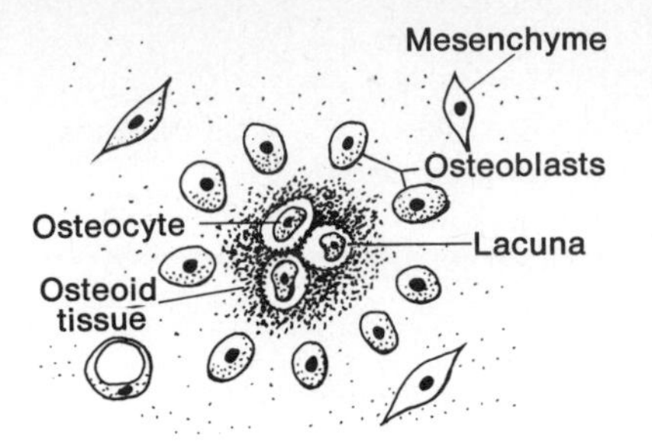

Fig. 2-34

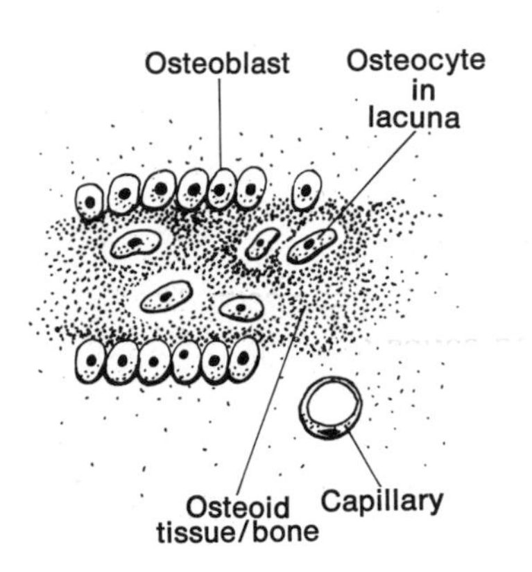

Fig. 2-35

Figure 2-36: Periosteum has formed on both sides of developing flat bone. The major tissue here is now bone—the osteoid tissue is becoming mineralized. Note the osteocytes in their lacunae. Large spaces at the sides and in the center are marrow cavities containing blood vessels—arterioles and veins—osteoblasts, and osteoclasts.

Intramembranous bone, even if fully developed, is resorbed by osteoclasts and replaced by osteoblasts throughout life. The osteoclasts and osteoblasts are probably derived from undifferentiated mesenchymal cells (UMC) and fibroblasts of the inner layer of periosteum. The material broken down by the osteoclasts is absorbed into the blood—calcium levels of the blood are maintained in this way. The marrow cavities—once occupied by mesenchyme—are filled with hemopoietic tissue and blood vessels.

Process of endochondral ossification

In the early stages of embryonic development, skeletons of cartilage are formed from mesenchyme, and these cartilaginous models have the basic shape of the adult bones

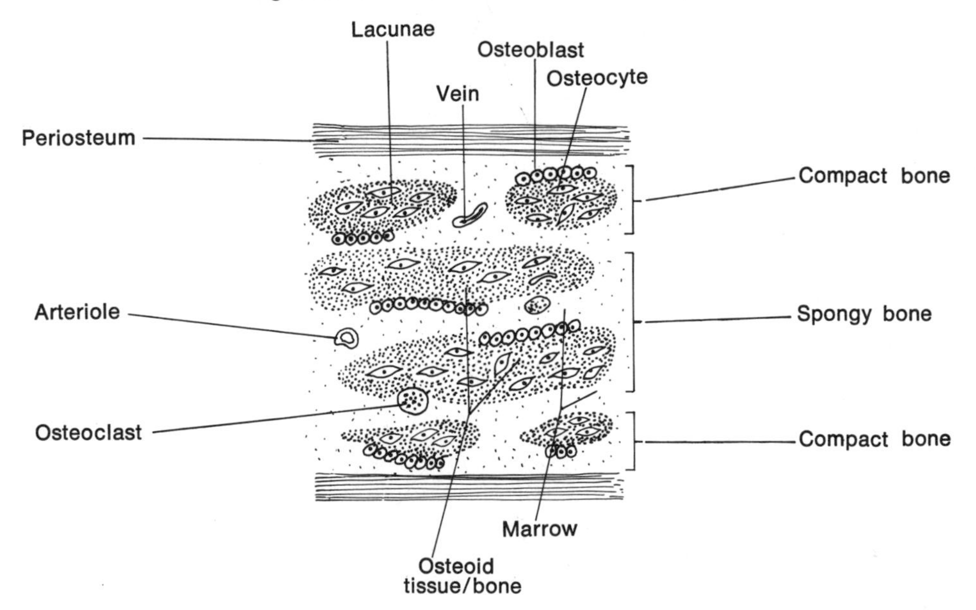

Fig. 2-36

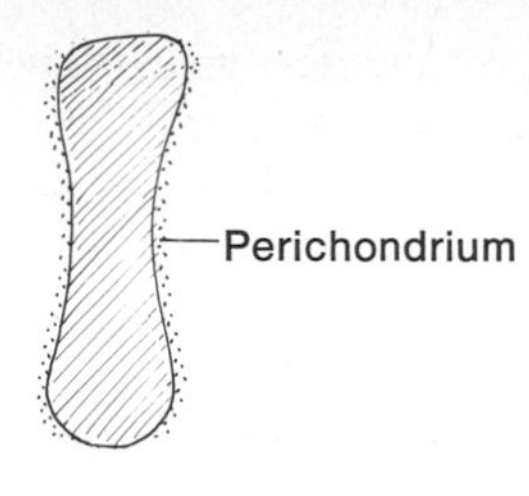

Fig. 2-37

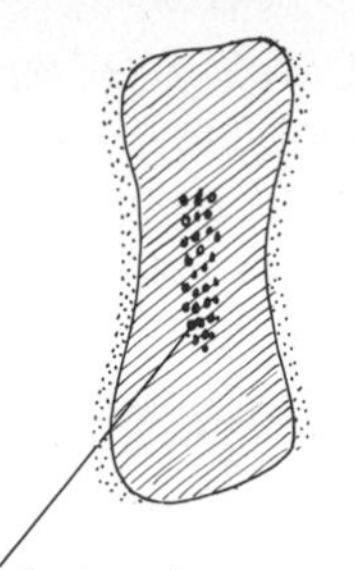

Fig. 2-38

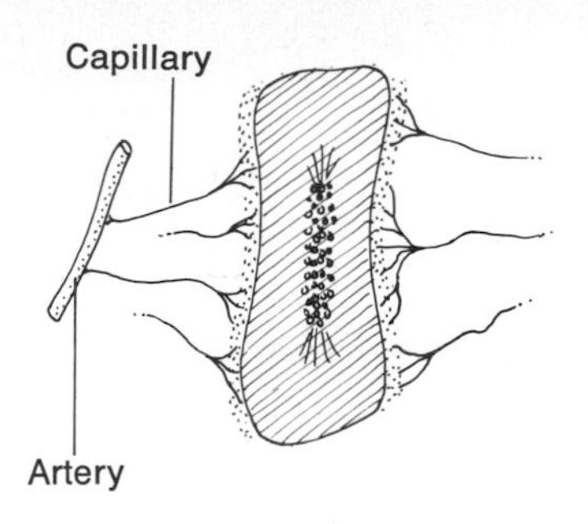

Fig. 2-39

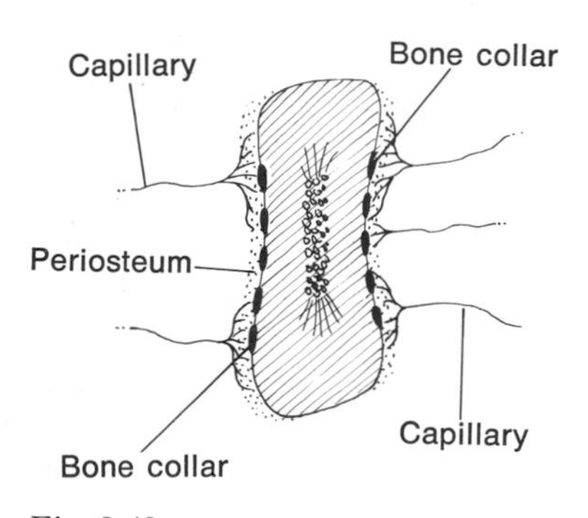

Fig. 2-40

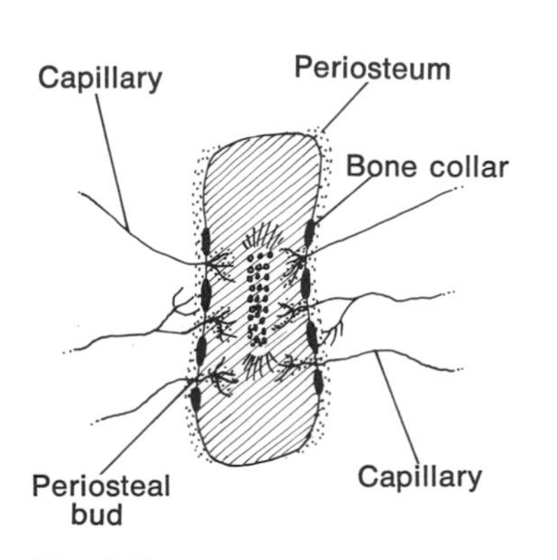

Fig. 2-41

which later develop. Bone cells arise in a separate line of development from mesenchyme cells; therefore in endochondral ossification bone cells must and will replace the cartilage cells. Once mesenchymal cells become cartilage cells they lose the ability to change into bone cells. The bone to be developed in the following series of illustrations might be the bone of the arm—the humerus.

Figure 2-37: The perichondrium is the source of chondroblasts (transformed from mesenchymal cells) which add to the growing cartilage by apposition (appositional growth).

Figure 2-38: The chondrocytes within the model mature and enlarge. As they do, the matrix thins out and the chondrocytes lose much of their source of nutrition, particularly in the center of the shaft. They enlarge. And die. The thin spicules of the matrix begin to absorb calcium salts (calcify). Calcification of cartilage is, in a sense, a pathological process. In this case, it is necessary in order that bone may replace it.

Figure 2-39: While the spicules of the central zone of cartilage calcify, the perichondrium is invaded by capillaries. This event influences the mesenchymal cells of the inner layer of perichondrium to differentiate into osteoblasts (as well as chondroblasts). The perichondrium now becomes known as periosteum and a layer of bone is secreted along the sides of the central zone of the shaft. This layer is the bone collar (Fig. 2-40).

Figure 2-41: The bone collar supports the shaft of the cartilage model while the central zone, now filling with calcium crystals, disintegrates. Capillaries of the periosteum grow into this central zone. These are periosteal buds and carry with them osteoblasts and undifferentiated mesenchymal cells.

Figure 2-42: The central zone of calcified spicules (trabeculae), periosteal bud, osteoblasts, and UMC constitute a

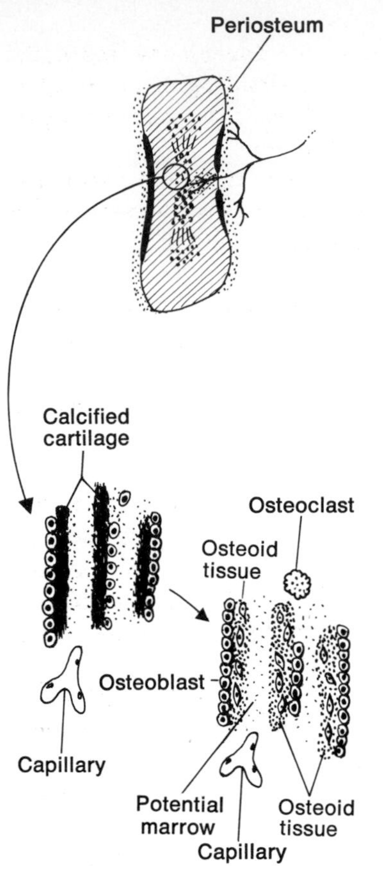

Fig. 2-42

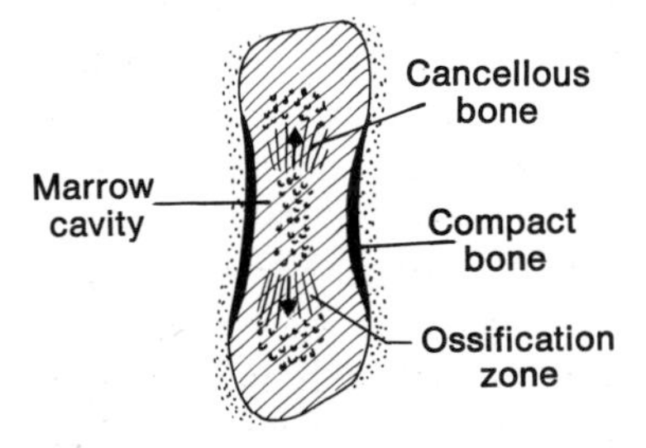

Fig. 2-43

diaphyseal (shaft) center of ossification. The osteoblasts line up around trabeculae of calcified cartilage and secrete osteoid tissue on them. The calcified cartilage disintegrates and is absorbed into the capillaries. Osteoid tissue remains and is subsequently mineralized to become bone. Note the osteoclast already removing new bone.

Figure 2-43: Note the progressive ossification along the sides of the shaft, which is to be compact bone, and in the central zone, which is to be cancellous bone. As the bone collar increases in length and thickness, the interior cancellous bone dissolves to a large extent, leaving a cavity—the marrow cavity.* At each end of the marrow cavity, ossification about trabeculae of calcified cartilage continues.

Figure 2-44: The process of chondrocyte hypertrophy and calcification of cartilage begins in the epiphyses (ends of the long bone) as it occurred previously in the diaphysis (shaft). Capillary invasion brings in periosteal buds and starts a new epiphyseal center of ossification. The process of ossification spreads out.

The epiphyseal ossification continues short of the outer limit of the cartilage model, which is the articular surface. The cartilage here will remain throughout life as articular cartilage. Between the epiphyseal and diaphyseal centers of ossification is a plate of growing cartilage—the **epiphyseal plate** (Fig. 2-45).

I hope you are still with me in this discussion of endochondral ossification. You should, at this point, understand the following about bones that develop by this method:

1. Bones are preceded by cartilage models.
2. The bone collar surrounds and supports the shaft while the cartilage of the central zone is disintegrating.
3. The periosteal bud brings into the central zone the cellular elements necessary for ossification.
4. The key feature of the ossification process is replacement of calcified cartilage by bone.
5. The formation of epiphyseal centers of ossification makes possible the mechanism for bone growth (lengthening).
6. The epiphyseal plate is a zone of cartilage between the epiphyseal and diaphyseal centers of ossification which is in a losing race with those centers in terms of rate of mitosis. As a consequence, bone lengthens.

* The marrow cavity becomes filled with undifferentiated mesenchymal cells, ground substance and blood vessels. This will become a center for development of blood cells (hemopoiesis).

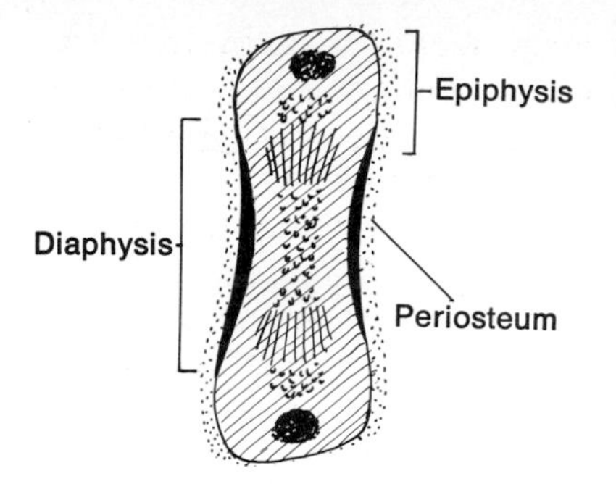

Fig. 2-44

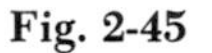

Fig. 2-45

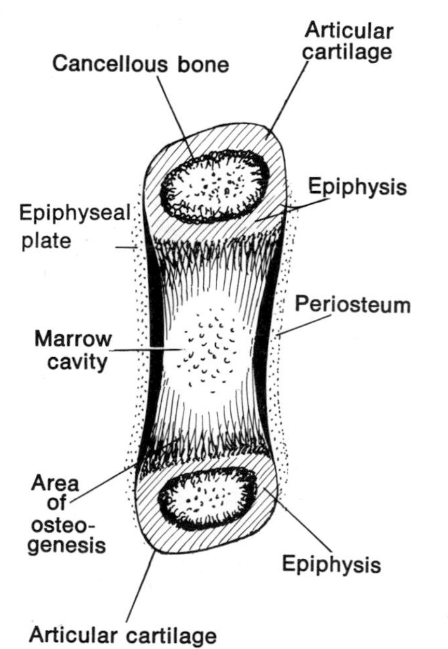

Bone growth

Assuming you understand the above phenomena, the pathway has been cleared for an understanding of the mechanism of bone growth—as seen in a 30-in.-tall 2-year-old growing to become 6 ft tall in about 16 years.

The secret to long bone growth lies in the epiphyseal plate. Once the epiphyseal ossification centers have appeared, the lengthening of a developing long bone will commence. See how in Fig. 2-46.

The epiphyseal plate separates epiphyseal bone from diaphyseal bone and, in fact, because of its continued growth, pushes the epiphysis away from the diaphysis.

The epiphyseal plate, however, does not become larger or thicker. It becomes smaller and thinner! The rate of proliferation of chondrocytes in the epiphyseal plate is, in a sense, in a race with the rate of proliferation of developing bone on the diaphyseal side of the plate. The proliferation of cartilage in the plate makes the plate thicker. The proliferation of developing bone on the diaphyseal surface of the plate (at the expense of the epiphyseal plate cartilage) makes the plate thinner.

This contest of rates of growth goes on for 18 to 25 years after birth in humans. The rate of diaphyseal bone formation exceeds the rate of cartilage growth, and the epiphyseal plate ultimately thins to a faint line not easily observable in adult bone.

Even after growth of bone ceases, reabsorption and replacement of bone continues throughout life (Fig. 2-47). These activities provide for possible change in the architecture of cancellous bone to compensate for certain stresses, e.g., increased body weight, increased muscle pull, nutritional deficiences, postural changes, and fractures. As a result of the reabsorption (resorption) of the mineral elements of bone, calcium is set free as ions for absorption by the blood—minimum levels of calcium in the blood are essential for life.

Multinucleated osteoclasts (probably originating from fused osteoblasts) are responsible for the erosive breakdown of bone and the resorption of the calcium elements into the blood. The subsequent replacement of bone takes place through the activities of osteoblasts. While compact bone is replaced in definitive architectural patterns, the Haversian systems (Fig. 2-47), cancellous bone is replaced in an irregular pattern of layers. The Haversian system allows for the continuing nourishment of osteocytes in their lacunae through Haversian canals, Volksmann's canals, and canaliculi (Fig. 2-32B).

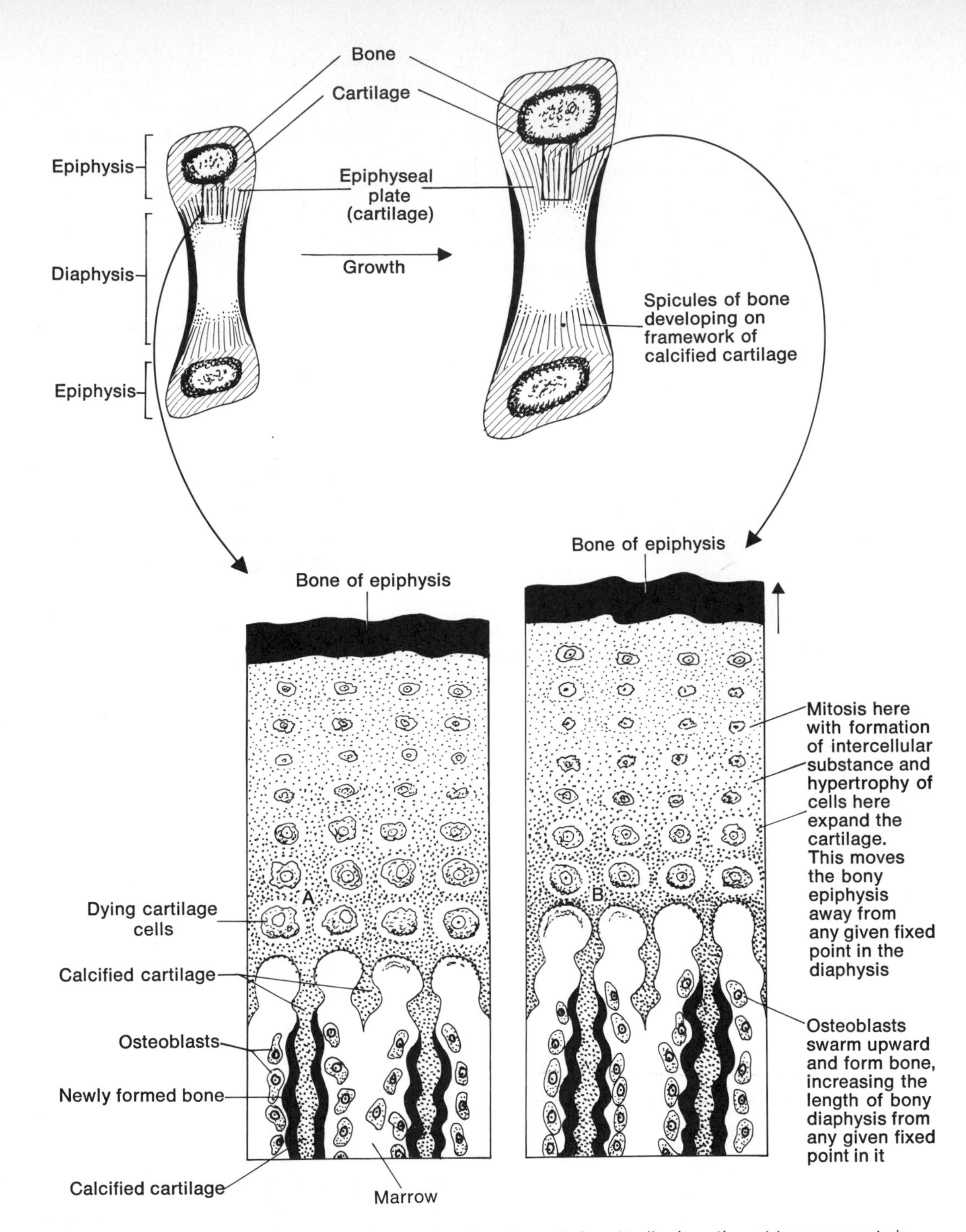

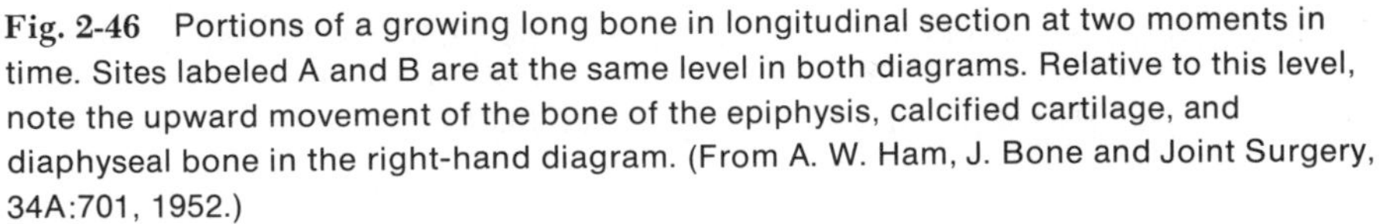

Fig. 2-46 Portions of a growing long bone in longitudinal section at two moments in time. Sites labeled A and B are at the same level in both diagrams. Relative to this level, note the upward movement of the bone of the epiphysis, calcified cartilage, and diaphyseal bone in the right-hand diagram. (From A. W. Ham, J. Bone and Joint Surgery, 34A:701, 1952.)

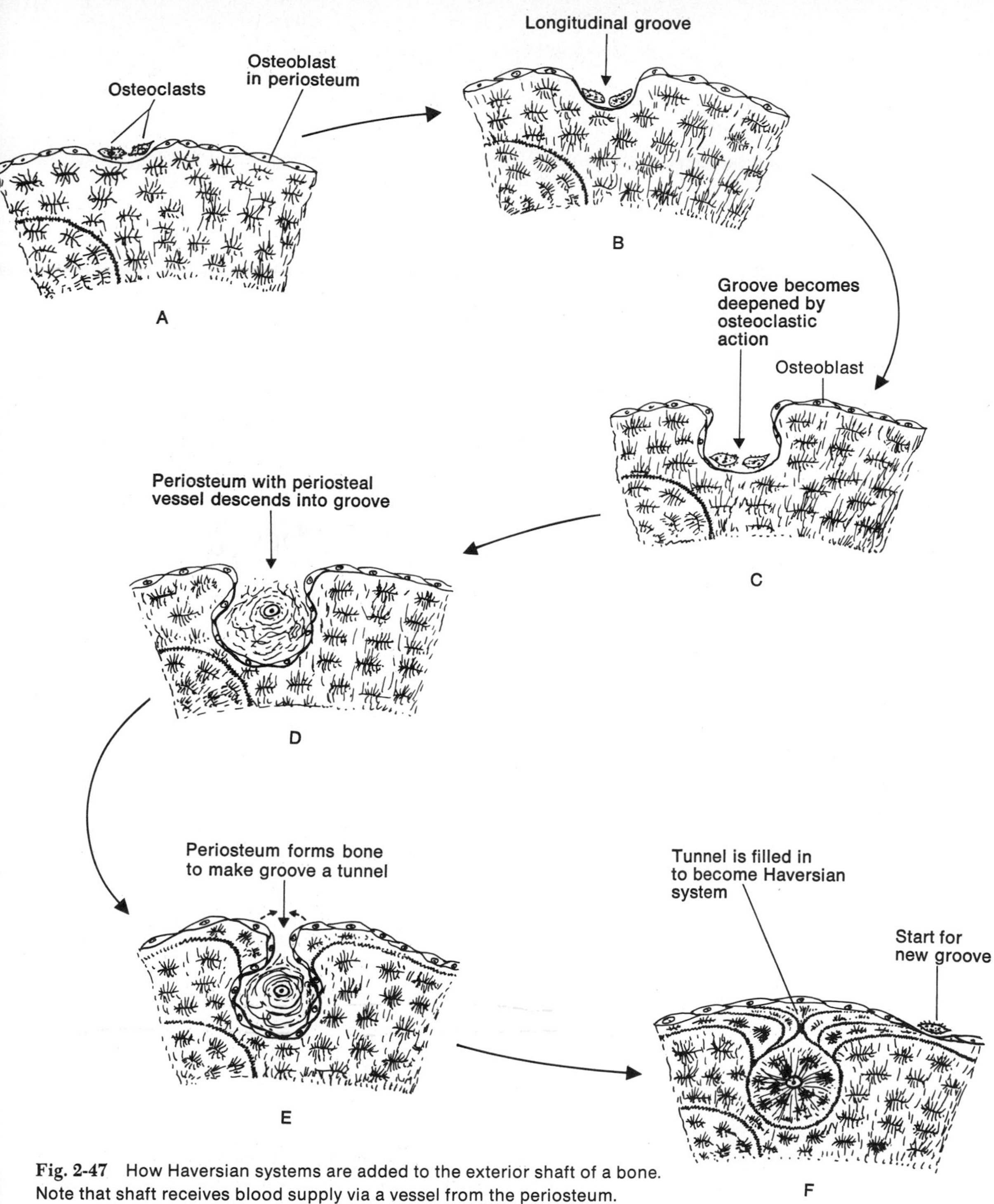

Fig. 2-47 How Haversian systems are added to the exterior shaft of a bone. Note that shaft receives blood supply via a vessel from the periosteum. (From A. W. Ham, *Histology*, 6th ed., J. B. Lippincott, Philadelphia, 1972.)

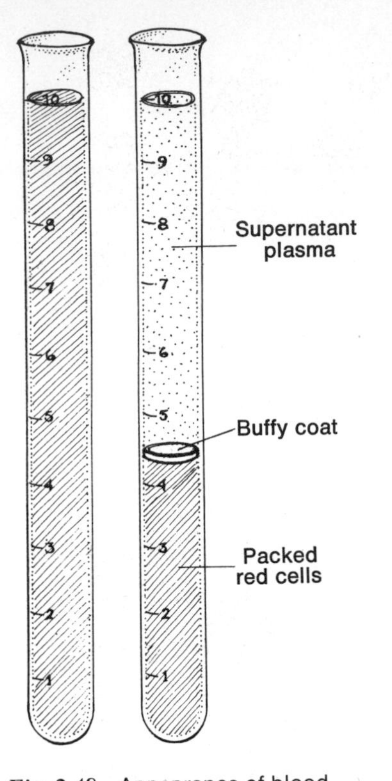

Fig. 2-48 Appearance of blood before and after sedimentation. The marks on sides of the tubes refer to percent. The volume of packed RBC's is about 44 percent; the leukocytes and platelets (buffy coat) make up about 1 percent; the rest (supernatant) is plasma.

The macroscopic nature of bones will be considered in more detail in Unit 3.

QUIZ

Now that you are knowledgeable about bone formation and the microscopic anatomy and function of bones, can you score 100 percent on this quiz? Indicate whether true (T) or false (F)*:

1. Bone is developed through a process of cartilage cells changing into bone cells (F).
2. Bone is a living tissue which undergoes resorption and replacement throughout the life of the individual (T).
3. Haversian systems of bone deposition may be found in both compact and cancellous bone (F).
4. Bone is about 65 percent mineral by weight (T).
5. The process of bone development for the flat bones of the cranium is called intramembraneous ossification (T).

BLOOD AND LYMPH

Blood and lymph are a type of connective tissue consisting of a fluid as well as a cellular phase.

The fluid phase of blood is called **plasma.** Plasma is a kind of extracellular fluid (ECF) and makes up about 5 percent of the body weight.† Plasma contains an unknown but great number of ions, molecules—both inorganic and organic—which easily communicate with other fluid spaces of the body. Digested foodstuffs of molecular dimensions, drugs, hormones, and a variety of protein enzymes, antibodies, and other substances regularly pass into and out of the plasma. Infective agents also employ the plasma as a vehicle for spreading to other areas.

If plasma is allowed to stand in air, it will clot, and the fluid remaining after the clot is **serum.** Since the clot is the result of a complex series of reactions involving the plasma proteins, serum differs from plasma in having fewer such proteins as well as an absence of other clotting substances.

If blood is centrifuged, or is allowed to stand without clotting, the plasma separates from the cellular portion. The plasma phase is straw-colored (Fig. 2-48). Plasma and serum

* Answers: F, T, F, T, T.

† Plasma makes up about one-third of the total body ECF volume and is in communication with the other two-thirds (which largely bathes the cells) by processes of diffusion and active transport. It is interesting to note that ECF and intracellular fluid (ICF) make up 60 percent of the body weight.

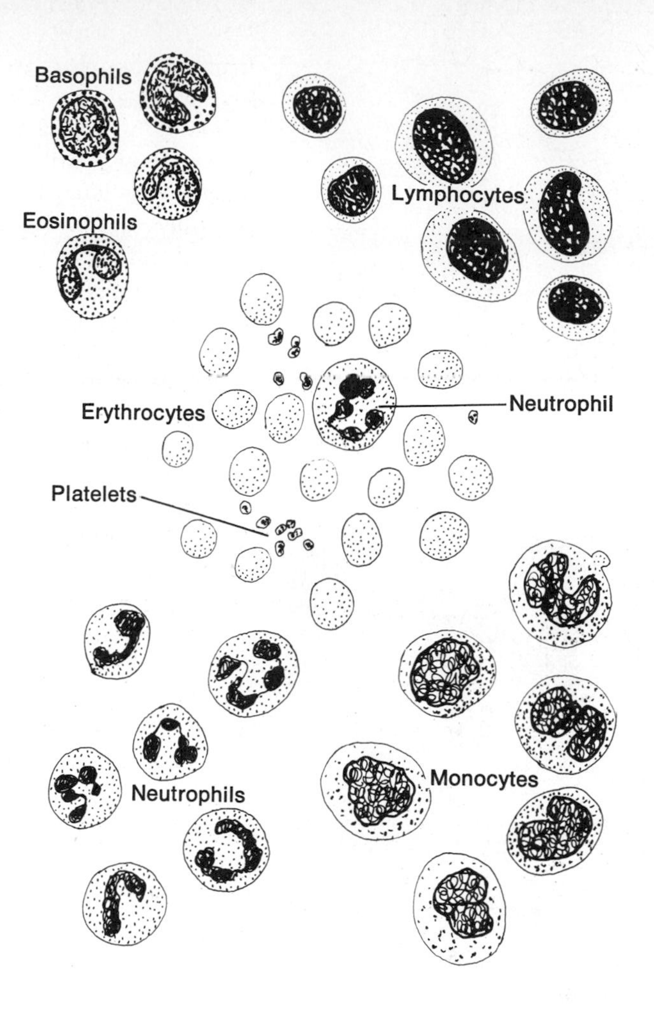

Fig. 2-49 Formed elements of the blood. (From R. O. Greep, *Histology*, 2d ed., McGraw-Hill, New York, 1966.)

isolated from blood samples taken from patients may reveal symptoms of a disease or destructive process underway on being tested in the clinical laboratory by the medical technologist.

The cellular phase of blood makes up about 45 percent of the total blood volume. The constituents of this phase are called **formed elements** (Fig. 2-49) and they consist of:

1. **Erythrocytes,** or red blood corpuscles (RBC), 44 percent by volume of blood.
2. **Leukocytes,** or white blood cells (WBC), 1 percent by volume of blood.
3. **Platelets,** less than 1 percent by volume of blood.

All of these elements develop from stem cells in the bone marrow or lymphatic organs.

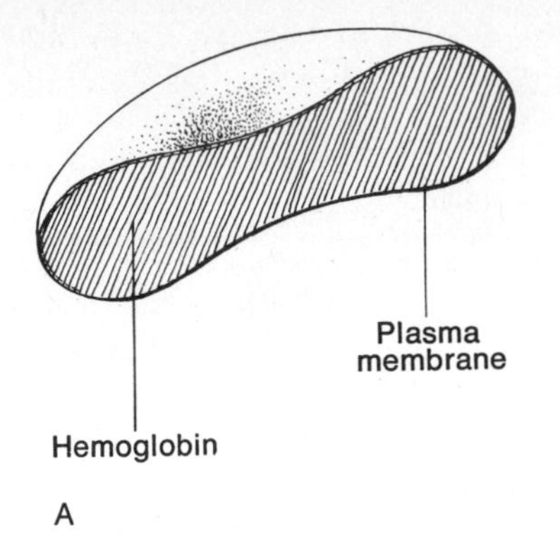

B

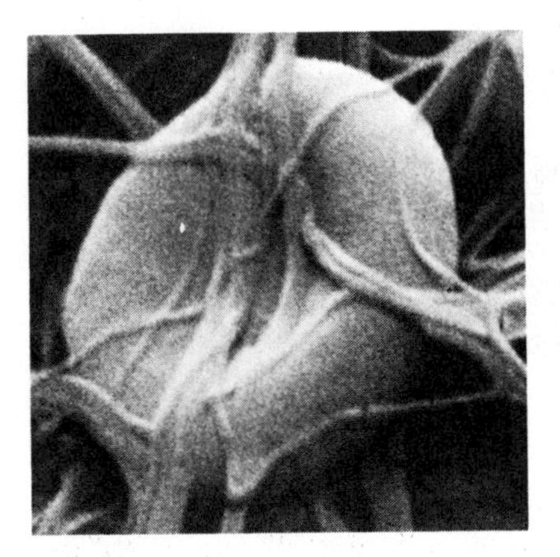

Fig. 2-50 A. An erythrocyte bisected. Note the disc shape. B. Electron micrograph of an erythrocyte enmeshed in fibrin. (Courtesy Emil Bernstein and Eila Kairinen, Gillette Company Research Institute, Rockville, Md.)

Erythrocytes (RBC's)

Erythrocytes (Fig. 2-50) are the most commonly seen elements in the blood; there are about 5,000,000 per cubic millimeter. They are the remains of cells which extruded their nuclei in the latter phases of development, hence they are not correctly termed cells; corpuscles or elements would be more accurate.

An erythrocyte is a biconcave disc (Fig. 2-50) measuring about 8 micrometers in diameter. In essence, each is a sac of hemoglobin enclosed by a plasma membrane. The primary functions of the erythrocytes are: (1) to combine with *oxygen* inhaled into the lungs (hemoglobin + oxygen = oxyhemoglobin) and (2) to release that oxygen at the tissue level (by a process of diffusion). A reduction in quality or quantity of RBC's in the blood, for any reason, is termed anemia. An excess of RBC's in the blood constitutes polycythemia. The former is much more common than the latter.

Leukocytes (WBC's)

Leukocytes (Fig. 2-49) are the truly cellular elements of the blood. They make up about 1 percent of the total blood volume—but a vital 1 percent they are! They average about 8,000 cells per cubic millimeter of blood. A significantly decreased number of leukocytes constitutes leukopenia (*-penia,* poverty) while an increased number is leukocytosis. An increased number of a particular cell type due to a malignant process (neoplasm) is generally called leukemia (*-emia,* in the blood).

Leukocytes are generally larger than RBC's (10 to 16 micrometers in diameter) and are not at all concerned with oxygen transport. The leukocytes of the blood are as follows, with approximate percentage of total leukocytes given:

1. **Neutrophils,** 65 percent.
2. **Eosinophils,** 3 percent.
3. **Basophils,** 1 percent.
4. **Lymphocytes,** 30 percent.
5. **Monocytes,** 1 percent.

The first three listed (neutrophils, eosinophils, and basophils) are collectively termed granular leukocytes because there are obvious granules in their cytoplasm. The last two leukocytes are termed agranular (nongranular) leukocytes. **Neutrophils** are so called because they stain with a neutral (not acidic or basic) stain. They are highly mobile and can migrate out of the circulation into the connective tissue spaces to help combat an infection by phagocytosing bacteria. They may also be involved in immune reactions to foreign substances where specific antibodies play a major role.

Eosinophils are so called because their cytoplasmic granules stain with eosin, an acid dye. Their function is unknown. However, they are believed to be associated with allergic reactions, as they may be seen in great numbers in the connective tissues of mucous membranes involved in such reactions. Thus they are known to be motile and they apparently cross the vascular boundaries with ease.

Basophils are so called because their cytoplasmic granules stain with a basic dye. Their function is not known but they are known to contain significant titers (concentrations) of heparin, histamine, and serotonin. In this sense, they seem to be related to the mast cells of the connective tissues.

Lymphocytes arise from lymphatic tissue (found in the spleen, lymph nodes, tonsils, and nodules of the mucous membranes). They are generally nongranular when seen under the light microscope but occasionally are seen to contain sky-blue (azurophilic) granules. Lymphocytes come in large, medium, and small sizes. Only the latter two are found in the blood. Large lymphocytes are found in lymphatic tissue and are believed to be the forerunners of many of the lymphocytes responding to an infection. Lymphocytes are involved in immune responses, i.e., they are known to secrete antibodies which counteract the effect of antigens (foreign substances in the bloodstream).

Fig. 2-51 A lymphatic vessel. Note lymphocytes and lymph within the lumen.

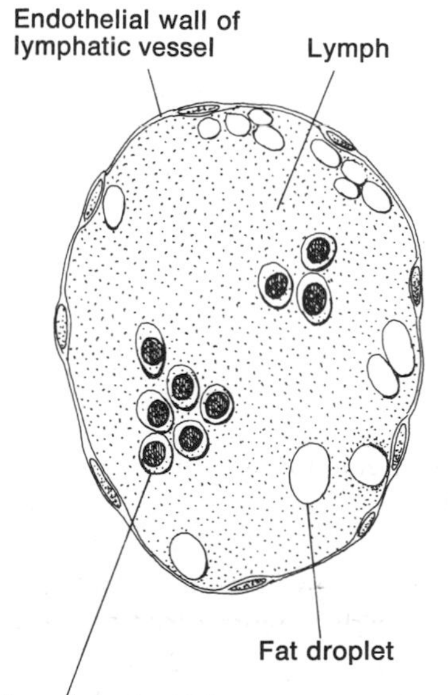

Monocytes are the largest cells of the blood and are believed to be phagocytic. They are capable of passing from the blood into the connective tissues and of transforming into wandering macrophages in response to an infection. Monocytes are often difficult to distinguish from lymphocytes, but their larger size and kidney-shaped nucleus will often identify them.

Platelets are pinched-off cytoplasmic bodies of megakaryocytes (cells found in the bone marrow); about 250,000 of them may be found in each cubic millimeter of blood. They are known to play a major role in blood clotting.

LYMPH

Lymph is a kind of ECF which is found in the lymphatic vessels of the body (Unit 3). It is a whitish milky fluid and there are large fatty molecules and small proteins as well as lymphocytes in it (Fig. 2-51). Lymph is generated in the tissue fluids and diffuses into lymph capillaries. As the lymph circulates through the lymph nodes lymphocytes are added. Ultimately, the lymph empties into one of two large veins draining into the heart.

QUIZ

Now, time for a brief quiz on the connective tissues:

1. Connective tissues generally function to bind, support, or link other structures in the body.
2. The three main categories of connective tissue are: connective tissue proper, supportive tissue and blood and lymph.
3. The most common of the first category (c.t. proper) is Loose c.t., filling all unoccupied spaces in the body.
4. Dense c.t. is characterized by its strength.
5. Dense irregular c.t. encapsulates bone (as periostium).
6. Dense regular c.t. forms tendons and ligaments.
7. Elastic tissue is found in the walls of large arteries.
8. Adipose c.t. is very important fat tissue.
9. The two most common cells of the connective tissue proper are fibroblasts and macrophages.
10. The three kinds of connective tissue fibers are: Collagenous, elastic, and reticular.
11. There are three kinds of cartilage in body: one kind is found in the nose (hyaline); one kind is found as the support of the external ear (elastic) and the other is found between the vertebrae as disks (fibrocartilage).
12. Cartilage is actively metabolic and comes under hormonal influences.
13. There are two types of bone based on structure: compact and cancellous.
14. In compact bone, the tissue is arranged in cylindrical haversian systems.

15. The fluid portion of blood is called plasma.
16. If blood is allowed to clot, the remaining fluid is called serum.
17. The "formed elements" of blood consist of erythrocytes, leukocytes, and platelets.
18. Erythrocytes transport oxygen.
19. Leukocytes make up about 1 percent of the total blood volume.
20. Lymph is derived from general lymphatic vessels and ultimately drains into one of two large veins.

MUSCLE TISSUE

Certainly if connective tissue takes up more of your body than any other tissue, muscle tissue must run a close second! For if *any part* of you moves—muscle moves it! Think about that for a moment. Several muscles are involved in just tapping your foot or scratching your head—not to mention running or walking. How about the twitches you get occasionally? Swallowing, breathing, digestion, defecation, urination? These movement-based processes all occur because muscle contracts. These are, of course, all obvious movements—there are many muscular contractions you are rarely aware of—contractions of the heart (the heart is practically all muscle), contractions of muscle in blood vessels so your blood pressure will rise when it's necessary; most of the muscular contractions of the digestive tract; tonic (sustained) contraction of postural muscles to keep you erect when sitting or standing.

By definition, muscle tissue is a contractile tissue composed of elongated cells often attached to one another through the medium of connective tissue fibers. Muscle cells are frequently called fibers because of their external appearance; however they are "living fibers" and not at all "fibrous" in the connective tissue sense of the word. Muscle tissues are classified on the basis of their microscopic appearance. Those muscle cells exhibiting cross striations throughout their full length are **striated** muscle; those muscle cells lacking cross striations are **smooth** muscle. Two types of striated muscle may be recognized; striated muscle associated with the bony skeleton of the body (skeletal) and that striated muscle making up the muscular component of the heart (cardiac). Smooth muscle is a constituent of visceral walls.

Skeletal muscle

Skeletal muscle fills out the body—provides it with form (Fig. 2-52). You might say that this is its passive function. Its active function is to move bones about joints.

The unit of structure of skeletal muscle is the muscle cell or fiber (Fig. 2-53).

Each muscle fiber is composed of:

1. A limiting membrane: the sarcolemma (L. *sarco*, flesh).
2. A cytoplasmic matrix: the sarcoplasm.
3. Common cell organelles as well as the highly specialized sarcoplasmic reticulum and contractile elements, the myofibrils.
4. Numerous nuclei.

The myofibrils are responsible for the longitudinal striations seen in skeletal muscle. The myofibrils are composed of myofilaments which are responsible for the cross striations seen with the electron microscope. The functional unit of myofilaments (between two Z discs) is a sarcomere (Fig. 2-54). These myofilaments apparently slide back and forth during muscular contraction and relaxation, and this action results in a short-

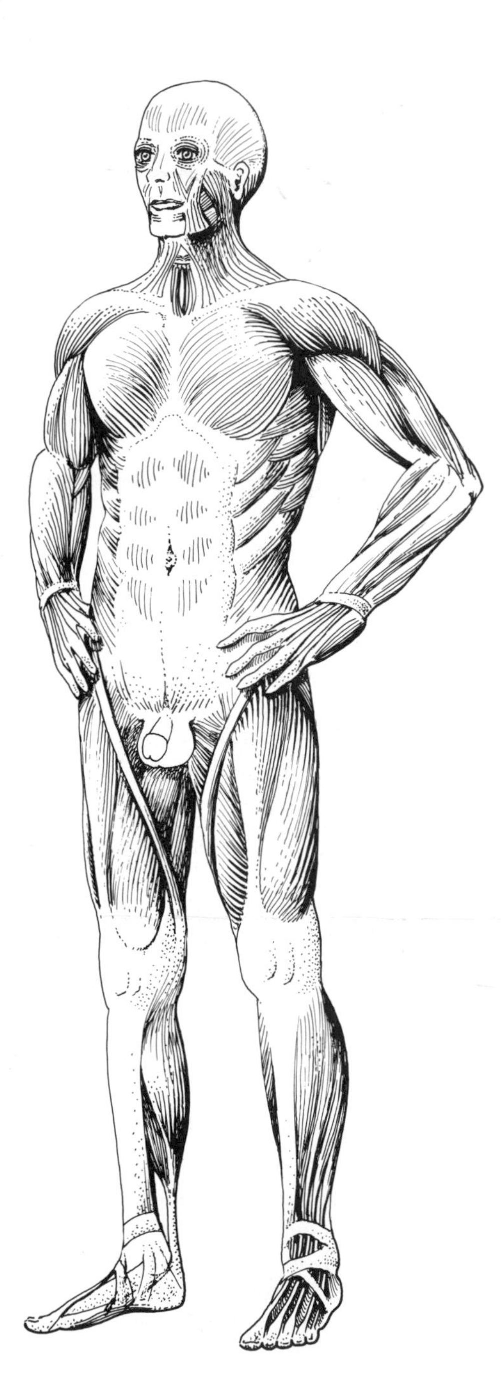

Fig. 2-52 The skeletal muscle mass creating the human form.

Fig. 2-53 Skeletal muscle in cross section (left) and longitudinal section (right). Photomicrographs. (Courtesy of James Runner.)

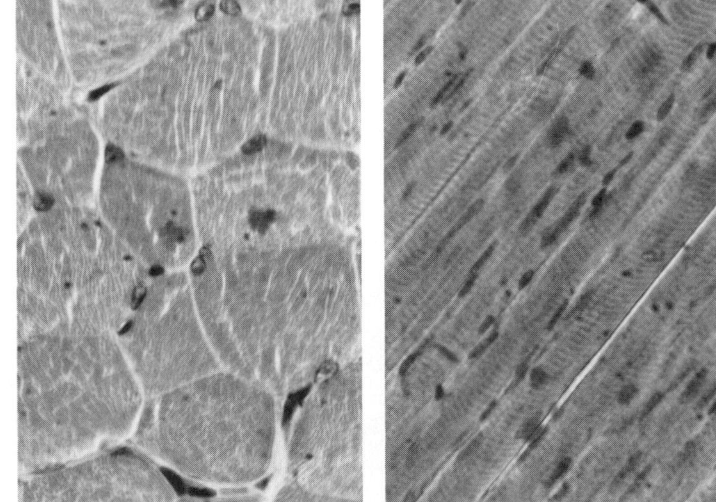

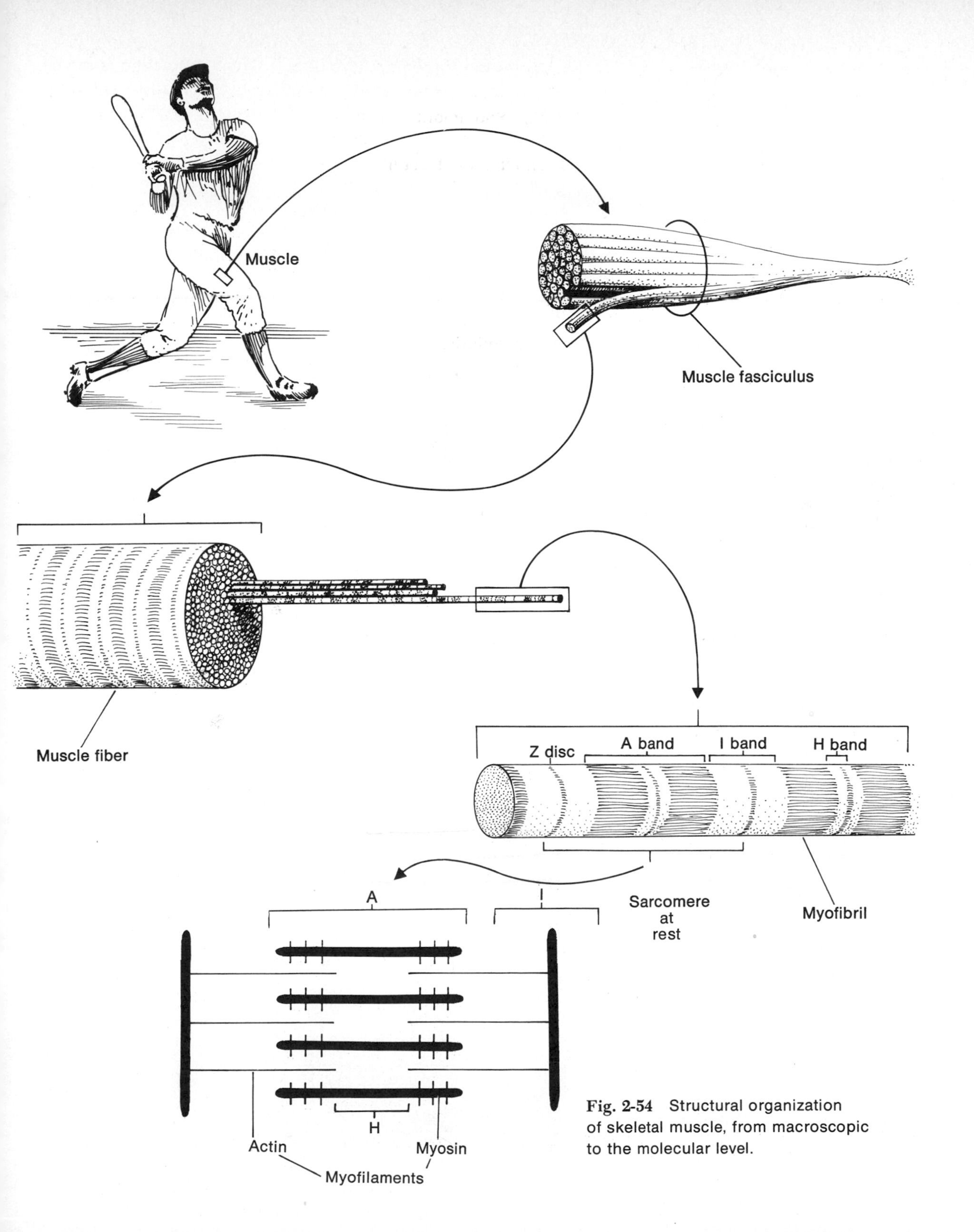

Fig. 2-54 Structural organization of skeletal muscle, from macroscopic to the molecular level.

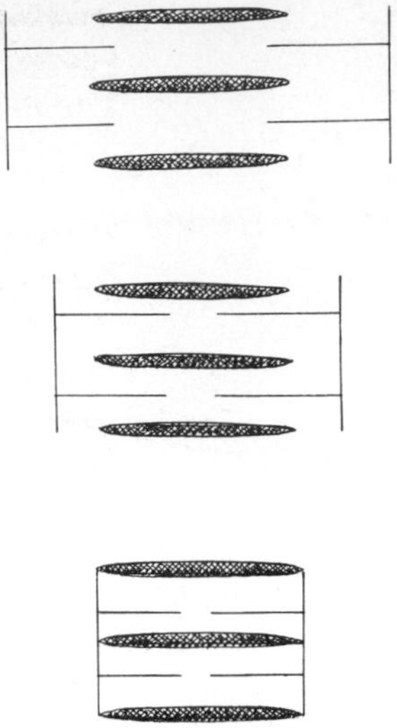

Fig. 2-55 The mechanism of skeletal muscle contraction (sliding filament hypothesis) in one sarcomere. Diagram.

ening of the sarcomere (Fig. 2-55). When this action is manifested throughout the whole muscle, the muscle shortens (contracts) by about one-third of its resting length.*

Now fibers in muscle tissue the size of the muscles of the arm, for instance, would fall apart like strings of spaghetti in a game of "pick up sticks" if they were not intimately bound together by connective tissue. By way of this connective tissue "packing," the muscle fibers receive their nerve and blood supply. Muscle tissue incorporating large amounts of this connective tissue is tough. In beef, it is this which makes certain cuts difficult to chew. The expensive, "you can cut it with a fork" meats are so because of a reduced amount of connective tissue. The anatomy of connective tissue investment of skeletal muscle can be seen in Fig. 2-56. Look at this figure as you read.

Each skeletal muscle fiber is delicately ensheathed in a thin film of reticular tissue termed **endomysium.** This endomysium blends with the sarcolemma.

A variable number of muscle fibers, so ensheathed, form a macroscopically visible bundle (fasciculus). Each fasciculus is bound by a layer of connective tissue called **perimysium,** which has continuity with the interlacing endomysium.

A variable number of fasciculi of skeletal muscle are bound together by a thick envelope of connective tissue, the

* For a more detailed discussion of this sliding filament hypothesis, see Bloom and Fawcett: *A Textbook of Histology* (see reference).

Fig. 2-56 The connective tissue investments of skeletal muscle.

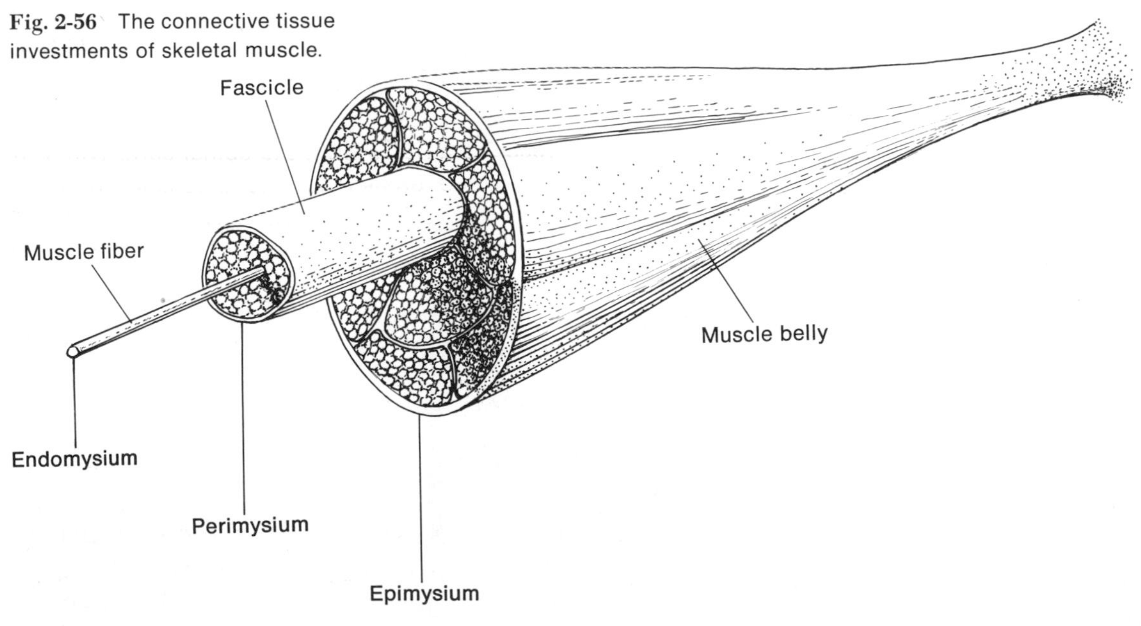

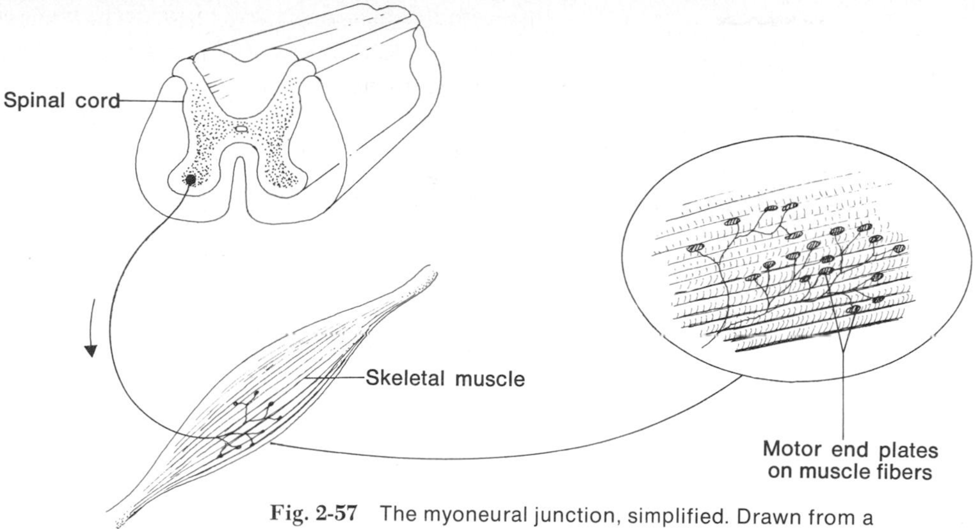

Fig. 2-57 The myoneural junction, simplified. Drawn from a photomicrograph.

epimysium, which is continuous with the finer fibers binding fasciculi and muscle fibers.

The epimysium merges at the extremities of the skeletal muscle mass to become continuous with tendons. Epimysium is generally indistinguishable from the deep fascia (see Unit 3) which is found throughout the musculoskeletal system.

Now it is a fact that skeletal muscle cannot contract unless it is innervated, i.e., served by nerve fibers, right down to single muscle cells. Further, should a muscle become denervated ("de-nerved"), it will atrophy and may largely disappear, leaving behind its fibrous tissue envelope.

Now look at Fig. 2-57 as you read the following:

A muscle receives a nerve from the spinal cord. Within the epimysium, the nerve breaks up into thousands of microscopically visible nerve fibers. *Each* nerve fiber entering a muscle will branch many times to innervate upwards of 150 muscle fibers, depending on the quality of movement of which the muscle is capable. The finer the movement, the fewer the muscle fibers managed by a single nerve fiber.

That area of a skeletal muscle fiber receiving a terminal motor nerve fiber is called a **motor end plate** or **myoneural junction.**

The nerve fiber and the muscle fibers which it innervates constitute a **motor unit.**

In the event of injury, muscle cells can regenerate, but slowly. Muscle cells are much more capable of enlarging (hypertrophy) as seen during postnatal growth and prolonged physical conditioning, or becoming smaller as in disuse atrophy, which is seen in chronic bedridden patients without benefit of physical therapy.

Smooth muscle

Smooth muscle is visceral muscle over which you have no voluntary functional control. It is innervated by a part of the nervous system whose functioning is also independent of your will. Smooth muscle is found in the walls of the digestive tract from esophagus to anus; in the walls of the urinary tract from kidney to urethra; in the walls of the reproductive tract from uterine tube to vagina and from the sperm duct (ductus deferens) to urethra; in the walls of the respiratory tract from pharynx to respiratory tissue (alveoli); in *all* arteries of the body and many large veins (Fig. 2-58). Smooth muscle permits you to change from far vision to near vision—it even permits you to cry! Remember the last time your hair stood on end? Smooth muscle erected those hairs. You have no control over the movements of these organs. Quite the contrary, they control you! Contraction or stretching of smooth muscle tells you when to eat, when to urinate, when to defecate, when to take a laxative, when to vomit, when you are going to give birth, and so on. Smooth muscle is decidedly different from skeletal muscle on a number of counts—both histologically and functionally:

1. Smooth muscle fibers are smaller than skeletal muscle fibers.
2. Smooth muscle fibers do not have more than one nucleus. (They are smaller than skeletal muscle fibers, and therefore do not require more than one nucleus.)
3. Nuclei of smooth muscle cells are centrally located in the cell; skeletal muscle nuclei are peripherally located.
4. Smooth muscle fibers lack a sarcolemma and a specialized sarcoplasmic reticulum.
5. Smooth muscle fibers do not each receive a nerve fiber.
6. The myofibrils of smooth muscle consist of very thin myofilaments which are not well organized. Hence cross striations are not apparent and thus its name, *smooth* muscle.

Fig. 2-58 (opposite) Smooth muscle in a small artery (A) and in the wall of the esophagus (B). C. Longitudinal section of smooth muscle. Photomicrograph. (Courtesy of James Runner.)

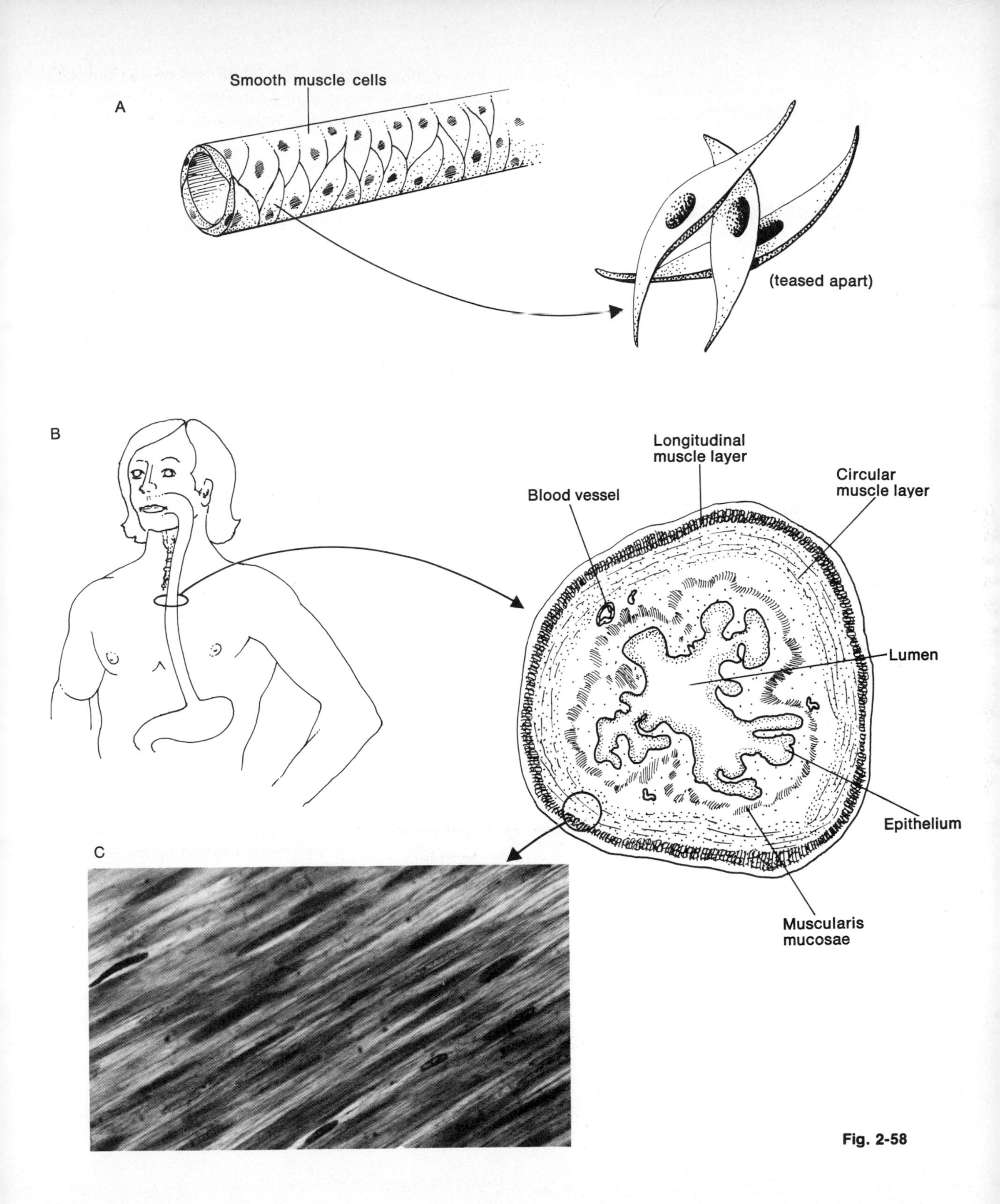
Smooth muscle cells
A
(teased apart)
B
Longitudinal
muscle layer
Circular
muscle layer
Blood vessel
Lumen
Epithelium
Muscularis
mucosae
C

Fig. 2-58

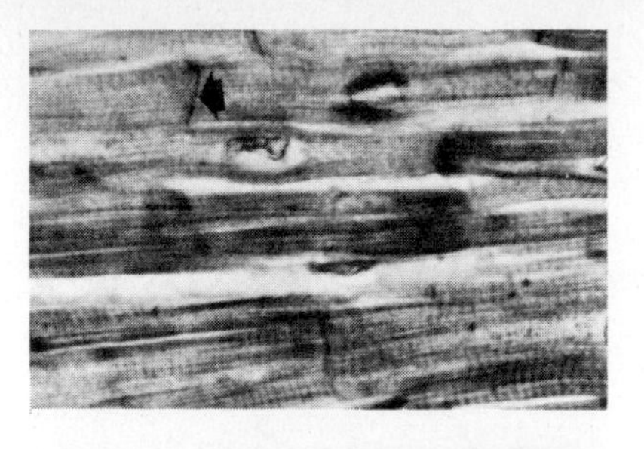

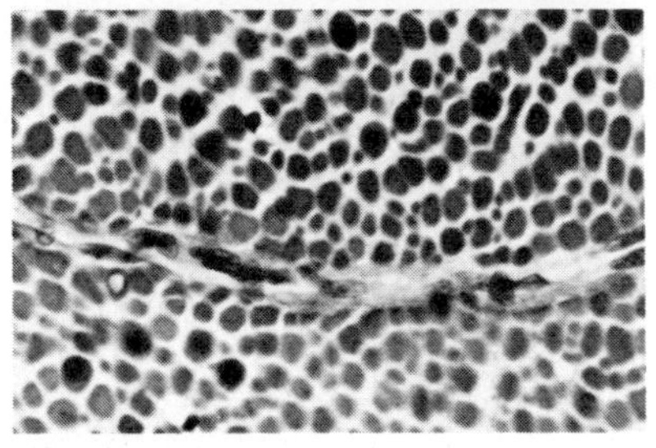

Fig. 2-59 Cardiac muscle. In longitudinal section (top), arrow points to intercalated disc. In cross section (bottom), note the capillary passing between fascicles. (Courtesy of James Runner.)

7. The contractions of smooth muscle are slower but can be sustained for longer periods of time.
8. Smooth muscle is more responsive to stretch.
9. Smooth muscle fibers may occur singly or in small groups; skeletal muscle fibers rarely do.

Smooth muscle may be induced into rhythmic contractions not only by nerve impulses but by hormonal stimulation or by stretching of the muscle itself. Motor end plates are not apparent. Since each muscle fiber does not receive a nerve fiber of its own, the impulses are probably transmitted from one cell to another via the cell membrane.

Smooth muscle can regenerate by mitosis. Also, since smooth muscle arises from mesenchyme, in the event of injury, cells of the mesenchyme may be able to replace the injured muscle cells.

Cardiac muscle

Cardiac muscle is the heart muscle (myocardium). In several ways cardiac muscle (Fig. 2-59) is physiologically similar to smooth muscle and structurally similar to skeletal muscle. Let's highlight these similarities:

1. Cardiac muscle, like smooth muscle, contracts rhythmically and the contractions are not voluntarily controlled.
2. Cardiac muscle contracts spontaneously, i.e., contracts without need of innervation.
3. Cardiac muscle is striated.
4. Cardiac muscle has unique dark bands called intercalated discs, which represent the junction between adjacent cells.
5. Cardiac muscle cells have single nuclei.

Cardiac muscle beats spontaneously at about 40 contractions (beats) per minute, but its rate of beat (contraction) is regulated by nerves functionally independent of your will. The myoneural junctions here are apparently not as complex as they are in skeletal muscle. Even though cardiac muscle beats without requirement of innervation, there must be a mechanism for ensuring synchrony of contraction among all the muscle cells in the wall of each of the chambers of the heart so that the heart may pump effectively. There is, in a specialized system of modified muscle cells (Fig. 2-60), the **cardiac**

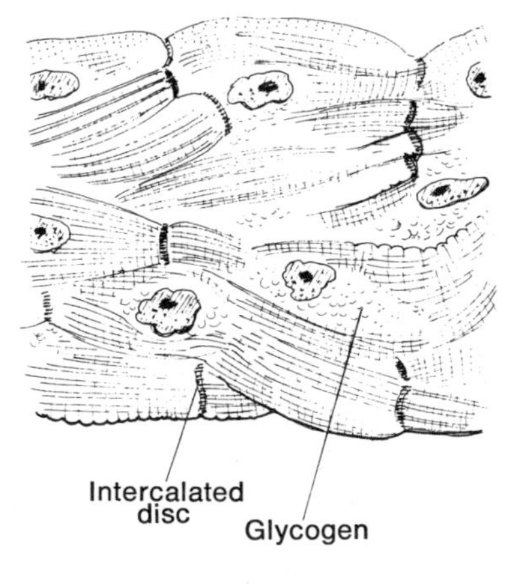

Fig. 2-60 Modified muscle cells constituting fibers of the cardiac conduction system. Note the glycogen surrounding the nuclei.

conduction system. It is made up of a couple of self-stimulating "electrical" impulse generators and a bundle of conducting muscle cells (fibers) distributed through the myocardium of the heart. These fibers, rich in glycogen,* are variably smaller and larger than cardiac muscle fibers but resemble cardiac muscle in that they have cross striations and intercalated discs.

Unlike smooth and somewhat like skeletal muscle, damaged cardiac muscle does not regenerate well at all, and following injury, the injured muscle cells, like skeletal muscle cells, will be replaced by connective tissue (scar tissue).

You might wonder how it is that cardiac muscle can beat continuously from 5 weeks' fetal age to an age often spanning 70 and sometimes 100 years. First, the heart (cardiac muscle cells) does *not* beat (contract) continuously from a cellular point of view. In fact, for every second of operation of a resting, beating heart, each muscle fiber rests about $^{6}/_{10}$ of that second. Second, the heart has a vast blood vessel network to provide needed oxygen. Third, the cardiac muscle fiber has a high content of glycogen on which to draw for energy.

NERVOUS TISSUE

Nervous tissue is excitable tissue; that is, it can generate nerve impulses as well as conduct them. All of your *sensations* (touching, hearing, seeing, etc.), all of your *thoughts* (rational and irrational, objective, and subjective, intellectual and emotional), and all of your *movements* (from the rate of beat of your heart to the wrinkling of your nose) are based on these two phenomena.

Nervous tissue is organized into many organs whose overall function may be likened to a vast communications network. The essential function of any communications system is to receive information and respond to that information in some manner. To do these things, a communications system must have:

1. *Receptors* to sense incoming information (stimuli).
2. A *processing center* to correlate, coordinate, integrate, and modify the information received.
3. *Conductors* to conduct incoming (sensory) stimuli to the processing center and to return evoked responses from the center to effectors which manifest the response.

* The presence of glycogen (a starch or multiple sugar) indicates that the structure is capable of a good deal of energy expenditure. The glycogen is the fuel for such expenditure.

and the nervous system has them! . . . in the following organization:

Central nervous system (CNS) (processing center and central conductors) consisting of a **brain** and **spinal cord** which process sensory information and generate appropriate responses (Fig. 2-61).

Peripheral nervous system (PNS) (receptors and conductors *outside* the brain and spinal cord) consisting of **spinal** and **cranial nerves.** Some of these take information to the CNS and others bring responses from the CNS to effector organs (Fig. 2-62).

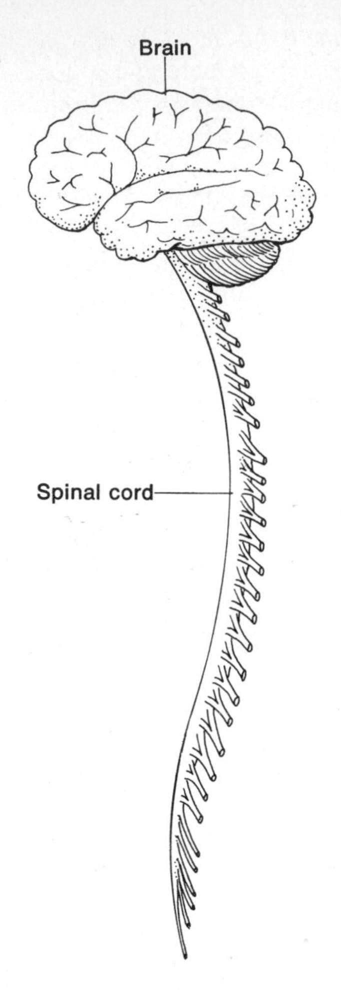

Fig. 2-61 The central nervous system, consisting of brain and spinal cord.

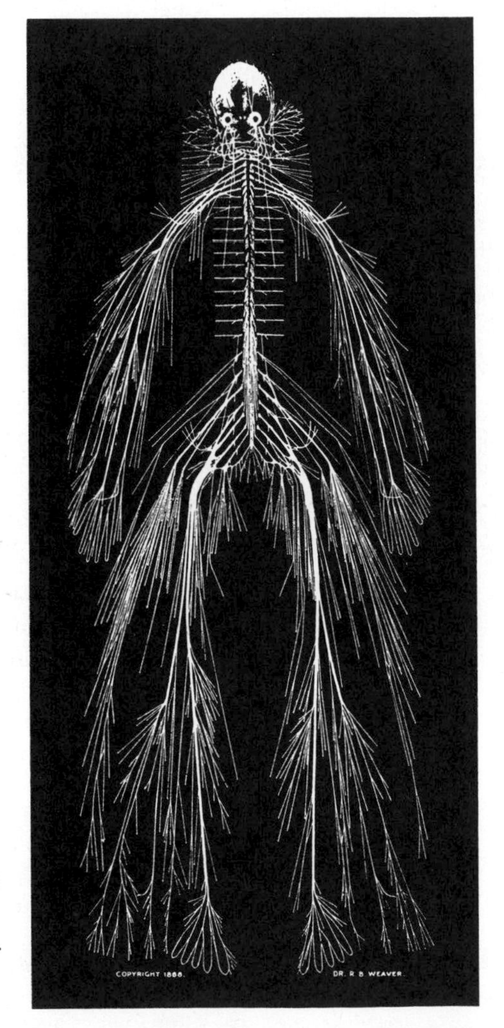

Fig. 2-62 The human peripheral nervous system. Photograph of a dissection of a female subject. She ("Harriet") was the janitress in the anatomy department at Hahnemann Medical College in the late 1800's and left her body for dissection by Dr. R. B. Weaver. (Courtesy of Dr. Peter S. Amenta, Hahnemann Medical College, Philadelphia.)

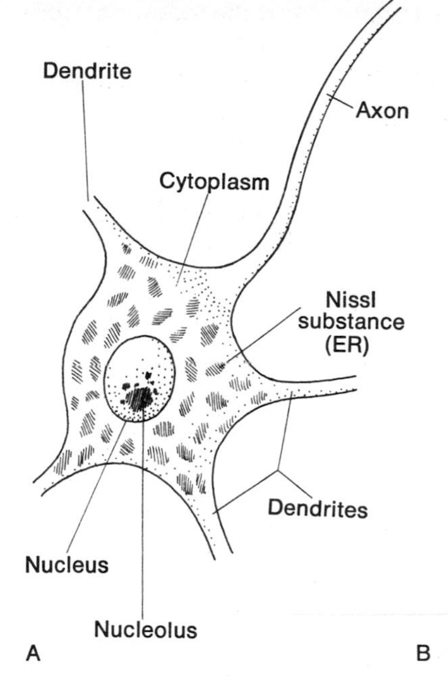

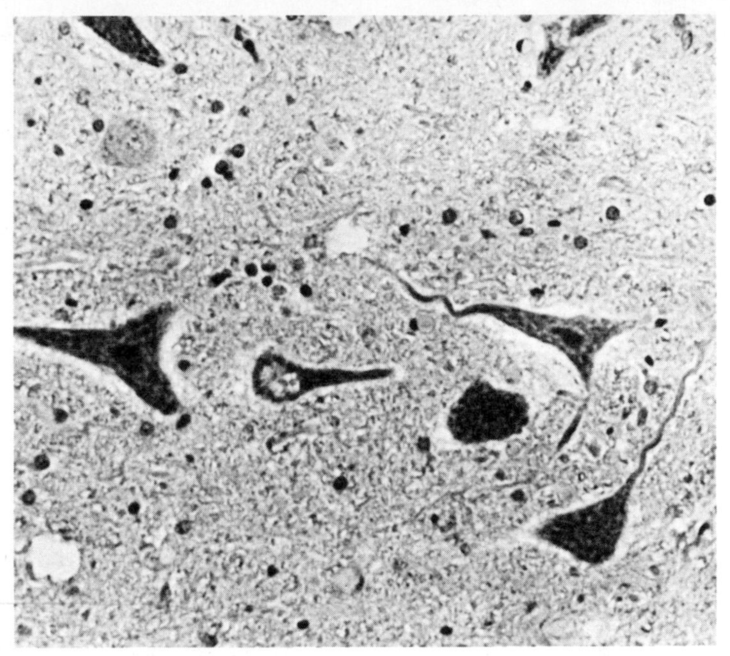

Fig. 2-63 A. The nerve cell body and stumps of its processes. B. Photomicrograph of several nerve cells in the spinal cord. (Courtesy of James Runner.)

The neuron

The neuron or nerve cell is the fundamental unit of structure and function in the nervous system. Two components of the neuron (Fig. 2-63) can be identified:

1. **Cell body:** the nucleus and its surrounding cytoplasm.
2. **Processes:** extensions of cytoplasm from the cell body.

The cell body (the nucleus, specifically) is, of course, the "administrative center" for the entire neuron. Frequently, cell bodies are congregated into groups in the nervous system. In the CNS these collections of nerve cell bodies are termed **nuclei;** in the PNS these collections of nerve cell bodies are called **ganglia.** Histologically, cell bodies are characterized by:

1. A large, centrally located, well-defined nucleus with prominent nucleolus.
2. Blocks of endoplasmic reticulum (ER) called *Nissl substance.*
3. Neurofibrils and other common organelles.
4. Plasma membrane.

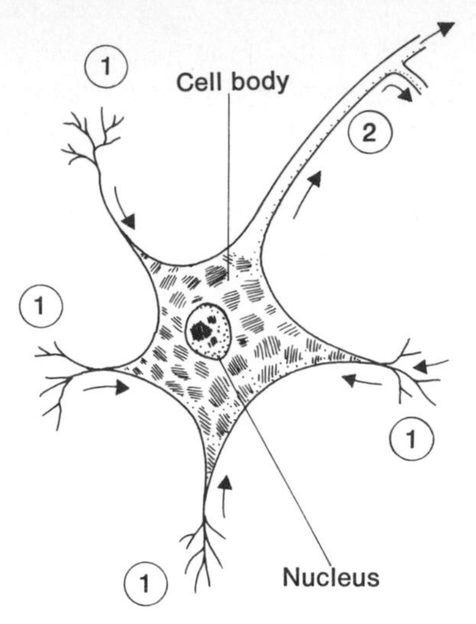

Fig. 2-64 Nerve processes. 1, dendrite; 2, axon.

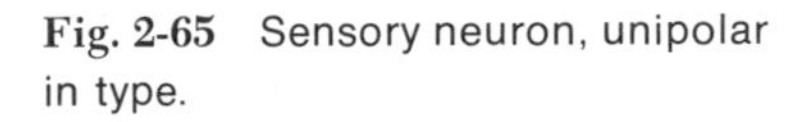

Fig. 2-65 Sensory neuron, unipolar in type.

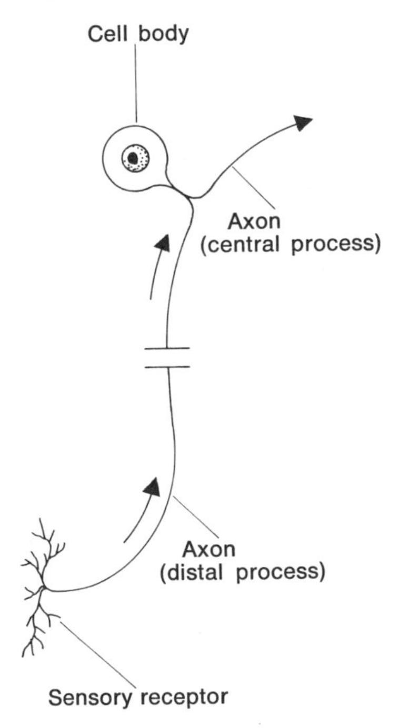

5. Inclusions (pigments, iron-containing granules, etc), particularly in neurons of older persons.

The shape of cell bodies varies from round to star-shaped. The shape of the cell is largely dependent upon the number of processes it has.

Neuronal processes are extensions of the plasma membrane and the cytoplasm of the cell body. They are often referred to as nerve fibers. These processes conduct nerve impulses to or from the cell body of origin. Neuronal processes conducting impulses toward their own cell body are called **dendrites** (Fig. 2-64). These processes are short, highly branched and receive many of the impulses coming into the cell body. They are not individually named since they do not form bundles of fibers. Their cytoplasmic constitution is like that of the cell body from which they originate.

Those processes transmitting impulses away from their own cell bodies of origin are called **axons** or axis cylinders (Fig. 2-64). Whereas there may be more than one dendrite per neuron, there is only one axon per neuron. In neurons with only one process, the process is generally referred to as an axon with distal and central branches. Axons usually travel in bundles, are often long (e.g., one group of axons extends from the spinal cord in the hip region and continues to the great toe without interruption), and may give off branches (collaterals).

Collections of nerve cell processes (mostly axons) within the CNS are called **tracts,** while collections of nerve cell processes (axons) in the PNS are called **nerves.** Most of these collections are named, e.g., optic tract, sciatic nerve.

Neurons may be functionally segregated into one of three classifications:

1. **Sensory neurons** (Fig. 2-65), whose axons conduct information from receptors (distal end of the neuron near the surface of the body, in viscera, or in muscles and blood vessels), to the CNS. Their cell bodies are usually located in sensory ganglia alongside (but outside) the CNS. The processes of sensory neurons are often termed afferent fibers. Generally, they belong to the PNS.
2. **Motor neurons** (Fig. 2-66) whose axons conduct movement-producing or gland-secreting impulses from the CNS to effector organs. Their cell bodies are located in the CNS (nuclei) or, in some cases, in motor ganglia outside the CNS. Processes of motor neurons are often called efferent fibers. Generally, these neurons belong to the PNS.

3. **Association neurons,** or **interneurons,** which remain entirely in the CNS and form a network between the motor and sensory neurons of the PNS (and the CNS). All of the coordination, correlation, integration, and modification in the CNS occurs by a complex interrelationship of millions of association neurons (Fig. 2-67).

Neurons may be structurally classified on the basis of numbers of processes.

Unipolar neurons (Fig. 2-65 and 2-68) are sensory neurons of the PNS in which only one process (pole) extends from the cell body. Cell bodies of unipolar neurons may be seen in collections alongside the brain and spinal cord (dorsal root ganglia). The single process of this neuron splits into central and distal branches.

Bipolar neurons (Fig. 2-69) are sensory neurons in which two processes (one axon, one dendrite) extend from the cell body. These neurons are associated with the organs of special senses (hearing, seeing, smelling, etc.).

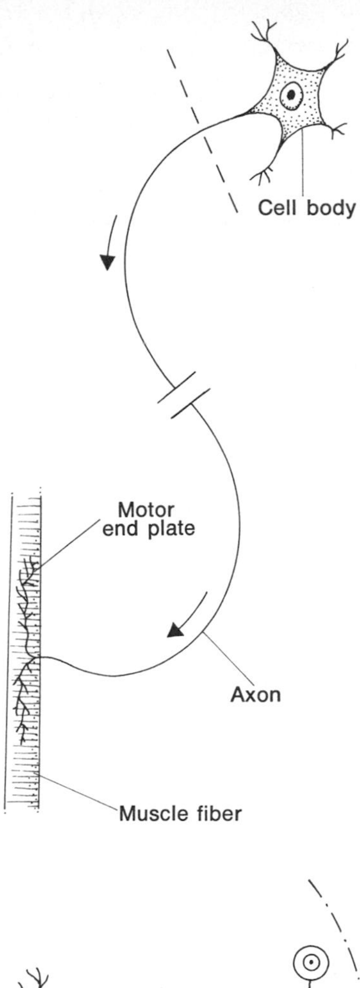

Fig. 2-66 Motor neuron.

Fig. 2-67 This schematic illustration shows how a motor response to pain stimulus might be influenced by memory area and other areas of the brain through a network of association neurons.

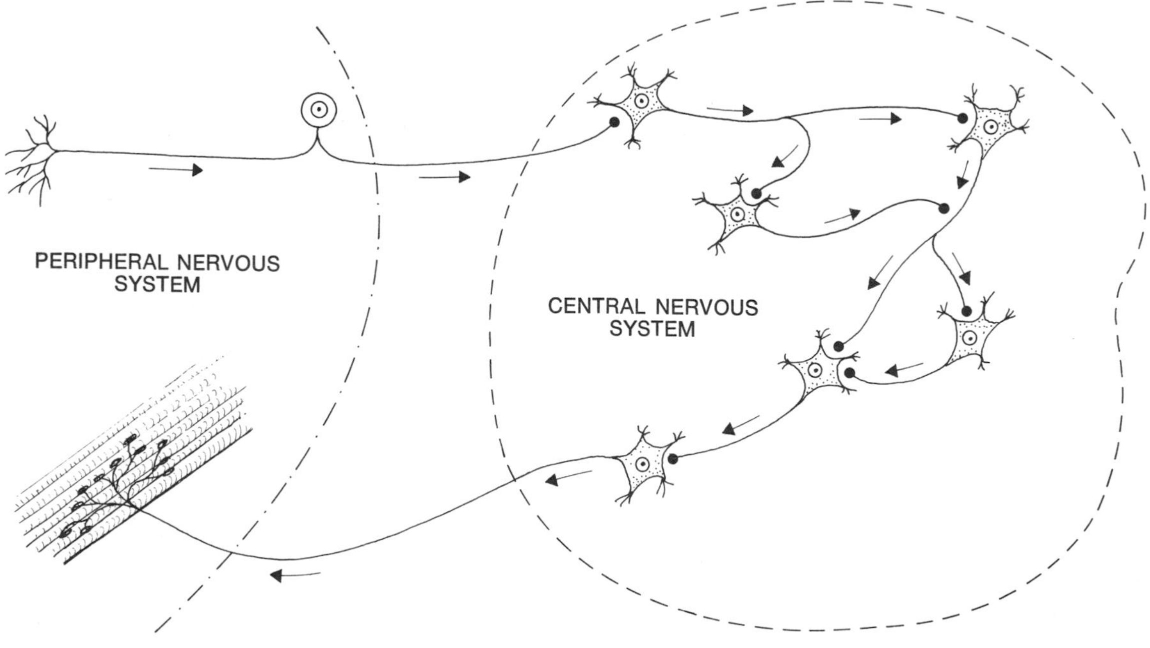

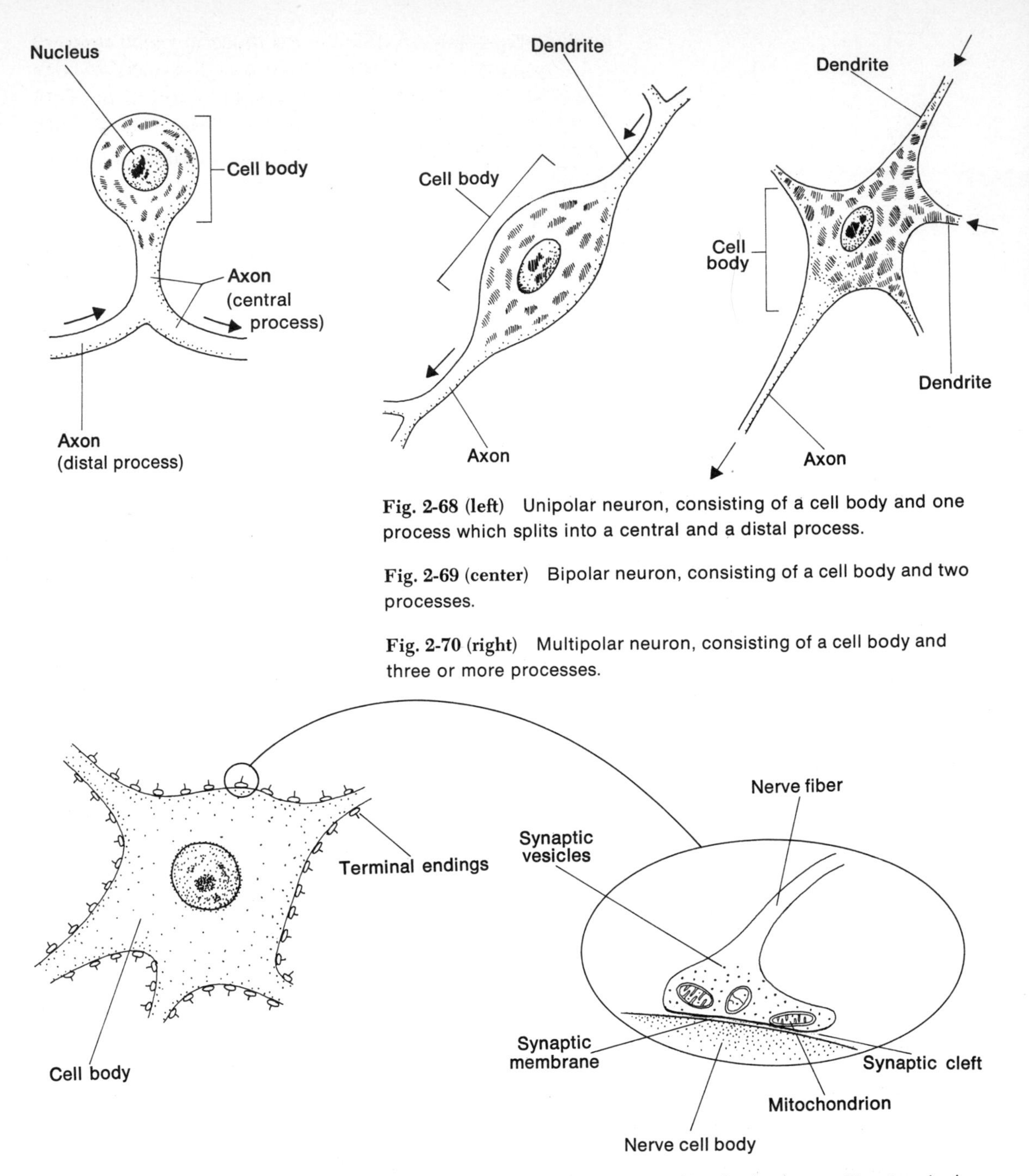

Fig. 2-68 (left) Unipolar neuron, consisting of a cell body and one process which splits into a central and a distal process.

Fig. 2-69 (center) Bipolar neuron, consisting of a cell body and two processes.

Fig. 2-70 (right) Multipolar neuron, consisting of a cell body and three or more processes.

Fig. 2-71 Anatomy of a synapse. Hundreds of nerve fiber terminal endings (boutons) synapse on a nerve cell body. When a nervous impulse arrives along the axon transmitter substance is released from synaptic vesicles into the synaptic cleft, where it stimulates the nerve cell body. Mitochondria supply energy for the transaction.

Multipolar neurons (Fig. 2-70) are those in which three or more processes (one of which is an axon) extend from the cell body. These neurons constitute a large part of the CNS and also some motor ganglia associated with the PNS. They may be motor, associative, or sensory.

Nerve connections and endings

It has been previously stated that two of the functions of the nervous system are (1) to correlate, coordinate, integrate, and modify information received and (2) to evoke appropriate responses. To do these things, there must be a free flow of information through innumerable circuits of neurons and this implies that neurons must interconnect. The connection between two neurons is called a **synapse** (Fig. 2-71). The structure of the various synapses is only now being appreciated with the invaluable electron microscope. The number of synapses is as great as the brain is complex—in fact, the complexity and efficiency of nervous communication is completely dependent upon the number of synapses. In short, the nervous system, like a politician, "has connections." The following are some characteristics of these connections:

1. One neuron may receive and make more than 1,500 synapses!
2. There is no structural contact between the two neurons involved in a synapse.
3. The intermediary between two neurons in a synapse is chemical.
4. The function of the chemical intermediary (transmitter) in a synapse is to excite the second neuron to generate an impulse or to inhibit its generation of an impulse.*

Nerve cell processes may end in several ways:

1. As a **terminal bouton,** the expanded end of an axon which synapses with the dendrite, cell body, or axon of another neuron (Fig. 2-72). Such synaptic endings are found in the CNS and ganglia of the PNS.

Fig. 2-72 Types of synapses. A. Axodendritic. B. Axosomatic. C. Axoaxonic.

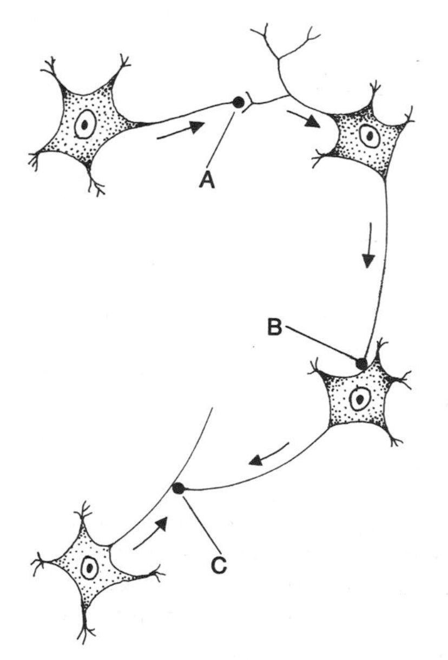

* A nervous impulse is often characterized as being "electrical." This is somewhat erroneous. An impulse is a wave of energy created by an unstable neuron plasma membrane. Normally the membrane is stable, but certain stimuli cause the membrane to become permeable to certain electrically charged chemical ions. This makes the membrane chemically and electrically unstable and this instability is transferred along the neuronal process. At a synapse, chemical transmitters are released when the rapidly moving membrane instability reaches the end of the process; these transmitters trigger membrane instability in the next neuron (in the same manner as before) and the "impulse" continues.

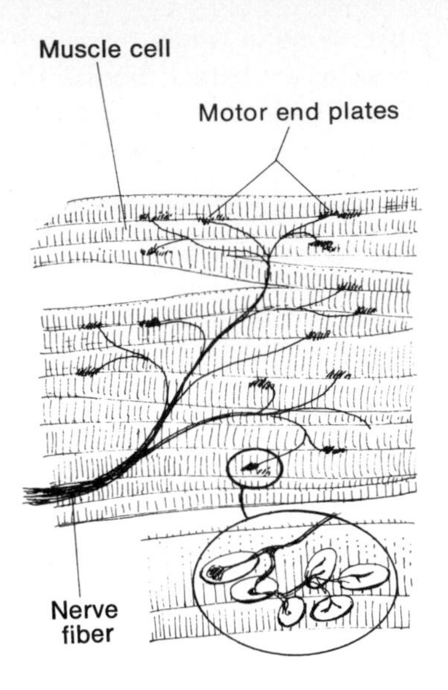

Fig. 2-73 Motor nerve endings on skeletal muscle.

2. As a **motor nerve ending,** the termination of a motor nerve fiber on a muscle or gland (Fig. 2-73).
3. As a **sensory nerve ending** (receptor), the distal termination of a sensory nerve fiber (Fig. 2-74); receptors are specialized for such modalities as pressure, touch, pain, hot, cold, vision, hearing, equilibrium, smell, taste, muscle sense, CO_2 concentration, and blood pressure.
4. As neuromuscular spindles, specialized muscle fiber–sensory receptor complexes in skeletal muscle. The receptors fire in response to muscle stretch (Fig. 2-75).

Nerve fiber coverings

Most neuronal processes of significant size, that is, axons, have one or more coverings, which they gather around themselves like cloaks (Fig. 2-76). The sheaths are of two kinds, myelin and neurilemma:

Fig. 2-74 A pressure receptor (Pacinian corpuscle) in longitudinal and cross sections. Cross section is a photomicrograph. (From W. F. Windle, *Textbook of Histology,* 4th ed., McGraw-Hill, New York, 1969.)

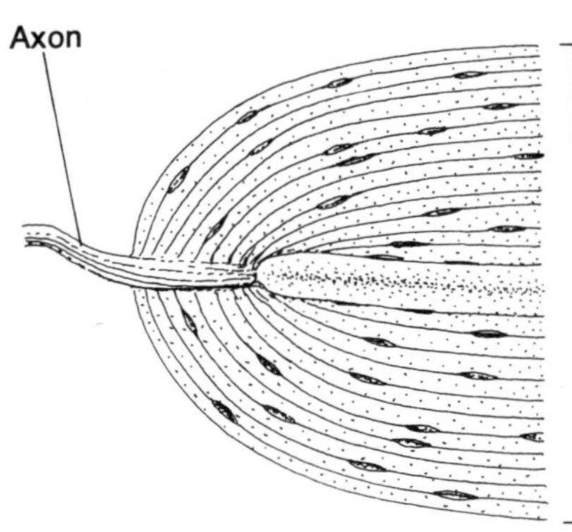

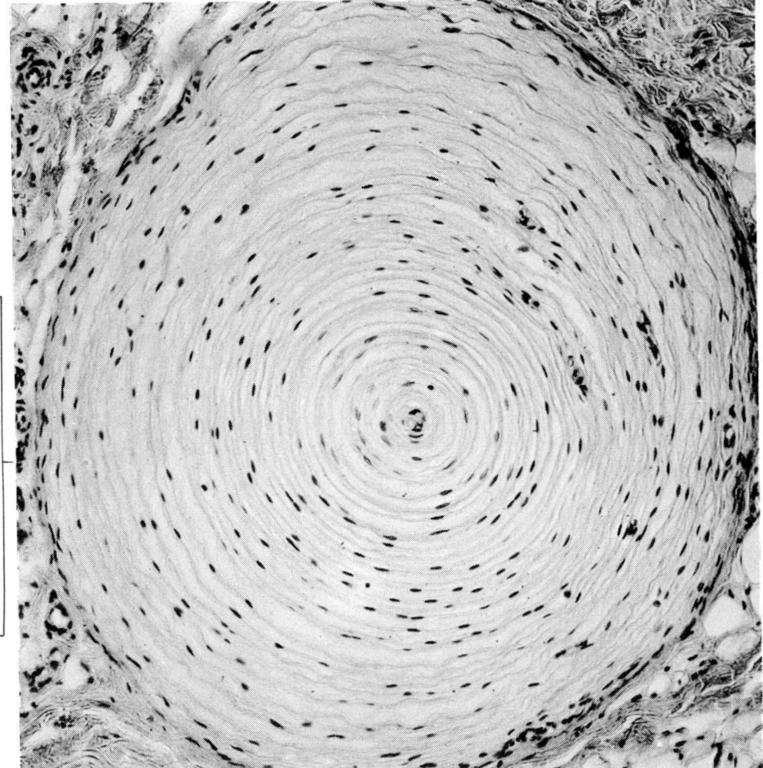

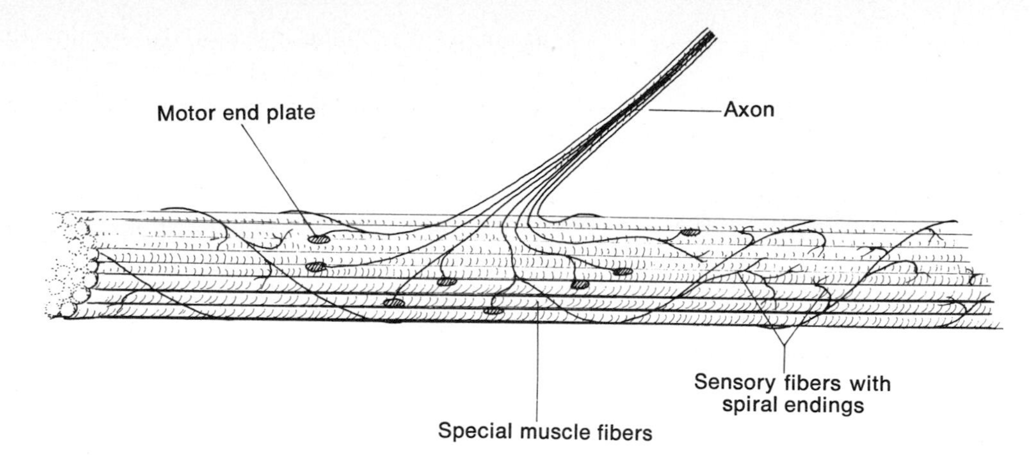

Fig. 2-75 A neuromuscular spindle. Sensory fibers entwine the special muscle fibers; motor fibers terminate at motor end plates.

Fig. 2-76 Development of coverings of an axon. A. Once invaginated, the axon appears to spiral or rotate, picking up layers of myelin from the Schwann cell plasma membrane. (Redrawn from B. O. Filho, based on data by J. D. Robertson.) B. A myelinated nerve fiber showing regular array of myelin layers. Electron micrograph. ×112,500. (From W. F. Windle, *Textbook of Histology,* 4th ed., McGraw-Hill, New York, 1969.)

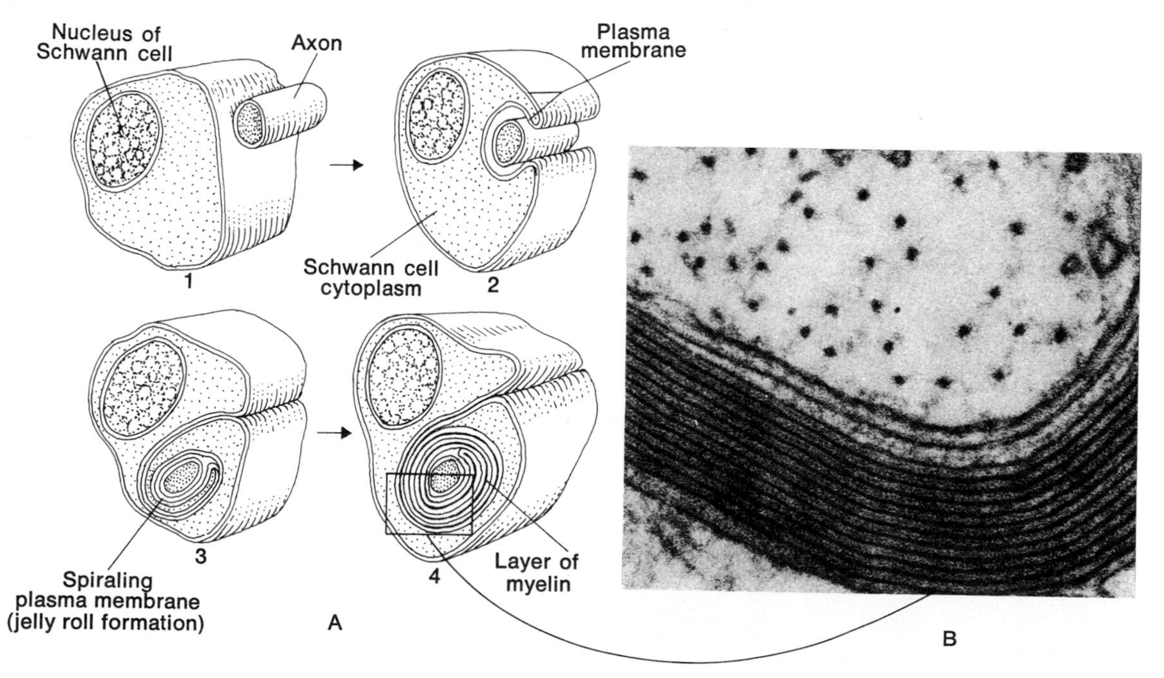

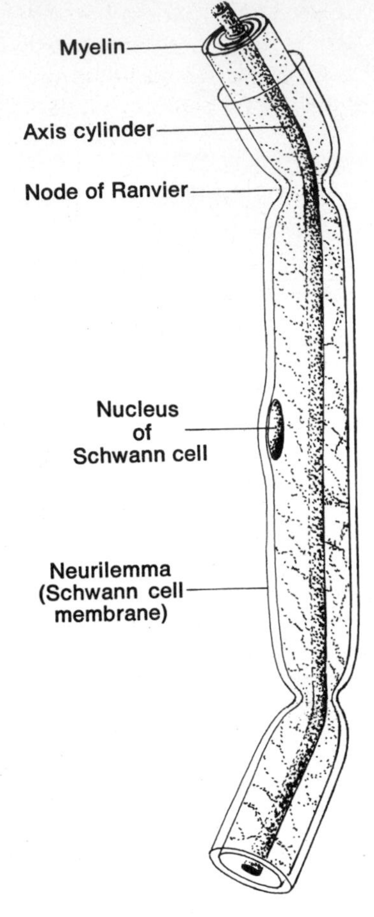

Fig. 2-77 Coverings of a peripheral nerve axon.

1. **Myelin** (Fig. 2-77) is a thick, fatty sheath found in the CNS and PNS. The tiny constrictions that can be seen along a myelinated nerve fiber are places where the myelin is interrupted and the axon is laid almost bare. These are nodes of Ranvier. They serve to facilitate the transmission of nervous impulses. Myelin is secreted by certain neuroglial cells in the CNS and by Schwann cells in the PNS.
2. The **neurilemma,** or sheath of Schwann (Fig. 2-77), is found on most fibers of the PNS. The neurilemma consists of cells (Schwann cells) wrapped around myelinated or nonmyelinated axons. The neurilemma is not found in the CNS since Schwann cells do not exist there. Neurilemma is apparently necessary for nerve regeneration, acting as a lattice for the growing nerve fibers. Hence CNS fibers cannot regenerate to any significant degree.

UPON REFLECTION

In this rather extensive unit, we have considered the basic structure and function of all the tissues in the human body. You will soon be studying the macroscopic (gross) structure of the body in some detail, and hopefully this introduction into the morphological microcosmos will make the former more clear and lucid. You should reflect now on how the four tissues can be combined to form organs prepared to perform the tasks of life—from the transport of blood to the conduction of air and from the storage of urine to the mechanical churning and digestion of ingested foodstuffs.

REFERENCES

1. Werner, Henry J.: *Synopsis of Histology,* 2d ed. McGraw-Hill Book Co., New York, 1967. (A good outline-style summary of the essentials of histology.)
2. Reith, Edward J., and Michael H. Ross: *Atlas of Descriptive Histology,* Harper and Row, Publishers (Hoeber Medical Division), New York, 1966. (Many good black-and-white photomicrographs with attendant descriptions. Should be a reference library item.)

3. Bloom, William, and Don W. Fawcett: *A Textbook of Histology*, 9th edition, W. B. Saunders Co., Philadelphia, 1968. (*The* treatise, complete and in depth. An outstanding *reference* for the nonmedical, undergraduate anatomy student.)
4. Ham, Arthur W.: *Histology*, 6th ed., J. B. Lippincott Co., Philadelphia, 1970. (This is *the* authority on bone. Highly recommended for bone and its development in detail. Also an excellent general histology treatise.)
5. Katz, Bernard: *Nerve, Muscle and Synapse*, McGraw-Hill Book Co., New York, 1966. (If you want a firm understanding of nervous impulses, synapses, and neuromuscular relationships, this is the paperback to which one should go!)
6. Peele, Talmage L.: *The Neuroanatomic Basis for Clinical Neurology*, 2d ed., McGraw-Hill Book Co., New York, 1961. (A thorough text on neuroanatomy.)
7. *Scientific American* offprints:
   - 20 The Nerve Impulse (Katz, 1952).
   - 51 White Blood Cells vs Bacteria (Wood, 1951).
   - 58 The Nerve Impulse & The Squid (Kernes, 1958).
   - 72 The Growth of Nerve Circuits (Sperry, 1959).
   - 79 Cilia (Satir, 1961).
   - 88 Collagen (Gross, 1961).
   - 155 Aging of Collagen (Verzar, 1963).
   - 158 Lymphatic System (Mayerson, 1963).
   - 176 How Cells Attack Antigens (Speirs, 1963).
   - 191 The Embryological Origin of Muscle (Konigsberg, 1964).
   - 196 Hemoglobin Molecule (Perutz, 1964).
   - 1001 The Synapse (Eccles, 1965).
   - 1007 The Sarcoplasmic Reticulum (Porter & Franzini-Armstrong, 1965).
   - 1011 The Physiology of Exercise (Chapman & Mitchell, 1965).
   - 1012 Evolution of Hemoglobin (Zuckerkandl, 1965).
   - 1014 The Flight Muscles of Insects (Smith, 1965).
   - 1018 The Production of Heat by Fat (Dawkins and Hull, 1965).
   - 1021 Electrical Effects in Bone (Basset, 1965).
   - 1026 The Mechanism of Muscular Contraction (Huxley, 1965).
   - 1038 The Nerve Axon (Baker, 1966).
   - 1064 Bone (McLean, 1955).
   - 1065 Sickle Cells and Evolution (Allison, 1965).
   - 1073 Small Systems of Nerve Cells (Kennedy, 1967).

clinical considerations

NERVE REGENERATION

Nervous tissue does not regenerate well at all. If and when it does, certain anatomical conditions must be met. Damage to the neurons of the brain and/or spinal cord is not normally followed by significant regeneration or replacement by new neurons. In other words, the damage is usually permanent, and it may be manifested by paralysis or behavioral changes. Damage to the nerves of the peripheral nervous system may be repaired with or without surgical assistance. The discussion in this section will be limited to nerve degeneration and regeneration in the PNS.

Hypothetical case: A person involved in an automobile accident suffers a deep transverse cut at the flexor side of the wrist. He is brought to the hospital. The resident surgeon notes that the arterial blood supply is still intact but that several tendons have been slashed as well as a prominent nerve to the hand—the median nerve. Sensory defects in and loss of certain motor capacities of the hand are noted, confirming the transection of the median nerve. The two cut ends of the nerve are approximated and united surgically and, following a period of several weeks, generally complete restoration of function and sensations in the hand will probably take place. What is the anatomical basis for this restoration of nervous activity in the hand?

Figure 2-78A diagrammatically demonstrates the microscopic appearance of the nerve immediately after the cut before any apparent cellular changes take place. All areas served by this nerve are affected, i.e., loss of muscular function and interruption in conduction of sensory information from affected area to the CNS. (Analogy: having the telephone wire cut while listening to another party on the phone.)

Subsequently, the following changes can be noted:

The nerve (axon) stump on the distal side* of the cut swells and degenerates (Fig. 2-78B). This occurs because these nerve processes have been separated from their cell bodies (located in the spinal cord and in ganglia alongside the cord) and therefore cannot survive. The myelin degenerates into

* *Proximal stump:* that fragment between the CNS and the cut; *distal stump:* that fragment between the cut and the effector organs/sensory receptors.

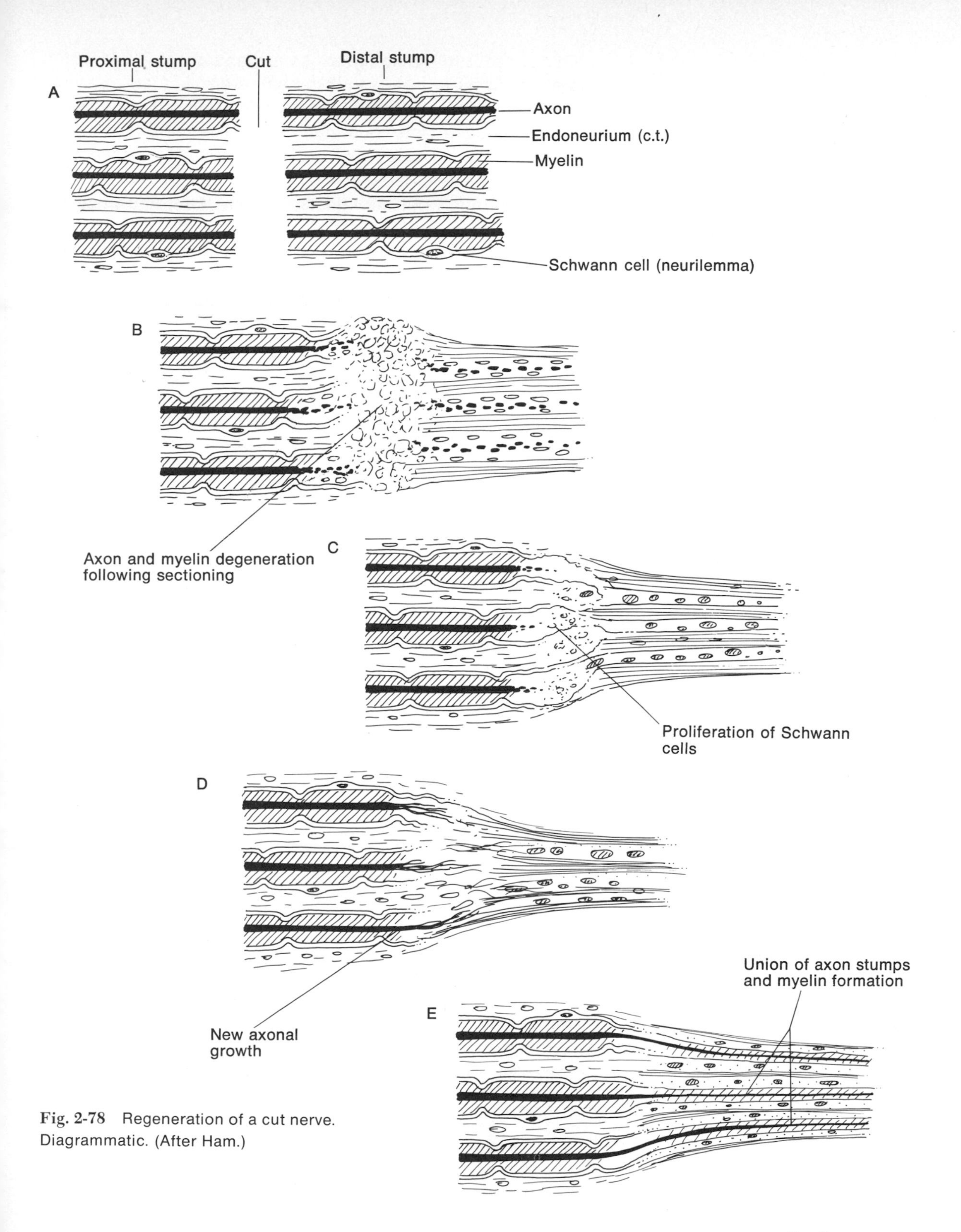

Fig. 2-78 Regeneration of a cut nerve. Diagrammatic. (After Ham.)

little "beads" following the axonal degeneration. This degeneration proceeds throughout the entire distal stump, leaving only a neurilemmal sheath (of Schwann). Unmyelinated fibers degenerate at a faster rate than myelinated fibers. The proximal stump will also degenerate to some degree, dependent on the proximity of the cut to its cell body. The closer the cut to the cell bodies of the affected nerves, the less chance of regeneration there is.

In the distal stump, macrophages from the neighboring connective tissues migrate in among the degenerated nerve fibers and myelin and phagocytose the debris (Fig. 2-78C). Meanwhile, Schwann cells from both the proximal and distal stumps proliferate and converge upon one another in the region of the cut—creating a tubular bridge between fragments. The fibroblasts from the endoneurium also proliferate, and connective tissue is laid down around the neurilemmal "bridge." During this time, it is *critical* that the original and developing neurilemmal tubes remain open.

Following the clearing of debris by the macrophages, the proximal, intact axon stumps start to grow distally, like the extension of so many fingers. This regeneration of nerve processes is dependent upon the existence of open neurilemmal sheaths—much as the growth of vines requires a lattice upon which to proliferate. Some of the growing nerve fibers manage to pass through the "bridge" of Schwann cells. These become enfolded longitudinally by the Schwann cells—in jellyroll fashion—as they progress distally.

The sprouting nerve fibers grow at a rate of about 1 to 2 mm per day (24.5 mm = 1 inch), about as fast as your fingernail grows. Thus for a cut at the wrist, the nerve fibers would progress down the neurilemmal sheaths to reach the effector or end organ in 6 to 10 weeks after the nerves began to grow. During this time, the Schwann cells secrete myelin as they envelop the axons (Fig. 2-78D). This myelin, a fatty substance, acts as an insulator for the nerve fibers. As each nerve fiber reaches its end organ, function is restored in that part (Fig. 2-78E). However, motor fibers which end at a sensory receptor and sensory fibers ending at a muscle end plate will not function. Eventually, the structure of the regenerated median nerve comes to appear much like it did before the transection.

The changes described in the foregoing account are classical—all conditions being ideal. There are a great many factors which influence nerve regeneration and which were not considered in this discussion, e.g., blood supply, absence of scar tissue, condition of the patient. This process of nerve regeneration is limited to the peripheral nervous system as Schwann cells (neurilemma) do not exist in the brain or spinal cord. But without open neurilemmal tubes the nerve fibers would grow in random directions and few, if any, would find their way across the "cut" to the end organs.

WOUND HEALING

One of the common phenomena of injury in which connective tissue plays an active part is *wound healing*. The spectrum of such wounds ranges from "sterile" surgical incisions to traumatic wounds involving damage to cartilage and the fracturing of bone. The discussion here will be limited to common flesh wounds (initiated by punctures and lacerations) in which the subcutaneous tissue is involved, specifically one caused by a nail passing through the skin into the underlying connective tissue.

Phase one (Fig. 2-79): As the nail tears through the epithelial membrane and enters the subcutaneous (loose) connective tissue, it unavoidably damages cells and rips open small capillaries and lymphatic vessels. Blood and lymph (fluid and cells) leak to the outside of the wound and are visible to the naked eye. Almost immediately, a clot (of fibrin and red and white blood cells) is formed, preventing further fluid loss and leakage. Depending on the condition of the nail and the flesh

Fig. 2-79 Trauma: initial phase of wound healing and inflammation.

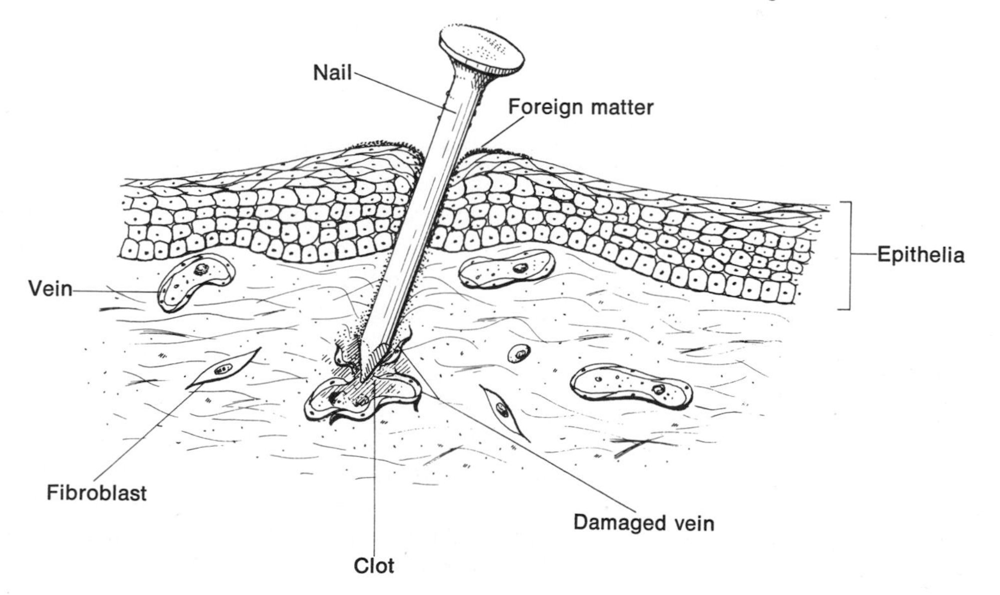

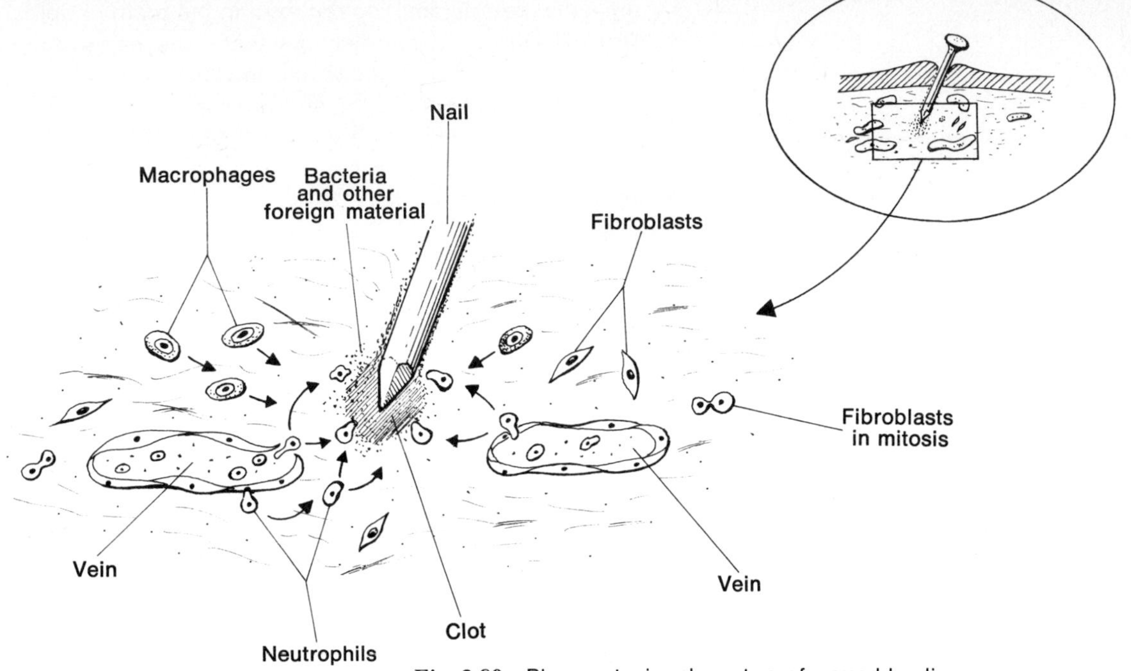

Fig. 2-80 Phagocytosis: phase two of wound healing.

through which it passed, variable numbers of bacteria start to proliferate about the clot.

Phase two (Fig. 2-80): Due to the toxins released by the bacteria, an acute inflammation develops (within 24 hr). The capillaries dilate (making the surface of the skin around the wound *red* and *warm*); the fluid of the blood (plasma) diffuses out into the connective tissue causing *swelling* around the wound. *Pain* accompanies these phenomena.

Due to chemical substances secreted by the damaged cells, certain phagocytic cells of the blood (neutrophils) and connective tissues (macrophages) are attracted to the clot. The neutrophils ingest the bacteria and clot components and subsequently die in the area; these dead white cells form visible masses called *pus*. The neutrophils will be subsequently phagocytosed by the macrophages, as will the remaining parts of the clot, and residual bacteria. All this will be accomplished in 48 to 72 hr in most cases.

Phase three (Fig. 2-81): After about 48 hr, changes in the connective tissue can be noted. Fibroblasts enlarge and enter into mitosis. As soon as the clot is cleared, fibroblasts "migrate" into the vacated area and start increasing their rates of protein synthesis. Within one week, increased numbers of re-

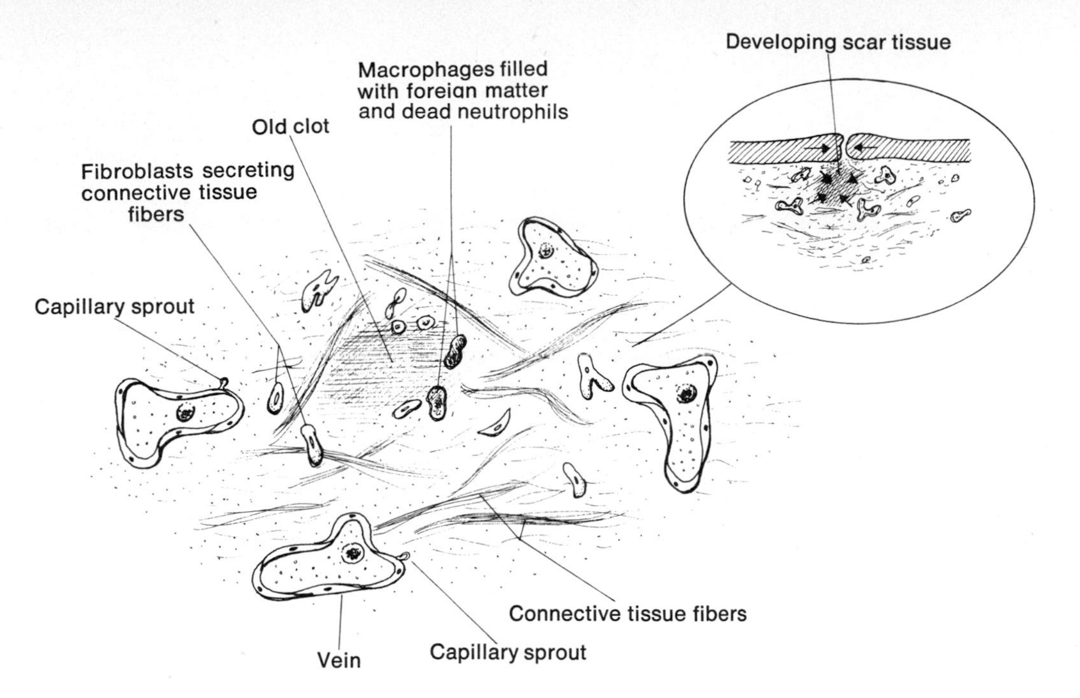

Fig. 2-81 Repair: phase three of wound healing.

ticular fibers, and then collagenous fibers, can be seen at the periphery of the wound area. By two weeks, a thick mass of collagenous fibers (dense, irregular connective tissue) solidly fills the area vacated by the clot. This mass is called **scar tissue** (a "scar"). While the scar tissue is forming, new blood vessels are formed (by sprouting from existing capillaries) and whole new capillary loops are formed within a few days. The excess fluid is reabsorbed by these capillaries and swelling is reduced. Within the developing scar tissue, capillaries are either pushed away or do not develop at all and the scar becomes generally avascular. Thus, a *reddish* wound becomes a *white* "scar," visible to the naked eye. As the scar tissue develops and matures, it contracts and the overlying edges of the torn epithelial membrane are approximated, and, through mitosis, the epithelial defect is closed.

Under normal conditions, the growth of connective tissue is very rapid—more rapid than any other tissue. When wounds cause large epithelial tears, the proliferating connective tissue may invade the epithelial defect and replace the epithelium "forever."

It is interesting to note that the proliferation of fibroblasts and blood vessels is retarded by the administration of cortisol (a well known adrenocortical hormone).

part three regions

unit 3 introduction to regional anatomy

The goal of this unit is to prepare you for study of the body as a functional complex of interrelated systems, organs, and tissues. Here you will get an overview (the broad-brush treatment) of some structural and functional characteristics of bones, muscles, nerves, blood vessels, lymphatics, and the skin. This overview should make your regional study more meaningful.

Generally, the basic *support* of the body consists of bone, to which many, many muscles are attached, providing the body with *form*. Since the various bones are connected in such a way that movement between most bones is permitted, it is convenient that muscles (which contract in response to nerve stimulation) cross these connecting points (joints), enabling bone *movement*. Now bone and muscle are living things, made up of collections of cells and their secretory products. Living things need oxygen and nourishment, and receive them by way of the blood, which is conducted from place to place in living, metabolically active vessels. Lymphatic vessels assist the blood vessels in maintaining the proper balance between ions and fluids in the body tissues. The skin and underlying fascia protect the body from the elements, house its exquisite sense receptors, and, finally, give the body the quality of smoothness and softness.

We will study the salient features of all these things.

introduction to osteology

SKELETON

The bones of the body make up a composite structure called the **skeleton.** Those bones oriented along the midline axis of the body constitute the **axial skeleton.** Those associated with the limbs or appendages constitute the **appendicular skeleton.**

Members of the axial skeleton (Fig. 3-1) include:

1. Skull
   a. Bones encasing the brain: **cranial bones** (8).
   b. Bones of the face: **facial bones** (14).
   c. Bones of the middle ear cavity: **ossicles** (6).
   d. Bone at the base of the tongue: **hyoid** (1).
2. Vertebral column (backbone)
   a. Supporting head and neck: **cervical vertebrae** (7).
   b. Supporting chest, neck, and head; articulates with ribs: **thoracic vertebrae** (12).
   c. Supporting abdomen, ribs, and chest (thorax); neck and head: **lumbar vertebrae** (5).
   d. Supporting the weight of the body less the lower limbs; articulates with and supports the hip bones: **sacrum** (five fused bones).
   e. The vestigial tail: **coccyx** (four fused bones).
3. Thorax
   a. The breast plate: **sternum** (1).
   b. Bones that surround the thoracic viscera and some of the abdominal viscera as well: **ribs** (24).

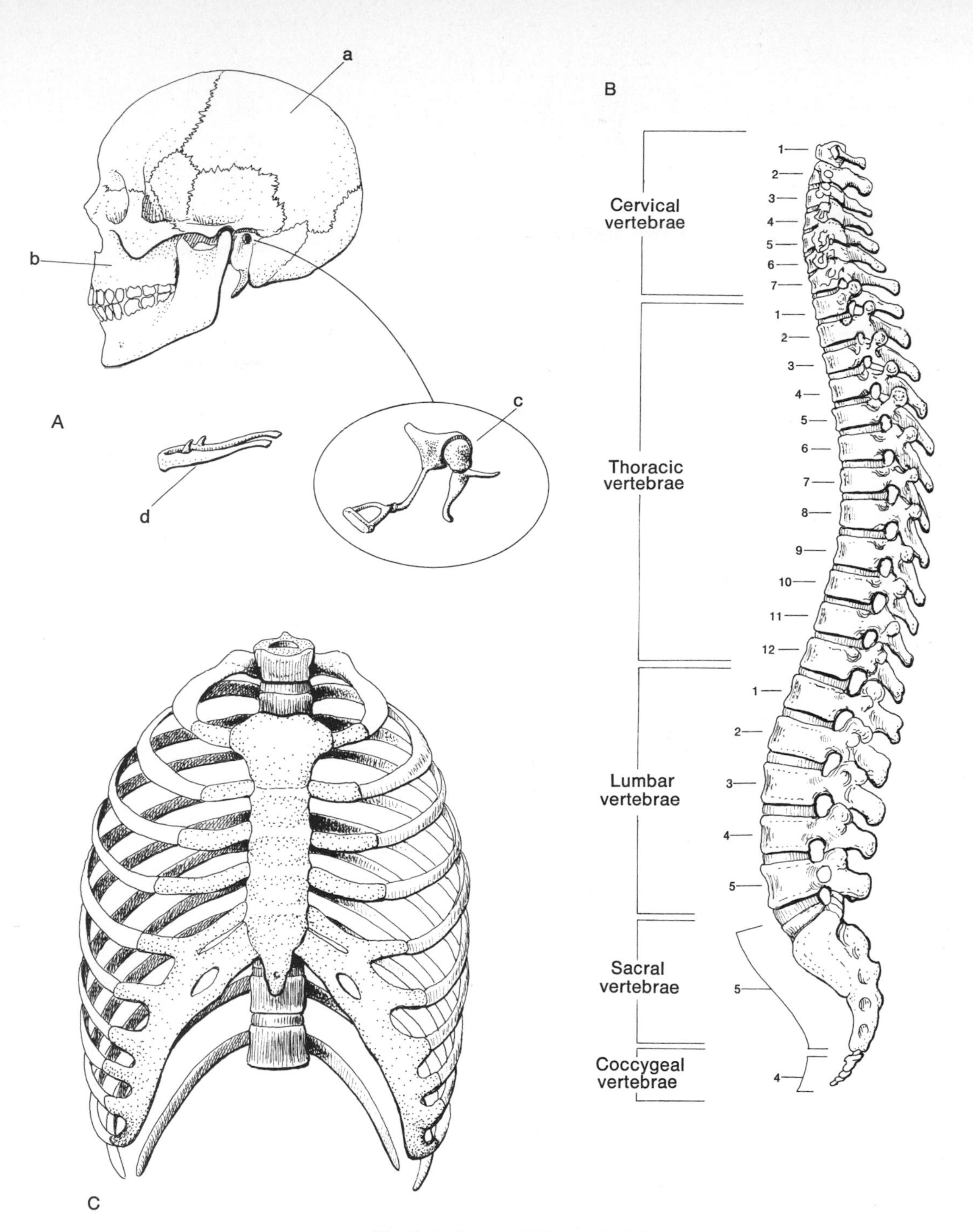

Fig. 3-1 Bones of the axial skeleton. A. Skull; a, cranial bones; b, facial bones; c, ear ossicles; d, hyoid bone. B. Vertebral column. C. Thorax.

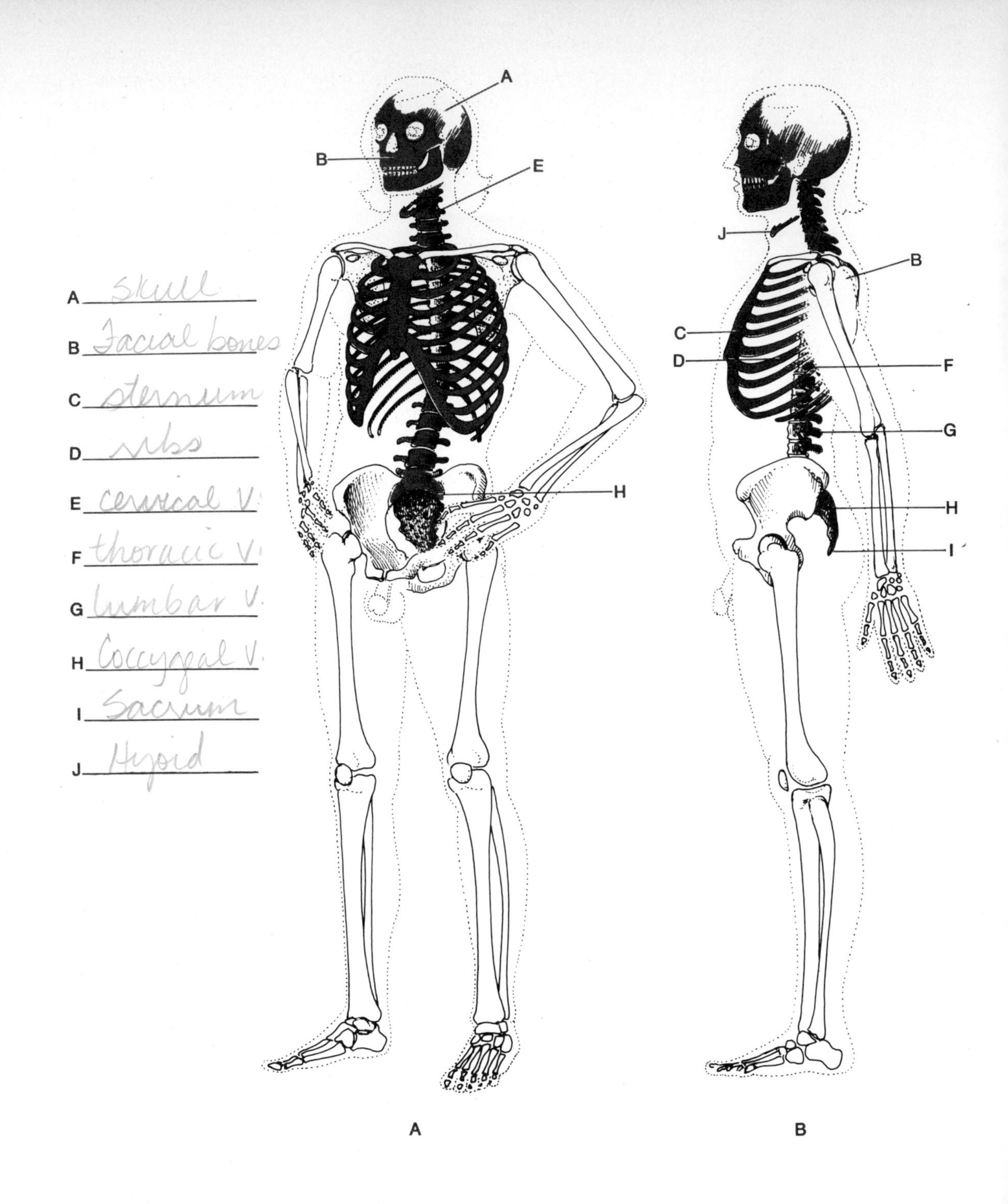

Fig. 3-2 Skeleton of the adult male. A. Anterior view. B. Lateral view. Axial skeleton has been blackened. Identify labeled bones.

So much for the axial skeleton—it may be reviewed in Fig. 3-2. How many of the bony regions can you remember without turning back to the text?

The appendicular skeleton consists of a pair of bony girdles and the limb bones to which they connect. In the case of the upper limb, the girdle surrounds the upper chest (pectoris) and is conveniently termed the **pectoral girdle.** The members of this girdle (Fig. 3-3) include:

1. The anterior components: **clavicle(s)** (2).
2. The posterior components: **scapula(e)** (2).

From the scapulae, the limb bones are fastened in the following sequence:

1. Bone of the arms: **humerus** (2).
2. Bones of the forearms: **ulna** (2) and **radius** (2).
3. Bones of the wrists: **carpals** (16).
4. Bones of the hands (and fingers): **metacarpals** (10) and **phalanges** (28).

The bones of the lower limb include the limb bones and the girdle of bone (the *hip* bones) encircling (and thereby forming) the pelvis—conveniently termed the **pelvic girdle.** From the **hip bones,** articulated in sequence, are:

1. Bone of the thigh: **femur** (2).
2. Bones of the leg: **tibia** (2) and **fibula** (2).
3. Bones of the foot: **tarsals** (14), **metatarsals** (10), and **phalanges** (28).

So much for the appendicular skeleton, your knowledge of which can be tested in Fig. 3-3.

Classification according to shape

All these bones of the skeleton, when sorted out, generally fall into about four classes according to shape (Fig. 3-4):

1. Those with a long axis—more length than width: **long bones.**
2. Those which are stubby—more or less cube-shaped: **short bones.**
3. Those which are platelike: **flat bones.**
4. All others: **irregular bones.**

Sesamoid bones

Among irregular bones is also a group of small sesame-seed-size bones **(sesamoid bones)** which develop in the tendons of certain muscles. Found in the foot, hand, and the knee, these

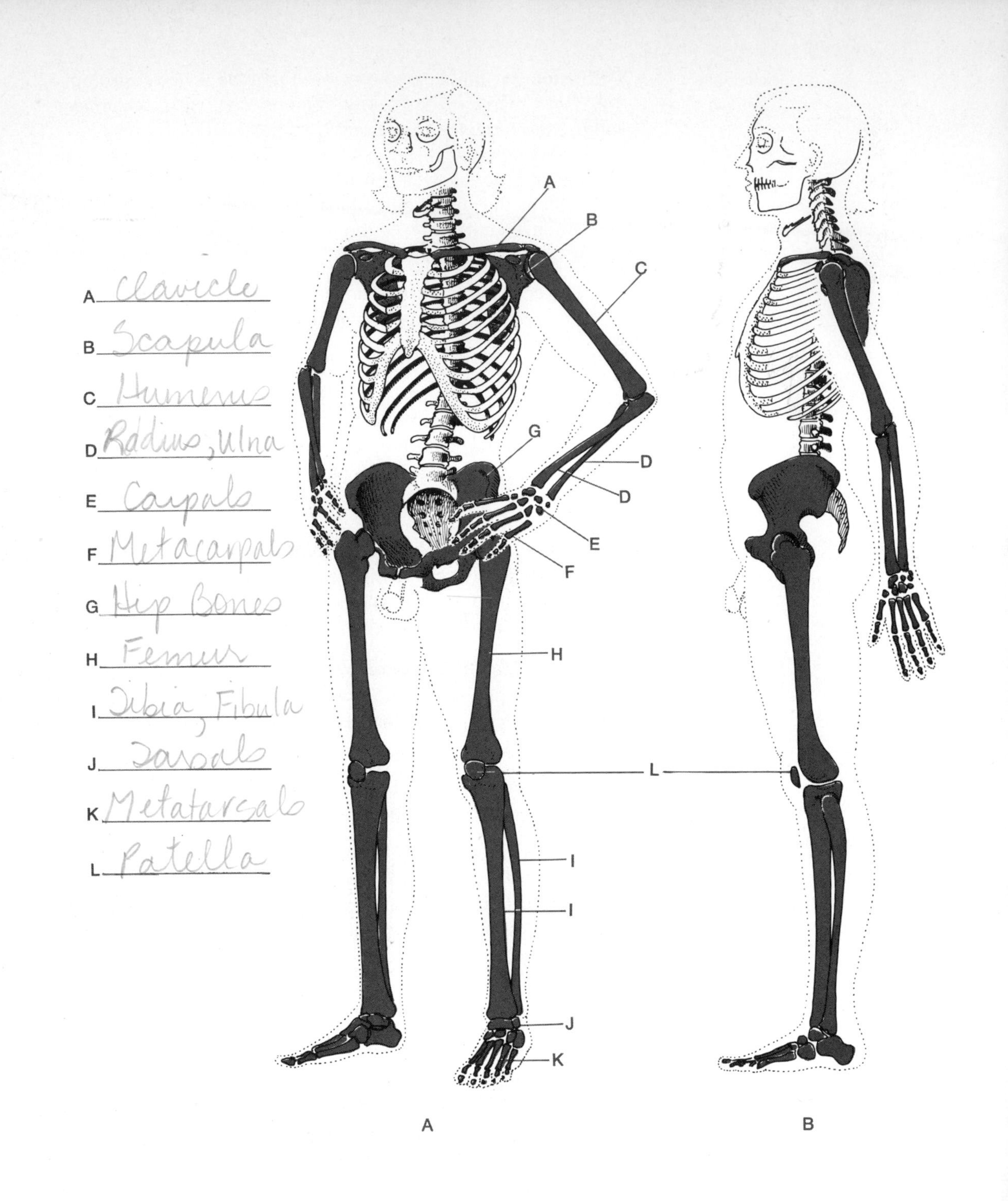

Fig. 3-3 Skeleton of the adult male. A. Anterior view. B. Lateral view. Appendicular skeleton has been blackened. Identify labeled bones.

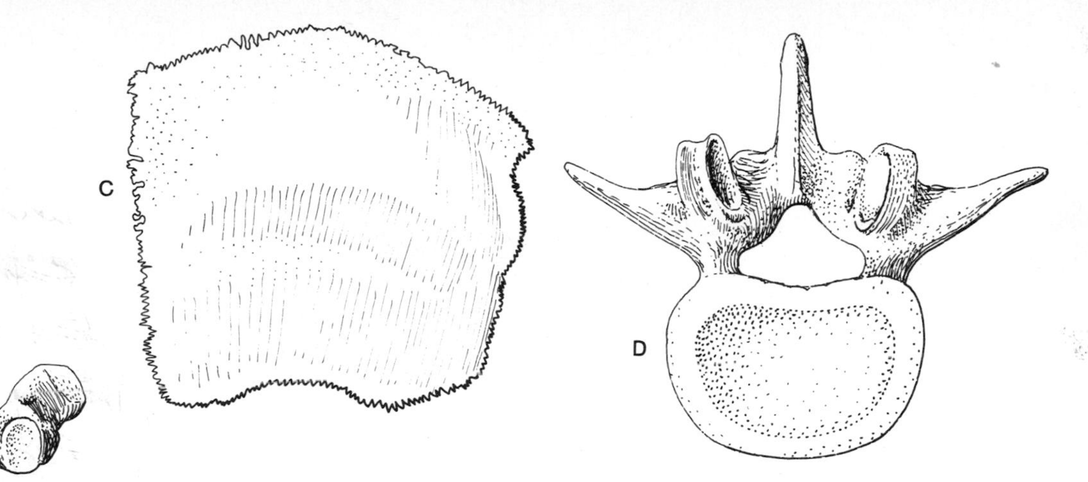

Fig. 3-4 Classification of bones according to shape. A. Long bone (phalanx). B. Short bone (wrist). C. Flat bone (skull). D. Irregular bone (vertebra).

bones apparently formed, in evolutionary terms, in response to friction generated by the tendons passing over bone. The largest (and most significant) sesamoid bone in our body is the **patella**—the knee cap (Fig. 3-3).

Characteristics of bone shape

During formation and growth, bones develop holes, projections, and depressions. The holes, or **foramina** (sing. foramen), are created by the passage of blood vessels or nerves into or out of bone. Ridges and grooves of bone are often created by nerves and/or vessels pressing on the soft developing bone (much as a paw-print is formed in newly laid cement) (Fig. 3-5). Many of the projections and depressions of bone are created by the pull of muscles and tendons (so-called *traction epiphyses*).

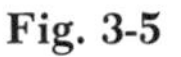

Fig. 3-5

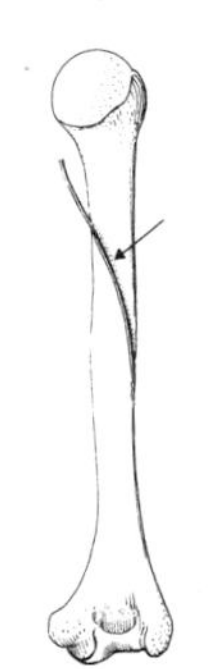

The terminology for the various foramina, ridges, grooves, projections and depressions involves the frequent use of synonyms. For example:

A hole in a bone
- if round, is a **foramen** (Fig. 3-6A),
- if slitlike, is a **fissure** (Fig. 3-6B),
- if tubelike, is a **canal** or **meatus** (Fig. 3-6C),
- if a hollow space, is a **cavity** (Fig. 3-6D).

A ridge on a bone
- if extensive, may be called a **crest** (Fig. 3-7A),
- if thin, may be called a **line** (Fig. 3-7B).

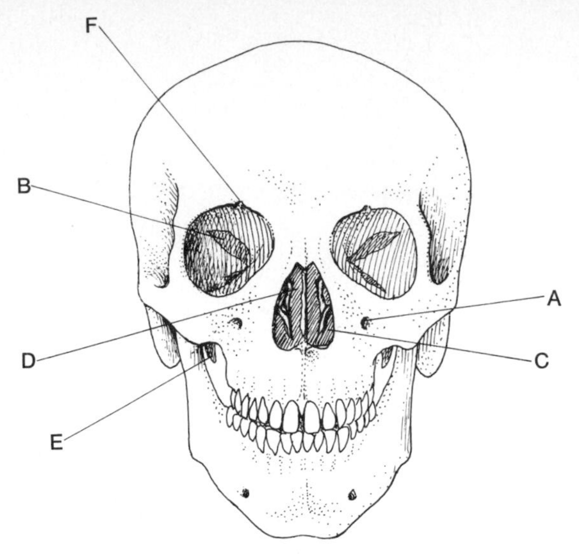

A foramen
B fissure
C canal or meatus
D cavity
E spine or styloid process
F notch

Fig. 3-6 Some foramina, grooves, projections, and depressions in the skull. Identify these characteristics in the blank spaces.

A groove on a bone (a depression with a long axis)
may be termed a **sulcus** or a **fissure** (Fig. 3-6B).

A projection on a bone
of large proportions, may be a **trochanter,** a **tuberosity,** or a **malleolus** (Fig. 3-7C),
if small, may be a **tubercle** or **process** (Fig. 3-7D),
if pointed or spear-like, may be a **spine** or a **styloid process** (Fig. 3-6E).

A depression in a bone
if small, is a **fovea** (Fig. 3-7E),
if larger and deeper, is a **fossa** (Fig. 3-7F)
if moon-shaped, may be a **notch** (Fig. 3-6F).

Whenever a bone articulates with another, the articular surface, capped with hyaline cartilage, may be
rounded—a **condyle** (Fig. 3-7G) or **head** (Fig. 3-7H), if
flat—a **facet.**

The bones of the skeleton serve at least one of a number of potential functions. Such functions include:

1. *Support* of the body or some part of it.
2. *Attachment* points for muscles and tendons.
3. *Protection* for internal viscera.
4. *Hemopoietic* (blood-forming) activity within the marrow cavity of long bones and others.
5. *A reservoir of calcium* for the fluids and cells of the body (mineral content of bone is 65 percent by weight).

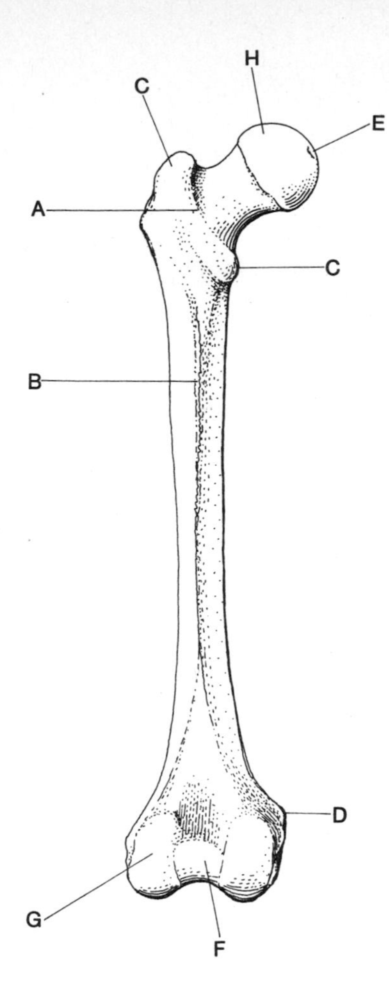

Fig. 3-7 Some projections and depressions on the femur. Identify these characteristics in the blank spaces provided.

A ____________ E ____________

B ____________ F ____________

C ____________ G ____________

D ____________ H ____________

QUIZ

Now, let's see if you have a good understanding of the skeleton:

1. The skeleton is responsible for the basic ____________ of the body.
2. The part oriented in the midline along the longitudinal axis of the body is the ____________ ____________.
3. The bones associated with the limbs are collectively titled the ____________ ____________.
4. In the former category, one finds the bones of the ____________, ____________ ____________, ____________
5. The limb bones—both upper and lower—are fastened to ____________ and ____________ girdles, respectively.
6. Holes in bone are generally called ____________ and exist for the ____________ of ____________ ____________ or ____________.
7. Trochanters and tuberosities are both large ____________ on a bone.
8. A fossa is an example of a ____________ in a bone.
9. The articulating surface of a bone may be called a ____________, ____________, or ____________.
10. Generally, the skeleton serves the functions of ____________, muscle ____________, ____________, ____________ activity and a calcium reservoir.

introduction to arthrology (joints)

An **articulation** describes the joining of two or more bones. There are various kinds of joints, classified according to type of movement permitted or what the joint is made of. Functionally, joints may be (1) *immovable,* (2) *slightly movable,* (3) *freely movable.* Structurally, joints are (1) *fibrous,* (2) *cartilaginous,* or (3) *synovial.* The latter classification is preferable for our purposes.

Now, fibrous and cartilaginous joints may be immovable or slightly movable, but they are never freely movable. However, all synovial joints are freely movable.

Fibrous joints are those in which the intervening substance between adjoining bones is fibrous (dense regular or irregular) connective tissue. Such joints may be seen between bones of the skull; these joints are often called **sutures** (Fig. 3-8). With aging, the fibrous tissue is replaced by bone and

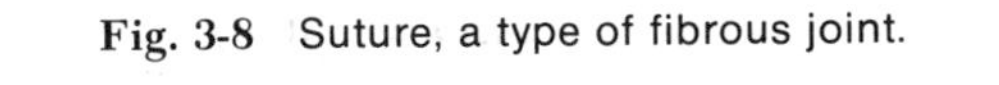

Fig. 3-8 Suture, a type of fibrous joint.

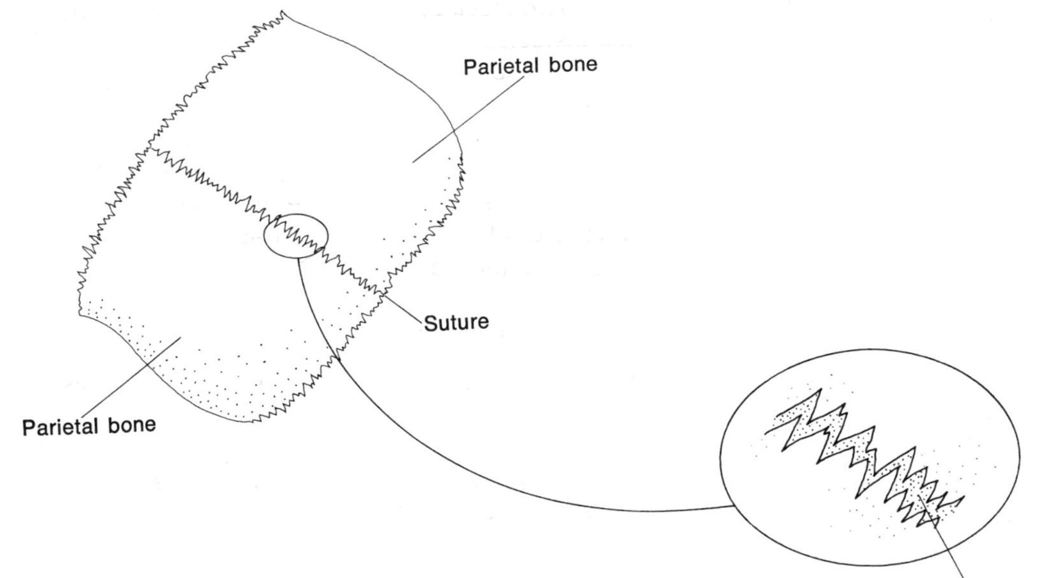

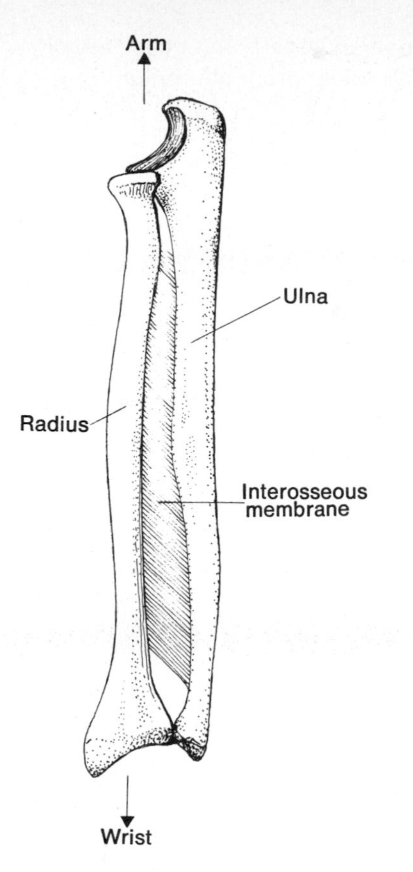

Fig. 3-9 Syndesmosis, a type of fibrous joint.

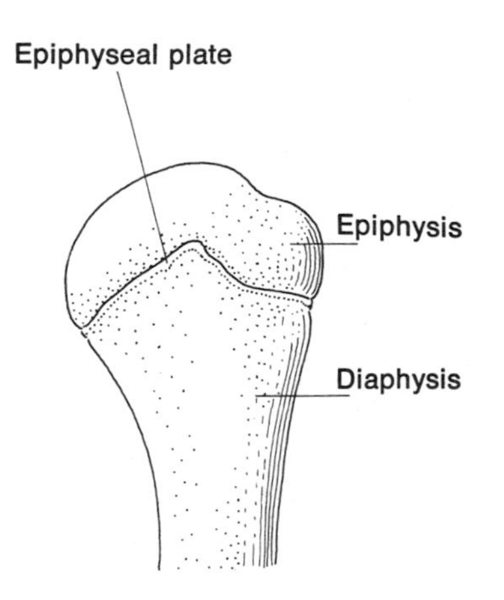

the suture becomes a *synostosis*—an immovable joint. Fibrous joints are also seen between two bones which are some distance apart. The intervening tissue is regularly arranged dense c.t. (ligaments). Seen between the bones of the forearm and bones of the leg, a joint of this type is often termed *syndesmosis* and is described as an interosseous membrane (Fig. 3-9). Syndesmoses are slightly movable.

Cartilaginous joints are those in which the intervening substance between adjoining bones is cartilage. Such joints may be seen between the epiphysis and diaphysis of developing bone. You will remember seeing these before. They were called epiphyseal plates. Recalling the descriptive terms synostosis and syndesmosis, how would you describe—in one word—the above joint? *Synchondrosis* (Fig. 3-10). Such a joint is immovable and in young adulthood becomes a synostosis. Cartilaginous joints are also found between bones where the intervening substance is fibrocartilage (in addition to the hyaline cartilage over the articular surfaces). This joint is slightly movable and may be seen at the interpubic joint, intervertebral joint, and between upper segments of the sternum. Such joints are called **symphyses** (Fig. 3-11).

Synovial joints are freely movable joints (Fig. 3-12) and are all characterized by:

1. A *fibrous articular capsule* enclosing the two articular surfaces, thus forming a joint cavity. These capsules, often reinforced by ligaments, arise from and merge with the periosteum on the sides of the bone. The capsules may be strong (as in the hip joint) or weak (as in the shoulder joint).
2. A *synovial membrane* lining the interior of the fibrous capsule. This loose connective tissue membrane (*a*epithelial) forms folds within the joint cavity but does not cover the actual articulating surfaces.
3. *Synovial fluid* secreted by the synovial membrane. It is a watery fluid, much like (and may be) plasma and acts to lubricate the joint and prevent overheating.

Synovial joints may also incorporate other structures which are necessary to joint stability. Such structures include articular discs or menisci (pads of fibrocartilage), rings of fibrocartilage around sockets, thus deepening the socket, or extra ligaments reinforcing the capsule.

Fig. 3-10 Synchondrosis, a type of cartilaginous joint. An epiphyseal plate of the developing humerus.

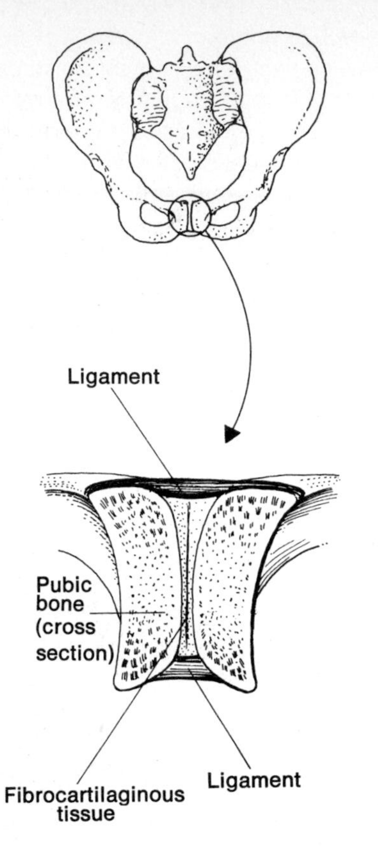

Fig. 3-11 Symphysis, a type of cartilaginous joint.

Synovial joints may be one of several varieties. Specific joints will be considered with the appropriate region, but for now, refer to Fig. 3-13:

1. **Gliding joints** (or plane joint): where the two articular surfaces are flat and slide across one another.
2. **Hinge joints:** where one articular surface is concave and the other is convex. Movement is limited to one axis. So in terms of movement this is a uniaxial joint.
3. **Pivot joints** (also uniaxial): where one bone rotates on or about another, much as a wheel rotates about its axle.
4. **Condylar joints:** where one bone is a socket and the other fits into the socket *but* the movements are limited to circumduction—*no rotation*. This is really a lesser degree of a ball and socket joint.
5. **Ball and socket joints** (multiaxial): characterized by articular surfaces shaped like a ball and a socket—rotation and circumduction are allowed.
6. **Saddle joint** (biaxial): where the concavity of one surface moves in the concavity of the other.

Fig. 3-12 Characteristics of a typical synovial joint.

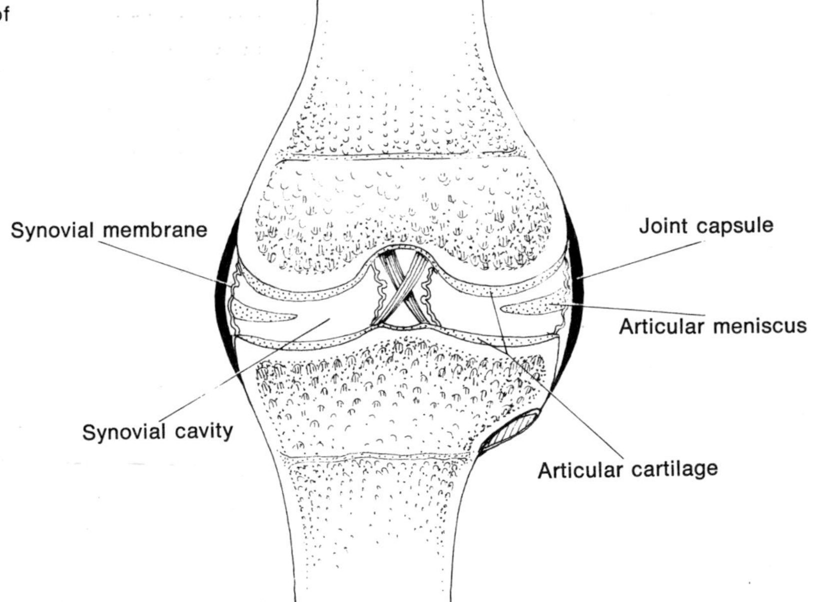

Fig. 3-13 (opposite) Some types of synovial joints.

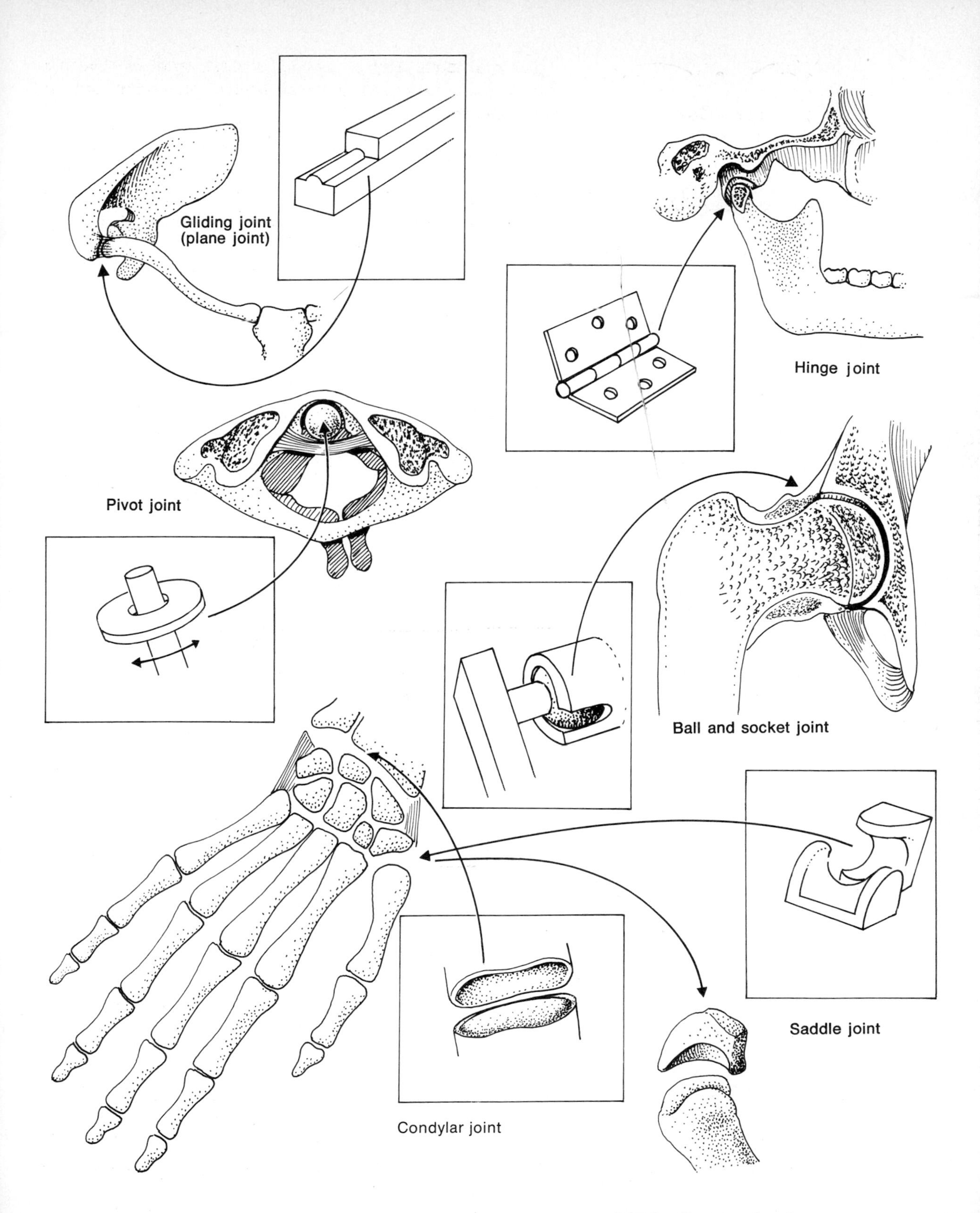
Gliding joint
(plane joint)
Hinge joint
Pivot joint
Ball and socket joint
Saddle joint
Condylar joint

introduction to skeletal myology

By now you are familiar with some general concepts and specific facts about the microscopic nature of muscle. Next we shall discuss skeletal muscle as an organized aggregation of organs which make up the *form* of the body as well as providing the latter with the ability to *move*.

GROSS ANATOMY

A skeletal muscle generally has two parts: the contractile unit, the **belly,** and the ends which attach to bones, **tendons** (Fig. 3-14). The muscle fibers themselves generally do not attach to bones but do so by means of connective tissue (tendon). If the tendon is flat and sheetlike in appearance, it is an **aponeurosis.** A tendon attaches to bone by merging with its periosteum or by inserting directly into the bony tissue. In some cases, e.g., the muscles of the face, tendons insert into and merge with the superficial fascia underlying the skin.

The individual skeletal muscle fibers do not generally extend throughout the whole length of a muscle. The muscle fibers are connected to fellow parallel fibers by connective tissue (endomysium) (Fig. 3-14).

Fig. 3-14 Anatomy of a skeletal muscle and its connective tissues.

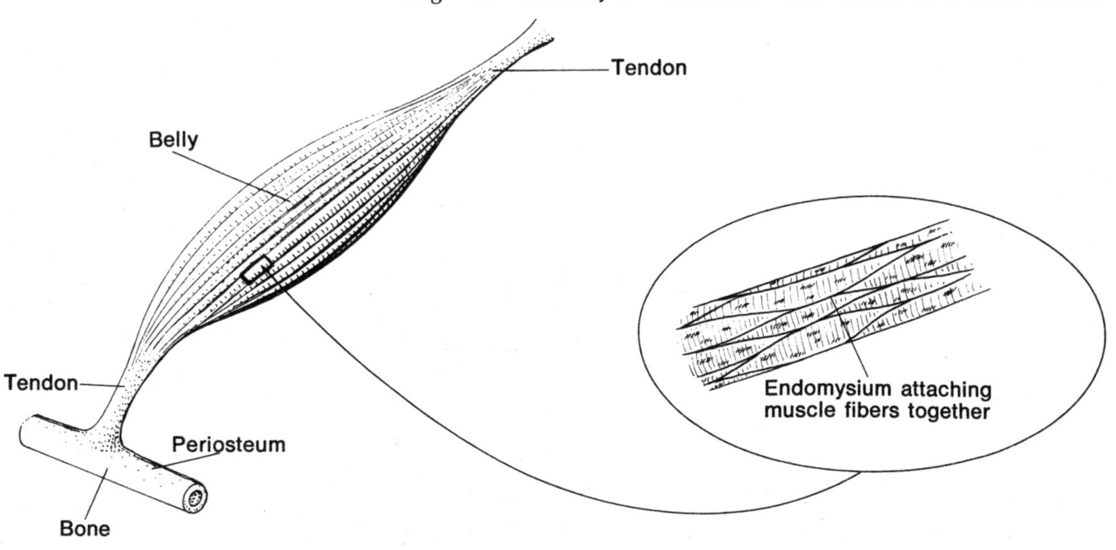

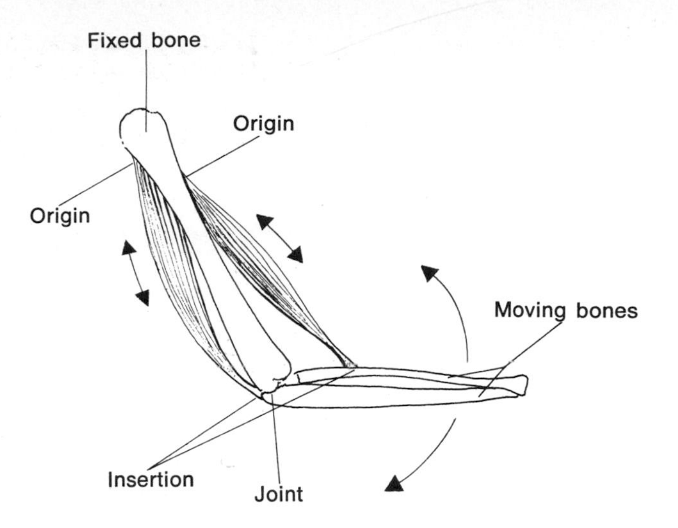

Fig. 3-15 Nomenclature of muscle attachment.

Fig. 3-16 General pattern of muscle fiber architecture.

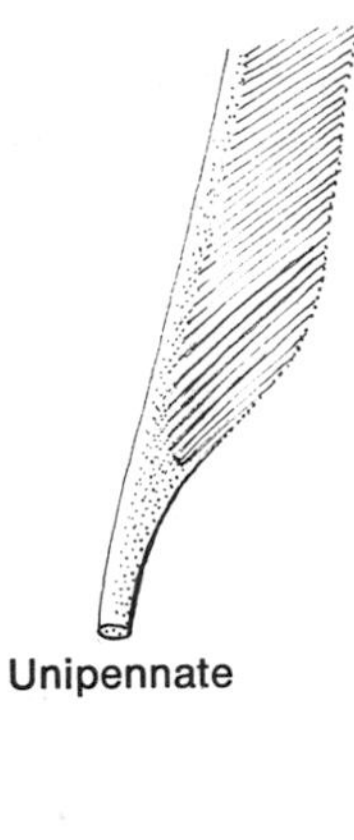

Bipennate

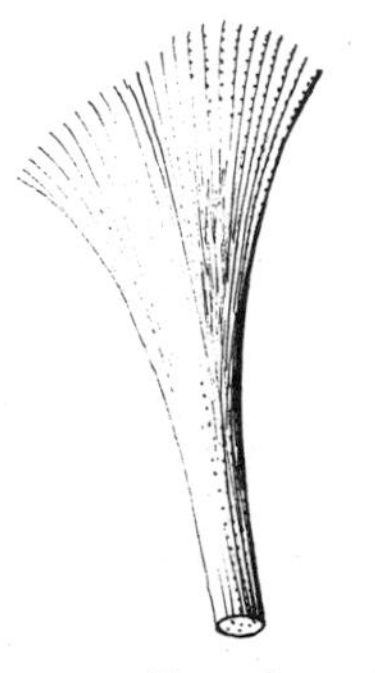

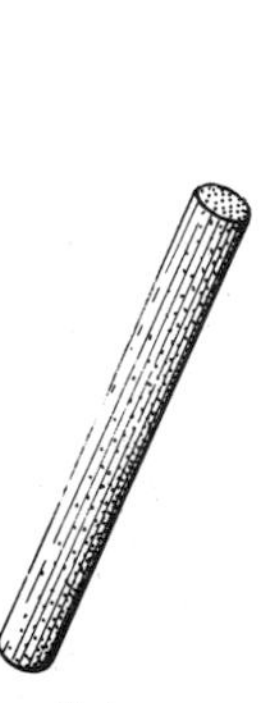

A muscle usually is attached (by tendons) to two bones that abut at a joint; one of those bones will move away from or toward the other when the muscle contracts, depending on the joint involved and the muscle's relation to that joint. It is important to differentiate the two attachments (Fig. 3-15) in order to appreciate the muscle's function:

1. The attachment to the more fixed (nonmoving) bone is the **origin.***
2. The attachment to the moving bone is the **insertion.**

The arrangement of fibers in skeletal muscle is variable but is frequently related to the muscle's functional capacity. The variation of fiber architecture is dependent on orientation of the tendons, presence of connective tissue septa, length of muscle, etc. The general patterns of fiber architecture may be appreciated in Fig. 3-16.

* Sometimes these definitions do not apply, for they are dependent upon the kind of movement taking place. In some cases, the origin may be that bone *closest* to the midline of the body or *closest* to the trunk of the body; the insertion may be the bone *farthest* from the midline of the body or *farthest* from the trunk of the body.

Multipennate

MUSCLE NOMENCLATURE

The skeletal muscles of the body may be named according to shape, location, attachment, size, or function. The following matching quiz is designed to aid in familiarizing you with the "strange" vocabulary of skeletal muscles. In many cases, the relationship is readily apparent. Muscles named according to *shape* include (match the definition to the proper term):

4 Teres	1. Kite-shaped
2 Deltoid	2. Triangular or delta-shaped
7 Quadratus	3. Trapezoid-shaped
5 Serratus	4. Round
1 Rhomboid	5. Saw-tooth attachments to bone
6 Latissimus	6. Widest or broadest
3 Trapezius	7. Square or rectangular-shaped

Answers: 4, 2, 7, 5, 1, 6, 3.

Muscles named according to *location* include:

2 Brachialis	1. Muscle of the ear
4 Pectoralis	2. Muscle of the arm
1 Auricularis	3. Muscle of the buttock
3 Gluteus	4. Muscle of the chest

Answers: 2, 4, 1, 3.

Muscles named according to *attachments* include:

3 Hyoglossus	1. Muscle attached to thumb (pollex)
5 Sternocleido-mastoid	2. Muscle of the cranium (frontal bone)
2 Frontalis	3. Muscle attached to hyoid bone and tongue
1 Pollicis	4. Muscle attached to great toe (hallux)
4 Hallucis	5. Muscle attached to sternum and mastoid process

Answers: 3, 5, 2, 1, 4.

Muscles named according to *size, orientation,* or *relative position* include:

6 Major	1. The long muscle
4 Magnus	2. The smaller of two muscles
3 Brevis	3. The short muscle
1 Longus	4. The great muscle
5 External	5. The outermost muscle
7 Superioris	6. The larger of two muscles
2 Minor	7. The most cephalic of two or more muscles
9 Rectus	8. An angular muscle (relative to the vertical)
8 Oblique	9. A straight muscle

Answers: 6, 4, 3, 1, 5, 7, 2, 9, 8.

Muscles named according to *function* include:

__4__ Sphincter	1. Narrows or constricts
__3__ Abductor	2. Straightens a joint
__1__ Constrictor	3. Takes away a bone from the midline
__5__ Depressor	4. A circularly arranged muscle which functions like a "purse string"
__2__ Extensor	5. Lowers a bone at a joint

Answers are apparent.

SIMPLE MACHINES OF THE MUSCULOSKELETAL SYSTEM

To appreciate the mechanics of skeletal muscle action, it is necessary to consider some basic mechanical principles. In order to act effectively, muscles employ the concept of *mechanical advantage.* They do this through the utilization of the principles of *simple machines.* A machine is merely a contrivance which modifies forces and movements so that work can be done more efficiently. The musculoskeletal system employs three of the six universally accepted simple machines* (Fig. 3-17): (1) the lever, (2) the wheel and axle, and (3) the pulley.

Levers

Levers require three basic components. See now Fig. 3-18:

1. Fulcrum (F): The axis or point in space about which a structure, e.g., bone, moves or turns. In the musculoskeletal system, such a point would be a *joint.*
2. Load (L): The structure, e.g., bone which will be moved

* The other simple machines include the wedge, inclined plane, and the screw.

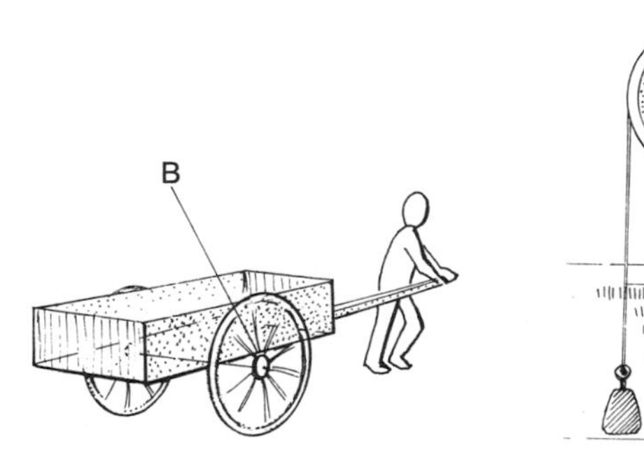

Fig. 3-17 Some simple machines as they apply to the musculoskeletal system. A. Lever. B. Wheel and axle. C. Pulley.

or turned about a fulcrum and which offers *resistance* to the force attempting to move it.

3. Power (P): The force applied to the load so that movement may occur. In the body, such a force would be derived from muscular work.

By rearranging the three components of the lever, it is possible to come up with three classes of levers (arbitrarily numbered I, II, III) of which II has little or no application in the musculoskeletal system.

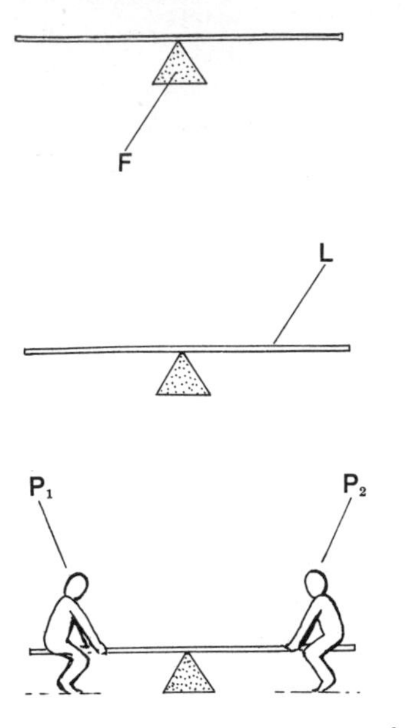

Fig. 3-18 The basic components of a lever machine. F, fulcrum; L, load; P, power.

An example of a class I lever can be studied in Fig. 3-19: Note the joint (F) is located between the skull (L) and the neck muscles acting on the skull (P).

Note the "seesaw" effect of alternately extending and flexing the skull. The "success" of this lever (in terms of functional efficiency) is related to the position of the fulcrum (atlantooccipital joint between the skull and the first cervical vertebra) relative to the load. If one wears a 10 lb weight on the back of one's head for a long period of time while still maintaining an erect head, one of the muscles (P) will begin to hurt from extended contraction. Which one, P_1 or P_2?* For the least output of energy, the skull should be balanced on the joint. If you study frequently with your head bent forward, you can now understand why the back of your neck often subsequently feels strained.

An example of a class III lever can be seen in Fig. 3-20: Note the point of power application (P) lies between the load (L) and the fulcrum (F). The resistance arm (distance between F and L) is significantly longer than the force arm (distance between F and P). Relative to a first-class lever where the converse of the above is true, the third-class lever requires more strength to lift a given load. This "deficit" is made up in the fiber pattern of the muscle employing a third-class lever. The deltoid muscle, shown in Fig. 3-20, is such a muscle. Its fiber architecture is multipennate (Fig. 3-16), in which many more muscle fibers play on any one tendon than can generally be demonstrated in the fiber pattern of muscles employing first-class levers. The result is considerably greater strength in the former. The great strength of the deltoid muscle can be demonstrated by standing in a doorway and forcefully abducting the upper limbs against the doorjambs for several seconds and then stepping out from the doorway. The tightly contracted deltoid muscles will effortlessly lift your arms into the air, momentarily overcoming the force of gravity.

* P_2.

Fig. 3-19 Anatomical example of a Class I lever. See text for explanation.

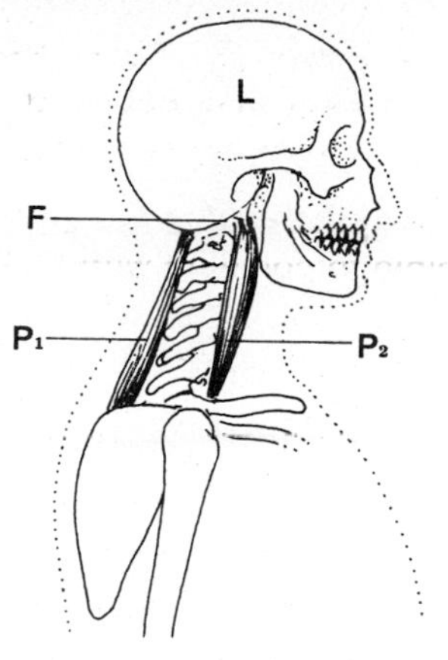

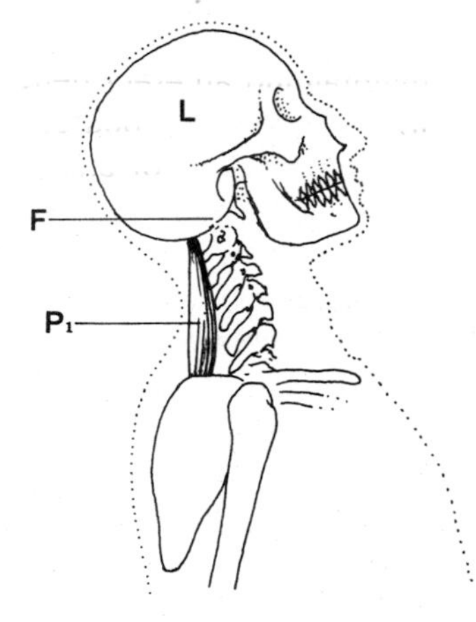

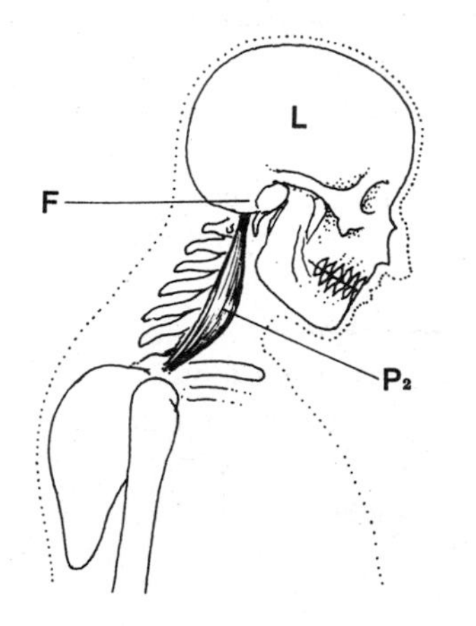

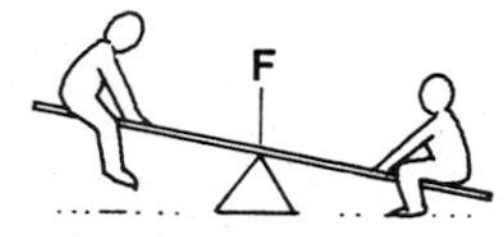

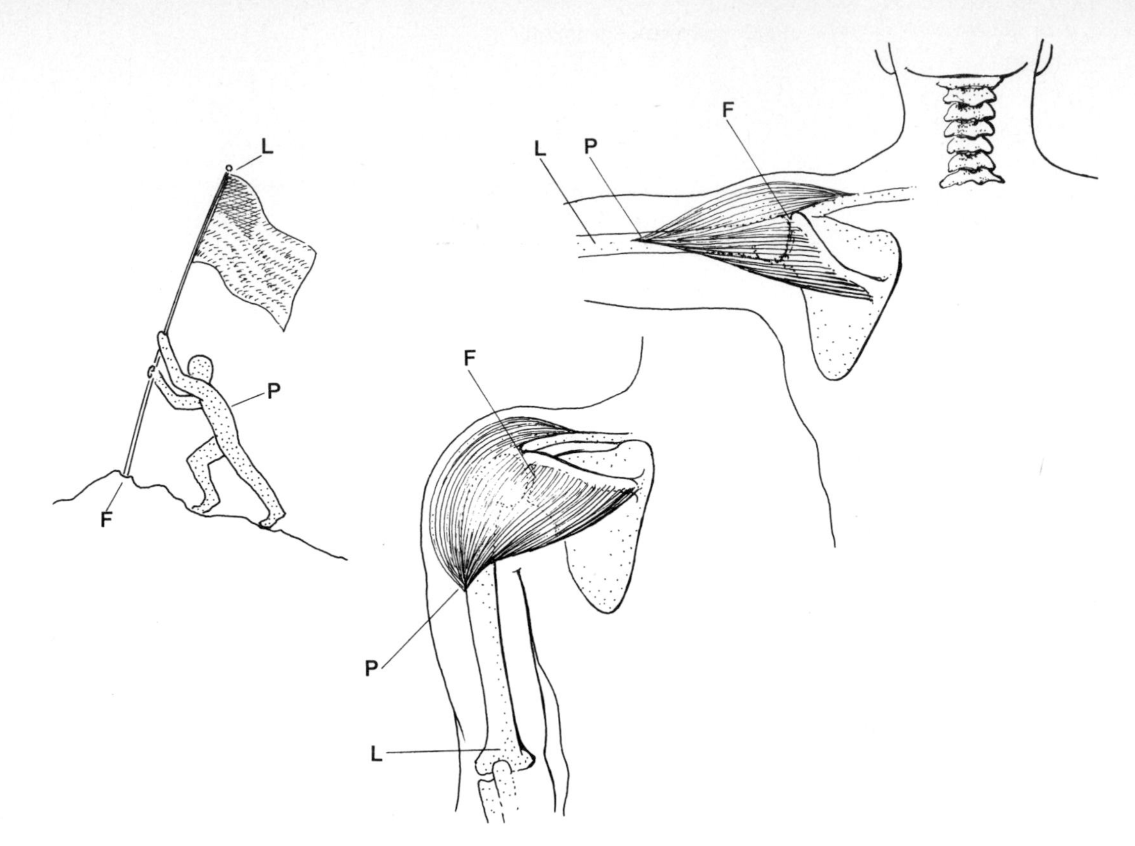

Fig. 3-20 Example of a Class III lever. See text for explanation.

An example of the wheel and axle machine may be seen in the head of the atlas (first cervical vertebra) pivoting about the axis (the second cervical vertebra), much as a wheel rotates about an axle.

An example of the pulley machine in the musculoskeletal system may be seen in Fig. 3-21.

In many cases, two or more muscles may act as parallel forces in carrying out a particular task. If these forces are directed oppositely, the effect is one of *rotation* and the forces are collectively termed a *force couple* (Fig. 3-22).

The mechanics of body movement, just superficially introduced here, are analyzed in more detail in the study of *kinesiology*.

INTEGRATION OF MUSCLE ACTION

In general, a number of muscles or muscle groups are active in any given movement. Rarely does one muscle act alone to

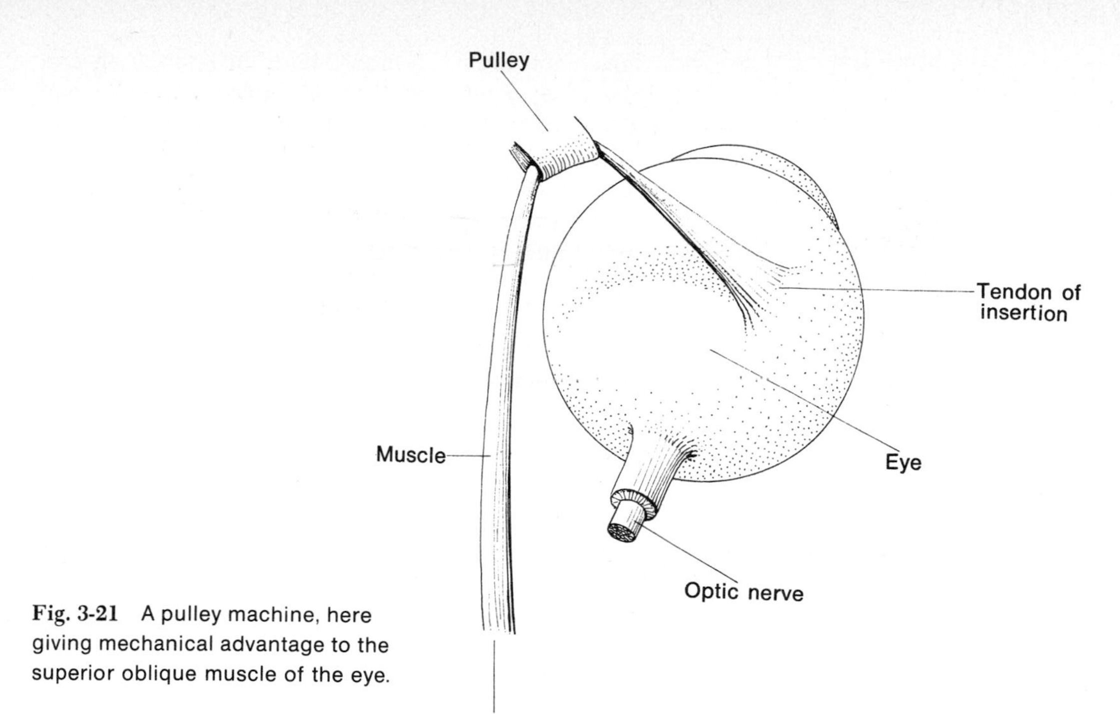

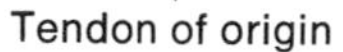

Fig. 3-21 A pulley machine, here giving mechanical advantage to the superior oblique muscle of the eye.

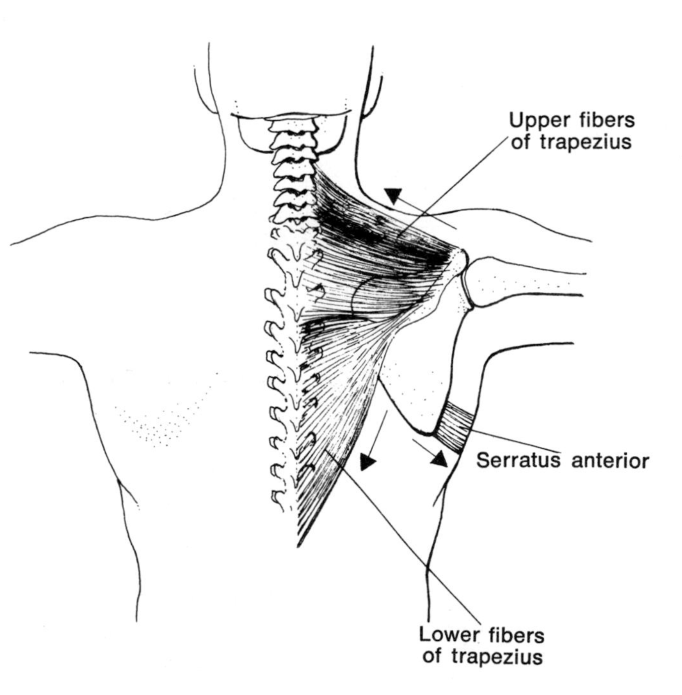

Fig. 3-22 A pair of force couples collectively acting to tilt the scapula up and outward. Upper trapezius and serratus anterior; upper and lower trapezius.

provide movement. In fact, a high degree of integration and coordination of muscular activity is often required. This can be demonstrated in the very simple activity of flexing the forearm at the elbow joint: Pick up a relatively heavy object (resistance) and flex your elbow joint with the forearm and hand pronated (palm down) (Fig. 3-23). Palpate the biceps brachii (arm) muscle and note its degree of contraction against the resistance. In this act of flexion, the biceps brachii (A) contracted and aided in flexing the forearm. It is therefore called a mover in this action. It is not, however, the principal mover. The principal actor was the brachialis (A′) and such a mover is called a **prime mover.**

The muscle opposing the act of flexion, an extensor, crosses the elbow joint on the opposite side of the biceps and the brachialis. This muscle, the triceps brachii (B) is termed an **antagonist** in the act of flexion. It may be a passive antagonist or it may come into active play as a check muscle in dampening excessively vigorous flexion movements* or in grading rates of flexion.

Now while the above acts of flexion and some negligible antiflexion took place, there were muscles (e.g., trapezius) stabilizing the upper extremity to the axial skeleton, compensating for the asymmetrical lifting of the weight. Such a muscle is called a **fixator** (C).

Finally, there are muscles which prevent undesired movements from one of the movers. For example, the biceps brachii is not only a flexor of the forearm, it is also a supinator (prime mover, in fact) of the forearm. If, in lifting a weight, it was desirable to keep the hand and forearm pronated (as you did when you lifted the weight in this example), it would be necessary for the pronators of the forearm (e.g., pronator teres, D) to prevent supination by the biceps. Thus, the pronators are **neutralizers** in this case.

The skeletal muscles of the body act to move one or more parts of the body. These movements are designated (as movements of a bone or bones):

______________ (taking away from midline)
______________ (bringing toward the midline)
______________ (straightening)
______________ (bending)
______________ (twisting)

If your recall of these terms is hazy, review the Introduction.

* This is a reflex phenomenon called *reciprocal inhibition.*

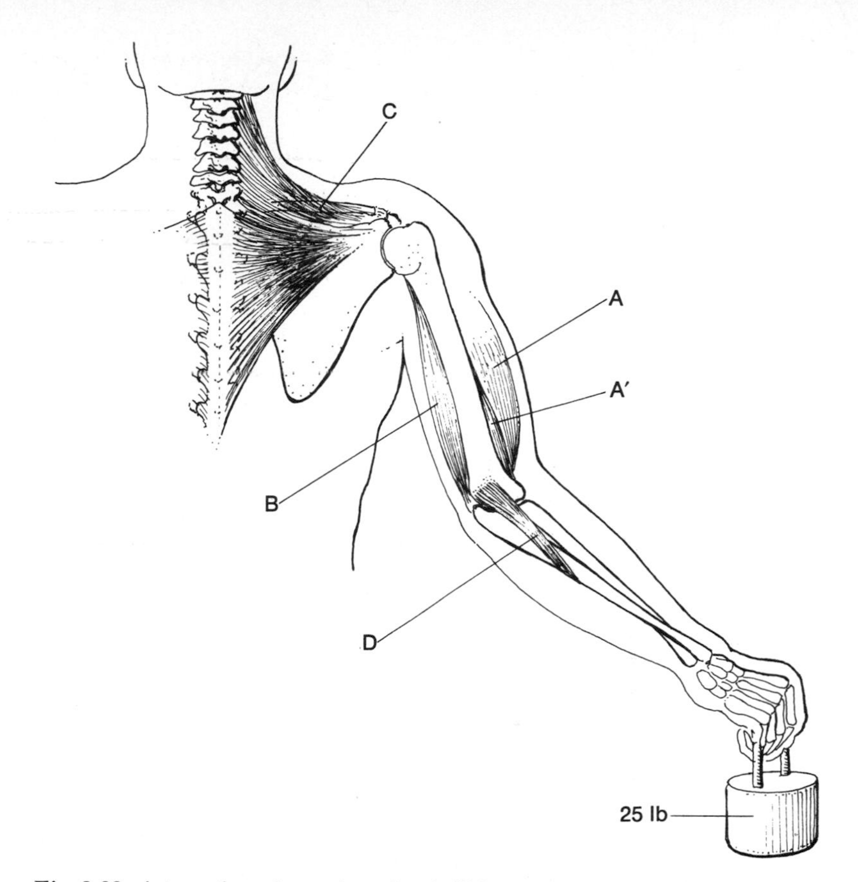

Fig. 3-23 Integration of muscle action in lifting a 25-lb weight when the forearm is pronated. A, biceps brachii; A′, brachialis; B, triceps brachii; C, trapezius; D, pronator teres. For explanation, see text.

In considering the action of a muscle and contemplating the amount of force required to complete the action, it is very important to consider the effects of gravity. In general, extensors act as antigravity muscles—they oppose the effect of gravity and act to keep the body erect. If you stand up straight, all limbs extended, your extensor muscles are acting in concert and your flexors are relaxed. If you suddenly collapse, your extensors relax and gravity influences your body to fall to the floor. Actually, your flexor muscles have little to do with gravity-effected flexion of the body joints, as in fainting. In general, most of the extensor muscles of the axial skeleton are located posterior to the midline frontal plane. This can be demonstrated in fainting, for when the muscles relax and tension is taken off the joints, the body tends to fold forward.

fasciae and bursae

Fascia (pl., fasciae) is a layer of loose or dense irregular connective tissue found throughout the body. In certain places it may be known by special terms (e.g., submucosa and lamina propria in viscera) but generally fascia is simply referred to as being either superficial or deep (Fig. 3-24).

The fascia just under the skin, variable in thickness and density, is *superficial fascia,* and is generally described as having an outer fatty layer (subcutaneous) and an inner fibrous layer. One or the other layer may be reduced or absent. The subcutaneous layers add a quality of smoothness to the form of the body—particularly in females* (Fig. 3-25). The

* Estrogen, a female sex hormone, influences the distribution of fat deposition at puberty and the result is that a woman's body is more curvaceous and her body form appears smoother than the muscular and angular form of the male.

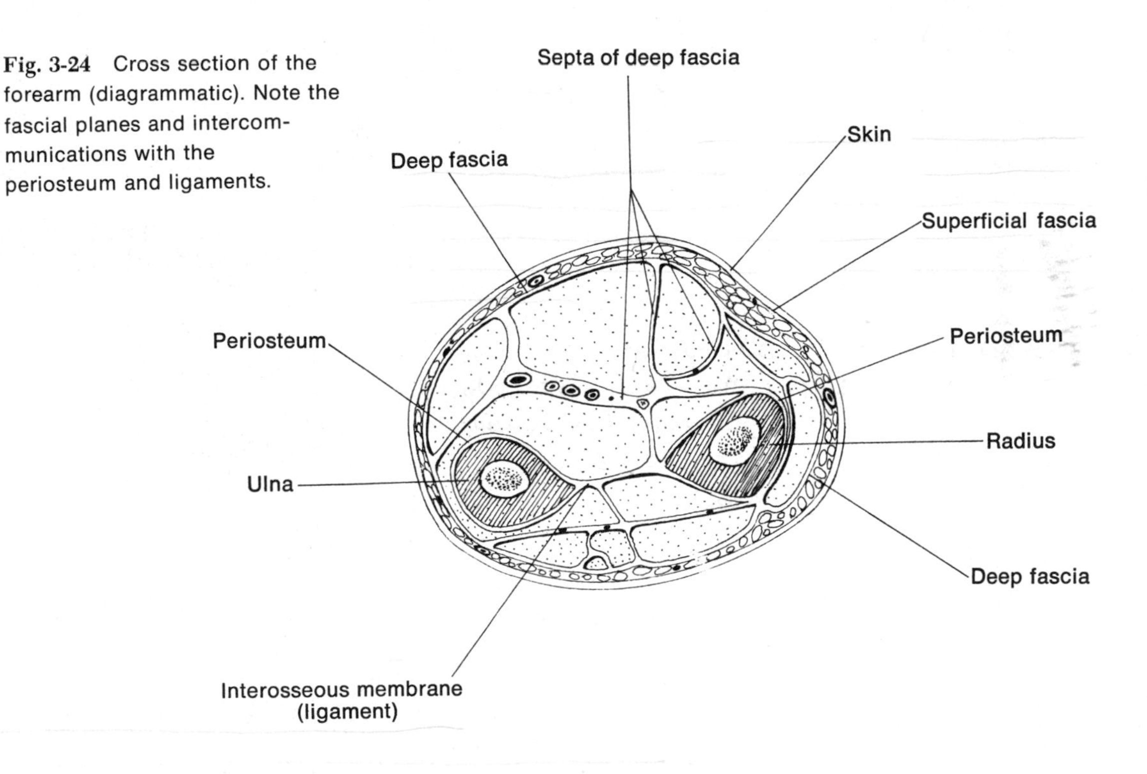

Fig. 3-24 Cross section of the forearm (diagrammatic). Note the fascial planes and intercommunications with the periosteum and ligaments.

Fig. 3-25 (opposite) Compare the smoothness of the male and female bodies. The deposition of fat in the superficial fascia is largely responsible for the differences. Notice the hard shadows and rough-hewn surface of the male.

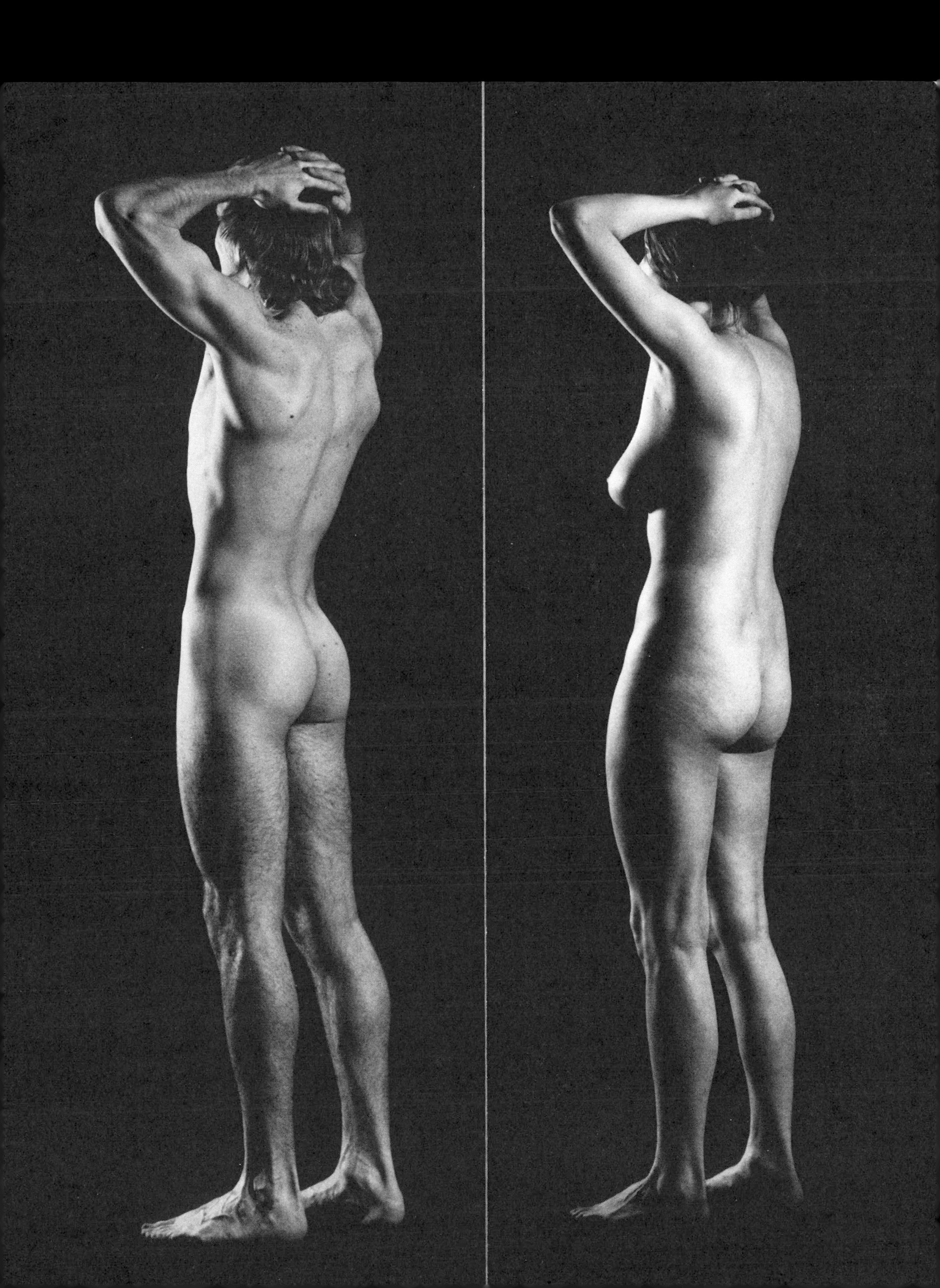

fibrous layer of superficial fascia has special significance for the surgeon, for its tough fibrous quality lends itself to stitching (suturing) wounds in surgical operations. In general, superficial fascia:

1. Insulates the body against cold.
2. Pads the space between skin and muscles.
3. Protects transiting nerves and vessels.
4. Acts as a reservoir of energy, for fat from fascia is quickly mobilized when the body requires it, as in dieting, and is used as a source of energy for metabolic activity; thus the body form changes as weight is lost.

Deep fascia (Fig. 3-24), in the conventional sense,* constitutes the connective tissue "envelope" bounding *skeletal muscle* (external to but contiguous with perimysium), and forming walls (septa) between muscles and muscle groups. It also contributes significantly to the construction of joints. Deep fascia fills out regions of deeper structures just as superficial fascia fills out the external body form. Deep vessels and nerves pass through and are protected by deep fascia. Deep fascia is often continuous with periosteum, ligaments, superficial fascia—providing a firm network of fibrous support for the muscles, viscera, vessels and nerves within.

A **bursa** (pl., bursae) is a partially collapsed connective tissue sac, like a partly filled balloon. Bursae are found in the superficial fascia and between tendons and bone, muscle and bone, muscle and muscle, or wherever two adjacent musculoskeletal structures rub together and create the heat of friction (Fig. 3-26). These bursae prevent the moving parts from inflaming one another due to contact irritation. The inner surface of the bursae is like a synovial membrane and secretes a fluid (synovia) which moistens the interior sufficient to allow one surface to rub on another frictionlessly. Not infrequently, bursae may communicate with joint cavities following chronic inflammation of adjacent tendons (tendinitis) or the bursa itself (bursitis).

Fig. 3-26 A pair of bursae in the shoulder region; the subacromial (A) and subdeltoid (B) (diagrammatic).

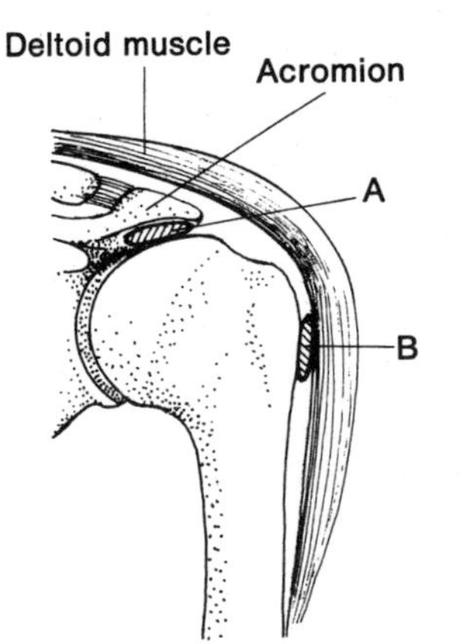

In places where tendons travel back and forth over bones, a tubular bursa, secured in place by a fibrous sheath, surrounds the tendon. Such a bursa is called a **tendon sheath** (Fig. 3-27); it allows frictionless movement of the tendons through the tunnels. These sheaths are particularly significant in the hand, as untreated infections of them can result in permanent functional impairment.

* Actually, deep fascia constitutes all the connective tissue associated with muscles and bone, including ligaments and periosteum.

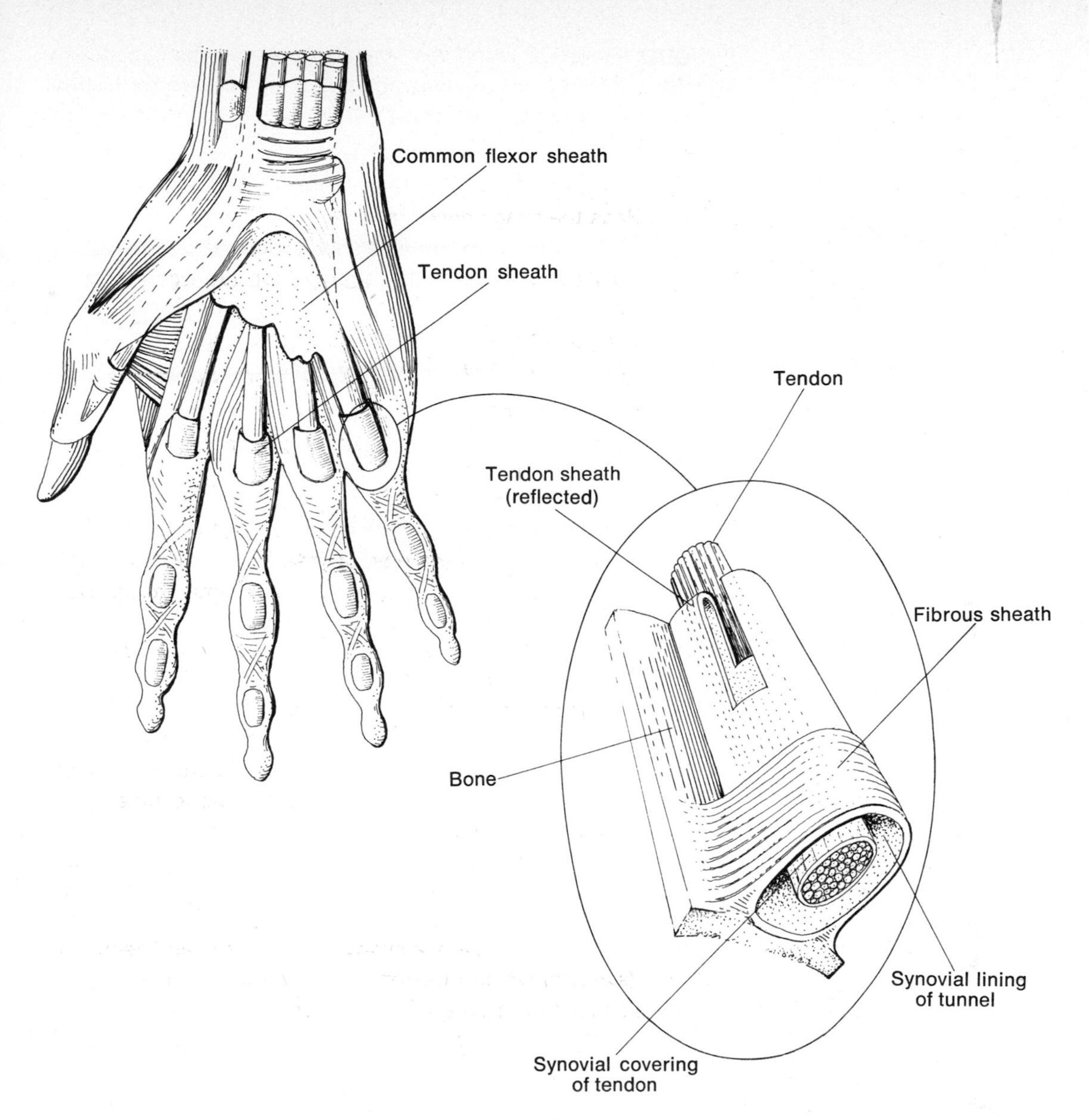

Fig. 3-27 Tendon sheathes of the palmar aspect of the hand and wrist. In the diagram of a tendon sheath, note how the synovial membrane is reflected upon itself. The fibrous sheath prevents the tendon from popping out of position when slack (bow stringing). (Adapted from J. V. Basmajian, *Primary Anatomy,* 6th ed., Williams and Wilkins, Baltimore, 1971.)

QUIZ

Time for a comprehension check—this one on joints, muscles, fasciae, and bursae:

1. The joining of two or more bones is called an articulation.
2. Structurally, joints are classified as cartilaginous, fibrous, or synovial.
3. Functionally, joints may be immovable, slightly moveable, or freely moveable.
4. A skeletal muscle, such as the biceps muscle of the arm, generally has two parts: the belly and the tendon.
5. The tendinous attachment to the more fixed bone is referred to as the origin; the attachment to the "moving" bone is called the insertion.
6. The three basic components of a lever-type machine include the fulcrum, load, and power.
7. If two muscles acting on one bone contract in parallel but opposite directions, the effect is one of rotation and the colective forces are termed a force couple.
8. A movement is the product of muscle action integration. Generally, a muscle in such a movement will play one of four roles: prime mover, antagonist, fixator, or neutralizer.
9. The layer of connective tissue just under the skin is superficial fascia. Deep fascia constitutes the connective tissue envelope of skeletal muscle.
10. Bursae may be found as partially collapsed connective tissue sacs, filled with fluid and existing between adjacent musculoskeletal structures which rub against one another.

vessels of the lymphatic and cardiovascular systems

The cardiovascular system (Fig. 3-28) is composed of a pump (heart) and a system of tubes (vessels) which (1) carry blood pumped away from the heart (arteries), (2) distribute and receive certain molecular material, both nutritive and waste, to and from the local tissues (capillaries); and (3) carry blood

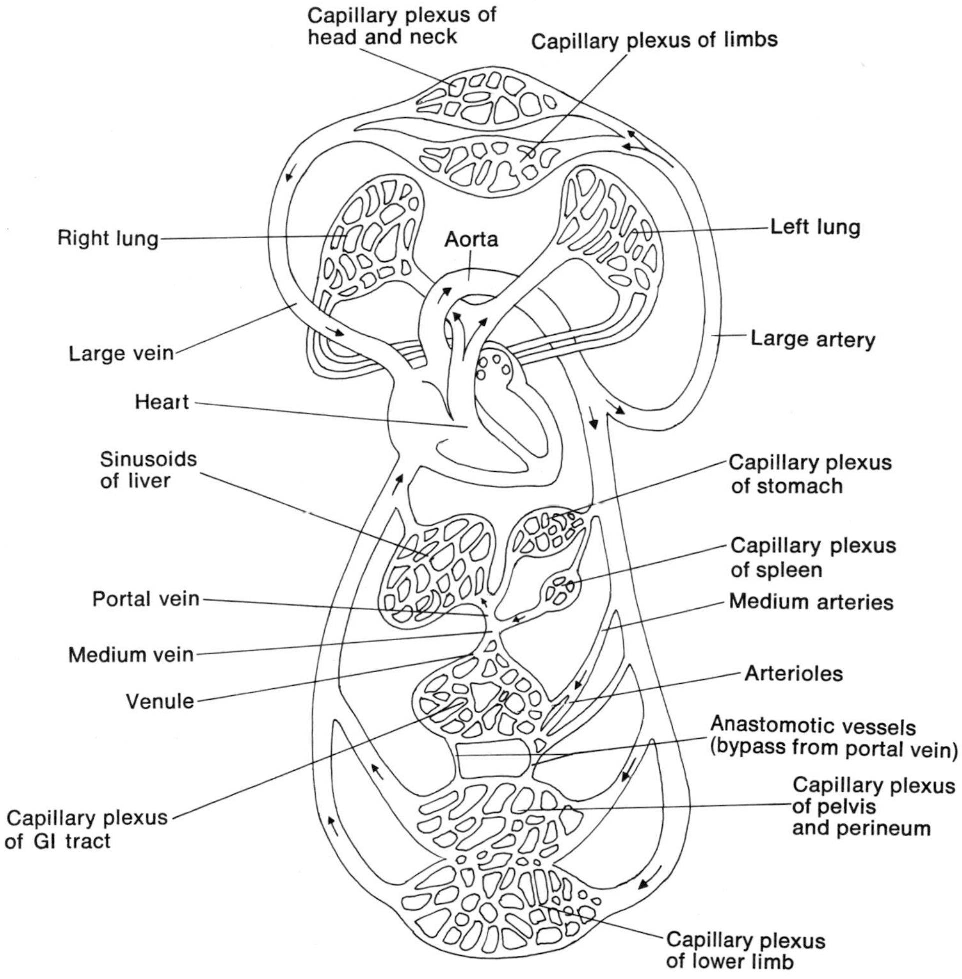

Fig. 3-28 The circulatory scheme of the body. To trace circulatory flow, start with the aorta.

to the heart from the tissues (veins). The heart will be described in Unit 8 along with other viscera of the thoracic cavity.

Within this circulatory system, one set of vessels is responsible for carrying oxygenated blood and nutritive material to the tissues and cells of the body and returning deoxygenated blood to the heart. This is the **systemic circuit.** Another set of vessels carries deoxygenated blood from the heart to the lungs for oxygen replenishment and returns the oxygenated blood to the heart. This is the **pulmonary circuit.** Oxygenated blood is reddish in color (due to the nature of the oxygen-binding hemoglobin molecules in red blood corpuscles) while deoxygenated blood is bluish (because of the reduced concentration of oxygen).

It should be appreciated that blood vessels are not simply a system of inert tubes but are metabolically active *organs* composed of epithelial, connective, muscle, and nervous tissues. The structure of the types of vessels described above varies significantly and is related to the task performed. The function of a particular vessel is related to the latter's proximity to the heart as well as to the amount of each of the tissues present in its walls. For instance, a vessel receiving blood from the heart (artery) at high pressure will certainly be structured differently from a vessel carrying blood at relatively low pressure toward the heart (vein); and a vessel composed primarily of elastic tissue certainly functions differently from a small epithelial capillary.

ARTERIES

Those vessels carrying blood from the heart to the tissues are called **arteries.** The arteries arising directly or almost directly from the heart are referred to as *large* arteries. As they spring out of the thoracic cavity and branch into smaller regional arteries they are called *medium* arteries. The smallest branches of these—communicating with the capillary networks (through which oxygen, nutritive elements, and wastes pass into and out of the neighboring tissues) are called **arterioles** or small arteries. Each of these rather arbitrary classes of arteries differs functionally and structurally.

Functionally, the large arteries are responsible for conducting blood from the heart to the regional vessels without sacrificing any significant pressure loss. They do this by stretching their walls during the blood ejection phase of the heart (systole) and recoiling during the relaxation phase

Fig. 3-29 Layers of a generalized muscular artery. Internal layer has simple squamous epithelial lining and loose connective tissue. Middle layer has smooth muscle cells circularly arranged. External layer has loose connective tissue merging with surrounding fascia. Note the internal elastic membrane and the more irregular external elastic membrane.

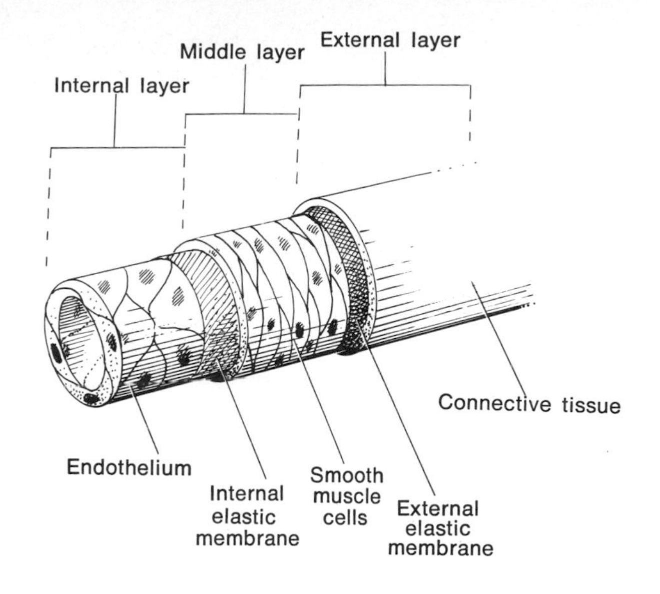

Fig. 3-30 An artery (solid arrow) with a vein (hollow arrow) folded over it. Note the prominent elastic membranes of the artery (small arrow) and the lack of distinct layering of the vein. Photomicrograph. (Courtesy of James Runner.)

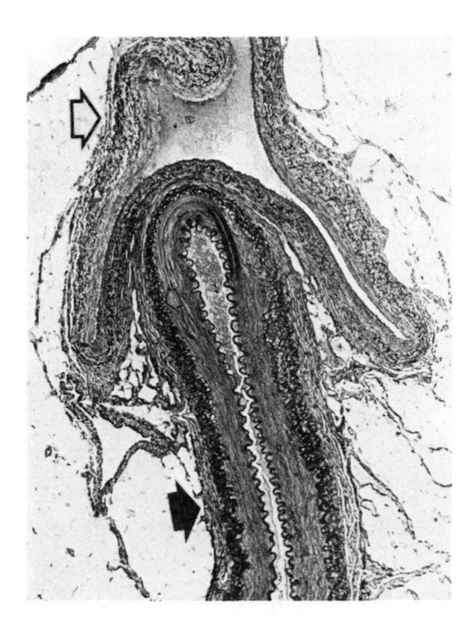

(diastole). What kind of tissue would you think predominates here? Elastic c.t. Hence, large arteries are often termed elastic or conducting arteries. You will study these vessels with thoracic viscera in Unit 8. Furthermore, large arteries all have small vessels perforating their walls to supply them with blood. Such vessels constitute **vasa vasorum.**

The medium-sized arteries—the most common of all arteries from the dissector's point of view—supply the various regions of the body. Thus, these arteries are often termed *distributing* arteries. Blood flow to one or more regions can be altered by varying the diameter of the appropriate distributing arteries. What kind of tissue, found in the walls of these vessels, is responsible for such action? *Smooth muscle.* Thus these arteries are also called muscular arteries (Figs. 3-29, 3-30).

The small arteries (including the arterioles) generally cannot be seen by the naked eye. Like muscular arteries, they contain one or more layers of smooth muscle and are responsible for *regulating* blood flow into the capillary networks. The greatest resistance to blood flow is produced here, and this is reflected in the blood pressure measurements of an individual, i.e., the greater the resistance the higher the pressure. Thus arterioles, as well as muscular arteries, are referred to as *resistance* vessels by the physiologist.

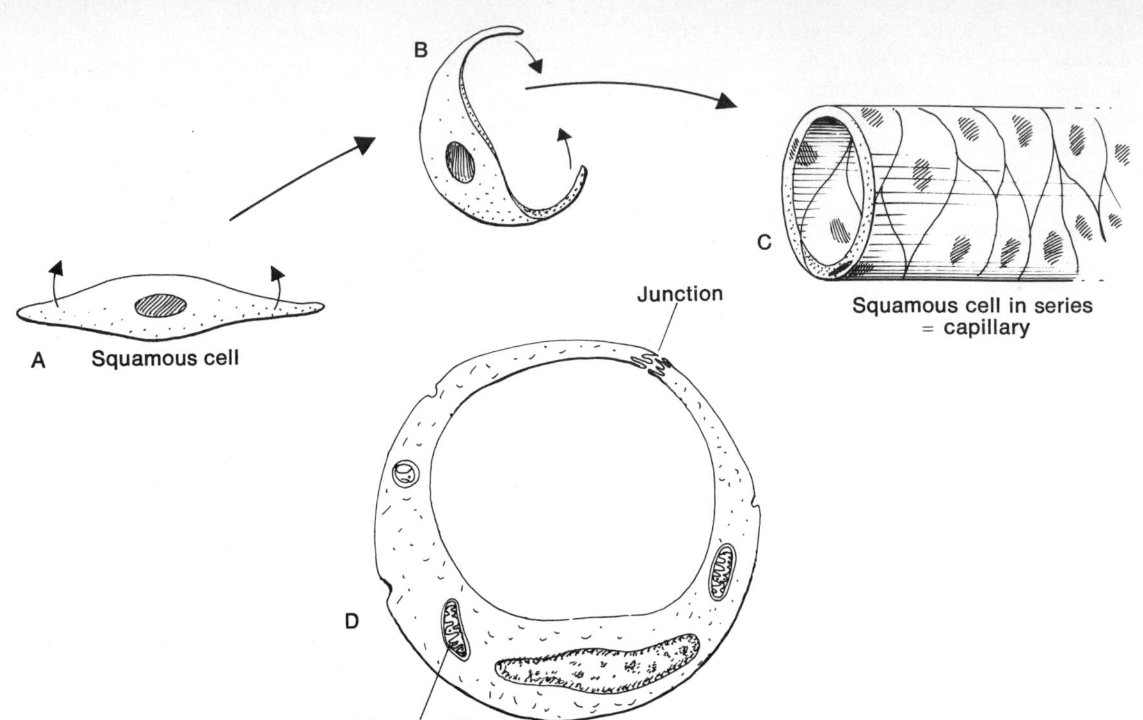

Fig. 3-31 Formation of a capillary with contiguous squamous cells (A-C). D. Capillary in cross section as viewed by an electron microscope.

CAPILLARIES

Capillaries are the thinnest and smallest of all the blood vessels—and also the most common. They usually cannot be directly seen by the naked eye, but their presence is reflected in the color of the nails, the skin, etc. It is by virtue of capillaries that nutritive elements and wastes pass into and out of the tissues, for capillaries are the only vessels thin enough for rapid transfer of small molecular materials. A capillary is a small tube composed of simple squamous (endothelial) cells arranged in series. The cross-sectional area is just large enough to accommodate the passage of one blood cell at a time (Fig. 3-31). The capillary is supported by a thin layer of connective tissue. Capillaries form interconnecting networks throughout most of the body (Fig. 3-28). When one becomes suddenly pale, as for example from shock, the change in skin color is due to a constriction of the arterioles feeding into the capillary network of the skin. Thus the blood flow through capillaries is diminished and the skin pales.

VEINS

Arteries conduct blood from the heart to capillaries from which cells of the body receive their nutrition. Waste* products are discharged by the cells and taken up by the capillaries which merge into vessels returning blood to the heart. Such vessels are **veins.** As veins get closer to the heart, they become larger, just as rivers become larger as they approach the sea. This is so because veins, like rivers, receive more and more tributaries as they near the heart (sea). The smallest veins are called **venules.** Arising from capillaries, venules are the tributaries† of medium veins (Fig. 3-30). These, in turn, are tributaries of the larger veins further downstream.

Venules are about the same size as arterioles but the lumen is larger and the walls are thinner in the former. The pressure of the blood within venules is about half that in arterioles and this fact is reflected in the thin walls of the former.

Veins have the capacity to distend and thus function as a reservoir of blood and are often called *capacitance* vessels (in contrast to small and medium arteries which are called *resistance* vessels). Larger veins, like arteries, are supplied by vasa vasorum.

In medium-sized veins of the extremities and other areas, **valves** may be found (Fig. 3-32). These structures function to prevent retrograde (back) flow in those veins where gravity has a significant influence. The valves, usually arranged in pairs, are cup-shaped endothelial pockets. The opening of the pockets always faces the heart. Such venous valves can often be seen in the neck of people yelling or straining.

Certain organs of the body include enlarged capillary-like vessels termed **sinusoids** in their structure, particularly where a great deal of molecular exchange takes place. Conducting blood toward the heart such venous sinusoids are found in the liver, bone marrow, and other "bloody" organs, and will be considered regionally. Sinusoids are often lined by phagocytic cells. Greatly enlarged veins within the skull and placenta are called **venous sinuses.**

* The term *waste* is a simplification, as what is waste to one cell may be life-sustaining to another.

† While veins have tributaries, arteries do not. Because the direction of blood flow in arteries is away from the heart, arteries are said to have branches.

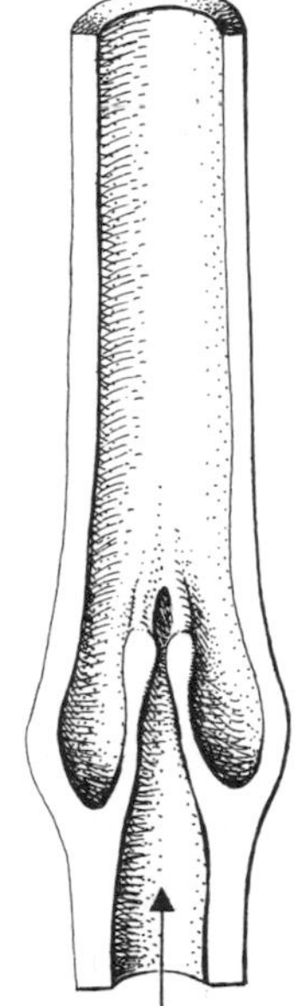

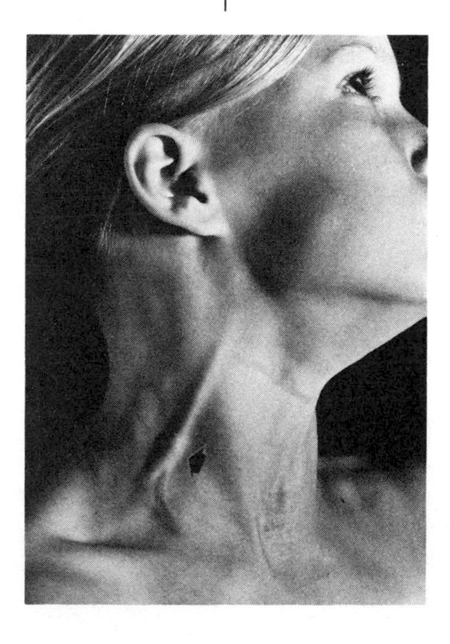

Fig. 3-32 A strip of vein cut longitudinally to illustrate a valve. In the photo an arrow indicates the blood-filled valve of the external jugular vein of an 11-year-old girl. Having her strain brought the valve into sharp relief.

GENERAL COMMENTS ON REGIONAL VESSELS

Medium-sized arteries and veins usually travel together as they pass through a region. This is not a chance meeting, as veins and arteries develop together from the same embryonic tissue. During development, arteries and veins appear at the same time and probably in the same number. At 5 weeks or earlier, the heart starts beating and the intravascular fluid ("blood") is forced through the developing vessels. Those vessels closest to the heart—the midline vessels—receive the greatest head of pressure and only they receive fluid (Fig. 3-33). It is suggested that these midline vessels become adult arteries while the others become obliterated. The pressure of blood in the veins is much reduced and most of the veins receiving blood become adult veins. Since blood pressure apparently does not influence development of venous patterns, veins are more numerous and more variable than arteries. In your dissections you will find this to be true.*

In the extremities and body walls, veins are usually superficial to their companion arteries. Thus following a mild laceration only capillaries and possibly veins would be involved. Except in certain places, deep cuts are required to sever an artery.

Particularly in the extremities, medium-sized veins traveling with a companion artery (and nerve) often travel in pairs **(venae comitantes)**—one on each side of the artery. This can be observed during dissection with the brachial veins, radial and ulnar veins, and similar veins of the lower limb.

CONCEPT OF COLLATERAL CIRCULATION

It should be appreciated that were it not for some sort of alternate routes of vascular circulation about the regions of the body, significant cuts and wounds involving ligation (tying) of distributing arteries would mean cutting off the blood supply to the part, with gangrene following. Cells must have a constant supply of oxygen (brought by blood). Interruption of that blood supply for any more than a few seconds would result in the death of many cells. Happily, there are usually a variety of vascular routes to any one region of the body and within that

* You will have little trouble distinguishing veins from arteries. Arteries hold their round shape and recoil when compressed or pinched because of the elastic tissue in the walls. Arteries rarely have blood in them and thus are a light flesh color. Veins, on the other hand, are larger than their fellow arteries, and collapse when compressed. They may appear beaded because of the dried blood within them. Because of that blood, veins are usually dark colored—blue or black. Veins are relatively easily torn, arteries are not.

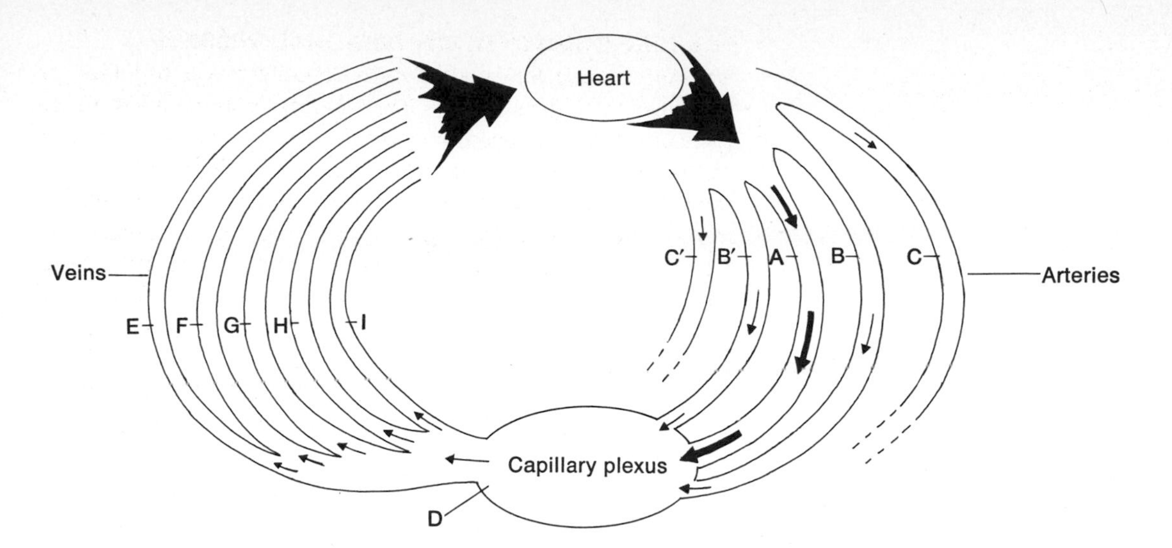

Fig. 3-33 A theoretical scheme of development and disintegration of embryonic blood vessels. Blood leaving the heart is under high pressure and is forced down the path of least resistance, i.e., the midline artery. Arteries B and B′ become collateral vessels while arteries C and C′ become obliterated because of insignificant blood flow. The blood leaving the capillary plexus (D) is at low pressure and trickles into any number of veins (E–I). The adult pattern of many, variably routed veins and fewer, straighter, and more consistently routed arteries is probably set by pressure patterns during vascular development.

region therc are interconnecting vessels. A connection between vessels is an **anastomosis.** Anastomoses provide alternate routes of blood circulation; such routes are called **collateral circulation.** Some organs are notably devoid of any significant collateral circulation, e.g., liver, kidney, heart.

PORTAL SYSTEMS

In most organs, the blood supply is channeled in by arteries, passes through a capillary plexus, and leaves the organ via veins (Fig. 3-28). In certain regions, e.g., the liver, GI tract, and the pituitary gland, the blood passes through *two* capillary networks before returning to the heart. Such a phenomenon is called a **portal system** (Figs. 3-28 and 3-34). Oxygen and nutritional exchange occurs in the first capillary plexus; the venous blood is then transported into a second capillary network, where the lining cells collect certain materials for processing and secrete other molecules into the venous blood.

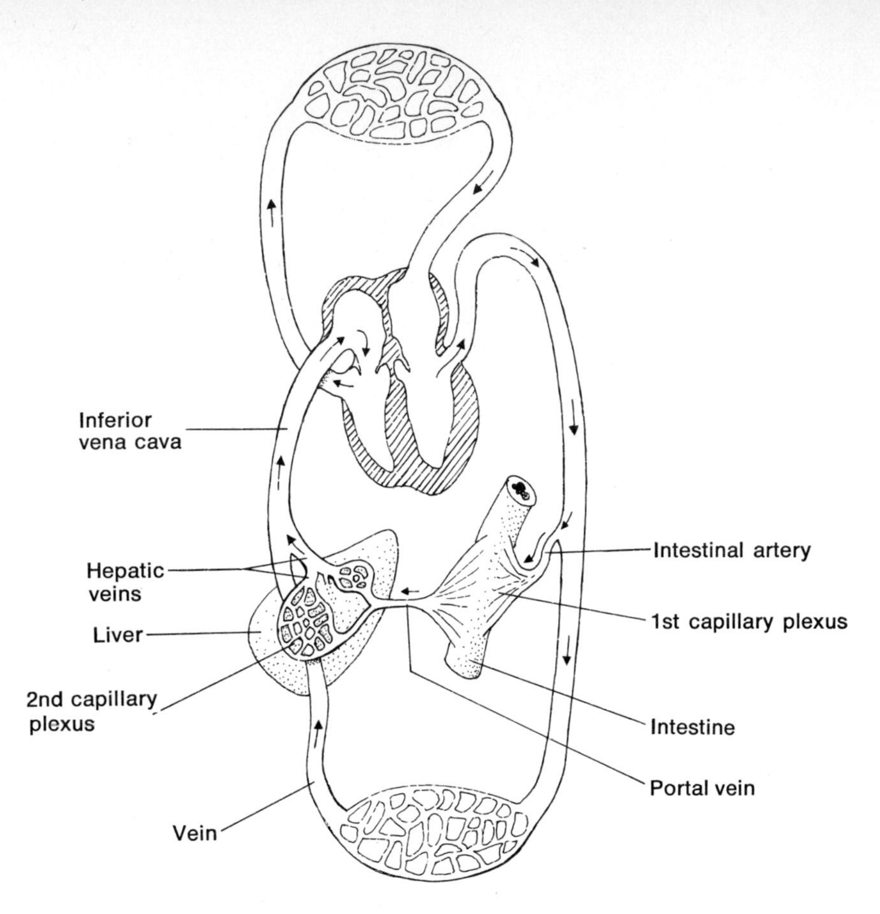

Fig. 3-34 Schematic of the body circulation emphasizing the hepatic (liver) portal system. Route of blood flow through the portal system: (1) intestinal artery, (2) capillary plexus of gastrointestinal tract, (3) portal vein, (4) sinusoidal plexus of the liver, (5) hepatic veins.

Reference will be made at appropriate times to the portal system.

LYMPHATIC VESSELS AND LYMPH NODES

The molecular concentrations of extracellular fluid (ECF) of the body are in equilibrium with molecules in the plasma (a type of ECF) of the blood. Functioning to help maintain a stable fluid and electrolyte balance among the cells, tissue spaces, and the blood (by drawing off excess fluid, fat, and small protein molecules) are the **lymphatic** vessels (Fig. 3-35). Lymphatic vessels make up a closed system of connective tissue–supported endothelial tubes, permeable to water and

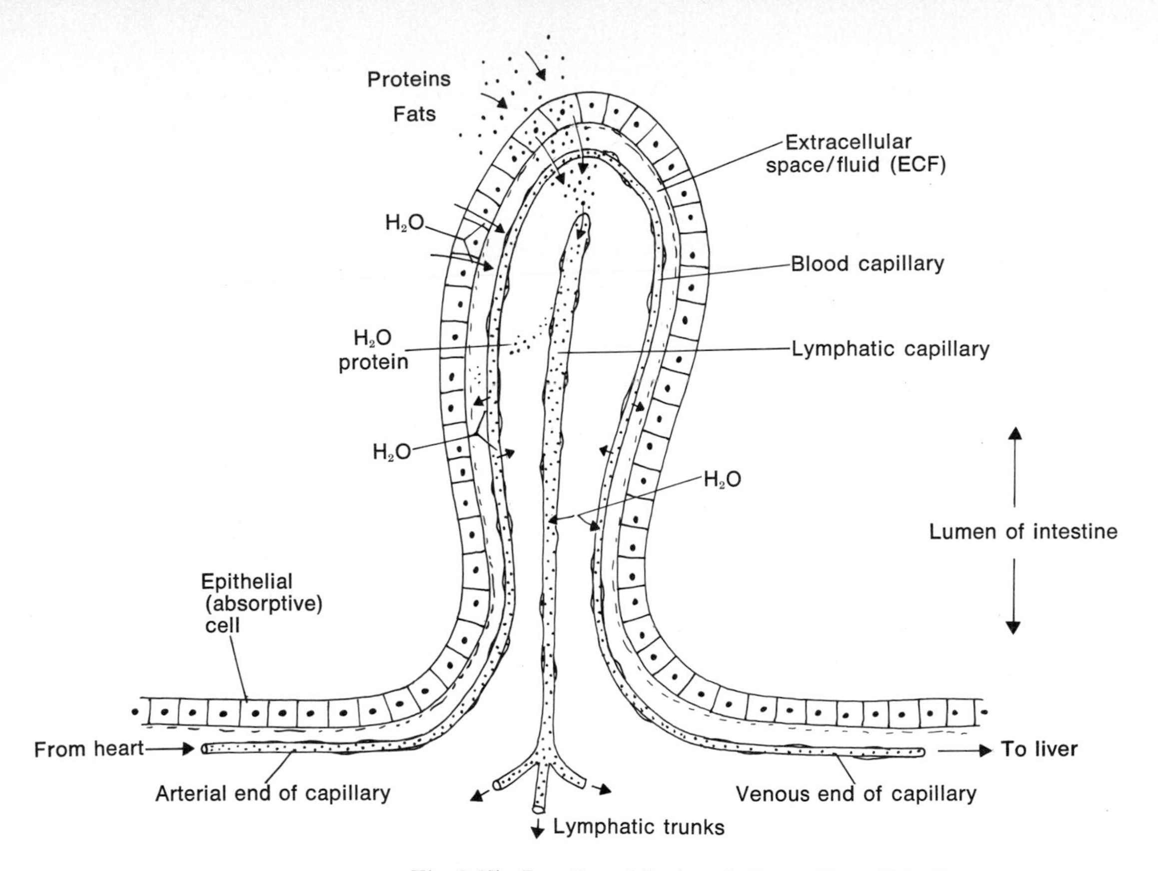

Fig. 3-35 Function of the lymphatic capillary. This diagram illustrates how proteins and fats which cannot be absorbed by blood capillaries pass into lymphatic capillaries.

molecules up to the size of small proteins, and found in most tissues throughout the body, with notable exceptions (e.g., brain). The smallest of the lymph vessels **(capillaries)** arise as blind tubes in the connective tissue. These anastomose to form the tributaries of larger lymph vessels.

The movement of lymph is created by local skeletal muscle movement (since no lymphatic "heart" exists) and made possible by valves in many of the lymph vessels. Lymph vessels, often running in company with neurovascular bundles, are tributaries of larger lymph channels, two of which drain into the venous system (Fig. 3-36). The major vessels will be considered regionally.*

In the mainstream of the lymph vessels, usually in specific locations about the body, one finds collections of **lymph**

* Lymph vessels are but one part of the lymphatic system. See Synopsis of Systems in the appendix.

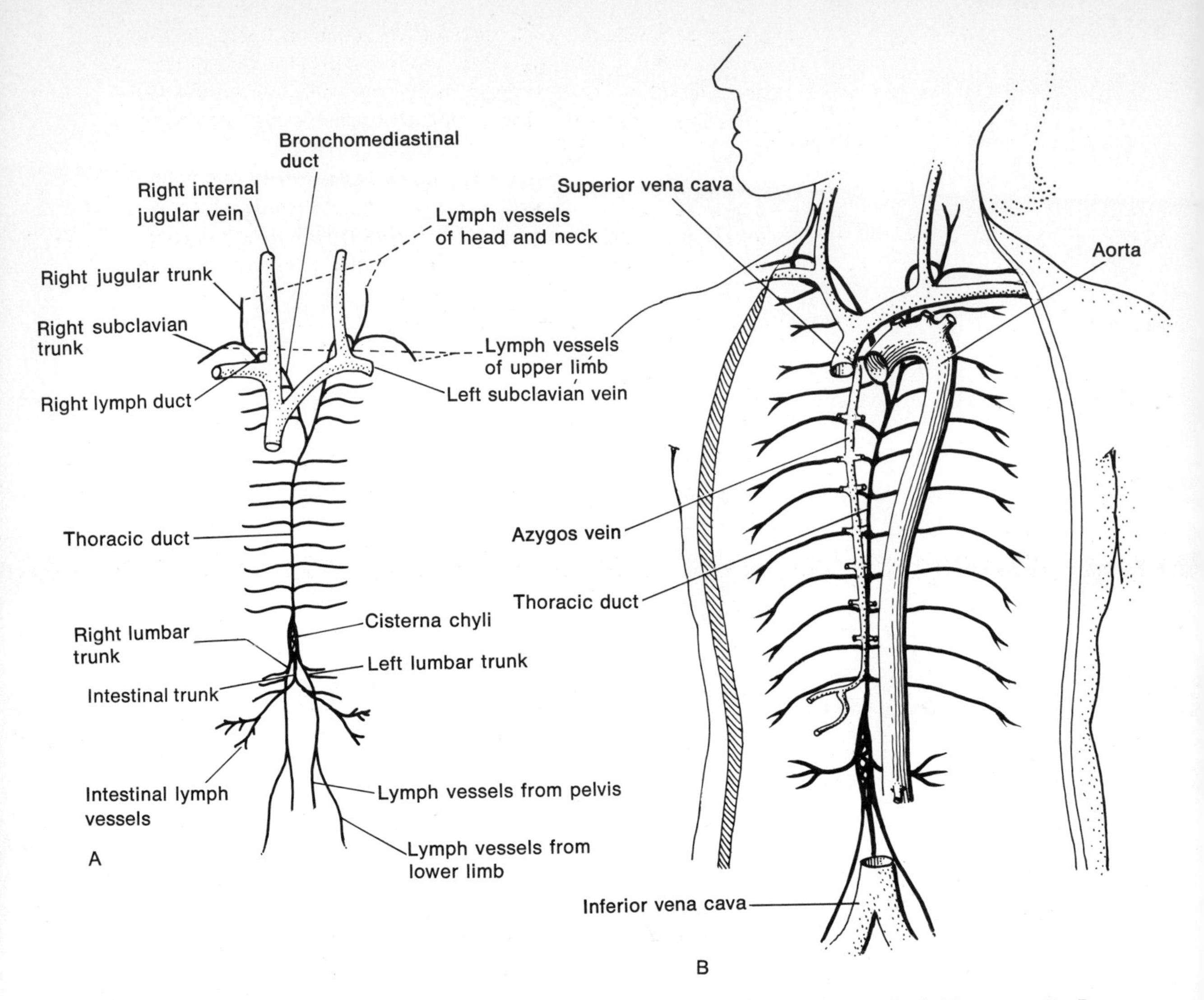

Fig. 3-36 The principal lymphatic vessels. A. Diagrammatic. B. Vessels superimposed on the posterior body wall.

nodes—variable sized, bean-shaped structures which filter the lymph fluid. Now refer to Fig. 3-37 as you read below:

Lymph nodes are encapsulated structures fed lymph fluid by one or more **afferent vessels** and discharging the strained lymph into one or more **efferent vessels** at the **hilus** of the node.

The outer one-third **(cortex)** of the node is populated by circular nodules of lymphatic tissue—masses of lymphocytes set in a dense but delicate nest of reticular fibers. In the center of the nodule **(germinal center)** a proliferation of lymphocytes often takes place, particularly when the node is involved in a defensive maneuver. Phagocytes may be found here too.

The more central part of the node **(medulla)** is characterized by irregular chains **(cords)** of lymphatic tissue through which vast numbers of interlacing lymph channels **(sinuses)** weave their course. These sinuses ultimately all arrive at the hilus to form the efferent lymph vessels. The sinuses are lined by phagocytic cells which play a starring role in the filtering

Fig. 3-37 The generalized pattern of superficial lymphatics in the body. One of the inguinal lymph nodes has been magnified and sectioned.

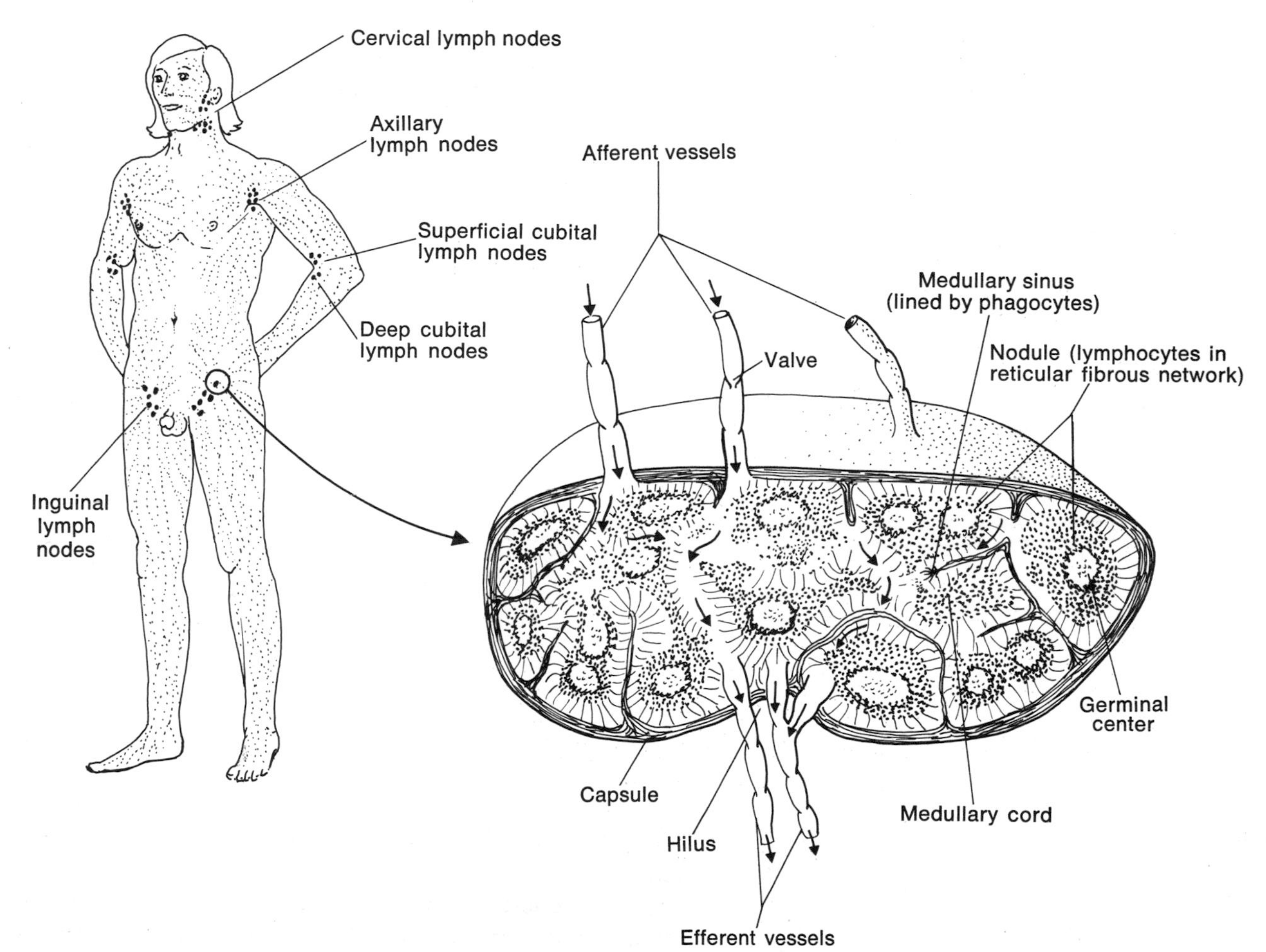

of foreign bodies (bacteria, etc.) from the circulating lymph.

Thus, lymph nodes not only "strain" the lymph but are important sources of lymphocytes. In times of adversity (infection), lymph nodes swell three or four times their normal size (head of a hat pin), a consequence of increasing numbers of lymphocytes within the node. These cells are often distributed peripherally via the lymphatic vessels to fight infection.

Aggregations of lymph nodes occur throughout the body, usually in association with deep or superficial veins. These will be mentioned regionally.

Superficial groups of lymph nodes can be palpated easily—particularly when they are enlarged following a local infection. Such groups include the **cervical lymph nodes** (felt on the side of the neck overlying the great sternocleidomastoid muscle—they can be rolled underneath the skin), **axillary lymph nodes** (in the "armpit" region), and **inguinal lymph nodes** (at the groin) (Fig. 3-37). These groups are employed by the physician as indicators of infection in regions drained by each group. Therefore, enlarged cervical nodes are expected during acute or chronic tonsillitis, the axillary nodes may trap metastasizing cells from carcinoma of the breast, and the inguinal nodes may be involved in an abscess of the knee.

introduction to the peripheral nervous system

The peripheral nervous system (PNS) (Fig. 3-38) is that part of the body nervous tissue organization which deals with (1) *receipt* of sensory information by receptors, (2) transmission of that information *to* the central nervous system (CNS), and (3) the transmission of motor commands *from* the CNS to the peripheral effectors (skeletal, smooth, and cardiac muscle; glands). The PNS is composed of:

1. **Spinal nerves:** consisting of *sensory* nerves, that is, nerves which transmit impulses to the spinal cord portion of the CNS from sensory receptors; and *motor* nerves, that is, nerves which transmit impulses from the spinal cord to the effector organs.
2. **Cranial nerves:** consisting of sensory nerves that pass along impulses to the *brain* portion of the CNS from sensory receptors, and motor nerves that carry impulses from the *brain* to the effector organs.
3. **Autonomic nervous system (ANS):** a division of the PNS with special responsibility for cardiac muscle and for the smooth muscle and glands of the viscera. The ANS will be discussed separately from the PNS in this unit.
4. **Ganglia:** collections of cell bodies of most sensory fibers of the PNS as well as certain motor fibers to be discussed with the autonomic division of the PNS later in this chapter.

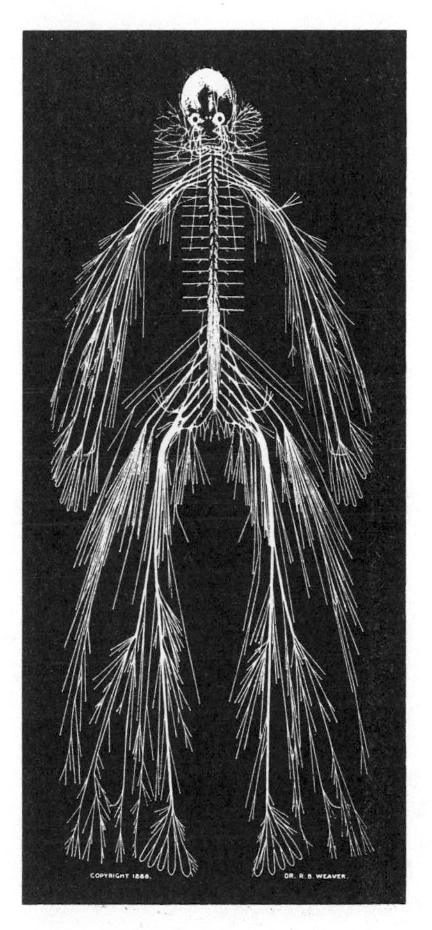

Fig. 3-38 The peripheral nervous system. Cranial nerves leave from the head and neck. Spinal nerves leave from the spinal column. (Courtesy of Dr. Peter S. Amenta, Hahnemann Medical College, Philadelphia.)

QUIZ

Before going on, to help you test your recall of some fundamental terms, try the following matching quiz (five right is a satisfactory level of understanding):

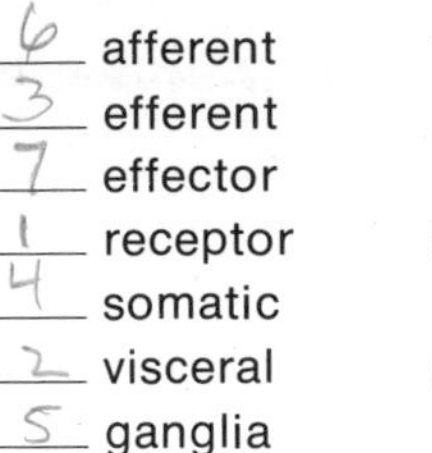

6 afferent
3 efferent
7 effector
1 receptor
4 somatic
2 visceral
5 ganglia

1. Structure that generates a nervous impulse following appropriate sensory stimulus.
2. Pertains to organs containing smooth muscle, glands, or cardiac muscle.
3. Relates to movement from the center to the periphery.
4. Pertains to skin, fascia, and the musculoskeletal system.
5. Collections of cell bodies in the PNS.
6. Relates to movement from the periphery to the center.
7. Muscle or gland that responds to efferent stimulation.

Spinal and cranial nerves, as well as tracts of the CNS, are composed of axons having one of several possible functions. In this respect, a nerve axon of the PNS is sensory (afferent) or motor (efferent) in function. If it is associated with viscera, it is designated visceral afferent or visceral efferent. If it is associated with the skin, fascia, or musculoskeletal system, it is designated somatic afferent or somatic efferent.

Can you organize the functional classification of axons, as given above, in the block diagram below?

Visceral — Afferent / Efferent

Somatic — Afferent / Efferent

SPINAL NERVES

A spinal nerve is composed of variably sized, functionally mixed axons of the PNS. Refer to Fig. 3-39 as you read the following:

A spinal nerve is formed from two spinal roots emerging from the spinal cord, sheltered by the bony vertebral column. One root, the sensory root, leaves the cord on its dorsal sur-

Answers: 6, 3, 7, 1, 4, 2, 5.

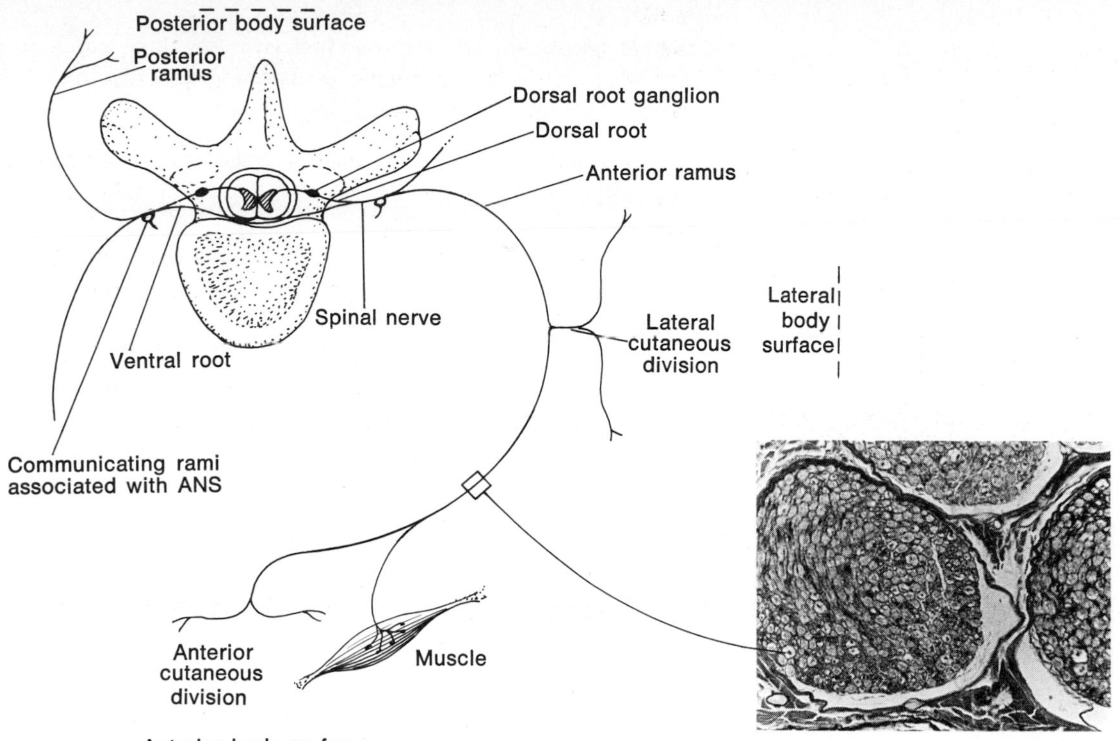

Fig. 3-39 Thoracic intercostal spinal nerves and rami. A ramus may travel with other fibers in bundles, or fascicles, as shown in the photomicrograph. (Courtesy of James Runner.)

face and so is called the **dorsal root.** The motor root leaves the cord on its ventral surface and is therefore called the **ventral root.** The two roots come together about 1 in. lateral to the cord to form a **spinal nerve.**

The spinal nerve then divides into an anterior ramus and a posterior ramus [pl., rami (branches)]. The posterior ramus, usually the smaller of the two, serves the musculoskeletal regions of the back, while the anterior ramus serves the anterolateral body wall and the limbs.

There are 32 pairs of spinal nerves exiting the spinal cord from skull to coccyx (Fig. 3-40). The nerves are divided into regions similar to the regional breakdown of vertebrae:

1. Cervical: 8 spinal nerves.
2. Thoracic: 12 spinal nerves.
3. Lumbar: 5 spinal nerves.
4. Sacral: 5 spinal nerves.
5. Coccygeal: 2 spinal nerves.

The precise arrangement and distribution of the nerves

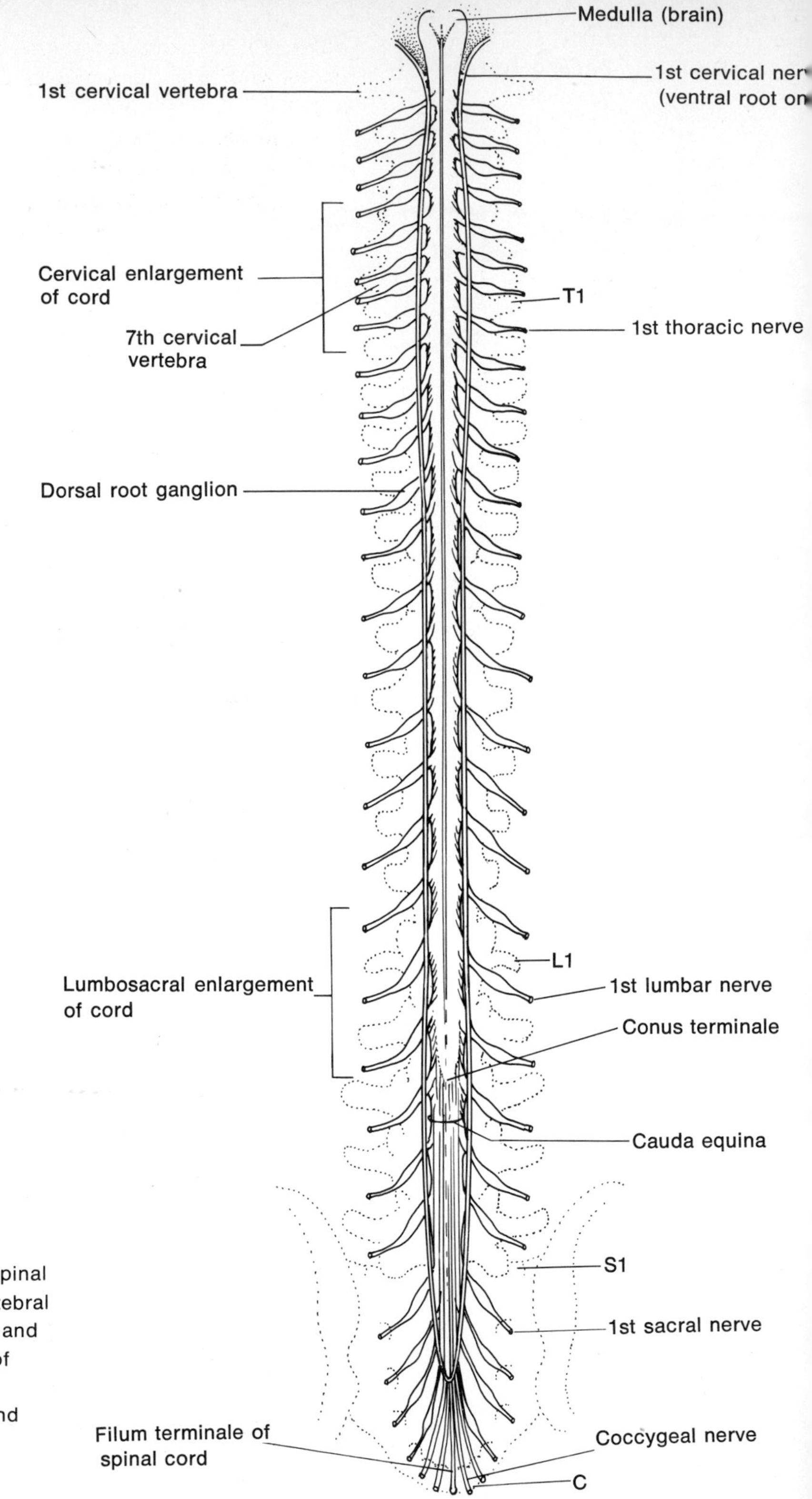

Fig. 3-40 The spinal cord and spinal nerve roots in relation to the vertebral column. Posterior view. Cervical and lumbar enlargements are areas of increased numbers of neurons relating to nerves of the upper and lower limbs.

differ from region to region; however, the thoracic spinal nerve is usually employed as the classical spinal nerve (Fig. 3-39).

In general, the spinal nerves form networks called **plexuses** (Fig. 3-41) before distributing out to peripheral organs. The major plexuses of peripheral nerves are formed by anterior rami. See if you can reason out the correct plexus or nerves

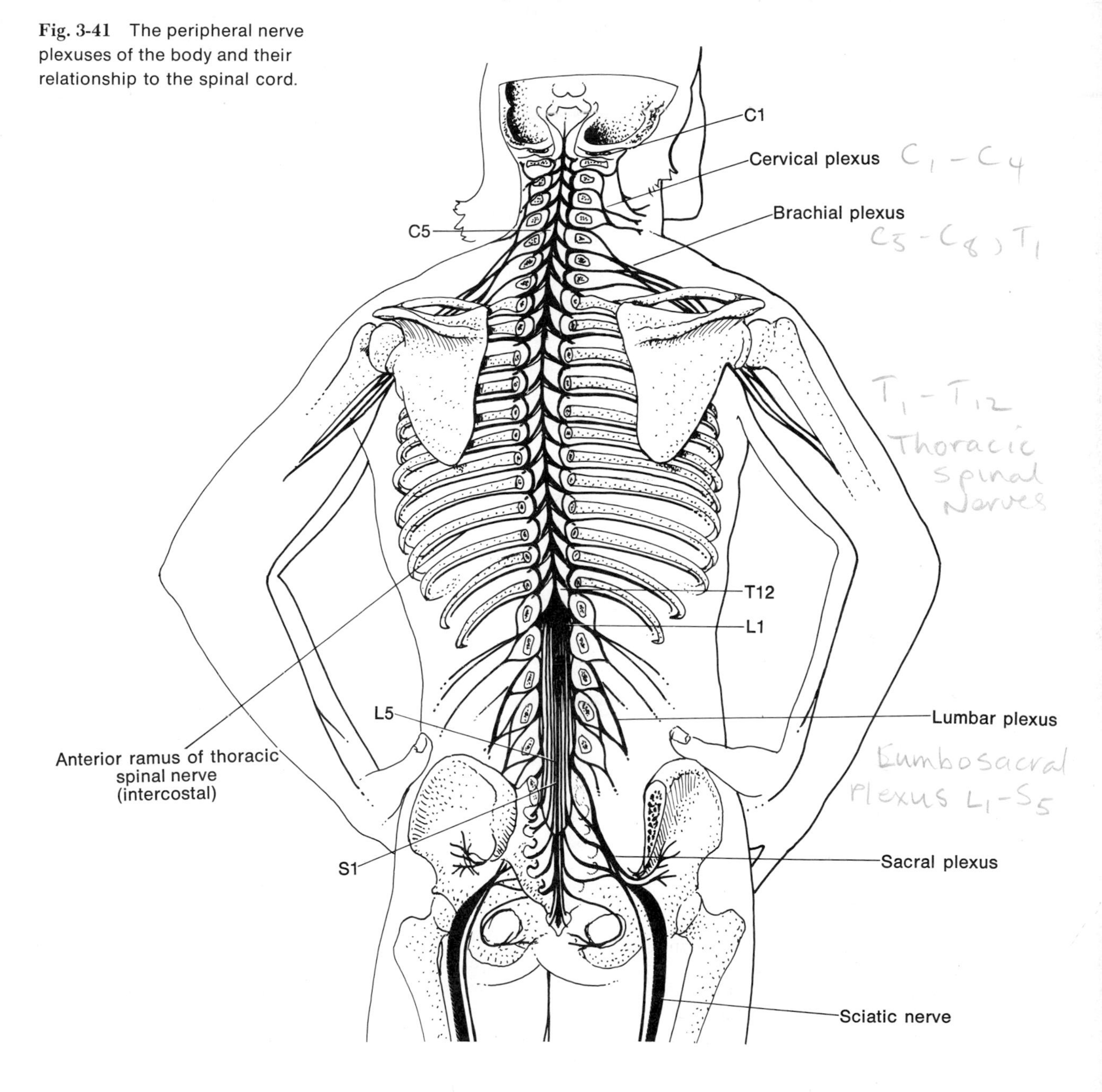

Fig. 3-41 The peripheral nerve plexuses of the body and their relationship to the spinal cord.

for each of the functions listed:

4 Cervical plexus (C1–C4)	1. Serves the muscles, etc. of the rib cage (intercostal) and the overlying skin and fascia.
2 Brachial plexus (C5–C8, T1)	2. Serves the muscles, skin, etc. of the upper limb.
1 Thoracic spinal nerves (T1–T12)	3. Serves the muscles, skin, etc. of the lower limb and pelvis.
3 Lumbosacral plexus (L1–S5)	4. Serves the anterior and lateral neck region, as well as the diaphragm, the principal muscle of respiration.

The nerves of each plexus will be considered regionally.

Each spinal nerve is distributed to a specific region of the body. The ventral root contributes efferent (motor) fibers, both somatic and visceral, to the muscles, glands, and viscera of that region. The dorsal root is composed of afferent (sensory) fibers coming in from pain, temperature, pressure, touch, and proprioception receptors of that region. The body-surface region served by one dorsal root of a spinal nerve is called a **dermatome** (Fig. 3-42). An appreciation of dermatomes is particularly important to the physician. A nerve dysfunction may be manifested by loss of some or all sensation in one or more dermatomes. Furthermore, pain in a dermatome may actually be caused by disorder not at the body surface but in one of the viscera. This is called **referred pain.** The dermatome affected is the one with sensory fibers entering the cord through the same dorsal roots as the afferent fibers from the injured organ.

CRANIAL NERVES

Structurally and functionally, cranial nerves are similar to spinal nerves. They are just related to different parts of the CNS. The 12 cranial nerves, with two exceptions, serve only the head region. Unlike spinal nerves, each cranial nerve is named, usually by function or destination (Fig. 3-43).

GANGLIA

The cell bodies of most sensory neurons of the peripheral nervous system are located in ganglia adjacent to but outside the brain and spinal cord (Fig. 3-44). The cell body, you will remember, is that part of the neuron containing the nucleus

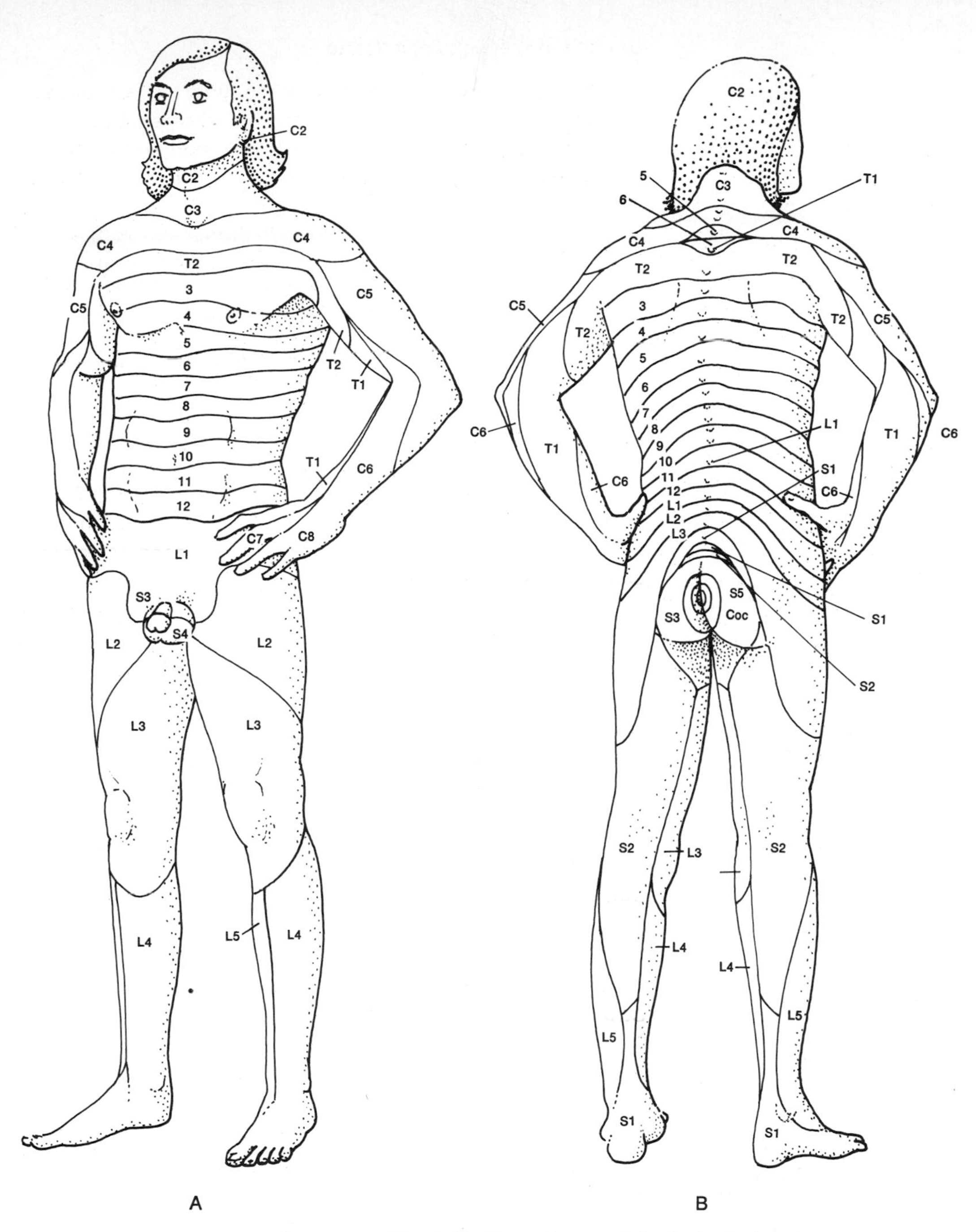

Fig. 3-42 Dermatomes of the body. A. Anterior aspect. B. Posterior aspect.

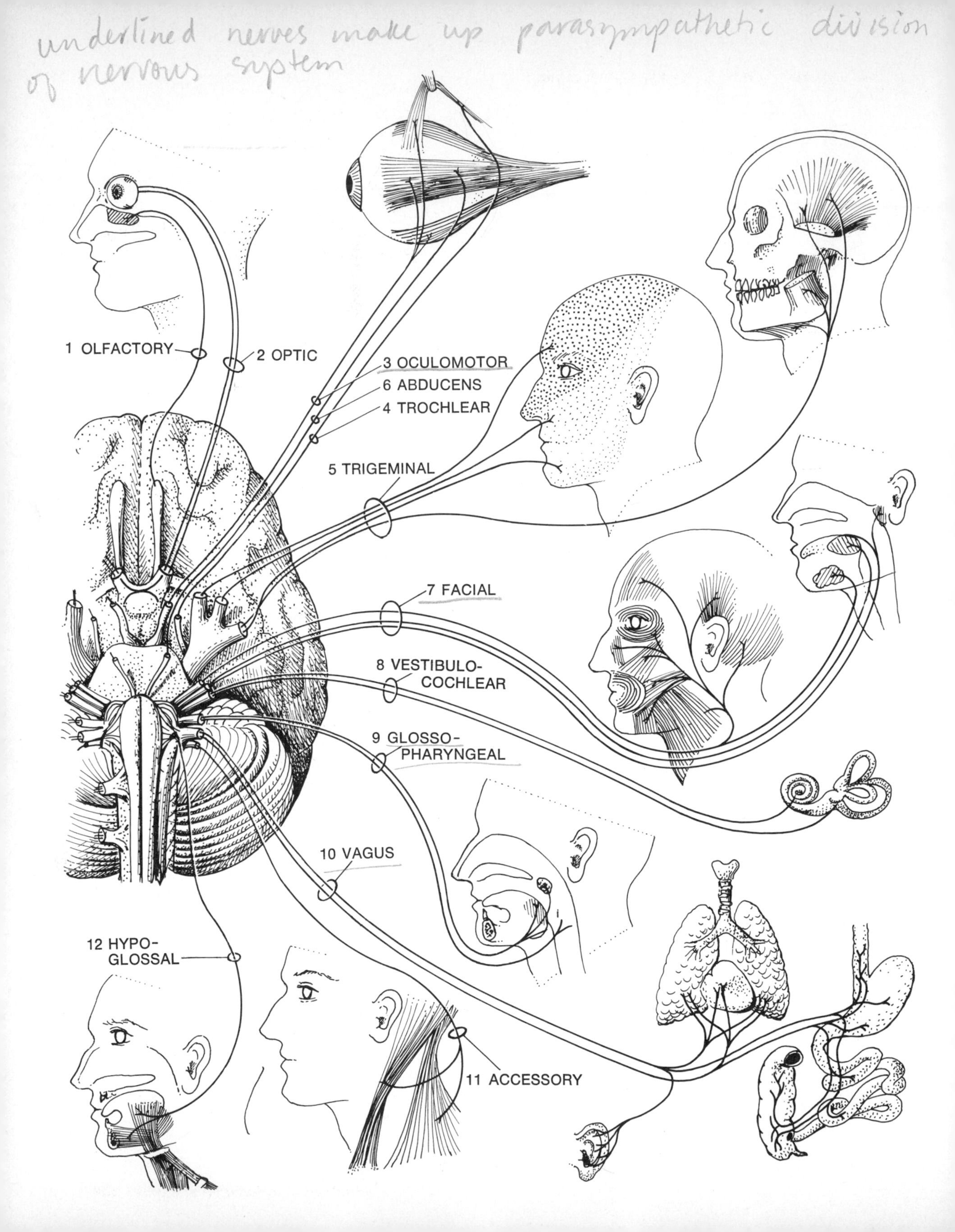
underlined nerves make up parasympathetic division of nervous system
1 OLFACTORY
2 OPTIC
3 OCULOMOTOR
6 ABDUCENS
4 TROCHLEAR
5 TRIGEMINAL
7 FACIAL
8 VESTIBULO-
COCHLEAR
9 GLOSSO-
PHARYNGEAL
10 VAGUS
12 HYPO-
GLOSSAL
11 ACCESSORY

Fig. 3-43 (opposite) The cranial nerves. (After Netter.)

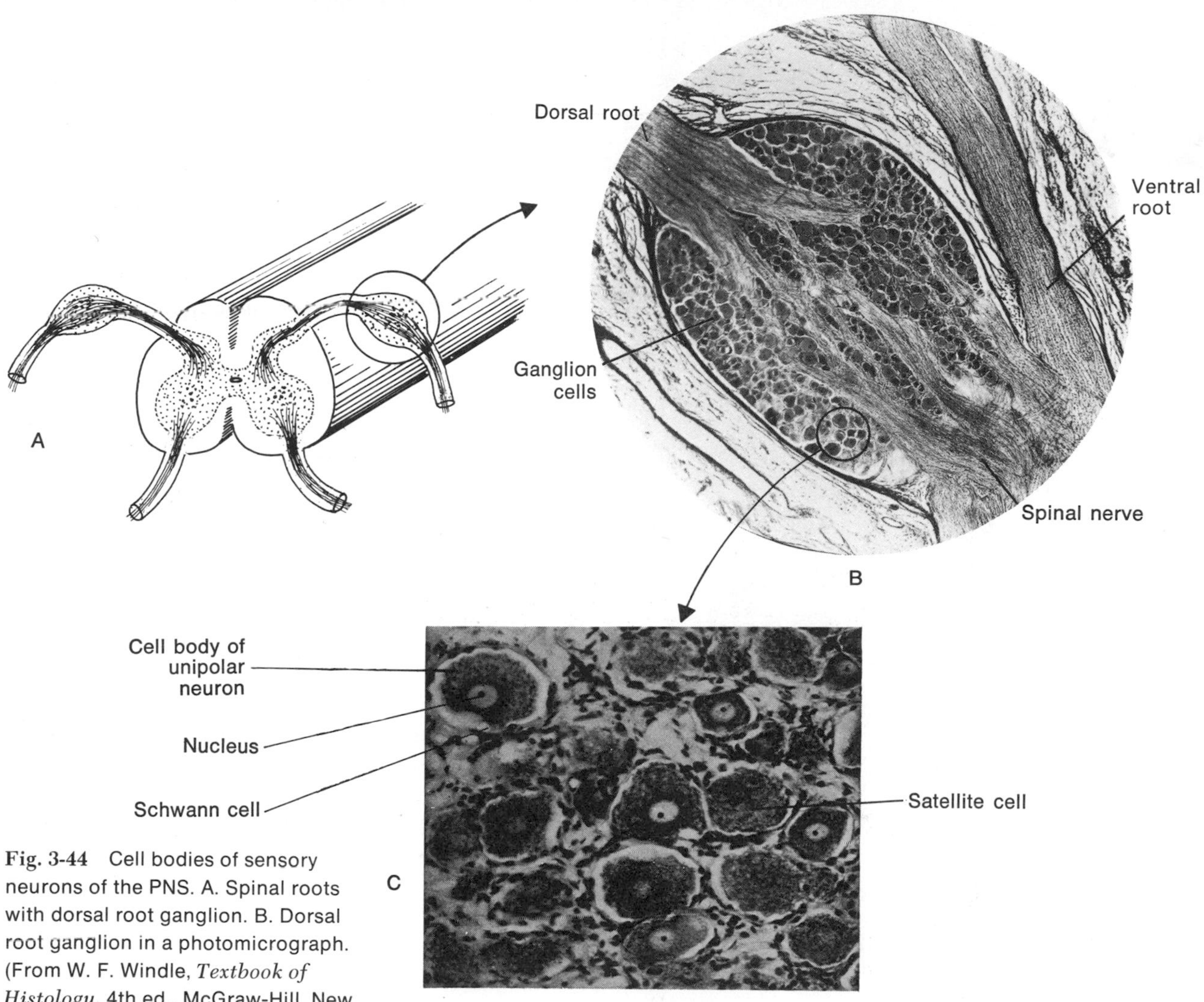

Fig. 3-44 Cell bodies of sensory neurons of the PNS. A. Spinal roots with dorsal root ganglion. B. Dorsal root ganglion in a photomicrograph. (From W. F. Windle, *Textbook of Histology*, 4th ed., McGraw-Hill, New York, 1969.) C. Dorsal root ganglion under higher magnification (×200). Photomicrograph. (From T. L. Peele, *The Neuroanatomic Basis for Clinical Neurology*, 2d ed., McGraw-Hill, New York, 1961.)

and some peripheral cytoplasm. The large masses of these cell bodies, their satellite cells, and axons produce a visible swelling along the dorsal root fibers within the enclosure of the intervertebral foramen.

Many of the cranial nerve sensory ganglia are comparable to spinal ganglia, i.e., they are unipolar. However, the ganglia of the olfactory, optic, and vestibulocochlear (I, II, VIII) nerves are bipolar cells located near the receptor organs. Motor ganglia, associated with the autonomic nervous system, will be discussed in the next section.

introduction to the autonomic nervous system (ANS)

A CONCEPT

The autonomic nervous system—responsible for distributing *motor* impulses to viscera—consists of visceral efferent (motor) fibers and their cell bodies. You are now familiar with the concept that the PNS is primarily a voluntarily activated system. The autonomic nervous system is predominantly an *involuntary* system capable, under certain conditions, of voluntary activation as well.

As your study of body structure progresses you will find that most of the viscera of the body have the capacity to move or secrete, or both. All of the viscera of the alimentary canal (stomach, intestines, etc.) have the capacity to churn their contents mechanically and to secrete enzymes as well in order to assist in the breakdown of these contents. They do this through the offices of smooth muscle and glands within their walls. You may note that this digestive process takes place without any willed (voluntary) action on your part.

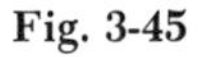
Fig. 3-45

Fig. 3-46

Likewise, the viscera of the respiratory system (lungs, etc.), cardiovascular system (heart, blood vessels), urinary system (kidneys, bladder, etc.), and reproductive system (glands and ducts), as well as the various glands throughout the body also have the capacity to move or secrete (or both) *and do*—for a lifetime—without any conscious effort on your part. Such visceral activity is the responsibility of the **autonomic nervous system.**

When "vegetation" and quiet relaxation are the order of the day (Fig. 3-45), the autonomic nervous system acts to maintain the internal environment accordingly—slower respiration rates, slower heart rate, increasing the digestive processes, etc. When a threat to one's survival is at hand (Fig. 3-46), the autonomic nervous system takes appropriate action, e.g., increasing blood flow to the skeletal muscles involved in "flight," increasing respiration rate, etc., designed to allow you to run faster than you thought you could or take other evasive actions.

The maintenance of a stable internal environment relative to outside environmental conditions is called **homeostasis,** and that, in a word, is the function of the ANS. So much for concepts.

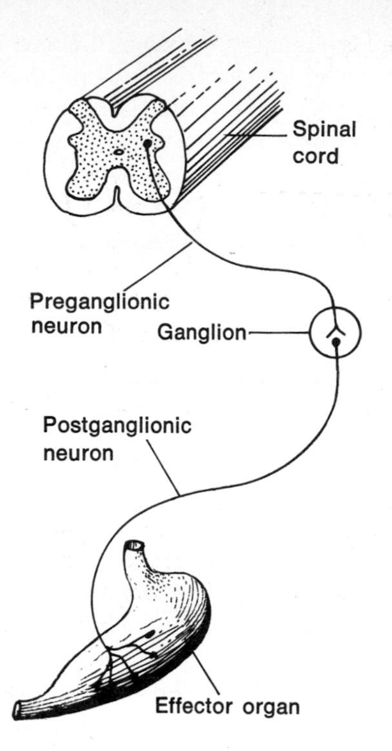

Fig. 3-47 The basic plan of autonomic innervation of the viscera.

ANATOMY OF THE ANS

You are now familiar with the pattern of a single motor neuron with cell body in the spinal cord whose axon exits via the ventral root and terminates at the motor end plate in skeletal muscle. However, ANS innervation of viscera is accomplished by *two* motor neurons stretching from CNS to effector organs.

Now study Fig. 3-47: The axon of the first neuron, with cell body in the CNS, passes out with other axons of motor neurons through the ventral root and terminates by synapsing

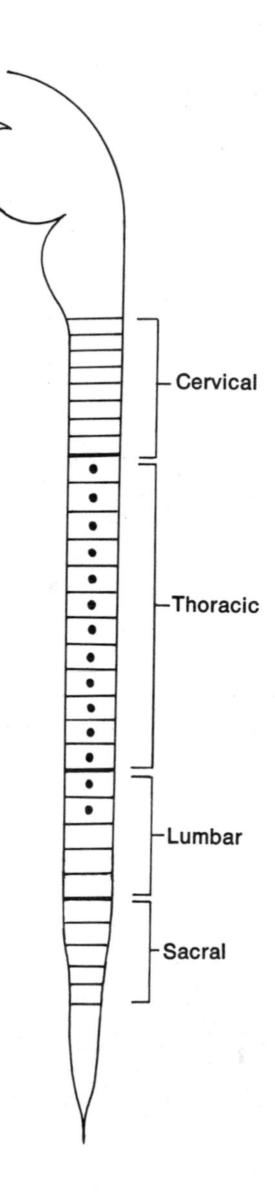

Fig. 3-48 Regions of the spinal cord (T1 to L2) in which cell bodies of thoracolumbar preganglionic neurons (represented by dots) are found.

with the cell body of the second neuron. Remember, collections of such cell bodies are called ganglia, except that ganglia you have met before, dorsal root ganglia, were composed of cell bodies of *sensory* neurons; the ganglia of the ANS are of cell bodies of *motor* neurons. The axon of the second neuron leaves the ganglion and ends eventually in smooth muscle, glands, or cardiac muscle. Since, in the sequence described, the first neuron is in front of the second neuron, more specifically, in front of its cell body, the first neuron is termed **preganglionic** and the second neuron **postganglionic.**

The autonomic nervous system can be subdivided structurally and functionally into two components (1) thoracolumbar (sympathetic) division and (2) craniosacral (parasympathetic) division.

Thoracolumbar division of the ANS

The **thoracolumbar,** or **sympathetic,** division is responsible for "flight-or-fight" activities (Fig. 3-46). It produces changes in appropriate visceral activity necessary for meeting such stresses as fear and panic and extended physical exertion. Such visceral activity includes increasing blood pressure and directing flow of blood to skeletal muscles, increasing heart rate, maintaining body temperature through sweating, increasing the blood sugar concentration in response to the greater demand by the body tissues, and so on.

The thoracolumbar division of the ANS is structurally characterized as follows:

All preganglionic cell bodies (multipolar) are located in the thoracic and lumbar regions of the spinal cord (Fig. 3-48).

The relatively short axons of these preganglionic neurons pass out of the cord via the ventral root and enter a chain of ganglia, **the sympathetic chain,** lying adjacent to the vertebral bodies (Fig. 3-49). The route of entry into the chain is the same for all preganglionic axons of this division: spinal cord → ventral root → spinal nerve → white* rami communicantes → sympathetic chain.

Having entered the sympathetic chain, the preganglionic axons may follow one of the three courses shown in Fig. 3-50 to synapse with postganglionic neurons.

The axons of the postganglionic neurons then take a relatively long journey to the effector organ. Note that in front of

* "White" because these preganglionic axons are myelinated (yielding a glistening white appearance); postganglionic axons are unmyelinated and therefore called "gray."

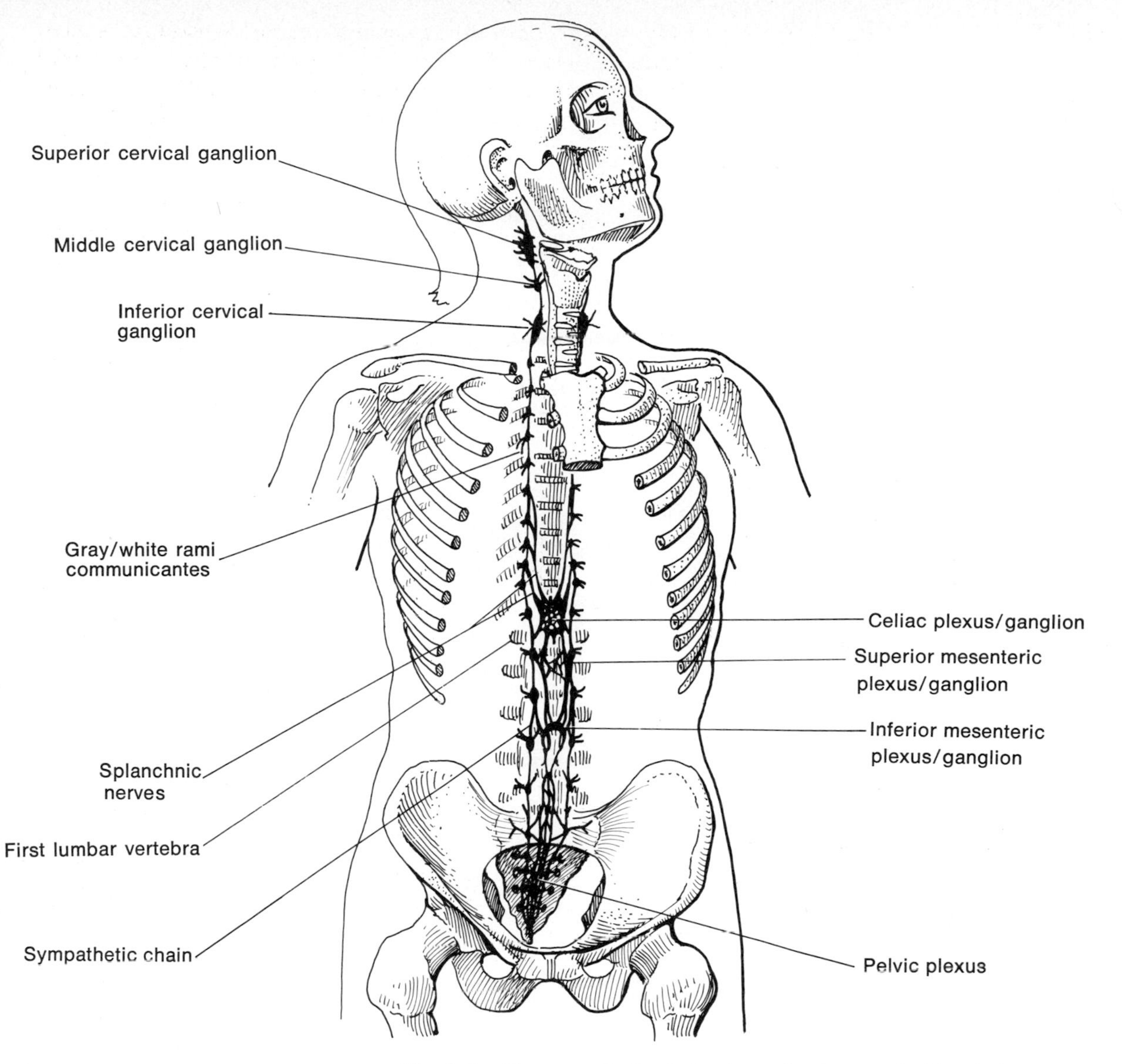

Fig. 3-49 The sympathetic chain of ganglia along each side of the vertebral column. Note the splanchnic (visceral) nerves passing to the prevertebral ganglia. (Adapted from E. L. House and B. Pansky, *A Functional Approach to Neuroanatomy*, 2d ed., McGraw-Hill, New York, 1967.)

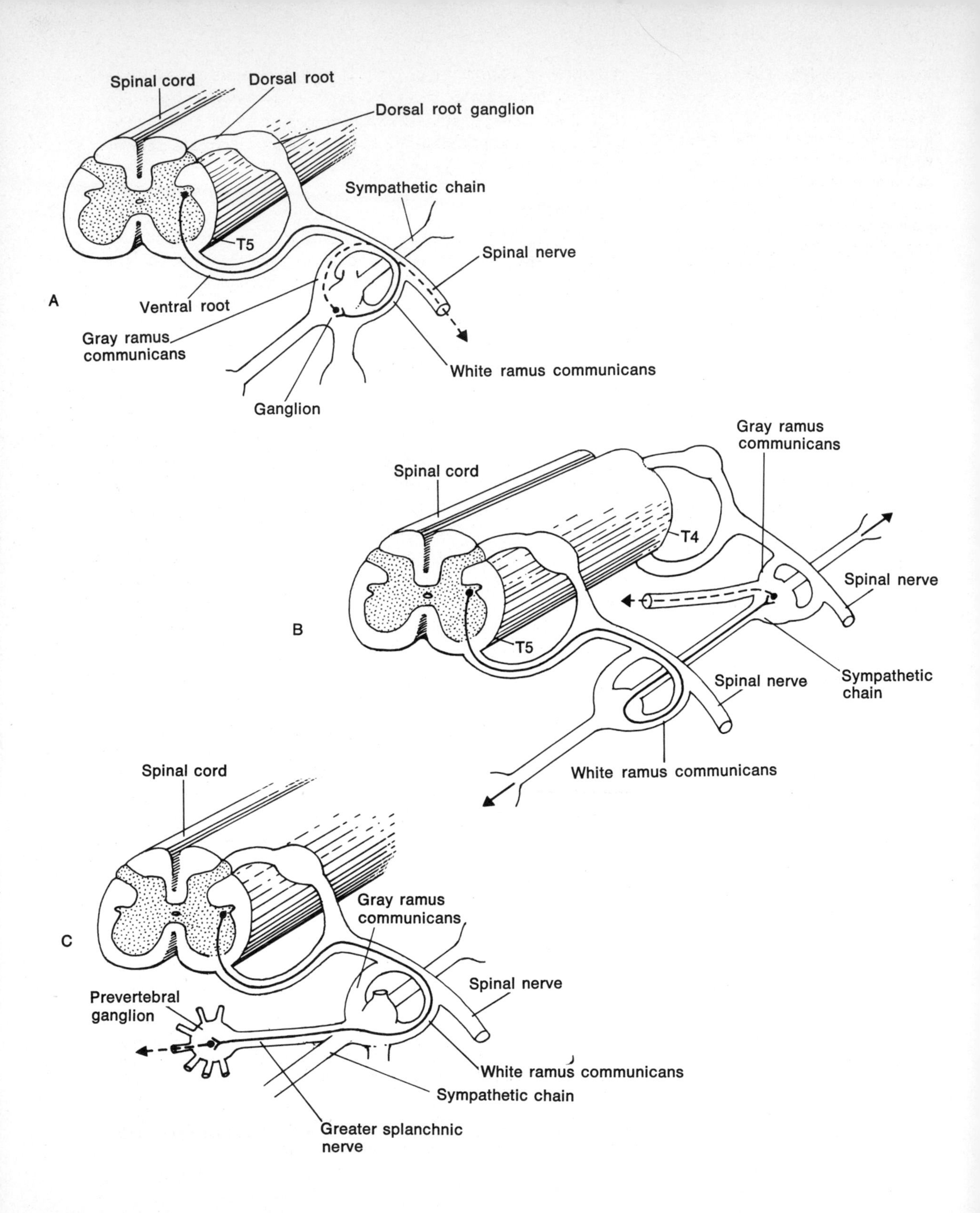
Spinal cord
Dorsal root
Dorsal root ganglion
Sympathetic chain
T5
Spinal nerve
A
Ventral root
Gray ramus communicans
White ramus communicans
Ganglion
Gray ramus communicans
Spinal cord
T4
Spinal nerve
B
T5
Sympathetic chain
Spinal nerve
White ramus communicans
Spinal cord
Gray ramus communicans
C
Spinal nerve
Prevertebral ganglion
White ramus communicans
Sympathetic chain
Greater splanchnic nerve

Fig. 3-50 (opposite) Three routes available to thoracolumbar fibers of the ANS. A. A fiber may follow spinal nerves to the body wall and limbs, where it innervates sweat glands or hair erector muscles. B. A fiber may travel up or down the chain before leaving to innervate viscera of the head, neck, or thorax. C. A fiber may pass through the chain without synapsing, to become part of a splanchnic nerve en route to the prevertebral ganglia.

the aorta and vertebral column in the abdominal cavity there are also three sets of ganglia known as **prevertebral ganglia** (Fig. 3-49). Because these ganglia are close to the organs they supply, the postganglionic neurons emerging are shorter than those from the sympathetic chain.

Such is the structure of the thoracolumbar division of the autonomic nervous system. Contemplation of Figs. 3-48 to 3-50 may ease the confusion many of you are doubtless suffering. When you think you understand matters this far, try Fig. 3-51.

Craniosacral division of the ANS

The **craniosacral,** or **parasympathetic,** division is responsible for "vegetative" functions (Fig. 3-45) involving such activities as the digestion of foodstuffs (mechanical churning and secretion of enzymes) and passing of digestive byproducts into the rectum for subsequent elimination. Through contraction of the smooth muscle of blood vessels (vasoconstriction), the flow of blood is shunted from other body regions to the abdominal viscera*. The heart rate and the force of the heart may or may not be decreased, depending on the time since the last meal. Thus, while the thoracolumbar division readies the body for action to maintain survival, the craniosacral division stimulates certain visceral ("vegetative") functions while somatic activity is generally relaxed.

All preganglionic cell bodies (multipolar) are located in the brainstem and the sacral regions of the spinal cord (Fig. 3-52). Hence the term *craniosacral.*

The longer axons of the cranial preganglionic neurons pass from the brain in company with certain cranial nerves (III, VII, and IX) and go to one of four ganglia in the head region (Fig. 3-53). There they synapse with the cell bodies of postganglionic neurons, whose axons terminate in the smooth muscle and glands associated with the eye, salivary and other head glands, and blood vessels of the head.

The remaining axons of the cranial preganglionic neurons make up the vagus nerve (X) and are quite long and extensive. These pass throughout the thoracic and abdominal cavities†, giving off branches to the heart, lungs, and many of the vis-

* The brain is extremely sensitive to this minute change in blood flow, and since the blood flow is slightly decreased to the brain, the individual "feels" sleepy following ingestion of a meal (effects of a mild hypoxia—low oxygen).

† In so doing, the main abdominal branches of the vagus nerve pass *through* the prevertebral ganglia without synapsing.

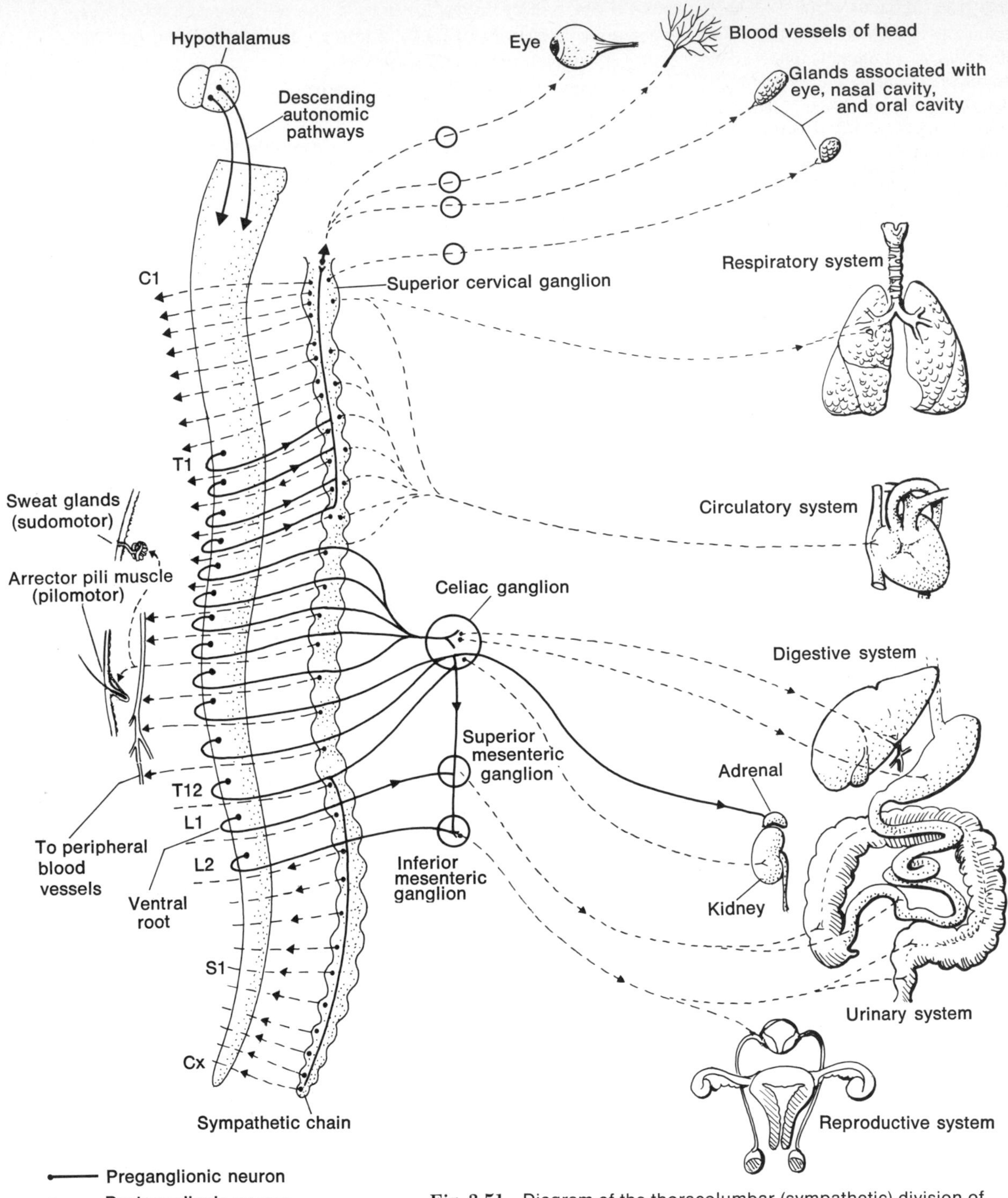

Fig. 3-51 Diagram of the thoracolumbar (sympathetic) division of the ANS.

Fig. 3-52 Regions of the brain and spinal cord in which cell bodies of craniosacral preganglionic neurons (represented by dots) are found. These regions include cranial nerves III, VII, IX, and X, and spinal levels S2 to S4.

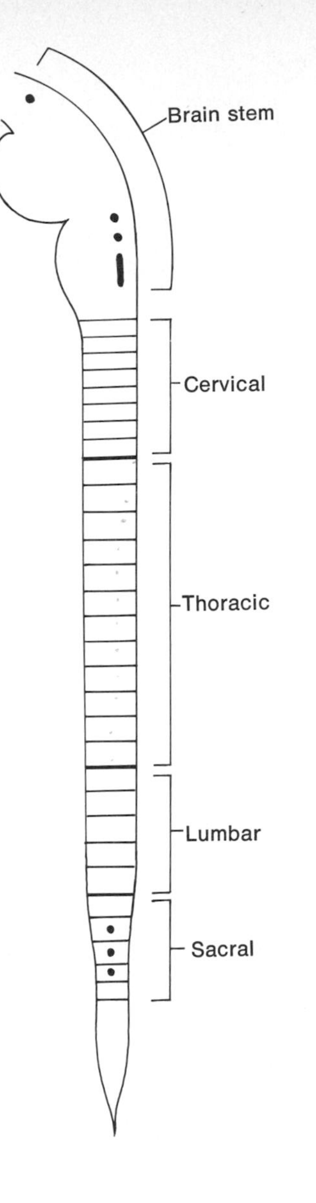

cera of the digestive and urinary systems (hence the name *vagus*, which means wanderer or vagabond). Within the walls of each of these organs, the preganglionic axons synapse with the cell bodies of postganglionic neurons. The very short axons of these neurons terminate in the smooth muscle layers and glands within the walls of that same organ (Figs. 3-53 and 3-54).

The long axons of the sacral preganglionic neurons pass out of the spinal cord via the ventral roots and join the sacral

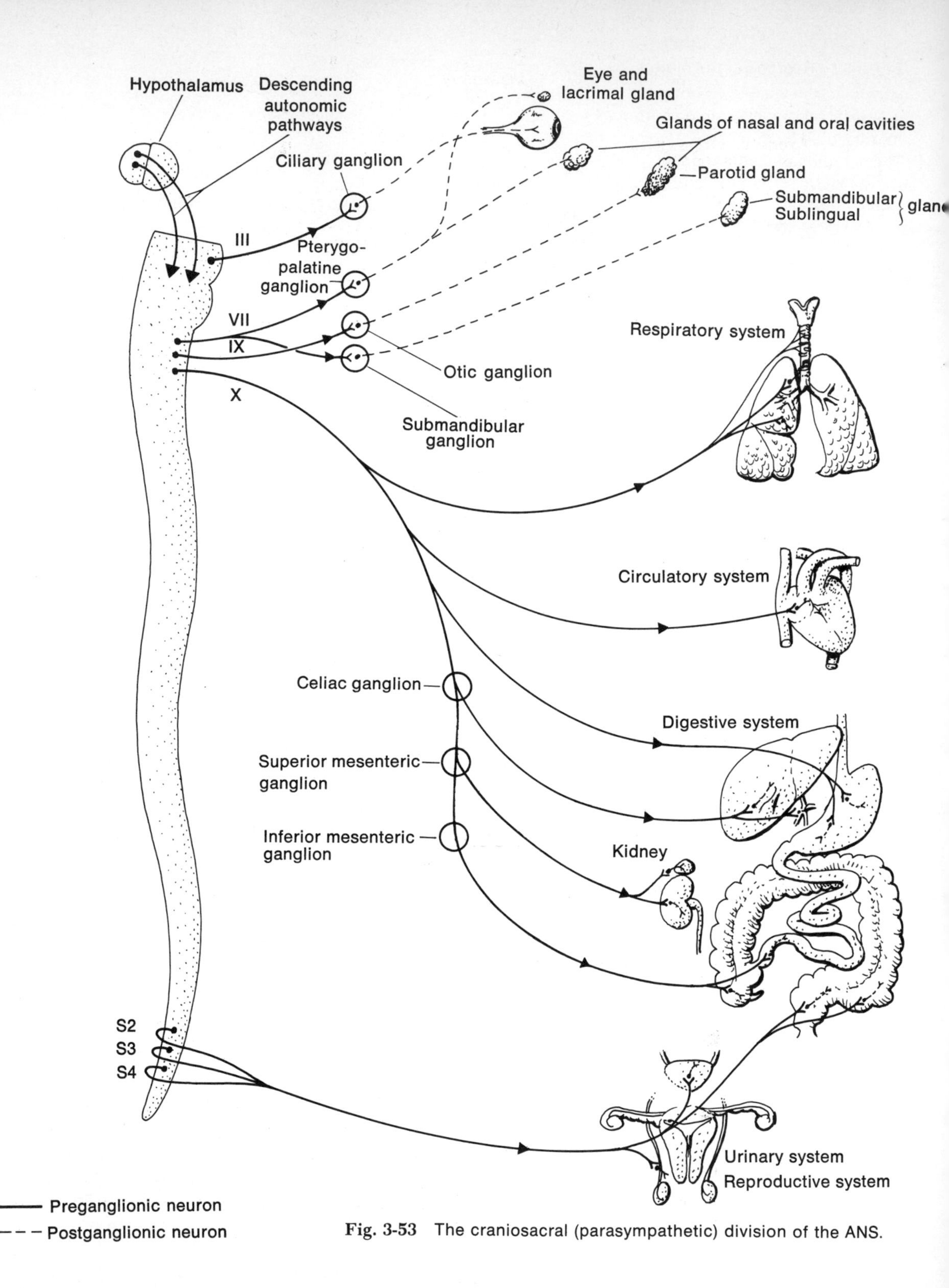

Fig. 3-53 The craniosacral (parasympathetic) division of the ANS.

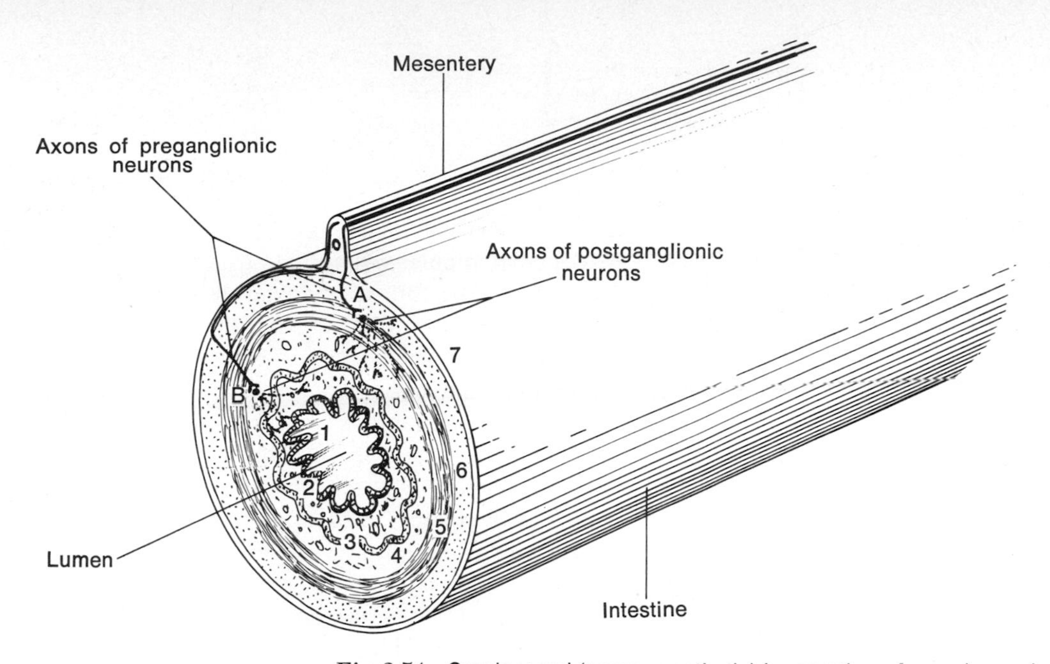

Fig. 3-54 Craniosacral (parasympathetic) innervation of muscles and glands in a section of intestine. The nerves pass between layers of the mesentery suspending the intestine. Nerve A supplies layers 4, 5, and 6. Nerve B supplies layers 2 and 3. The layers are 1, epithelium; 2, glands and connective tissue; 3, smooth muscle of the mucosa; 4, connective tissue with large vessels; 5 and 6, longitudinal and transverse layers of smooth muscle; 7, peritoneum.

spinal nerves (they do not pass through the sympathetic chain). Then they leave the somatic portions of the spinal nerves to form plexuses (Fig. 3-49). Next the axons pass into the walls of the large intestine, rectum, and anal canal; as well as those of the bladder and reproductive organs. Within these walls, the preganglionic axons synapse with the cell bodies of postganglionic neurons, whose very short axons terminate in smooth muscle and glands of those walls (Fig. 3-54).

Functionally, the two divisions of the autonomic nervous system work together to maintain homeostasis. They are not antagonistic—any more than rain is antagonistic to sunlight—but work, often simultaneously, to keep the internal body structure functioning relative to external and internal environmental conditions. It has been said, and quite accurately, that the visceral efferent nerves "operate" body activity as an organ makes beautiful music: "The sympathetics are like the loud and soft pedals, modulating all the notes together, while the cranial and sacral innervation are like the separate keys" (Cannon).

reflexes

In the functional sense, a reflex is an involuntary motor response to impulses generated by sense receptors. We demonstrate various reflexes constantly throughout the day and night. Reflexes are mechanisms by which our various body parts and organs can function without our conscious awareness. There are certain basic functions of the body which must go on, for example, when we are asleep: visceral movements during digestion, changes in heart and respiratory rates, musculoskeletal movement in response to itching, irritation , or pain, etc. These acts are accomplished reflexly. We reflexly move out of the way of a speeding object, or we reflexly set our bodies to pick up heavy weights—"without thinking about it" (in human relations, however, reflexes are no substitute for "thinking about it").

By definition, a reflex requires two structural components: an afferent limb and an efferent limb. The afferent limb starts with a sense receptor and ends in the CNS via an axon. The efferent limb starts in the CNS—directly or indirectly connected to the afferent limb—and ends at the effector organ via an axon. The CNS, then, is the key to all reflexes.

Reflexes are rarely simple; they usually involve many neurons (polysynaptic) (Fig. 3-55A), and this is important to keep in mind. However, simple two-neuron reflexes involving one synapse (monosynaptic) are extremely important, for these are the stretch (myotatic) reflexes (Fig. 3-55B). The significance of muscle contraction in response to stretch (i.e., muscle "tone") will be considered with pathways of the central nervous system (see Unit 6).

Stretch receptors are found in skeletal muscle (muscle spindles) and in tendons (Golgi tendon organs). Refer now to Fig. 3-55B as you read the following:

Skeletal muscle is sensitive to stretching. When stretched, the muscle will weakly contract (stretch reflex).

Fig. 3-55 Anatomy of a reflex. A. Polysynaptic reflex. The magnified cell body of a motor neuron shows, in the many synapses, the numerous influences converging on it from various sensory receptors as well as facilitatory and/or inhibitory centers in the brain. The response involves many muscles. B. Monosynaptic reflex. The axon of a stretch-sensitive sensory neuron found in skeletal muscle synapses directly with a motor neuron whose axon innervates the muscle affected by the stretching. See text for knee-jerk test.

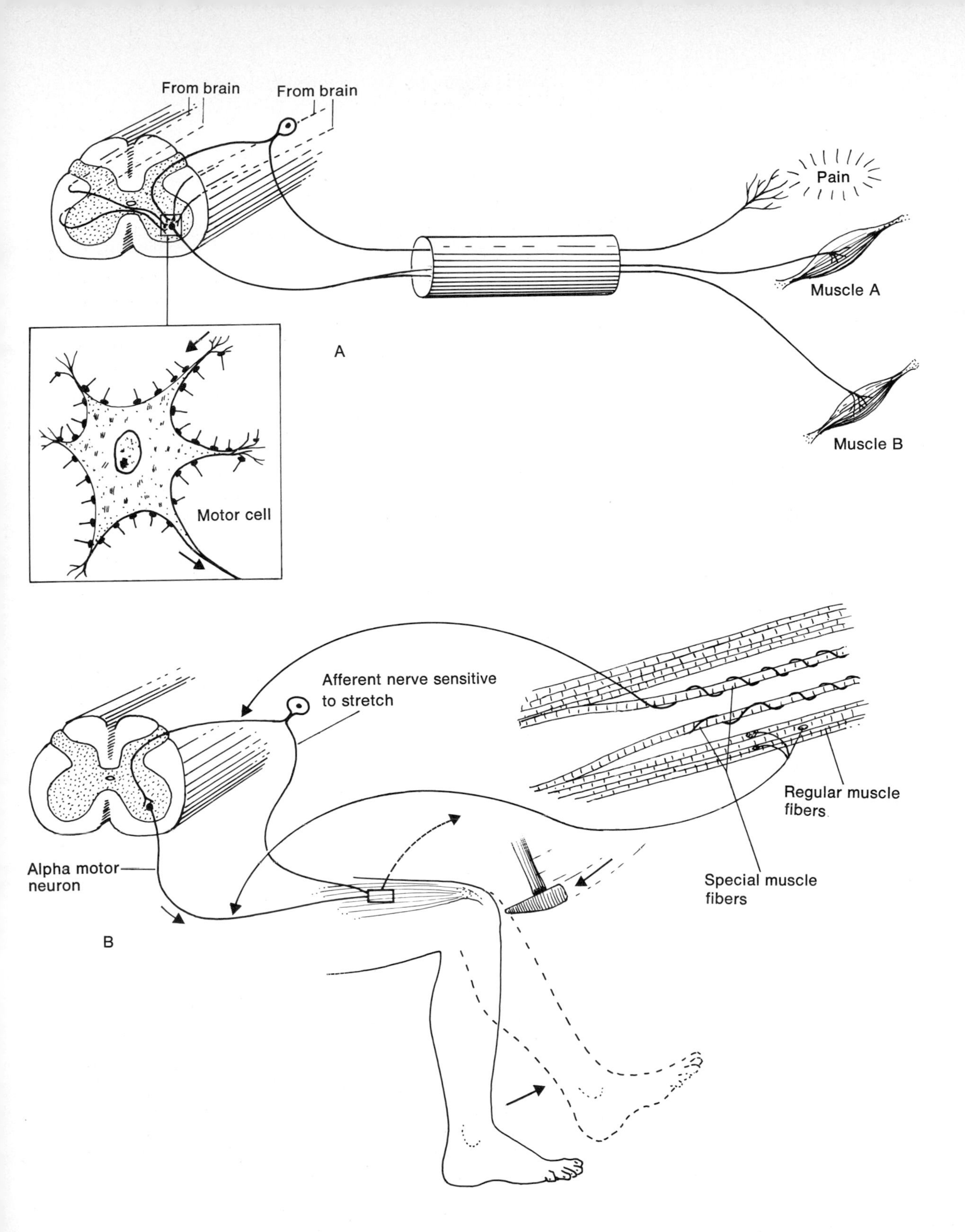
From brain
From brain
Pain
Muscle A
Muscle B
A
Motor cell
Afferent nerve sensitive
to stretch
Regular muscle
fibers
Alpha motor
neuron
Special muscle
fibers
B

The anatomical basis for the stretch reflex may be found in the **neuromuscular spindle** (Fig. 2-75).

The neuromuscular spindle consists, in part, of special muscle fibers within the mass of regular skeletal muscle. These fibers are entwined with a coil of nerve fiber sensitive to pulling/stretching.

When the entire muscle is stretched, the coiled stretch-sensitive afferent nerves fire off impulses to the spinal cord, or brain (in the case of cranial nerves).

The afferent nerve enters the cord via the dorsal root, passes to the anterior horn of the spinal cord, and synapses directly with a large (alpha) motor neuron.

The alpha motor neuron fires off impulses to the regular muscle fibers and the muscle contracts.

The method used clinically to test one of the stretch reflexes is illustrated in Fig. 3-55B (knee jerk). Try it yourself:

1. The tendon of the quadriceps femoris muscle passes over the kneecap to insert on the tuberosity of the tibia two fingersbreadth below the cap (feel this on yourself).
2. Cross one leg over the other and allow it to hang in relaxation. The tendon of the quadriceps muscle in the "crossing leg" is stretched, but not enough to reflexly contract.
3. Tap the tendon between the kneecap and the tibial tuberosity with the little-finger side of your hand.
4. If the muscle tendon is stretched sufficiently to "fire" the stretch receptors, the quadriceps femoris muscle will contract, the knee will jerk, while the leg partly extends.

The knee jerk is tested to check on certain functions of the spinal cord. If there are lesions in the CNS or the nerve being tested, they may be reflected in hyperstimulated or depressed stretch reflexes, representative of spastic or flaccid muscles, respectively.

The withdrawal reflex is an example of polysynaptic reflexes. The motor neurons generating the response (i.e., contraction of a flexor muscle necessary to withdraw from the pain or pressure) come under the influence of other neurons in the CNS as well as PNS. For example, if a cat steps on a piece of glass with its right foreleg, that foreleg will *flex* in response—the opposite foreleg will *extend,* as will the opposite hindleg. The cat will also cry out. Such is an example of a polysynaptic reflex. Reflexes may be purely somatic (somatosomatic), purely visceral (viscerovisceral) or mixed. All of the movements of our body parts which are not willfully induced are considered to be reflexive (involuntary).

skin

There is no magician's mantle to compare with the skin in its diverse roles of waterproof, overcoat, sunshade, suit of armor and refrigerator, sensitive to the touch of a feather, to temperature and to pain, withstanding the wear and tear of three score and ten, and executing its own running repairs.*

The truth here is as profound as the description is elegant. Further introduction would be redundant.

The skin of the body **(integument)** is an external surface lining continuous with the internal surface membrane of those body cavities open to the exterior. The smooth, uninterrupted junction of the external skin of the lip with its internal surface in the mouth is an excellent example of external-internal continuity and can be seen and felt on yourself. Note how dry the external portion of the lip is and how moist the internal portion is. The latter surface is a mucous membrane which continues caudally to line the entire gastrointestinal tract and is continuous with the skin of the anus as well.

The integument (Fig. 3-56) is arranged into two well-defined layers:

1. **Epidermis:** the outer, stratified epithelial layer.
2. **Dermis:** the connective tissue layer underlying the epidermis.

The dermis is bound to deeper structures (e.g., periosteum or deep fascia) by a variable layer of subcutaneous tissue, the superficial fascia. This layer may be thick or thin; for example, contrast on yourself (1) the depth or amount of fascia between the skin and periosteum on the anteromedial portion of your leg ("shinbone"); (2) the amount of fascia between skin and underlying muscle at the waist or hips just above the buttocks.

THICKNESS AND QUALITY OF SKIN

The thickness as well as the quality of the skin is significantly different in several regions of the body (Fig. 3-57). The thicker skin generally covers the posterior and/or extensor surfaces of the body, the thickest being found on the palms of the hands and the soles of the feet. The thinnest skin is prob-

* From Lockhart, Hamilton, and Fyfe, p. 3.

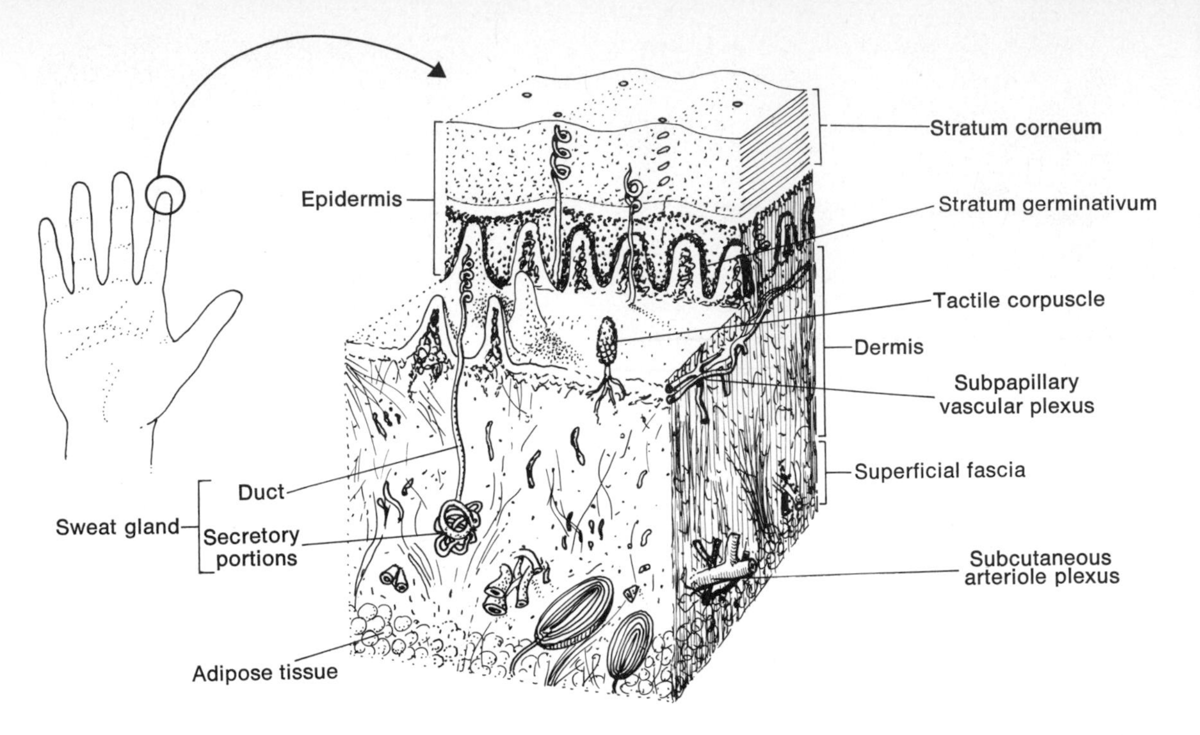

Fig. 3-56 Diagram of skin taken from a finger. (Adapted from R. D. Lockhart, G. F. Hamilton, and F. W. Fyfe, *Anatomy of the Human Body,* 2d ed., J. B. Lippincott Co., Philadelphia, 1965.)

ably that covering the eyelids. You can ascertain this on yourself. While you're at it, note (by pinching, or pressing, or just looking):

1. The difference in the thickness of abdominal skin and that over the middle or upper back.
2. The nature of the skin of the lips at the mouth in contrast to the skin of the anus—both skin-lined orifices are opposite ends of the same (digestive) tube.
3. The skin at the elbow in contrast to the skin of the forearm (compare both anterior and posterior surfaces).

The quality of skin is probably dependent upon the degree of moisture within. This, in part, is dependent, upon the number and kinds of glands in the skin. Aging tends to dry out the skin. The firmness and elasticity of youthful skin, created by an abundant, fatty superficial fascia thick with elastic fibers, is lost during the aging process as fluid is absorbed and adipose tissue disappears, causing the skin to sag and become loose

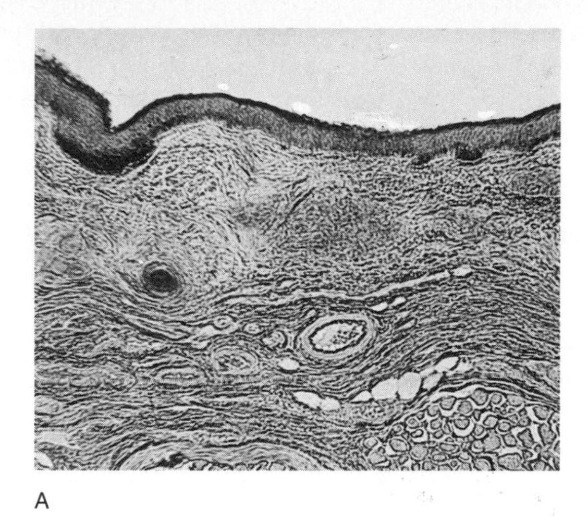

A

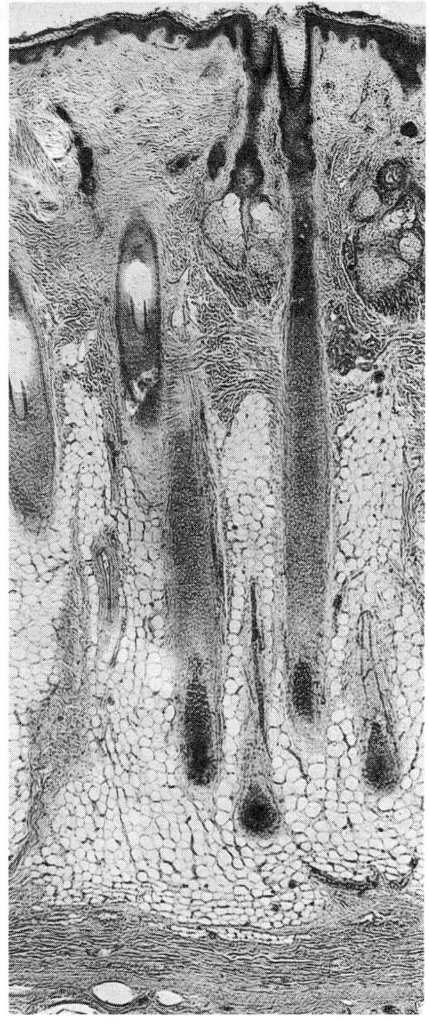

B

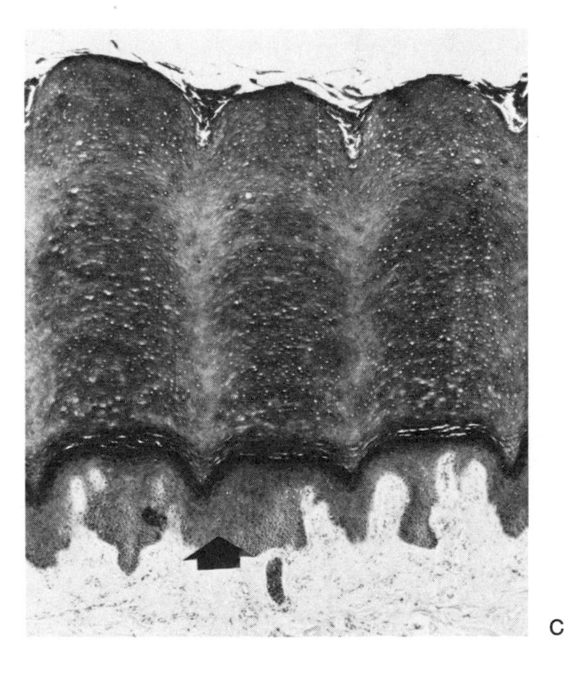

C

Fig. 3-57 Skin from three different body regions. A. Thin skin from the eyelid. B. Skin with hair. C. Thick skin. Note keratin layers at the center. Head of arrow is in stratified squamous epithelium, tail is in dermis. Photomicrographs. (A and B, from W. F. Windle, *Textbook of Histology*, 4th ed., McGraw-Hill, New York, 1969; C, courtesy of James Runner.)

and thin. Ultraviolet light (as from the sun) has a harmful effect on the elastic fibers of the dermis and superficial fascia, causing them to lose their resiliency over a span of several years. Thus, people who work largely out of doors in sunlight tend to have prematurely wrinkled ("weather-beaten"), highly creased skin. This is caused by a loss of elasticity. People who spend a good deal of time getting suntans over a period of years also tend to wrinkle prematurely. Interestingly, people of the dark races often look much younger than their years because the highly pigmented epidermis absorbs the ultraviolet rays and the elastic tissue remains "in good health."

SURFACE CHARACTERISTICS OF SKIN

The free surface of the integument is characterized by creases, wrinkles, various blemishes and, with regional variations, hair and color.

Consider the surface of your forearm at the wrist: note the many hairs exiting from their follicles through the skin—usually at intersections of the creases. Run your hand over the surface of the hair without touching the skin and note the sensation generated (there must be touch receptors associated with or near the hair follicles in the dermis). Note the difference in character among the hairs of your body: on your head, on your forearm, over your pubis, at the axilla ("arm pit").

Note the many creases at the back of the wrist and hand; compare with the front (anterior surface) of the arm, which has little or no creases. Why?

The surface of the palms of the hand and the plantar region (sole) of the foot is corrugated with ridges and furrows oriented in a general pattern—note these dermatoglyphics, as well as the flexure lines, on your own palm. Specific patterns are inherited, providing a rather unique method of individual identification. These ridges and furrows are believed to provide the "working" surfaces of the hands and feet with a friction-factor, enabling a more secure grasp.

SKIN COLOR

The pigment concentration in the skin differs among all people and all races. The major source of coloration of the skin is *melanin*—a pigment found in cells of the basal layer of epidermis. The more concentrated the pigment the darker the skin. Even in the same individual, there are differences in pigment dispersion; for example, contrast the difference in color between the scrotum and the palm of the hand, or the darkly pigmented smaller labia of the female genitals and the palm. Specific accumulations of pigment often occur as freckles in young people. Reddish or dark-brown spots (nevi, or moles) are commonly seen and may represent either concentrations of pigment or tiny tufts of capillaries pushed to the surface of the skin.

Aside from melanin pigment, the color of skin is also influenced by another pigment, carotene. Found in the skin of all races, it gives a somewhat yellow cast to the skin of Caucasians (white race).

Finally skin color is also a function of the color of blood in the vessels of the underlying dermis. The thinner the epidermis or the more diffuse the melanin, the more blood influences the color of skin. Physiological states may be identified on the basis of such influences; a person suffering from sunstroke may exhibit a very red face due to capillary dilatation in response to the heat, while a person in shock from fear will exhibit a white, pale face because of ANS-induced constriction of the underlying blood vessels. A person suffering from albinism—in which melanin pigment is lacking—often has a faint tinge of pink in skin color, again from the color of the blood in the vessels under the skin. In albinos this is particularly true in the iris of the eye, which lacks protective pigment, and is pink—reflecting the color of the blood in the vessels passing through the iris.

EPIDERMIS

The epidermis is simply stratified squamous epithelium. Slight differences in the character of the epithelium may be seen in any cross section, thus creating strata or layers.

Study Fig. 3-56 as you read the following:

The most basal (deepest) layer of cells of the epidermis is the germinating layer—columnar cells which divide constantly and whose progeny push up toward the free surface to replace those cells which have died and been sloughed off. These cells are closest to the underlying dermis and the capillary beds therein, thus their sustenance is assured.

The cells intermediately placed between the germinating layer and the surface layers of undernourished and dehydrated cells are of various sizes and shapes. They are literally pushed up by the cells being generated from below. As they get farther and farther away from the source of nutrition, relying on nourishment reaching them by diffusion, they begin to react, physically and physiologically.

The upper layer of dead cells—stratum corneum—make up the majority of the epidermis in thick skin. The outer cells are continuously sloughing off, as all people have at one time or another unconsciously appreciated: (1) the paint or stain you got on your hands and couldn't wash off—until, sometime later, you noticed the paint gone and figured it "somehow" got scraped off (it did, along with the cells on which it was imprinted); (2) the inevitable "peeling" following a painful sunburn.

An excessively thick stratum corneum over the base of the fingers of the palm is callus. It is this layer which is generally responsible for many of the functions of skin, e.g., protection from certain chemicals and ultraviolet light.

The material which gives the stratum corneum its characteristic quality is *keratin*, a kind of protein. It is also the basic constituent of hair and nails (and, in such animals as the bull or ram, horns*).

The epidermis is entirely without blood vessels and penetrated only by the free endings of pain receptors and the coiled ducts of sweat glands.

DERMIS

The dermis is the connective tissue layer underlying the epidermis; it is highly vascular and characterized by papillae (nipples) (Figs. 3-56, 3-57) which project up into the epidermis like so many cones. These papilla are characterized by capillary nets and certain sense receptors. Arising in the dermis are the hair follicles and coiled sweat glands, whose shafts and ducts, respectively, pass upward through the epidermis. In the dermis, associated with each hair shaft, is an oil or sebaceous gland and a small bundle of smooth muscle (arrector pili). The dermis is replete with sensory nerve fibers conducting sensory impulses from intradermal receptors and with ANS (sympathetic) motor fibers innervating the arrector pili muscles. The dermis is continuous below with the superficial fascia often without "tell-tale" distinction.† This fascia may be thick (hips) or thin (anteromedial leg), and may contain portions of sweat glands and hair follicles. It is usually continuous with deep fascia enveloping skeletal muscles or with the periosteum of rather superficial bones.

* But not antlers, which are something else again.

† Although the latter is frequently fatty, the former is not.

APPENDAGES OF THE SKIN

The appendages of the skin include: (1) hairs, (2) nails, (3) glands (sebaceous, sweat, mammary).

Hair

Most of the human body is covered with hair, even where this is not immediately apparent. In certain regions of the body, hair takes on a special character as in the axillary and pubic regions. In other regions (parts of external genitalia, palms, soles, lips, and nipples) hair is absent. It would seem that hair on people is largely vestigial, but in other mammals, hair plays an important role in protection and heat insulation. In people, there may be significance in the fact that touch receptors are intimately associated with hair shafts.

A hair is an elongated shaft of dried, keratinized cells and a root arising from its surrounding follicle (Figs. 3-57 and 3-58). The follicle, formed from an invaginated bud of surface epithelium in the embryo, is the germinating component of the hair. The hair grows lengthwise by mitosis occurring at the root, pushing the keratinized cells of the hair upward. The hair exits onto the skin surface through the "passageway" created

Fig. 3-58 A. Diagram of a hair follicle and related sebaceous gland and arrector pili muscle. Relate to Fig. 3-57B. B. Electron micrograph of cut hair follicles. (Courtesy of Emil Bernstein and Eila Kairinen, The Gillette Co. Research Institute.)

B

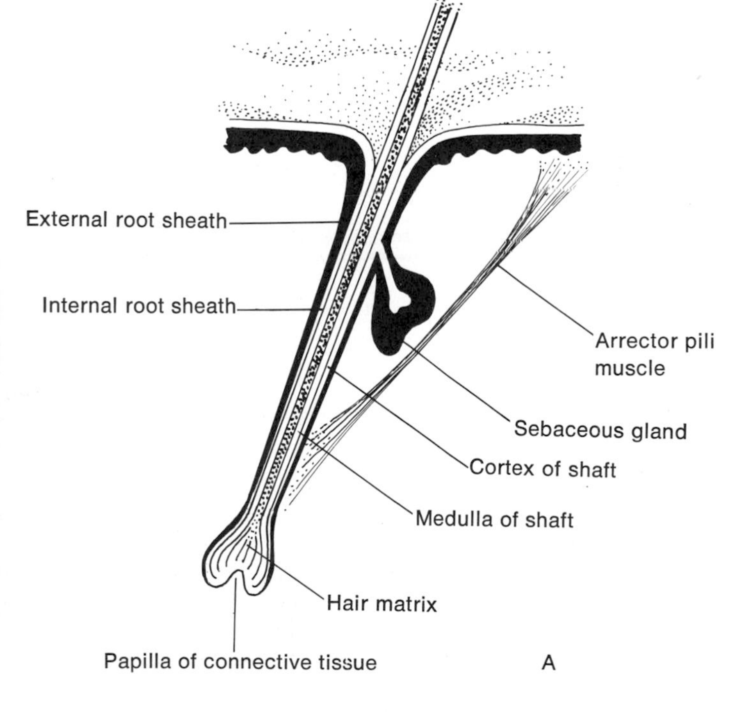

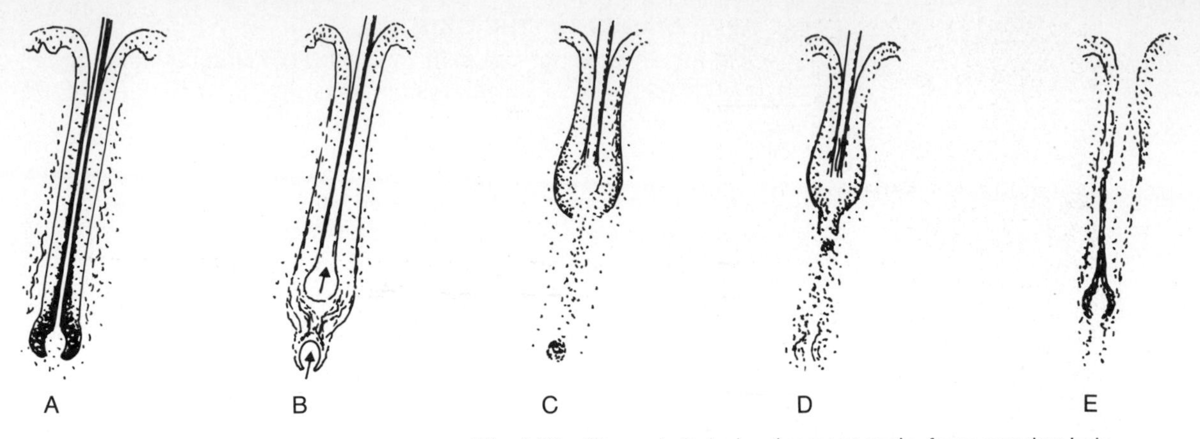

Fig. 3-59 Stages in hair development cycle, from growing hair (extreme left) to hair loss (extreme right). In men who tend toward baldness, there is a sharp increase of hairs that are in the resting club hair phase (second from right). These hairs are easily removed. (Adapted from M. M. Wintrobe et al., eds., *Harrison's Principles of Internal Medicine,* 7th ed., McGraw-Hill, New York, 1974.)

earlier by the invaginating epidermal bud. As long as the follicle is not injured or removed, the hair shaft will grow regardless of how many times it is cut—a truth to which many adult males will (wearily) attest! Actually, a hair will stop growing only when its follicle has ceased mitotic activity, a phenomenon under genetic control. Periodically, the major portion of the follicle dies and vanishes. Some of the remaining cells then form a new follicle and a new hair is generated, forcing the old (club) hair out (Fig. 3-59). Apparently the development, growth, and recession of certain hairs comes under the influence of sex hormones (testosterone in the male and estrogen in the female). When these hormones appear in significant concentrations in the blood (at puberty) development of the coarse hairs of the axillary, pubic, and (in the case of the male) other regions commences. Loss of scalp follicles or diminution in their activity in the male, causing balding (alopecia), may also be related to the onset of puberty. This tendency apparently is inherited.

Nails

Nails are composed of modified cells of the stratum corneum and, as you can see, are at the tips of the fingers and toes on the posterior surface. They are diminutive forms of claws and hoofs of man's distant ancestors. The nail has its nail root in the nail groove and is attached on its underside to the nail

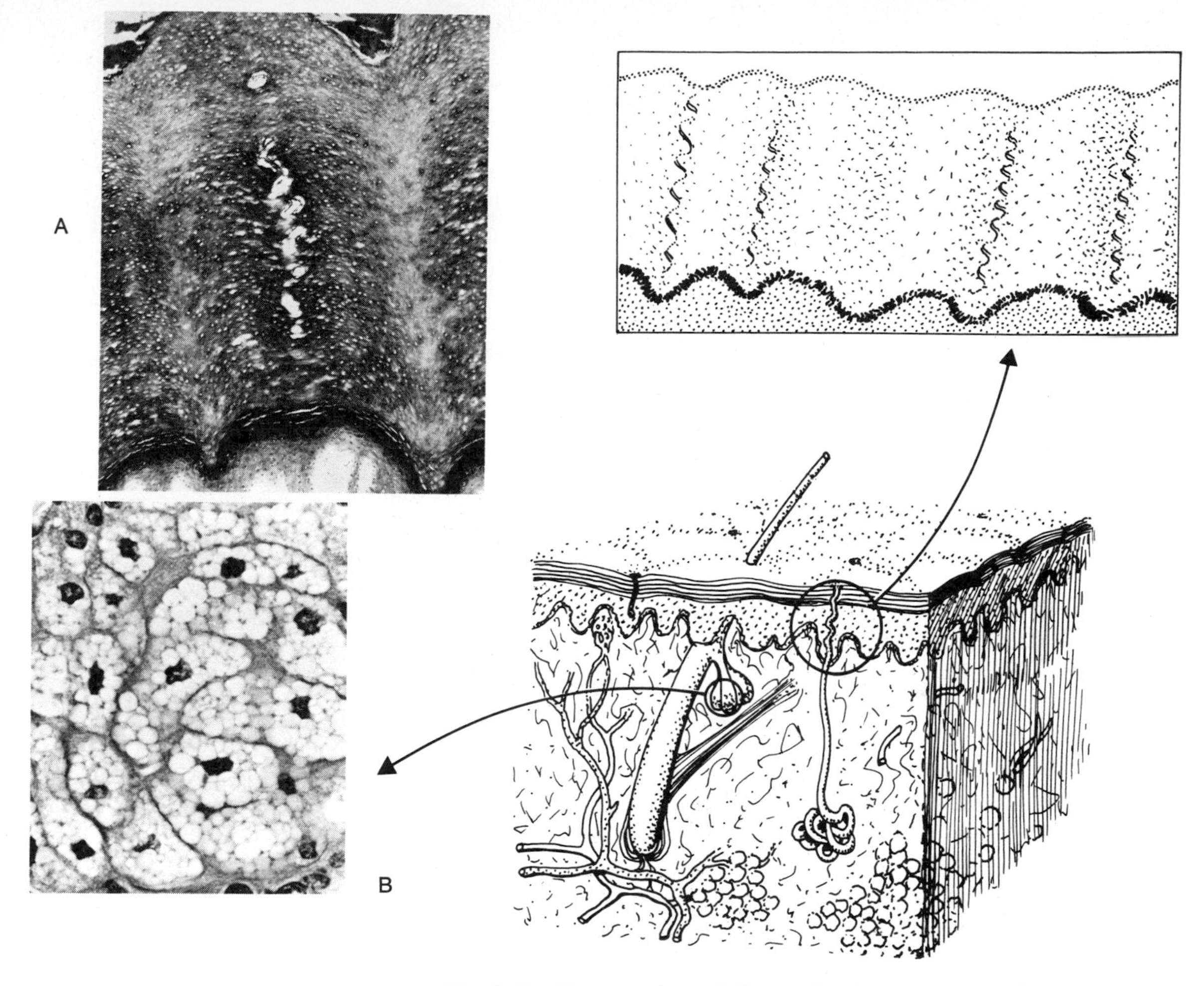

Fig. 3-60 Diagram of hair follicle with related glands. A. Ducts of sweat glands coiling up through the stratum corneum of the epidermis. B. Cross section through a sebaceous gland. (A and B, courtesy of Dr. C. Bordelon.)

bed or epidermis. The nail root contains a mass of actively proliferating cells which generate nail growth. The skin overlying the nail root (cuticle, eponychium) often harbors a good deal of cellular debris—to the delight of the beautician.

Glands

The development of exocrine glands has been presented in Unit 2. In general the skin contains two kinds of glands:

1. **Sweat glands,** which are simple, coiled glands located usually in the deep dermis (Fig. 3-60).

2. **Sebaceous glands,** which are simple, sac-shaped exocrine glands of the dermis whose ducts open into the distal sheaths of hair follicles.

The coiled portions of sweat glands are secretory, composed of pyramid-shaped columnar cells and closely related myoepithelial cells. The latter cells are innervated by fibers of the sympathetic division of the autonomic nervous system. When stimulated, the processes of these cells, oriented around the secretory cells, "squeeze" the secretions from the cells into the ducts. This salty, watery fluid is transported up through the corkscrew duct of each gland and deposited onto the skin surface. The general secretory product of a sweat gland is perspiration. The secretory activity of sweat glands, as a way of releasing heat, plays an important role in the body's temperature-regulating mechanism.

Sebaceous glands (Fig. 3-60) are absent in regions without hair. Each sac (termed an alveolus) consists of a mass of epithelial cells; the lining cells of this alveolus divide frequently and their progeny (offspring) are pushed into the center. These cells ultimately break down and become the oily secretory product (sebum) of the alveolus. Excretion of sebum onto the skin surface occurs through the action of the arrector pili muscles (which are smooth muscles) attached to the sheath of each hair. There are specializations of these glands in the eyelids (tarsal glands). Their general function is unknown but may be related to waterproofing of the skin.

VASCULAR SUPPLY AND DRAINAGE OF SKIN

Everybody is aware that a mere nick of the skin yields an expression of blood to one degree or another. However, one rarely has to fear that a pulsating artery has been cut because, generally, except in the scalp and face, arteries of significant size are not subcutaneous. They usually travel in deep fascia, often deep to or surrounded by muscle. A brief study of the dorsum of your own hand will illustrate the above is *not* true for veins or, of course, capillaries. These are found in *both* superficial and deep compartments.

Refer to Fig. 3-61 and note the arteriovenous distribution in the dermis: The direct connections between small arteries and veins in the skin (arteriovenous shunts) play an important role in regulating body and local regional temperatures. To conserve body heat, blood can be shunted, by-passing the capillaries and preventing loss of heat by convection to the skin's surface. In response to fear, for example, sympathetic-

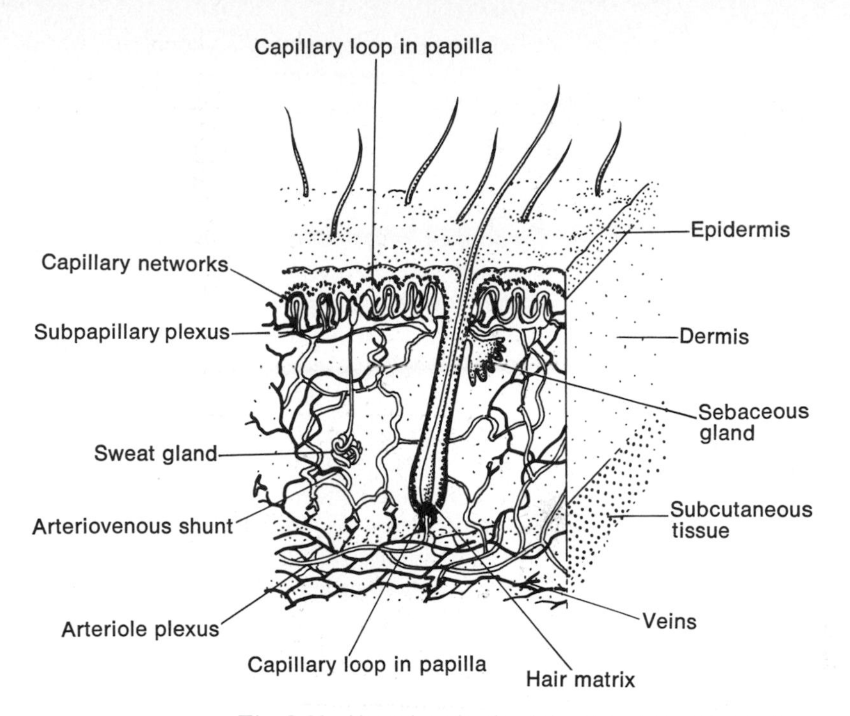

Fig. 3-61 Vascular supply of the skin. Note the arteriovenous anastomosis (shunt) and the absence of vessels in the epidermis.

induced vasoconstriction of the arteriole plexuses reduces blood flow to the dermis, and the skin whitens ("blanches"). In contrast, blushing is a manifestation of vasodilatation of these plexuses.

GENERAL SENSORY RECEPTORS

General sensory receptors may be differentiated from special receptors in that the former are more diffusely distributed throughout the body near or in the skin. Special receptors are located in the head region and are sensitive to distant as well as local stimuli. General receptors are more dependent upon physical contact with the immediate environment. Further, special receptors are generally more complex in organization, and will be studied in Unit 6.

Sensory receptors may be responsive to: (1) stimuli external to the body (exteroceptors), (2) internal stimuli (interoceptors), or (3) changes in equilibrium in the muscle, tendon, or joint (proprioceptors). Sensory receptors may be associated with viscera (visceral) or related to the surface of the body (somatic).

A number of receptors found in the skin or underlying fascia have been functionally differentiated:

Reception of pain stimuli (Fig. 3-62A) is accomplished by free nerve endings which ramify under and within the epithelial layer(s) of the skin and mucous membranes. These free, unencapsulated endings are the distal termination of sensory neurons which transmit painful stimuli to the spinal cord and/or brain. Awareness of pain and a reflex movement away from the source of pain usually follow. Pinch your skin to generate a sensation of pain and contemplate what is going on in your body in response.

Reception of cold stimuli (Fig. 3-62B). Although a relative sensation, a feeling of coldness is apparently generated by encapsulated corpuscles (of Krause)–naked nerve endings wrapped in a ball-like covering of connective tissue. Such receptors are generally found in the dermis of the skin. Impulses generated here travel to the CNS by way of the axons of sensory neurons. Place your hand around a piece of metal (on your chair, desk, etc.) and contemplate the sensation of coldness and how you are made aware of it.

Reception of warmth stimuli (Fig. 3-62C). These sensors are also encapsulated and are located in the dermis of the skin. The impulses generated by these receptors pass to the CNS by way of the axons of sensory neurons. Touch the side of your neck or other warm area of your body and contemplate the sensation and how it is transmitted to the CNS.

Reception of touch stimuli (Fig. 3-62D) is implemented by one of a variety of encapsulated and uncapsulated receptors (discs, corpuscles, follicular endings) located in the dermis, usually just below the epithelial layer, and around the shafts of hair follicles. Gently brush your finger over the hair of your forearm without touching the skin. What particular touch receptors are being stimulated? (Fig. 3-62). Touch your lip and then touch the palm of your hand. Although the sensations (and neural "signals") of touch are identical, two structurally different receptors are responsible for generating these sensations. As with other sensory stimuli, touch sensations are transmitted to the CNS via sensory neurons whose cell bodies are located in dorsal root ganglia.

Reception of pressure stimuli (Fig. 3-62E). To generate a feeling of pressure, one must push relatively hard against an area of the body. Try it on your forearm—touch the hairs (one kind of receptor), then the skin (another kind of touch re-

Fig. 3-62 (opposite) A section of skin and its sensory receptors (schematic).

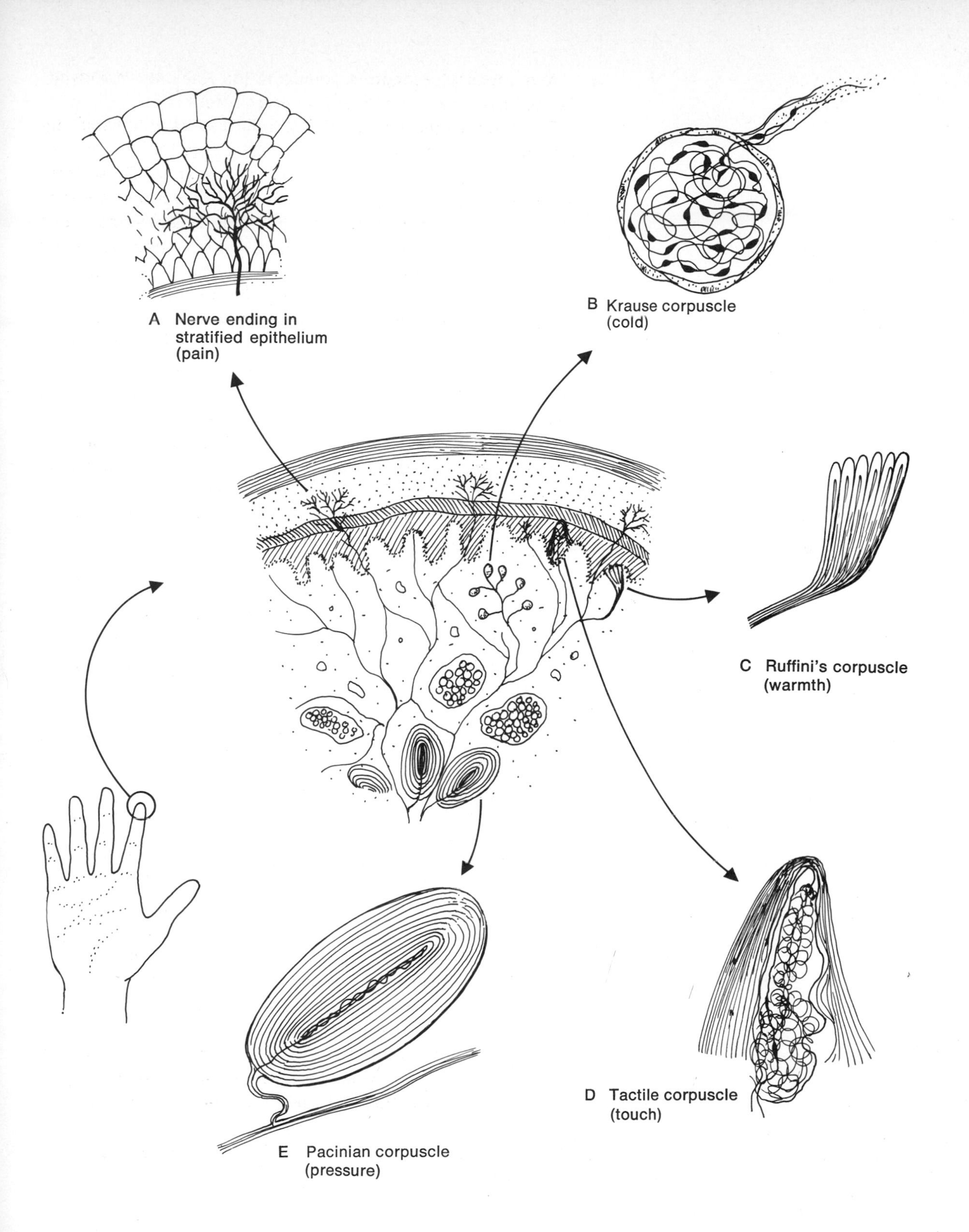
A Nerve ending in stratified epithelium (pain)
B Krause corpuscle (cold)
C Ruffini's corpuscle (warmth)
D Tactile corpuscle (touch)
E Pacinian corpuscle (pressure)

ceptor), then push hard to stimulate the Pacinian corpuscle (pressure receptors) located in the deep dermis or underlying superficial fascia. The sensations are conducted to the CNS as before.

Of all the axons of sensory neurons conducting sensations of touch, pressure, pain, and temperature, those conducting pain are apparently the smallest in diameter. This phenomenon is believed to be the basis for the effectiveness of topical analgesics (agents which relieve or prevent pain on the surface of the skin). Certain anesthetic agents (e.g., lidocaine) block impulse transmission first on the small pain fibers and later (or not at all) on the larger axons conducting other sensations. The axons of larger motor neurons are not involved in this therapy, as they are much larger in diameter.

Finally, it should be mentioned that there are temperature, pressure, and pain receptors in the walls of viscera as well as in the skin. Obviously, tactile receptors are absent. Apparently, the visceral general sensory receptors (interoceptors) are not as sensitive as the exteroceptors, or are fewer in number per unit area, except, perhaps, pain receptors.

It can be appreciated that the skin and its array of sense receptors are the first contact that your body makes with the external environment. Whether it be hot or cold, fluid or solid, painful or not, it keeps you informed as to your environment so that in the event of possibly harmful changes in that environment, you can respond—voluntarily or *in*voluntarily.

UPON REFLECTION

One of the most important concepts that you should have firmly in mind as you approach a regional study of the body is the general structural and functional relationships of body components. The body is *not* a *bag* of skin *filled* with isolated systems for digestion, respiration, circulation, etc.

Since all of the components of the body are basically cellular and therefore living, a constant supply of nutrition is required. This means that all parts of the body must have a blood supply. In order for the blood to get there, blood vessels are necessary, and fascia is required to bind down and support these vessels as they traverse the body. At the point of blood distribution to the "consumer organs," vast capillary networks are required for a fair and equitable distribution of oxygenated fluid and nutritive elements. And of course, vessels transporting used (venous) blood from organs *to* the heart are required and they are usually twice as numerous as the arteries transporting blood to the organs *from* the heart. *Every organ, no matter how small, contains blood vessels necessary for its sustenance.*

All organs of the body require some sort of nerve supply—either as a source of stimuli for function or as sensors of information often required for survival. Therefore, traveling with arteries and veins throughout the body, are nerves—all bound together in "fascial packing" to prevent damage or injury during body movements. Just as capillaries occur in extensive networks, nerves terminate by dividing into great numbers of fine branches within the muscular or connective tissue walls of an organ. There is also an accessory circulation system, the lymphatics, aiding the blood vascular system in draining the organs and tissues of the body. These numerous, small vessels ramify about most organs and their larger vessels may be found packaged in with arteries, veins, and nerves.

You are also aware now that connective tissue binds and supports the multitudes of organs so that they (1) do not fall apart and (2) do not bounce unnecessarily during body movement.

As you dissect and/or contemplate this marvel of structural and functional continuity called the body, do think in terms of collectiveness; each organ being a collection of functional units (epithelium, connective tissue, blood vessels, nerves, and muscle). As you approach each organ for study, ask yourself:

1. What function is carried out by this structure? With what system is it associated?
2. How does the architecture of this organ lend itself to the accomplishment of its task(s)?
3. What tissues are involved here and how are they structurally related? (In other words, *what* am I looking at?)
4. How is this organ supported?
5. What is its blood supply?
6. What is its nerve supply?

REFERENCES

1. Anson, B. J. (ed.): *Morris' Human Anatomy,* 12th ed., McGraw-Hill Book Co., New York, 1966.
2. Basmajian, J. V.: *Muscles Alive,* Williams and Wilkins Co., Baltimore, 1962.

3. House, E. L., and B. Pansky: *A Functional Approach to Neuroanatomy*, 2d ed., McGraw-Hill Book Co., New York, 1967. (Well-organized text written for medical students but can help the advanced undergraduate.)
4. Noback, C. R., and R. J. Demarest (illustrator): *The Human Nervous System*, McGraw-Hill Book Co., New York 1967. (Difficult text but outstanding illustrations for beginning students.)
5. Scientific American (reprint) references:
   - 64 Microcirculation of the Blood (Zweifach, 1959)
   - 158 The Lymphatic System (Mayerson, 1963)
   - 1003 The Skin (Montagna, 1965)
   - 1032 Adaptions to Cold (Irving, 1966)
   - 1093 The Venous System (Wood, 1968)
   - 1117 The Wonderful Net (Scholander, 1957)
6. Wells, K. F.: *Kinesiology*, 4th ed., W. B. Saunders Co., Philadelphia, 1966.

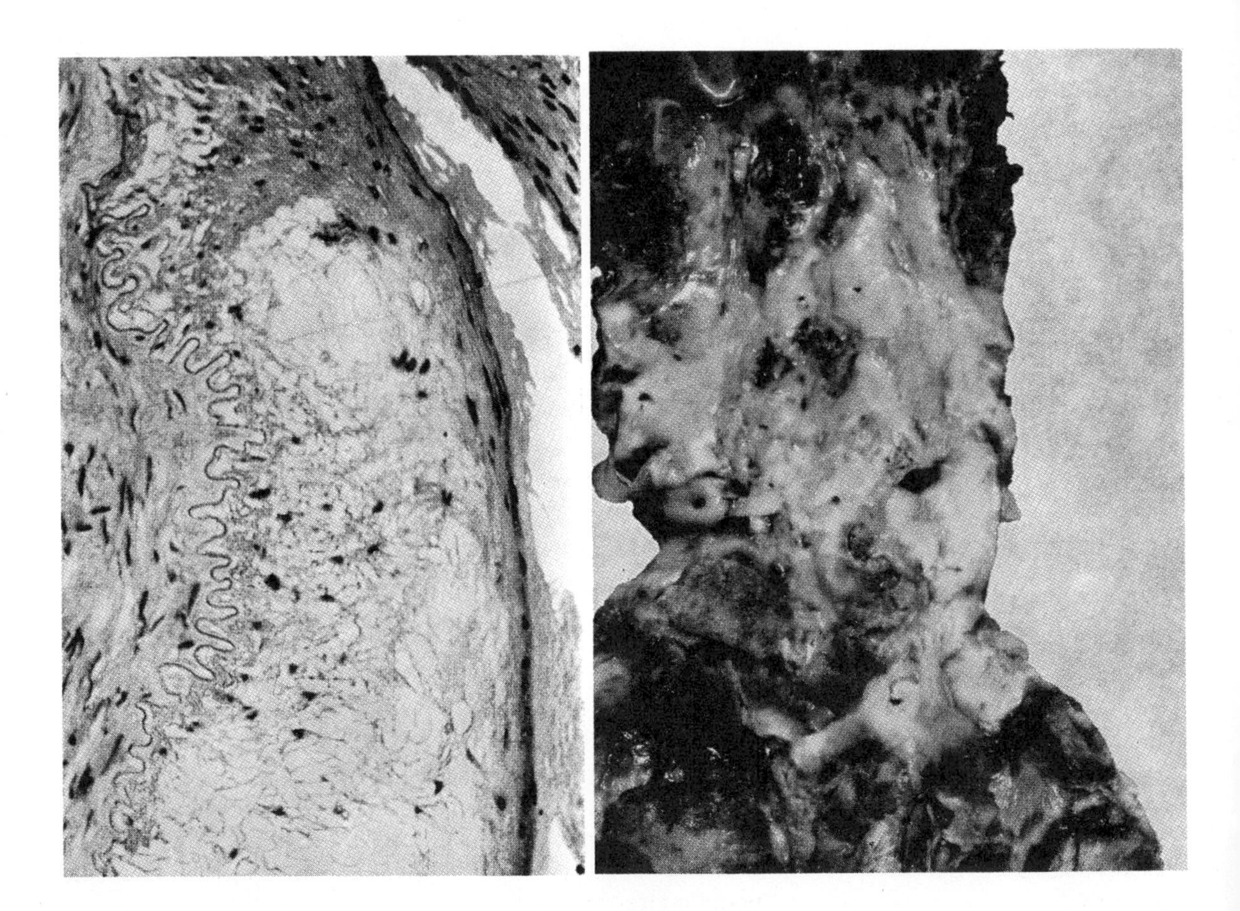

clinical considerations

Fig. 3-63 (opposite) Atherosclerosis. The photomicrograph on the left shows a small atherosclerotic plaque in the internal layer of an artery. On the right is a macroscopic view of an artery, split longitudinally, on which several atherosclerotic plaques may be seen. (From R. P. Morehead, *Human Pathology*, McGraw-Hill, New York, 1966.)

ATHEROSCLEROSIS

Atherosclerosis (Gk., *athero*, gruel or porridge; *sklerosis*, condition of hardening) is a condition in which the innermost layer of one or more arteries accumulates fat, fibrous tissue, calcium, and other deposits, which are collectively termed "plaques" (Fig. 3-63). The flow of blood through such vessels may be subsequently altered. In time, these plaques invade the middle layer, destroying the internal elastic membrane and adjacent smooth muscle cells. In advanced cases, cholesterol and calcium deposits are seen in the middle and internal layers.

Atherosclerosis usually manifests no signs or symptoms unless the formed plaques are broken off and pass downstream. Such a free-flowing plaque is an *embolus*. An embolus may block the flow of blood through a vessel. If the vessel is a critical one, for which there is little or no collateral circulation, cells in the area are immediately deprived of oxygen and within seconds suffer permanent injury or death. The area is then called an *infarct*. A myocardial infarct is commonly the result of occlusion of a coronary artery. A cerebral infarct results from occlusion of one of the cerebral arteries. The consequences of both may be serious, even fatal.

The cause of atherosclerosis is unknown but is believed to be related to high intake of dietary saturated fat. Atherosclerosis is probably not related to arteriosclerosis, a disease involving degenerative changes of the middle layer of medium-sized arteries.

unit 4 the upper limb

The upper limb is truly a marvelous set of structures. Consider some of the tasks we can accomplish with it: we can climb trees using our upper limbs as stabilizers and as hoisters as we go from one level to another. We can throw a baseball with amazing speed and accuracy and loft it 100 yards with the business end of a bat. The knife-thrower at the circus depends (as does the throwee) upon the delicate neuromuscular coordination of his upper limb. In golf we work the shoulder and wrist joints with powerful muscular movements to effect a 300-yard drive, and ask for subtle nuances of upper limb movement for that 6-in. putt. And who can tell a story without appropriate gestures of the hands?

The anatomical basis for these and like things is before you.

osteology

GENERAL

You will remember that the bony framework of the body is called the skeleton. That part of the skeleton along the principal axis of the body is the axial skeleton, while the bones of the limbs and their bony girdles comprise the appendicular skeleton (Fig. 3-3).

The upper limb is attached to the axial skeleton by the **pectoral girdle.** The pectoral girdle is composed of two clavicles anteriorly and two scapulae posteriorly (Fig. 4-1).

As you can see in Fig. 4-2:

- The clavicle articulates with the scapula laterally at the acromioclavicular joint.
- The arm bone, or humerus, articulates with the glenoid fossa of the scapula.
- The bones of the forearm (radius and ulna) articulate above with the humerus and below with the wrist bones.
- The wrist, or carpus, is composed of eight small bones of

Fig. 4-1 The pectoral girdle in relation to the thorax. Viewed from above.

Fig. 4-2 (opposite) Bones of the right upper limb.

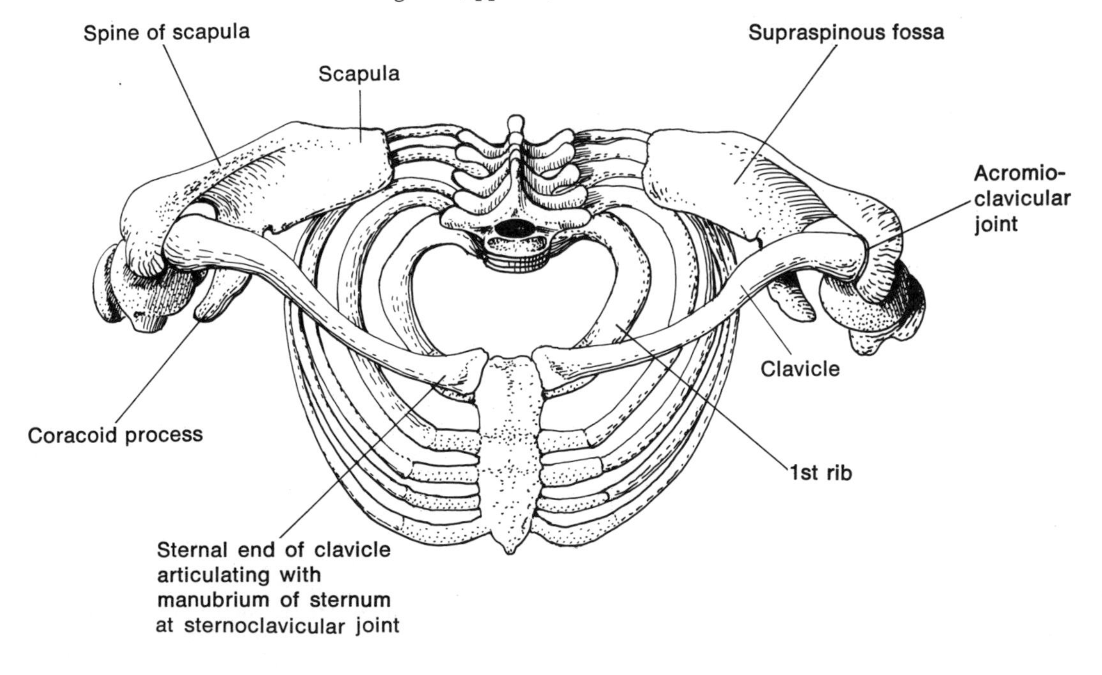

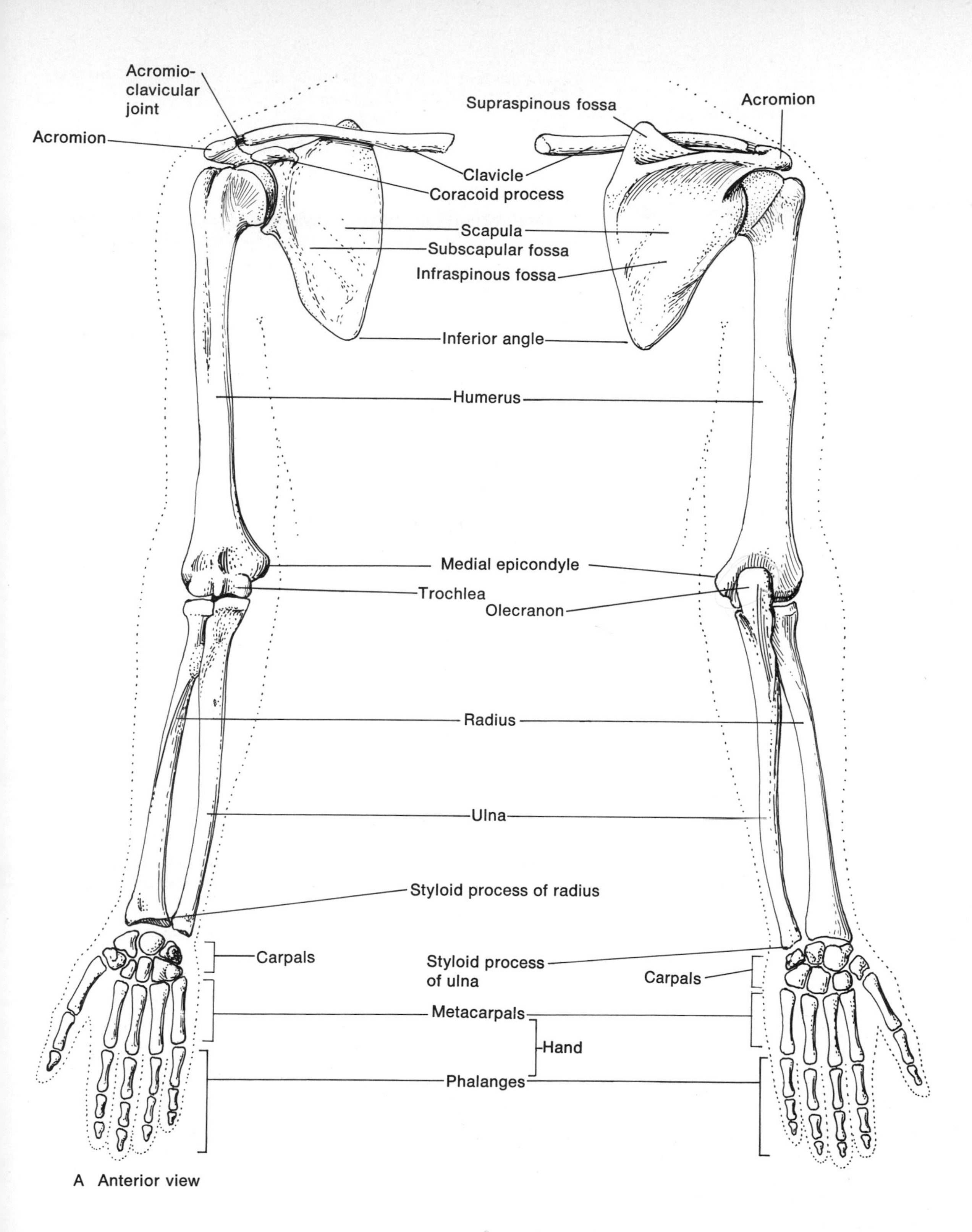

A Anterior view

a variety of assorted shapes, and the bones are named according to these shapes.

- The distal row of the carpal bones articulates with the five metacarpal bones which make up the skeleton of the palm and dorsum of the hand.
- Each metacarpal articulates with a proximal phalanx of a digit. There are three phalanges in each digit of the hand, with one exception, the thumb, which has two phalanges.

CLAVICLE (collarbone)

The **clavicle** (L., small key) is the anteriormost bone of the pectoral girdle and is often referred to as the *strut* of the upper limb, for it forces the scapula backward and laterally away from the chest wall (Fig. 4-1). The region about the clavicle-scapula articulation is the shoulder. Medially the clavicle articulates with the sternum (sternoclavicular joint). Note the relationship of the clavicle to the scapula and the humerus in Figs. 4-1 and 4-2. You can now understand why, in falls on the hands or violent blows against the shoulder, the clavicle (strut) may snap . . . and frequently does.

Laterally the clavicle articulates with the acromion of the scapula at the acromioclavicular joint. The acromioclavicular ligaments reinforcing the joint are rather weak, and thus shoulder separations at this joint are not infrequent.

Much of the clavicle, on its upper surface, is subcutaneous and is easily palpated. *Feel it yourself.* Place your fingertips on the more distal part of the clavicle and move your arm about. You can feel it move, especially during abduction of the arm and elevation of the tip of the shoulder. Now feel the entire clavicle—from its attachment at the manubrium to the tip of the shoulder (acromion). As you draw your fingers laterally from the medial attachment, you will note that the bone first curves forward and then posteriorly to the acromion. This first curve is quite significant, as it allows passage underneath for the important plexus of nerves from the neck to the limb, as well as for the subclavian artery and vein.

The clavicle gives attachment to important muscles of the back, the shoulder, the neck, and the chest.

SCAPULA (shoulder blade)

The **scapula,** the posteriormost bone of the shoulder girdle, is a flat bone with a number of surfaces and borders (Fig. 4-3). These offer points of attachment for many muscles—17 in all.

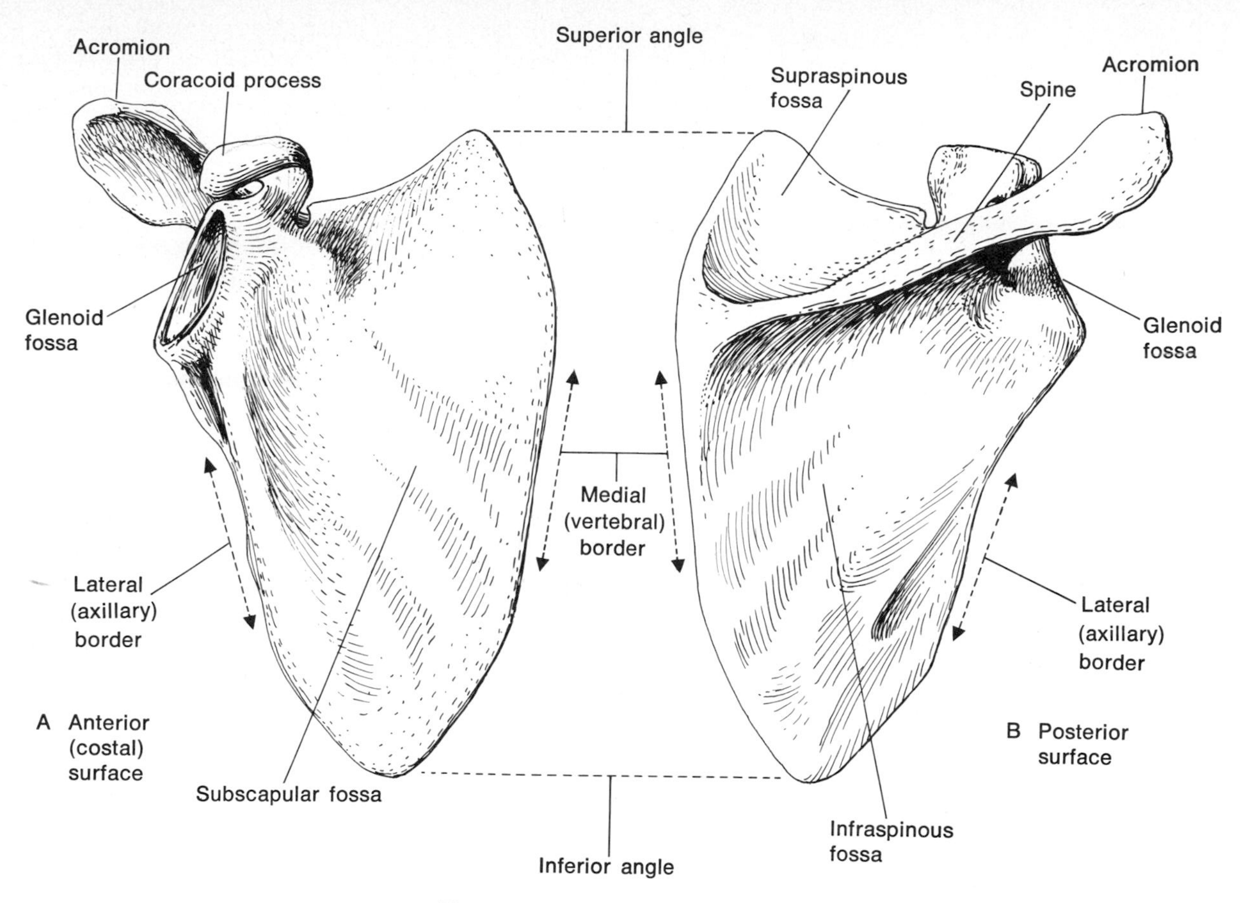

Fig. 4-3 The right scapula.

This brings to light a rather important concept for the pectoral girdle: the scapula has no direct bony connection with the axial skeleton (only indirectly, *via* the clavicle). Yet the humerus (arm bone) articulates with the scapula, which seems to imply that the integrity of the whole upper limb is dependent upon the flimsy sternoclavicular joint. Obviously it is not, as any one who has done "chin-ups" will testify, for if only the sternoclavicular joint secured the upper limb, the pectoral girdle would literally slip off the thorax with every "chin-up." So some other tissue must be involved, and there is: muscle.

It is largely because of the muscle securing the scapula to the axial skeleton that the scapula is capable of a wide range of movement (look ahead to Fig. 4-11). This is important, for if the scapula were not capable of rotation, elevation, or depression, freedom of movement of the upper limb would be severely restricted.

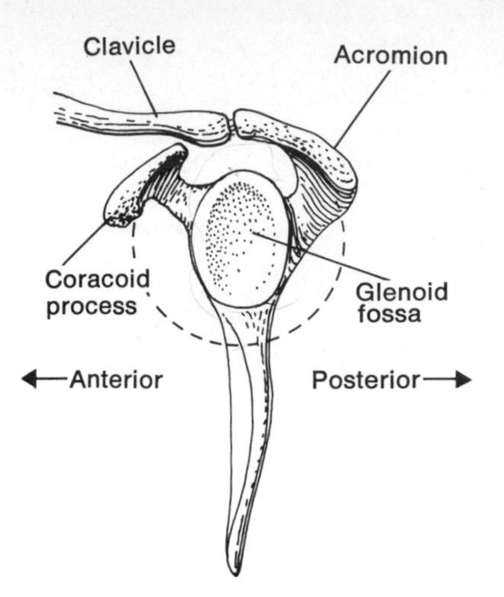

Fig. 4-4 Lateral view of the scapula demonstrating the shallow glenoid fossa. Circumference of the humeral head is designated by the dotted line.

The security of the scapula-arm (glenohumeral) joint is deficient in one way: the fossa of the scapula that receives the head of the humerus is shallow (Fig. 4-4). Again, muscles as well as ligaments reinforce this joint (see Fig. A-1, Appendix, for x-ray view).

On your own scapula (and referring to Fig. 4-3) palpate the acromion at the tip of the shoulder; now follow the bone down the back obliquely—the acromion becomes the spine of the scapula. By wrapping your arm around your thorax just below the axilla (armpit) you can feel the lateral border of the scapula, in part. It is easier to distinguish as you abduct your arm, because the scapula rotates upward during abduction, and as it does you feel the lateral border of the scapula move into your hand. Bring one hand around in back as if you were trying to scratch your upper back or scapula. In so doing, the scapula is forced away from the thorax posteriorly, presenting the medial border and the inferior angle for observation (in a mirror) and palpation (Fig. 4-5).

Fig. 4-5 Retracted scapulae. Note the "winged" vertebral borders and the prominent inferior angles (arrows).

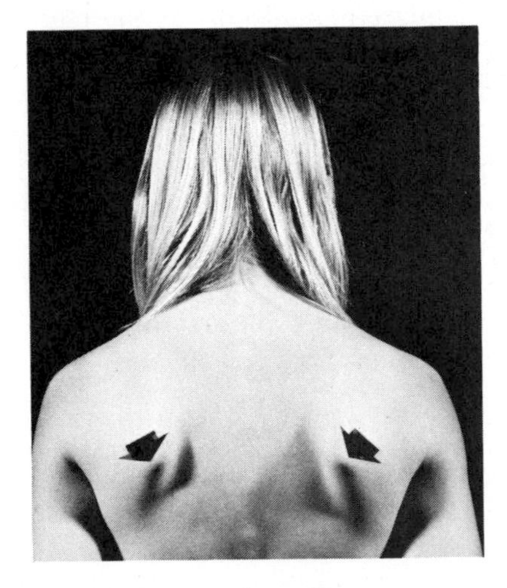

HUMERUS

The **humerus** is the bone of the arm and is easily identified by its "bald" head above and the pulley-shaped articulating condyle (trochlea) distally (Fig. 4-6). The humerus articulates with the glenoid fossa of the scapula superiorly and with the two bones of the forearm below, the ulna and radius. The humerus can be palpated in various places throughout the arm; (refer to Fig. 4-6 while doing so): about two fingersbreadth below the acromion anteriorly one can feel a small projection of bone. This is the lesser tubercle, which receives the attachment of an important muscle. The greater tubercle (three muscles insert here) can be felt at the tips of the shoulder just below the acromion laterally. The shaft of the bone can be felt throughout from side to side, including the prominent deltoid tuberosity (a bump about halfway down the lateral aspect of the arm) where the deltoid (shoulder muscle) inserts. At the elbow, the medial epicondyle is very prominent and feeling its posterior aspect may result in tingling sensations in the fourth and fifth digits, which means that you have pressed the ulnar nerve, to be discussed later.

The humerus is subject to fracture and displacement following falls and violent motions, such as in throwing a ball. The displacement of the two broken ends occurs because of muscular pull. Dislocation of the shoulder joint can also

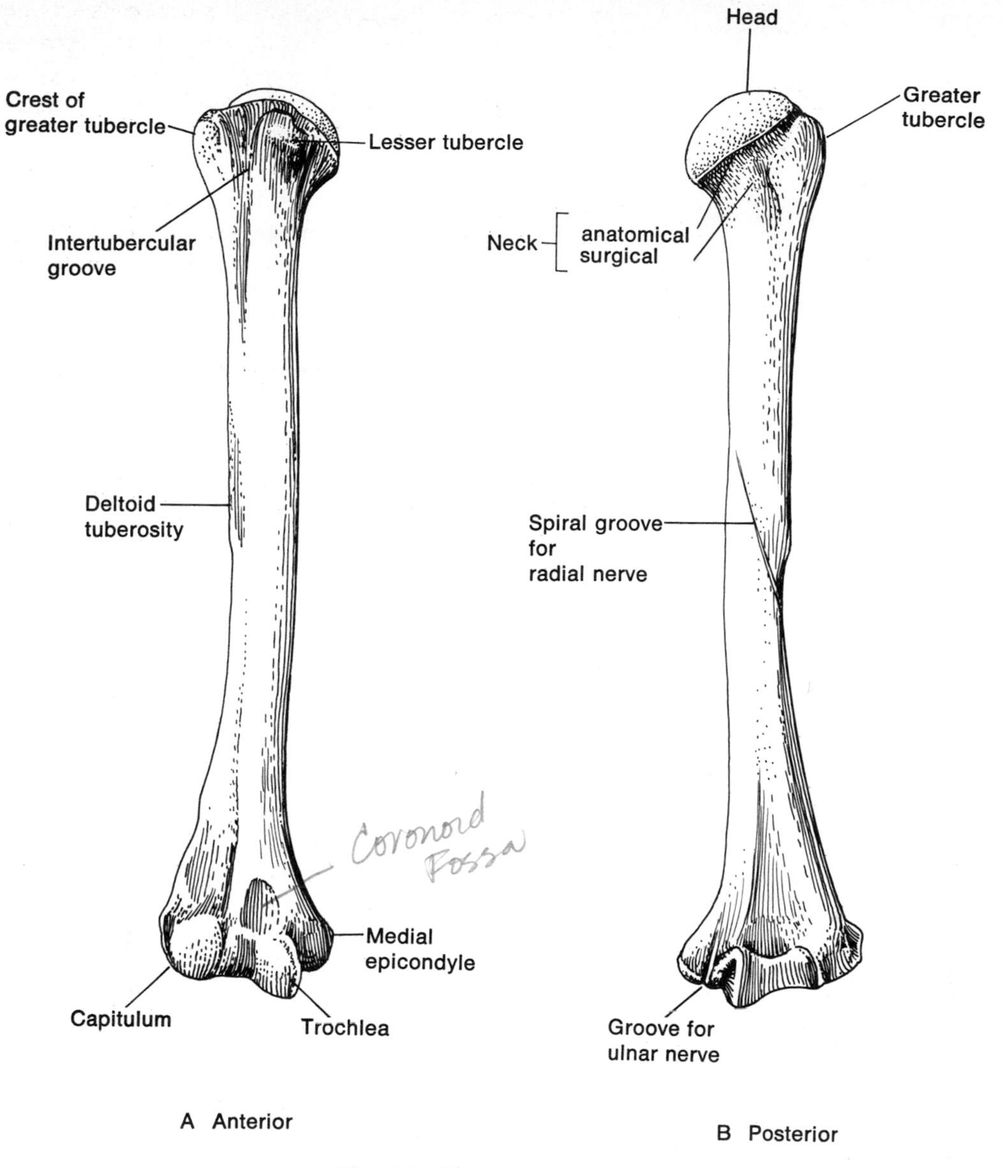

Fig. 4-6 Right humerus.

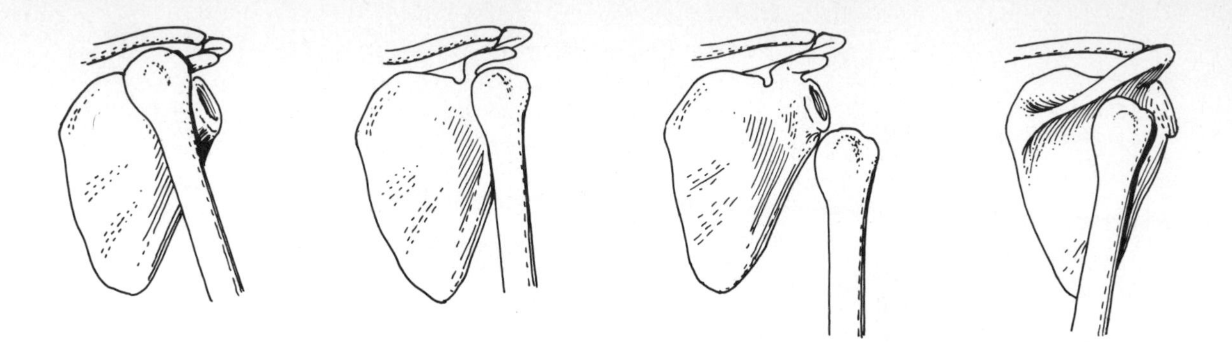

Fig. 4-7 Varieties of shoulder (glenohumeral) dislocations.

occur (Fig. 4-7). The head of the humerus usually moves inferiorly (why not superiorly?) and then forward or backward depending on the force.

ULNA

The **ulna** is one of the two bones of the forearm (Fig. 4-8; see Fig. A-2, Appendix, for x-ray view). In the anatomical position, it occupies the medial (little-finger) side of the forearm (Fig. 4-2). The ulna is large above, articulating with the trochlea of the humerus, and slender and small below, where it articulates with the radius directly and the wrist indirectly via an articular disc.

The ulna is further characterized by an unusual C-shaped process superiorly—the olecranon (olek-ran-on). The olecranon can best be palpated when the elbow joint is flexed. Note that the olecranon does not move when you rotate the forearm from pronation to supination and back. Try it! Thus the humeroulnar joint is a true hinge joint and the ulna does *not* rotate. How then does the forearm rotate? We'll see soon.

The posterior aspect of the ulna can be felt almost throughout down to the wrist. Here it ends as a rounded projection of bone, easily felt on the posterior aspect of the forearm just above the wrist, medially. This is the styloid process when the forearm is supinated—it is the head of the ulna when the forearm is pronated (Fig. 4-8). With your forearm pronated, place a finger on the head of the ulna (a significant anterior bump at the little-finger side of the wrist). Now rotate laterally—the projection largely disappears, and what you feel is the styloid process of the ulna. What's happening? It's a fact that the ulna does *not* rotate, but movement of skin and tendons uncover the head of the ulna as the forearm is pronated.

RADIUS

The radius is that bone of the forearm occupying the lateral or thumb side (Fig. 4-8; see Fig. A-2, Appendix, for x-ray view). A rounded or tabletop-shaped head at the proximal end is its distinguishing characteristic. In contrast to the ulna, the radius thickens from above to below to form a stout distal extremity which articulates with two bones of the wrist, as well as the ulna. Proximally the radius articulates with the capitulum (head-like process) of the humerus, as well as the radial notch of the ulna. The radius is largely covered with muscle except at its distal extremity, where its prominent lateral border (remember the anatomical position) can be felt down to the lateral border of its styloid process.

Note that when you rotate the forearm, the radius follows the thumb. So extend the elbow joint with the limb supinated

Fig. 4-8 Right radius and ulna.

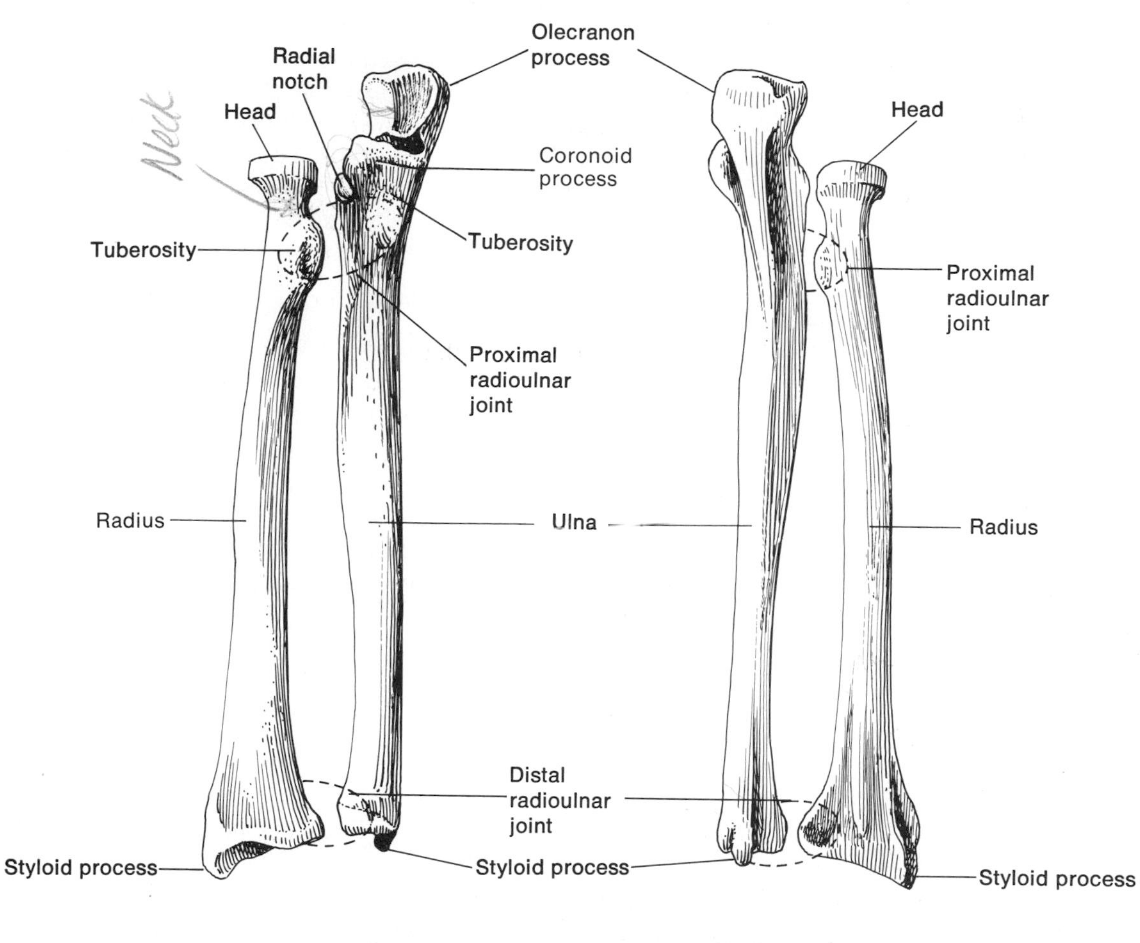

(anatomical position) and note that the ulna is definitely medial to the radius throughout the forearm. Now, keep your fingers on the olecranon of the ulna and slowly pronate (medially rotate) the *hand* (without rotating your arm), keeping your eye on the distal, lateral border of the radius. You can see that the *radius* crosses over the distal two-thirds of the ulna in the act of pronation. The ulna, as you can feel, does not rotate. Is the radiohumeral joint adapted for rotation? It sure is, for the tabletop head of the radius pivots under the capitulum of the humerus. This is a pivot type of synovial joint.

As you may know from personal experience the radius is a common site of fracture. What part of the radius is usually involved in such a fracture? The distal portion. This is usually caused by falling on the outstretched hands.

CARPUS (see Fig. A-3, Appendix, for x-ray view)

The carpus, or wrist, is the proximal part of the hand and is composed of eight bones. The carpal bones are arranged in two rows of four each (Fig. 4-9). Each carpal bone has its own distinguishing characteristics and is named accordingly:

Proximal row:

scaphoid	lunate	triquetrium	pisiform
(boat-shaped)	(moon-shaped)	(three-sided)	(pea-shaped)

Distal row:

trapezium	trapezoid	capitate	hamate
(little table)	(table-shaped)	(headlike)	(hook)

There is little value, if any, in studying the individual wrist bones. If they are to be studied, they should be studied collectively as an entity. The scaphoid and lunate articulate with the radius while the triquetrium articulates with an articular disc distal to the ulna. Of all the bones of the wrist, the scaphoid seems to be most subject to fracture.

METACARPALS (see Fig. A-3, Appendix, for x-ray view)

Metacarpals are the long bones of the palm of the hand (Fig. 4-9). There are five of them—one for each finger. Their bases articulate with the distal carpal row proximally (carpometacarpal joints) and their heads with each proximal phalanx distally (metacarpo-phalangeal joints). Got that? These bones are more easily felt on the posterior surface. Their heads are felt distally as knuckles in a clenched fist. Can you palpate each metacarpal?

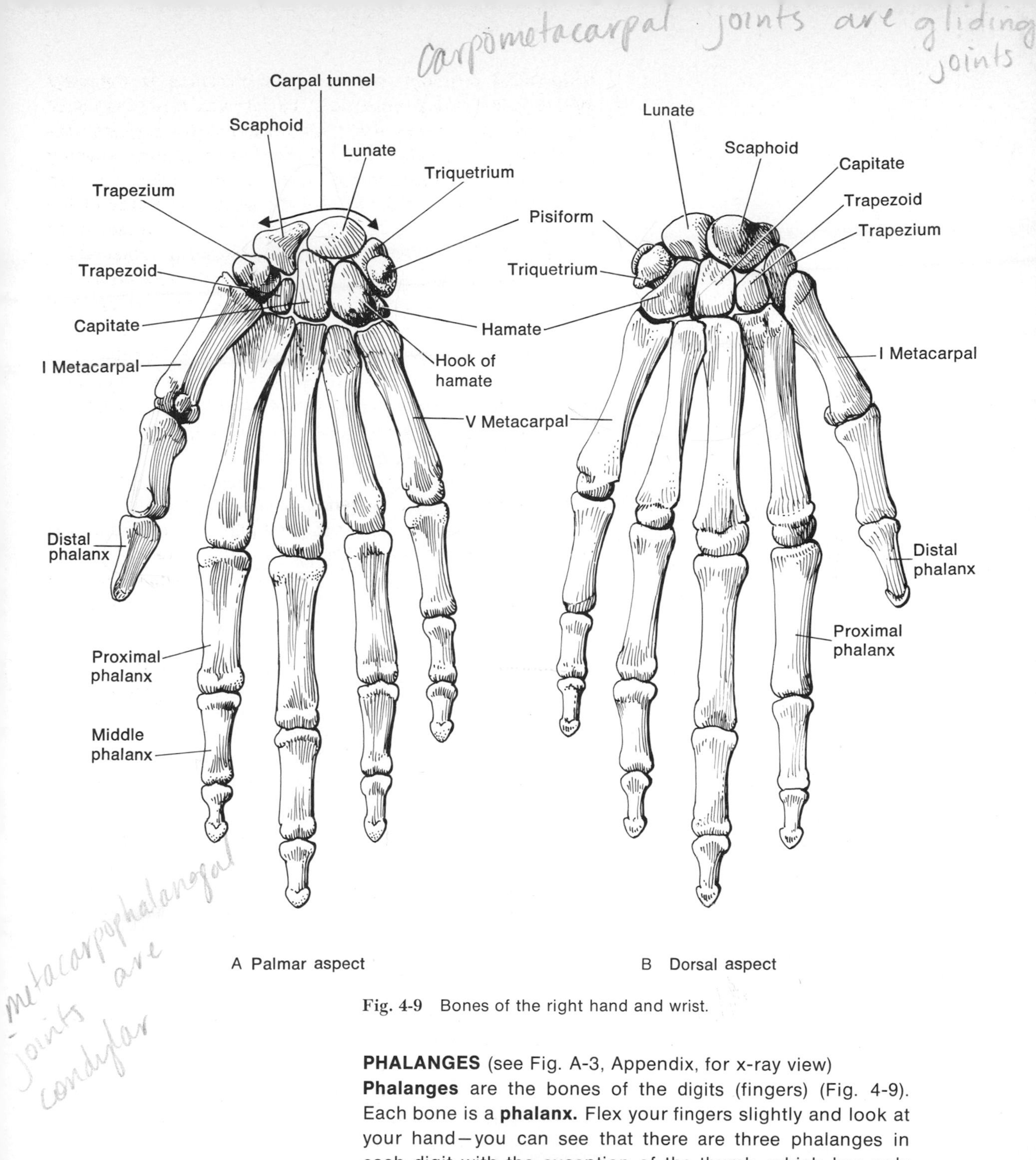

Fig. 4-9 Bones of the right hand and wrist.

PHALANGES (see Fig. A-3, Appendix, for x-ray view)
Phalanges are the bones of the digits (fingers) (Fig. 4-9). Each bone is a **phalanx.** Flex your fingers slightly and look at your hand—you can see that there are three phalanges in each digit with the exception of the thumb, which has only two. By experimenting, you will find that the interphalangeal joints are of the hinge variety.

myology

MUSCLES "MOORING" SCAPULA TO AXIAL SKELETON

trapezius
rhomboids
levator scapulae
pectoralis minor
serratus anterior

The scapula, supported by muscle from the axial skeleton, is capable of a number of movements (Fig. 4-10) and these are usually associated with simultaneous movements of the arm. You can demonstrate to yourself some of the movements of the scapula: Sit or stand erect, shoulders back—now move your shoulders forward and medially (as if trying to keep warm) and at the same time place your fingers on one of your clavicles. Notice how the clavicle's lateral extremity is pushed forward and upward. Since the scapula articulates with the lateral end of the clavicle, it follows that if the clavicle is moved upward and forward, the scapula must have moved with it. Now continue moving your shoulders forward and backward while observing the tip of your shoulder (acromion of the scapula). If you concentrate, you can "sense" the scapula sliding medially and laterally over the thorax during the above movements. The muscles which caused the scapular movements are some of the mooring muscles of the scapulae. *They all arise from the axial skeleton and they all insert on the scapula.* To securely moor the scapula, it would seem reasonable to assume that the muscles inserting on the scapula must arise laterally, medially, superiorly, anteriorly and inferiorly to it, and they do (Fig. 4-11).

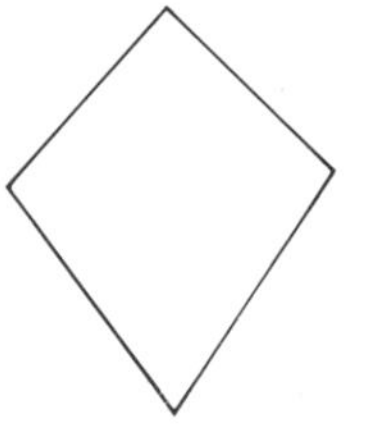

Of the five muscles supporting the scapula, the largest is a superficial, trapezoid-shaped muscle which provides the slope of the shoulder from the neck and gives both the back of the neck and the upper back their form (Fig. 4-10). Obviously then, this muscle arises superiorly and medially to the scapula. It also arises inferiorly to its scapular attachment (skull to T12). This is the **trapezius** muscle (Fig. 4-11). Study the illustration of this muscle and contemplate its attachments and its probable actions*. Place your left hand over the slope of your right shoulder and, with your right hand, pull up on the right edge of your chair seat. What function did the trapezius serve by contracting when you pulled up on the chair?

* Remember—muscles never push—they always pull!

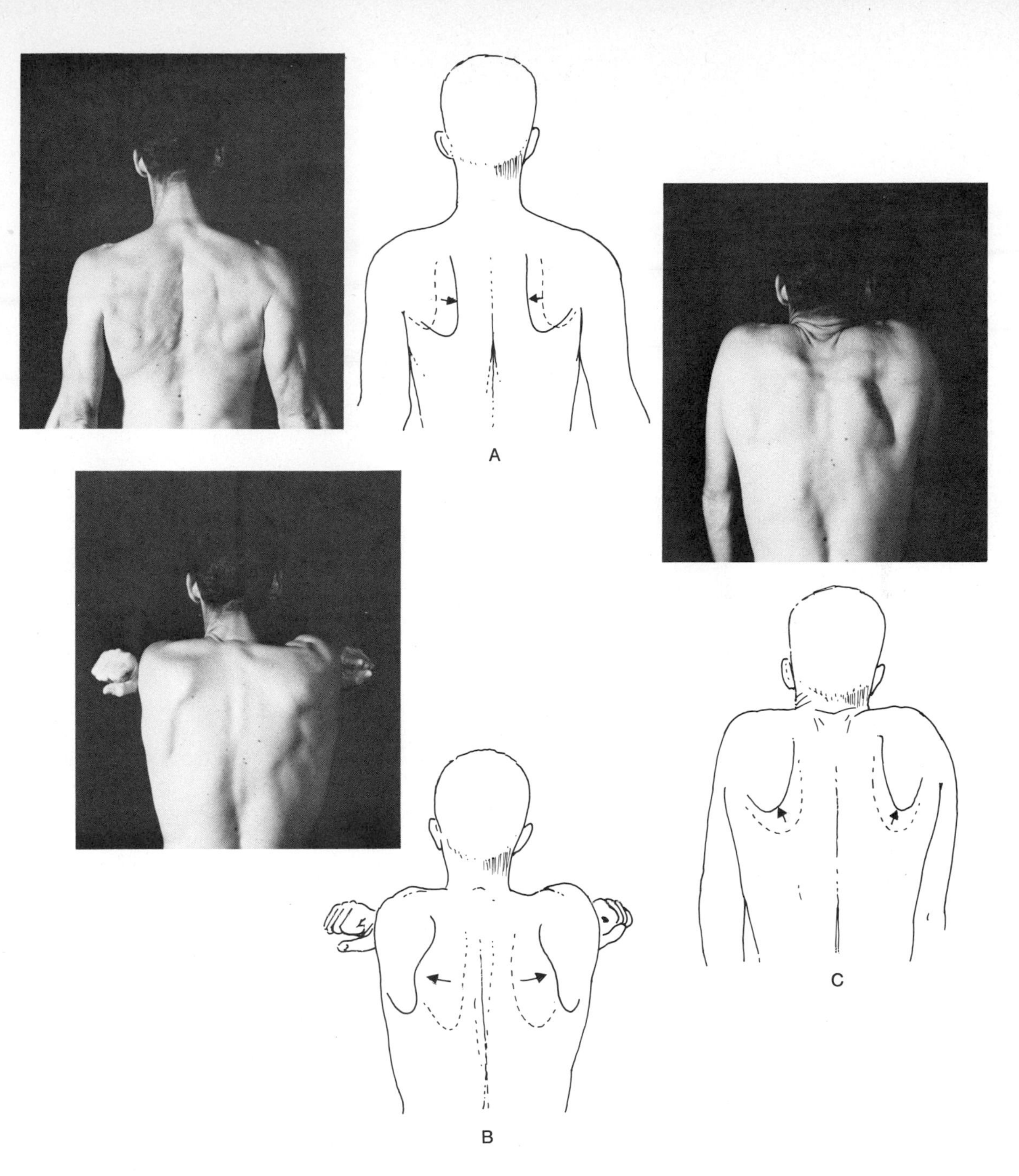

Fig. 4-10 Movements of the scapula as seen on the body surface. A. Retracted scapulae. B. Protracted scapulae. C. Elevated scapulae. D (turn page). Upward tilt or rotation. E. Depressed scapulae.

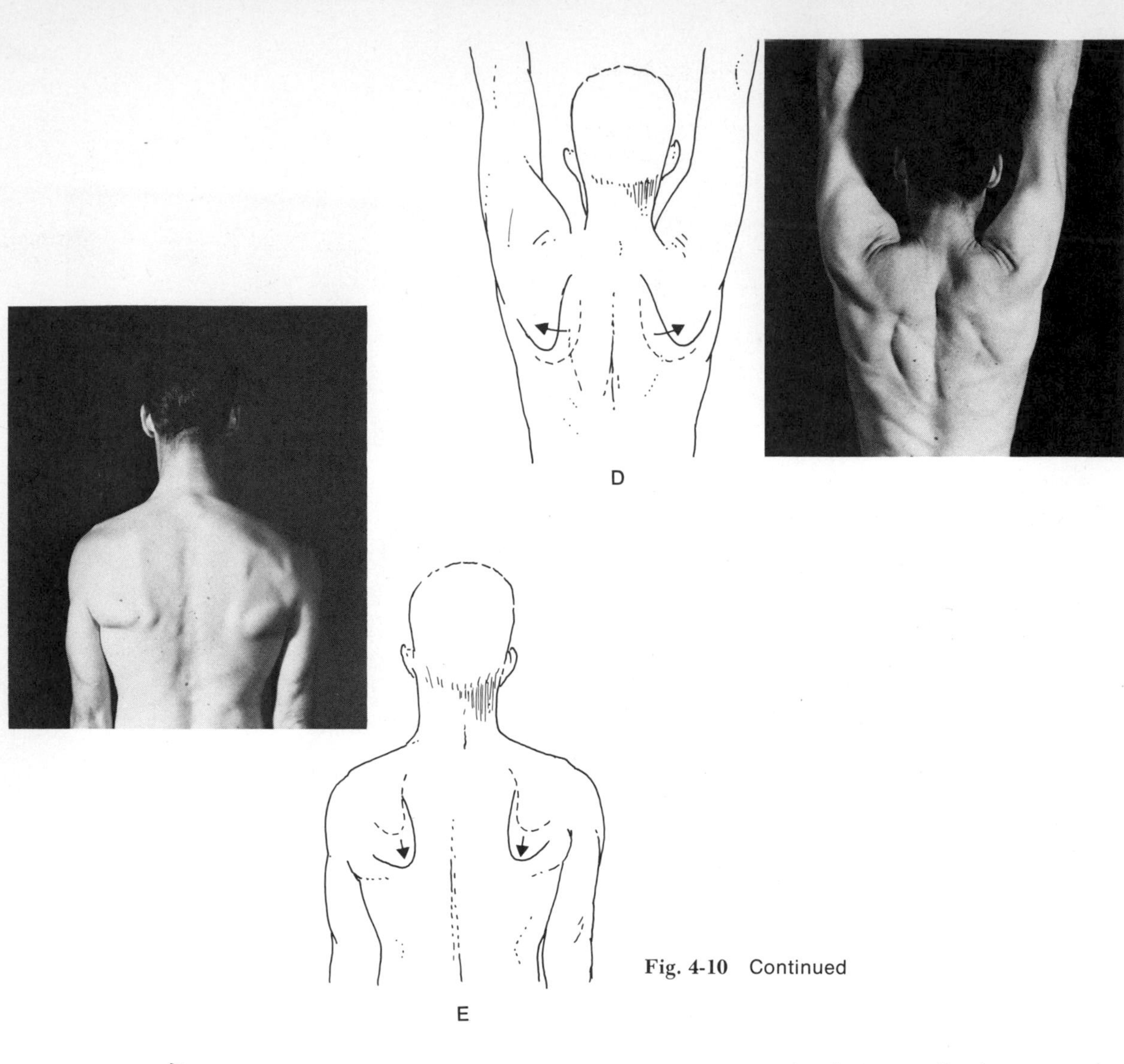

Fig. 4-10 Continued

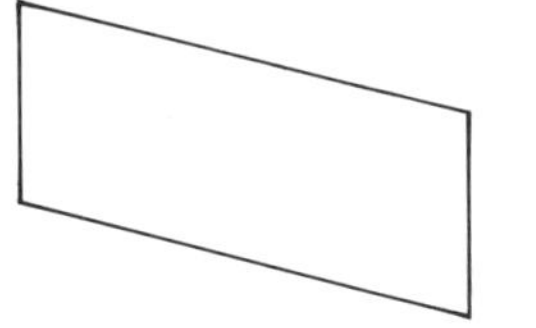

Immediately deep to trapezius between the two scapulae there is a rhomboid-shaped muscle on each side of the vertebral column attaching to the vertebral border of the scapula. These muscles reinforce the action of the middle fibers of trapezius by retracting the scapula. These muscles are called the **rhomboids** (Fig. 4-11). They cannot be felt.

Also hidden under the cover of the trapezius immediately superior to the rhomboids is a straplike muscle inserting at the superior angle of the scapula and arising from the upper vertebrae of the neck. It serves as a superior mooring muscle of the scapula as well as acting to elevate it—hence its name **levator scapulae** (Fig. 4-11).

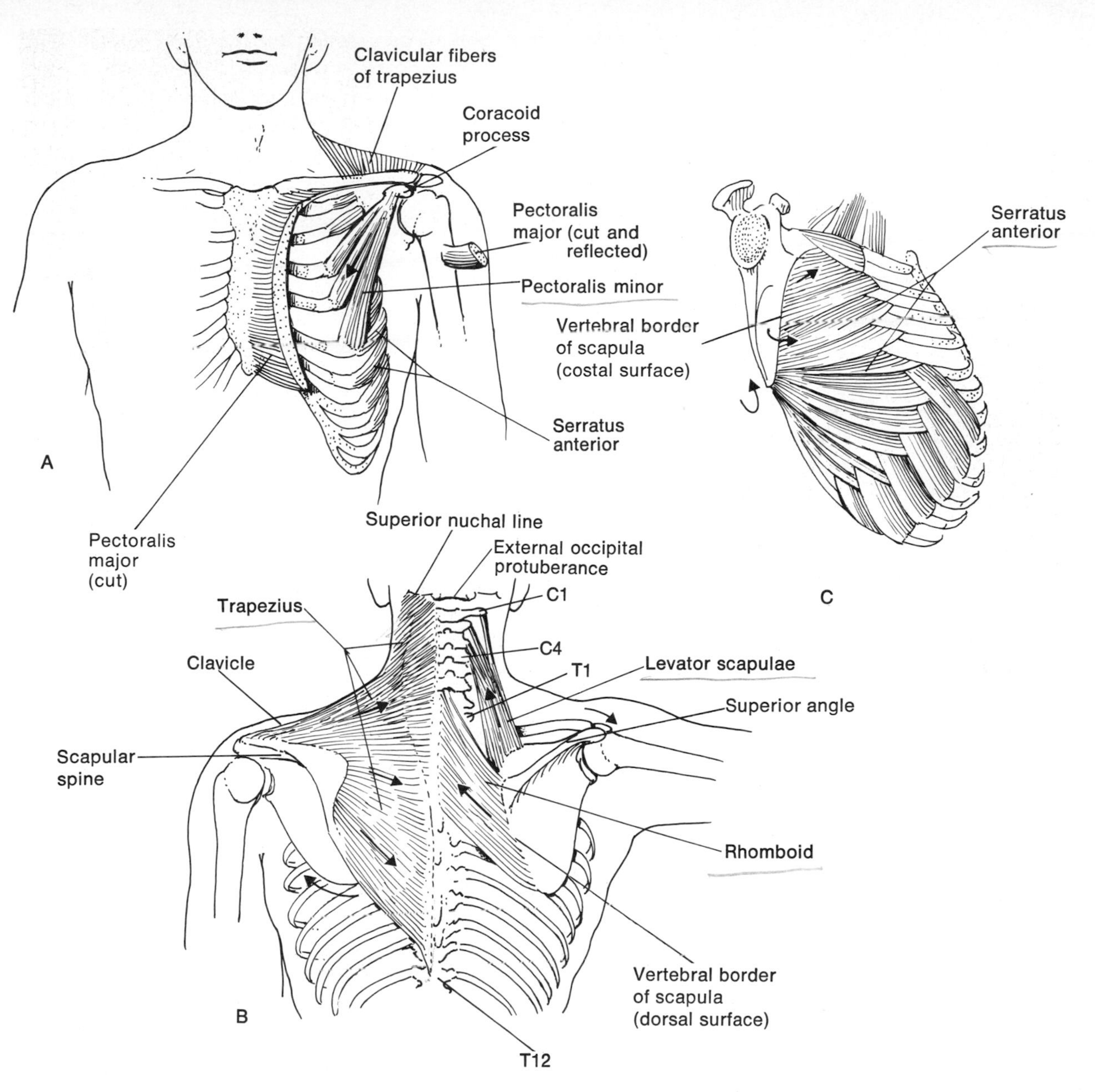

Fig. 4-11 The muscles stabilizing and moving the scapula about the posterior thoracic wall. Arrows show their direction of pull on the pectoral girdle. A. Anterior view. B. Posterior view. C. Lateral view.

So much for the medial, superior, and inferior mooring muscles.

The lateral mooring muscle of the scapula is one which arises from the lateral aspect of ribs 1 to 8 and in so doing, appears saw-toothed or serrated on the body surface (Fig. 4-12). Being the anteriormost of two such serrated muscles, you might suspect its name: **serratus anterior.** Serratus anterior can frequently be felt contracting one and one-half handbreadths below the axilla when the arm/shoulder is abducted against resistance, for the muscle is principally a protractor of the scapula (pulls scapula laterally about the rib cage) (Fig. 4-11).

Fig. 4-12 Surface features of some muscles of the upper limb.

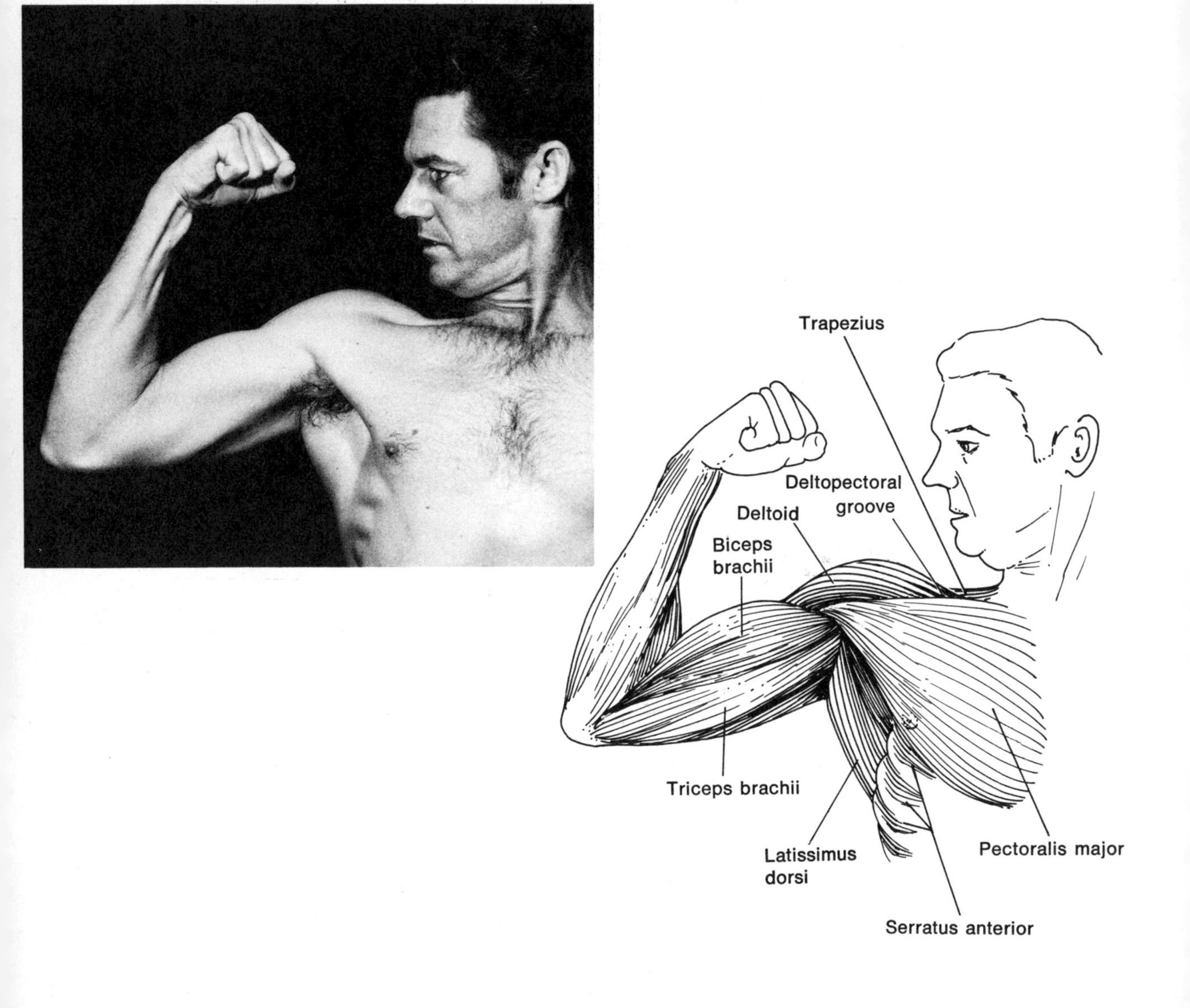

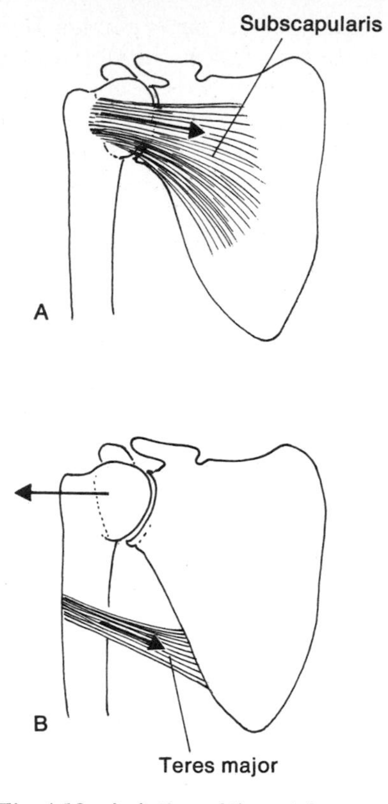

Fig. 4-13 A. Action of the rotator cuff muscles on the glenohumeral joint resulting in reinforcement of joint security. B. A muscle which crosses well below the plane of the joint actually forces the head out of the socket.

The anterior mooring muscle of the scapula arises from the anterior chest wall (pectoral region) and is the smaller and deeper of two such muscles in this region. From this you might suspect its name: **pectoralis minor** (Fig. 4-11). This muscle inserts on the coracoid process of the scapula, but it cannot be palpated because it lies under the cover of its larger fellow, pectoralis major. Pectoralis minor functions in depression and protraction of the scapula.

In review then, consider the illustrations on Fig. 4-11 and the surface anatomy photographs (Fig. 4-10 and 4-12). Remember, these muscles are collectively responsible for *mooring the scapula to the axial skeleton.* Can you relate the muscle illustrations to the same muscles on your own body? By use of a mirror and the appropriate exercises (contracting against resistance) you can probably see your own trapezius and serratus anterior muscles.

MUSCLES SECURING HEAD OF HUMERUS TO SCAPULA

subscapularis　　supraspinatus　　infraspinatus
teres minor

At this point, you might acquaint yourself with the glenohumeral (shoulder) joint. You can see that in order for the head of the humerus to be retained in the glenoid fossa, the muscles doing the retaining must insert about (that is, in the same plane as) the head of the humerus and *not* below it. Muscles inserting on the upper shaft but below the head have the effect of forcing the head out of the fossa when contracting (Fig. 4-13). You should understand this before going on.

All four of the muscles holding the head of the humerus in the glenoid fossa of the scapula arise from the scapula and insert on the lesser or greater tubercles of the humerus (which are in the same plane as the glenohumeral joint).

Refer now to Fig. 4-14 and note:

1. One muscle inserts on the anterior aspect of the humerus (lesser tubercle): **subscapularis.**
2. One muscle inserts at the crown of the greater tubercle: **supraspinatus.**
3. Two muscles insert on the posterior aspect of the humerus (greater tubercle): **infraspinatus** and **teres minor.**

A cuff is thus formed—from the lesser tubercle anteriorly over the top (crown) of the greater tubercle and down the posterior part of the greater tubercle—*a musculotendinous*

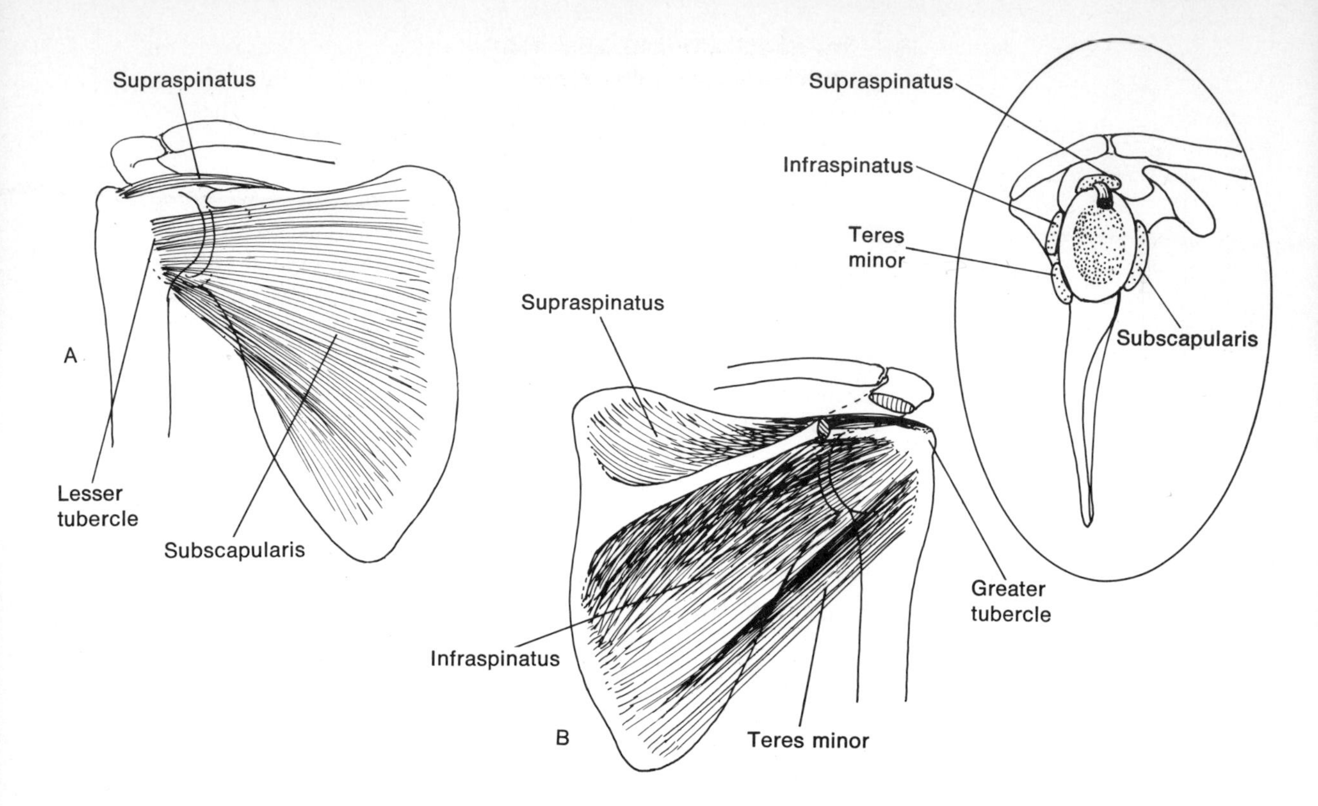

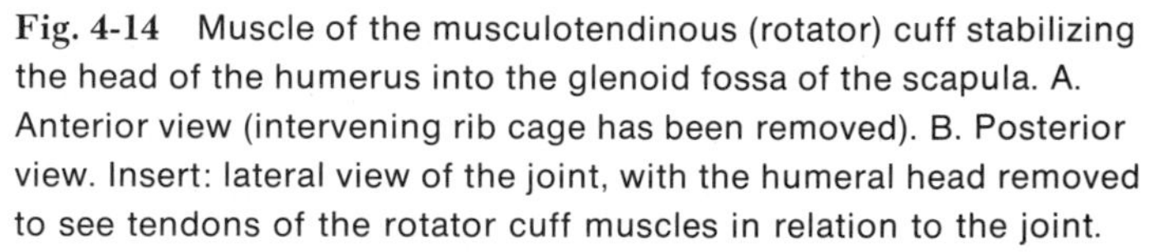

Fig. 4-14 Muscle of the musculotendinous (rotator) cuff stabilizing the head of the humerus into the glenoid fossa of the scapula. A. Anterior view (intervening rib cage has been removed). B. Posterior view. Insert: lateral view of the joint, with the humeral head removed to see tendons of the rotator cuff muscles in relation to the joint.

cuff. When these four intrinsic muscles of the shoulder contract in unison, the head of the humerus is held securely in the glenoid fossa. The origins of the cuff muscles are made obvious by their names, with the exception of teres minor. By studying Fig. 4-14 you can see that the cuff muscles inserting on the posterior aspect of the humerus are lateral rotators, and the muscle inserting anteriorly is a medial rotator; hence the group is often called the **rotator cuff.** The supraspinatus, as you can see, is an abductor. Of the muscles of the cuff, usually only the infraspinatus can be visualized on the surface of the back, and then only in well-developed males. Can you "place" these muscles in your own body?

MUSCLES MOVING THE ARM*
(less musculotendinous cuff)

pectoralis major   deltoid   latissimus dorsi
teres major

Pectoralis major (Fig. 4-15) is an anterior muscle of the chest, the larger of two, having three heads of origin: (1) from the clavicle (clavicular head), (2) from the sternum and ribs (sternocostal head), and (3) from the aponeurosis of an abdominal muscle (abdominal head). Attaching to the upper, anterior aspect of the humerus, pectoralis major *en masse* acts to adduct and medially rotate the humerus. You can feel this on yourself: place your arm (with forearm extended) on the table and turn your body to the side so your arm is abducted. Now adduct your arm against the resistance of the table while placing your hand over your chest. You can probably feel the whole lower border of pectoralis major (in front of the axilla) as well as the upper fibers contract. Now put your closed fist on top of the table (while facing the table) and push down (extending the arm against resistance from a flexed position) and note (by palpation) that the sternal portion of pectoralis major contracts and the clavicular portion (just below the medial clavicle) does not. Thus, the sternal head of pectoralis major acts to extend the arm.† Now place your fist *under* the table and lift *up*. The clavicular portion contracts and the sternal head does not, demonstrating that the clavicular head of pectoralis major acts to flex the arm,‡ especially when the arm is already partly flexed. Now see Fig. 4-16. Yet pectoralis major—one of the most powerful muscles of the body—is not the primary flexor or extensor of the arm.

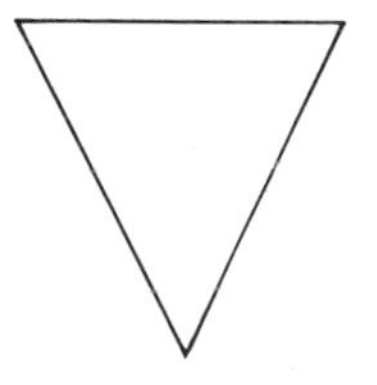

The muscle of the shoulder has the shape of the Greek letter delta, but a delta turned upside down. Thus the name of this muscle is the **deltoid** (Fig. 4-15). The fibers of this exceptionally strong muscle may be divided into anterior, middle, and posterior segments. You can see that the anterior fibers of the deltoid flex the arm, the posterior fibers extend the arm, and the middle fibers abduct the arm. In fact, the deltoid muscle is (1) *the primary flexor of the arm,* (2) *one of two primary extensors of the arm,* and (3) *the primary abductor of the arm.* Flex your arm against resistance while touching the anterior portion of the deltoid. Next, place your left hand over

* Coracobrachialis, and the long heads of biceps brachii and triceps brachii are also weak movers of the arm. The latter two will be considered with muscles of the arm.

† The sternal head also acts to adduct and medially rotate the humerus.

‡ The clavicular head also acts to medially rotate the humerus.

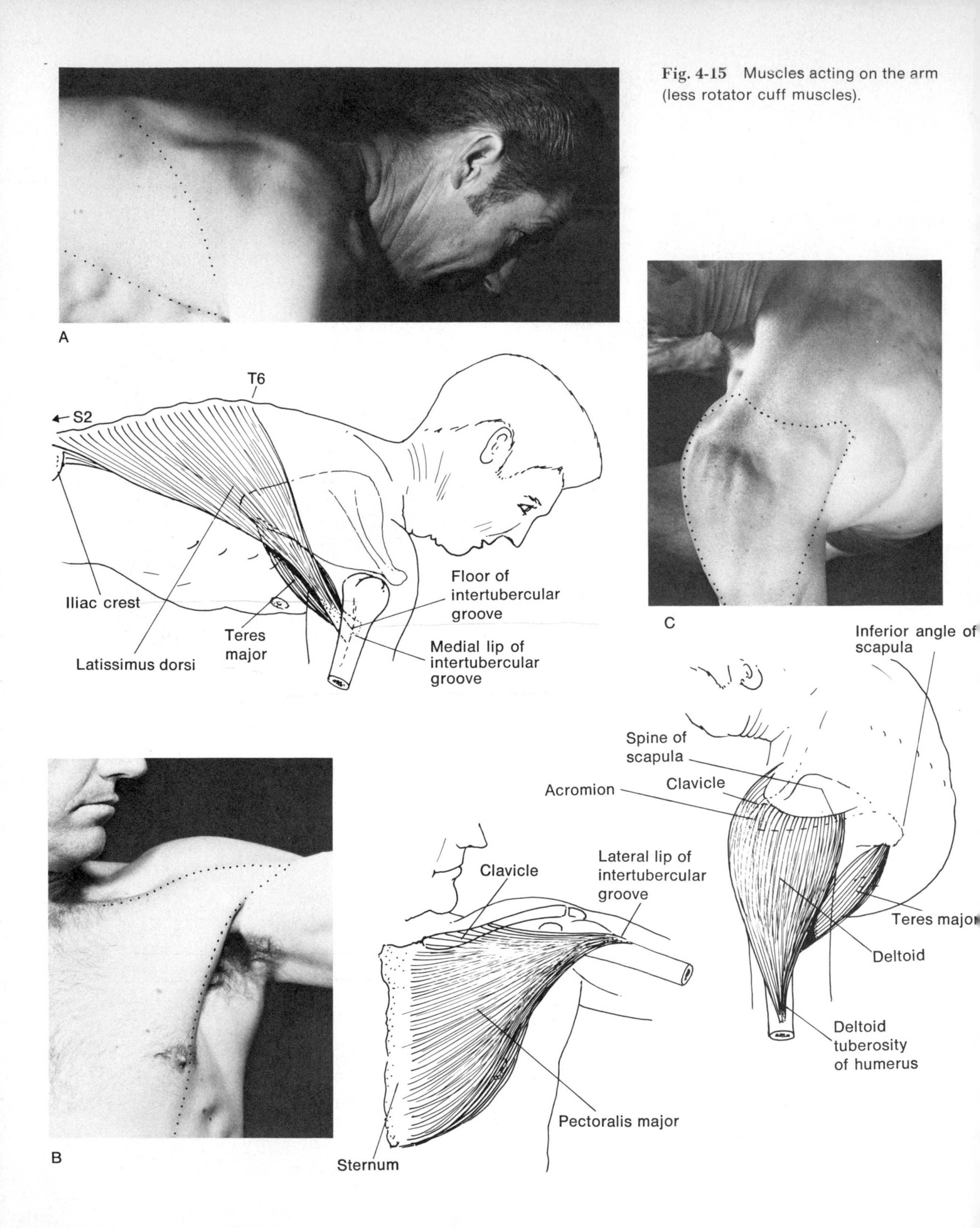

Fig. 4-15 Muscles acting on the arm (less rotator cuff muscles).

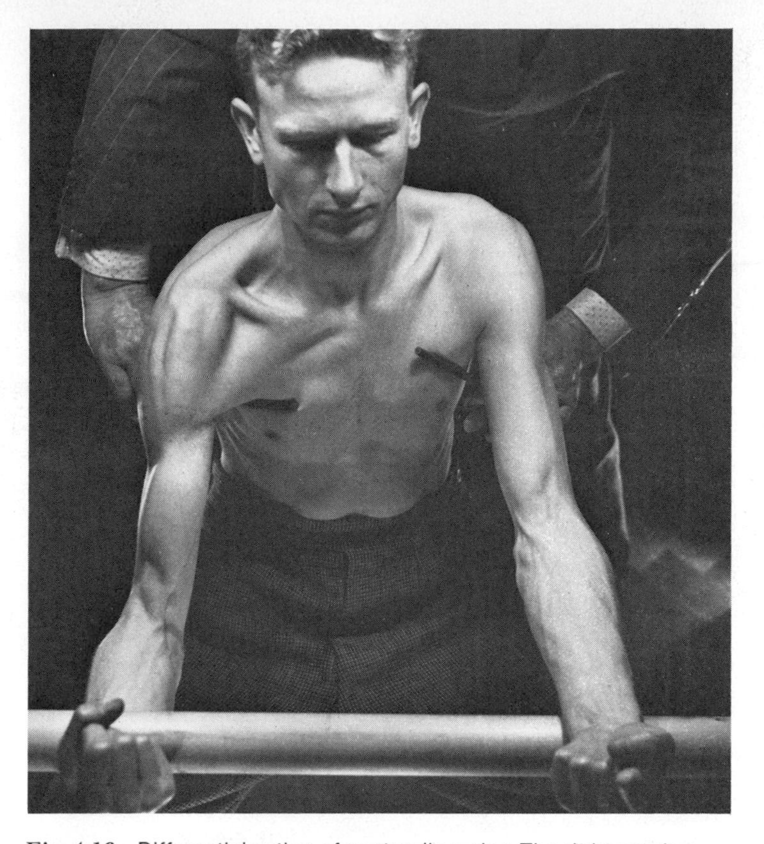

Fig. 4-16 Differential action of pectoralis major. The right arm is being raised (flexed) and the left arm is being depressed (extended), both against resistance. The clavicular head of pectoralis major is active in flexion (right arm); the sternal head is active in extension (left arm). (Courtesy of R. D. Lockhart, *Living Anatomy*, 6th ed., Faber & Faber, Ltd., London.)

the point of the shoulder with your fingers palpating the posterior portion. At the same time, with your right hand reach back toward your hip. The muscle bulging into your left hand is the deltoid. As a final example of the power of the deltoid, stand in a doorway and abduct your limbs against the door jambs until your arms hurt. Then step out smartly and feel your arms abduct effortlessly! The attachments of the deltoid may be seen in Fig. 4-15.

The broadest muscle of the back is the **latissimus dorsi** (Fig. 4-15). This muscle is a powerful extensor and adductor of the arm. The lateral border of this muscle can be seen on yourself in a mirror when adducting your arm against resistance from an abducted or extended position. Bring your arm around your chest about a handsbreadth below the axilla so that your fingers contact the mid-back—then extend as instructed above—in so doing, you will feel latissimus dorsi bulge into your hand. This muscle is particularly important in resisting superior displacement of the whole pectoral girdle when a person hangs by the arms or uses crutches for support of the body. The broad origin and narrow insertion of this muscle can be appreciated in Fig. 4-15.

The only other muscle (appropriate to this section) left for consideration is a round muscle, the larger of two: **teres major** (Fig. 4-15). This muscle is an adductor and a medial rotator. Be careful here! Teres *minor* is a lateral rotator. What fundamental difference is there in the distal attachments of these two muscles that forms the basis for their antagonistic actions? This muscle can best be seen in a well developed male who adducts or medially rotates the arm against resistance (Fig. 4-10E). The attachments of teres major can be seen in Fig. 4-15.

MUSCLES OF THE ARM

- biceps brachii
- triceps brachii
- coracobrachialis
- brachialis

The muscles of the arm (Fig. 4-17) arise from the humerus or from the scapula immediately adjacent to the humerus, and all but one cross the elbow joint to insert on the ulna or radius. The humeroulnar joint (elbow) is a hinge joint, hence all the arm muscles on the anterior aspect of the arm are flexors of the forearm and the muscle on the posterior aspect of the arm is an extensor of the forearm. In Fig. 4-17 note the attachments of the principal *flexors* of the forearm, **biceps brachii** and **brachialis;** and the *extensor* of the forearm, **triceps brachii.**

Now flex your forearm against resistance and feel/see the bulge of the biceps brachii on the anterior surface of the arm. Unless your forearm is fully flexed, note how the triceps brachii on the posterior surface is also contracted. Why? Now extend your forearm against resistance and feel the triceps brachii contract.

The surfaces features of the arm may be seen in Fig. 4-17.

Fig. 4-17 (opposite) Muscles clothing the arm acting principally on the elbow joint. A. Anterior view, superficial and deep dissections. B (turn page). Posterior view. Try to correlate the drawings with the photographs of the arm as well as your own.

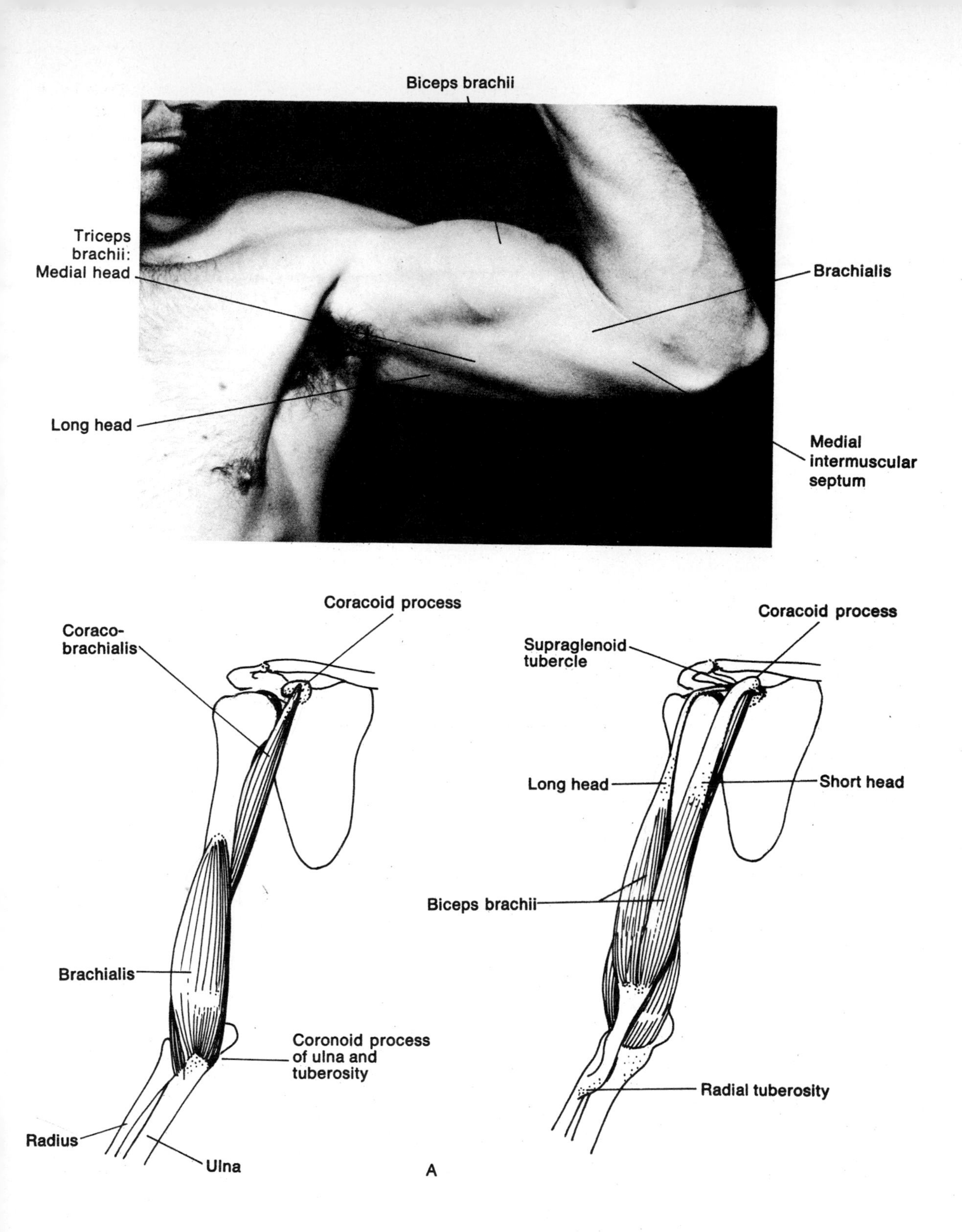

Biceps brachii
Triceps brachii: Medial head
Brachialis
Long head
Medial intermuscular septum
Coracoid process
Coraco-brachialis
Brachialis
Coronoid process of ulna and tuberosity
Radius
Ulna
Coracoid process
Supraglenoid tubercle
Long head
Short head
Biceps brachii
Radial tuberosity
A

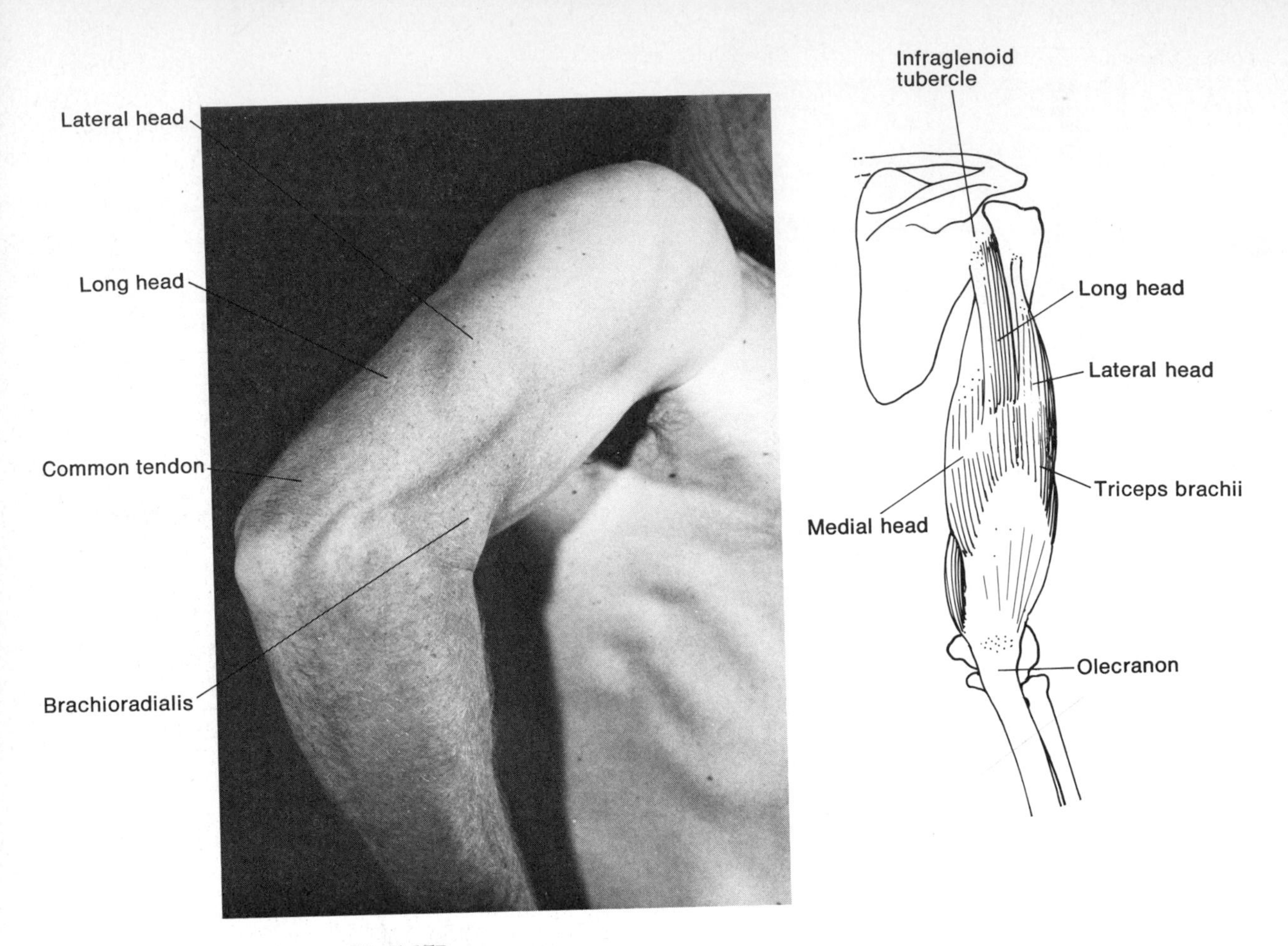

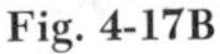
Fig. 4-17B

MUSCLES OF THE FOREARM

There is no shortage of muscles in the forearm—or lack of things for them to do. Consider your own forearm:

The bony arrangement of the forearm is such that two compartments are created: anterior and posterior (with the limb in the anatomical position).

Move your hands at the wrist joint and move your fingers. The muscles doing the work, as you can see, are largely in the forearm. Further, the muscles pulling on the anterior aspect seem to be generally involved in flexing the wrist or fingers—thus the anterior compartment might be called the **flexor compartment** (Fig. 4-18). Similarly, the posterior com-

Fig. 4-18 (opposite) Muscles of the anterior right forearm. A,B. Superficial layers. C. Middle layer. D. Deep layer. Try to correlate the drawings with the photograph of the forearm and with your own.

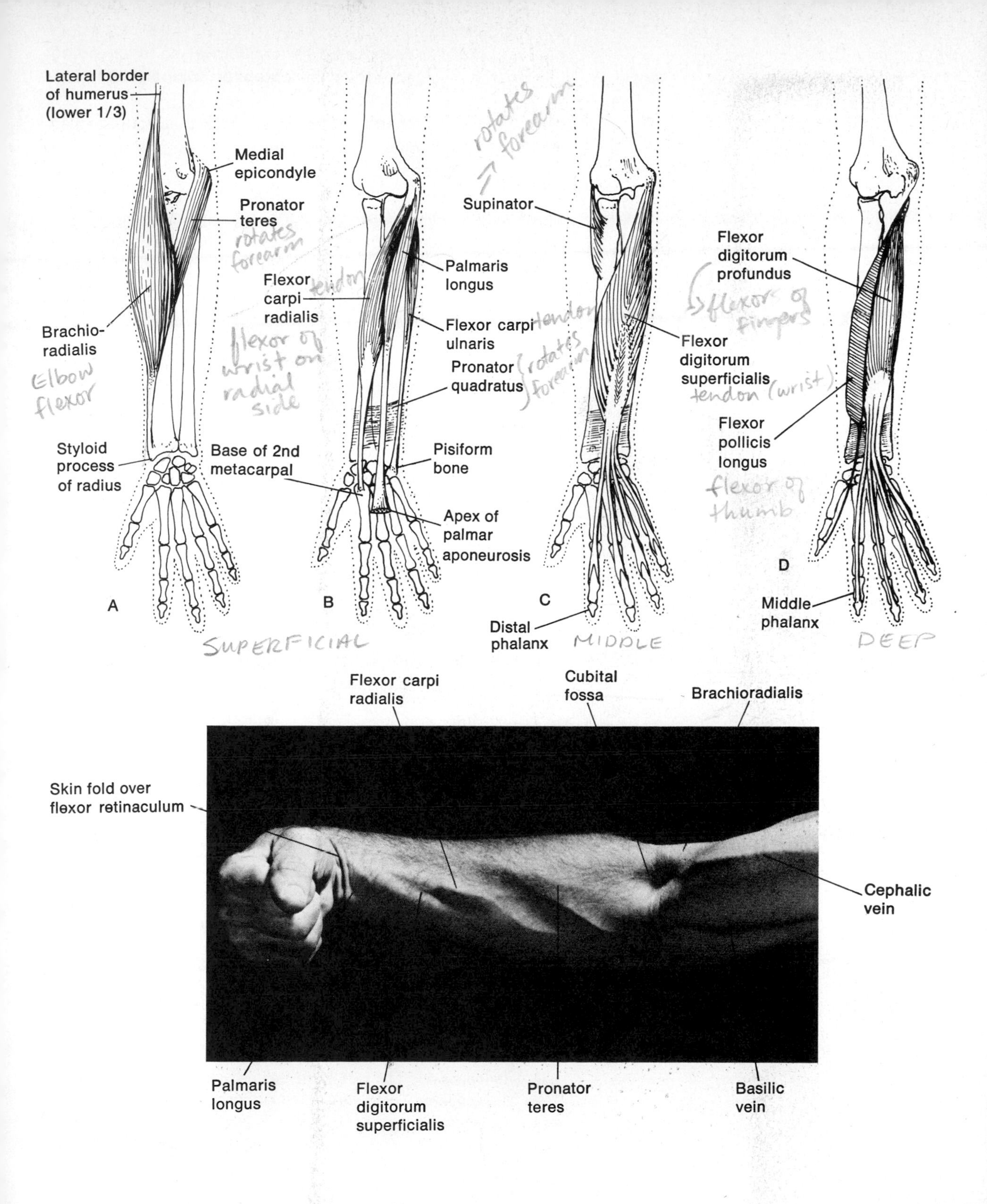

Lateral border of humerus (lower 1/3)
Medial epicondyle
Pronator teres
Brachio-radialis
Styloid process of radius
Flexor carpi radialis
Palmaris longus
Flexor carpi ulnaris
Pronator quadratus
Base of 2nd metacarpal
Pisiform bone
Apex of palmar aponeurosis
Supinator
Flexor digitorum superficialis
Distal phalanx
Flexor digitorum profundus
Flexor pollicis longus
Middle phalanx
A
B
C
D
Flexor carpi radialis
Cubital fossa
Brachioradialis
Skin fold over flexor retinaculum
Cephalic vein
Palmaris longus
Flexor digitorum superficialis
Pronator teres
Basilic vein

partment muscles seem to be generally involved in extending the wrist and fingers—thus the **extensor compartment** (Fig. 4-19).

Now move your hand at the wrist. Note that you can flex and extend the wrist. Therefore, there must be a flexor of the wrist and an extensor of the wrist.

Now note that you can also adduct and abduct the wrist. These actions are also often described as *radial deviation* (abduction or movement of the hand toward the radius or thumb side) and *ulnar deviation* (adduction or movement of the hand toward the ulna or little finger). A flexor and an extensor of the wrist arising in the forearm and inserting near the base of the thumb could (and do) collectively cause abduction of the wrist. Alternately, a flexor and an extensor of the wrist arising in the forearm and inserting near the base of the little finger could (and do) collectively cause adduction of the wrist.

Therefore, the principal actors of the wrist found in the forearm are:

1. A flexor of the wrist on the ulnar side: **flexor carpi ulnaris.**
2. A flexor of the wrist on the radial side: **flexor carpi radialis.**
3. An extensor of the wrist on the ulnar side: **extensor carpi ulnaris.**
4. Extensors of the wrist on the radial side—a long one and a short one: **extensor carpi radialis longus** and **brevis.**

In connection with these names, remember that the wrist is the carpus (Latin genitive form: *carpi).*

Now just flex and extend your fingers without moving your wrist—it is apparent that some muscles responsible for moving fingers (L., *digits;* genitive form: *digitorum*) are also in the forearm:

1. Flexors of the digits **(flexor digitorum)** found in a superficial layer **(superficialis)** and deep layer **(profundus).**
2. Extensors of the digits **(extensor digitorum).**

The thumb is one of the body's really special structures. It is highly mobile—move it around and see. As you circumduct your thumb (moving it in a circle), note the tendons "pop out" proximal to its base. Note they cross the wrist, and therefore

Fig. 4-19 (opposite) Muscles of the posterior forearm. A. Superficial layer. B. Deep layer. Try to correlate the drawing with the photograph of the forearm.

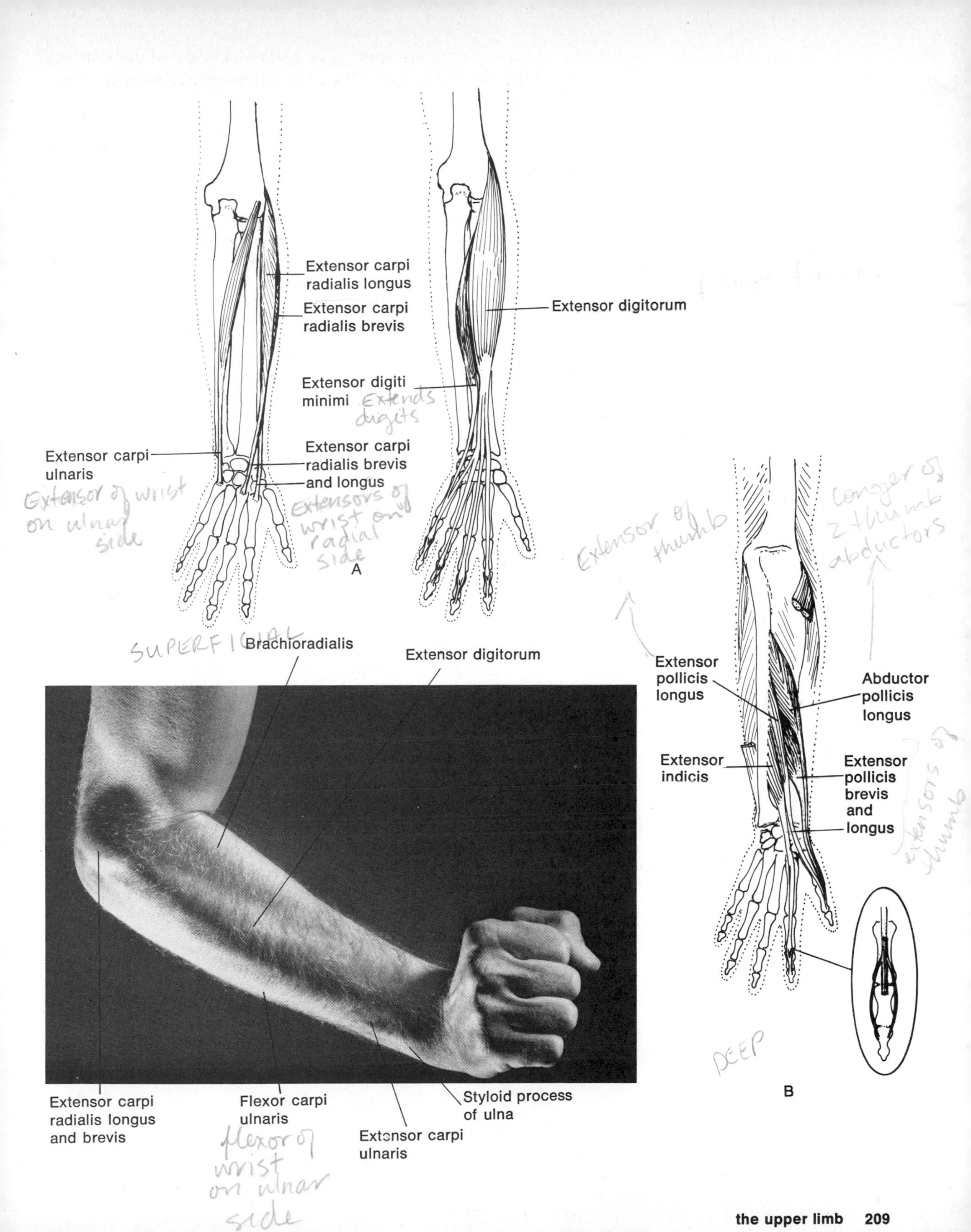
Extensor carpi radialis longus
Extensor carpi radialis brevis
Extensor digitorum
Extensor digiti minimi
Extensor carpi ulnaris
Extensor carpi radialis brevis and longus
A
Brachioradialis
Extensor digitorum
Extensor carpi radialis longus and brevis
Flexor carpi ulnaris
Extensor carpi ulnaris
Styloid process of ulna
Extensor pollicis longus
Abductor pollicis longus
Extensor indicis
Extensor pollicis brevis and longus
B

must come off the forearm. Thus (Figs. 4-18 and 4-19) there are:

1. A flexor of the thumb—longer of two (short one is in the thumb): **flexor pollicis longus.**
2. Extensors of the thumb—a long one and a short one: **extensor pollicis longus** and **brevis.**
3. An abductor of the thumb—longer of two (the short one is in the thumb): **abductor pollicis longus.**

See how you can rotate your forearm (pronation and supination). The principal movers for these functions are also found in the forearm and are reasonably labeled:

1. **Supinator.**
2. **Pronators teres** (the round one) and **quadratus** (the rectangular one).

Now for the finishing touch: Neutralize your forearm and hand between pronation and supination (you should be looking at the radial border of your forearm). While restraining your wrist with your other hand, attempt to flex your elbow joint. As you do, look for a bulge just below the arm on the upper forearm (lateral surface). That bulge is the **brachioradialis** muscle, an effective flexor of the elbow. This muscle is nice to have around in case you injure the nerve (musculocutaneous) supplying brachialis and biceps; this is so because the brachioradialis is innervated by another nerve (radial).

The precise arrangement of the above muscles is important for medical students and kinesiology students. If it is of significance for you, the muscles may be studied in the laboratory or in one of the classical anatomical treatises cited in the reference section of this unit. A general arrangement is discussed later in this unit.

Now see the flexor side of the wrist (Fig. 4-18). By flexing the wrist against resistance, you can visualize and palpate the tendons of:

1. **Palmaris longus** (in the middle—a functionally insignificant muscle, missing in 10–15 percent of the population).
2. **Flexor carpi radialis.**
3. **Flexor carpi ulnaris.**
4. **Flexor digitorum superficialis** (best seen in those who play racket sports).

The other tendons of the wrist are covered by a thick fascial bandage and cannot easily be felt.

MUSCLES OF THE HAND

Look at the palm of your hand. The muscular pad just below the thumb is the **thenar eminence.** The pad along the little finger side of the palm is the **hypothenar eminence.** In between these pads are small muscles between and anterior to the metacarpals.

Refer now to Figs. 4-20 and 4-21 and note:

The thenar pad is composed of:

1. A short abductor of the thumb: **abductor pollicis brevis.**
2. A muscle opposing thumb to 5th digit: **opponens pollicis.**
3. A short flexor of the thumb: **flexor pollicis brevis.**

The hypothenar pad is composed of:

1. A flexor of the little finger: **flexor digiti minimi.**
2. An abductor of the little finger: **abductor digiti minimi.**
3. A muscle opposing 5th digit to thumb: **opponens digiti minimi.**

The other intrinsic muscles of the hand include: finger adductors (palmar interossei), finger abductors (dorsal interossei), an adductor of the thumb (adductor pollicis), assistant extensors of the interphalangeal joints (lumbricals, interossei) and assistant flexors of the metacarpophalangeal joints (lumbricals, interossei).

The movement that sets us and primates on a somewhat higher anatomical plateau than our more primitive animal friends with fingers is *opposition of the thumb and fifth finger*—a rotation of the thumb and flexion of the 5th digit such that the pads of the two fingers come into opposition. This fundamental (but not spectacular) neuromuscular feat is involved in many of our precise grasping motions.

QUIZ

It might be advisable here to check your mental absorption on the significant facts revealed in this unit so far. For example:

1. The upper limb is attached to the axial skeleton by the pectoral girdle (shoulder).
2. The clavicle functions as the Strut of the upper limb.
3. The clavicle is attached to the axial skeleton at the sternoclavicular joint. It is attached to scapula laterally at the acromioclavicular joint.
4. The scapula is supported primarily by muscle.

(Text continues on page 214)

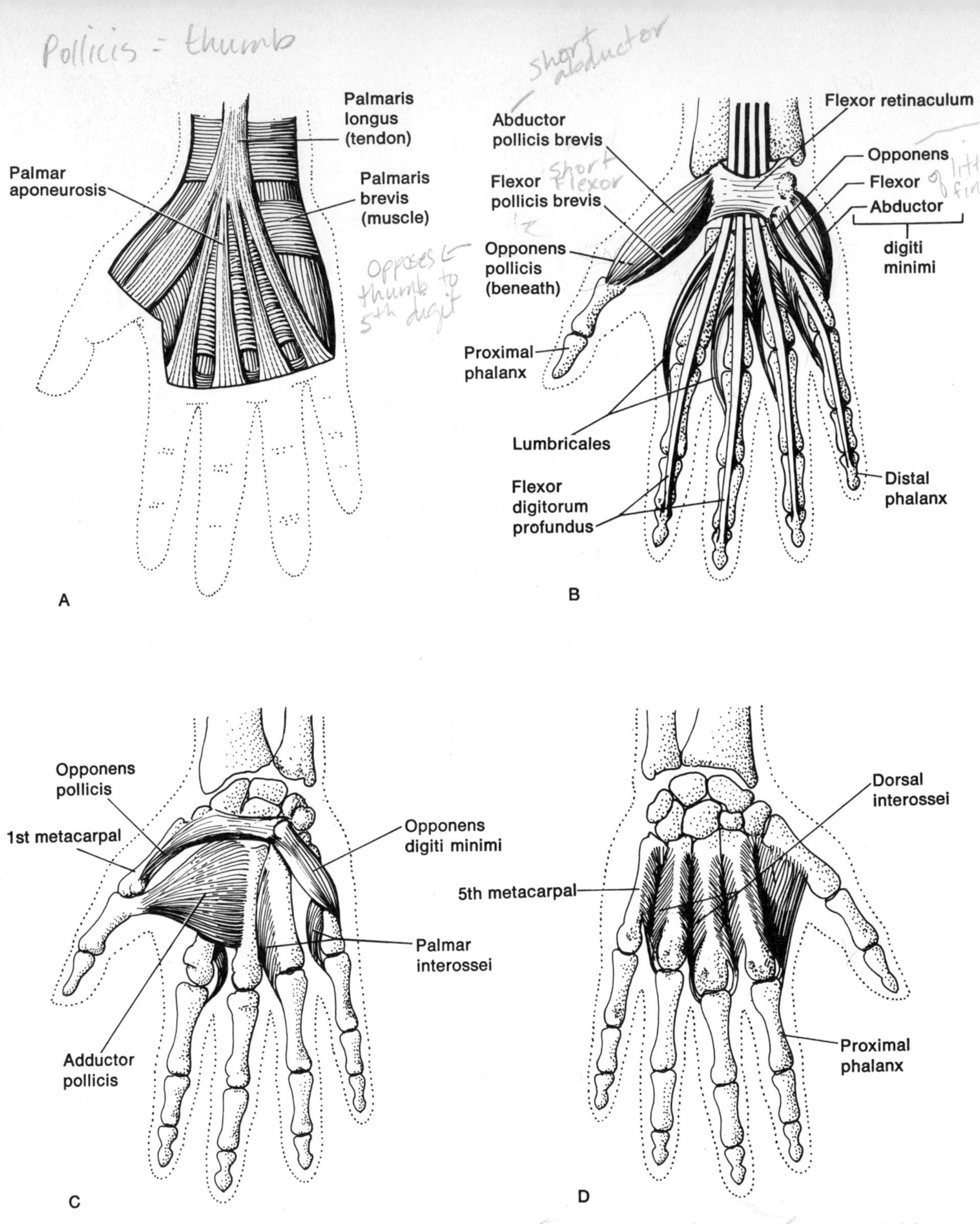

Fig. 4-20 Muscles of the hand. A–C, palmar view. A. superficial layer. B. Middle layer. C. Deep layer. D. Dorsal interossei muscles as seen on the dorsum of the hand.

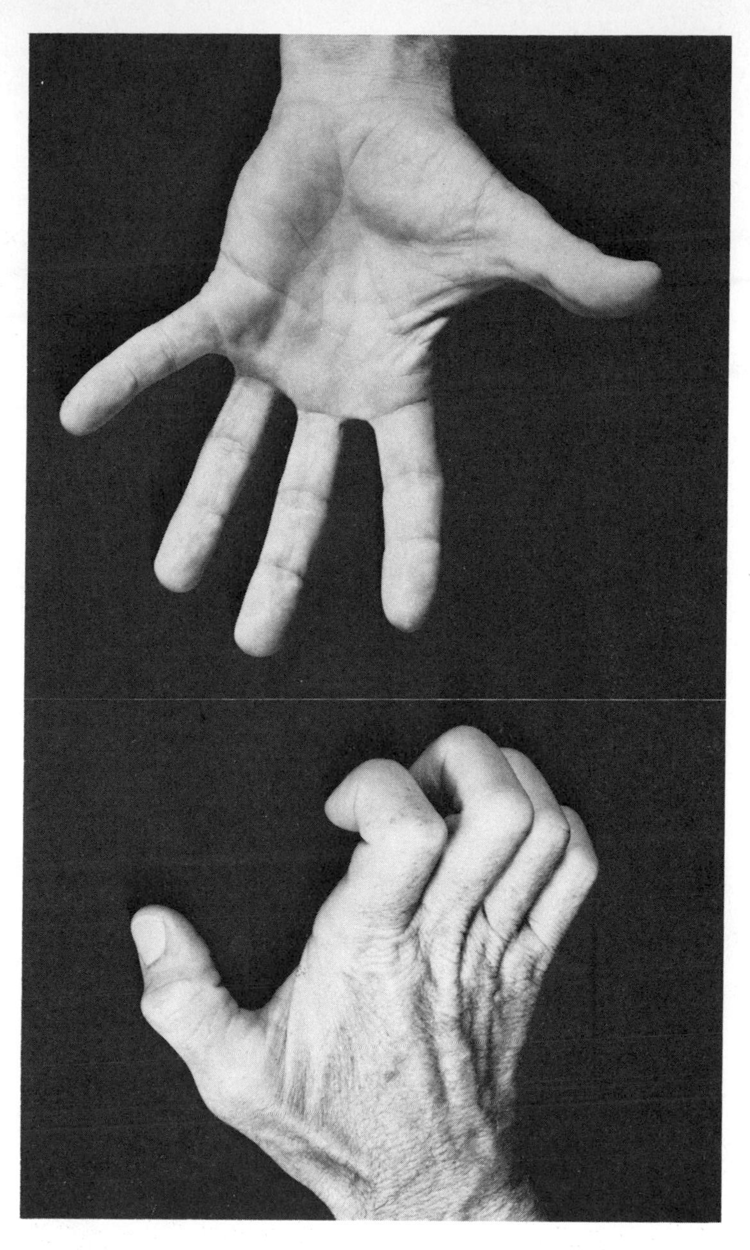

Fig. 4-21 Surface features of the hand. (Top) Palmar surface. (Bottom) Dorsum.

5. The bone of the arm is called the Humerus.
6. In rotation of the forearm, the ulna does not move; the radius rotates about it.
7. Muscles which secure the scapula to the axial skeleton are characterized by arising (origin) from the axial skeleton and inserting on the scapula.
8. The shallow glenoid fossa provides a poor socket for the head of the humerus. This joint is reinforced by 4 muscles which cross the joint in the same plane as the joint.
9. The deltoid muscle is the primary flexor, abductor, and extensor of the arm.

Subscapularis
Supraspinatus
Infraspinatus
Teres Minor

regional and neurovascular considerations

AXILLA

The axilla is a space created by the relations of the scapula, the clavicle, and the humerus to the lateral aspect of the rib cage (area of 1st and 2nd ribs) (Fig. 4-22; A-1, Appendix). You will remember that the clavicle forces the scapula laterally and posteriorly. The chest wall is curved laterally and its girth (circumference) decreases from the level of the 5th rib upward. Thus, under the protection of the acromion, a space is created. It is an area of particular interest to the anatomist, clinician, and the surgeon for several reasons, not the least of which is the important neurovascular bundle and its branches/tributaries projecting out to the limb from the neck and chest (Fig. 4-23). Disease of or injury to the axillary struc-

Fig. 4-22 Bony boundaries of the axilla, viewed from above.

Fig. 4-23 Muscular boundaries of the axilla in relation to the neurovascular bundle passing through. Seen in cross section through the axilla. (Adapted from R. D. Lockhart, G. F. Hamilton, and F. W. Fyfe, *Anatomy of the Human Body,* 2d ed., J. B. Lippincott, Philadelphia, 1965.)

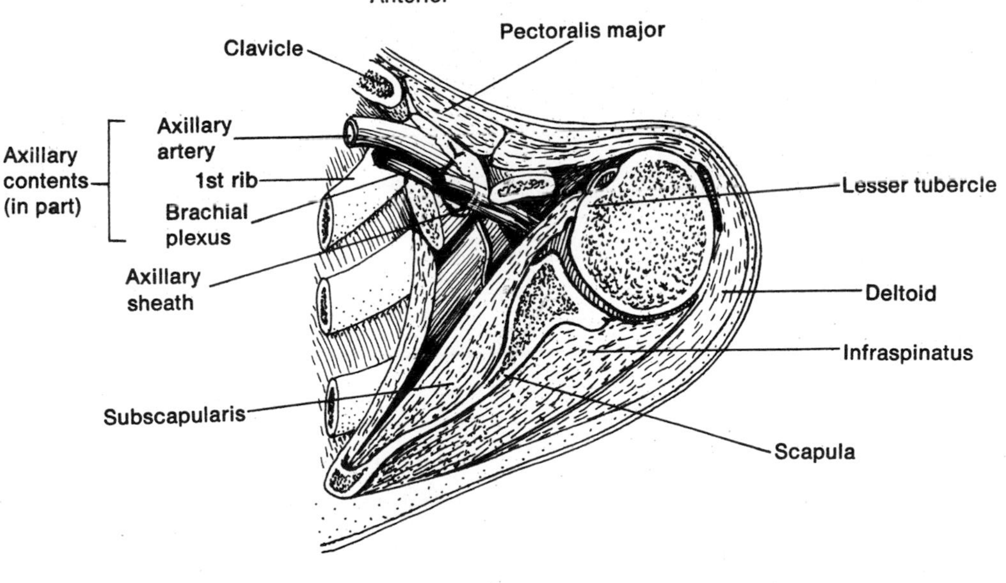

Pectoral region = breast region

tures often involves the scapular region, the breast, pectoral muscles, shoulder, arm, forearm, wrist and/or hand. Cancer of the breast is often reflected in enlarged lymph nodes in the axilla. In few regions is there such a complex array of structures involving so many areas. Such a fascinating area warrants some investigation.

The axilla is a dome-shaped compartment bounded by:

1. An *anterior wall* created primarily by the pectoralis major muscle crossing from the chest to the humerus.
2. A *posterior wall* created by the scapula clothed by the subscapularis, plus latissimus dorsi and teres major muscles, which cross the axillary space just lateral to the scapula en route to the humerus.
3. A *medial wall* formed by ribs 2 to 6 and intervening intercostal musculature—all of which is covered by the serratus anterior.
4. A *lateral wall* consisting of the thin intertubercular groove of the humerus between the pectoralis major anteriorly and teres major posteriorly.
5. The *apex* represented by the acromion.
6. The *base* created by a suspensory ligament, which gives the axilla a bell shape and is composed of fascia and skin.

p. 198

Now refer to Fig. 4-12 and relate it to your own axillary region: the anterior wall can be easily felt as can latissimus dorsi of the posterior wall. Press your fingers medially up into the axilla and feel the serratus anterior muscle blanketing the ribs. Now move the back of your fingertips laterally and feel the intertubercular groove (Fig. 4-6) between the tendons of the pectoralis major and the latissimus dorsi. You can grasp the base of the axilla easily, demonstrating that it is indeed skin and fascia. Feel and see the apex of this space at the top of the shoulder. It is the bony acromion.

Within the space, and bound securely in deep fascia, are:

1. The brachial plexus of nerves.
2. The axillary artery and vein.
3. Branches and tributaries of the above.
4. The axillary lymph nodes.
5. Adipose tissue.

The **brachial plexus** (Fig. 4-24) is a network of nerves derived from spinal nerves C5 to T1. It passes out of the neck partly under cover of the trapezius, deep to the clavicle but over the first rib, and into the axilla. Within the axilla the plexus breaks up into an interesting pattern around the ax-

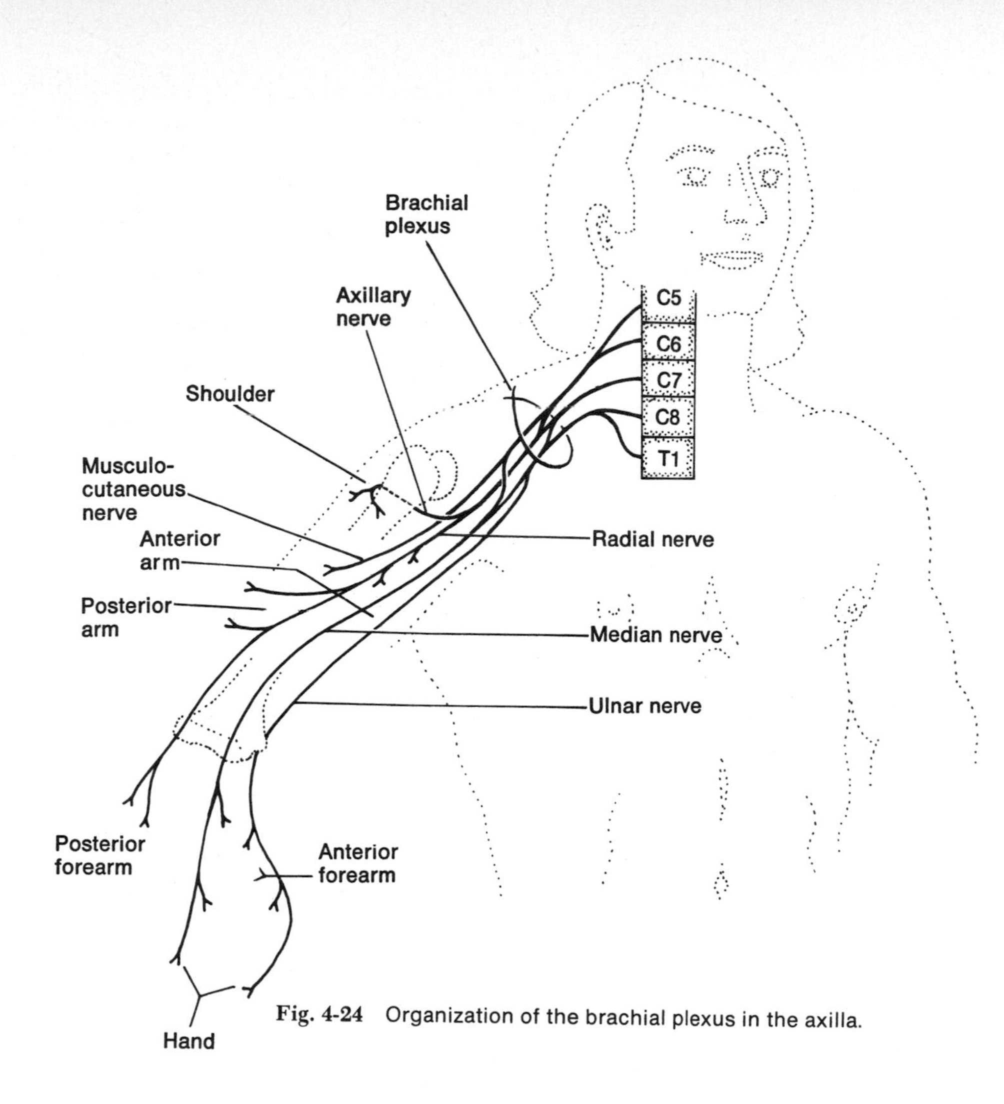

Fig. 4-24 Organization of the brachial plexus in the axilla.

illary artery. From within the axilla, terminal branches develop which supply the entire upper limb with motor and sensory innervation. Refer now to Fig. 4-25 and note:

1. The roots of the plexus are the anterior rami of spinal nerves C5 to T1.
2. These roots soon combine to form three trunks which divide into three cords.

Because the cords are closely wrapped around the axillary artery they are named according to their relation to that artery: lateral, medial, and posterior. Look closely at the cords and you will see them break up into the final branches which

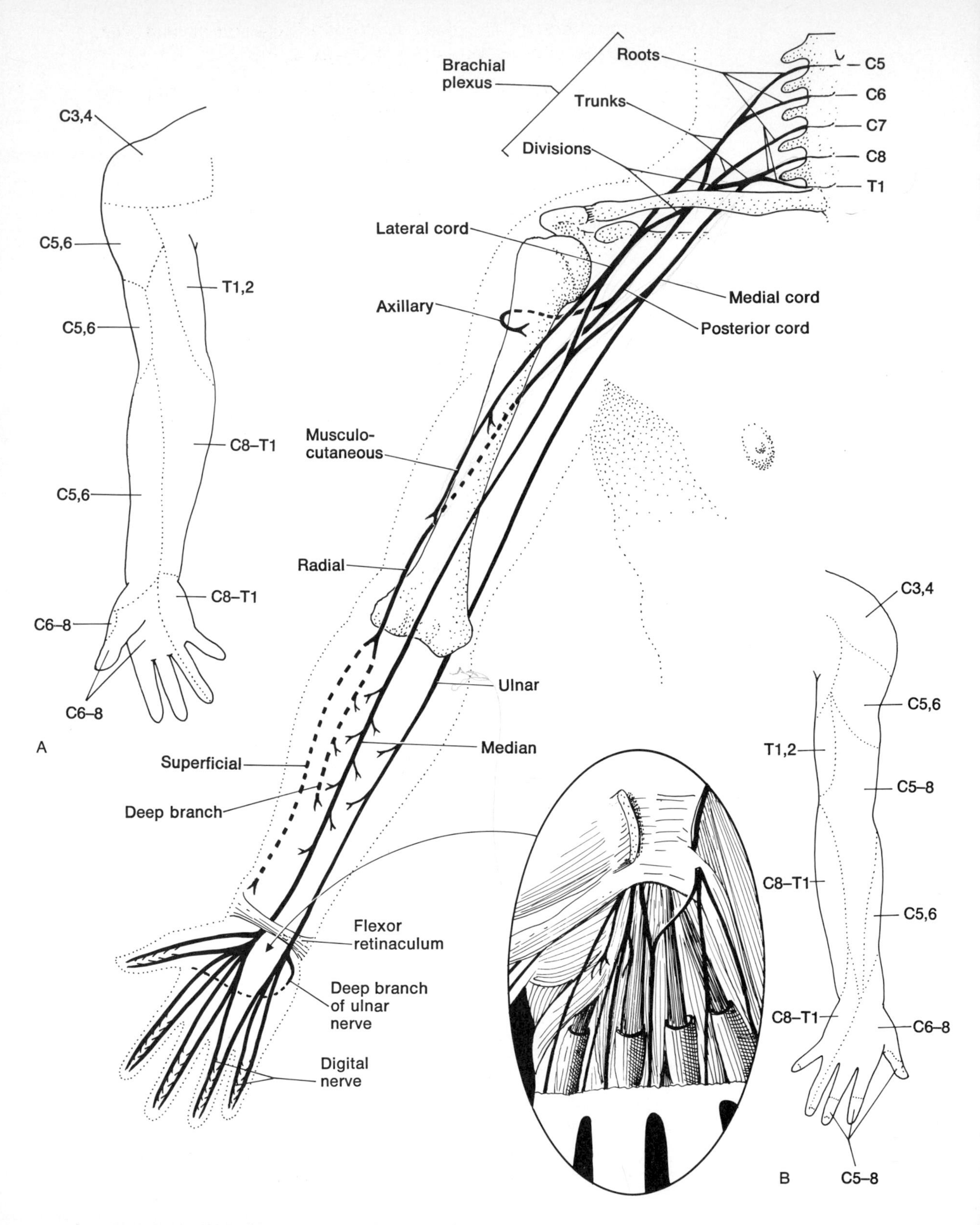

Brachial
plexus
Roots
Trunks
Divisions
C5
C6
C7
C8
T1
Lateral cord
Medial cord
Posterior cord
Axillary
Musculo-
cutaneous
Radial
Ulnar
Median
Superficial
Deep branch
Flexor
retinaculum
Deep branch
of ulnar
nerve
Digital
nerve
C3,4
C5,6
T1,2
C5,6
C8–T1
C5,6
C8–T1
C6–8
C6–8
A
C3,4
C5,6
T1,2
C5–8
C8–T1
C5,6
C8–T1
C6–8
C5–8
B

pass through the upper limb and supply muscle and skin. These terminal nerves (Fig. 4-25) are as follows:

1. **Axillary nerve** (posterior cord) to the deltoid and teres minor muscles and related cutaneous areas.
2. **Radial nerve** (posterior cord) to the triceps brachii and posterior forearm muscles and related cutaneous areas.
3. **Musculocutaneous nerve** (lateral cord) to flexor muscles of the arm and the lateral cutaneous region of the forearm.
4. **Ulnar nerve** (medial cord) to the anterior forearm and hand.
5. **Median nerve** (from both medial and lateral cords), supplying the lion's share of the anterior forearm and hand.

Each nerve will be considered with the appropriate region. Keep in mind that each nerve has sensory as well as motor components. As you read about the regions of the upper limb you may want to return to Fig. 4-25 from time to time to maintain your grasp of the "big picture."

The **axillary artery** (Fig. 4-26) is a continuation of the subclavian artery, which itself is a direct branch of the brachiocephalic artery coming off the arch of the aorta. The branches of the axillary artery supply the shoulder region (including the joints), the pectoral region, part of the thorax, as well as the proximal arm. The axillary artery continues into the arm to become the brachial artery.

The **axillary vein** drains into the subclavian vein, which drains into the brachiocephalic vein entering the heart. The tributaries of the axillary vein may be seen in Fig. 4-27.

There are a number of groups of lymph nodes in the axillary fascia and adipose tissue (Fig. 4-28) which are of particular significance: first, because they are the first important group of nodes to receive lymph drainage from the entire limb. Second, because they drain the lymphatic vessels of the pectoral region, i.e. the breast. Thus, any infection of the breast or any part of the arm, forearm, wrist, or hand will pass into the axillary nodes via the various lymphatic vessels and cause their enlargement*. Neoplasms of the breast are usually followed by a spreading of the malignant cells (metastasis) into the

* When infective agents or neoplastic cells pass into a lymph node, the large lymphocytes start dividing rapidly and the node swells and enlarges visibly. Thus swollen nodes, of which several sets are palpable, imply a disease state.

Fig. 4-25 (opposite) Nerves of the right upper limb, anterior view. A. Anterior dermatomes. B. Posterior dermatomes.

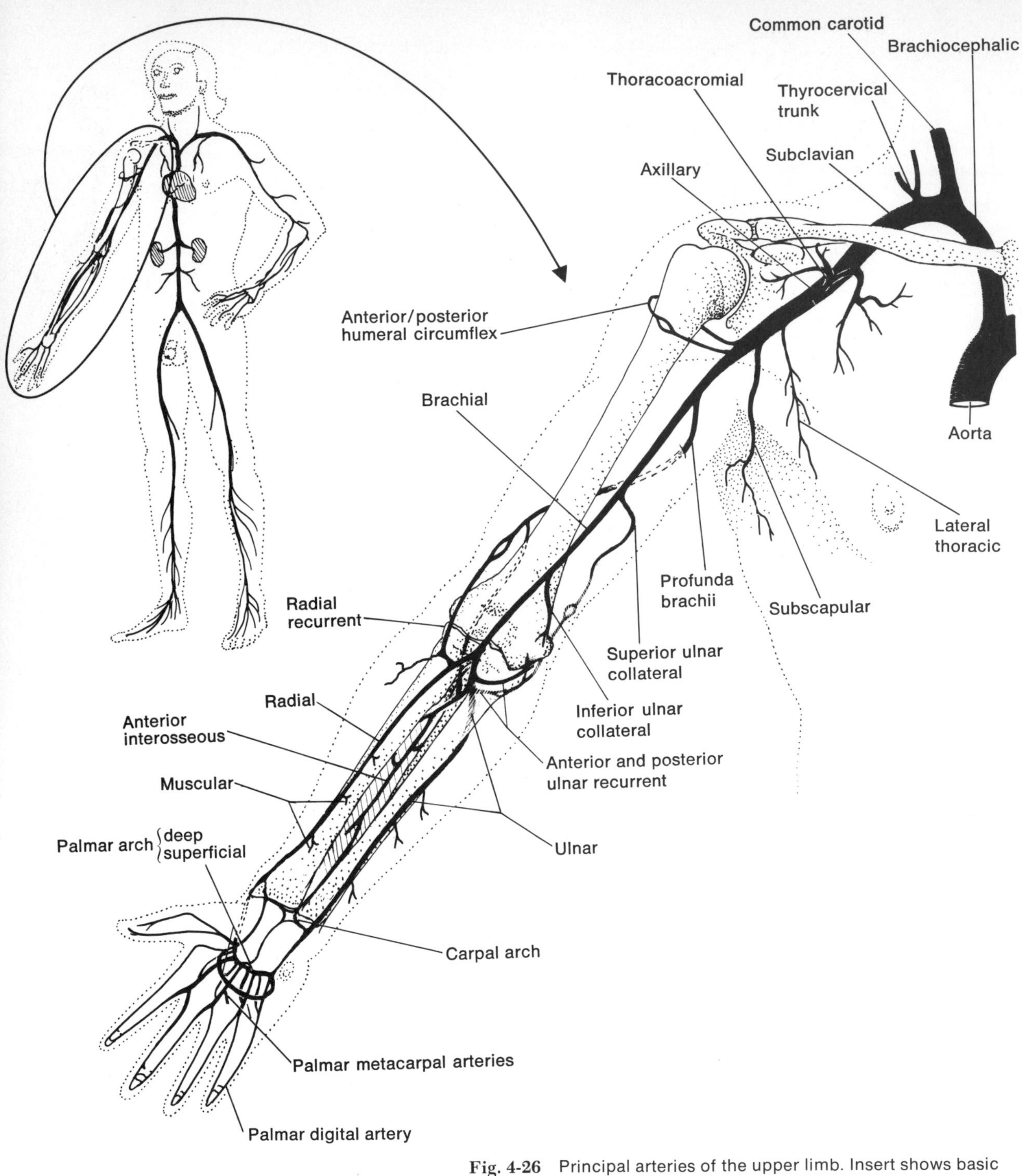

Fig. 4-26 Principal arteries of the upper limb. Insert shows basic body arterial plan.

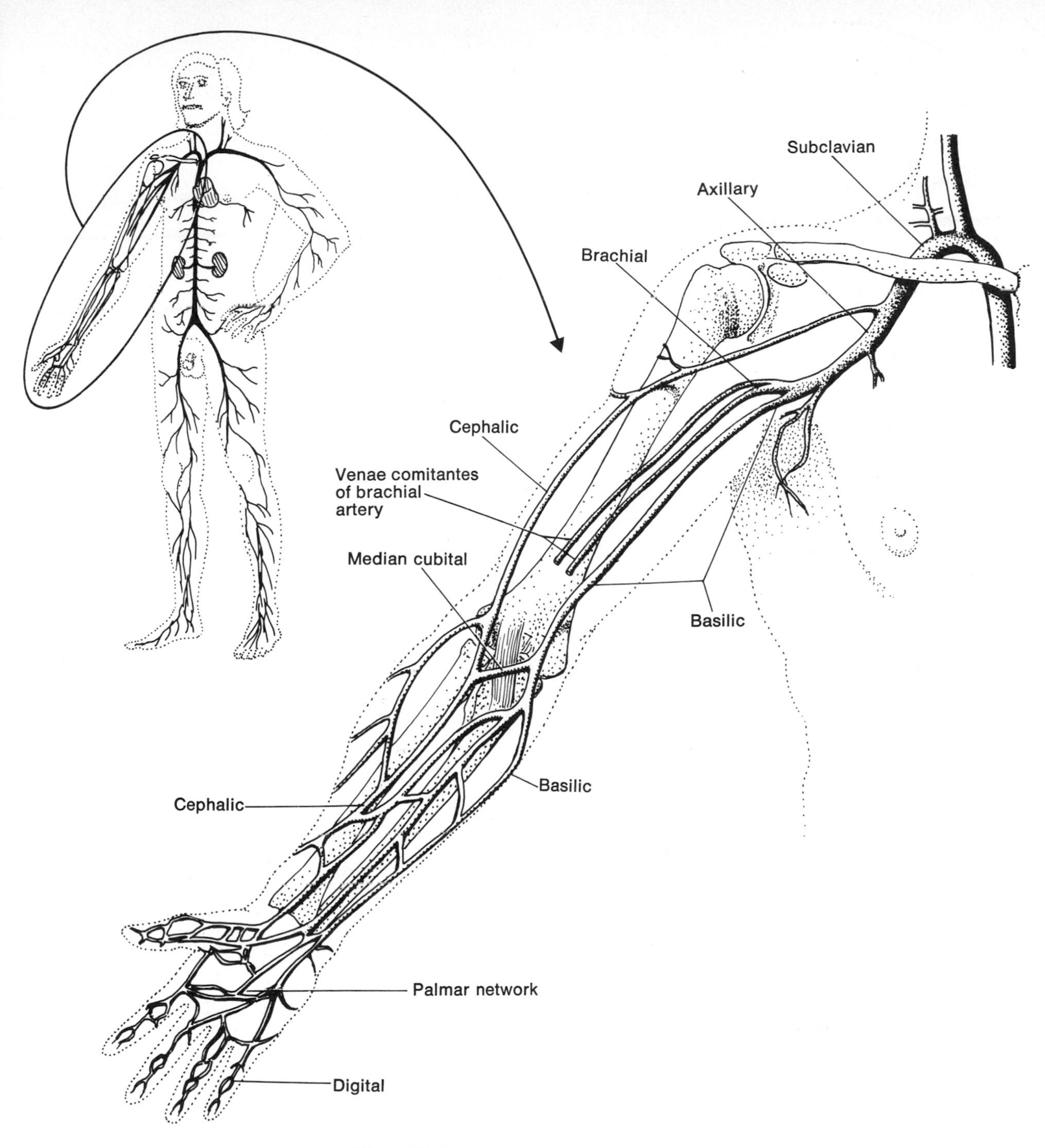

Fig. 4-27 Principal veins of the upper limb. Some of the deep veins (venae comitantes) following the arteries are not shown. Insert shows the basic body venous plan.

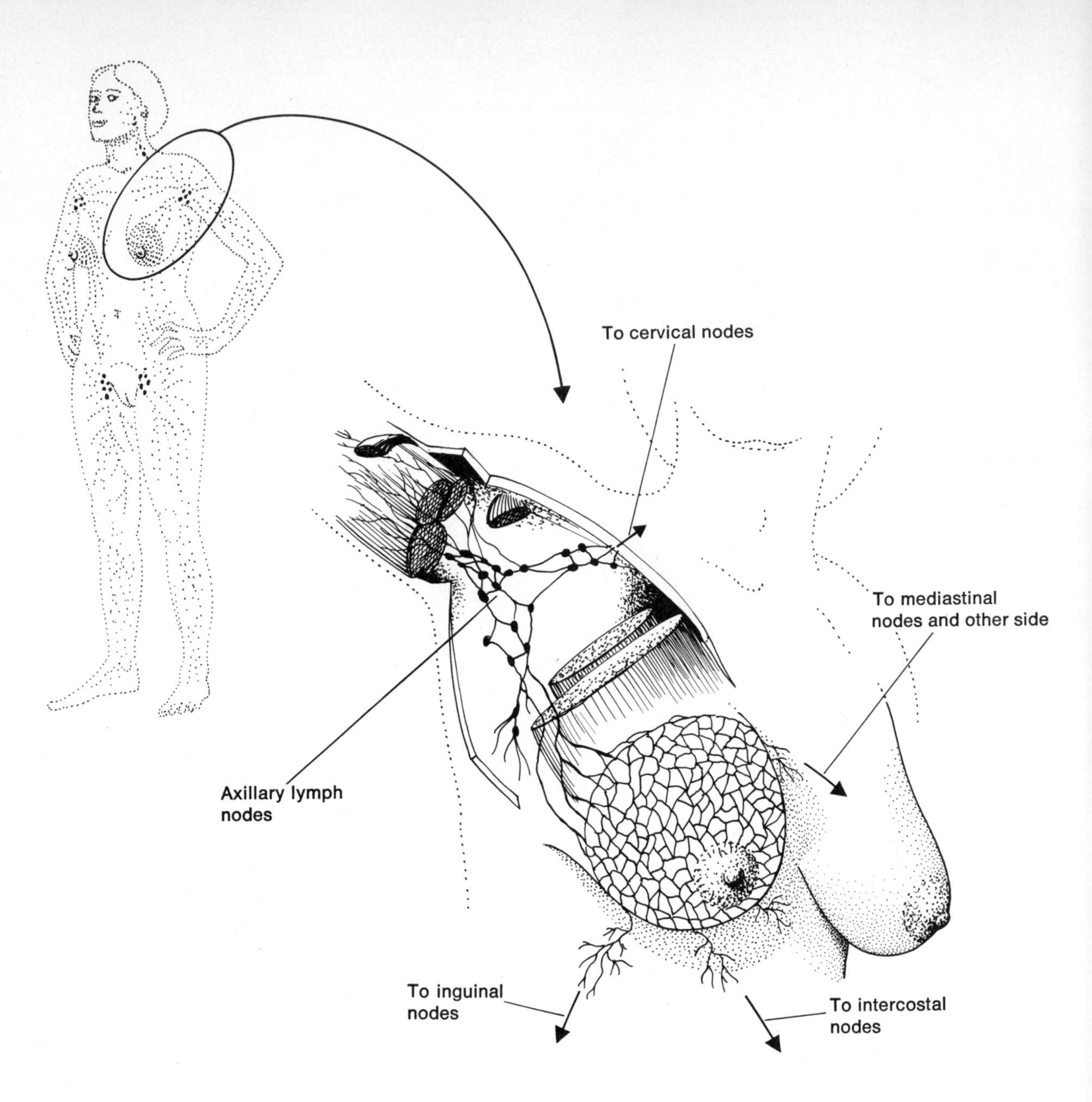

Fig. 4-28 Lymphatic circulation about the axilla and breast. Insert shows the basic body plan of superficial lymphatic vessels and lymph nodes.

lymphatic vessels, which transmit these cells to the axillary lymph nodes (as well as to others). These nodes act as a "filter" in an attempt to prevent further metastasis. Thus early diagnosis of carcinoma of the breast is critical. Palpation of the axillary nodes is commonly performed by physicians during routine physical examinations.

SCAPULAR REGION

The scapular region, superficial to the rib cage and deep back muscles, is both bony and muscular, consisting largely of the scapula and related muscles, that is, trapezius, rhomboid, levator scapulae, infraspinatus, supraspinatus, and latissimus dorsi (Figs. 4-11 and 4-14). Site of the common "slap on the back," this region boasts a thick but mobile hide. The scapula, trapezius, and latissimus dorsi are normally quite visible and easily palpated in the living person (Figs. 4-10 and 4-15A). Most of the above muscles (mooring and rotator cuff muscles) are innervated by nerves of the brachial plexus, principally the posterior cord branches. The trapezius is served by the accessory nerve (XIth cranial nerve).

The most significant feature of this region are the extensive vascular anastomoses arranged about the scapula (Fig. 4-29).

Fig. 4-29 Scapular anastomoses. A. The vessels around the scapula (arterial circle) can receive blood from a variety of different arteries. Note the ligature. With the blood in the axillary artery damned up at that point, consider the alternate routes by which blood can get to the brachial artery (highly schematic). (Adapted from Otto C. Brantigan, *Clinical Anatomy*, McGraw-Hill, New York, 1963.) B. The arrangement of arteries about the scapula.

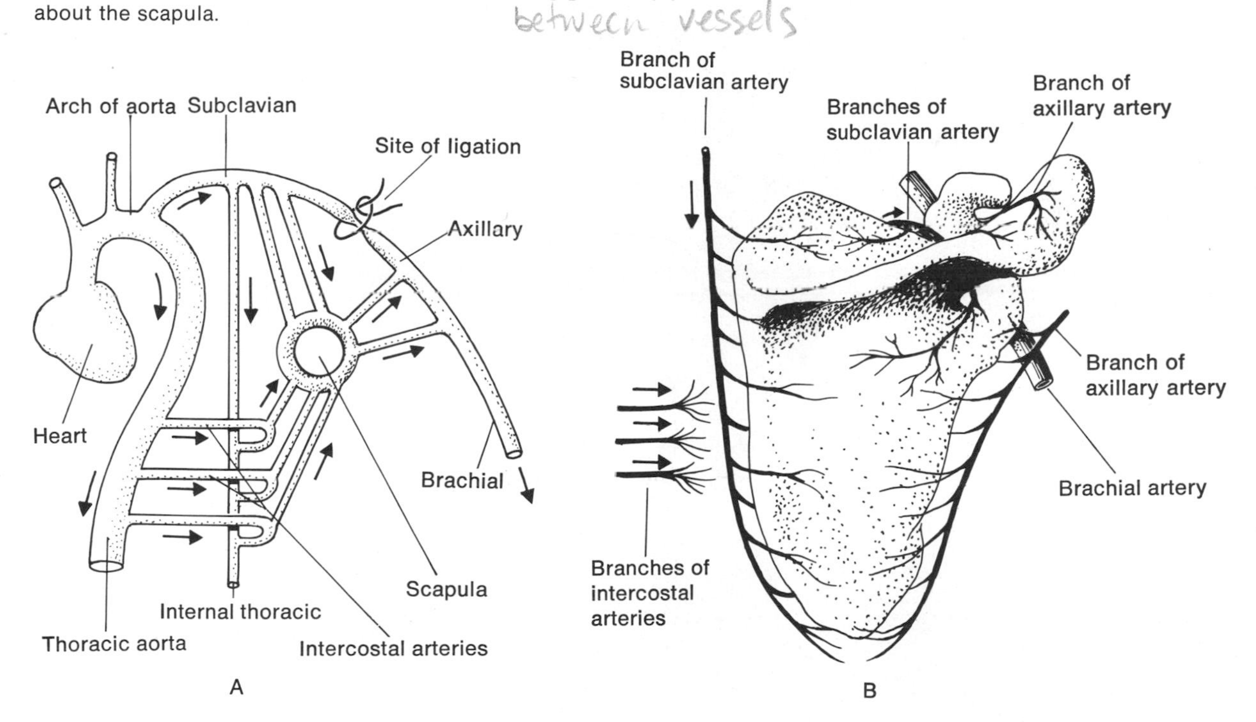

Should the subclavian or axillary artery be ligated following trauma, blood can still reach the most distal part of the limb by these interconnecting vessels. The source of blood to the scapular region is largely branches of the subclavian artery.

PECTORAL REGION

Anterior counterpart of the scapular region, the pectoral region, consists primarily of the pectoralis muscles (major and minor) overlying the thoracic cage (Figs. 4-11 and 4-12). Like the scapular muscles, they are concerned with movements of the upper limb. In both males and females, pectoralis major is both easily seen and palpated, and is largely responsible for the chest form, even in females. The pectoral muscles are supplied by pectoral nerves from the brachial plexus (Fig. 4-25) and receive vascular branches from the axillary artery (Fig. 4-26). The skin of this region is significantly

p. 218

Fig. 4-30 The adult female breast. The dissected view is a sagittal section.

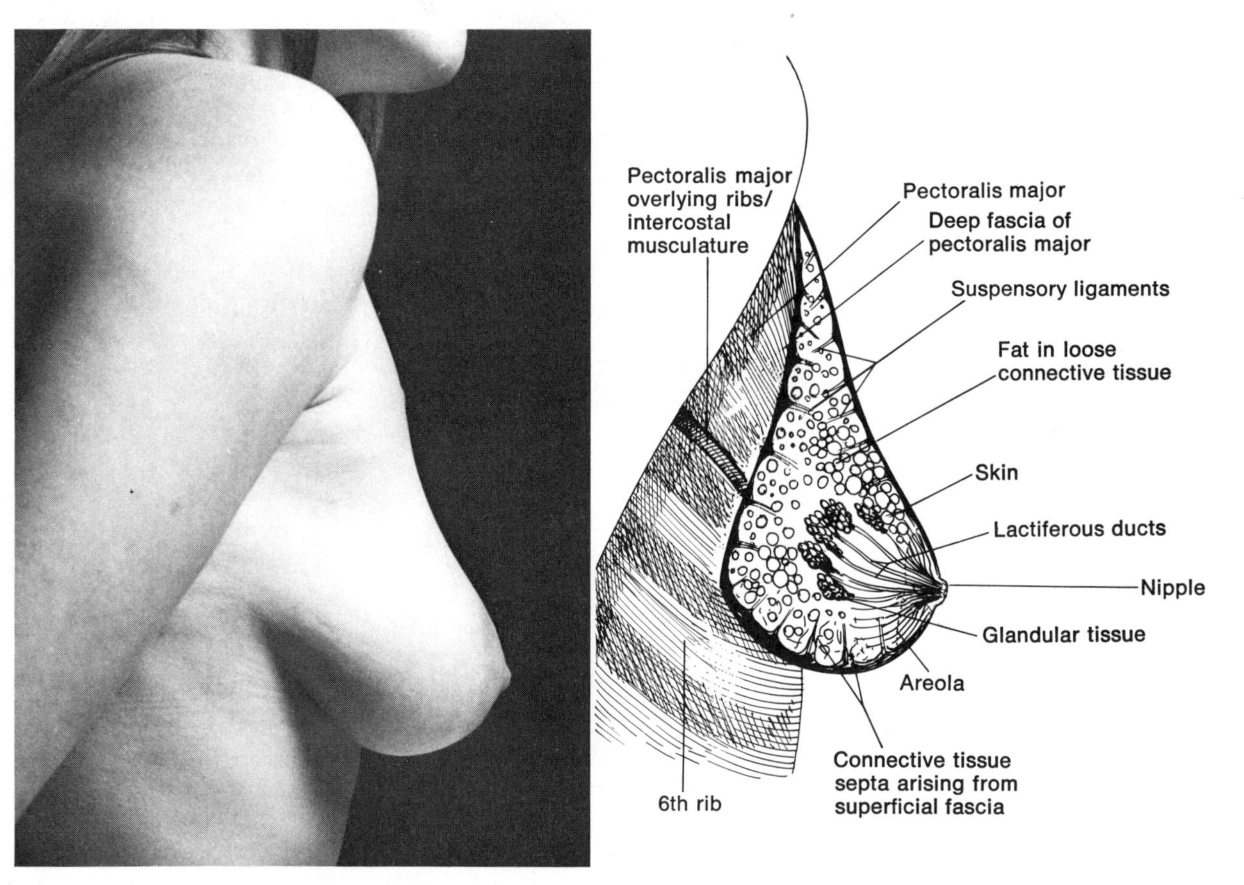

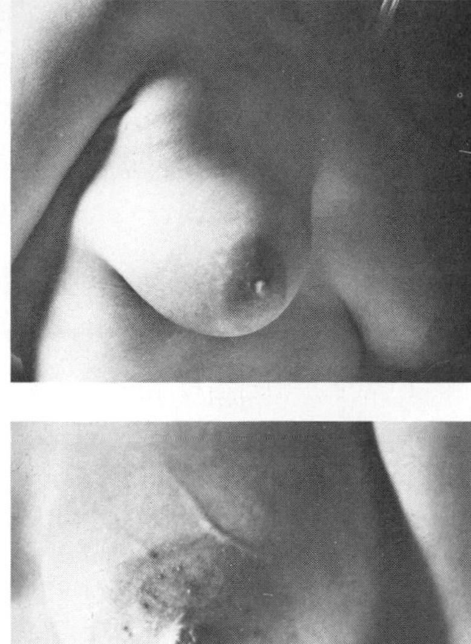

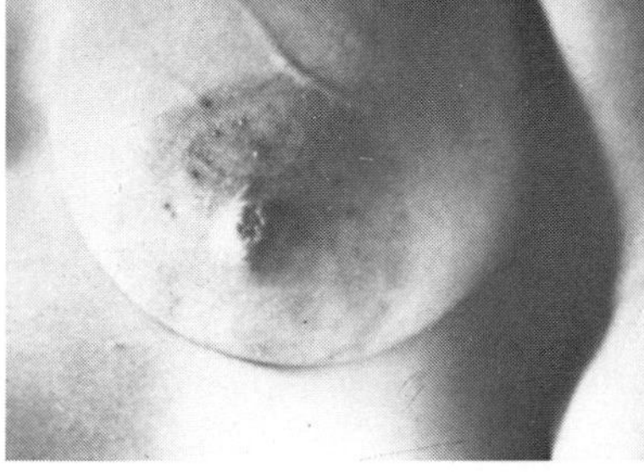

Fig. 4-31 Surface features of the breast in various functional states. A. Anterior view of non-gravid breast of 21-yr-old woman. B. Lactating breast. (B, courtesy of R. D. Lockhart, *Living Anatomy*, 6th ed., Faber & Faber, Ltd. London.)

thinner than the skin of the upper back, and its underlying superficial fascia is characterized by the presence of mammary glands.

The breast

In spite of its cultural significance, a breast is simply a group of modified sweat glands, packed in fat, supported by fibrous ligaments borrowed from local fascia, and covered with delicate skin. Undeveloped in the immature (prepubescent) girl, the breasts begin to enlarge in puberty under the guidance of estrogen, the female sex hormone. The increase in breast size is principally due to a pronounced increase in deposition of adipose tissue. Full development of the glandular and duct components does not take place until pregnancy.

It should be understood that the breast is a subcutaneous structure—superficial to the deep fascia of pectoralis major. This can be quickly verified by contracting the great pectoral muscle and noting how easily the breasts move over the muscle mass. A typical, well-developed breast extends from about the second to the sixth ribs and from the lateral border of the sternum to (and often into) the axilla.

Each breast contains about 15 to 20 glandular **lobes,** each packaged in generous quantities of fat (Fig. 4-30). The **ducts** of the lobes converge toward the apex of the breast, reaching the surface by way of a raised area of skin, the **nipple.** Supported by connective tissue and smooth muscle, the nipple is capable of erection (like the hairs of the skin) when touched or exposed to cold. The circular pigmented area of variable size about the nipple **(areola)** is characterized by a number of underlying sebaceous glands (Fig. 4-31). These are active in nursing (lactation) during periods of sucking and function to prevent painful cracking of the skin.

Strands of fibrous connective tissue infiltrate the breast between superficial and deep fasciae (Fig. 4-30), helping to suspend it. In time these ligaments are stretched and the breast tends to lose its shape. The wearing of brassieres is believed to prevent premature stretching of these ligaments.

The breasts function to provide nourishing milk for the newborn child. Thus it is not until pregnancy that the breasts fully mature, during which they may double in mass because of the proliferation of glands and ducts. The lymphatic and blood vessels increase in number, the areola darkens, and the breast becomes turgid with milk waiting to be expelled. After birth, the breast responds to the instinctive sucking of the infant, and milk is ejected by means of a straight-forward neuroendocrine reflex. Following the lactation period the breast returns to its normal size.

Blood is transported to the breast by branches of the subclavian, axillary, and intercostal arteries (Fig. 4-26). Drainage occurs by way of the corresponding veins. The pattern of lymphatic vessels of the breast is quite extensive, as a glance at Fig. 4-28 will verify. The lymph from these vessels flows into nodes principally located in the axilla. Every woman should be aware of the ever-present danger of malignancy in the breast and that prolonged enlargement of the axillary nodes when there is no infection present is a cause for concern. Gynecologists recommend routine palpation of the breasts for detection of abnormal growths or cysts. Because of the tremendous vascularity of the breast and its responsiveness to estrogen (a hormone which, among other things, stimulates cell division), metastasis may quickly follow development of a malignancy.

The nerve supply to the breast is derived from thoracic (intercostal) nerves 2 to 6. These are, of course, primarily sensory nerves, for the breast, particularly the nipples, is exquisitely sensitive to touch, which facilitates the expulsion of milk during lactation.

SHOULDER REGION

The shoulder region includes the acromion and neighboring scapular spine and lateral clavicle, the acromioclavicular and glenohumeral joints, and related muscles (see Fig. A-1, Appendix, for an x-ray view). The glenohumeral joint, related bursae, and nerves and vessels of the region will be discussed here, as well as the surface features.

The **glenohumeral joint**—a multiaxial, ball and socket articulation—is one of the most flexible joints of the body (Fig. 4-32). Its loose ligaments invest a fibrous joint capsule; the socket (glenoid fossa of the scapula) is, as we have mentioned, insecure (look back to Fig. 4-14). A lip of fibrocartilage about the perimeter of the socket helps retain the humerus. And tendons of a number of muscles further reinforce the connection. With numerous muscles approaching and leaving the humerus near the joint at various angles it is no wonder that a great deal of friction is generated in vigorous arm movements. The fact is that numerous bursae may be found in and around the shoulder joint where muscles tend to rub against one another (Fig. 4-33). These function to reduce the wear and tear of local muscles and tendons due to friction. Inflammation of these bursae (bursitis) and associated

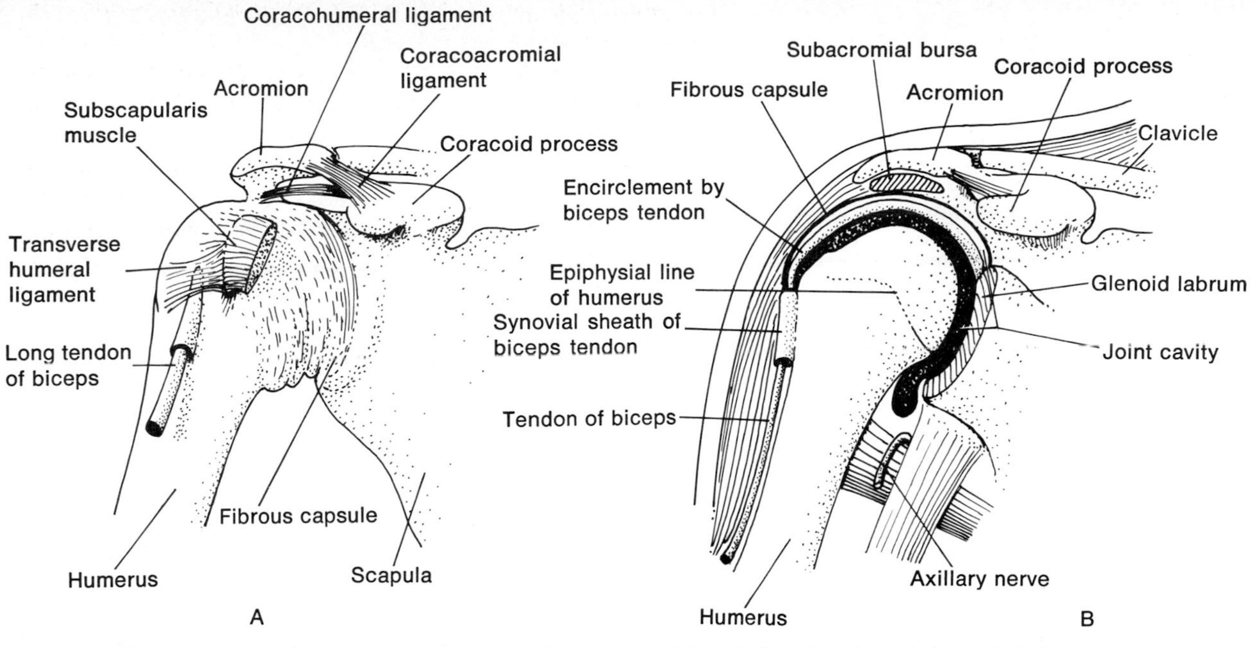

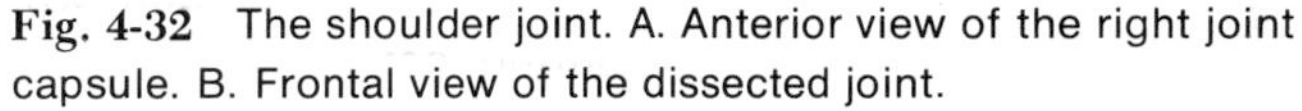

Fig. 4-32 The shoulder joint. A. Anterior view of the right joint capsule. B. Frontal view of the dissected joint.

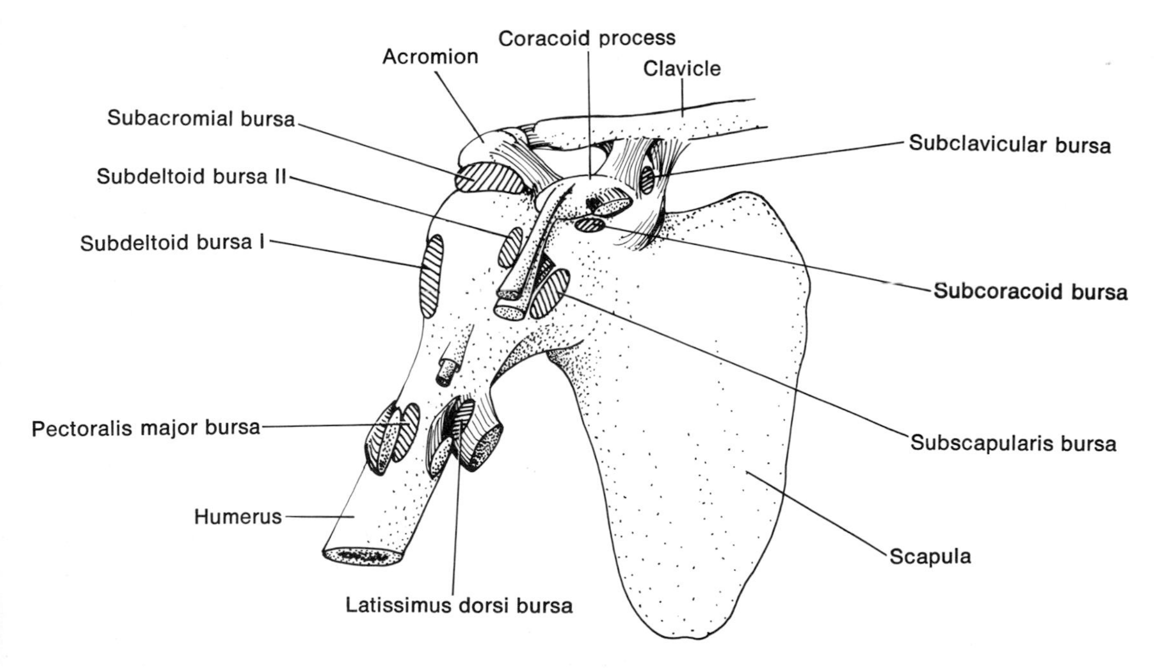

Fig. 4-33 Bursae around the shoulder joint.

tendons (tendonitis) frequently occurs with strained muscles among laborers and others who constantly put their shoulder joints under stress. In chronic cases it is not unusual to find torn tendons and passageways along which inflammation from bursae has spread to joint cavities. Then movement of the arm is very painful and permanent dysfunction may result. The "once-a-year-ball-game-at-the-high-school-reunion" athlete can often avoid muscle strain, bursitis, and torn tendons simply by "warming up" beforehand. Warm-up gives the circulatory vessels a chance to increase the blood flow to the muscles and tendons before they undergo the unaccustomed stress. How do you think prevention of bursitis may be related to increased blood flow in such instances?

Now feel your shoulder and note the bony prominence of the acromion and its proximity to the scapular spine and clavicle. Feel the full extent of the deltoid muscle and its neighbors, pectoralis major (anteriorly) and trapezius (posteriorly) (Fig. 4-15). Next press your fingers into the furrow between the pectoralis major and the deltoid about two fingersbreadth below the lateral aspect of the clavicle and dig hard for the coracoid process of the scapula (Fig. 4-2), which receives the tendons of the coracobrachialis, the short head of biceps brachii, and the pectoralis minor. Inflammation of these tendons is not uncommon. Traveling in the deltopectoral groove (Fig. 4-12) is a blood vessel—the **cephalic vein**—often large, sometimes absent, which is at times the site of a catheter insertion. This vein is one of two large superficial veins draining the hand, forearm, and arm. It will be discussed shortly.

About 1½ in. below the acromion on the deep surface of the deltoid, the **axillary nerve** and circumflex humeral artery/vein cross from posterior to anterior. Denervation of the deltoid (cutting of the axillary nerve) results in atrophy of the deltoid muscle mass and, in time, a visible shrinking of the shoulder, indicating the extent to which the deltoid gives it form.

REGION OF THE ARM

The **arm** is the region between the shoulder and elbow joints. In addition to the humerus it comprises an anterior compartment of principally elbow flexors, a posterior compartment of principally elbow extensors, and a medial neurovascular bundle. All of this can be felt on yourself (Fig. 4-17). Anteriorly, the biceps bulges into the hand. Just deep to the bellies of the biceps, lying atop the brachialis is the **musculocutaneous nerve.** This nerve supplies the muscles of the anterior compartment. Posteriorly, the more slender triceps muscle can be palpated all the way down the common tendon to the olecranon of the ulna. About one-third of the way down the humerus, the **radial nerve** wraps around the posterolateral humerus (Fig. 3-5) in a downward spiraling course en route to the brachioradialis muscle, under which it lies (Fig. 4-25). Thus in arms broken midway up the shaft or higher the radial nerve may suffer injury. In the lower, lateral aspect of the arm, the brachioradialis muscle can be palpated when the forearm is in the neutral position and the elbow joint is flexed against resistance (Fig. 4-19). Medially, at the upper arm, the coracobrachialis and short head of the biceps muscles converge toward the humerus. Immediately adjacent and posterior to the muscles is a neurovascular bundle consisting of the brachial artery and veins, and median, musculocutaneous, and ulnar nerves. The musculocutaneous nerve ducks between biceps and brachialis, and the **ulnar nerve** heads for the medial epicondyle (look back to Fig. 4-2) about one-third of the way down the arm. With your fingertips you should be able to roll the ulnar nerve over the medial epicondyle (the "funny bone").

The source of blood to the arm comes primarily from the brachial artery (whose pulse can be felt in the upper, medial aspect of the arm) and its branches (Fig. 4-26). The paired brachial veins course with the artery (Fig. 4-27).

The concavity at the anterior aspect of the elbow is known as the **cubital fossa** (Fig. 4-18). It is created by the lateral brachioradialis and medial common flexor muscle mass of the forearm and deepened by the tendons of biceps and brachialis diving for their bony attachments (check these things on yourself and in Fig. 4-18). The cubital fossa embraces the brachial vessels and median nerve en route to the forearm (Fig. 4-34). Unfortunately because of the fossa's aponeurotic roof—the bicipital aponeurosis—they cannot be palpated. This roof is an extension of the tendon of the biceps. It merges with the fascia overlying the flexor muscle mass of the upper forearm (Fig. 4-18). Flex your elbow against resistance and you will be able to feel the bicipital aponeurosis and the tendon of the biceps in the fossa quite well. You may also note a vein or two passing over the fossa in the superficial fascia. The one crossing from lateral to medial is probably the **median cubital vein,** a structure of some clinical significance as a site for intravenous injection. This vessel is easily tapped because it overlies the bicipital aponeurosis, which makes a nice firm platform for it when the elbow is extended.

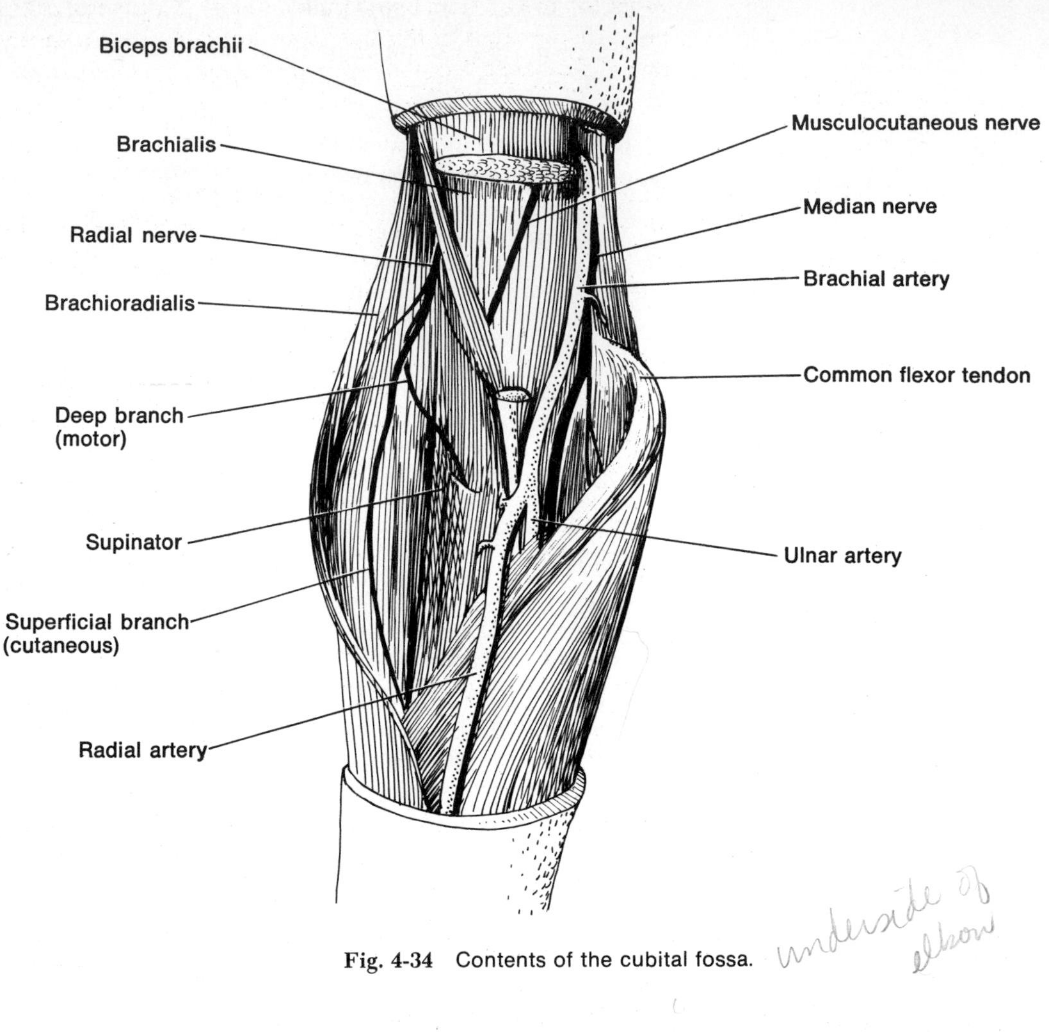

Fig. 4-34 Contents of the cubital fossa.

The **elbow** or humeroulnar joint is a simple uniaxial hinge joint between the olecranon of the ulna and the olecranon fossa of the humerus (Fig. 4-35). The medial and lateral epicondyles are easily felt on either side of the joint. Lateral to the humeroulnar joint the radius articulates with the capitulum of the humerus in a pivot joint, and forms another pivot joint, with the ulna called the proximal radioulnar joint (Fig. 4-8). These three joints share a common fibrous joint capsule, often in communication with local bursae. Pain and swelling in the elbow region are frequently signs of inflammation of these bursae—a common plight of miners and players of racket sports.

The neck of the radius is bound firmly to the radial notch of the ulna by an annular ligament (Fig. 4-35A), which also prevents the radius from being pulled out of its humeral articulation. The strength of this ligament is sometimes exceeded by overenthusiastic mothers who yank balky children from the candy counter by hand or wrist.

REGIONS OF THE FOREARM AND HAND

The **forearm** consists of two bones (radius and the more medial ulna) attached to one another at three places (proximal, intermediate, and distal radioulnar joints), packed tightly with muscle bellies and long sinuous tendons. The bones and intervening interosseous membrane divide the forearm into anterior (flexor) and posterior (extensor) compartments.

The anterior compartment consists of three muscular layers, with the larger vessels (ulnar and radial arteries and veins) and nerves (median and ulnar) coursing between the middle and deep layers. The taut deep fascia makes specific muscles difficult to feel, and nerves and vessels are too deep for palpation. At the anterior wrist, the ulnar artery's pulse can

Fig. 4-35 The elbow joint. A. Medial view showing important ligaments. B. Sagittal section through the humero-ulnar joint.

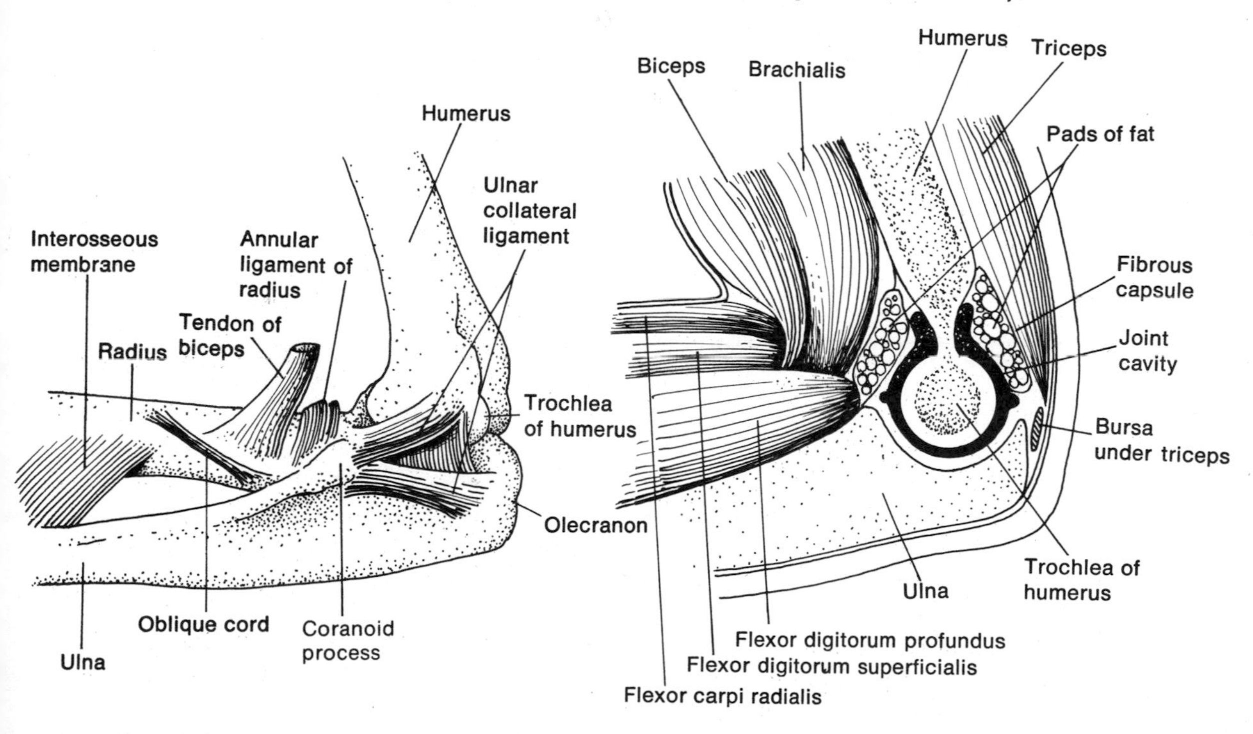

be felt on the ulnar side. Lying adjacent to the ulnar artery is the ulnar nerve. These structures pass through the forearm under cover of the flexor carpi ulnaris muscle (the ulnar nerve innervating that muscle and one-third of flexor digitorum profundus as well) and enter the palm of the hand superficial to the flexor retinaculum (Fig. 4-25). The median nerve surfaces from under the flexor digitorum superficialis to pass into the palm deep to the tendons of the flexor carpi radialis and palmaris longus (Fig. 4-25). With the exception of the two muscles innervated by the ulnar nerve, all muscles of the flexor compartment are served by branches of the median nerve.

The posterior compartment consists of two layers of muscle with the principal neurovascular elements lying between and deep to the two. The superficial layer of muscle is directed to the wrist and digits while the deep layer of muscle is concerned with the thumb and index finger (Fig. 4-19B). The **radial nerve** enters the forearm under the brachioradialis, slips under the supinator and terminates in the extensor muscles. Since there are no extensor muscles in the hand, it follows that the radial nerve will not continue into the hand except for cutaneous branches.

The radial nerve and its muscles are critical elements in normal functioning of the hand, for the wrist must be extended if the fingers are to be utilized with any significant strength. See this for yourself: Note the difference in the strength of your grip (fingers curled) with the wrist flexed and extended. Loss of the radial nerve at the elbow or higher results in "wrist drop" with an inability to extend the wrist as when showing your fingernails to an interested party. The blood supply to the extensor compartment is derived from the radial artery and the common interosseous artery (Fig. 4-26), the former passing toward the wrist under the brachioradialis, the latter providing branches on either side of the interosseous membrane. The radial artery can be located by feeling its pulse at the wrist, flexor aspect, radial side (just medial to the tendon of the brachioradialis—Fig. 4-18A).

The tendons at the wrist cannot be easily felt on the extensor surface but they can be palpated easily on the flexor surface. Just distal to the wrist, the tendons projecting to the thumb on the extensor surface are easily seen, particularly when the thumb is abducted and extended (Fig. 4-21).

A triangular hollow (the "anatomical snuffbox") can be seen between the long and short tendons of extensor pollicis (Figs. 4-19 and 4-21). It is said that this hollow made an excellent depository for snuff about to be inhaled. The floor of

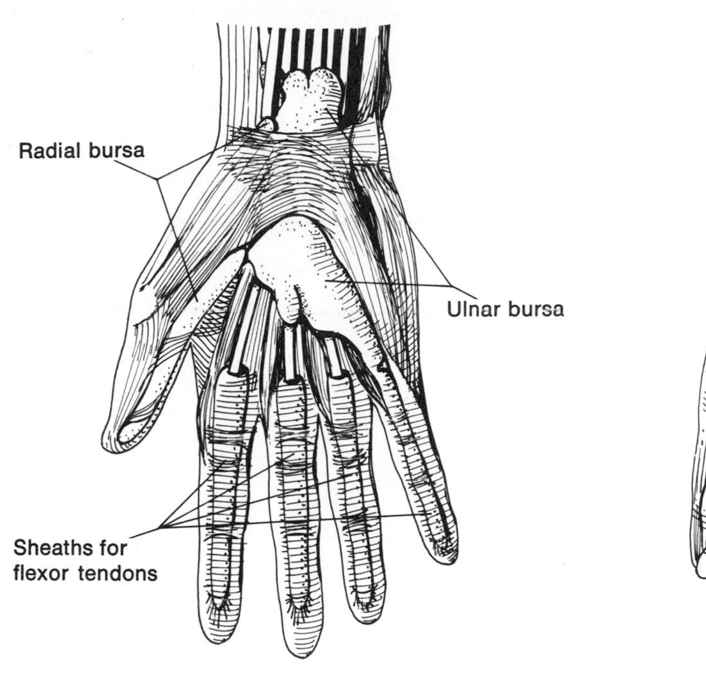

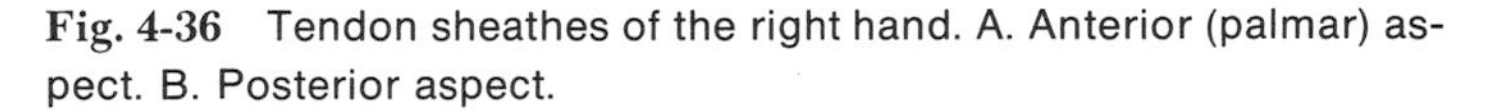

Fig. 4-36 Tendon sheathes of the right hand. A. Anterior (palmar) aspect. B. Posterior aspect.

the "snuffbox" accommodates the radial artery in its trek to the extensor surface of the hand. Since the artery dives deep here, it cannot be palpated without difficulty. Circumduction of the thumb allows you to see the tendons in this area.

As the tendons of the forearm cross the very mobile wrist, they pick up synovial sheaths as a defense against the trauma of friction when sliding over the carpal bones (Figs. 3-27 and 4-36). From a clinical standpoint, the anterior sheaths are more significant than the posterior ones. This is so because penetration wounds occur more frequently on the palmar surface and because of the extensive synovial tunnels about the digital flexor tendons. Not infrequently these long sheaths communicate with the larger radial and ulnar bursae; should infective material be introduced at any point in these sheaths, a generalized infection of the whole hand may follow, resulting in functional impairment of the hand.

UPON REFLECTION

The ultimate anatomical tool is the hand; note the precision with which it conducts its business. Magnificently manipulated by the nervous system, the hands provide the means by which we play piano concertos of Rachmaninoff, create beautiful woodwork, tool a precision valve from a hunk of metal, an perhaps best of all—make new friends with a simple handshake.

The scapula, clavicle, humerus, radius, ulna, carpus, and their associated musculature are the servants of the hand; their architectural arrangement you can now appreciate. In the days before you, heady with the knowledge just learned, reflect on how the structure of your upper limb lends itself to the task it performs.

REFERENCES

1. Anson, B. J. (ed.): *Morris's Human Anatomy,* 12th ed., McGraw-Hill Book Co., New York, 1966.
2. Goss, C. M. (ed.): *Gray's Anatomy of the Body,* 28th ed., Lea and Febiger, Philadelphia, 1966.
3. Lockhart, R. D.: *Living Anatomy,* 6th ed., Faber & Faber, Ltd., London. (A singular atlas of photographs of muscular activity and bony landmarks.)
4. Royce, J.: *Surface Anatomy,* F. A. Davis Co., Philadelphia, 1965. (Excellent photographs of the surface anatomy of the body. Shows changes in body proportion with age.)
5. Wells, K. F.: *Kinesiology,* 4th ed., W. B. Saunders Co., Philadelphia, 1966. (Outstanding source of information on muscle action, both academically and in relation to physical activity.)

clinical considerations

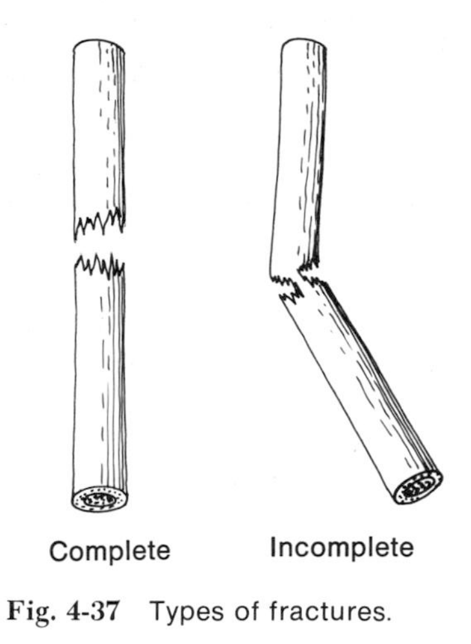

Fig. 4-37 Types of fractures.

FRACTURES OF THE CLAVICLE AND HUMERUS

A *fracture* is simply a break in a periosteum-lined bone. The physician usually called upon to reduce* a fracture is an orthopedist†. Fractures are caused by trauma or disease (cancer, infection, or metabolic disorders). Complications from fractures are generated when surrounding soft tissues or neurovascular structures are penetrated. Fractures which break the skin are called compound or open; and simple when the skin is not broken. Some fractures involve complete breaks in the bone (complete fracture); some are breaks in the bone with the bone still in one piece (incomplete fracture) (Fig. 4-37). Within each category just noted, there are variations in the appearance of the fracture—some are impacted (telescopeal), others are comminuted (broken into several fragments), etc.

Following a fall or accident in which a fracture occurs, the surrounding tissues are torn and bleeding (hematoma) and leakage of tissue fluid occurs. The former is indicated by a blue-black area appearing over the fracture (ecchymosis; G. *chymos*, juice). The region concerned becomes swollen with fluid (edema). With movement, there may be pain and a grating sound of the broken ends (crepitation). At such times it is important to restrain movement in order to diminish further soft-tissue injury. X-rays will usually positively identify the type of fracture. Knowing the soft-tissue anatomy about the fracture will help you judge the severity of injury.

In the case of the bones of the upper limb, as well as others, muscles will pull on each of the fragments (angulation) and often cause overriding (shortening). In such instances, neurovascular structures can become stretched or torn with further movement, if they have not been damaged already.

Fracture of the clavicle

Assume a clean break (transverse, simple fracture) has been confirmed by x-ray. Characteristically, the proximal fragment will be pulled upward by the sternocleidomastoid muscle of

* *Reduction*, to restore to normalcy.

† *Ortho*, straight; *pedic* from *pais*, child. An orthopedist formerly attended the musculoskeletal deformities of children but now cares for all diseases and injuries of the musculoskeletal system.

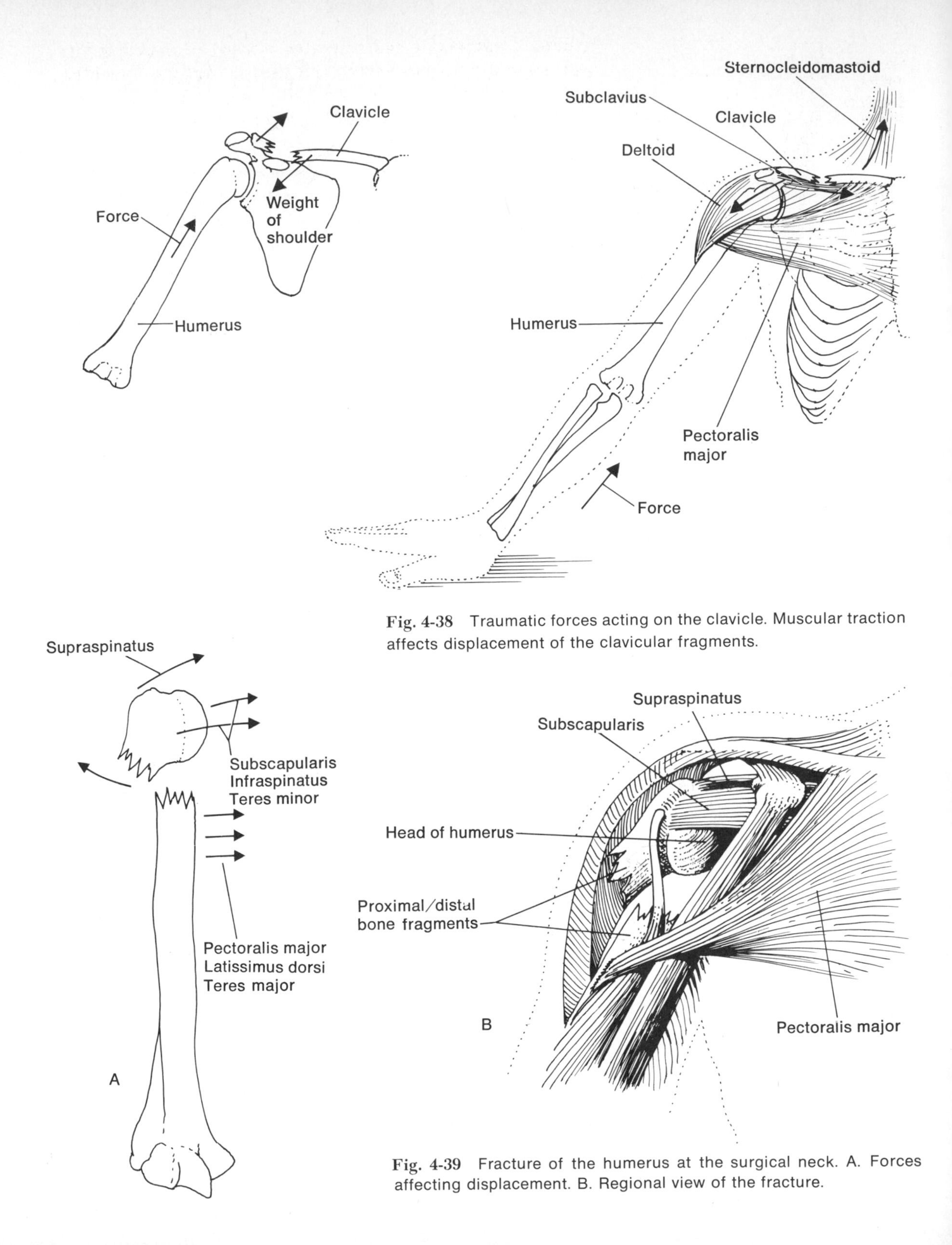

Fig. 4-38 Traumatic forces acting on the clavicle. Muscular traction affects displacement of the clavicular fragments.

Fig. 4-39 Fracture of the humerus at the surgical neck. A. Forces affecting displacement. B. Regional view of the fracture.

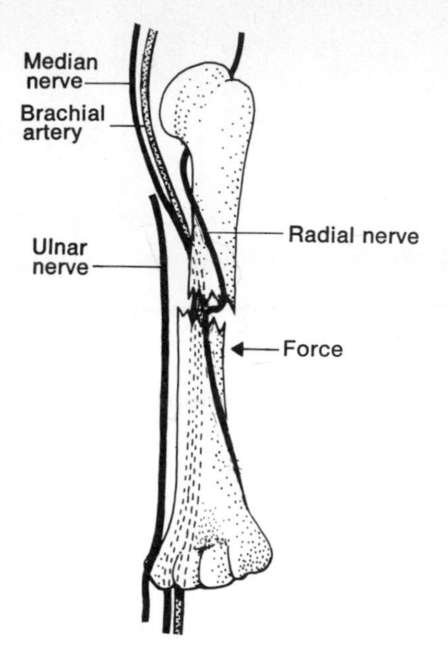

Fig. 4-40 Course of neurovascular structures along the posterior and anterior shaft of the humerus. Note how the radial nerve is endangered at the fracture site. Intervening muscle buffers the brachial artery and the median and ulnar nerves against the fracture.

the neck, apparently overriding the traction of the clavicular portion of the pectoralis major. (Turn your head and feel the anterolateral neck on the opposite side—this is the sternocleidomastoid.) The distal fragment will be pulled downward by the deltoid (Fig. 4-38). What prevents this distal fragment from rupturing the subclavian vessels under the clavicle? A small, rather insignificant muscle, the subclavius, which cushions the vessels and absorbs piercing fragments.

Fracture of the humerus

In general, the fragments of a fractured humerus assume characteristic positions because of muscular traction. If the fracture is at the surgical neck, the proximal segment will be lifted up and out (abducted) by the rotator cuff muscles (Fig. 4-39). The distal fragment will tend to be pulled medially by the muscles inserting just below the break (latissimus dorsi, teres major, and pectoralis major). What neurovascular structures might be involved in such a break?* How would such involvement be manifested?

Consider a fracture of the humerus along the shaft at almost any point and the question of radial nerve involvement becomes relevant (Fig. 4-40). If the nerve was severely damaged, one sign would be a loss of extensor tone at the wrist, with wrist drop. Consider what muscles would affect displacement of the humeral fragments if the fracture took place at:

1. The upper third of the humerus above the deltoid tuberosity and below the intertubercular groove.
2. The level of the deltoid tuberosity.

In the case of a fracture at the supracondylar ridge (Fig. 4-41), the neurovascular structures entering the cubital fossa may be damaged. How would such damage show up?

* Consider: the radial nerve, cephalic vein, circumflex humeral arteries, etc.

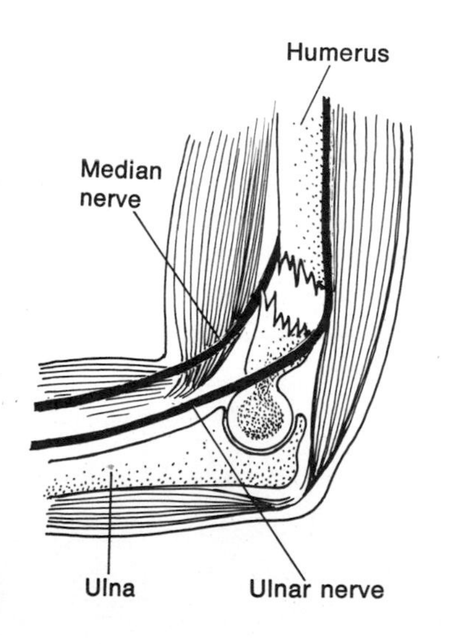

Fig. 4-41 Involvement of ulnar and median nerves and brachial artery in a supracondylar fracture.

unit 5 the lower limb

osteology

The bones of the lower limb are the rigid framework for the hips, thighs, legs, and feet. They are part of the appendicular skeleton. Many have counterparts, or homologues, in the upper limbs. But, unlike the bones of the upper limbs, these bones must support the entire body and provide it with a means of locomotion. Their structure reflects these tasks. Compare the general structure of lower limb bones and upper limb bones in Fig. 5-1. Try to identify the differences resulting from the difference in function.

The homologue of the pectoral girdle is the **pelvic girdle,** the incomplete circle of bone which supports the vertebral column directly and the body trunk, head, and upper limbs indirectly. The pelvic girdle is composed of two hip bones (right and left) which articulate with one another anteriorly and with the sacrum posteriorly (Fig. 5-3). The pelvic girdle and the sacrum with coccyx make up the circular wall of the

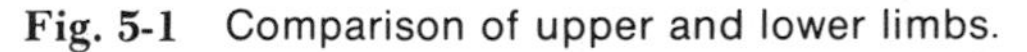

Fig. 5-1 Comparison of upper and lower limbs.

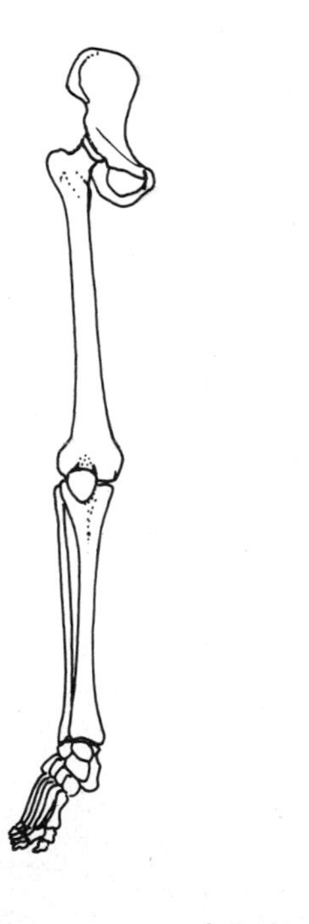

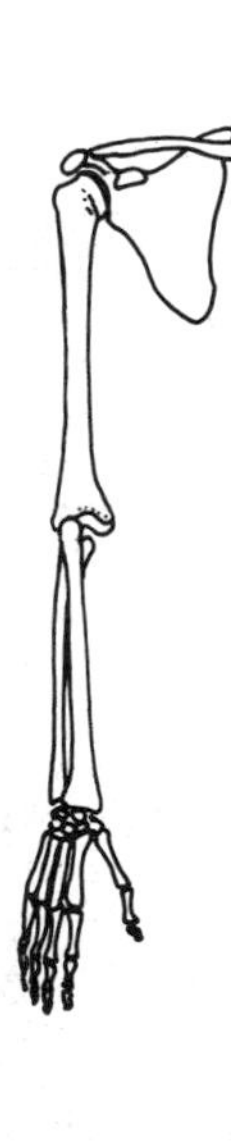

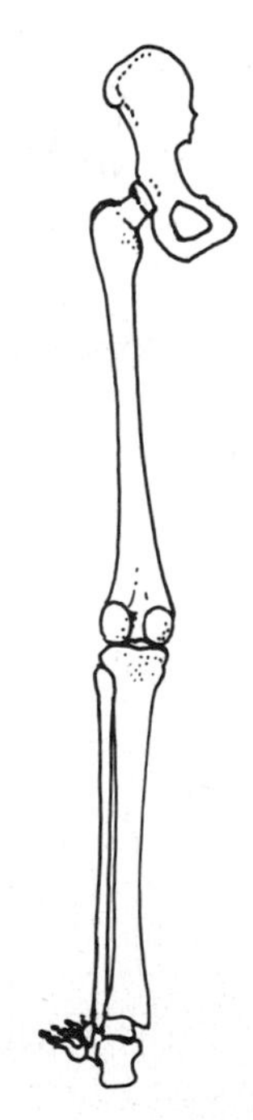

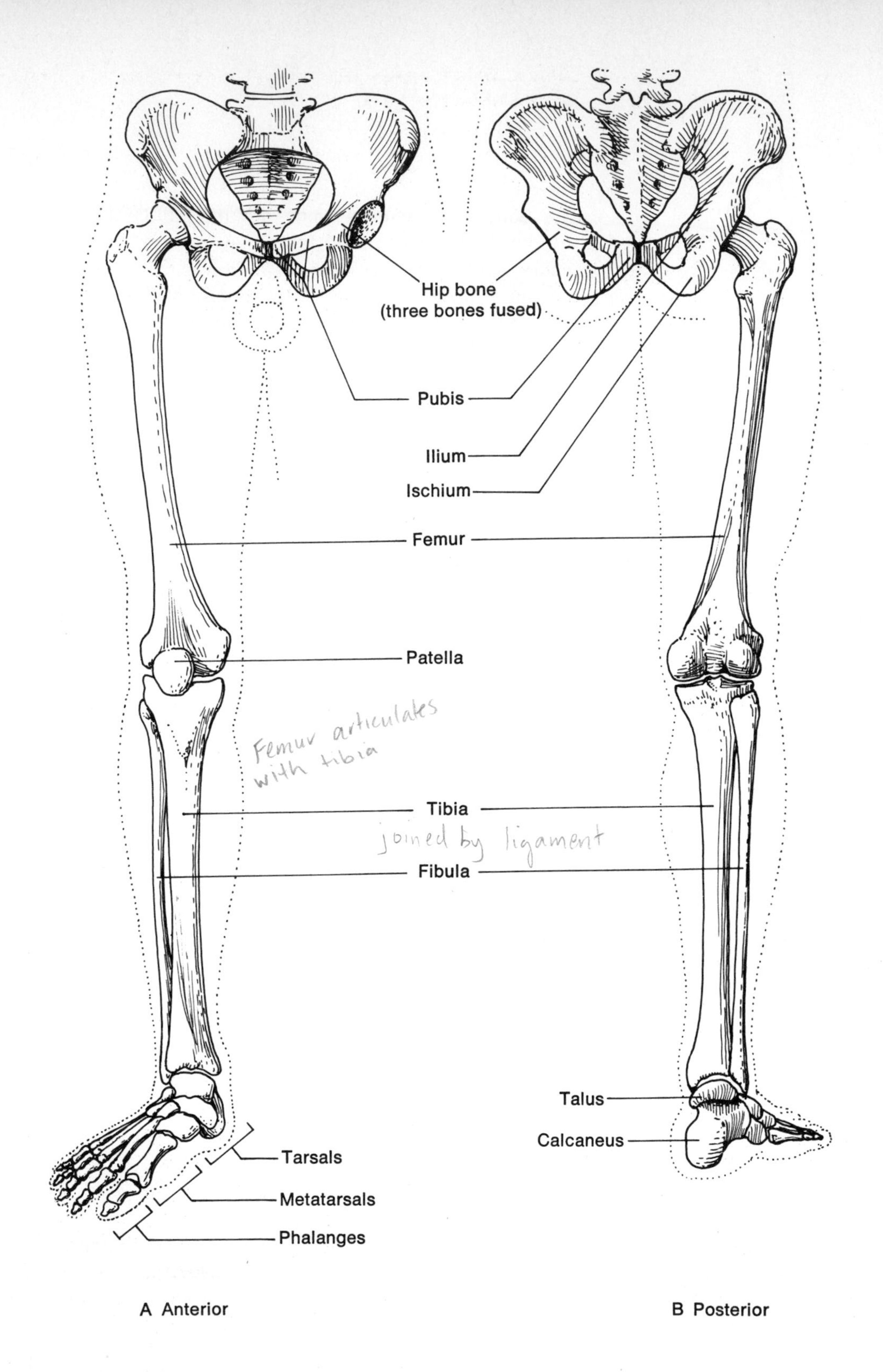
Hip bone
(three bones fused)
Pubis
Ilium
Ischium
Femur
Patella
Femur articulates
with tibia
Tibia
joined by ligament
Fibula
Talus
Calcaneus
Tarsals
Metatarsals
Phalanges
A Anterior
B Posterior

pelvis, which it may help to think of as a basin which retains pelvic viscera.

Study Fig. 5-2 as you read:

The pelvis, via the sacrum, articulates with the 5th lumbar vertebra above and the coccyx below. Anterolaterally, each hip bone articulates with a thigh bone, the **femur.**

The femur articulates with the larger bone of the leg, the **tibia,** as well as the largest sesamoid bone of the body, the **patella.** This complex three-bone arrangement among the femur, patella, and tibia is the **knee joint.**

The tibia articulates with the femur above, the talus below, and the **fibula** at both ends and, by means of a ligament, throughout its length. The tibia and fibula support the soft structures of the leg; the latter bone is long and slender and attaches to the tibia above and talus below.

The **tarsus,** composed of seven bones, supplies the bony support for the ankle and posterior half of the foot. It articulates with the tibia and fibula (via the talus) proximally and the five metatarsals distally.

Each metatarsal articulates with a phalanx distally. As in the hand, there are three phalanges in each digit, except for the first digit, which has two.

HIP BONE (see Fig. A-4, Appendix, for x-ray view)

Each hip bone (L., *coxa,* angle, or hip) articulates with the sacrum posteriorly at the sacroiliac joint and with the other hip bone anteriorly at the pubic symphysis (Fig. 5-3).

Until now we have discussed bones which generally lie in a vertical or horizontal plane. A quick look at the skeleton might fool you into thinking that the pelvis is oriented along these two planes the way the main axis of the body is in a standing position. But Fig. 5-4 shows you that the plane of the upper opening, that is, the **pelvic inlet,** is tilted about 60° from horizontal from posterior to anterior. An appreciation of this orientation is fundamental to an understanding of the following points about the pelvis and its neighbors:

1. The abdominal and upper pelvic cavities are without a bony anterior wall.
2. The pelvic tilt compensates for the lumbar curvature of the vertebral column.
3. The pelvis can handle the weight of the body more efficiently with this orientation.
4. There is a tendency for the 5th lumbar vertebra (or its disc) to slip forward on the upper surface of S1. When this happens the condition is called spondylolisthesis.

Fig. 5-2 (opposite) Bones of the pelvis and right lower limb.

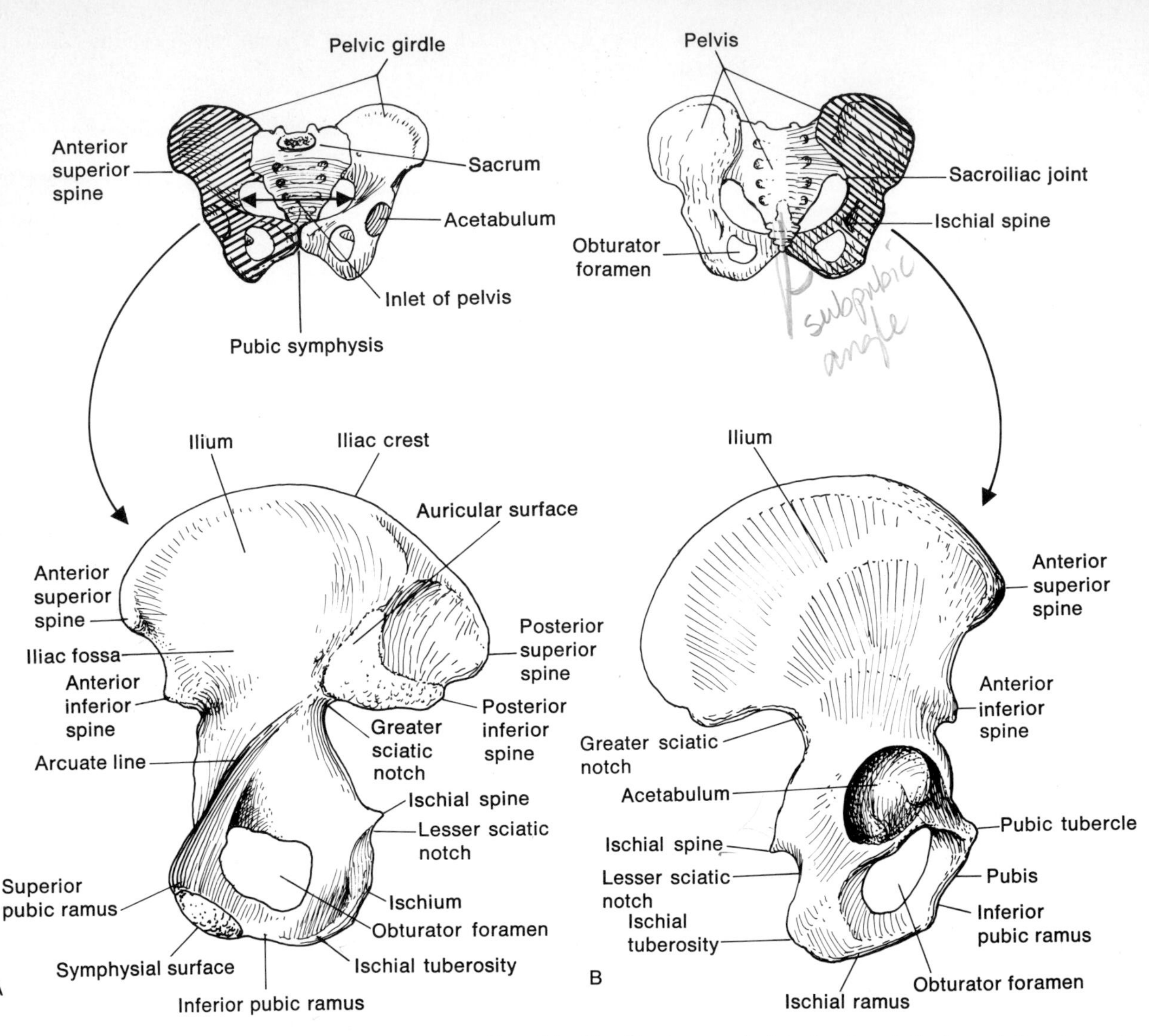

Fig. 5-3 The hip bones. A. Interior (medial) view of the right hip bone. Note its position relative to the articulated pelvic girdle. B. Exterior (lateral) view of the right hip bone. Note its position relative to the articulated pelvic girdle above.

Each hip bone is actually a fusion of three separate bones, the union not taking place until about the age of 14 (Fig. 5-5). The uppermost bone, characterized by a broad ala (wing) is the **ilium.*** It articulates with the other two bones to form the

* "Ilium" sounds like "ileum," but there the resemblance ends. The ileum is part of the gastrointestinal tract.

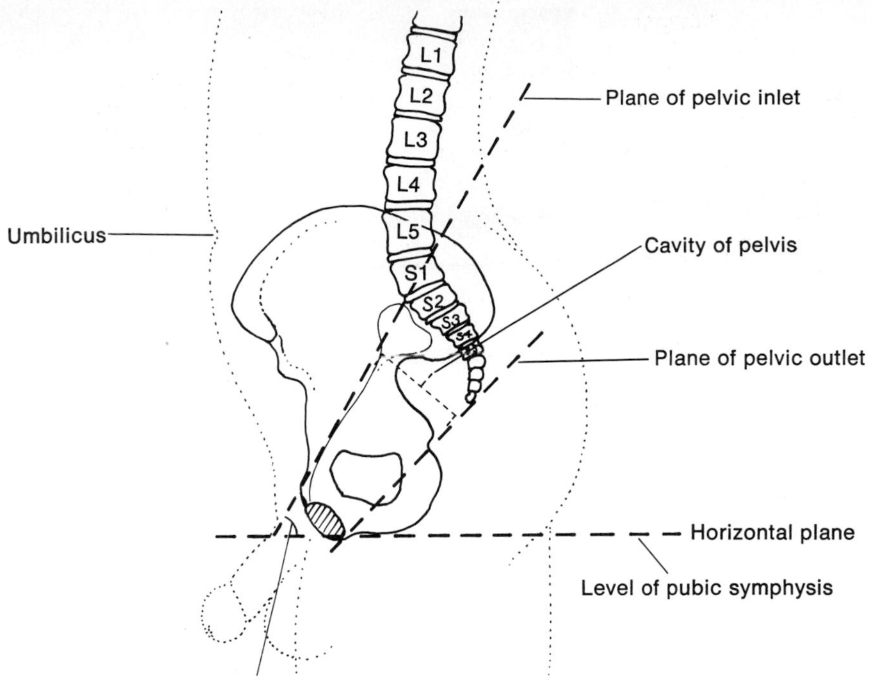

Fig. 5-4 Orientation of the pelvis. The angle of the pelvic inlet from the horizontal plane is about 60°.

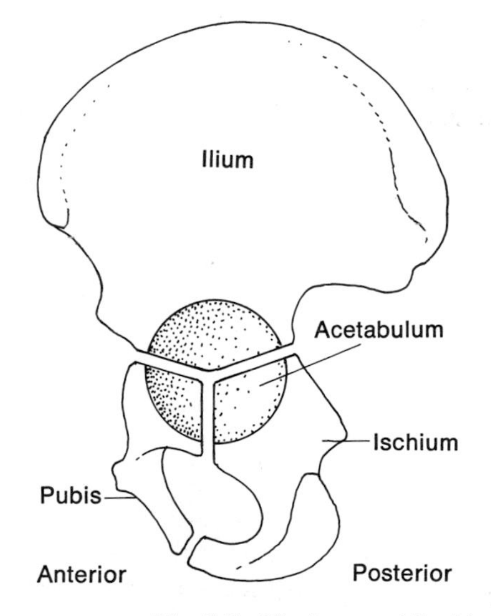

Fig. 5-5 The bones of the hip and their site of fusion, the acetabulum (shaded).

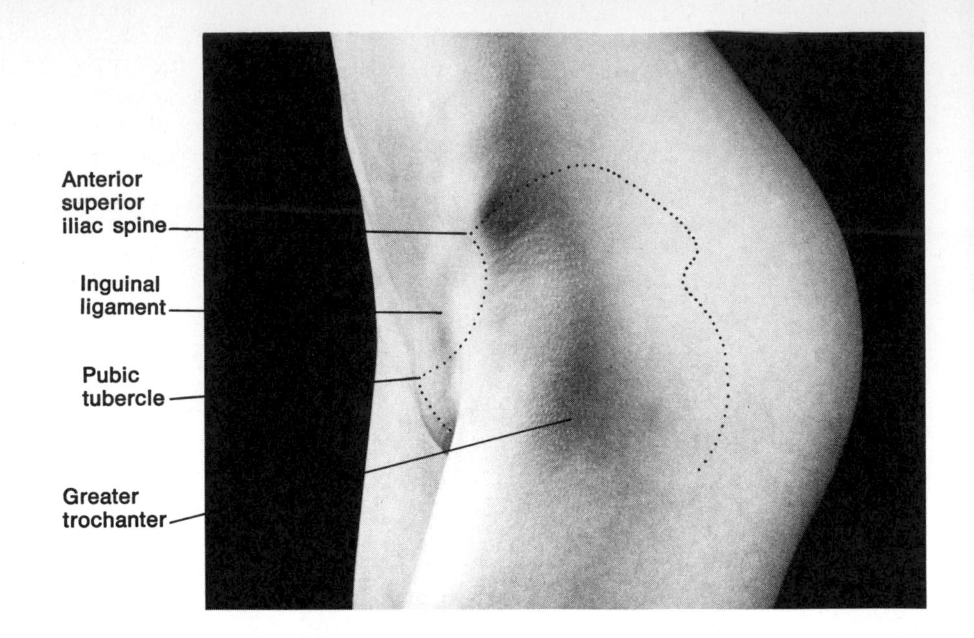

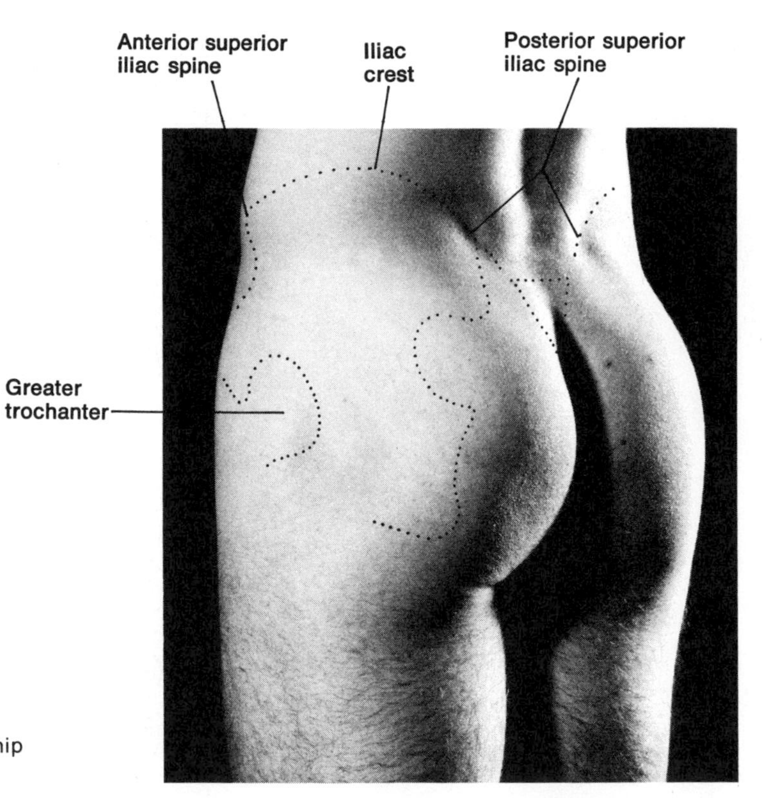

Fig. 5-6 Surface features of the hip bones.

socket, or **acetabulum,** for the head of the femur. The anteriormost of the two is the **pubis;** posteriormost is the **ischium.** Once the centers of ossification within the pubis, ischium, and ilium join, the bone becomes one, just as the union of the diaphysis and two epiphyses of the humerus make one bone.

Most of the pelvic girdle is covered with muscle, but some features can be palpated. Place your hands on your hips and you will feel a ridge of bone, the **iliac crest.** (Refer to Figs. 5-3 and 5-6 as necessary.) Trace this crest anteriorly to the point of a drop-off. This point is the **anterior superior iliac spine,** an important attachment point for anterior thigh muscles. Inhale and hold your breath; now you may be able to see the anterior superior spines. Next trace the iliac crest posteriorly to the **posterior superior iliac spine.** Using a mirror, you can probably see these spines as dimples in the skin, at about L5. Now trace around to the front of your body at the L5 level and note that these posterior spines are as high or higher than the anterior spines. The iliac crests are at about L4, about the level of your navel, or umbilicus.*

Now feel laterally and posteriorly just below the iliac crests. The gluteal muscles covered with fatty superficial fascia and skin may be felt there.

Next place your fingers on an anterior superior iliac spine and trace horizontally to the anterior midline of your body. Press in and note: no bone. Now move your finger inferiorly until you touch bone. This is the pubic bone. Note its relationship to the anterior superior spines of the ilium. Now refer to Fig. 5-4, where the tilt of the pelvis may be seen. With some difficulty the pubic bone can be traced by hand to the ischium. Here, deep to the buttock, the bone becomes enlarged on each side in an ischial tuberosity. It is on these tuberosities that you sit (Fig. 5-7). An extensive bursa is associated with each tuberosity, and bursitis in the area is not uncommon in people who spend much time sitting. (Strangely, students seldom suffer from it!) Again with difficulty, the ischial tuberosities can be traced posteriorly to the sacrum, deep to the crease of the buttocks, the gluteal crease. This region around the ischial tuberosity and the sacrum contains important muscles and ligaments and structures passing into the thigh from the pelvis.

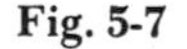

Fig. 5-7

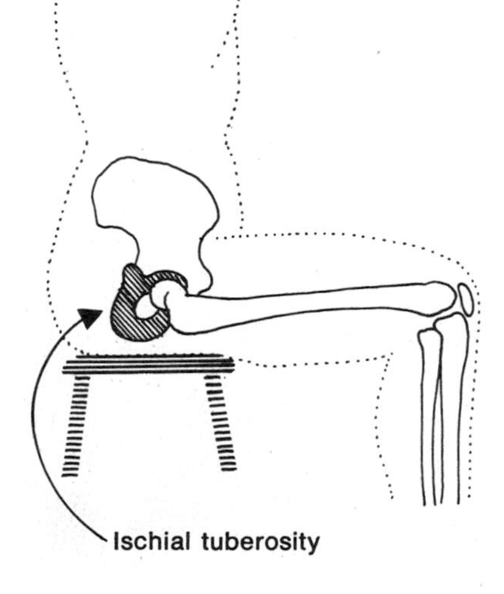

* The relationship of the navel, the iliac crest, and the anterior superior spine is clinically significant. A line drawn from the navel to the right anterior superior iliac spine may cross the appendix about two-thirds of the way. In appendectomies the surgeon will often make the incision 90° to this line at that point.

The pelvic girdle provides support for the pelvic viscera (bladder, rectum, internal organs of reproduction, and associated vessels, nerves, and ducts) and the perineal structures as well (the perineum lies between the thighs and includes the anus and external reproductive organs). The pelvic girdle of females, therefore, must be able to accommodate the developing child during the first 9 months of its life *and* allow its exit from the uterus within the frame of the pelvic

Fig. 5-8 Surface differences in hip structure between males and females. Note in the female: (1) accentuation of the ilia, resulting in wider hips, (2) more acute angle between neck of femur and shaft as manifested by a greater inward slant of thighs from hips to knees. Note in the male: the greater length of the lower limbs prevents him from lowering his head over his knees. The length of both trunks appears the same.

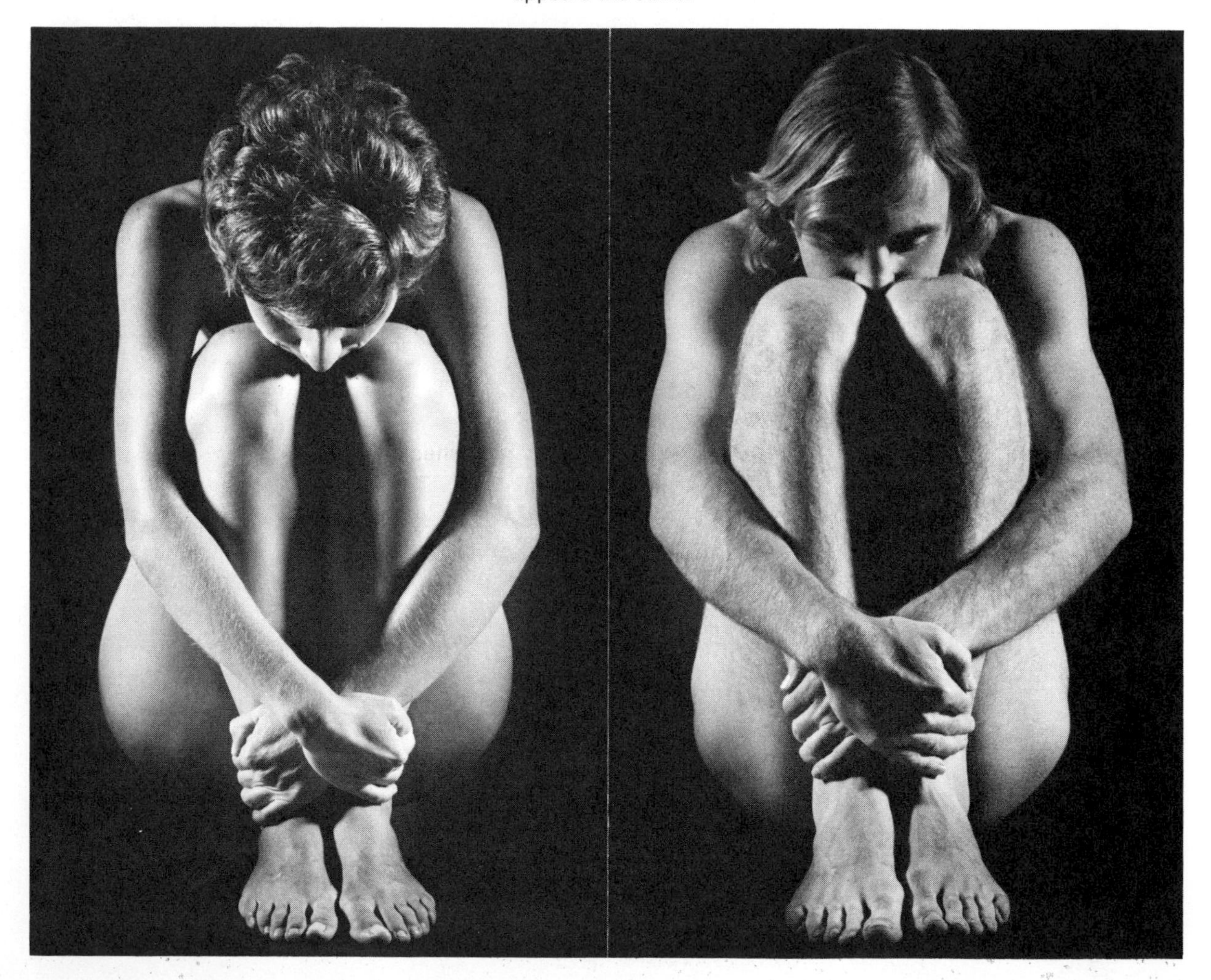

outlet. Since males are without the essential apparatus for such activity, you might suspect that male pelves differ from female pelves. These differences are of course reflected in differences in "typical" male and female body form about the hips (Fig. 5-8), as "girl watchers" and "boy watchers" know intuitively.*

The sacrum and coccyx will be considered with the vertebral column (Unit 7).

FEMUR

The **femur** (L., thigh bone) is the longest bone in the body. It is the bony framework of the thigh and articulates with the acetabulum above and the tibia and patella below. Some of the characteristic features of the femur (Fig. 5-9) can be felt on your own body. Note the unusual neck protruding laterally from the head. The neck ends laterally at a large protuberance, the **greater trochanter,** a term implying that there must be a **lesser trochanter** . . . and there is. Now place the palms of your hands on your iliac crests (lateral aspect) with fingers extending downward. Your distal phalanges should be touching a bony prominence just under the skin and fascia—the greater trochanter (Fig. 5-9). The distance between left and right greater trochanters is greater in females than males.† When one is standing, a horizontal line passing medially from the greater trochanter goes through the center of the hip joint, a ball and socket, synovial joint.

The shaft of the femur is completely clothed in muscle and cannot be palpated. At the distal extremity of this bone the adductor tubercle can be felt medially just above the medial condyle of the femur (Fig. 5-9). The adductor tubercle is an important attachment point for one of the adductors of the thigh. Feel also for the condyles of the femur and note that they can be palpated only medially and laterally. Why not anteriorly?

The large neck of the femur forces the upper shaft of the bone laterally, and so the femur is angled medially from above to below when one stands erect, with knees together. The angle between the neck and the shaft of the femur differs

* Differences in distribution of body fat in superficial fascial spaces are also important factors (see Fig. 3-25).

† For an interesting, simple experiment each member of the class should measure this intertrochanteric distance and bring the data to the instructor the next day. The average measurements for males and for females should be figured and compared. What significance does this measurement have?

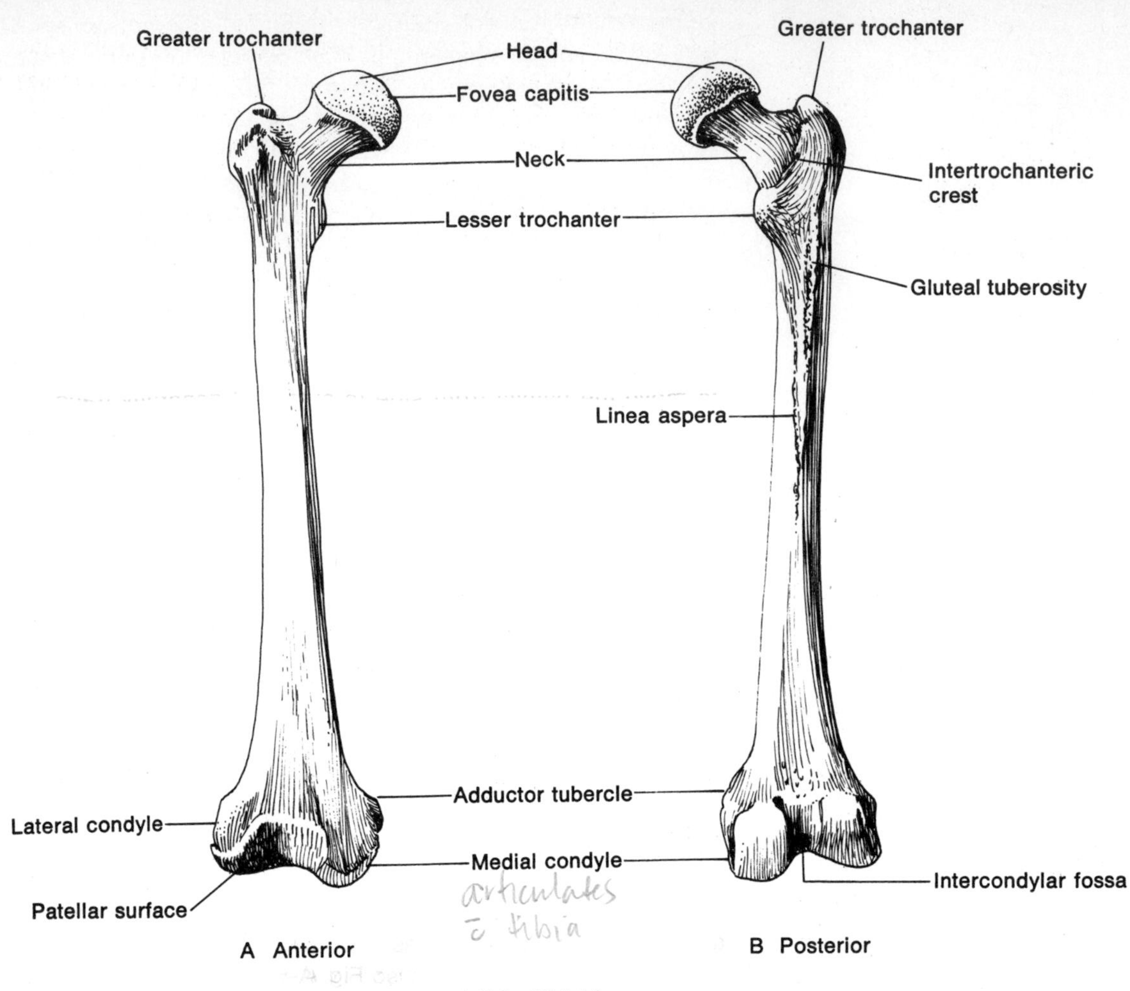

Fig. 5-9 Right femur.

from person to person. In some people in whom the angle is abnormally large, an unusual manner of walking, or gait, is the result. Stand erect with feet together and run your hands downward from the lateral hips to the lateral aspect of the knee joints. The angle formed between the neck and the shaft of the femur is less in females than males (Fig. 5-8). Why? How does this show up in a woman's gait?

After studying Figs. 5-1 and 5-9 name some of the like and unlike features of the following structures, and explain them in functional terms:

1. The head of the humerus and the head of the femur.
2. The neck of the humerus and the neck of the femur.
3. The condyles of the humerus and the condyles of the femur.

PATELLA (see Fig. A-5, Appendix, for x-ray view)

The **patella** (L., small pan), or knee cap, is the largest sesamoid bone of the body (Fig. 5-10). Early in fetal development cartilaginous tissue is formed in the tendon of the quadriceps femoris, which crosses the knee joint to insert on the tibia. By puberty (age 12 to 14 years), the patella has ossified. In the evolutionary sense the patella probably developed in response to the trauma of friction generated by the movement of the tendon over the femur (and tibia). The patella articulates only with the femur.

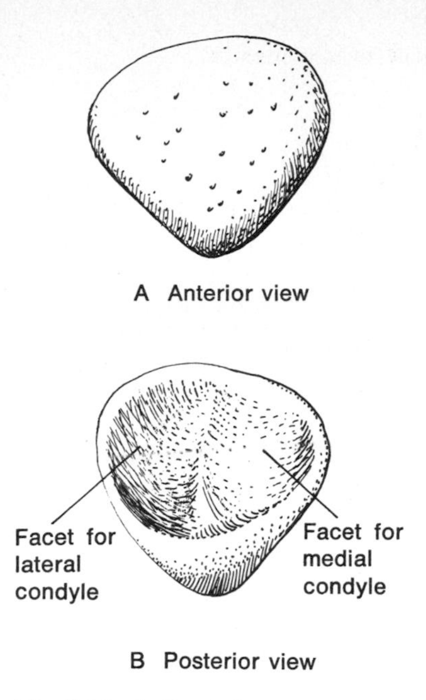

Fig. 5-10 Left patella.

You can easily feel the patella. While sitting extend your knee joint, resting your foot on the floor. You should be able to move the patella from side to side and generally trace its entire anterior surface. Note the loose mass just distal to the patella. Keep your hand on it as you slowly flex your knee. That "cord" which takes shape as you stretch it is the patellar ligament, the distal part of the tendon of the quadriceps femoris. Notice that the patella is immobile when the knee is flexed as well as when it is extended against resistance, because the quadriceps femoris—whether stretched or contracted—holds it in place. The patella gives added mechanical advantage to the quadriceps in its job of extending the knee.

TIBIA

The **tibia** (L., musical pipe), popularly called the shinbone, is the second largest bone of the body and corresponds to the ulna of the forearm. The tibia articulates with the medial and lateral condyles of the femur above, with the fibula throughout on the lateral aspect, and with the talus of the ankle below (Fig. 5-11; see also Fig. A-5). The stoutness of the tibia is related to its function, support of the body.

Much of the anterior tibia can be palpated. With the knee flexed, the **medial condyle** can be felt, as can the entire anteromedial surface of the shaft. Note how subcutaneous, that is, how close to the surface, the tibia really is. At the upper end of the shaft you can feel the **tuberosity of the tibia,** which receives the patellar ligament. The anteromedial surface of the tibia terminates distally as the very prominent **medial malleolus.** What function might this projection (and its fellow on the lateral side) serve for the tendons from the leg that pass around them and into the foot? Your answer should be: "a pulley."

FIBULA

The **fibula** is a narrow sticklike bone occupying the lateral side of the leg and corresponding to the radius of the forearm

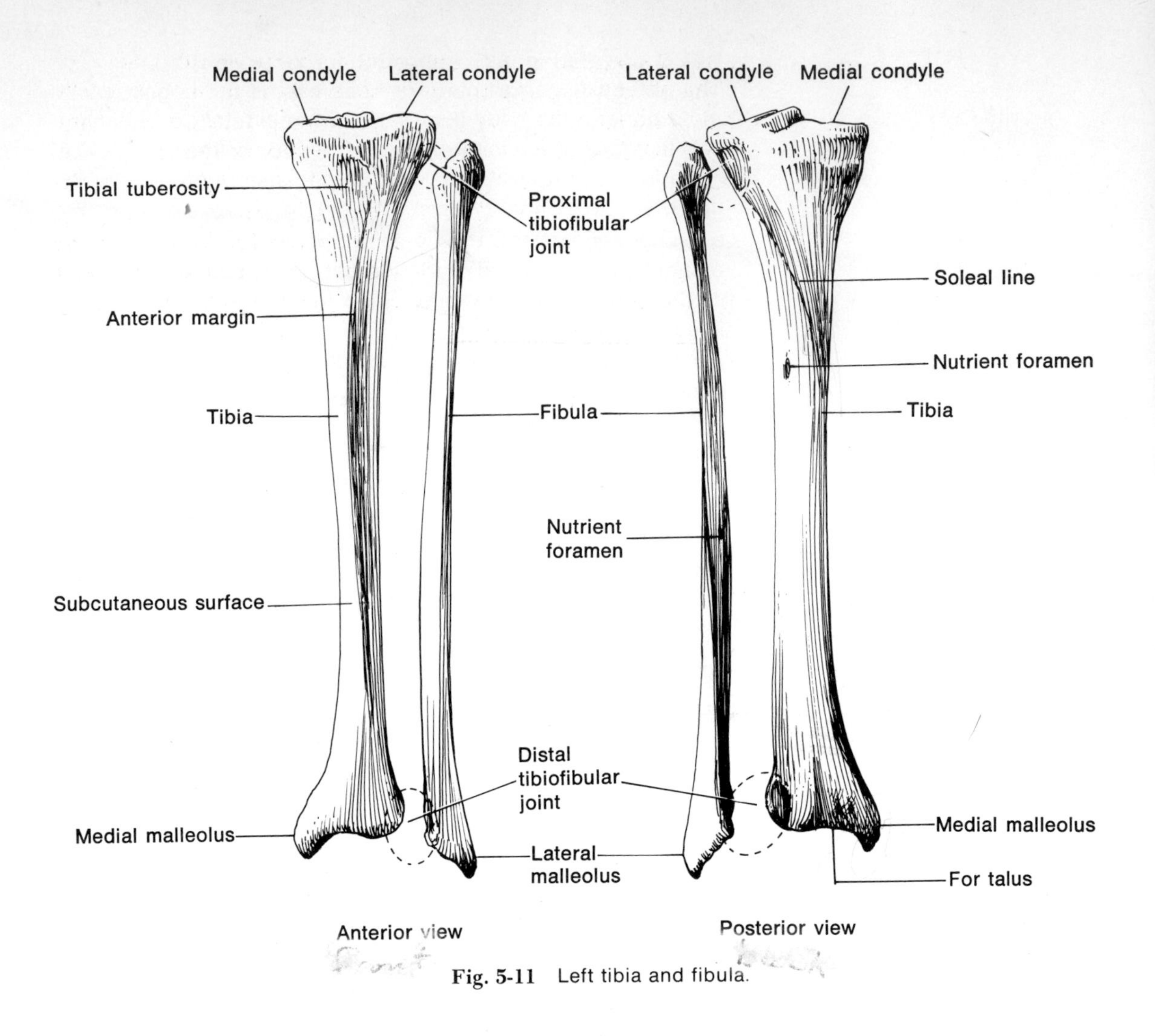

Fig. 5-11 Left tibia and fibula.

(Fig. 5-11). The fibula not only serves as an important attachment point for muscles but acts as a kind of strut in support of the ankle joint.

BONES OF THE FOOT (see Fig. A-6, Appendix, for x-ray view)
If you place your hand palm down next to the corresponding foot you will see that the bones of the two structures are roughly alike. The bones of the foot consist of the tarsal bones, the metatarsals, and the phalanges. Approximately one-half of the posterior foot is supported by the tarsal bones (Fig. 5-12), and you can identify some of these by palpation.

The **talus** may be felt about three fingersbreadth below the medial malleolus by approaching the ankle from behind, with your thumb palpating laterally and your fingers palpating medially. Moving forward from below the malleolus about an inch, you will feel a projection on the medial surface: the tuberosity of the **navicular** bone. Feel also the bone of the heel, the **calcaneus.** Study the relationships of these three bones in Fig. 5-12. The remaining four tarsal bones, all in a line from lateral to medial, are the **cuboid** and three **cuneiforms.** Distally they articulate with the metatarsals, the latter being easily palpated from the dorsal surface.

ARCHES OF THE FOOT

While sitting down with shoes and socks off, you should easily be able to feel an arch on the medial side of the foot between the floor and the plantar surface (Fig. 5-13). This is one of the components of the longitudinal arch of the foot,

Fig. 5-12 Skeleton of the right foot.

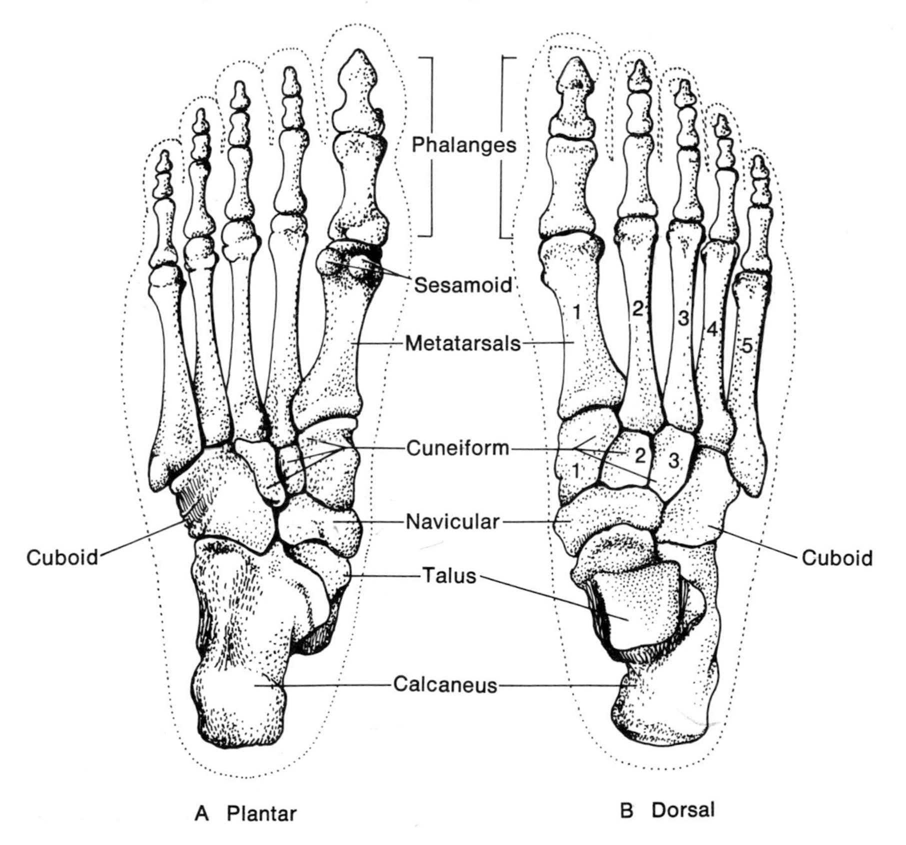

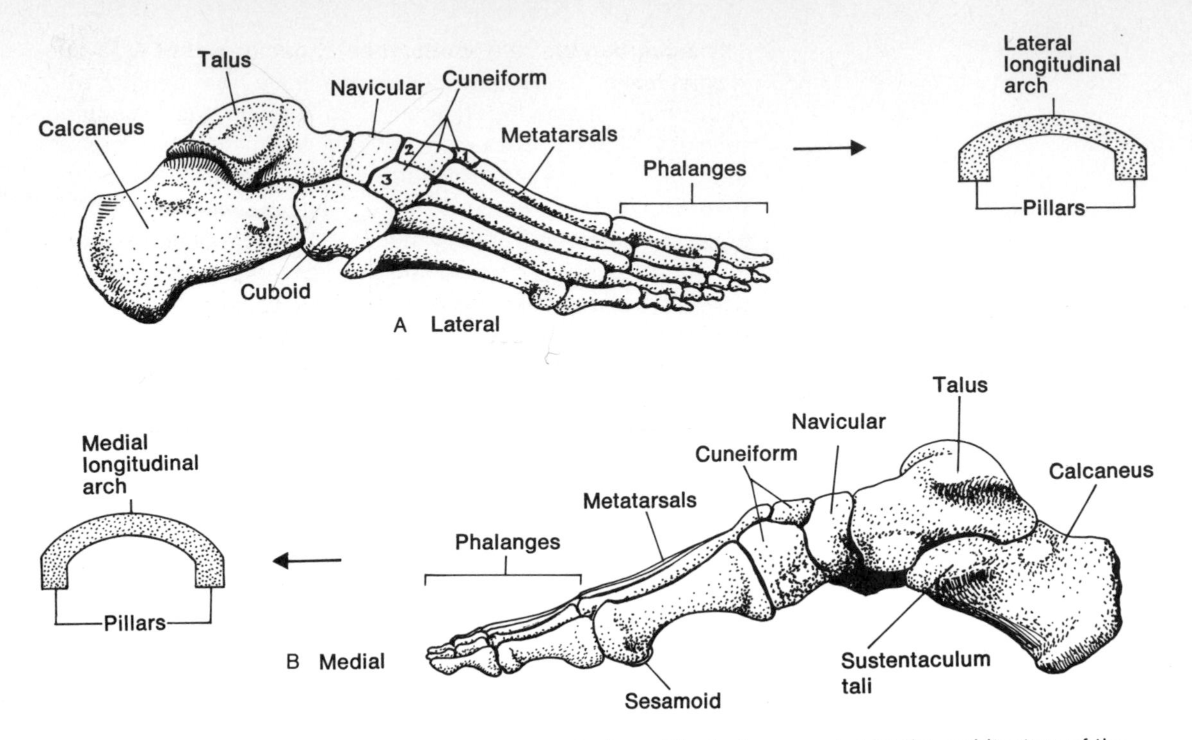

Fig. 5-13 Arches of the feet are created by the architecture of the bones. Note the larger medial arch.

specifically the **medial longitudinal arch.** There is also a less obvious **lateral longitudinal arch** on the lateral side. The longitudinal arch extends from the calcaneus to the heads of the metatarsals. The medial component of this arch (Fig. 5-13B) is formed by:

1. Calcaneus.
2. Talus.
3. Navicular.
4. Three cuneiforms.
5. Metatarsals 1, 2, and 3.

Can you confirm this on yourself? If your arch tends to be low, you may not be able to do it. Low arches are quite normal unless they are severe and create a handicap in walking, a condition called pes planus ("flat foot"), which is frequently caused by abnormal bone structure or joint defects. The height of the longitudinal arch differs from person to person—look at the wet footprints around a pool (although the sole sometimes masks the arch).

The components of the lateral longitudinal part of the arch include the:

1. Calcaneus.
2. Cuboid.
3. Lateral two metatarsals.

Since the lateral longitudinal arch is the more flat and less mobile one, it is the primary bearer of weight and support, whereas the elastic and flexible medial arch functions in locomotion as a shock absorber as well as an aid to balance. Both arches are involved in standing and walking. They are maintained by ligaments.

Aside from the longitudinal arch in each foot there is a transverse arch created when the two feet are placed side by side. This arch flattens when one is standing and adds to the overall flexibility of the foot.

myology

The muscles of the lower limb are numerous but not unmanageable when sorted into groups. Happily most of them fit neatly into anatomical compartments; further, their names often suggest their function, and their relationships to the joints literally tell the task they perform.

To some degree the lower limb muscles can be correlated with those of the upper limb. However, the correlation breaks down somewhat, partly because lower limb musculature is concerned with *support* and *locomotion,* the upper limb with *mobility.*

The muscles of the lower limb are organized as follows:

1. Muscles of the buttock, which work the hip joint.
2. Muscles of the thigh, which work the hip and/or knee joints.
3. Muscles of the leg, which work the knee, ankle, and/or foot joints.
4. Muscles of the foot, which work the joints of the digits.

MUSCLES OF THE BUTTOCK (gluteal region)

gluteus maximus	deep lateral rotators
gluteus medius	tensor fasciae latae
gluteus minimus	

The muscles of the buttock correspond nicely to those of the shoulder, in a functional sense. Thus lateral and medial rotators, abductors, a flexor, and an extensor can be seen in both regions (Fig. 5-14).

Although there are some 10 muscles of the buttock, much of the prominence of the buttock in the muscular male is created by the massive gluteus maximus. In others, for better or worse, adipose tissue plays a starring role.

The arrangement of the gluteal muscles can be seen in Fig. 5-15:

1. The **deep lateral rotators** occupy an area immediately posterior to the hip joint.
2. The **gluteus minimus,** flush against the iliac fossa, lies in the same plane as the lateral rotators—but above them.
3. The **gluteus medius** overlies the minimus.

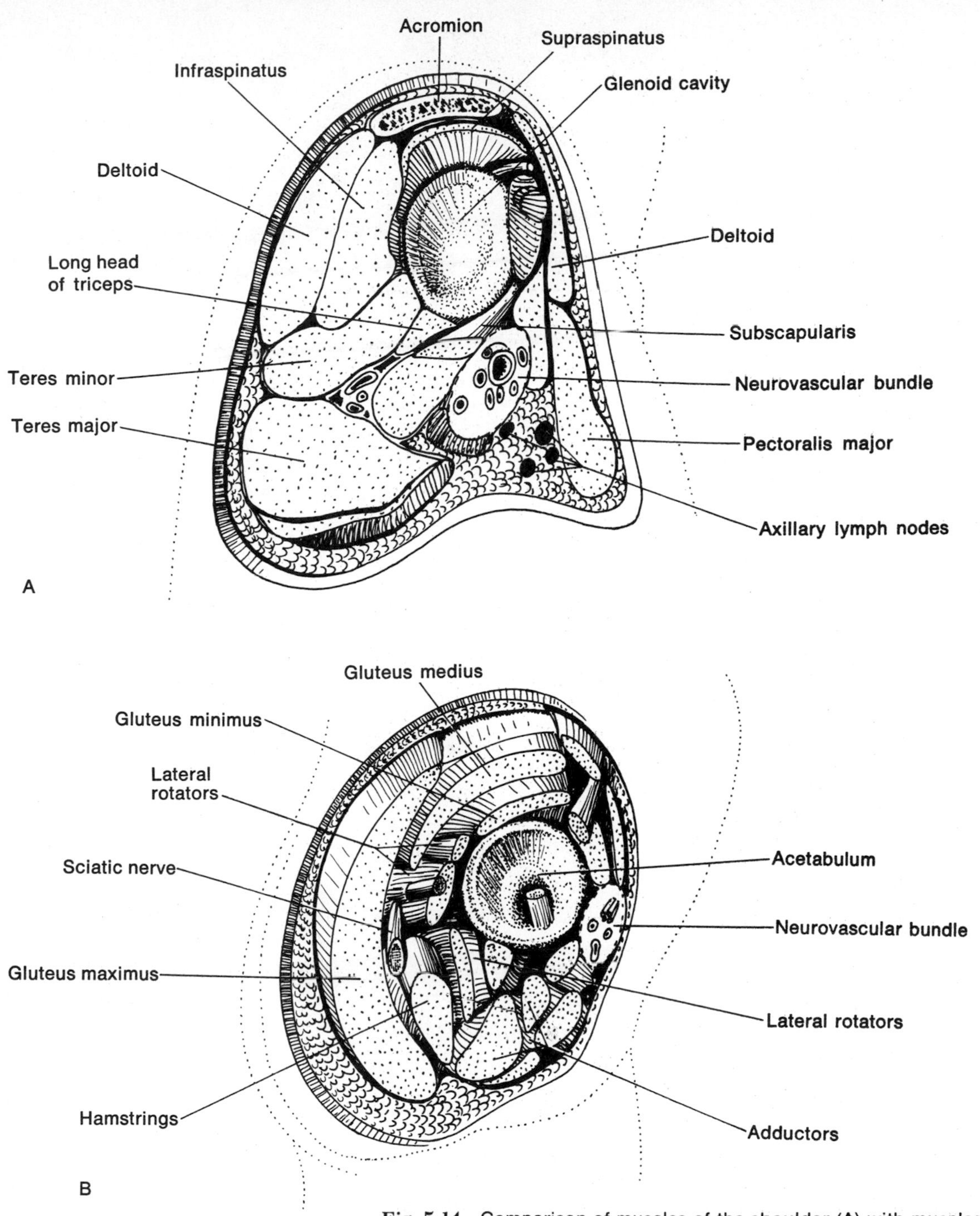

Fig. 5-14 Comparison of muscles of the shoulder (A) with muscles of the buttock (B) in cross section. (Adapted from R. D. Lockhart, G. F. Hamilton, F. W. Fyfe, *Anatomy of the Human Body*, 2d ed., J. B. Lippincott, Philadelphia.)

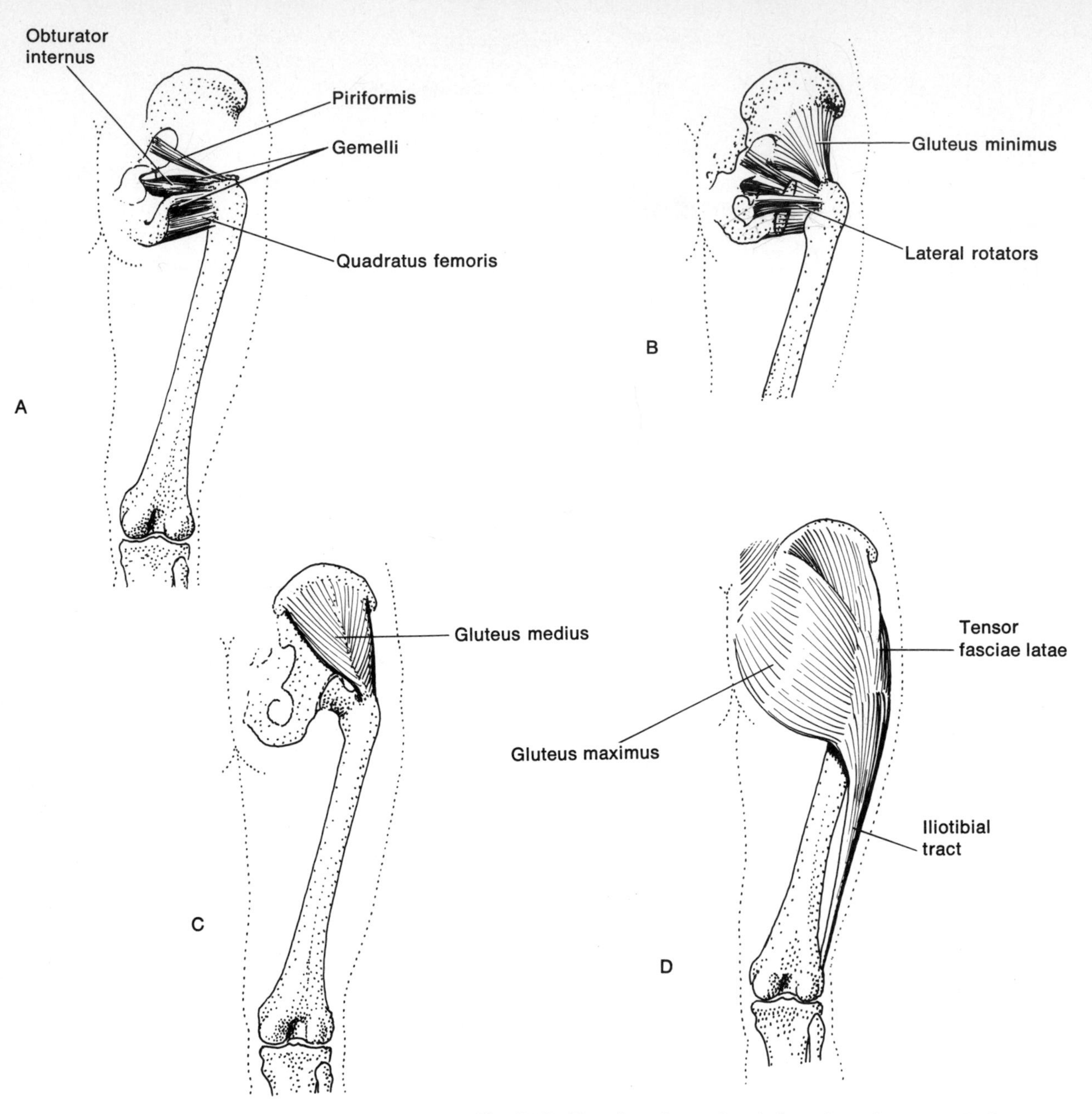

Fig. 5-15 The gluteal muscles. A. Deep lateral rotators. B. Gluteus minimus. C. Gluteus medius. D. Gluteus maximus.

4. The thick **gluteus maximus**—most superficial of all—overlies all but a portion of the medius. Compare this arrangement with the orientation of the posterior shoulder muscles shown in Fig. 5-14A.

The **deep lateral rotators** arise from internal and external surfaces of the pelvic girdle, cross the hip joint posteriorly and insert about the medial aspect of the greater trochanter (Fig. 5-15). These muscles are often difficult to see completely in dissection. They are all lateral rotators of the femur, but, depending on their orientation and specific attachments, they may also be secondary abductors or adductors; e.g., those that insert above the level of the joint would probably be abductors. This group of muscles is important in reinforcing the hip joint and it is analogous in this respect with the rotator cuff muscles of the upper limb.

The **gluteus minimus** is the deepest of the three gluteus muscles arising from the lateral surface of the ilium. It is primarily an abductor, as you might suspect from its attachments (Fig. 5-15B). It is too deep to be palpable.

Gluteus medius, on the other hand, is palpable, just above the greater trochanter (Fig. 5-16). By observing its attachments (Fig. 5-15C), you can see it is an *abductor* of the femur; attempt to abduct your leg against a table or wall while palpating the gluteus medius and you will feel it contract. Gluteus medius is an important postural muscle. A study of Fig. 5-17 will show you how it functions to keep the hips level during walking. As one foot is moved forward, the weight of

Fig. 5-16 Surface features of the gluteal region (buttock). The evident contractures of the gluteus maximus and medius were created by resisting the attempts of the subject to abduct and extend the thighs.

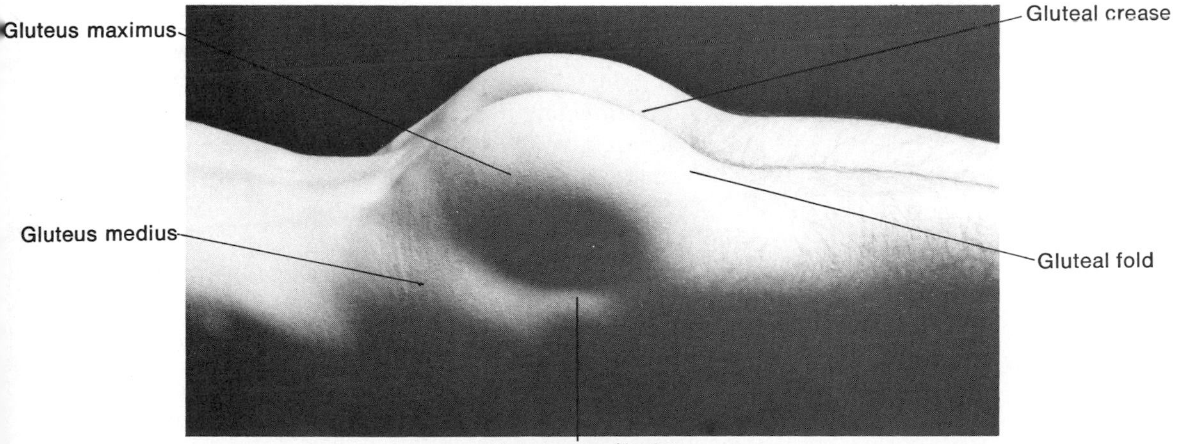

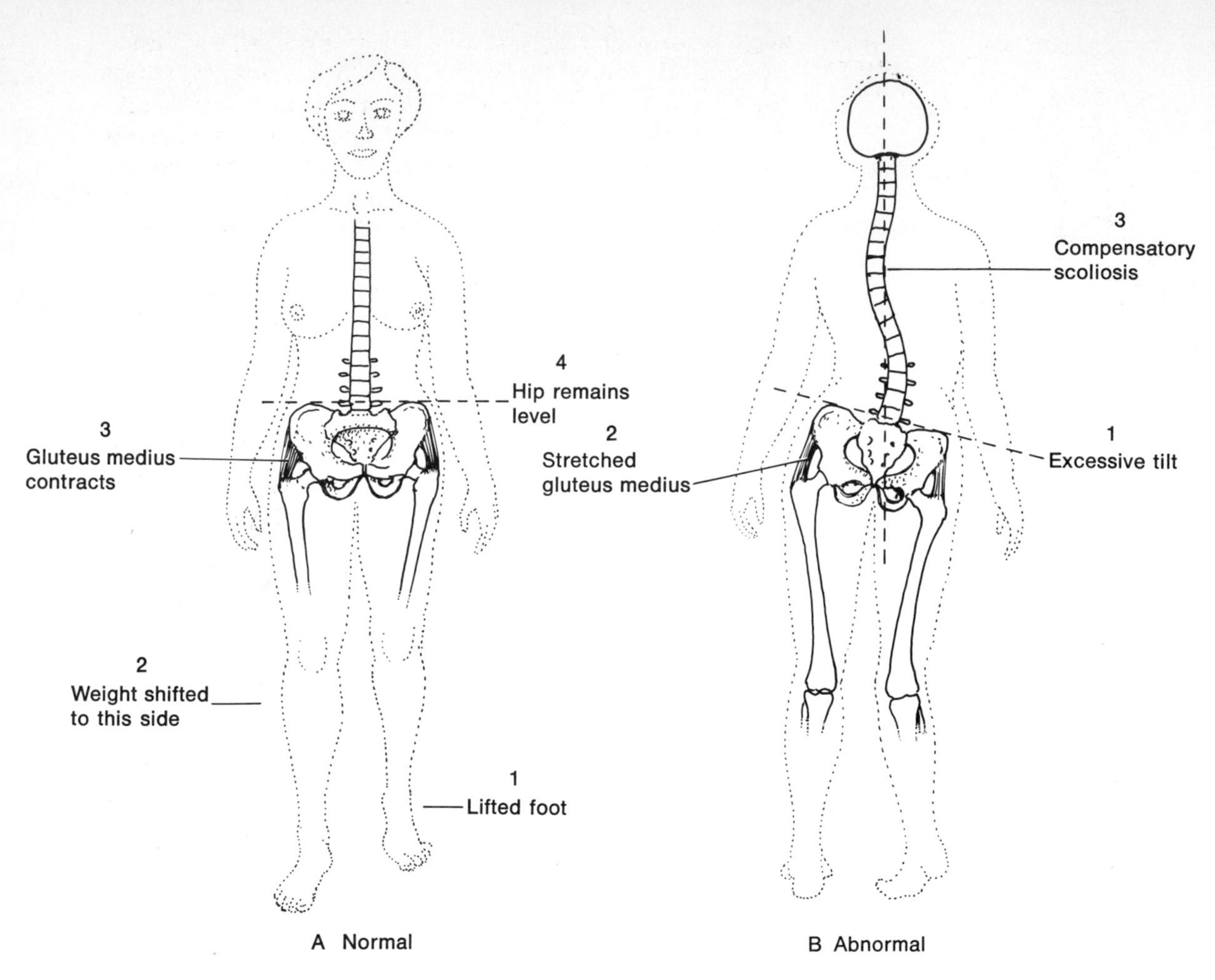

Fig. 5-17 Influence of gluteus medius in controlling the lateral pelvic tilt during walking.

the body is shifted to the opposite limb. See this on yourself. As the weight shifts the gluteus medius contracts on that same side, abducting the hip from the femur making the step. By this action, the hips remain level. In association with other weak abductors as well as a loose iliofemoral ligament, an unrestrained gluteus medius accounts for excessive lateral tilt of the pelvis ("hip swinging") during walking.* Muscle strain could be a consequence of such excessive swinging.

Gluteus maximus can be easily palpated (Fig. 5-16). By virtue of its attachments (Fig. 5-15D), it is the great *extensor* of the thigh. Thus gluteus maximus can best be felt when the

* It may draw attention but it plays havoc with good posture and compromises stability of the vertebral column.

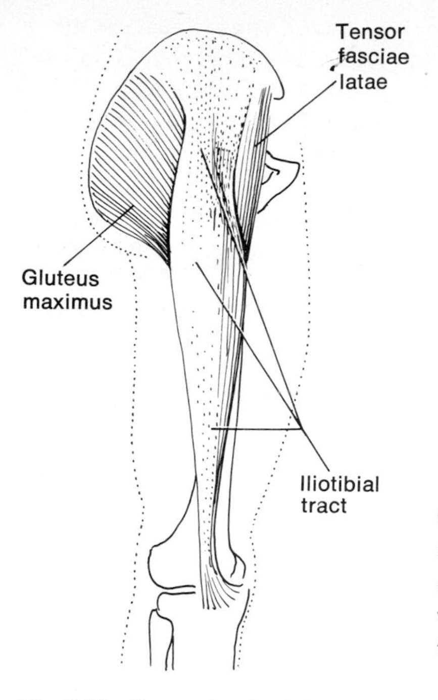

Fig. 5-18 Tensor fasciae latae muscle in the iliotibial tract.

thigh is extended against resistance; e.g., attempt, while standing, to extend the whole limb by pressing one heel against a wall. Because of the broad insertion of this muscle relative to the hip joint, the upper fibers tend to abduct the femur and the lower fibers tend to adduct it. Prove this on yourself.

The thigh is wrapped in a tight, dense stocking of deep fascia (fascia lata). Laterally, the fibers of this fascia are oriented vertically, in a band extending from the ilium to the tibia, which is therefore named the **iliotibial tract** (Fig. 5-15). Arising from the anterior part of the iliac crest, a muscle passes down with and inserts into the iliotibial tract. Aside from acting to flex and abduct the femur, it also puts tension on the fascia lata and is thus named the **tensor fasciae latae** (Fig. 5-18). It can be felt about four fingersbreadth anterior to the greater trochanter when abducting or flexing the thigh against resistance. Classically, the tensor fasciae latae is included with the gluteal muscles because of its similar innervation. In functional respects, it is more a muscle of the anterior thigh.

MUSCLES OF THE THIGH

The muscles of the thigh are remarkably oriented into three compartments, each having its own nerve supply:

1. The **anterior compartment,** supplied by the femoral nerve.
2. The **medial compartment,** supplied by the obturator nerve.
3. The **posterior compartment,** supplied by the sciatic nerve.

Now if you refer to your own thigh while standing in the anatomical position, note that the hip joint is a multiaxial joint—the thigh can be adducted, abducted, flexed, extended, rotated, and circumducted at that joint. Try it! Furthermore, the knee joint is operationally a hinge joint, for most intents and purposes. The leg can be flexed and extended. In moving your leg, note that flexion is 180° out of phase with flexion of the forearm at the elbow joint, i.e., the leg is drawn up on the posterior side of your body, the forearm on the anterior side.

Standing in the anatomical position, you can reason out what movers are in what compartments by testing your own thigh movements, as follows:

Anterior compartment of the thigh: These muscles must generally flex the thigh (if they cross the hip joint) and extend the leg (if they cross the knee joint). And so they do.

Posterior compartment of the thigh: These muscles must generally extend the thigh (if they cross the hip joint) and flex the leg (if they cross the knee joint). And so they do.

Medial compartment of the thigh: These muscles must adduct the thigh, if they cross the hip joint. Since the knee is largely a hinge joint, adduction of the leg is not possible. Thus only one muscle of the medial compartment crosses the knee joint and it does so behind and to the side of the joint, playing a role in flexion of the leg.

MUSCLES OF THE POSTERIOR THIGH

semimembranosus   biceps femoris
semitendinosus

These three muscles make up the bulk of the posterior thigh compartment and may be palpated on yourself extensively. While standing, flex your leg (at the knee) against resistance, while placing your hand under the gluteal fold (Fig. 5-16). Alternately flex and relax. This is the origin of the three muscles, the fibers attaching at the ischial tuberosity. The muscle mass can be palpated throughout the posterior thigh. Just above the knee, feel the mass seem to split into two cords (tendons). The medial tendon is longer and takes up one-half the muscle, which incidentally may help you to remember its name: **semitendinosus.** Deep to this muscle is the **semimembranosus.** The tendon palpated on the lateral side is the tendon of a two-headed muscle of the thigh, hence its name: **biceps femoris** (Fig. 5-19).

Collectively these muscles constitute the "hamstrings," and act together as secondary extensors of the thigh (they cross the hip joint), as well as primary flexors of the leg (they cross the knee joint). These muscles have little slack (are short) as is demonstrated when trying to touch your toes without bending your knees. As the flexion of the hip is increased (e.g., a high kick) the "hams" involuntarily flex the knees. See this on yourself. In Fig. 5-19, note the space created behind the knee by the diverging tendons of the hamstring muscles. This is the popliteal fossa, an area to be discussed shortly.

MUSCLES OF THE MEDIAL THIGH

adductor magnus   gracilis
adductor longus
adductor brevis

These muscles make up the major mass of the medial thigh compartment (Fig. 5-20). As a group they may be palpated

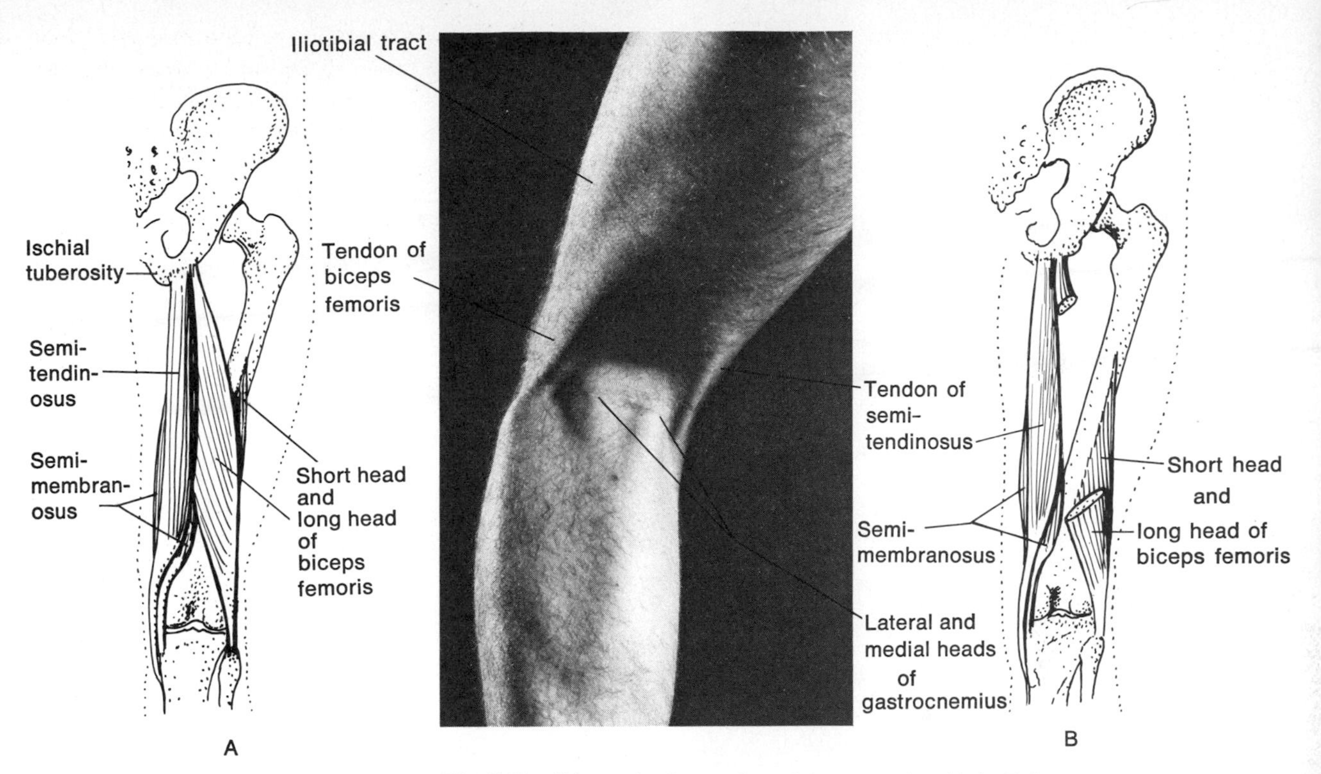

Fig. 5-19 "Hamstring" muscles of the posterior thigh. Relate drawings to the surface photograph.

when the thigh is adducted against resistance. While standing, spread your legs about 2 ft apart and hook one foot around the leg of a piece of heavy furniture. Feel now for the inferior pubic ramus and ischial ramus on one side and adduct the limb on that side against resistance. The origin fibers of the adductors may be felt to contract. Feel the extent of this muscle mass inferiorly. This is mainly the **adductor magnus.** While sitting, spread your thighs and then adduct them against resistance and you can feel as well as see the adductor mass contract. The insertions of these muscles range along almost the entire posterior shaft of the femur. Only the **gracilis** inserts on the tibia, thus it can be a flexor of the knee as well as an adductor of the femur.

MUSCLES OF THE ANTERIOR THIGH

iliopsoas	vastus medialis	quadriceps femoris
pectineus	vastus intermedius	
sartorius	vastus lateralis	
	rectus femoris	

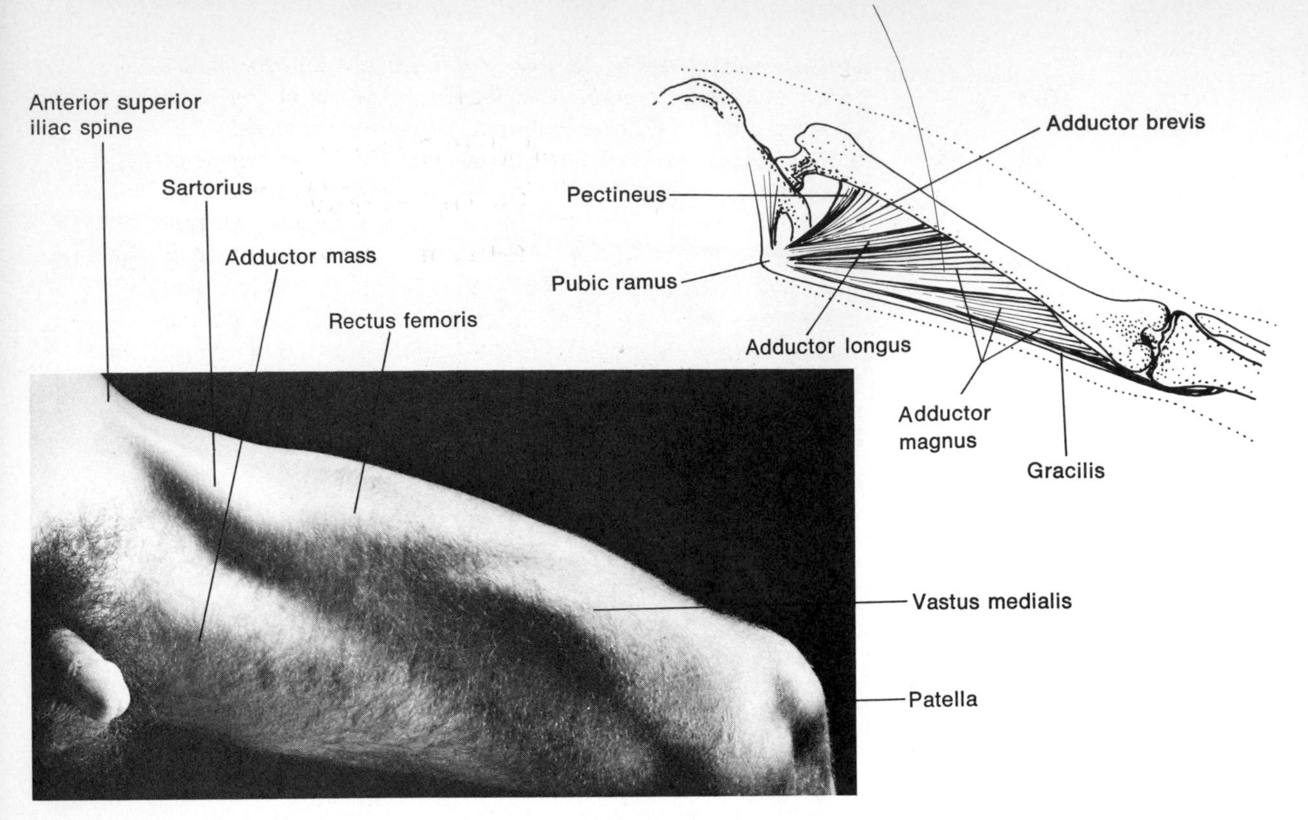

Fig. 5-20 Muscles of the medial and anterior thigh.

As was said before, the muscle mass of the anterior thigh may be functionally subdivided into:

1. Muscles that flex the femur at the hip.
2. Muscles that flex the femur at the hip and extend the leg at the knee.
3. Muscles that extend the leg at the knee.

The principal flexors of the thigh are a pair of muscles which arise from the iliac fossa (the iliacus) and along the vertebral column (the psoas) and join to insert on the lesser trochanter of the femur (Fig. 5-21A). Collectively they form a structural and functional unit—the **iliopsoas.** They can be palpated only with difficulty. Besides flexing the thigh, the iliopsoas is an important postural muscle.

The pectineus, adjacent to the iliopsoas (Fig. 5-21A), is a secondary flexor and adductor of the femur.

A pair of the longest muscles of the body arise from the anterior iliac spine, cross the hip joint and the knee joint, and insert on the tibia. One of the pair can be palpated directly on the anterior aspect of the thigh while the knee is extended against resistance—try it; this is the straight muscle of the femur or **rectus femoris** (Fig. 5-21B). It inserts at the tibial tuberosity via the patellar ligament. Therefore it is an extensor of the knee as well as a flexor of the hip. The other muscle of the pair crosses the thigh obliquely from anterior to medial as it descends to the tibia. This is the "tailor's muscle" or **sartorius** (Fig. 5-21B). It is best felt at its origin when the hip joint is flexed. Note its attachment on the tibia. Thus it is also a flexor of the knee joint. Because of its oblique course across the thigh, the sartorius is a lateral rotator of the thigh. Thus, to assume the tailor's posture at work (sitting cross-legged), flex and laterally rotate the thighs at the hip joints, flex the knee joints, and let gravity have its way.

The three **vasti** closely hugging the femur all join with the more superficial tendon of rectus femoris to insert on the tibial tuberosity. Thus the foursome is known collectively as

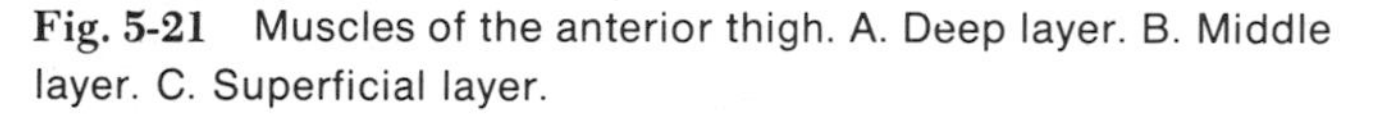

Fig. 5-21 Muscles of the anterior thigh. A. Deep layer. B. Middle layer. C. Superficial layer.

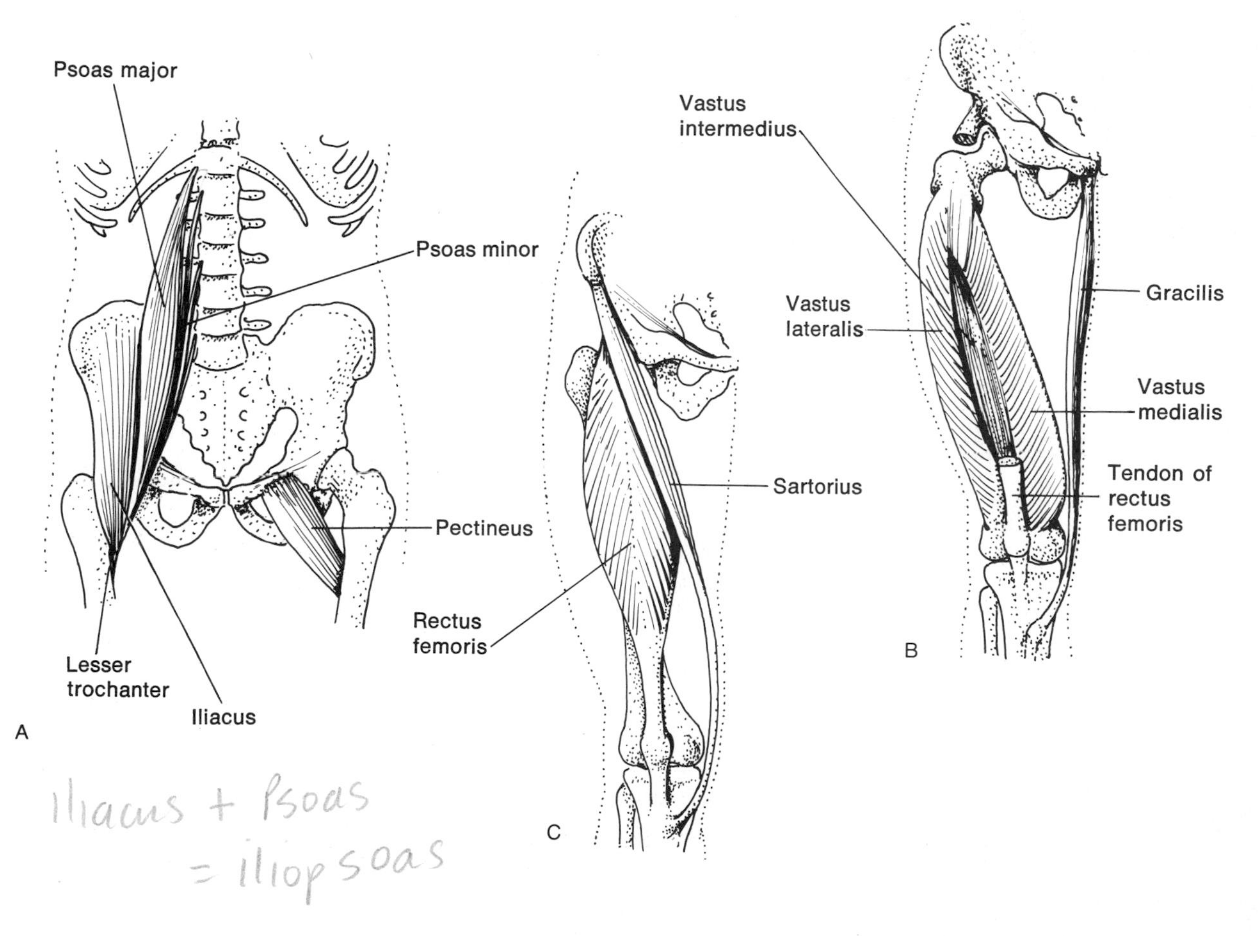

the **quadriceps femoris** (Fig. 5-21). They function as extensors of the knee joint. **Vastus lateralis** may be palpated just lateral to rectus femoris when the knee is extended against resistance. **Vastus medialis** may be palpated likewise medially; **vastus intermedius** cannot be felt.

MUSCLES OF THE LEG

The muscles of the leg are reasonably named and are therefore easily remembered. You will recall the forearm incorporated musculature which manipulated the digits as well as the wrist. So it is with the leg. Further, there are muscles in the forearm that operate the thumb (pollex). Thus there are muscles in the leg that move the great toe (hallux). Like the forearm, the leg contains anterior and posterior muscle compartments. It also has a lateral compartment. The function of

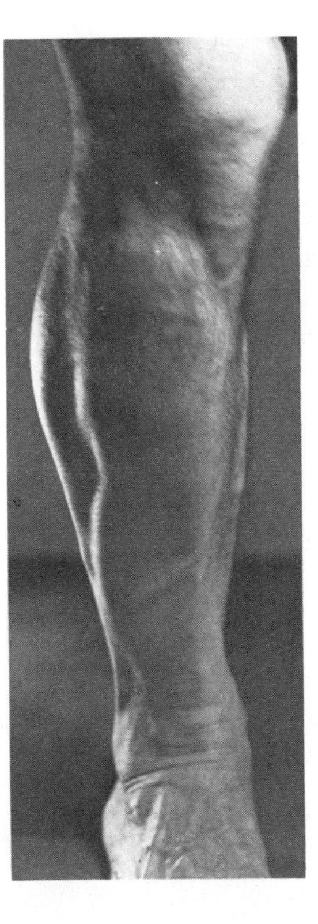

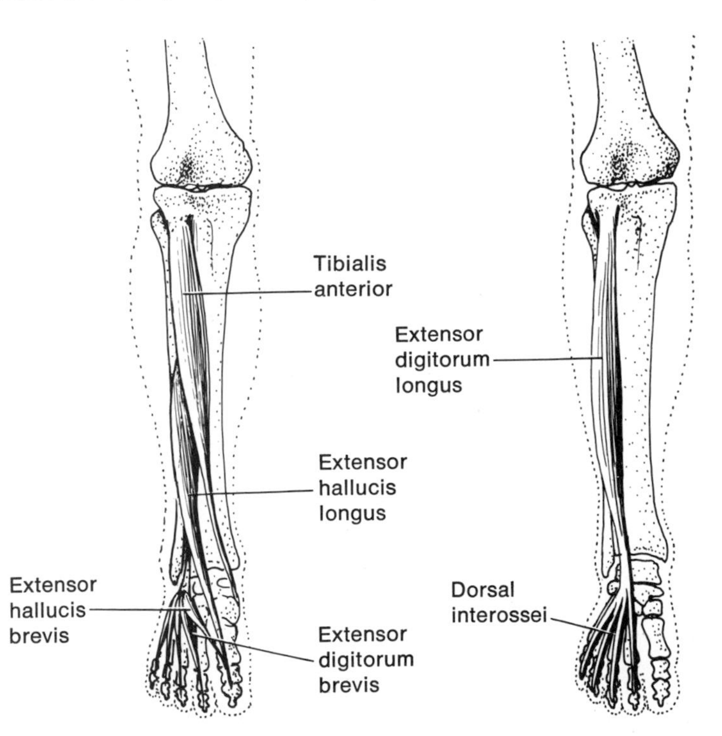

Fig. 5-22 Anterior leg (crural) muscles. Extensor hallucis lies deep to the extensor digitorum. Relate the drawings to the surface photograph. (Courtesy of R. D. Lockhart, *Living Anatomy*, 5th ed., Faber and Faber, Ltd., London, 1959.)

these groups is straightforward and is related, in some cases, to the names of the individual muscles.

However, to correlate muscle groups of the leg with those of the forearm, it is necessary to supinate the forearm and hand (as opposed to pronation as in the anatomical position). Place your hands as if they were feet. In this way, you can see and remember that:

1. Extensors of the foot and toes are in the anterior compartment. These extensors are also called dorsiflexors.
2. Flexors of the foot and toes are in the posterior compartment. These flexors are also called plantarflexors.
3. Muscles (of the anterior and posterior compartments) whose tendons pass to the medial aspect of the foot are inverters/adductors* of the foot.
4. Muscles (of the lateral compartment) whose tendons pass to the lateral aspect of the foot are everters/abductors of the foot.

MUSCLES OF THE ANTERIOR COMPARTMENT OF THE LEG

tibialis anterior	extensor digitorum
	extensor hallucis longus

The three muscles of the anterior leg (Fig. 5-22) all cross the ankle joint and function to dorsiflex (extend) the foot. The tendon of one of these muscles passes into the medial longitudinal arch and acts on the tarsal joints to effect inversion and adduction (Fig. 5-23). Try these movements on yourself. Now see Fig. 5-22 again and try to reason out the tasks of each of the muscles indicated from its attachment (more than one answer may apply):

____ Extensor hallucis longus	1. Dorsiflexion (extension) of the foot at the ankle joint
____ Extensor digitorum longus	2. Extension of the great toe
____ Tibialis anterior	3. Extension of toes 2 to 5
	4. Adduction/inversion
	5. Abduction/eversion

* Inversion/eversion = a turning inward/outward of the plantar surface of foot at the subtalar joints; adduction/abduction = bending of front of foot medially and laterally, respectively. These movements are made in concert with inversion/eversion, respectively.

Answers: EHL: 1, 2; EDL: 1, 3; TA: 1, 4

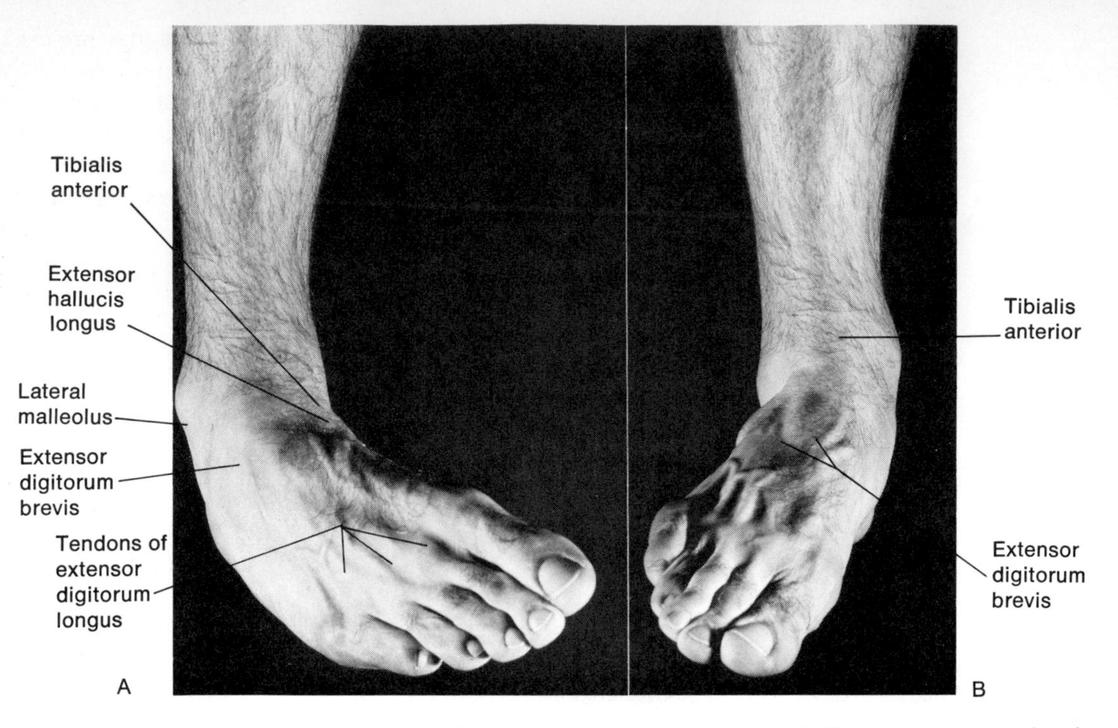

Fig. 5-23 A. Inversion and adduction. B. Eversion and abduction. In A, the great toe is extended.

Tibialis anterior is easily palpated in the anterolateral compartment while the foot is dorsiflexed against resistance (Fig. 5-24) and inverted (Fig. 5-23). While one is standing, tibialis anterior pulls the leg forward on the ankle joint (as in leaning forward) to assist balancing the body on the foot.

Deep and slightly lateral to the tibialis anterior is the common extensor of the toes: **extensor digitorum longus.** It cannot be palpated in the leg; however, if you extend your toes, its tendons can be seen and felt as they emerge from under the extensor retinaculum and pass into the foot.

The extensor of the great toe, **extensor hallucis longus,** is buried deep to the two muscles just discussed, but its tendon may be palpated while the great toe is extended (Figs. 5-22 and 5-23).

MUSCLES OF THE POSTERIOR COMPARTMENT OF THE LEG

gastrocnemius	popliteus
soleus	tibialis posterior
	flexor hallucis longus
	flexor digitorum longus

Stand on your toes while palpating the back (posterior compartment) of the leg. If you weigh 200 lb, that prominent mass you feel just lifted 200 lb and somewhat precariously set it on the heads of the metatarsals and on your toes. That mass consists of the superficial muscles of the posterior compartment: **gastrocnemius** and **soleus** (triceps surae) (Figs. 5-25 to 5-27).

Gastrocnemius (Gk., belly of the leg) and soleus (named after the sole fish, whose shape it resembles) are the principal flexors (plantarflexors) of the ankle joint. They literally pull the heel (calcaneus) upward when you stand on your toes, as you have just demonstrated. The muscles narrow distally to form a thick tendon (of Achilles) inserting on the calcaneus (tendo calcaneus). The attachments of these muscles may be seen in Fig. 5-25. "Gastroc," larger of the two muscles and an important stabilizing muscle of the ankle joint, also crosses the knee joint and is, therefore, a secondary flexor of the leg.

The deep members of the posterior compartment may be studied in Fig. 5-25. Covered by the superficial muscles just discussed, they cannot be felt except for their tendons, which pass just below the medial malleolus onto the plantar surface of the foot. As in the case of the anterior compartment muscles, the muscles here are named reasonably with respect to function.

See now Fig. 5-25. Can you reason out the tasks of each of the muscles of the deep posterior compartment of the leg

Fig. 5-24 Surface features of the tendons of tibialis anterior and extensor digitorum longus.

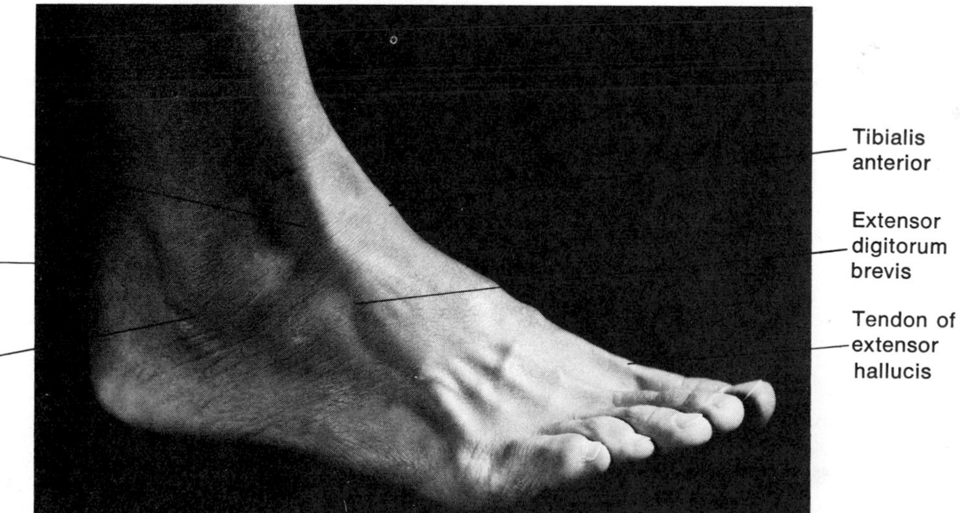

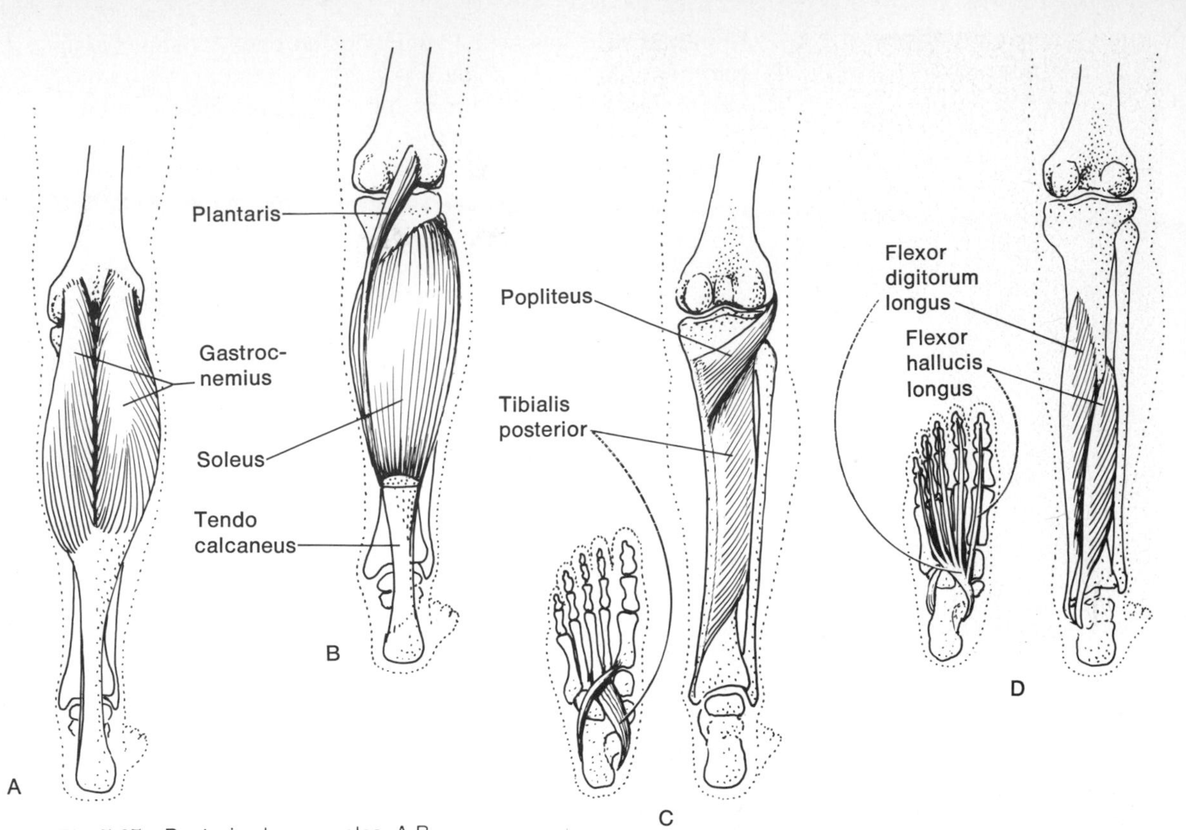

Fig. 5-25 Posterior leg muscles. A,B. Superficial layers. C,D. Deep layers. Note sites of insertion on plantar surface of foot.

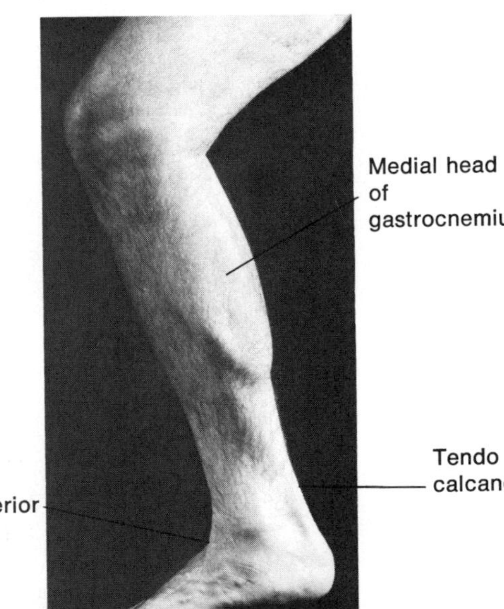

Fig. 5-26 Surface features of the posteromedial leg.

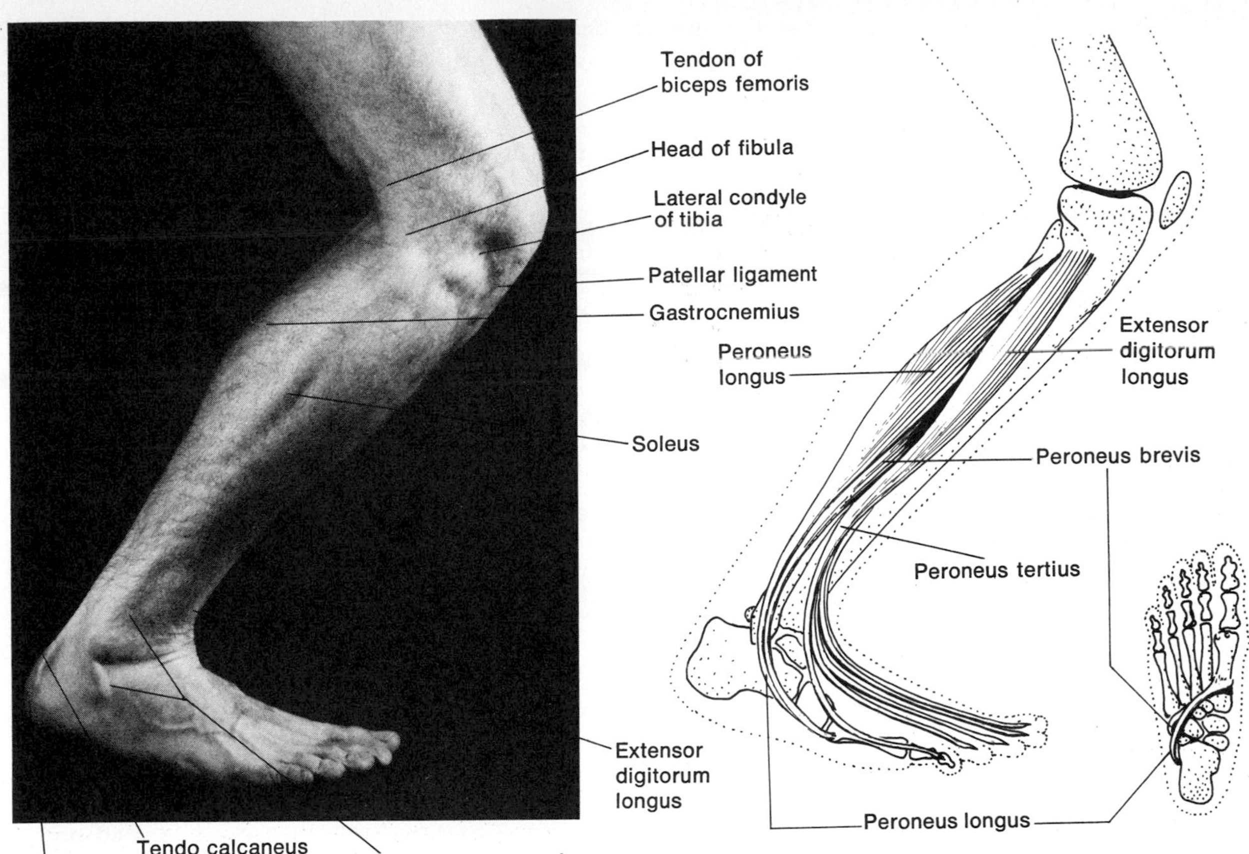

Fig. 5-27 Lateral leg muscles.

based on their attachments/names? (warning: more than one answer may apply):

1, 2, 4 Flexor digitorum longus
1, 3, 4 Flexor hallucis longus
1, 4 Tibialis posterior

1. Plantarflexion of the foot at the ankle.
2. Flexion of toes 2 to 5.
3. Flexion of the great toe.
4. Adduction/inversion.
5. Abduction/eversion.

Answers: FDL: 1, 2, 4; FHL: 1, 3, 4; TP: 1, 4.

The **popliteus,** corresponding to the pronator teres in the forearm, acts to flex and medially rotate the leg. It also helps to stabilize the knee joint. It cannot be felt, as it is covered by the soleus and the gastrocnemius.

The **tibialis posterior** is an antagonist of the tibialis anterior in balancing the leg on the foot. Since they both insert on the medial aspect, plantar surface of the foot, they act as inverters/adductors of the foot.

Flexors hallucis and **digitorum longus** are the antagonists of extensors hallucis and digitorum in movement of the great toe and toes 2 to 5, respectively.

MUSCLES OF THE LATERAL LEG

peroneus longus   peroneus brevis

Peroneus longus is the most superficial of the two and can be seen and palpated on the lateral aspect of the leg when the foot is everted against resistance (e.g., against the leg of a table). You can readily see (Fig. 5-27) that these muscles, by virtue of their insertions, are plantarflexors of the ankle and foot. However, their role is minimal. The soleus is 95 percent responsible for plantarflexion. The course of their tendons around the lateral aspect of the foot also tells you that these muscles are (principally) abductors/everters of the foot.

INTRINSIC MUSCLES OF THE FOOT

By definition, the intrinsic muscles of the foot arise and insert within the foot (Fig. 5-28). They are actively contracting during movement of the toes. Feel for a soft muscular mass (extensor digitorum brevis) just anterior to the lateral malleolus. Now extend your toes and feel and see this mass contract, forming a prominence (Fig. 5-23).

All but two of these intrinsic muscles* are located on the plantar surface of the foot deep to the thick plantar aponeurosis and aid in flexion of the toes. They should be thought of as a functional unit—not memorized individually. They are *not* significant actors in maintaining the arches of the foot; this task is left to the appropriate ligaments. Relate them to the intrinsic muscles of the hand.

QUIZ

1. Structurally, the bones of the upper and lower limbs are reasonably similar. Functionally, the bones and joints of the upper limb are related to mobility while those of the lower limb are related to ________ and ________.
2. The homologue of the pectoral girdle in the lower limb is the ________ ________.

* Extensor digitorum brevis and extensor hallucis brevis. On the basis of Fig. 5-22, can you differentiate the two by palpation?

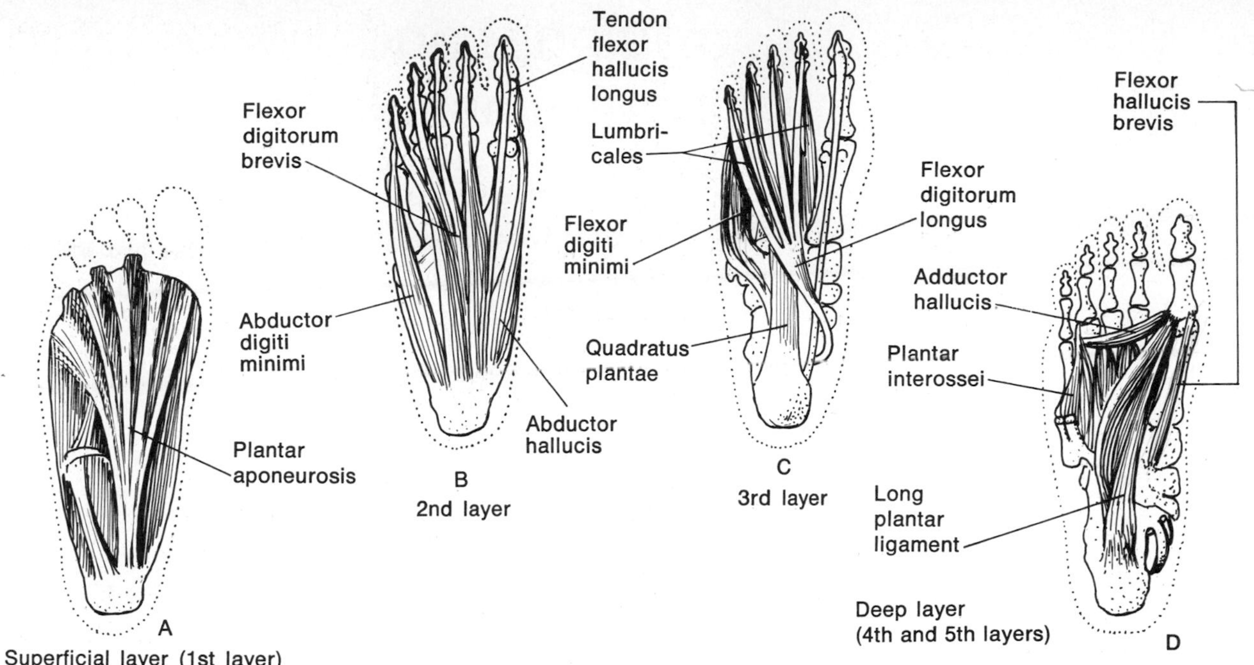

Fig. 5-28 Intrinsic muscles of the foot, plantar aspect. For muscles on the dorsum, see Fig. 5-22.

3. Each hip bone is composed of three bones: the ___ilium___, ___ischium___, and the ___pubis___.
4. The shaft of the ___tibia___ is characterized by having its anteromedial surface entirely subcutaneous.
5. The muscles of the lower limb corresponding to the musculotendinous cuff of the upper limb are largely ___lateral___ ___rotators___ (functionally speaking).
6. The great extensor of the thigh is ___gluteus___ ___maximus___.
7. Excessive hip swinging during locomotion could cause spasms or strain in the muscle ___gluteus___ ___medius___.
8. The deep fascia of the thigh is called ___fascia___ ___lata___. p. 259
9. The adductors of the thigh occupy the ___medial___ thigh compartment.
10. The support of the arches of the foot ~~is~~/is not (cross out one) provided by the intrinsic muscles of the foot.

regional and neurovascular considerations

GLUTEAL REGION

The gluteal region is that area of the posterolateral hip bordered by the lumbar region (flank) above and the posterior thigh below. The prominence of the region is the **buttock,** which is often striking in people who have huge deposits of fat in the superficial fascia. The left and right buttocks are separated from one another by a gluteal crease; they terminate inferiorly at the gluteal fold (Fig. 5-16).

The arrangement of structures in the gluteal region can be seen in Fig. 5-29. Study this figure as you read the following:

The bony and ligamentous framework of the gluteal region consists of the iliac fossa, the ischial spine and tuberosity, the sacrum and coccyx, the greater and lesser sciatic foramina (created by the presence of the sacrospinous and sacrotuberous ligaments), and the greater trochanter of the femur. The great sciatic nerve is shown in relation to these structures. It leaves the pelvis via the greater sciatic foramen.

Note carefully the piriformis muscle, most prominent of the deep lateral rotators, projecting through the greater sciatic foramen. Vessels and nerves exiting or entering the pelvic interior do so via the greater sciatic foramen either above piriformis (for instance, superior gluteal vessels and nerve) or below it (for instance; inferior gluteal vessels and nerve, sciatic nerve, pudendal vessels, and pudendal nerve to the perineum).

Study the gluteus medius, overlying the deeper gluteus minimus, inserting about the greater trochanter. The gluteus maximus, superficial to all, is somewhat inferior to gluteus medius and more horizontally oriented.

Finally, discover on yourself that the gluteus medius and maximus can be palpated against resistance (Fig. 5-16).

The muscles of the gluteal region are all movers of the thigh in abduction, rotation, extension, and, in one case, adduction. Secondarily, the deep lateral rotators reinforce the security of the hip joint, as the rotator cuff muscles stabilize the shoulder joint. In fact, there is a good deal of structural similarity between the two areas (Fig. 5-14).

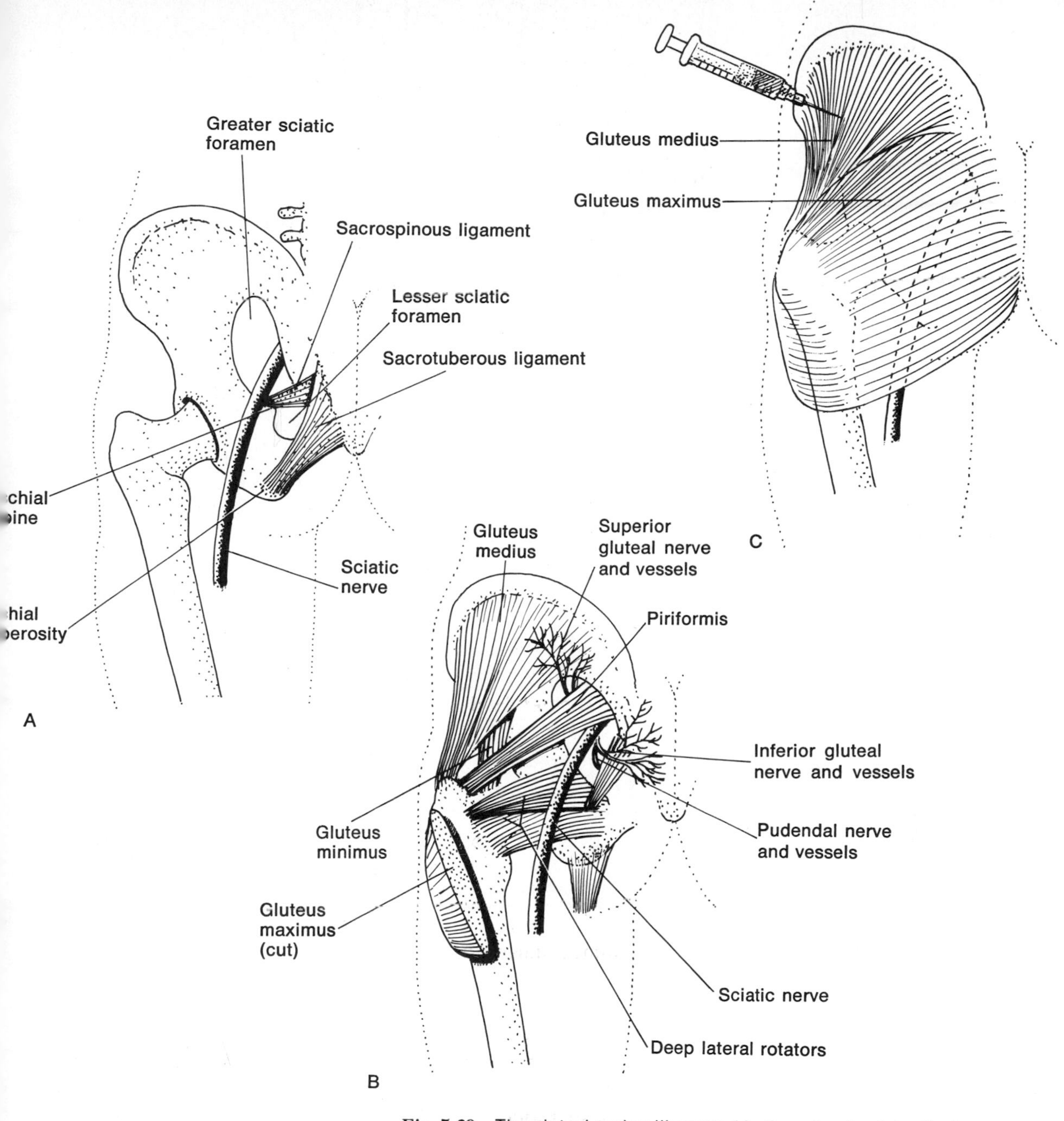

Fig. 5-29 The gluteal region illustrated in three levels of depth. A. Bony and ligamentous floor of the gluteal region. B. Middle layer with vessels and nerves superimposed. C. Superficial muscle layer. The preferred site of needle penetration is the upper, outer quadrant of the buttock.

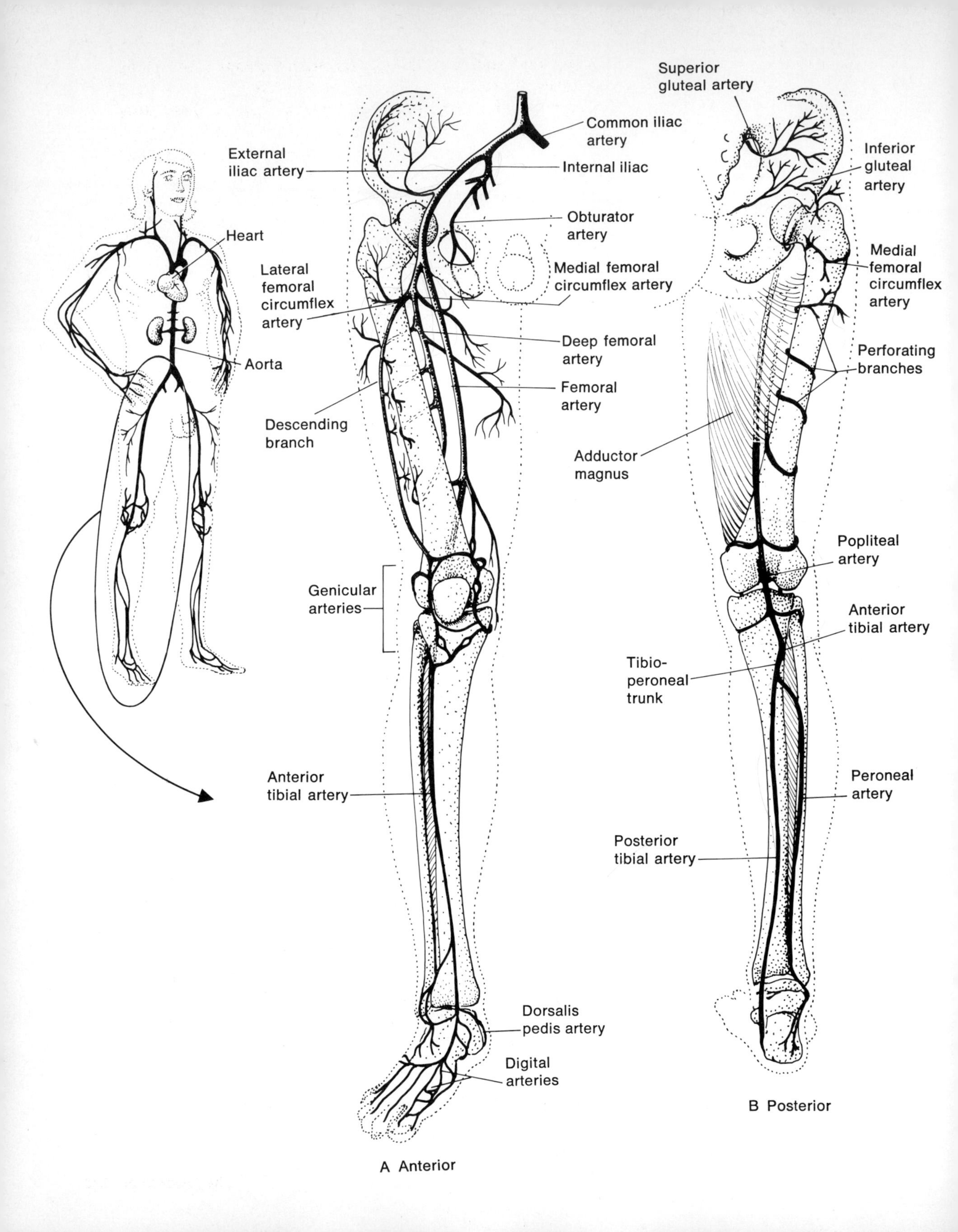

Heart
Aorta
External iliac artery
Common iliac artery
Internal iliac
Obturator artery
Lateral femoral circumflex artery
Medial femoral circumflex artery
Deep femoral artery
Femoral artery
Descending branch
Genicular arteries
Anterior tibial artery
Dorsalis pedis artery
Digital arteries
A Anterior
Superior gluteal artery
Inferior gluteal artery
Medial femoral circumflex artery
Perforating branches
Adductor magnus
Popliteal artery
Anterior tibial artery
Tibio-peroneal trunk
Peroneal artery
Posterior tibial artery
B Posterior

The blood supply to the gluteal region comes from within the pelvis. Refer now to Fig. 5-30 as you read:

The abdominal aorta bifurcates into the common iliac arteries at about the level of the fourth lumbar vertebra.

The internal iliac artery (a branch of the common iliac) enters the pelvis and supplies the pelvic viscera (see Unit 10).

The superior and inferior gluteal arteries spring directly off the internal iliac, pass through the greater sciatic foramen, and enter the gluteal region to supply the structures there.

If the internal iliac artery is ligated routes of collateral flow may be visualized.

The veins draining the gluteal region (Fig. 5-31) ride alongside the arteries, are similarly named, and drain into the inferior vena cava.

The nerves to the gluteal region arise from the lumbar and sacral segments of the spinal cord via the **lumbosacral trunk** (Fig. 5-32). The nerves are appropriately named according to the muscles they innervate. Thus gluteus medius/minimus are innervated by the superior gluteal nerve; gluteus maximus by the inferior gluteal nerve; piriformis by nerve to piriformis.

Passing through the gluteal region is the longest and largest nerve of the body, the **sciatic nerve,** which will innervate the structures of the posterior thigh, the entire leg, and the foot (Fig. 5-33). In its trek through the gluteal region, the sciatic nerve is superficial to all muscles of the gluteal region except the gluteus maximus. Thus, an injection directed into the gluteal region may have serious consequences if the needle should strike or release its contents about the sciatic nerve or other vessels and nerves. Symptoms may range from a sharp pain to paralysis of a group of muscles. For reasons such as these, intramuscular injections of the buttock are introduced in the upper and outer quadrant in the region of the gluteus medius muscle (Fig. 5-29).

HIP JOINT

The hip joint is where the acetabulum of the hip bone articulates with the head of the femur. It is a synovial joint of the ball-and-socket variety. The capsule of the joint is thick and tough, yet loose enough to permit multiaxial movements. The capsule is reinforced by three stout ligaments on all surfaces (Fig. 5-34). Since each ligament arises from one of the bony contributions to the hip, and all attach on the femur, you will find the names of these important ligaments quite reasonable.

Fig. 5-30 (opposite) Principal arteries of the lower limb.

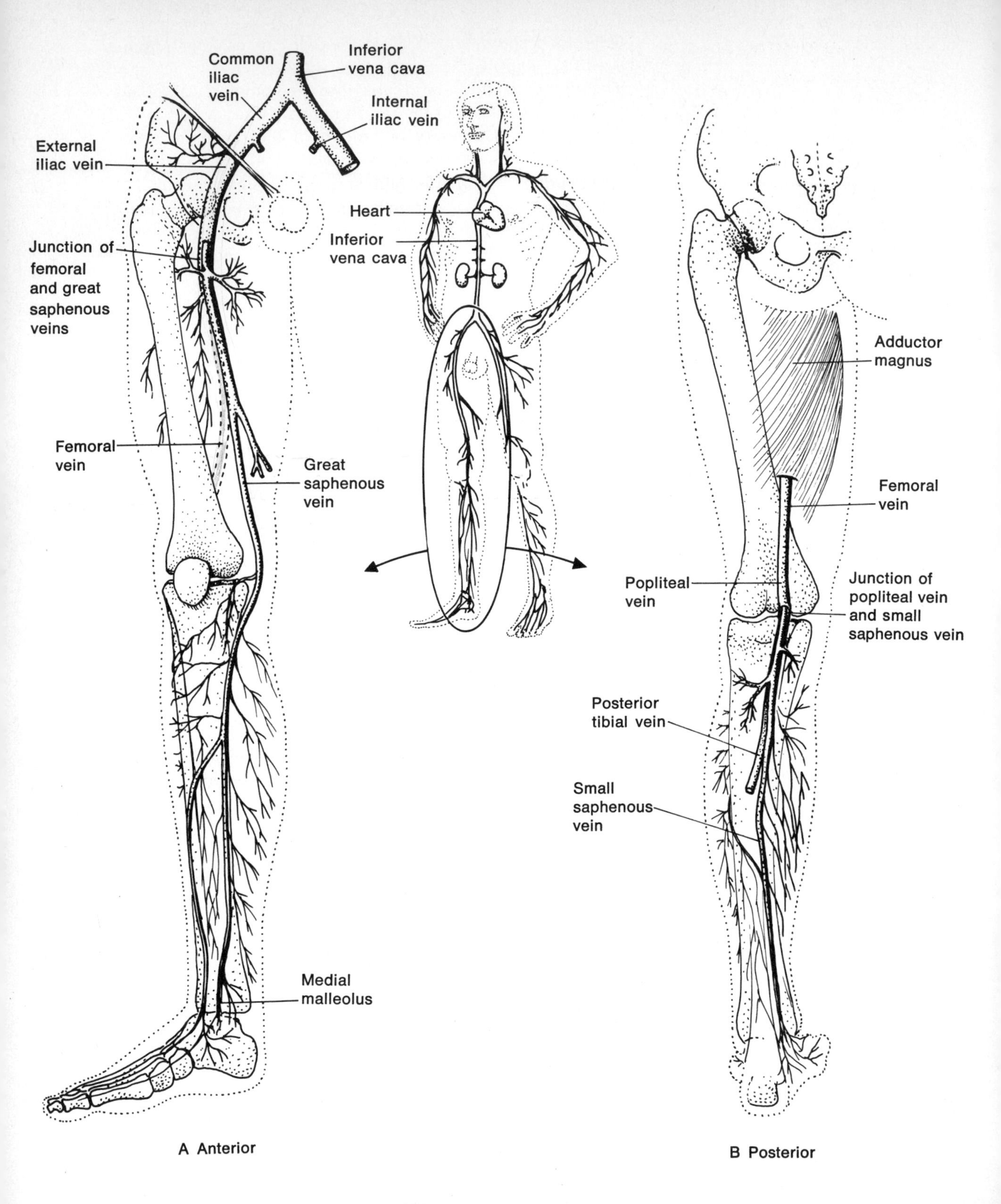

Common iliac vein
Inferior vena cava
Internal iliac vein
External iliac vein
Junction of femoral and great saphenous veins
Femoral vein
Great saphenous vein
Medial malleolus
Heart
Inferior vena cava
Adductor magnus
Femoral vein
Popliteal vein
Junction of popliteal vein and small saphenous vein
Posterior tibial vein
Small saphenous vein
A Anterior
B Posterior

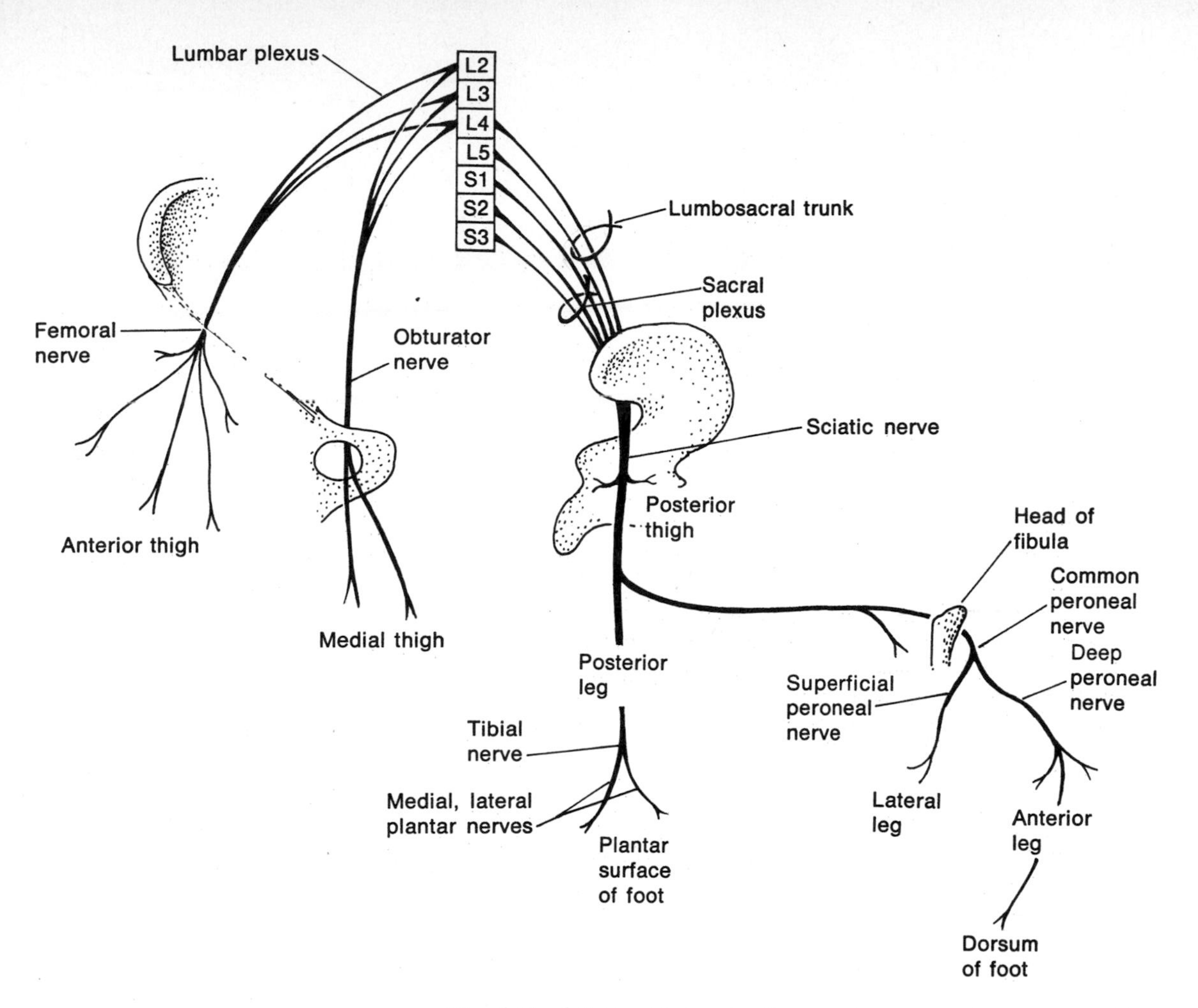

Fig. 5-32 Scheme of the distribution of major motor nerves to the lower limb. (Adapted from J. C. B. Grant, *Atlas of Anatomy*, 6th ed., Williams and Wilkins, Philadelphia, 1971.)

Fig. 5-31 (opposite) Principal veins of the lower limb. Insert: basic body venous plan.

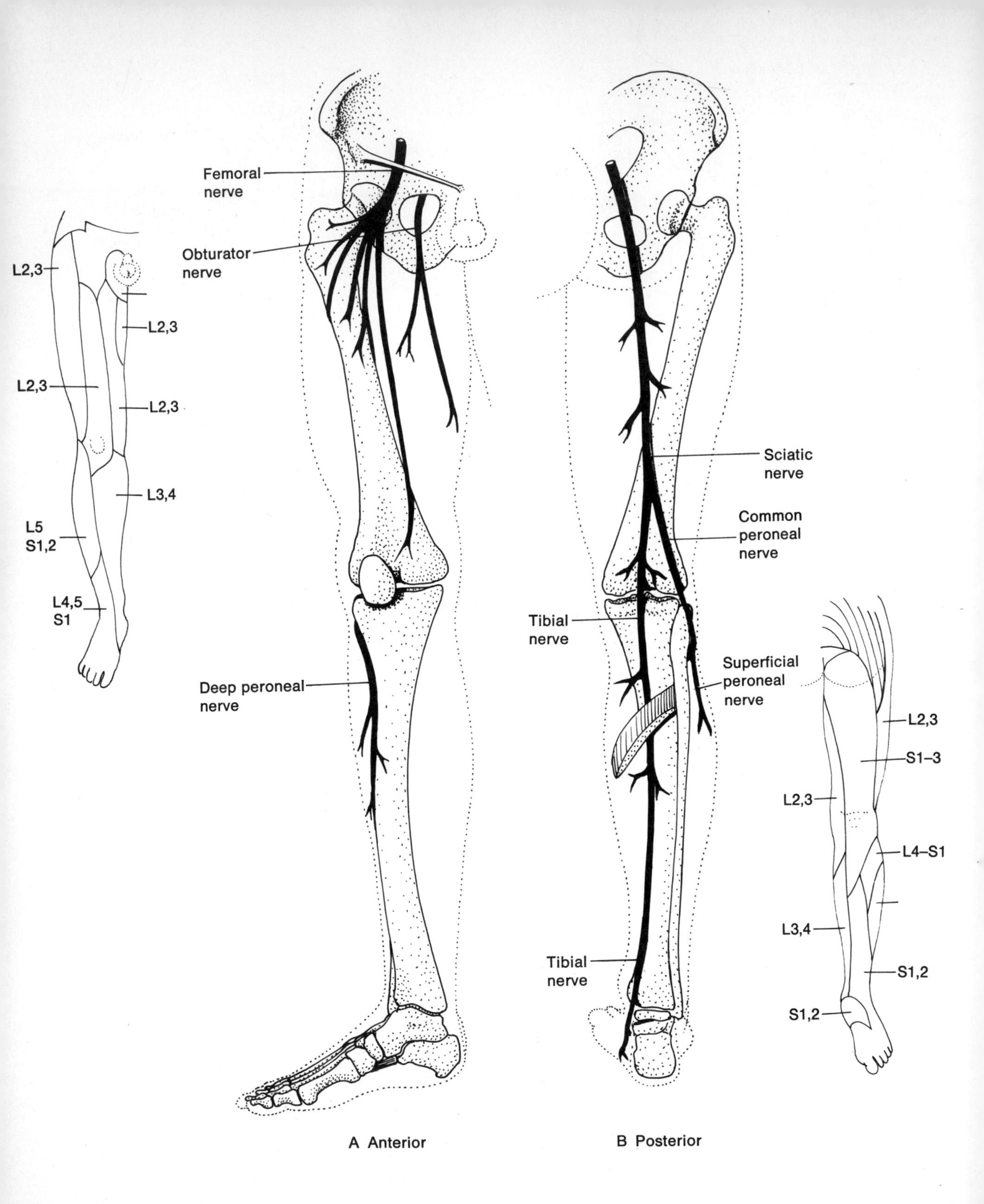

A Anterior

B Posterior

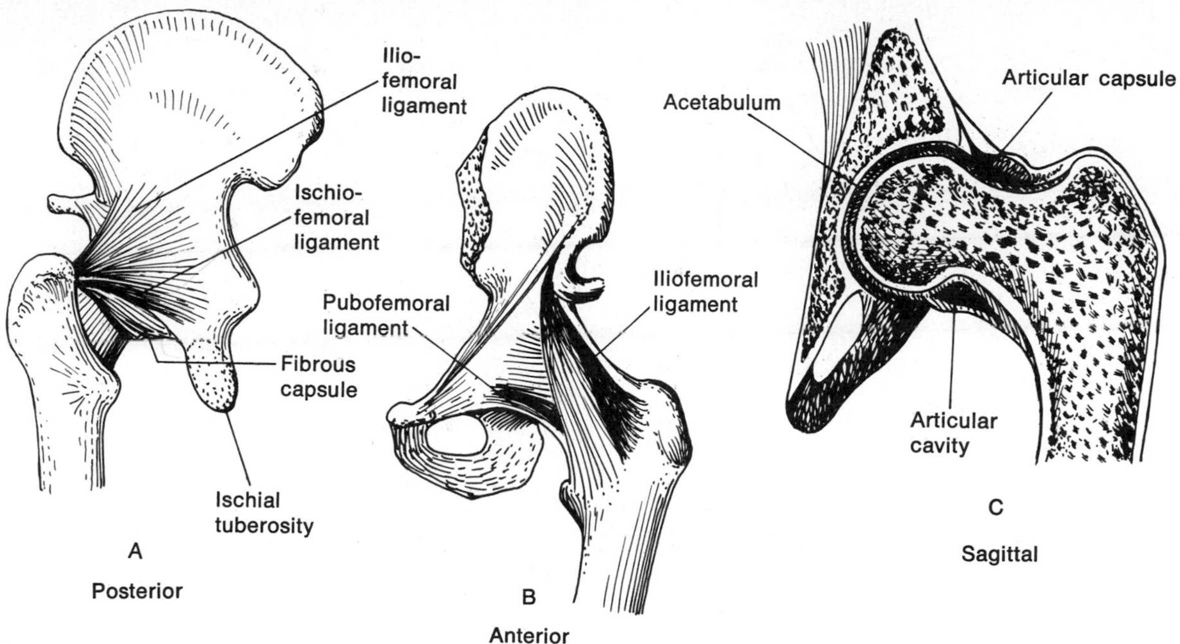

Fig. 5-34 The hip joint.

These ligaments are responsible for checking excessive movement at the joint.

The interior of the joint capsule is lined with synovial membrane and is partially filled with fat.

The vascular and nerve supply to the joint is by several small twigs from neighboring vessels and nerves.

The joint, in addition to the ligaments just discussed, is reinforced by several muscles of the hip, the most effective being the deep lateral rotators of the hip joint (Fig. 5-29).

FEMORAL REGION

The skin and superficial fascia of this region are continuous with those of the buttock. The amount of fat in the superficial fascia of the thigh is considerably more than in the arm but less than that of the buttock. You can discover this on yourself by rubbing the skin of each of these regions over the underlying deep fascia and mentally "measuring" the depth of fat between the two (see also Fig. 5-35).

The deep fascia of the thigh (fascia lata) is strong and taut. Superiorly it is continuous over the gluteus maximus and is

Fig. 5-33 (opposite) Nerves and dermatomes of the lower limb.

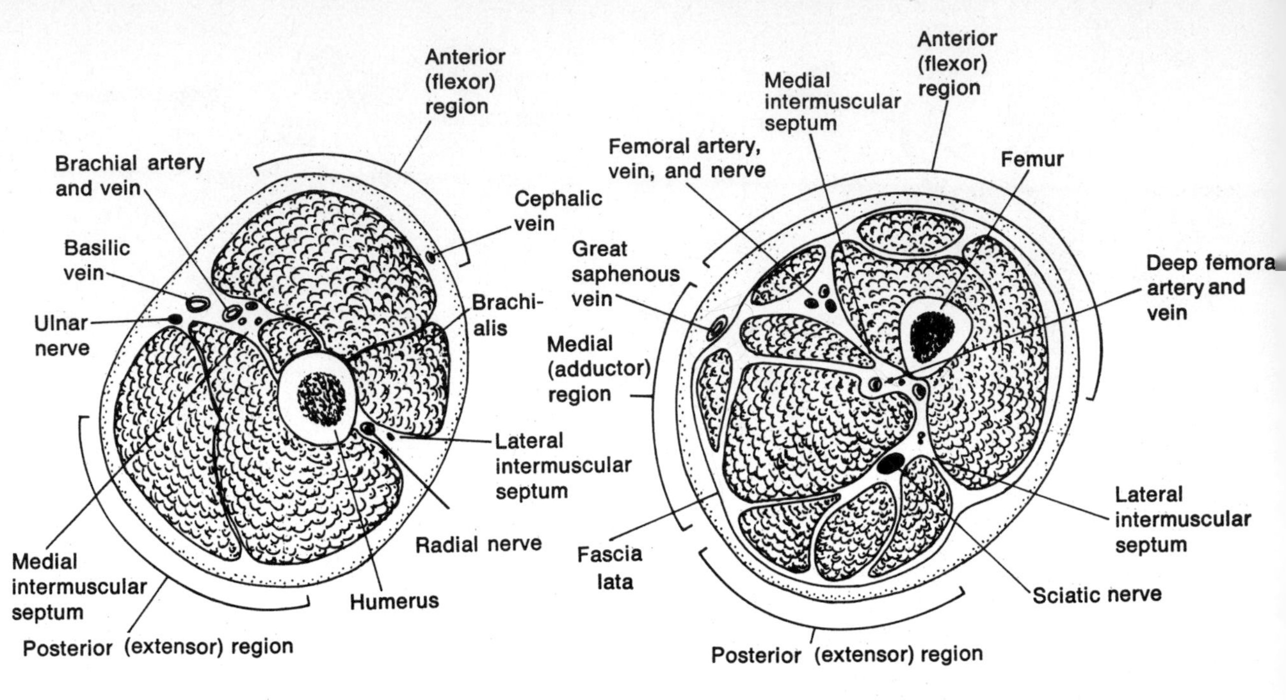

Fig. 5-35 Comparison of the mid-arm and mid-thigh in cross section.

strengthened laterally as the iliotibial tract. It is continuous distally with the capsule of the knee joint. Now feel on yourself the lateral, posterior, anterior, and medial thigh compartments and concentrate on the tautness of the deep fascia covering the superficial musculature of each region. Which region has the most condensed deep fascial layer? As in the arm the thigh incorporates medial and lateral intermuscular septa derived from the deep fascia. In Fig. 5-35 note that the medial septum separates the anterior muscle compartment from the medial muscle group, while the lateral septum walls off the posterior thigh muscles from the anterior compartment. The nerves and vessels to the thigh, except those destined for cutaneous regions, generally travel in one or more of the layers of deep fascia.

You will recall that the muscles of the thigh are arranged into anterior, posterior, and medial compartments. In general, each of the three nerves supplying the thigh pass from the pelvic interior directly into its related compartment (Figs. 5-32 and 5-33). The medial thigh compartment (adductors) is served by the **obturator nerve;** the anterior thigh compartment (flexors) is served by the **femoral nerve;** the posterior

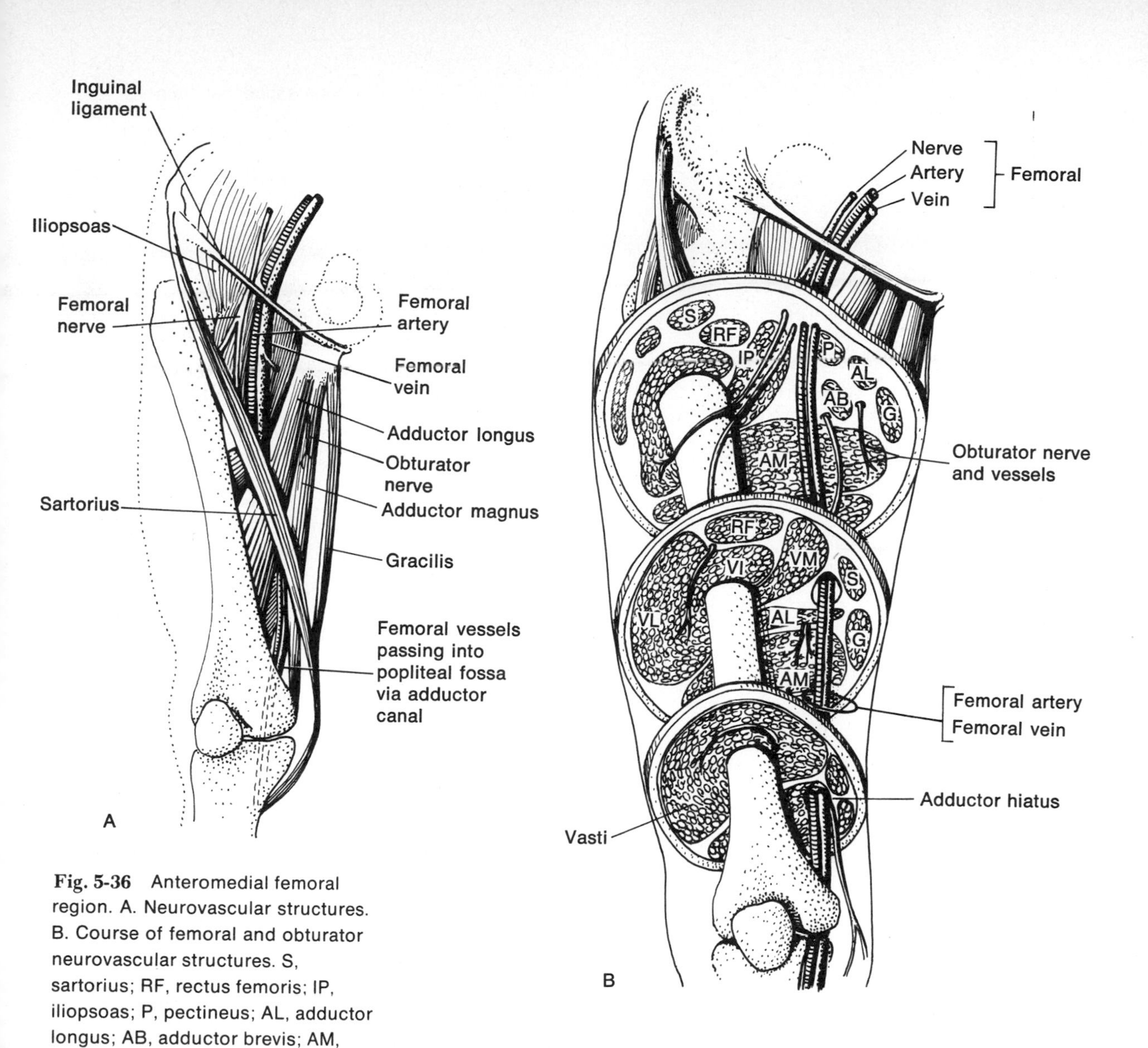

Fig. 5-36 Anteromedial femoral region. A. Neurovascular structures. B. Course of femoral and obturator neurovascular structures. S, sartorius; RF, rectus femoris; IP, iliopsoas; P, pectineus; AL, adductor longus; AB, adductor brevis; AM, adductor magnus; G, gracilis; VI, VL, VM, vastus intermedius, lateralis, medialis. (Adapted from J. E. Healey, *A Synopsis of Clinical Anatomy*, W. B. Saunders, Philadelphia, 1969.)

thigh compartment (extensors of the thigh, flexors of the knee) is supplied by the **sciatic nerve.**

The arterial blood reaches the thigh principally from the **femoral artery,** the distal extension of the external iliac artery (Fig. 5-30). Venous drainage is handled by the **femoral** and **saphenous veins.** The former becomes the external iliac vein as it passes under the inguinal ligament (Fig. 5-31). This area of the thigh sports a muscular triangle created by the obliquely descending sartorius and adductor longus muscles (sides) and the inguinal ligament (base). This **femoral triangle** (Fig. 5-36) may be palpable on yourself, and, if you are lean

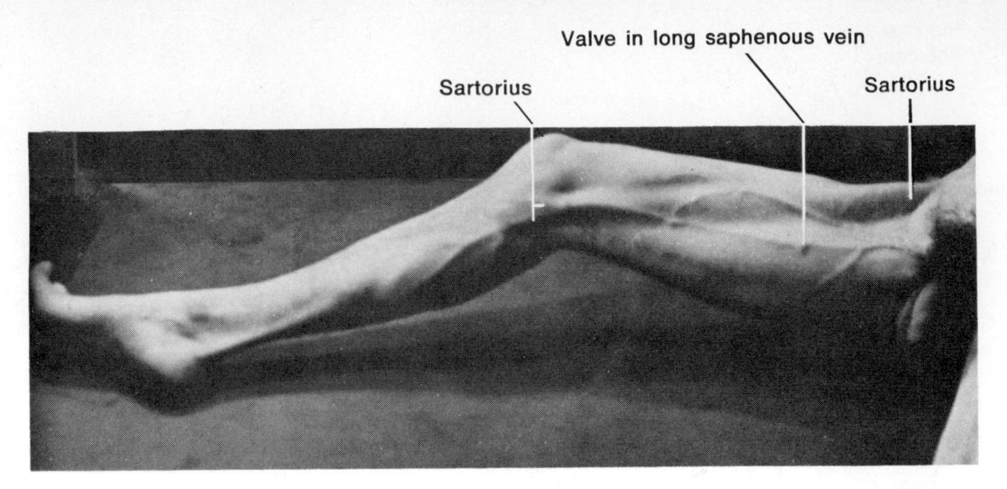

Fig. 5-37 The course of the great saphenous vein across the medial aspect of the thigh and leg. (Courtesy of R. D. Lockhart, *Living anatomy*, 5th ed., Faber and Faber, Ltd., London, 1959.)

and muscular as the subject is in Fig. 5-20, you may be able to visualize it as well. This triangle-shaped area is roofed only by fascia and skin, and the deep-lying iliopsoas and pectineus are its floor. The triangle accommodates the transiting femoral artery, vein, and nerve; thus it is here that the pulse of the femoral artery may be felt and here that pressure may be applied in the event of arterial hemorrhage. It is also here that the femoral vein receives the great saphenous vein through an opening in the fascia lata—a good site for catheterization. The great saphenous vein and its tributaries (Fig. 5-37) throughout the thigh, leg, and foot are of clinical interest due to their tendency to become varicosed in many people.

Lymph flowing through the dense lymphatic network of the thigh and lower limb ultimately drains into the superficial and deep sets of **inguinal lymph nodes** (Fig. 5-38). These nodes are drained primarily by deep lumbar trunks in the abdomen which pour into the thoracic duct (Fig. 3-36). The quite palpable superficial inguinal nodes are employed by the physician during physical examinations, for swollen nodes can be a sign of one of any number of disease processes.

KNEE JOINT (see Fig. A-5, Appendix, for x-ray view)

The knee joint—the largest and most complex joint of the body—is the combined articulations of the femur with the patella and the tibia (at both of its condyles) (Fig. 5-39). All three

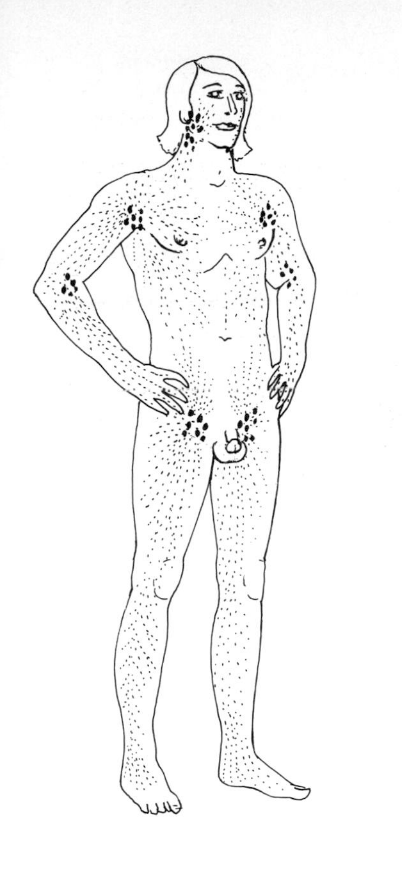

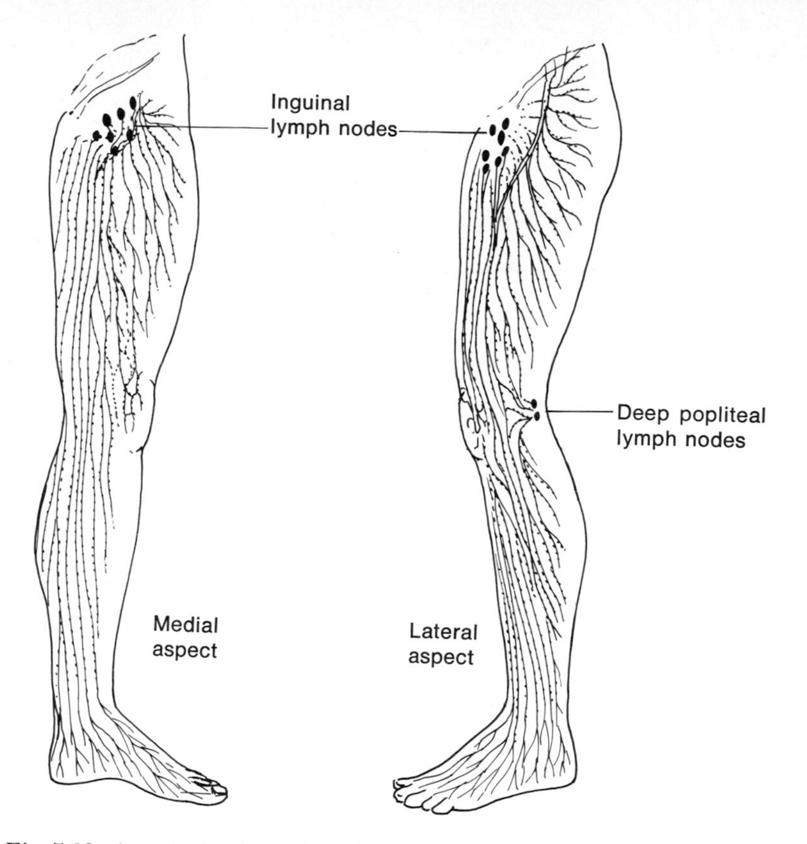

Fig. 5-38 Lymphatic circulation of the lower limb. Insert: basic plan of superficial lymphatics.

articulations are synovial in character—the patellofemoral articulation is the gliding type and the tibiofemoral articulations are both essentially of the hinge type.

The knee joint supports the weight of the body. The center of gravity in the erect position passes through the condyles of the femur (Fig. 5-40). Furthermore, it acts as a fulcrum for two of the longest bones in the body. The joint is strengthened by the presence of many ligaments and tendons as well as muscles crossing the joint. Refer now to Fig. 5-39 as you read the following:

The capsule of the tibiofemoral (knee) joints is reinforced medially and laterally by the **tibial** and **fibular collateral ligaments,** respectively. The tibial collateral ligament is considered by most authorities to be the primary stabilizing ligament of the knee. It is also subject to tearing under excessive side loads, that is, forced abduction of the tibia, occurring frequently in such sports as football.

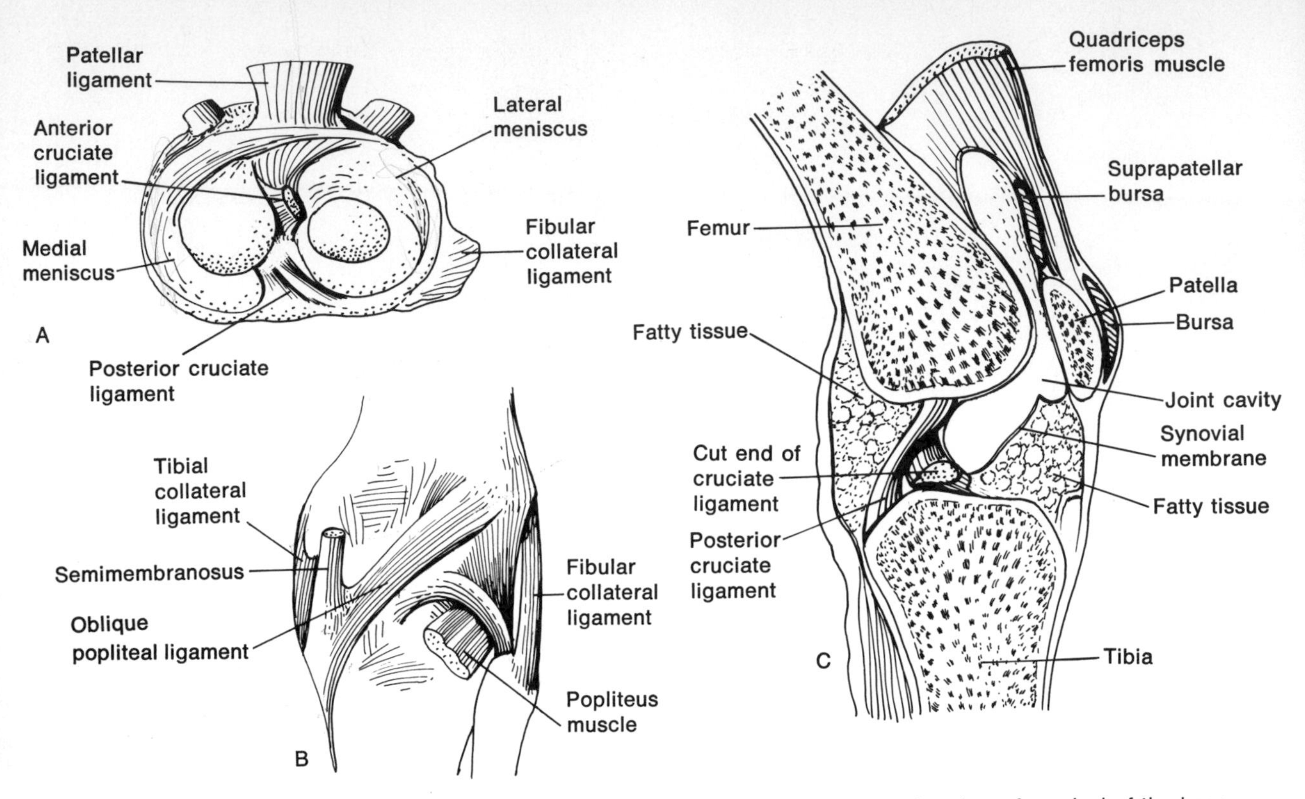

Fig. 5-39 The knee joint. A. Superior view of menisci of the knee joint. B. Posterior view. C. Sagittal section.

The posterior capsular wall of the joint is reinforced by ligaments which act to resist hyperextension ("bending backward") of the joint. The tendons of the gastrocnemius and the "hamstring" muscles reinforce the posterior aspect of the joint, also acting to check hyperextension.

The tendons of the gracilis, the sartorius, and the iliotibial tract reinforce the joint medially and laterally.

The anterior capsular wall is interrupted by the patella.

Proximal to the patella, the capsule is reinforced by the tendon of the quadriceps femoris. Distally, the capsule is strengthened by the patellar ligament. These are palpable.

Note that the condyles of the tibia and femur do not interlock precisely. Fibrocartilaginous lateral and medial **menisci** create a more agreeable union by deepening the concavity of the tibial condyles.

The anterior and posterior **cruciate** ("cross-shaped") **ligaments**—the internal ligaments of the joint—act to steady the tibiofemoral articulations during movement.

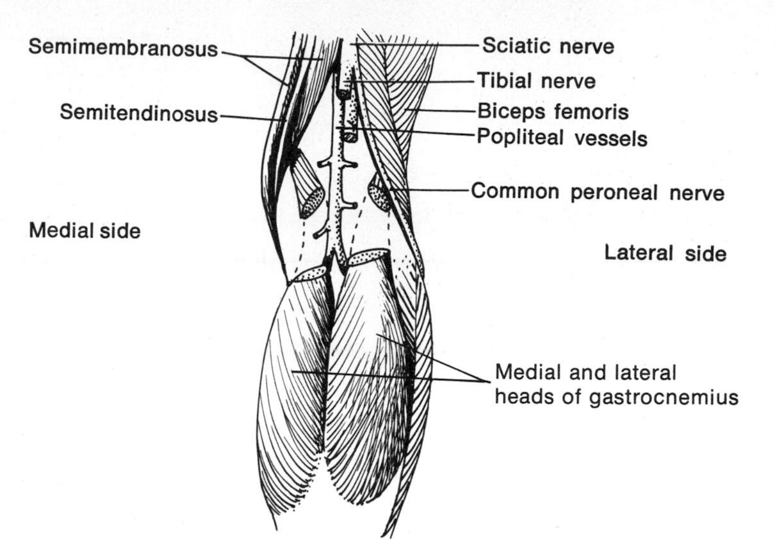

Fig. 5-41 Popliteal fossa, deep dissection.

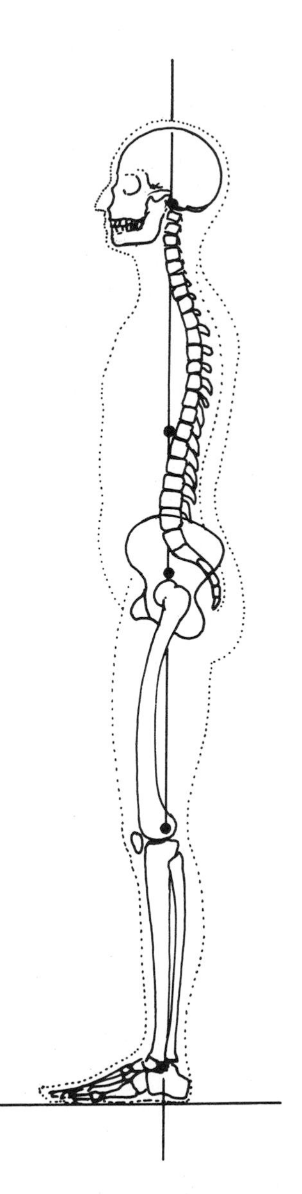

The joint is serviced by twigs from the three great nerves of the lower limb and by branches of the arterial anastomoses about the knee (Fig. 5-30).

POPLITEAL FOSSA

The popliteal fossa (Fig. 5-41) is created by (1) the popliteal surface of the femur and the posterior wall of the knee joint capsule (both forming the floor of the fossa) and (2) the "hamstrings" above and the gastrocnemius below (these are the fossa's sides). The fossa is covered by muscle (in part), fascia lata, superficial fascia with variable amounts of fat, and skin. The walls of the fossa are easily palpable (Fig. 5-19): sit on a chair with your knee flexed about 45° and place your hands around the knee such that your fingers are in contact with the sides of the fossa. Now alternately flex and relax your leg and feel, laterally, the biceps femoris tendon, and medially, the tendons of semitendinosus and semimembranosus (the tendon of the former is more posterior and more prominent when the leg is flexed).

If you will also lift your heel off the floor (extending your toes and plantarflexing the ankle and foot) while feeling the

Fig. 5-40 Line of gravity passes through the femoral condyles in normal erect position.

lower part of the fossa you can often feel the two heads of gastrocnemius contract.

The principal structures in transit within the popliteal fossa may be seen in Fig. 5-41. They are as follows:

1. The distal continuation of the femoral artery, the popliteal artery. Branches of this artery join with more proximal branches of the femoral and deep femoral arteries to form a collateral route of circulation around the knee.
2. The popliteal vein, the proximal continuation of the small saphenous, anterior, and posterior tibial veins.
3. The tibial nerve, one of two terminal branches of the sciatic nerve.
4. The common peroneal nerve, the other terminal branch of the sciatic nerve.

All of these structures are, of course, bundled in deep fascia and adipose tissue and may be difficult to palpate because of this. The popliteal fossa is a pressure point in the event of arterial bleeding in the leg or below. A number of popliteal lymph nodes are also found within the fossa receiving distal afferent lymphatic vessels and sending off proximal efferent vessels to the inguinal nodes.

CRURAL REGION

The leg (L., *crus*) is that region between the knee and the ankle. It is further divisible into anterior, posterior, and lateral compartments. By palpating the leg (Figs. 5-26 and 5-27) you will come in contact with structures already familiar to you (Fig. 5-42). Proximally, the head of the fibula on the lateral side and the tibial tuberosity on the anterior side are well-known landmarks. The entire anteromedial aspect is taken up by the subcutaneous surface of the tibia. Anterolaterally, the tibialis anterior can be felt, as can the peroneal muscles laterally. Posteriorly, the gastrocnemius can be easily felt, as can its tendon, tendo calcaneus. Look now at the anterior aspect of your leg. The skin and superficial fascia are relatively thin here so tributaries of the great saphenous vein may often be seen ascending obliquely toward the great vein, which may itself be seen on the anteromedial aspect of the leg and thigh (Fig. 5-37).

A careful study of Fig. 5-42 will reveal a more detailed consideration of the neurovascular structures and relations of the leg. Study this figure as you read.

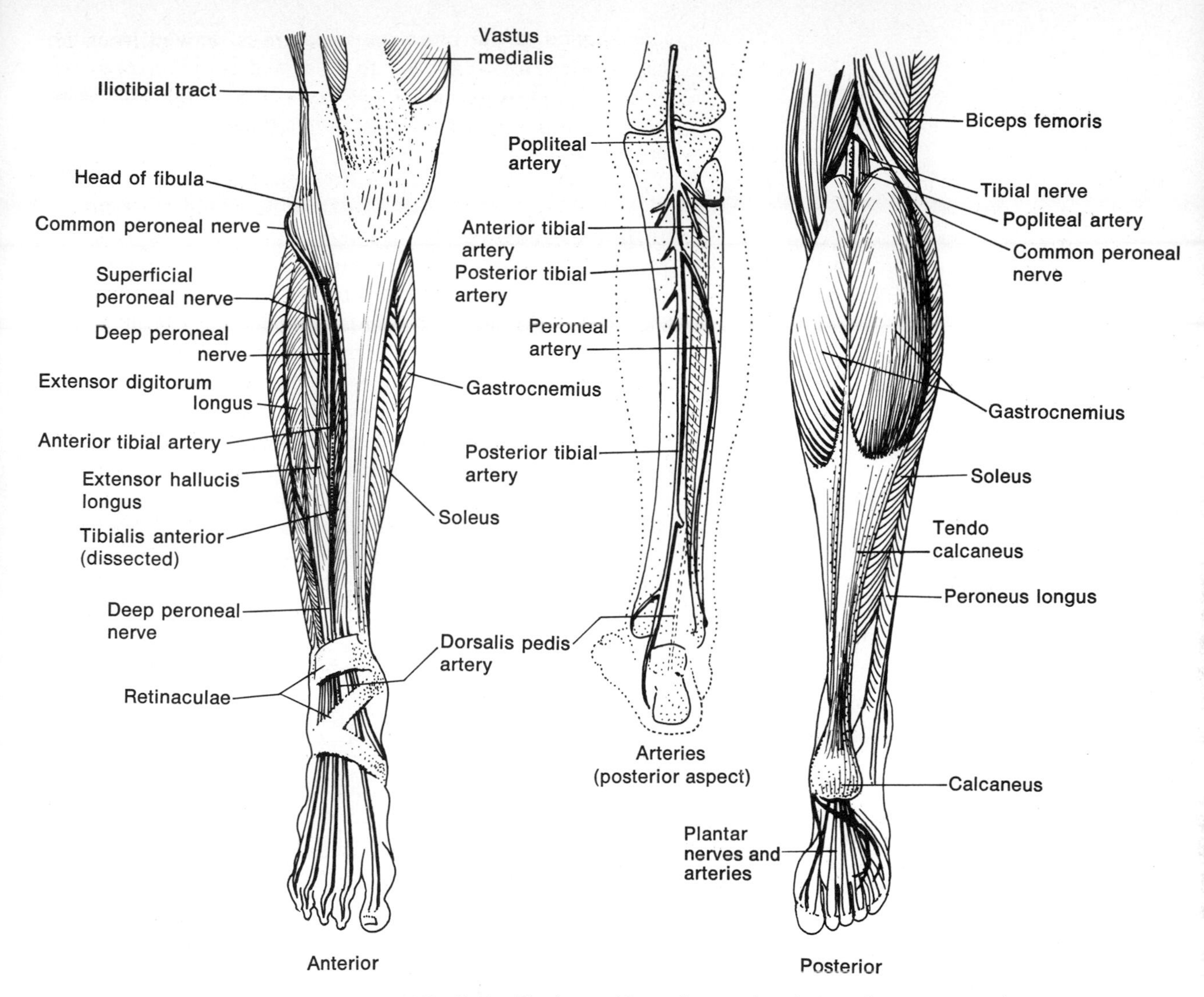

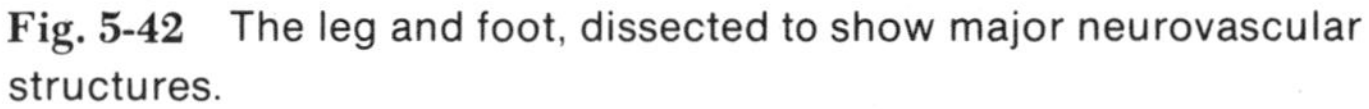

Fig. 5-42 The leg and foot, dissected to show major neurovascular structures.

The major source of arterial blood to the leg is the **popliteal artery,** which divides into an **anterior tibial artery** and a **posterior tibial artery.**

The anterior tibial artery courses distally on the anterior aspect of the interosseous membrane, sending muscular branches to the crural extensor muscles.

The posterior and lateral compartment of the leg is served by the larger posterior tibial artery whose largest branch, the peroneal artery, helps supply both flexor and extensor musculature.

Innervation to the crural musculature is derived from the sciatic nerve, which splits into tibial and common peroneal nerves just above the popliteal fossa (Fig. 5-33). The **tibial nerve** supplies the posterior compartment; the **common peroneal nerve** supplies the anterior and lateral regions.

REGION OF THE FOOT

The region of the foot is generally described as that part of the lower limb distal to the malleoli. The surface features of this region have been considered earlier.

You are already familiar with the structure of and functional concepts regarding the arches of the foot; thus you may remember that the primary support for these arches is derived from ligaments. Secondary support is achieved by muscles and the architecture of the bones themselves.

By referring to Fig. 5-43 you may study the primary ligaments most effective in supporting the longitudinal arches.

The plantar region of the foot is innervated by branches of the tibial nerve; the extensor region is served by the terminal

Fig. 5-43 Some important ligaments of the foot. A. Medial aspect. B. Lateral aspect. C. Plantar surface.

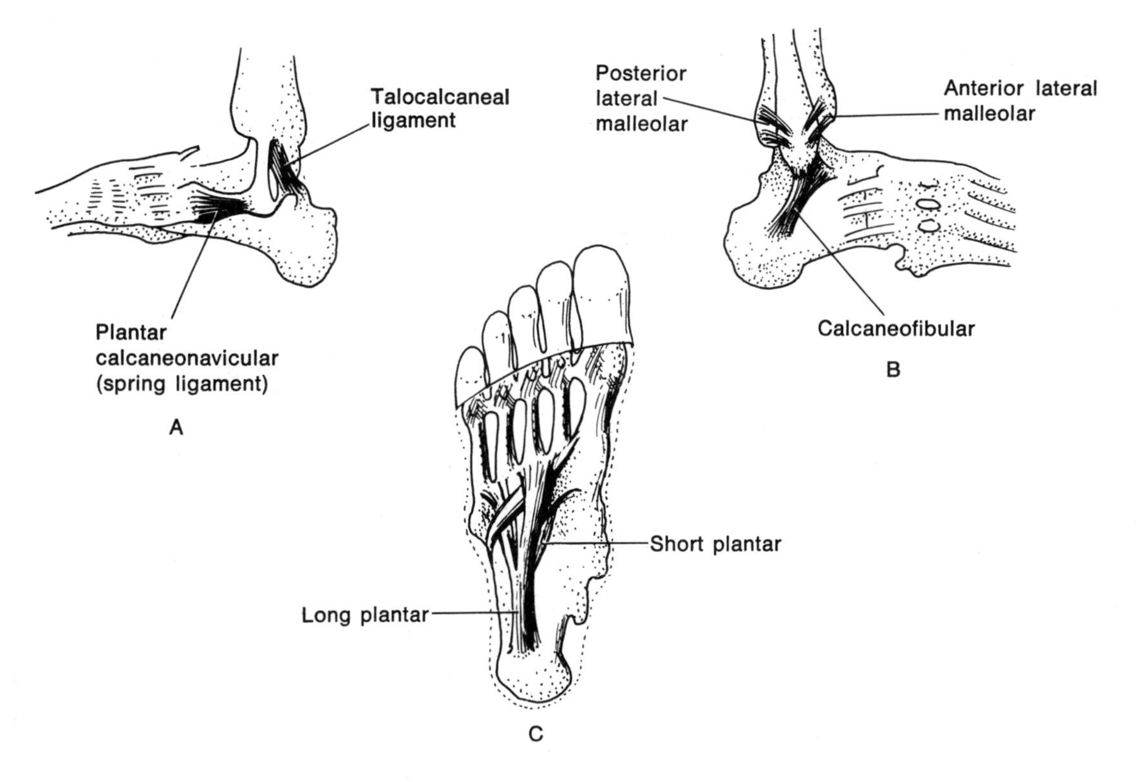

fibers of the deep peroneal nerve. The arterial supply and venous drainage as well as the nerve supply may be studied in Fig. 5-42.

The tendons of the foot are enclosed in tunnel-like tendon sheaths at critical points but they lack the interconnections of their fellows in the hand and they are not as subject to injury.

UPON REFLECTION

You have now considered both upper and lower components of the appendicular skeleton, the muscles that clothe them, the supplying arteries and nerves, and the draining veins and lymphatics. You should now reflect on their structural and functional similarities and disparities. One could philosophize at length as to the significance of these similarities, e.g., are the likenesses related to our humble evolution in that we at one time used our upper limbs for support during walking? Whatever, there are enough similarities in morphology and function to make one wonder. Are you aware of these similarities?

REFERENCES

1. Anson, B. (ed.): *Morris' Human Anatomy,* 12th ed., McGraw-Hill, New York, 1966.
2. Wells, K. F.: *Kinesiology,* 4th ed. W. B. Saunders Co., Philadephia, 1966.
3. Healey, J. E.: *A Synopsis of Clinical Anatomy,* W. B. Saunders Co., Philadelphia, 1969.
4. Lockhart, R. D., G. F. Hamilton, and F. W. Fyfe: *Anatomy of the Human Body,* 2d ed., J. B. Lippincott Co., Philadelphia, 1965.
5. Royce, J.: *Surface Anatomy,* F. A. Davis Co., Philadelphia, 1965.
6. Lockhart, R. D.: *Living Anatomy,* 5th ed., Faber and Faber, Ltd.,London, 1959.
7. Scientific American (reprint) references:
   1070 The Antiquity of Human Walking (Napier, 1967).
   1114 How Animals Run (Hildebrand, 1960).

clinical considerations

In this section, injuries to the major nerves of the lower limb will be examined.

Injury of the femoral nerve can be caused locally or may result from trauma to the spinal roots (L2, 3, 4) or the central nervous system itself.

Knowing the muscles innervated by the femoral nerve, you would probably suspect (and rightly so) that, following severe injury or cutting of this nerve, flexion of the hip is weakened while active extension of the knee is nearly impossible. Such paresis (weakening) and paralysis are manifested in walking, so that the leg must be swung forward into extension. Walking uphill or under any condition in which there is resistance to extension of the knee is not possible.

Injury of the obturator nerve (obturator palsy) is characterized by the affected limb swinging laterally in an arc during walking. The decreased action of the adductor muscles causes this. Paralysis of the adductors is rarely complete, as a portion of adductor magnus is innervated by the sciatic nerve.

Injury to the sciatic nerve is more common, probably because of its extensive origin and travel through the lower limb. Causes of such injury include diseases of the CNS, a herniated intervertebral disc pressing on roots of the nerve, stretching of the nerve as in combined flexion of the hip and full extension of the knee, and other pathological conditions of the pelvis or lower limb. If the injury is proximal to the nerves to the "hamstring" muscles, movement of the lower limb would be reduced to flexion, adduction, and possible rotation of the thigh. All movements and sensations in the leg and foot would be lost.

Injury to the tibial or common peroneal nerve may show up in one's gait. If you can recall the muscle groups served by each nerve you may be able to predict the deviations from normal gait following transection of each nerve. Study Fig. 5-44 as you read the following:

1. A normal gait is shown in A.
2. A "shuffling" gait (B) follows transection of the right tibial nerve, leading to paralysis of plantarflexors of the ankle and foot and flexors of the toes. If the nerve is cut at the ankle, the foot may assume a "claw" appearance

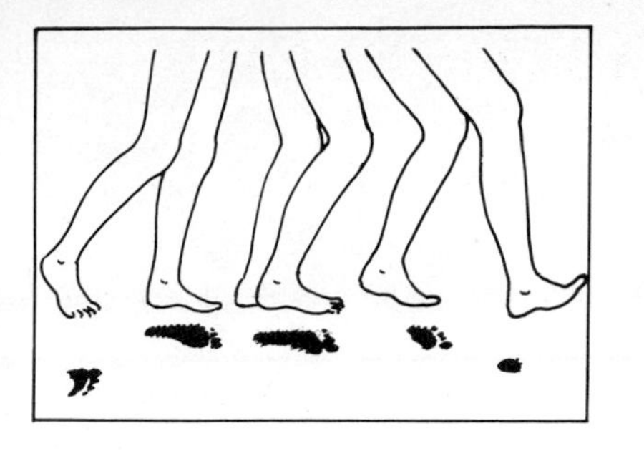

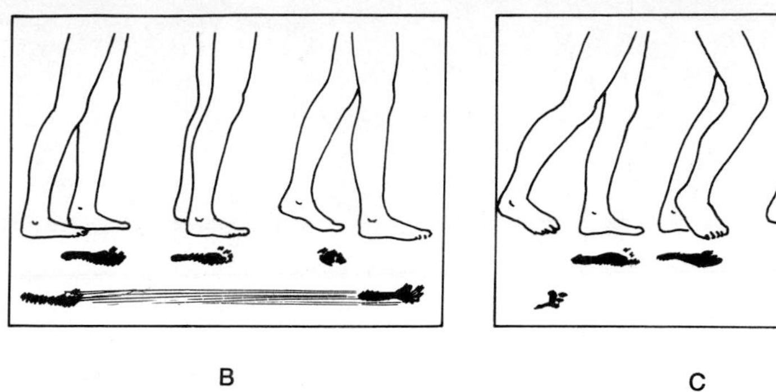

A B C

Fig. 5-44 A. Normal gait. Left: right heel is raised and foot actively plantarflexed for push off. Middle: right knee slightly flexed, leg carried forward mainly by gravity. Foot is dorsiflexed to clear ground. Right: right knee extended at end of step. Heel reached ground first. B. The "shuffling" gait of tibial nerve paralysis on the right side. Left: no push off since right foot cannot be plantarflexed nor heel raised. Middle: right leg dragged forward; no free swing as the foot does not leave the ground. Right: right foot still flat on the ground. It receives transferred weight. C. The "high step and flop" gait seen in paralysis of the common peroneal nerve on the right side. Left: the push off is mainly from the lateral border of the foot, inverted by the unopposed action of tibialis posterior. Middle: right knee flexed more than normal to compensate for the dropped foot, the toes and lateral border of which drag. Right: right foot flops flat to ground, ball of the foot first. (Courtesy of R. D. Lockhart, G. H. Hamilton, and F. W. Fyfe, *Anatomy of the Human Body,* 2d ed., J. B. Lippincott, Philadelphia, 1965.)

(like the "claw hand" caused by median and ulnar nerve injury).

3. The "high step and flop" gait in C is brought about by transection of the right common peroneal nerve, which causes the loss of the extensors of the right foot and toes (foot drop or "flop") and the unopposed inversion of the foot. Relative to the latter sign, the everters/adductors of the foot are all paralyzed.

unit 6 the head and neck

In a manner of speaking, the body is the instrument of the head and the neck is the thoroughfare between the two. It is in the head that all the decisions are made. It is also there that the air we breathe is first received. It is in the head that we verbalize our thoughts and our feelings, and with some of its organs that we take in food and process it for ultimate digestion. Further, the head supports those special sense receptors related to vision, hearing, equilibrium, taste, and smell, which report directly to the brain. The anterior surface of the head, the face, is so structured with musculature that a myriad of emotional expressions can be created. It is by way of the neck that orders of movement in response to decisions of the head are carried out, and that the air inhaled reaches the lungs and the food eaten passes to the stomach.

Certainly these are reasons enough to investigate the structure and function of our head and neck.

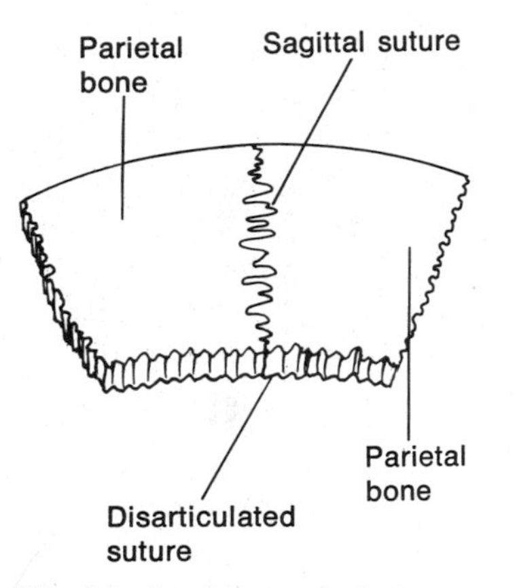

Fig. 6-1 A suture, typical of the joints between cranial bones.

bones of the skull

The bones making up the framework of the head are collectively termed the **skull.** That part of the skull forming a vault for the brain is the **cranium.** The vault itself is the **cranial cavity.** The upper, smooth part of the cranium is the **calvaria,** or skullcap. Its bones are easily palpated—try it! What remains are the bones of the face, which too are easily felt on the anterior part of the skull. These facial bones are of various shapes, unlike the cranial bones, which are all *flat*. The bones supporting the neck (cervical vertebrae) will be considered with Unit 7.

Most of the bones of the head are interconnected by fibrous tissue, the articulating surfaces being uneven like two pieces of a puzzle (Fig. 6-1). These joints, you may remember, are immovable and are termed *sutures*. With age, the fibrous tissue is replaced by bone (synostosis). In the fetus and newborn infant, the edges of the flat bones of the calvaria have not yet ossified, and diamond-shaped fibrous membranes make up the defect. These "soft spots" of the skull are termed *fontanelles*, and are eventually replaced by bone (Fig. 6-2).

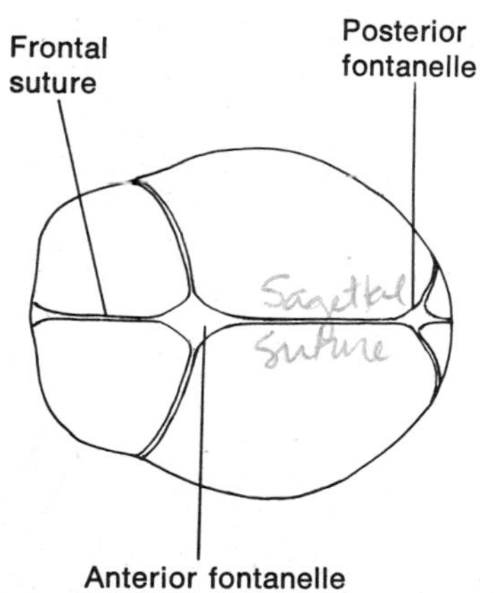

Fig. 6-2 The fontanelles as seen in the skull of the newborn. Superior view.

THE SKULL AS A WHOLE

While palpating your own skull and reading the following paragraphs, refer to Fig. 6-3:

The calvaria is relatively smooth, with small rises and depressions being more or less recognizable.* The anterior border of the calvaria may be felt as the upper ridges of the orbits (bony sockets for the eyeballs); laterally the calvaria disappears deep to the arch (zygomatic arch) palpable at the sides of the face; posteriorly it disappears deep to and below the muscles of the posterior neck encroaching upon the occipital bone.

Feel as much of the extent and borders of the orbits as you can—seven different bones join together to form the walls of these cavities.

The bony nasal cavities cannot be palpated because of the presence of that cartilaginous protuberance (the nose) lengthening the cavities anteriorly. However, the upper segment of the nose (nasal bones) can be felt.

The oral cavity is a space open to the outside when the lower jaw (mandible) is lowered (depressed). Note that it is bordered anterolaterally, in the closed condition, by the teeth.

The articulation of the skull with the cervical vertebrae is lost to the touch in the thick and muscular posterior neck.

THE SKULL: ANTERIOR ASPECT

(See Fig. A-7, Appendix, for x-ray view)

The anterior aspect of the skull is the framework for the face. This area is characterized by the two orbits, the paired nasal cavities, the teeth, and several foramina issuing forth some of the arteries, veins, nerves, and lymphatics supplying the facial structures. Refer again to your own skull and Fig. 6-3 while reading.

The bone of the forehead, conveniently termed the **frontal bone,** is the anteriormost bone of the calvaria and protects, in part, the frontal lobes of the brain. The lower part of this bone forms the variably prominent supraorbital ridge, and then curves inferiorly and posteriorly to become the roof of the orbit on each side—parts of which can be felt on yourself. Palpate an eyebrow (overlying a supraorbital ridge) and feel for a slight notch on the ridge which may or may not be

* Some of these "bumps" (like the one you might have suffered from banging your head against a low overhead) are not bony but are swollen areas of the scalp. The pseudo-science (false science) of relating intelligence to the topography ("bumps") of the skull is called phrenology.

present—this is the **supraorbital foramen,** accommodating the supraorbital vessels and nerves to the scalp. Within the anterior part of this bone is a cavity, the **frontal sinus,** which is in communication with the nasal cavity.

The bones taking up most of the facial area and contributing to the orbit and nasal and oral cavities are the **maxillae** (maxilla, sing.). You can feel (1) the infraorbital margin, to which the maxilla contributes, (2) part of the floor of the orbit, (3) the medial part of the prominent bony buttresses just below the orbits, (4) the bone surrounding the nasal cavities, (5) the alveolar processes with sockets for the maxillary teeth, and (6) the anterior three-fourths of the palate by placing a finger against the roof of the mouth.

The maxilla is further characterized by a cavity, the **maxillary sinus** (in communication with the nasal cavities), and a foramen below the orbit (infraorbital foramen) through which pass vessels and a nerve of obvious title (infraorbital) to the surrounding face.

The relatively small facial bone separating the maxilla and frontal bone anterolaterally is the **zygomatic.** This bone contributes to the walls of the orbit, as do the frontal bone, maxilla, and others. The frontal process—a portion of the zygomatic bone—curves up along the outer margin of the orbit to

Fig. 6-3 The skull. Anterior view.

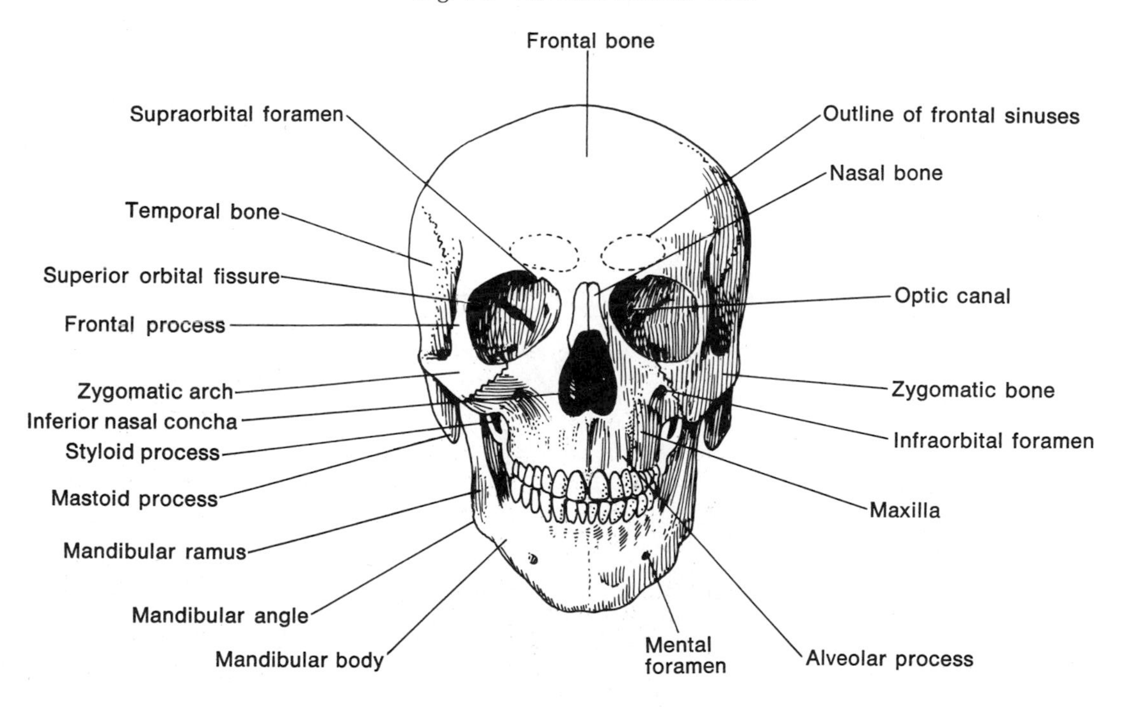

support the frontal bone. This process forms the major prominence of each cheek.

Two small adjoining bones form the anteriormost bony part of the roof of the nasal cavity. These are the **nasal bones** and are palpable throughout their anterior surface.

The lower jaw bone or **mandible** is easily felt. Start at the "tip" of your jaw (mental protuberance) and feel upward for the alveolar processes accommodating the lower teeth; note the body of the mandible; feel laterally toward the angle of the mandible. The ramus branching upward from the body can be palpated except where the massive masseter muscle interrupts. (Lock your jaws to feel the masseter.) Note the mental foramen midway along the body of the mandible. Can you name the structures in transit through this foramen?

Note that the surpraorbital, infraorbital, and mental foramina all lie on the same vertical plane. Is this by chance or do you think there may be some developmental significance behind it?

Fig. 6-4 The skull. Lateral view.

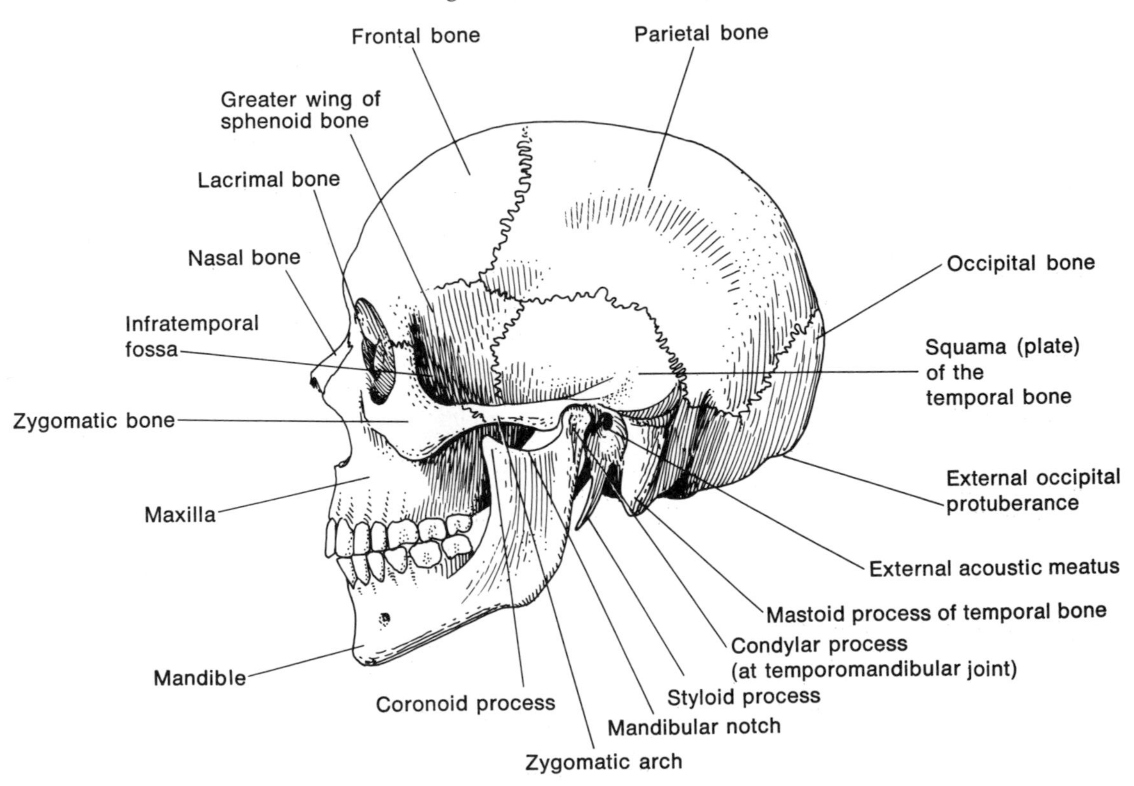

THE SKULL: LATERAL ASPECT

(See Fig. A-8, Appendix, for x-ray view)

Refer to Fig. 6-4 and move your fingers over your skull as you read the following.

The side of the skull is made up of the frontal and parietal bones, a large part of the temporal bone, parts of the sphenoid and zygomatic bones, as well as the ramus of the mandible.

The **parietal bones,** making up the major portions of the calvaria, protect the parietal lobes of the brain as well as a good part of the frontal lobes.

The **temporal bone** (to be considered in some detail shortly), occupies a major segment of the lateral aspect of the skull. The squamous portion, covered with fascia and muscle, is inferior to the parietal bone and is quite thin. Projecting anteriorly from the squamous portion is the zygomatic process, articulating, appropriately enough, with the zygomatic bone. Feel for the middle of the squamous part of your own temporal bone and trace inferiorly until you feel a definite protuberance. Trace forward across the zygomatic arch (temporal and zygomatic bony segments). In a manner of speaking, the arch acts as a strut or buttress to prevent the zygomatic bone from being driven into the cranium* as might occur upon falling on one's face. Feeling back to the protuberance at the posterior end of the arch, open and close your jaw and note the proximity of the temporomandibular joint. Move your finger posteriorly and inferiorly from the protuberance and it will enter the external auditory meatus, a canal in communication with the middle ear cavity. Behind the flap of the ear, you will feel the mastoid process of the temporal bone. Feel this process throughout. Its very existence is a result of the pull (traction) of the great muscle of the neck: sternocleidomastoid.

Just behind the frontal process of the zygomatic bone is a portion (greater wing) of the sphenoid bone (not palpable, as it is covered with fascia and muscle). It slopes down and inward with the squama of the temporal bone. The projection of the zygomatic and temporal bones (zygomatic arch) skirt around these bones, creating a deep fossa (infratemporal) which is filled with an array of structures—to the anatomist's delight—and will be considered shortly.

The rest of the lateral aspect of the skull is taken up by the ramus of the mandible. Note that the upper border of the

* An interpretation of J. C. B. Grant in *An Atlas of Anatomy,* 6th ed., Williams and Wilkins Co., Baltimore, 1970.

ramus (unpalpable) ends as two processes. The posterior one is the condyle for the joint; the anterior one is the coronoid process for muscular attachment.

THE SKULL: POSTERIOR ASPECT

Most of the posterior aspect (Fig. 6-5) can be palpated on yourself—including the major part of the occipital bone, as well as the contributions of the temporal and parietal bones. In the middle of the occipital bone, at the upper border of the hollow of the posterior neck, feel for a projection of bone—this is the **external occipital protuberance.** Extending laterally (on each side) from this point are the superior nuchal lines, which may not be palpable, as several muscles of the neck and back insert in this area.

Heretofore you have been able to palpate most of the structures discussed. In the following section, the structures are not within touch of the fingers and their consideration more properly belongs in the laboratory. However, simplified diagrams and brief discussions of key structures are presented here to make the laboratory experience more meaningful.

Fig. 6-5 The skull. Posterior view.

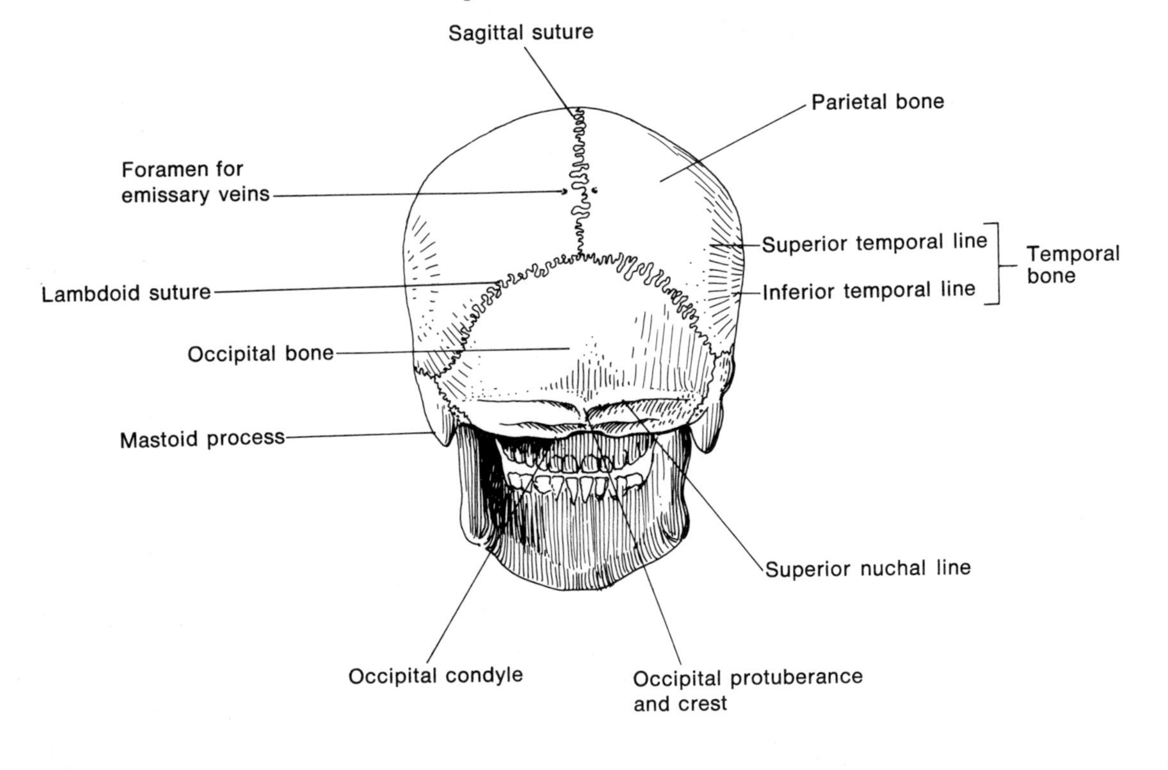

INTERIOR OF THE SKULL

The natural contours of the skull interior provide for a simple sorting of the structures within into three cranial fossae: (1) **anterior,** (2) **middle,** and (3) **posterior** (Fig. 6-6).

The **anterior cranial fossa**—supporting primarily the *frontal lobes* of the brain—is a high plateau relative to the other fossae. Refer to Fig. 6-7 and note that the general topography is more or less flat with the exception of the central part, which consists of a spear-like projection, and its neighboring depressions (the superior aspect of the ethmoid bone). The posterior aspect of the anterior fossa ends abruptly and the middle fossa begins well below the level of the former. The anterior fossa is composed of parts of three bones:

1. Frontal bone (roof of the orbit).
2. Sphenoid (lesser wing).
3. Ethmoid (roof of nasal cavities).

The **middle cranial fossa**—supporting the *temporal lobes* of the brain and the pituitary gland (hypophysis)—has a central saddle-shaped zone and two lateral cavernous depressions. Many of the nerves and vessels leaving or entering the cranium to and from the neck do so by way of foramina or canals in this fossa.

Refer to Fig. 6-6 as you read the following paragraphs:

The anterior wall of this fossa is sphenoid bone, punctuated by foramina and fissures giving passage to nerves and vessels to and from the orbit. These openings can be seen best in Fig. 6-11 (at least until you can get into the laboratory and see the actual skull). This anterior wall is also part of the posterior wall of the orbit.

The middle "Turkish saddle" **(sella turcica)** has a back (dorsum sellae) with two anterior and two posterior processes. The term *clinoid* is used to describe these processes. The Greek word *clinoid* means "like a bed," and indeed the sella and its processes can be likened to a four-poster bed. (It even has a membranous canopy.) The pituitary gland occupies the saddle. The sella and its processes belong to the sphenoid bone.

The floor of the middle cranial fossa is created from sphenoid and temporal bone. The posterior boundary of the fossa consists of a dense ledge of bone, the **petrous** (rocky) **portion** of the temporal bone, which also serves as the anterior boundary of the posterior cranial fossa and houses the fragile internal and middle ear structures.

The **posterior cranial fossa**—the deepest of the three fossae—supports the cerebellum of the brain. It is formed

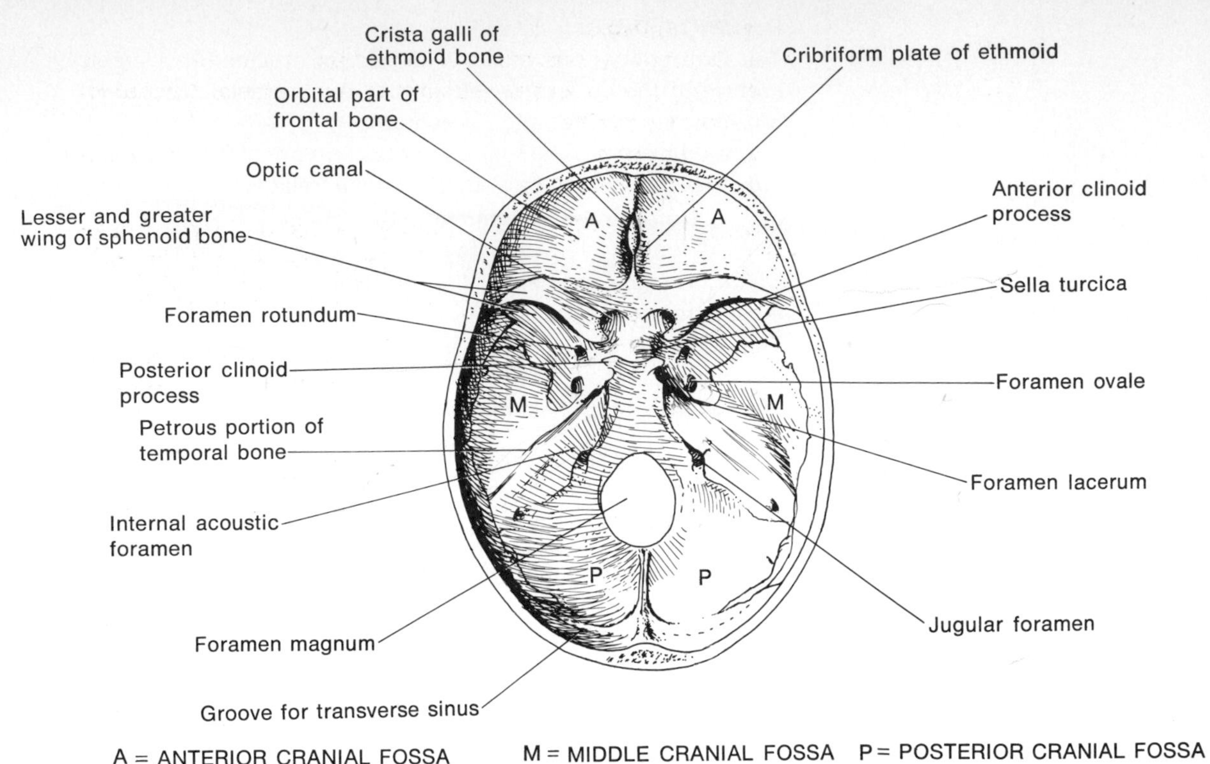

Fig. 6-6 Floor of the cranial cavity.

primarily by the occipital bone with an anterior contribution by the petrous portion of the temporal bone. In this fossa is the great foramen **(foramen magnum)** which marks the boundary between brain and spinal cord.

In the laboratory you will see that there are a variety of foramina at the floor of the skull. Consideration of these will be reserved for the laboratory experience. However note that they are there and that they serve to transmit vessels and nerves to and from the interior of the cranium.

A rough estimate of the geography of the interior bony structures on yourself can be made by considering some surface landmarks of the skull.* Concentrate on Figs. 6-3 and 6-4 and try to visualize the positions of the following structures along the lateral aspect of your own skull:

- Contents of the anterior cranial fossa.

* Walls (roof, floor, etc.) of the orbit, frontal processes, nasal bones, external auditory meatus, squamous portion of temporal bone, mastoid process, external occipital protuberance.

- Sella turcica.
- Contents of the middle cranial fossa.
- Internal and middle ear structures (petrous portion of temporal bone).
- Cerebellum.
- Foramen magnum.

A sagittal view of the skull affords a different approach to studying the interior of the cranial vault. Refer now to Fig. 6-7 and read the following description, moving from anterior to posterior.

The **frontal sinus** in the anterior wall of the anterior cranial fossa (frontal bone) drains into the nasal cavity (middle meatus).

The thin floor of the **anterior cranial fossa** has contributions by frontal, sphenoid, and ethmoid bones.

The **crista galli** is a projection of the ethmoid bone between the perforated **cribriform plates** in the anterior cranial fossa.

Fig. 6-7 The skull. Sagittal section. Internal view of the left half.

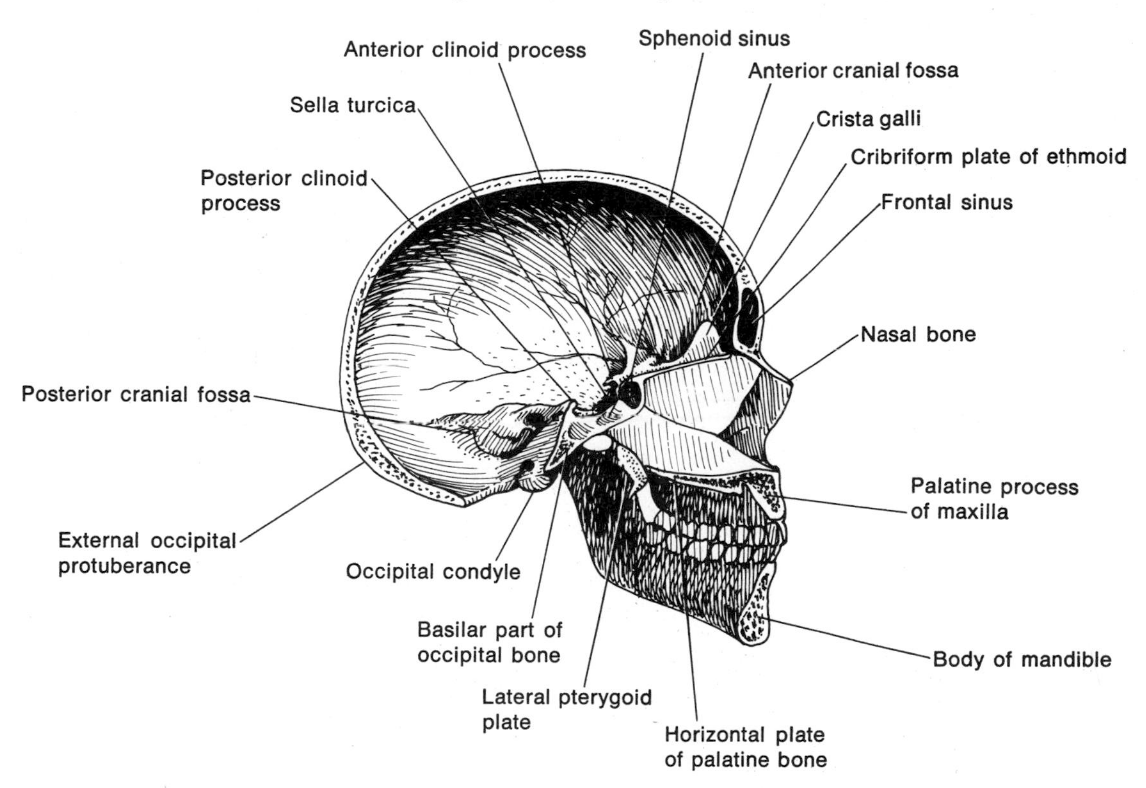

The **sphenoid sinus** of the sphenoid bone just anterior to the sella turcica drains into the uppermost chamber of the nasal cavity (sphenoethmoid recess). How might the sinus be surgically drained if infected or how might a surgeon approach the pituitary gland?*

The base of the occipital bone blends with the posterior surface of the sphenoidal body.

The **septum** divides the nasal cavity in two. The upper part of the septum is the perpendicular plate of the ethmoid; the lower is the vomer; the anterior part is cartilaginous.

The roof of the oral cavity as well as the floor of the nasal cavity is the **palate,** which consists of a pair of bones (three-fourths the maxilla, one-fourth the palatine) on each side of the midline.

EXTERNAL BASE OF THE SKULL

A quick look at Fig. 6-8 will reveal what appears to be a complex and unorganized mass of projections, depressions, arches, holes, and foramina. This, happily, is only partly true. Look again and relate it to the scheme provided in Fig. 6-9:

The area about the **foramen magnum** represents the occipital bone. This bone has a pair of condyles lateral to the foramen magnum which articulate with the first cervical vertebra (the atlas). Muscles supporting the head on the neck and providing movement of the head about the neck attach here.

The region between the base of the occipital bone and palatine portion of the palate, bordered laterally by the pterygoid plates, is taken up by the **pharynx** and its complex muscular attachments. The pharynx receives air from the nasal cavities, the bony housing of these cavities terminating at the posterior nasal apertures, and the air then passes into the pharynx. The oral cavity also communicates with the pharynx just beyond the posterior edge of the palate.

The region lateral and somewhat posterior to the pharynx is related to a pair of significant foramina on each side. The jugular foramen transmits the great internal jugular vein and the vagus nerve, along with other structures. The carotid canal transmits the internal carotid artery.

The region anterior to the posterior edge of the palatine bone between the upper left and right quadrants of teeth is the bony roof of the oral cavity, the **palate.** The foramina in

* Through the nasal cavity.

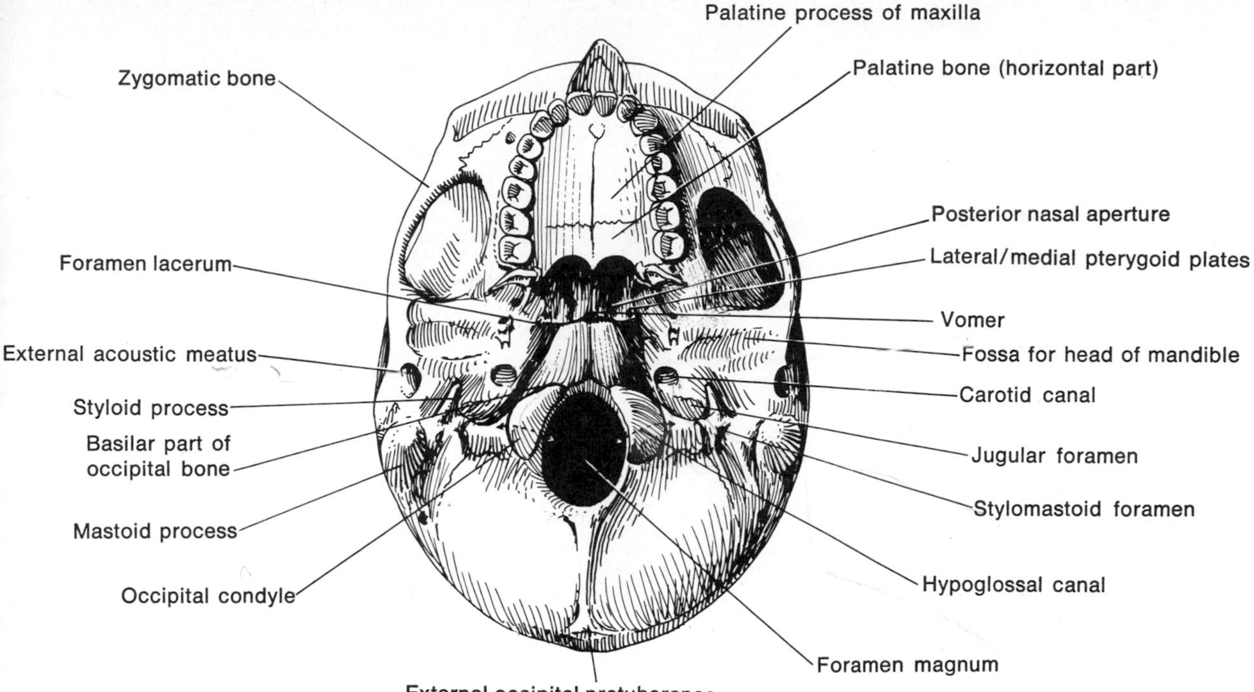

Fig. 6-8 The skull. Inferior view of the base.

this area conduct vascular and nerve structures to the mucous membrane of the palate and neighboring areas.

Finally, on the external surface of the base of the skull, there is the **infratemporal fossa,** a space lateral to the attachments of the pharynx and bordered externally by the zygomatic arch. The infratemporal fossa is stuffed with muscles of mastication, muscles relating to the middle ear, and blood vessels and nerves.

BONES OF SPECIAL SIGNIFICANCE

These are bones whose structure will provide meaningful insight as to the organization of the skull. As you read on each bone, study the appropriate illustration and concentrate on its relations to the others (Fig. 6-10).

Sphenoid bone

This is the "key" to the skull (Fig. 6-11). It occupies a central position, taking part in the walls of (1) the orbit, (2) the nasal

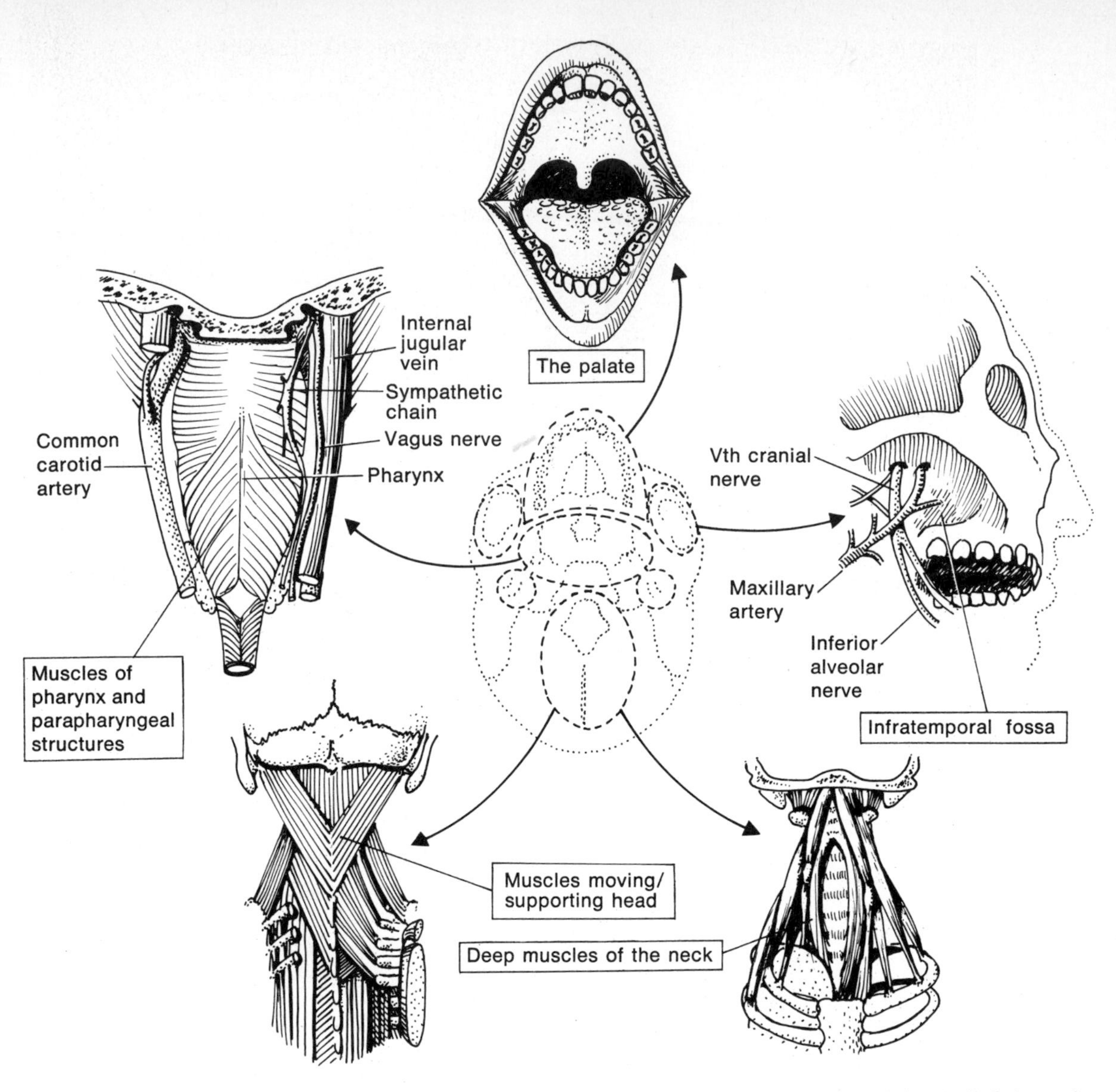

Fig. 6-9 Structural relationships of the base of the skull. Schematic.

cavity, (3) the outermost two of the three cranial fossae, and (4) the infratemporal fossa. As you can see, the sphenoid bone is the shape of a bat.

It has a body which is hollow **(sphenoid sinus),** articulating with the frontal and ethmoid bones anteriorly and joining posteriorly with the occipital bone (see Fig. 6-6).

It has a **greater** and **lesser wing** on each side, the lesser wings forming a part of the anterior cranial fossa floor, the greater wings constituting a large part of the posterior wall of

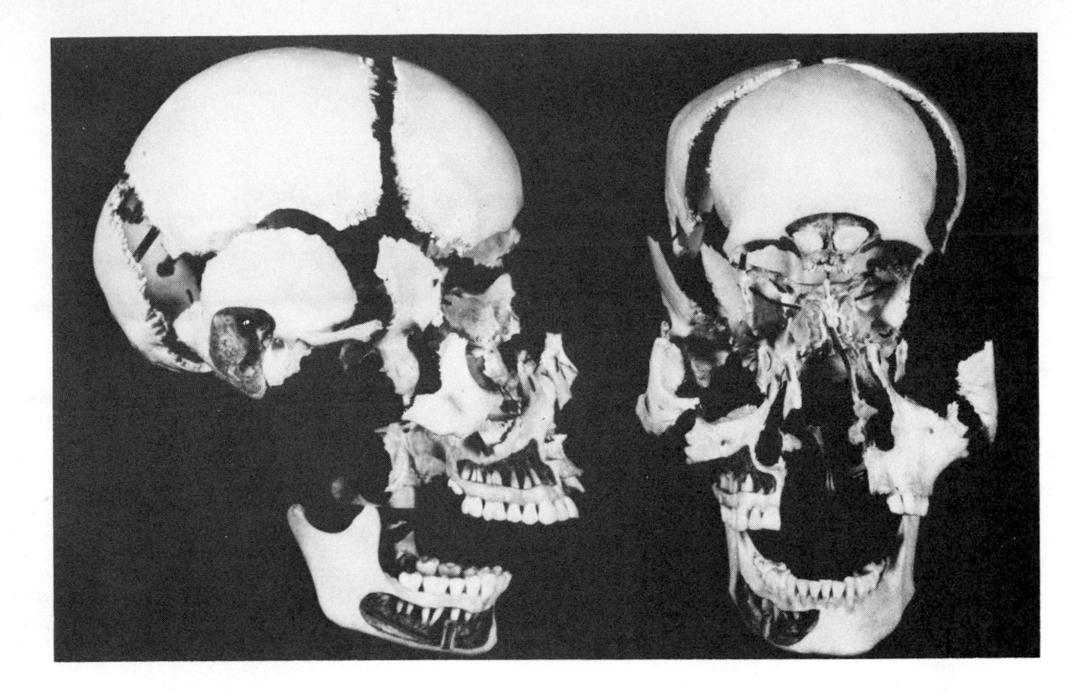

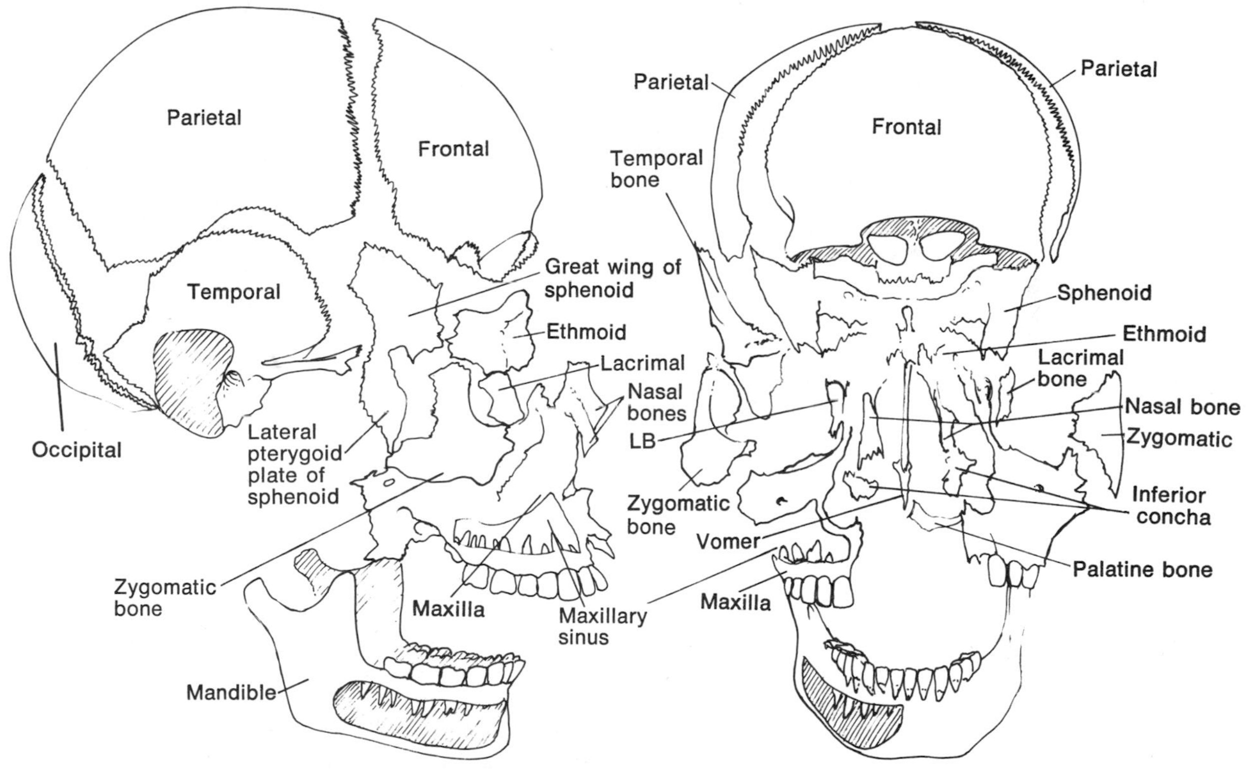

Fig. 6-10 A disarticulated skull (Beauchene skull).

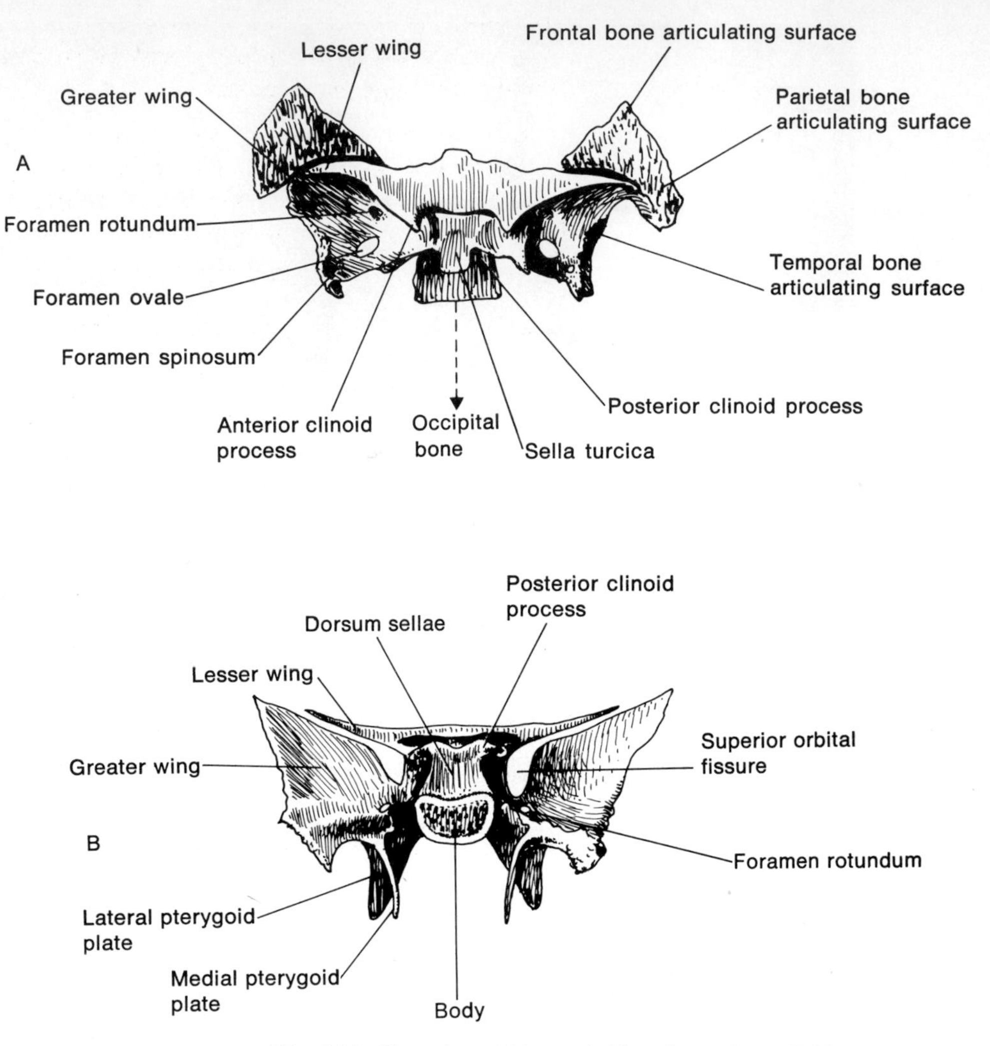

Fig. 6-11 The sphenoid bone. A. View from above. B. View from behind.

the orbit and medial wall of the infratemporal fossa. The space between the two wings is the **superior orbital fissure** (Fig. 6-11B), transmitting vessels and nerves to the orbit.

It has a double pair of "legs," the medial and lateral **pterygoid plates.** A muscle of the pharynx and a pair of masticating muscles attach on these plates.

If you will study the relationship of the sphenoid to its immediate neighbors you will have gone a long way toward understanding skull structure.

Temporal bone

The temporal (Fig. 6-12) is a complicated bone which has three basic embryologic parts:

1. A flat **squamous** part, which forms a major part of the lateral wall of the skull. The zygomatic process projects anteriorly from it. Under the posterior aspect of the process is the mandibular fossa, receiving the condyle of the mandible. The temporalis muscle of mastication arises from the squamous part.
2. A rocky **petrous** part, separating the middle and posterior cranial fossae. It houses the internal ear (containing vestibular and hearing sense organs) and middle ear structures. The very important internal carotid artery and facial nerve course through this bone. Laterally, the petrous bone terminates as the breast-shaped mastoid process.
3. A **tympanic** part, which houses the tympanic membrane (ear drum) and contributes to the external auditory meatus.

To truly appreciate the orientation of the various passageways within the temporal bone is a Herculean task; the above superficial description should be sufficient to convey the significance of the temporal bone in the general structure of the skull.

Fig. 6-12 The left temporal bone, lateral view. A. The disarticulated bones. B. The whole bone.

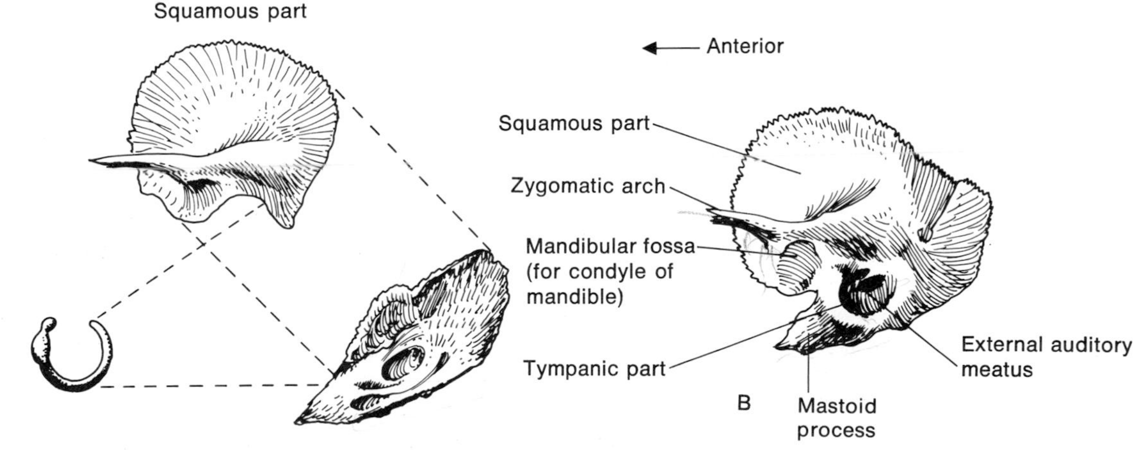

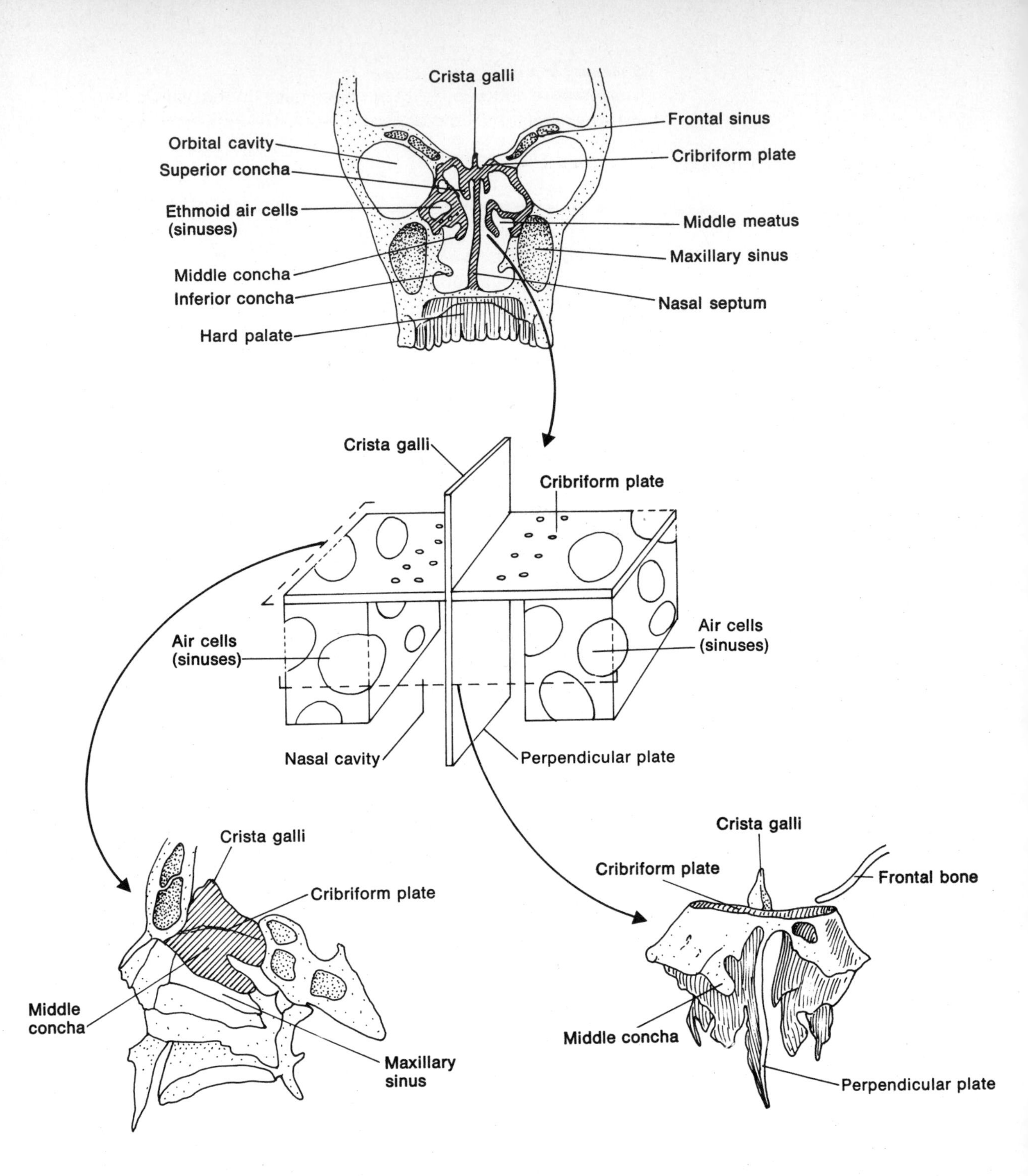
Crista galli
Frontal sinus
Orbital cavity
Cribriform plate
Superior concha
Ethmoid air cells
(sinuses)
Middle meatus
Maxillary sinus
Middle concha
Inferior concha
Nasal septum
Hard palate
Crista galli
Cribriform plate
Air cells
(sinuses)
Air cells
(sinuses)
Nasal cavity
Perpendicular plate
Crista galli
Cribriform plate
Middle
concha
Maxillary
sinus
Crista galli
Cribriform plate
Frontal bone
Middle concha
Perpendicular plate

Ethmoid bone

The ethmoid is quite delicate because of its thin plates and many air cells. It is a complicated-looking bone. However, if you concentrate on Fig. 6-13, a bit of light may be shed. The ethmoid can be likened to a box. The sides of the box are the medial orbital walls; the top of the box is formed by the cribriform plates; the box has a middle partition composed of a perpendicular plate which above is the crista galli and below is the septum partitioning the nasal cavity into two chambers. Each half of the box interior is again divided into a lateral series of spaces (air cells) and a medial passageway (nasal cavity).

Now look again at Fig. 6-13 and note:

- The **crista galli** (separating, in part, the frontal lobes and, specifically, the olfactory bulbs, where olfactory sensations are dispatched to the brain).
- The **cribriform plate** (perforated and through which the fine olfactory nerve fibers pass into the nasal cavity).
- Groups of **air cells** (anterior, middle, posterior) bordered by the nasal cavity medially and the medial orbital wall laterally and communicating with the nasal cavity.
- The midline **perpendicular plate** forming the posterior and superior component of the nasal septum and continuous above with the crista galli.
- The **middle concha** and **superior concha** arising from the wall of an air cell.
- The middle meatus under cover of the middle concha. Orifices of this meatus drain the frontal and maxillary sinuses and anterior and middle ethmoidal air cells.
- The central passageway of the nasal cavities—continuous anteriorly with the outside (via the anterior nasal apertures) and posteriorly with the nasopharynx (via the posterior nasal apertures, or choanae).

Fig. 6-13 (opposite) The ethmoid bone. At the top is a frontal section through the ethmoid and related bones. A schematic view of the ethmoid bone is in the center. At the lower left is a lateral view. A posterior view is at the lower right.

QUIZ

1. The bones of the head are collectively termed the ______.
2. That part of the skull enclosing the brain is termed the ______.
3. All cranial bones are ______ bones.
4. The joints between bones of the head are called ______.
5. Diamond-shaped fibrous membranes representing the unossified edges of cranial bones in the fetus and newborn child are termed ______.
6. The frontal, sphenoid, maxillary, temporal, and ethmoid bones all contain ______.
7. The following foramina are aligned on a vertical plane and transmit vessels and nerves of the same name: ______, ______, and ______ foramina.
8. Feel for a bony prominence on the back of the head (on the occipital bone); this is the ______ ______ ______
9. The natural contours of the skull interior create three cranial fossae. The anterior one supports the ______ ______ of the brain; the middle supports the ______ ______ of the brain; the posterior supports the ______ of the brain.
10. The bone especially important for an understanding of the skull is the ______.

regional organization according to function

REGIONS AND VISCERA RELATED TO PHONATION AND RESPIRATION

There are basically four regions of the head and neck devoted to the generation of intelligible (and unintelligible) noise, and to the preparation of atmospheric air inhaled for oxygenation of the blood by the lungs. These are:

1. Nose (including nasal cavities).
2. Sinuses of the skull.
3. Pharynx, specifically, the nasopharynx.
4. Larynx.

Nose

The nose consists of the external nose and the nasal cavity. The nasal cavity begins at the nostrils, or **nares,** and extends back to the **choanae,** the apertures that represent the entrance to the nasopharynx (Fig. 6-14).

External nose. The external nose is a skin-lined cartilaginous affair hung from a roof of nasal and maxillary bone. Feel your own nose for its pliability; note that the hard nasal bones descend from the frontal bone (the bridge of the nose). One-third to one-half of the way down the roof, the bone is replaced by cartilage. This cartilage on each side is continuous with the cartilage of the nasal septum, forming a T-shaped structure; feel this on yourself. This is significant for people who have suffered "broken" noses, for not infrequently the upper cartilage is torn and the septum of the nasal cavity bent to one side. This deviated septum is a factor in some upper respiratory problems. The flare of the nostrils is largely due to the cartilage framework.

Nasal cavity. On the basis of its microscopic structure, the nasal cavity is generally arranged into a vestibule, respiratory region, and upper olfactory region. Behind the nares is the **vestibule** (Fig. 6-14), where there are stout hairs (vibrissae) and "outdoor" skin gives way to "indoor" mucosa, that is, stratified squamous epithelium is replaced by pseudostratified columnar epithelium with cilia and goblet cells—typical respiratory epithelium.

The lateral walls of the cavity are characterized by three mucous-membrane-lined **conchae** (Fig. 6-14). The upper and middle conchae belong to the ethmoid, while the lower concha is "its own bone." The underlying **meatuses** are perforated by passageways from the sinuses. The conchae serve to increase the surface area of the nasal cavity and to create air tunnels so the warming and humidifying of air is more effective. The floor of the nasal cavity is the roof of the oral cavity—the palate, consisting of the palatine process of the maxilla in front and the horizontal plate of the palatine bone posteriorly (Figs. 6-8 and 6-9). The narrow roof of the cavity has the perforations of the cribriform plate of the ethmoid bone (Fig. 6-13).

The nasal cavity and its extensions (the sinuses and their ducts) are all carpeted with a mucous membrane (Fig. 6-14). The epithelium appears to be stratified but in fact is columnar. This pseudostratified epithelium furthermore contains secretory goblet cells and cilia. These structures are of paramount importance in that the secretions aid in trapping minute particulate matter and the cilia sweep the mixture toward the oral pharynx to be swallowed.

The lamina propria contains mucous and serous glands, particularly on the inferior concha. The tissue here is so vascular it is suggestive of the erectile tissue of the penis and clitoris. Typically, a sexually aroused individual has a "stuffy" nose, because the engorged tissue blocks the nasal passageway. It is interesting to consider the evolutionary aspect of this phenomena, for the sense of smell has long been known to be related to sexual activity in lower animals.

The many vessels of the lamina propria release a good deal of heat, which percolates up through the epithelium and heats the air passing into the nasopharynx. The lamina is contiguous with the local periosteum and perichondrium, there being no muscular layer.

The microscopic nature of the nasal cavity varies from the transitional nature of the vestibule to the classical respiratory epithelium of the respiratory region to the yellow sensory mucosa of the olfactory region at the roof.

In summary then, the nasal cavity functions to:

1. Warm the passing air (blood vessels).
2. Humidify the passing air (glands and heat).
3. Trap particulate matter (mucous cells and glands).
4. Sweep the debris to the oral pharynx (cilia).
5. Enable the sense of smell (olfactory region).

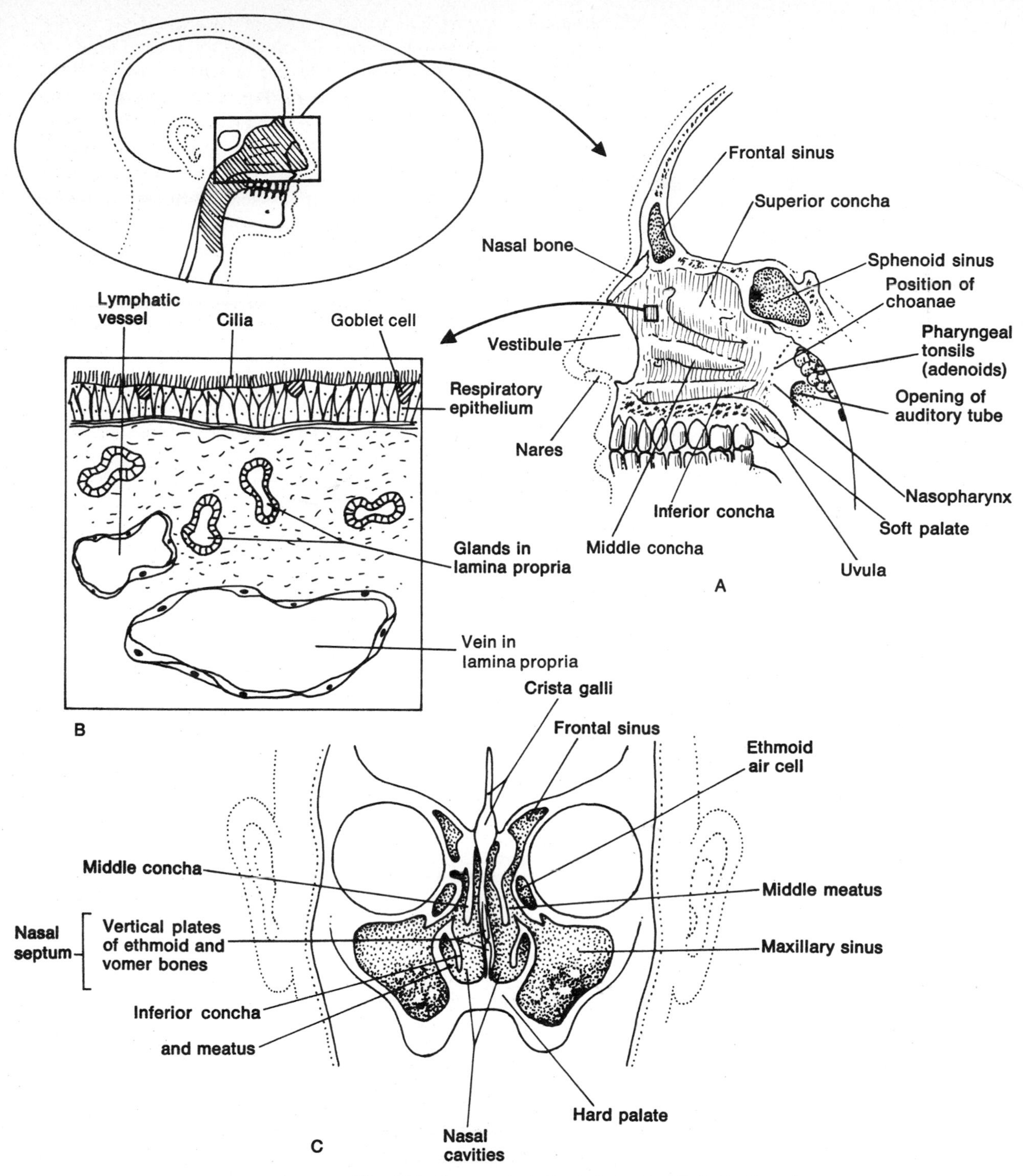

Fig. 6-14 The nasal cavity. A. Lateral wall. B. Section of mucosa taken from the nasal cavity. C. Frontal section of nasal cavity.

Sinuses

The sinuses of the skull may be found in the frontal bone, the sphenoid bone, the maxillary bones, and the ethmoid bone. These are the **paranasal sinuses** (Fig. 6-15). There is also a collection of air cells in the mastoid process of the temporal bone. All of these are in communication with one another by virtue of the nasal cavity and therein lies their greatest significance. The mucosal carpet of the nose continues into each of the orifices in the three meatuses and goes on to cover the walls of each of the sinuses; thus infected sinuses drain into the nasal cavity. Or, perhaps more significantly, a "common

Fig. 6-15 Paranasal sinuses and their relationship to the nasal cavity.

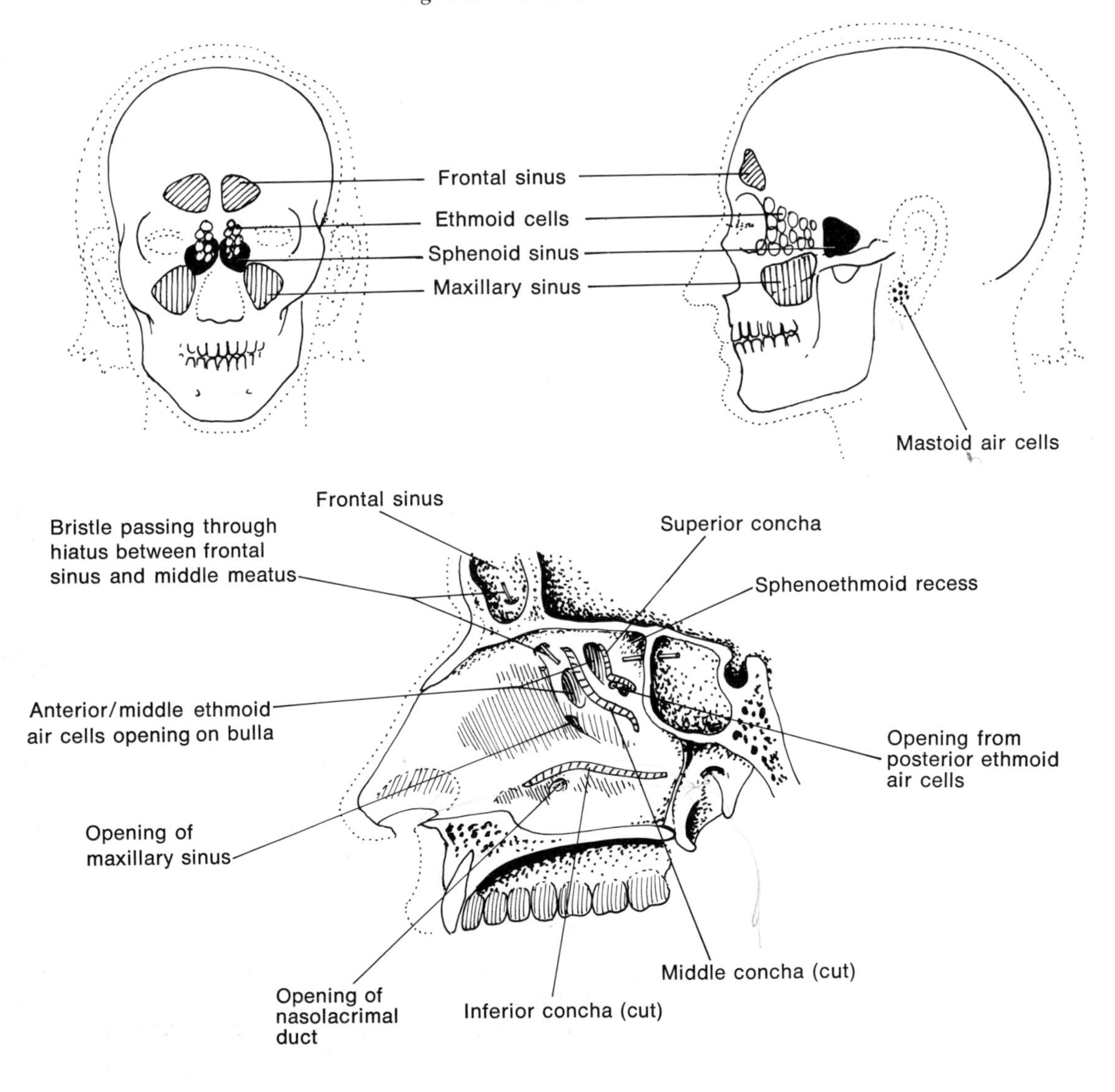

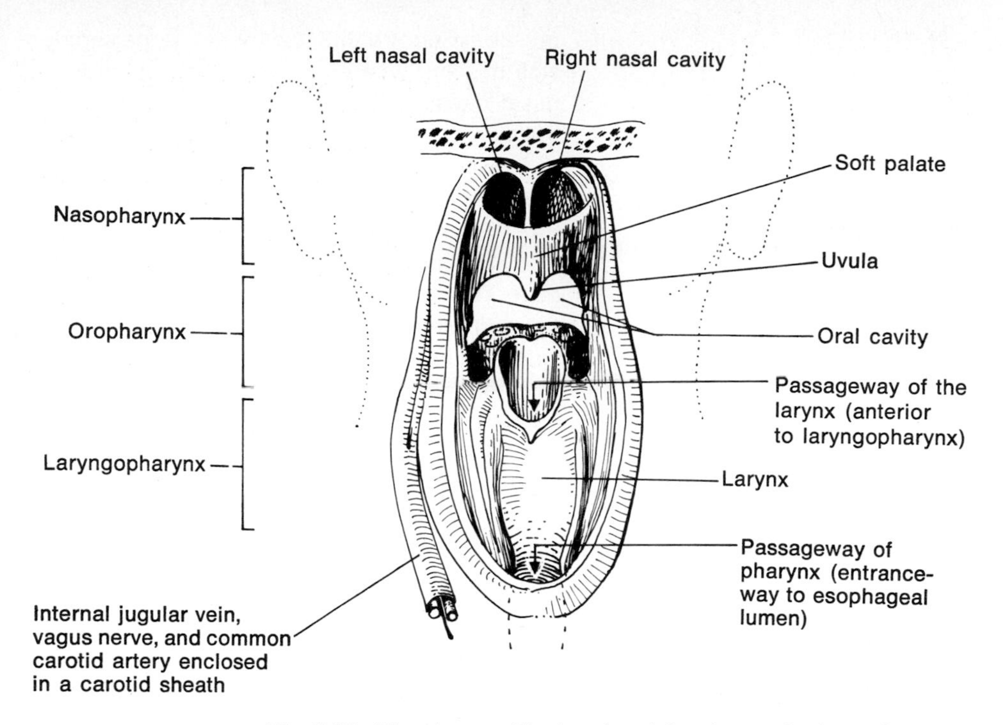

Fig. 6-16 The pharynx. The interior of the pharynx is shown from behind with the posterior wall removed.

cold" can infect the sinuses, the nasopharynx, the pharyngotympanic tube, and the middle ear cavity—and often does.

Other than a site of frequent congestion and infection, and a cause of frontal headaches, sinuses certainly influence the quality of the voice, as anyone who has ever had a "code in da head" knows. The sinuses serve to lighten the load of the head, but any other sophisticated significant functional features have yet to be uncovered.

Pharynx

The pharynx (Fig. 6-16) is divided into three regions:

1. That part associated with the nasal cavity, the **nasopharynx,** which is continuous with . . .
2. That part associated with the oral cavity, the **oropharynx,** which is in turn continuous with . . .
3. That part related to the esophagus and larynx, the **laryngopharynx.**

The latter two regions more properly belong in a discussion of swallowing, and will be presented a bit later. As an introductory note, it should be stated that the pharynx is a fibromuscular, contractile tube with a large segment of its anterior wall missing where it is in communication with the oral and nasal cavities. It is attached to the base of the skull, like a hollow bag, receives the effluent from nose and mouth alike, and discharges each into the appropriate "receptacle"—with infrequent error—at the level of the C6 vertebra. Feel for the most prominent bony bump at the back of your lower neck (midline), draw an imaginary line to the front and there you have—roughly—the distal termination of the pharynx. What is the approximate length of your pharynx, from the level of the nasal cavity to that of the esophagus? About 5 in. The posterolateral walls of the pharynx are muscular and lined with mucous membrane. The anterior wall is nonexistent, for the pharynx is open to the oral and nasal cavities here, except at the laryngopharynx, where the anterior wall is the larynx itself.

The nasopharynx commences at the choanae, thus it is truly an extension of the nasal cavities. See now Fig. 6-14 as you read the following:

The floor of the nasopharynx is the muscular soft palate and uvula.

The roof is concave, composed of sphenoid and occipital bones, as well as the first and second cervical vertebrae. The carpet of mucous membrane incorporates crypts of lymphatic tissue (tonsils), called the adenoids.

The lateral walls of the nasopharynx are pierced by the orifices of the paired pharyngotympanic (auditory) tubes, which pass to the middle ear cavities.

The mucosa of the nasopharynx merges with oral mucosa (characterized by stratified squamous epithelium) at the naso-oral pharyngeal junction.

Larynx

The larynx is a cartilaginous, ligamentous, and muscular apparatus in the neck which functions to:

1. Generate sound at various frequencies and volumes (phonation).
2. Guard the lower respiratory tract by means of the cough reflex.

The larynx is related to the laryngopharynx posteriorly (Fig. 6-16), the carotid sheath and its contents laterally, and the trachea below. Anteriorly, the larynx is superficial and may be palpated on yourself freely. Try it. Above, the larynx is continuous from the oropharynx. The larynx takes up the space between vertebral levels C3 and C6.

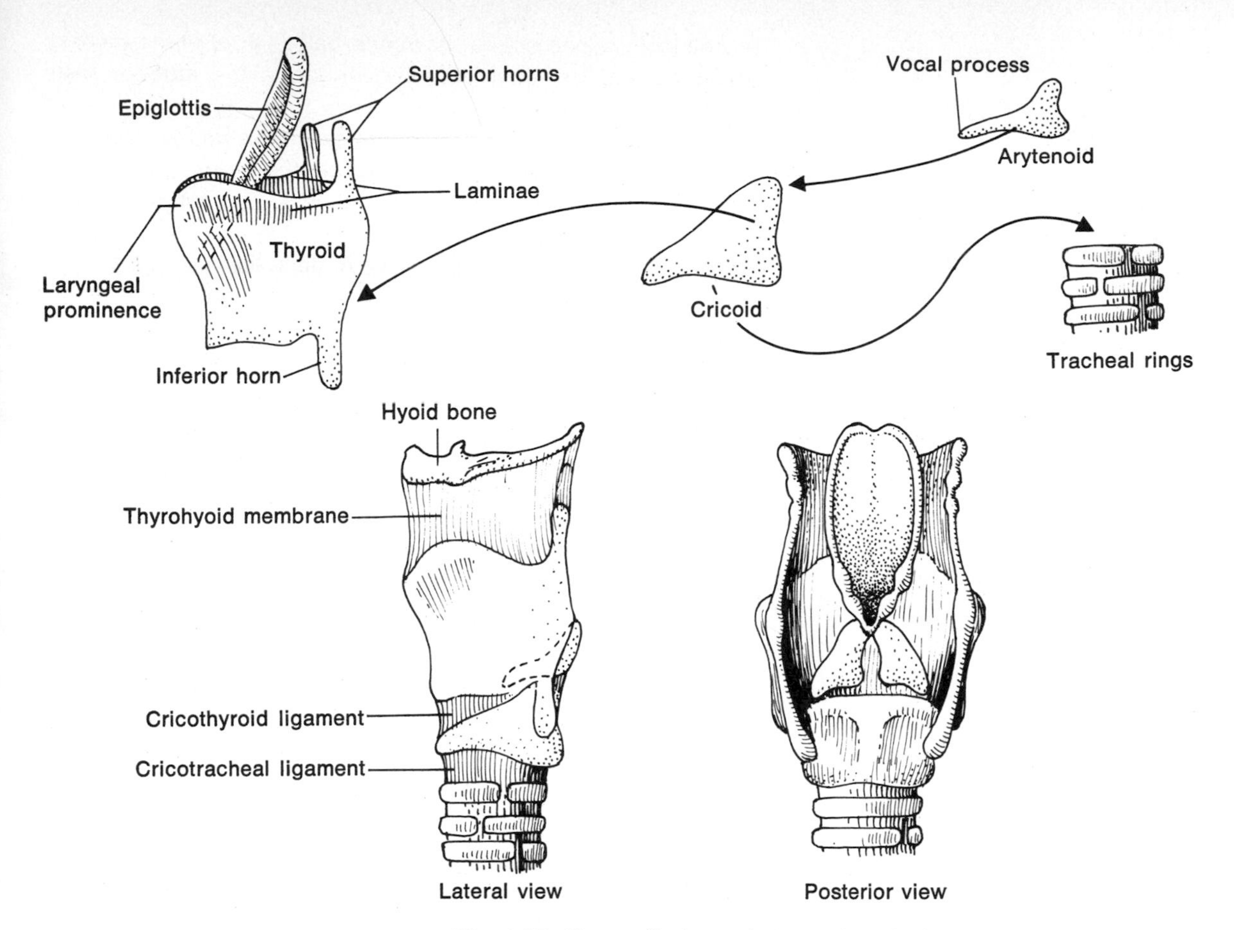

Fig. 6-17 The cartilaginous framework of the larynx.

Cartilages. The larynx has a cartilaginous framework; pieces are stacked more or less on top of one another. Consider Fig. 6-17 as you read below:

The **cricoid** (ring-shaped) **cartilage** fits on the trachea via an intervening ligament. It articulates with two cartilages:

1. The single **thyroid** (shield-like) **cartilage** laterally.
2. The paired **arytenoid cartilages** above.

The inferior horn of the thyroid articulates with the cricoid in a hinge joint. Thus to some degree the thyroid can tilt back and forth on the cricoid like a visor on a knight's armored headpiece.

The arytenoids sit atop the lamina of the cricoid and rotate in a pivot joint, with the vocal processes directed anteriorly (toward the thyroid cartilage).

The leaf-shaped epiglottis, is one-half in front of the thyroid cartilage and one-half above it. It forms the anterior wall of the laryngeal inlet.

Now, how much of the laryngeal cartilages can you feel on yourself? Place your fingers on the tracheal rings just above the sternum—in the midline of the neck. Extend your neck slightly.

Walk up the trachea with your fingers till you come to a ring-shaped cartilage somewhat bigger than the ones below it. This would be the **cricoid.** Feel its lateral extent.

Continuing superiorly, note the larger **thyroid cartilage** characterized by a median (laryngeal) prominence. This prominence (the "Adam's apple") is merely the line of fusion of the two plates (laminae) of the V-shaped thyroid cartilage. It is typically larger in males because the angle between the two plates is more acute, a function of increased male sex hormone secretion at puberty, and directly related to the voice's lower pitch acquired at about that time.

Now continue upward to the angle of the neck (where the neck takes a turn toward the chin) and with your thumb and index finger grasp the sides of the larynx and swallow. That bone you felt move up and down was the **hyoid.** Relate it to the laryngeal cartilages in Fig. 6-17.

So much for the cartilages and joints of the larynx.

Membranes and ligaments. The walls of the larynx, supported by the cartilages just studied, have fibroelastic membranes connecting the various cartilages. The external membranes, seen in Fig. 6-17, lie between:

1. Thyroid and hyoid (thyrohyoid).
2. Cricoid and thyroid (cricothyroid).
3. Cricoid and trachea (cricotracheal).

There are also internal membranes connecting the cartilages, and it is these membranes that create the "vocal cords" (Fig. 6-18), in the following way: An elastic membrane (conus elasticus) joins the thyroid, arytenoid, and cricoid cartilages on each side. The upper edge of this membrane is free (unattached), thickened, and ligamentous. This ligament—attached to the vocal process of the arytenoid cartilage posteriorly and the thyroid lamina anteriorly—constitutes the **vocal ligament** (cord). With a neighboring muscle, both of which are covered with mucous membrane, the ligament is retitled the **vocal fold** (Fig. 6-19).

Another membrane, in the same plane as the conus elasticus, connects the sides of the epiglottic cartilage to the ary-

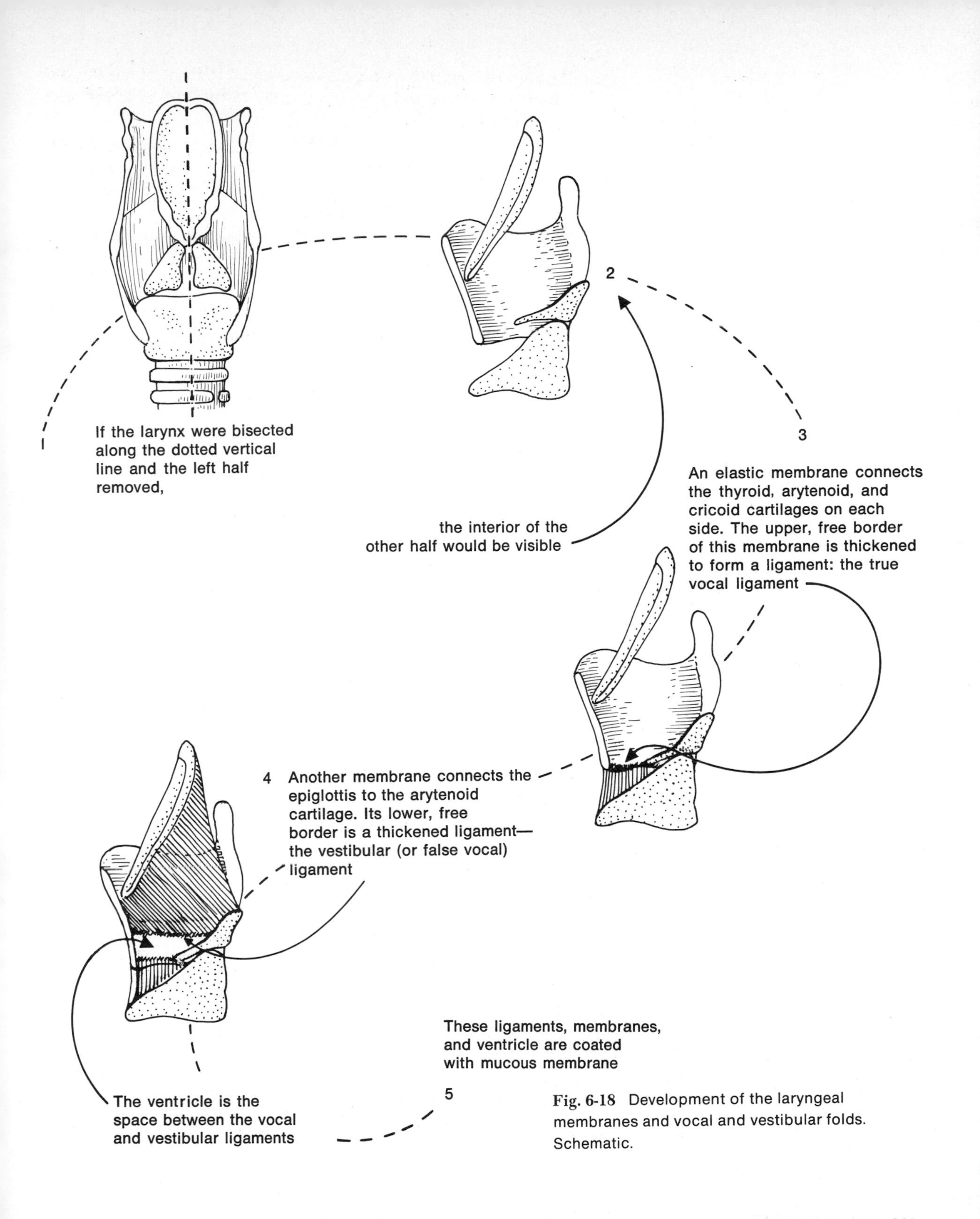

Fig. 6-18 Development of the laryngeal membranes and vocal and vestibular folds. Schematic.

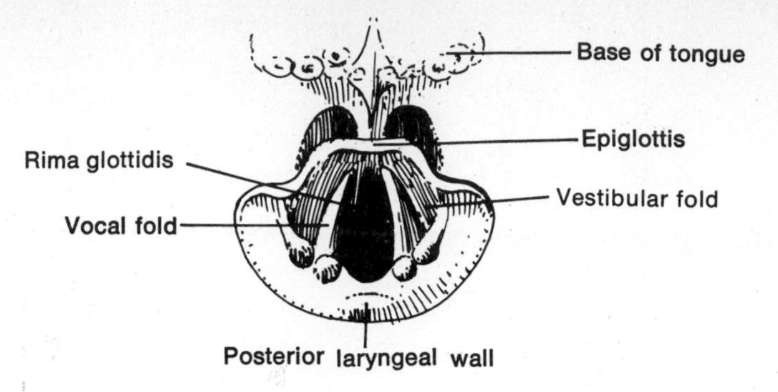

Fig. 6-19 Interior of the larynx. Superior view of larynx as seen through the laryngoscope during inspiration.

tenoid cartilages. The upper free border of this membrane constitutes the aryepiglottic fold on each side, thus creating the inlet of the larynx. The lower border is also unattached. It is ligamentous in nature and as the false vocal ligament, covered with mucous membrane, it is called the **vestibular fold.** The space between the two folds—vocal and vestibular—constitutes the **ventricle.**

With muscles (yet to be discussed) in place, and the entire laryngeal interior carpeted with mucosa, we can see that the larynx has:

1. An inlet created by the aryepiglottic folds.
2. A cavity above the vocal folds called the **vestibule.**
3. A cavity below the vocal folds called the **infraglottis** which is continuous below with the lumen of the trachea.
4. A variable space between the two muscular/ligamentous vocal folds called the **rima glottidis.**
5. A space between the two connective-tissue–filled vestibular folds (rima vestibuli). This space can become closed off by edema in certain instances of laryngeal trauma—an often fatal event.

The mucosa of the larynx reflects the functional activity of the organ: overlying much of the walls of the vestibule, and in particular, the vocal folds, the mucosa is characterized by a stratified squamous epithelium. These folds are traumatized by the violent gales created by coughing, yelling, and passing particles of pollutants. The rest of the mucosa is lined with typical respiratory epithelium.

Muscles and function. The larynx carries out its function by virtue of a set of intrinsic muscles which generally act on the vocal folds, directly or indirectly, to control phonation (voice). The muscles work around one of two joints: the cricothyroid and the cricoarytenoid. With a couple of exceptions, the muscles are named accordingly; see for yourself in Fig. 6-20:

1. The muscle attaching to the cricoid and the inner aspect of the thyroid lamina, crossing the cricothyroid joint, is the **cricothyroid muscle.** It tilts the thyroid on the cricoid (or vice versa?) and consequently stretches the vocal ligaments—tensing them and adducting them. This action—seen during phonation—may assist in achieving high-pitched sounds.*
2. There are two muscles attaching the arytenoids to the cricoid—one descends laterally from the arytenoid **(lateral cricoarytenoid);** the other descends posteriorly **(posterior cricoarytenoid).** The former muscle adducts the two cords, the latter abducts them.

These three muscles just discussed are the principal muscles acting in phonation: one abductor and two adductors of the vocal folds, resulting in opening and closing the rima glottidis, respectively.

3. The **vocalis** muscles, in concert with the more lateral thyroarytenoid, varies the tension on the vocal folds.
4. The ary- and thyroepiglottic muscles—located in the aryepiglottic folds—tend to close the inlet of the larynx during swallowing.

Speech or sounds are created by expelling air which passes between the vibrating vocal folds and "manipulating it" by lowering the mandible and moving the tongue and lips. The final result is influenced by the nasal cavities and sinuses as well as the pharyngeal and oral cavities. The relative position and tension of the vocal folds influence the volume of sound as well as the pitch.

REGIONS AND STRUCTURES RELATED TO PREDIGESTION

The structures in the head and neck related to digestion are specifically in the food processing business and may be summarized in this functional sequence:

1. Food is taken into the body through the service of the **oral cavity** with the aid of the **lips.**

* Think of a guitar: the greater the tension on the strings, the higher the pitch; the less the tension, the lower the pitch.

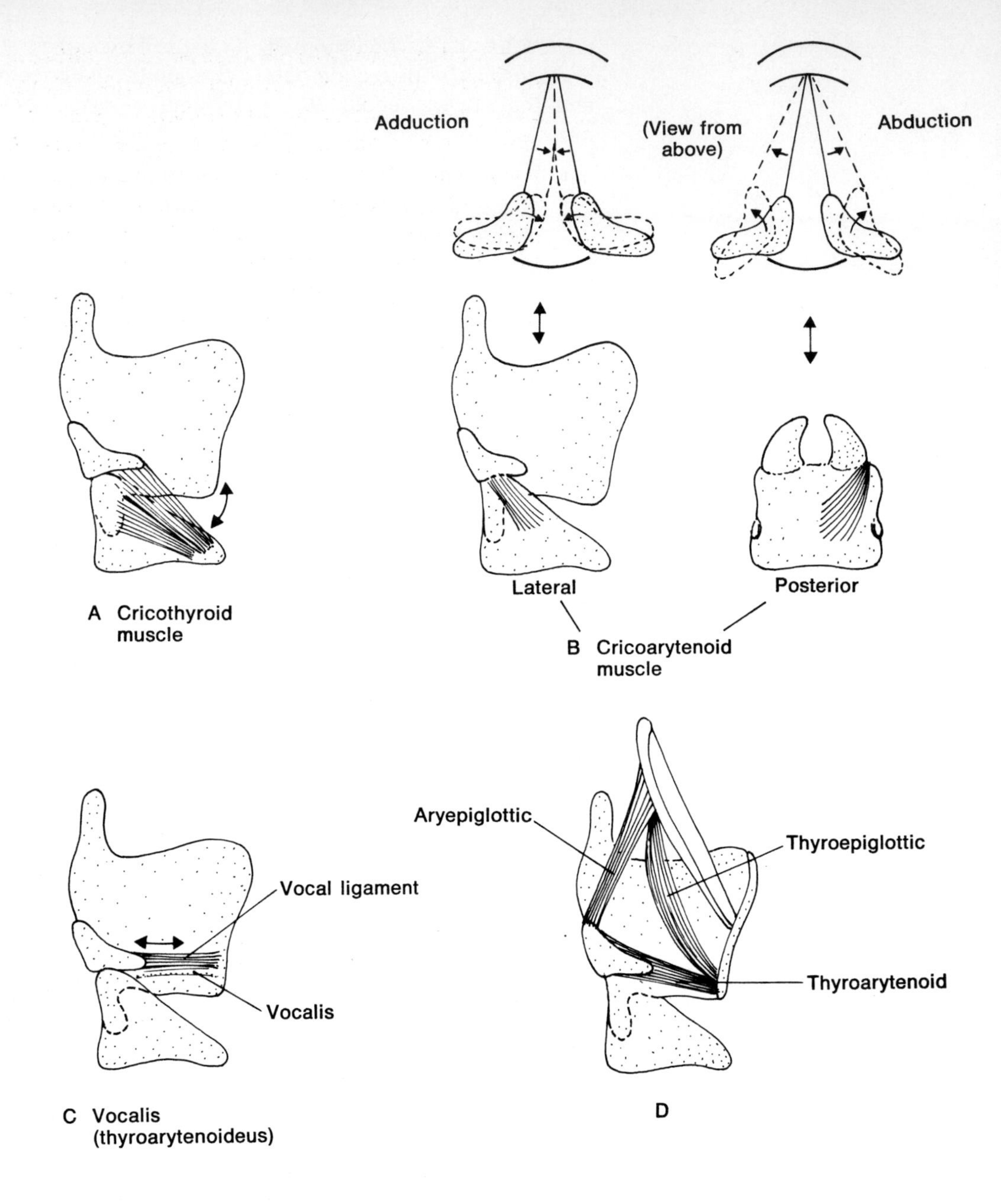

Fig. 6-20 Muscles of the larynx.

2. The food is mechanically processed by chewing. This phenomenon requires: (*a*) a joint about which the mandible moves, the **temporomandibular joint,** (*b*) **muscles of mastication** to move the mandible, (*c*) **teeth** to do the grinding, and (*d*) a **tongue** to do the tossing.
3. The food is enzymatically processed by the secretions of the three **salivary glands** and the many mucous and serous glands of the oral cavity.
4. The initially processed food is ejected into the oropharynx by the tongue and swallowed (a rather complex affair involving the palate, the oro/laryngopharynx, and the esophagus).

Oral cavity

Use a mirror and your sense of touch as you read the following paragraphs:

The lips border the entrance to the mouth. The vermilion free border (exterior) of the lips represents a transition from skin to mucous membrane. The mucosa contains no glands, hence the lips are dry. Internally, the lips are glandular and moist. They are highly mobile organs, for some 12 muscles insert about their muscular base, the **orbicularis oris.** See on yourself how mobile your lips really are. The redness of the lips is due principally to the superficial vascular patterns characteristic of the mucosa.

The space between the lips and the gums or teeth **(vestibule)** is characterized by a highly glandular mucosa and a midline fold between lip and gum (frenulum)—easily felt on yourself, with tongue or finger.

The oral cavity proper is within the frame of the teeth and their gums. Probe with your finger as you read the following:

The roof of the cavity is the **hard palate** (lined with a glandular, stratified squamous mucosa) and posterior to this, the muscular **soft palate,** terminating posteroinferiorly as the conical uvula. You can feel the transition from hard palate to soft with your tongue. You can also "wiggle" your uvula with your tongue.

The floor of the cavity is largely filled by the tongue, whose voluminous muscle base is circumscribed by the mandible. The mucosal surface of the floor blankets the sublingual and submandibular glands, whose ducts open into the mouth on the sublingual fold (Fig. 6-21). The muscular base of the floor is provided by the suprahyoid muscle group. This can be felt under the chin—the thickness of the floor can be measured by placing the index finger under the tongue and the thumb

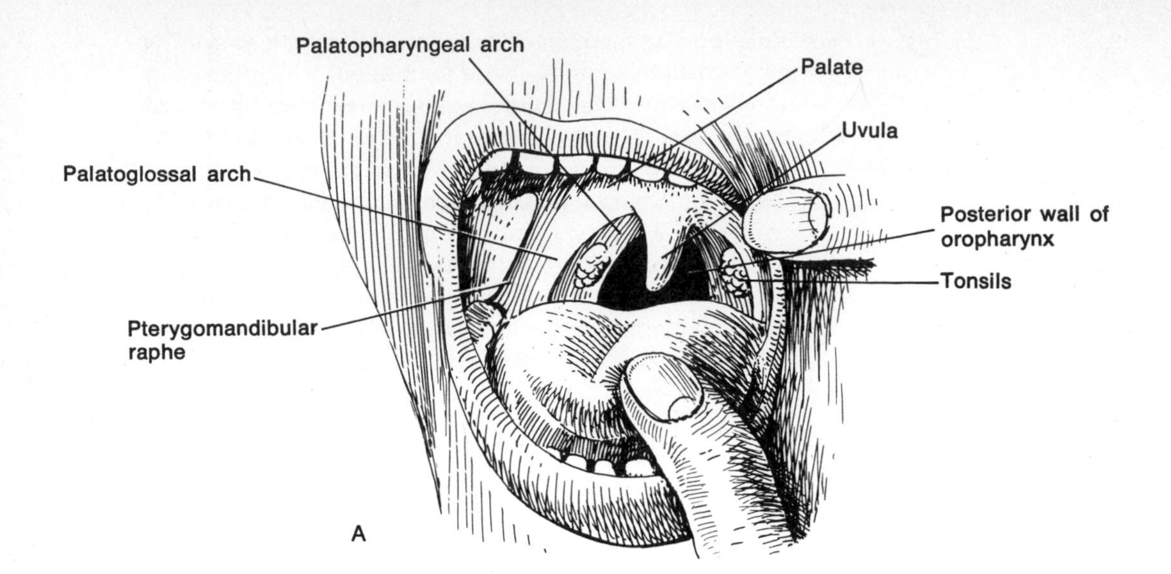

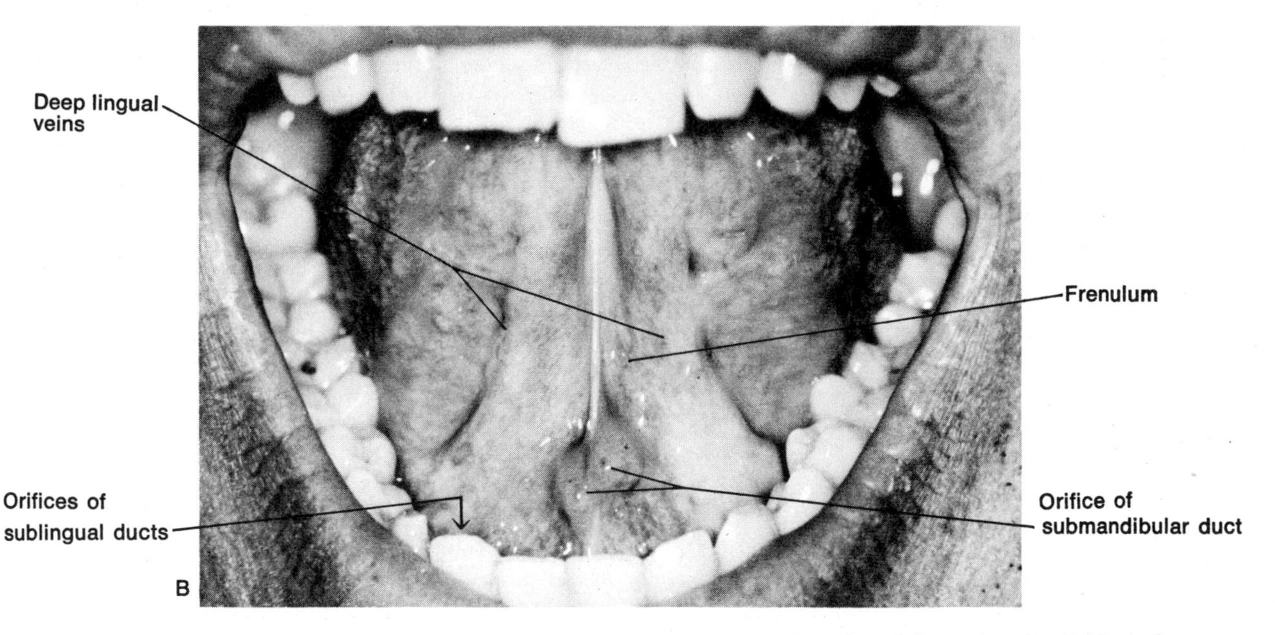

Fig. 6-21 Oral cavity. A. Drawing of the oral cavity. B. View of underside of tongue and floor of mouth.

of the same hand under the chin. The muscle group can be made to contract by jutting the chin forward.

The lateral walls of the cavity are the cheeks, consisting of the mucosa-lined **buccinator muscle,*** covered externally with a fat pad (giving a youthful look to the cheek) and skin. Place your tongue on the inside of your cheek and contract the buccinator (as if you were whistling). Next move your tongue up opposite the upper second molar and feel for a small papilla—this is the opening of the duct from the parotid gland. Now feel back with your thumb to the vertical ramus of the mandible. The soft covering there is the **masseter muscle** (of mastication) (Fig. 6-22). Keep your thumb there and clench your jaw and feel the masseter contract.

Posteriorly, the oral cavity is open to the oropharynx, hailed by the paired great arches of the mouth (Fig. 6-21). The fossae created by these arches house the palatine tonsils.

Mastication. The act of chewing is permitted by the complex **temporomandibular joint**† (Fig. 6-22), which is the articulation of the head of the mandible (condyloid process) with the mandibular fossa of the temporal bone. Feel the joint movements by placing a finger in the opening of the ear while chewing. The lower jaw is capable of:

1. **Depression** (dropping the jaw down as in opening your mouth).
2. **Elevation** (closing the mouth, bringing the teeth into occlusion).
3. **Protraction** or jutting of the chin forward. Here the mandible glides forward somewhat in its socket.
4. **Retraction** (the reverse of protraction).
5. **Lateral** displacement (side-to-side movements).

These movements are generated principally by the muscles illustrated in Fig. 6-22. The four muscles of mastication are all innervated by branches of the mandibular division of the trigeminal (Vth) cranial nerve (look ahead to Fig. 6-42).

The **temporalis** muscle may be felt at the sides of the temporal bone while biting down and the **masseter** has been palpated at the angle of the mandible. Feel the entire extent of the mandible. Sometimes the duct of the parotid gland can be

* This muscle is helpful in preventing food from spilling into the vestibule when chewing. Dogs and cats have no such muscle, hence when chewing they lose much of the food unless they gulp it!

† A double synovial joint, commonly classified as hinge but really capable of side-to-side and anterior-posterior slippage—as you can feel on yourself.

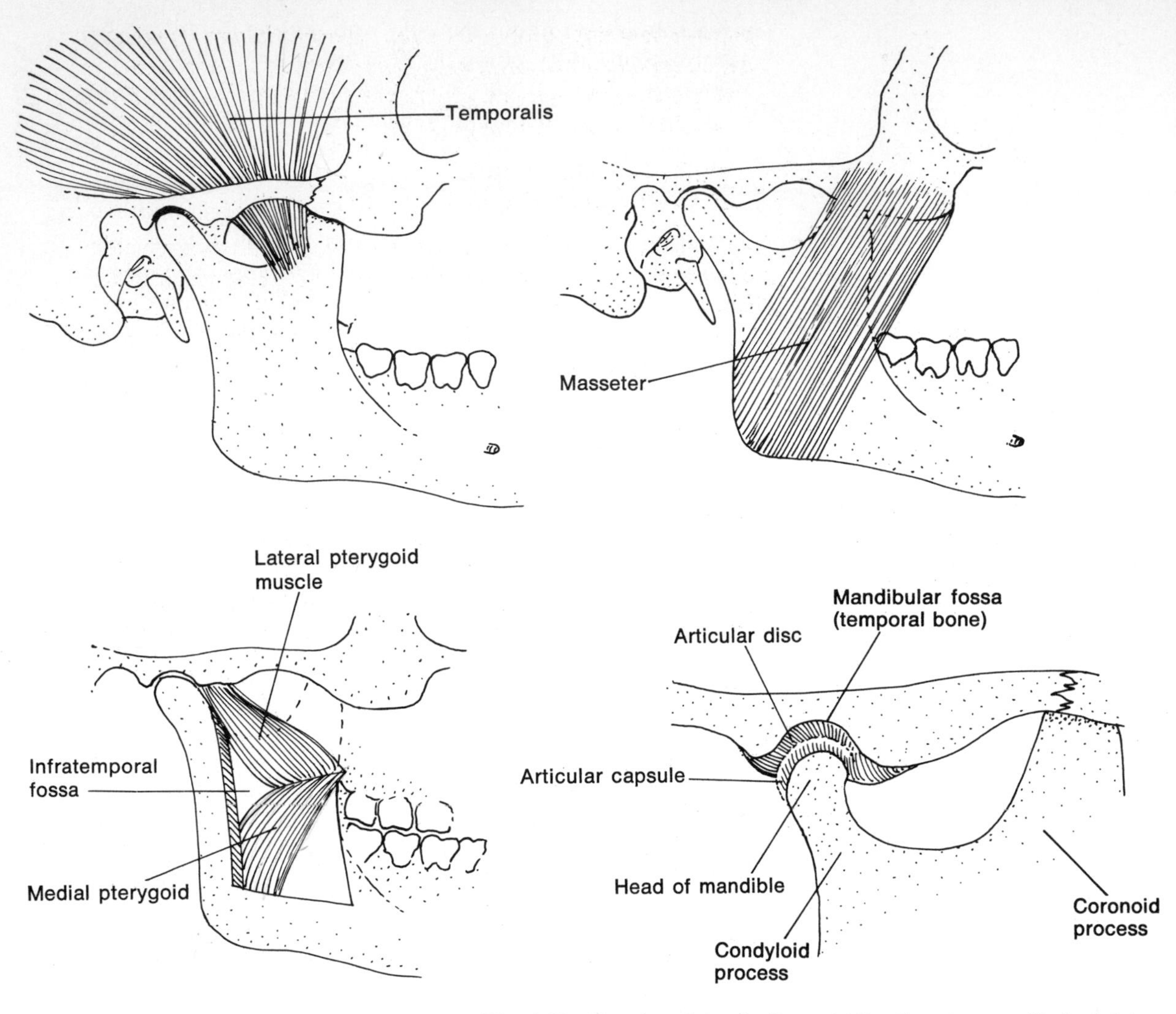

Fig. 6-22 Muscles of mastication and the temporomandibular joint.

rolled on the medial edge of the contracted masseter muscle a fingersbreadth below the zygomatic arch. The **pterygoid muscles,** residents of the infratemporal fossa, cannot be palpated.

Teeth

The teeth are primary agents in processing food which needs to be bitten off (incised) or pulverized. The teeth which fulfill the former function are at the front of the dental arches, while those which grind are found more posteriorly in the arches. See Fig. 6-23 as necessary, but use your own teeth as the primary learning tool and note: There are two U-shaped dental arches: upper and lower (maxillary and mandibular). If an imaginary median line bisected each arch into halves there would be four quadrants of teeth:

- Upper right and left (2).
- Lower right and left (2).

Classically, there are 32 teeth in the adult, or 8 in each quadrant. The dental constitution of each quadrant is identical.

Four kinds of teeth can be easily distinguished in each quadrant (Fig. 6-23):

1. Those that incise or nip: **incisors** (2).
2. Those that tear: **canines** (1).
3. Those with two cusps that grind: bicuspids or **premolars** (2).
4. Those with four cusps that grind: **molars** (3).

Since there are the same number and sequence of teeth in each quadrant, the arrangement of teeth (front to back) can be expressed by type: 2123, a simplification of

$$\frac{2123}{2123} \quad \frac{2123}{2123}$$

This is the **dental formula** in the adult.

Using a mirror, note how each type of tooth has its characteristic shape at the exposed position (crown) and how these shapes are related to their function. Each tooth consists of several parts (Fig. 6-23):

1. **Crown:** that part projecting above the gum, the surface of which is **enamel.** Enamel is 96 percent calcified tissue—the hardest in the body. It is not replaced or repaired naturally after having once formed, and therefore tends to erode with age.
2. **Neck:** the constricted part of the tooth where the gum attaches. The gum is a mass of thick, dense fibrous tissue covered with a vascular, pain-sensitive, mucous membrane. Clinically, each gum is called a **gingiva;** hence inflammation of the gums is gingivitis.
3. **Root:** that part of the tooth embedded in the alveolar socket of bone, and covered by a layer of fibrous tissue (cement) adhering the tooth to the neighboring bone. The cavitated core of the root and the crown **(pulp cavity)** contains a highly vascular and pain-sensitive loose connective tissue **(pulp).** Pulp secretes a calcified tissue **(dentine)** which lines the pulp cavity and makes up most

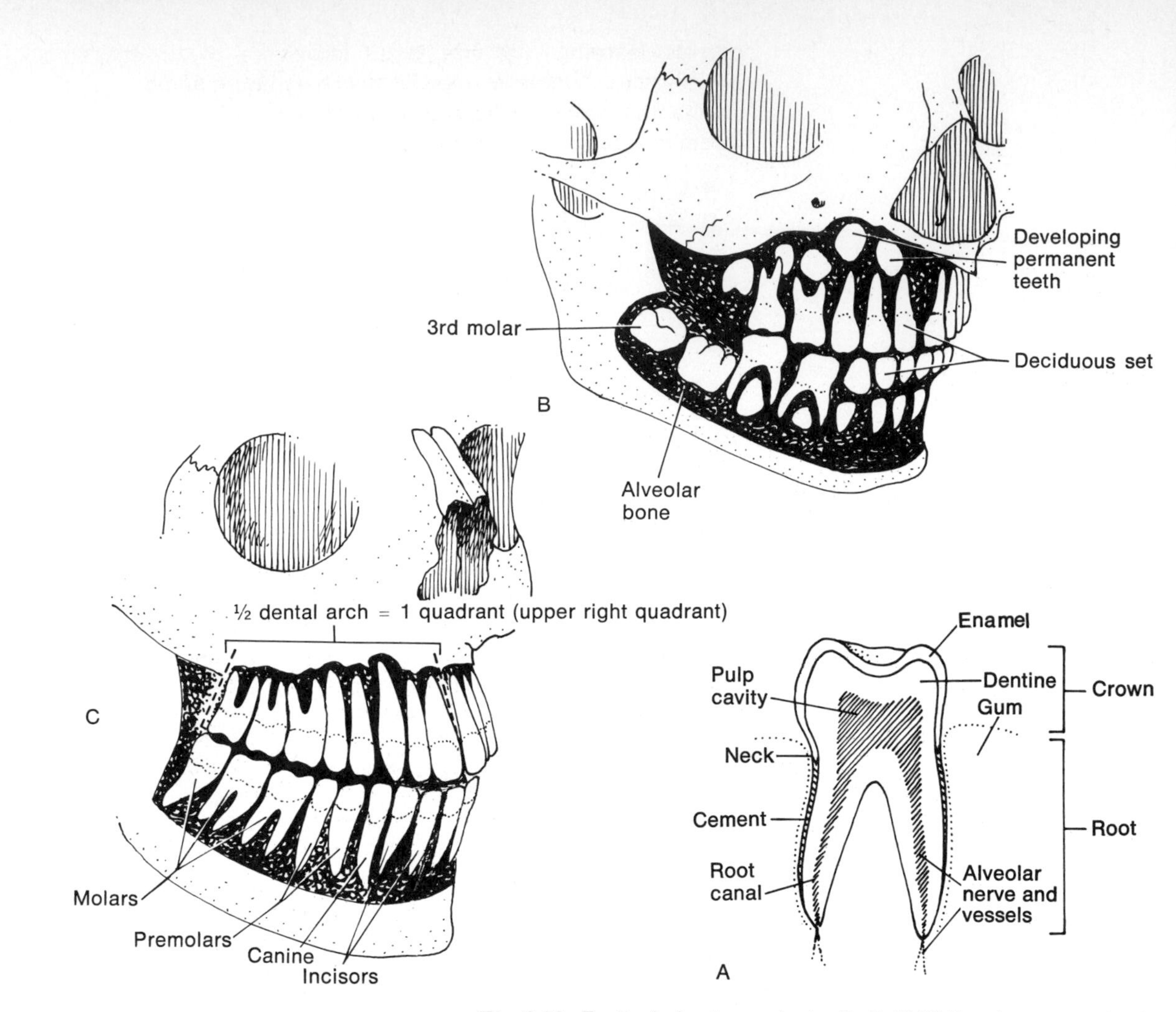

Fig. 6-23 Teeth. A. Anatomy of a tooth. B. "Milk" and permanent sets in a child's skull. C. Permanent teeth.

of the tooth mass. Dentine is replaced throughout life, unlike enamel.

4. **Root canal:** continuous with the pulp cavity, transmits nerves and vessels to the pulp and adjacent tissues.

At birth and during the first few months afterward, the teeth have not sufficiently developed to the point where they are visible in the oral cavity—but they are there, masked by the gums. These are the first set (**deciduous** or "milk" teeth) and generally make their appearance (erupt) according to the following schedule:

1. Lower central incisors (2) at 6 months.
2. Upper central and lateral incisors (4) at 8 months
3. Lower lateral incisors (2) at 15 to 20 months.
4. Canines (4) at 15 to 20 months.
5. First molars (4) at 15 to 20 months.
6. Second molars (4) at 20 to 24 months.

The dental formula for the deciduous set, then is 212 which, when fully spelled out, is

$$\frac{212}{212} \quad \frac{212}{212}$$

for a total of 20 teeth. What "adult" teeth are missing in the deciduous set?

While the 20 milk teeth are erupting, the second **(permanent)** set of teeth are developing deep to the first. These include the 20 replacing the milk teeth plus 8 molars and premolars (three in each quadrant) which had no predecessors.

The milk teeth begin to shed at about the sixth year. The roots are absorbed (disappear) and the enamel caps, held to the gums at the neck, break off, a painless and usually bloodless event. The milk teeth have usually dropped out by 12 years. The permanent set of dentition usually makes its appearance according to the following schedule:

	Age, years	No. of teeth
First molars	6	4
Central incisors	7	4
Lateral incisors	8	4
First premolars	9	4
Second premolars	10	4
Canines	11	4
Second molars	12	4
Third molars	18 to 25	4 (often impacted in the gums)
Total		32

It has been said by dentists everywhere that if people took care of the rest of their body as they did their mouths, we would all be dead in our twenties. The good condition of the teeth and gums is a function of inheritance, nutrition, and continuous dental care. A lack of one or more of these factors constitutes a basis for the premature and often needless cavitation, ulceration, and rejection of permanent teeth. Lesions of the enamel or dentine often show up on x-rays as "ulcerations" in the tissue, characterized by an abnormal cavity within the tooth frequently invisible to the naked eye. These spots of decay are called caries (L., decay). Filling them with a nonreactive metal such as silver or gold may prevent further erosion of the tooth (Fig. A-8, Atlas of X-rays).

Tongue

The tongue is a muscular organ, largely enveloped in mucosa and characterized on its dorsal surface by papillae, the largest of which support taste buds. The tongue is stabilized by a number of **extrinsic muscles** which form its root. All of these move and position the tongue in speech and during swallowing. The mass of **intrinsic muscles** makes up the body of the tongue.

Roughly two-thirds of the tongue is related to the oral cavity—the posterior one-third is vertically oriented and related to the pharynx (Fig. 6-24). Whereas the former is characterized by papillae (Fig. 6-94), the pharyngeal part contains lymphatic follicles termed **lingual tonsils.**

The vessels (lingual) and nerves (lingual, hypoglossal) to the tongue reach the tongue via its root. The most superficial of the veins can be seen on yourself on either side of the median frenulum on the underside of the tongue. Properly coated medication can be taken sublingually for rapid absorption into the blood because of the superficial location of these veins (Fig. 6-21).

The muscles of the tongue are innervated by the hypoglossal (XIIth) nerve. Taste buds on the anterior two-thirds of the tongue are served by the facial (VIIth) nerve while general sensations (e.g., pain) are transmitted by way of the lingual branch of the trigeminal (Vth) nerve; both taste buds and gen-

Fig. 6-24 The tongue. Sagittal section showing root of the tongue. Note attachments to mandible and hyoid.

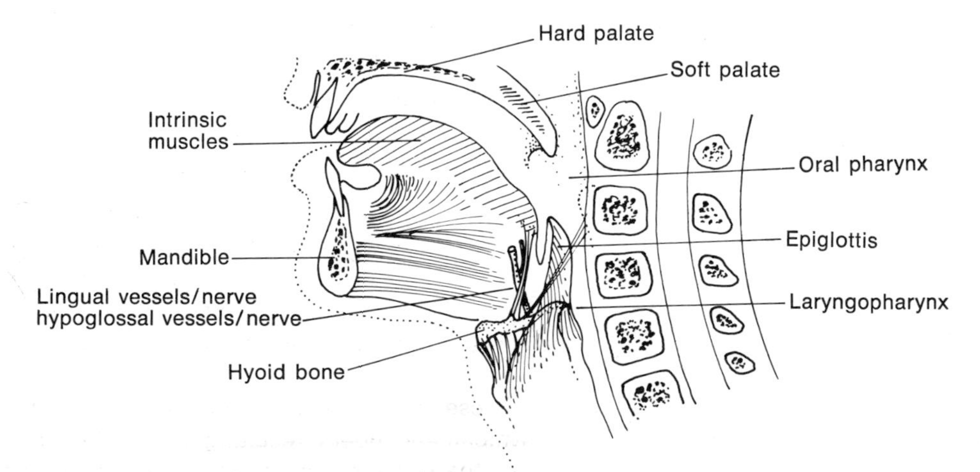

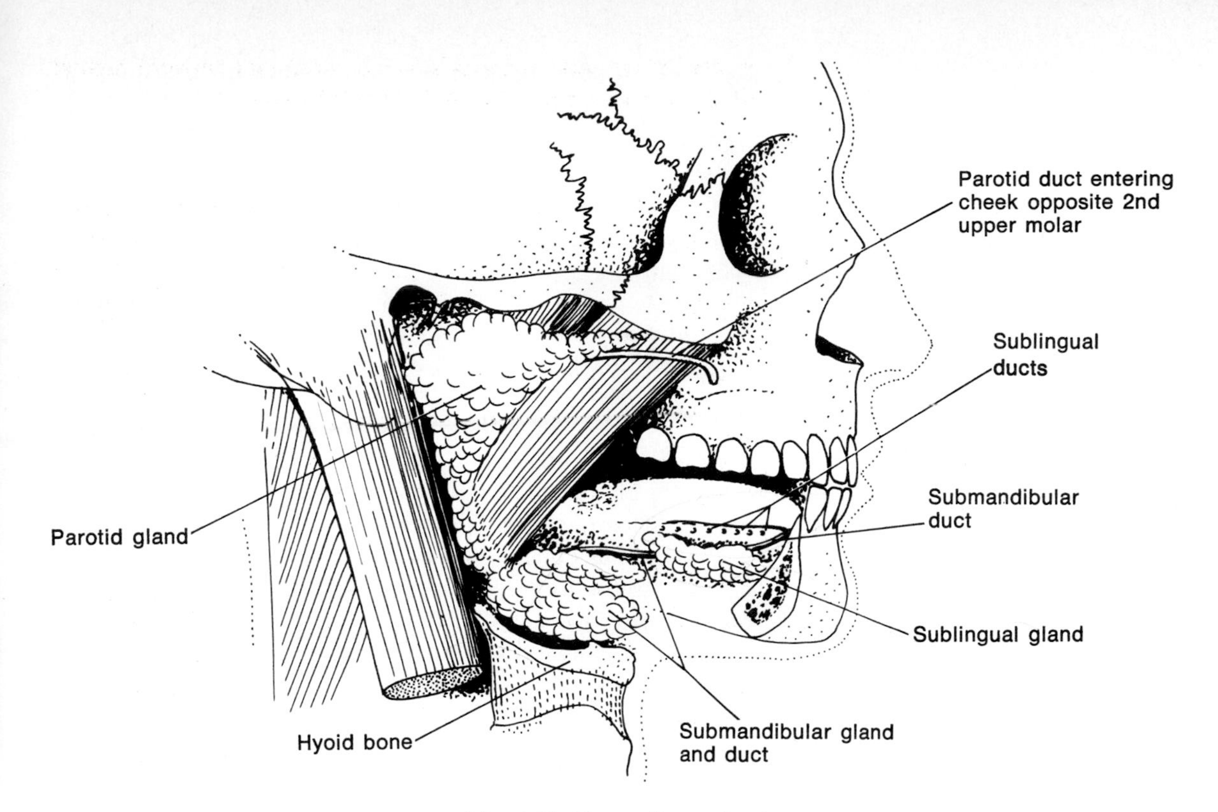

Fig. 6-25 The salivary glands.

eral sensations on the posterior two-thirds of the tongue are served by the glossopharyngeal (IXth) nerve.

Salivary glands

The salivary glands are responsible for the enzymatic processing and wetting of food prior to swallowing and digestion. There are three and they lie adjacent to or near the oral cavity (Fig. 6-25):

1. **Parotid:** in the parotid region.
2. **Sublingual:** under the tongue.
3. **Submandibular:** bent around the mylohyoid muscle.

The duct of the parotid muscle enters the oral cavity opposite the upper second molar and can be felt there. The ducts of the submandibular and sublingual glands open at the floor of the mouth along the sublingual papilla and fold (Fig. 6-25). These can be seen on yourself with the aid of a mirror (Fig. 6-21).

These glands are innervated by both sympathetic and parasympathetic fibers resulting in a reflex secretion of water and the enzyme ptyalin (which works on carbohydrates), in response to the ingestion (or the thought of ingestion) of food.

Palate, pharynx, and deglutition

The hard palate serves as the roof of the oral cavity and is continuous posteriorly with the soft palate. The soft palate is seemingly supported by the arches of the oral cavity and terminates posteroinferiorly as the uvula. Highly vascularized, the soft palate consists of a mucosa-covered muscular complex including levators and tensors of the palate as well as muscles which block off the nasopharynx from the oral pharynx. These muscles work in coordination with the muscles of the pharynx during swallowing.

We are interested now in the muscular constitution of the oral and laryngopharynx. Refer to Fig. 6-26 as you read.

The muscles constituting the walls of the pharynx are principally the **constrictors—superior, middle,** and **inferior.** These constrictors are lined internally with mucosa (stratified squamous epithelium) and externally with fascia related posteriorly to the vertebral column.

Laterally, the pharynx is a confluence of many muscles, with contributions from the suprahyoid group, the palate, etc.

Inferiorly, the laryngopharynx merges with the muscular tunic of the esophagus.

Swallowing (or deglutition) is accomplished by the concerted action of the tongue, muscles of the palate and pharynx, and the esophagus. Study Fig. 6-27 and swallow several times while palpating the hyoid bone. Swallowing is divided into two stages:

First stage: the mouth must be shut. The food mass (bolus) lies on the dorsum of the tongue. Inspiration of air ceases. The tongue drives the bolus down through the arches into the oropharynx. The sensory receptors of the glossopharyngeal (IXth) nerve sense the bolus's presence in the pharynx and fire off messages to the brain which reflexly orders the palatopharyngo-esophageal muscle complex to action, principally via the vagus (Xth) nerve.

Second stage: the hyoid bone and, therefore, the larynx rise with contraction of the suprahyoid muscles. The walls of the pharynx move anteriorly; the soft palate lifts up, and the nasopharynx is closed off. These actions are created by the levators and tensors of the palate and the lifters of the pharynx. Now the oral cavity is sealed off from the pharynx by the tongue pressing against the soft palate, and the bolus has nowhere to go but down! The larynx is elevated sufficiently to close off the inlet of the larynx—the epiglottis probably

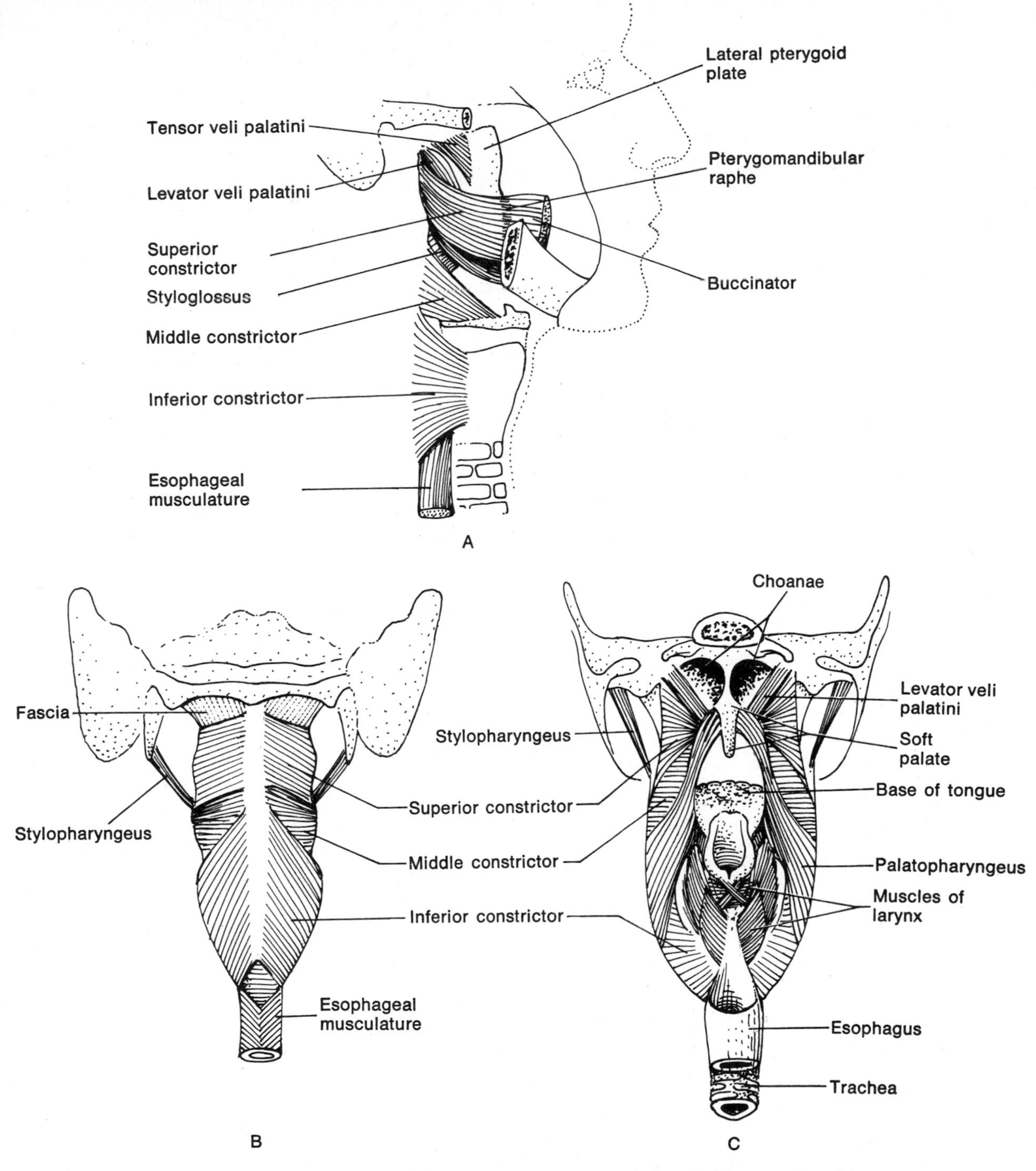

Fig. 6-26 The pharynx. A. Lateral view. B. Pharyngeal wall from behind. C. Pharynx with posterior wall removed.

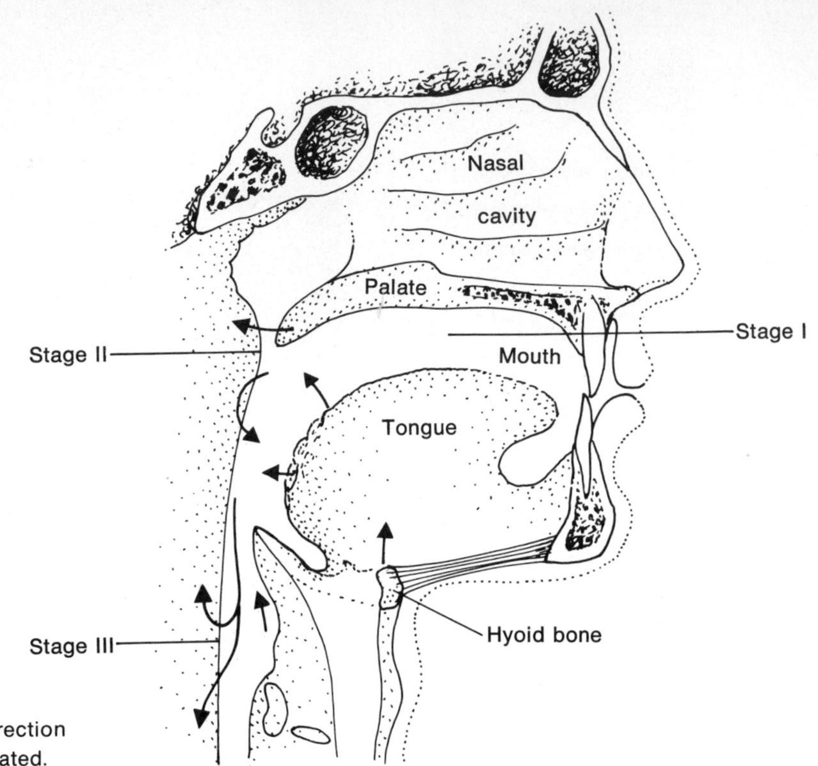

Fig. 6-27 Phenomenon of deglutition. Arrows indicate direction of movement of muscles indicated.

playing only a small role in this closure, since people who have had their epiglottis removed swallow nicely and securely.

Third stage: The constrictors of the pharynx move the bolus by peristalsis to the esophagus.

Should a portion of the bolus slip into the vestibule of the larynx, the vagus triggers a reversal of the just-explained events—a phenomenon familiar as the **cough reflex.** The glottis closes tightly, abdominal muscles contract, intrathoracic pressure builds up, and air is expelled explosively as the glottis suddenly opens. The bolus reverses direction!

Tonsils. Tonsils are masses of lymphatic tissue embedded in the mucosa of the pharynx. They are oriented about the entrance to the oropharynx in an incomplete ring. The constituents of this ring (Fig. 6-28) include:

1. The **palatine tonsils.**
2. The pharyngeal tonsil or **adenoids.**
3. The lingual tonsils.
4. The tubal tonsils.

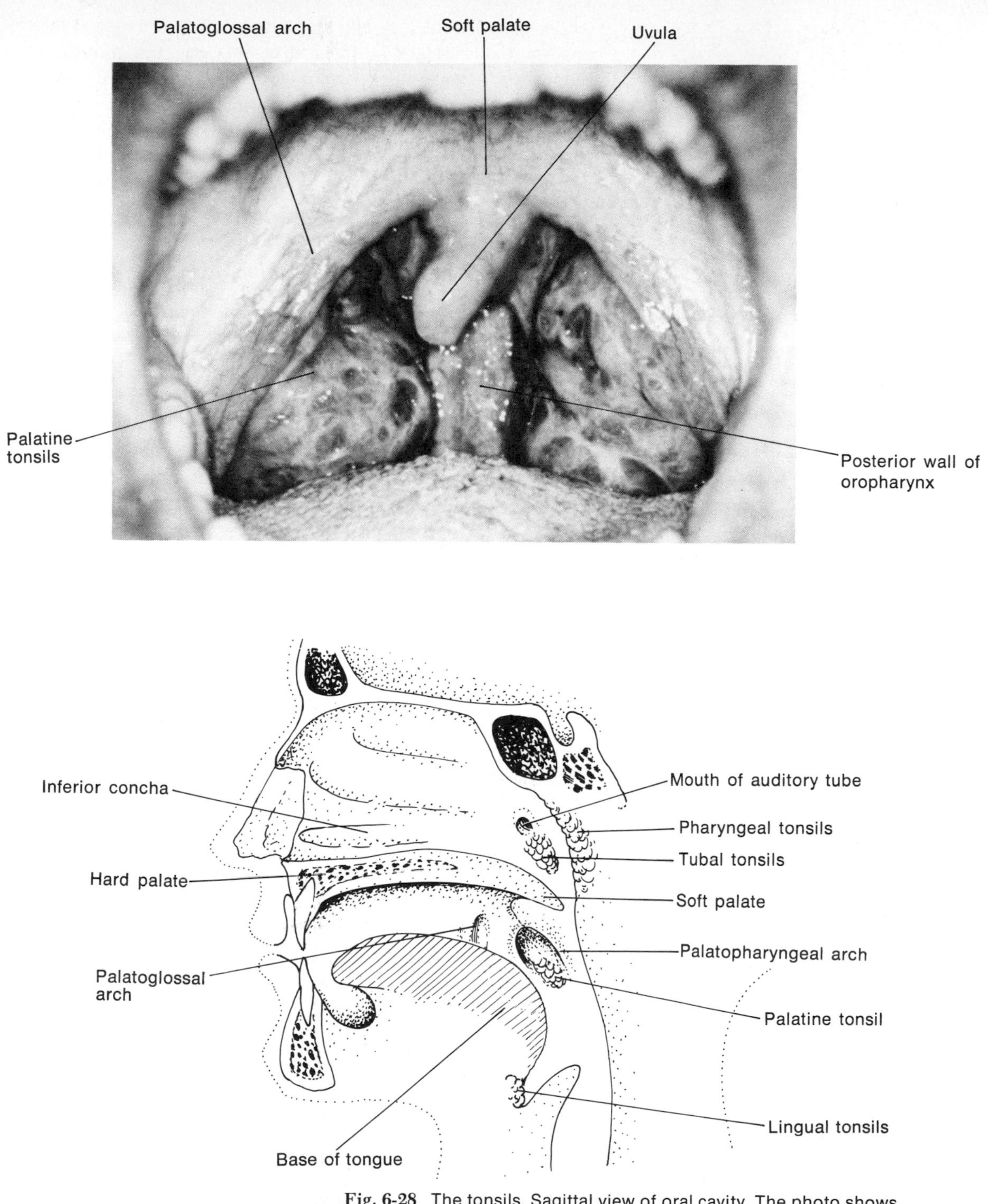

Fig. 6-28 The tonsils. Sagittal view of oral cavity. The photo shows greatly enlarged palatine tonsils in a 24-year-old male.

Once thought of as an inconsequential defense system against bacteria and viruses brought into the mouth, the tonsils are now a matter of controversy. They may be part of the antibody cell–seeding phenomenon in the young person. After puberty tonsils generally atrophy. During adolescence, they frequently become inflamed and consequently hypertrophy. In so doing, they can occlude the airway, cause gagging, and release pus into the middle ear cavity. In such cases, removal of the tonsils **(tonsillectomy)** is often in order.

VISCERA RELATED TO NEUROENDOCRINE SECRETION

You will recall that endocrine glands are ductless epithelial structures which secrete hormones* into the vascular system. The endocrine glands are generally directly or indirectly influenced by certain nuclei of the hypothalamus which secrete releasing factors. The endocrine system is so tied up with the central nervous system that the functional complex deserves the title neuroendocrine system. Components of the neuroendocrine system of the head and neck to be discussed here are the

1. Hypophysis.
2. Thyroid.
3. Parathyroids.

Hypothalamus

Certainly the hypothalamus is a critical component of this system, but its role in the economy of the endocrine system more properly is a topic of physiology, and will not be covered here in any detail. It is sufficient to say that the hypothalamus contains secretory neurons some of whose axons terminate in the posterior lobe of the hypophysis. The hormones are apparently synthesized in the hypothalamus and released in the posterior lobe, from where they enter the general circulation. These will be discussed shortly. Other secretory neurons of the hypothalamus are related to a plexus of veins through which their secretory products pass into the anterior lobe of the hypophysis and stimulate (or inhibit) the release of anterior hypophyseal hormones.

* Hormone (Gr., to arouse). A chemical agent secreted by a group of cells into the vascular system or body fluids. Hormones are usually effective at one or more "target" organs located at some distance from the source. Their effects are varied, widespread and, in some instances, life-sustaining, but generally result in the coordination of the parts of an organism (Bern).

Hypophysis (pituitary gland)

The hypophysis rides the Turkish saddle (sella turcica) of the sphenoid bone and is connected to the hypothalamus by a stalk. The hypophysis (Fig. 6-29) may be seen to have (1) an anterior lobe, (2) an intermediate part, and (3) a posterior lobe.

The anterior lobe is known to secrete growth hormone, adrenocorticotropic hormone, thyroid-stimulating hormone, gonadotropins, and prolactin.

Growth hormone (GH) is a protein which stimulates the lengthening of bone in young people by accelerating cartilage deposition at the epiphyseal plate (see Unit 2). Thus excessive secretion of GH results in gigantism, and chronic diminished secretion of GH (and sex-stimulating hormones) results in infantilism.

Adrenocorticotropic hormone (ACTH) is a protein that is directed toward the rind (cortex) of the adrenal gland atop the kidney. It stimulates the cortex to secrete a number of hormones called glucocorticoids.

Thyroid-stimulating hormone (TSH) is a protein which influences the thyroid gland in the neck to secrete thyroxine, a hormone which increases the basal metabolism in most normal tissues.

Gonadotropins are a pair of hormones which stimulate the ovary to develop follicles (follicle-stimulating hormone or FSH), and to ovulate (expel the ovum or egg) and transform postovulatory follicles into fatty hormone–secreting bodies (luteinizing hormone or LH). In the male, FSH stimulates spermatogenesis and LH stimulates the secretion of testosterone (male sex hormone) by the testis.

Prolactin is a hormone which stimulates the production of milk in pregnant (gravid) women.

The posterior lobe (Fig. 6-29) is known to release oxytocin and antidiuretic hormone. **Oxytocin** is a hormone which acts on myoepithelial cells of the mammary glands in response to sucking of the lactating breast, causing an ejection of milk from the glands and ducts into the mouth of the infant. Oxytocin also stimulates uterine muscle contraction at the time of birth (labor).

Antidiuretic hormone or ADH acts primarily on the kidney tubules to retain water, as when one is in the desert and conservation of body water is essential to maintain life.

The function of the intermediate lobe is not well understood in humans.

Because the hypophysis has such a widespread influence throughout the body directly and indirectly via its effect on other endocrine glands, e.g., the thyroid, hypopituitarism

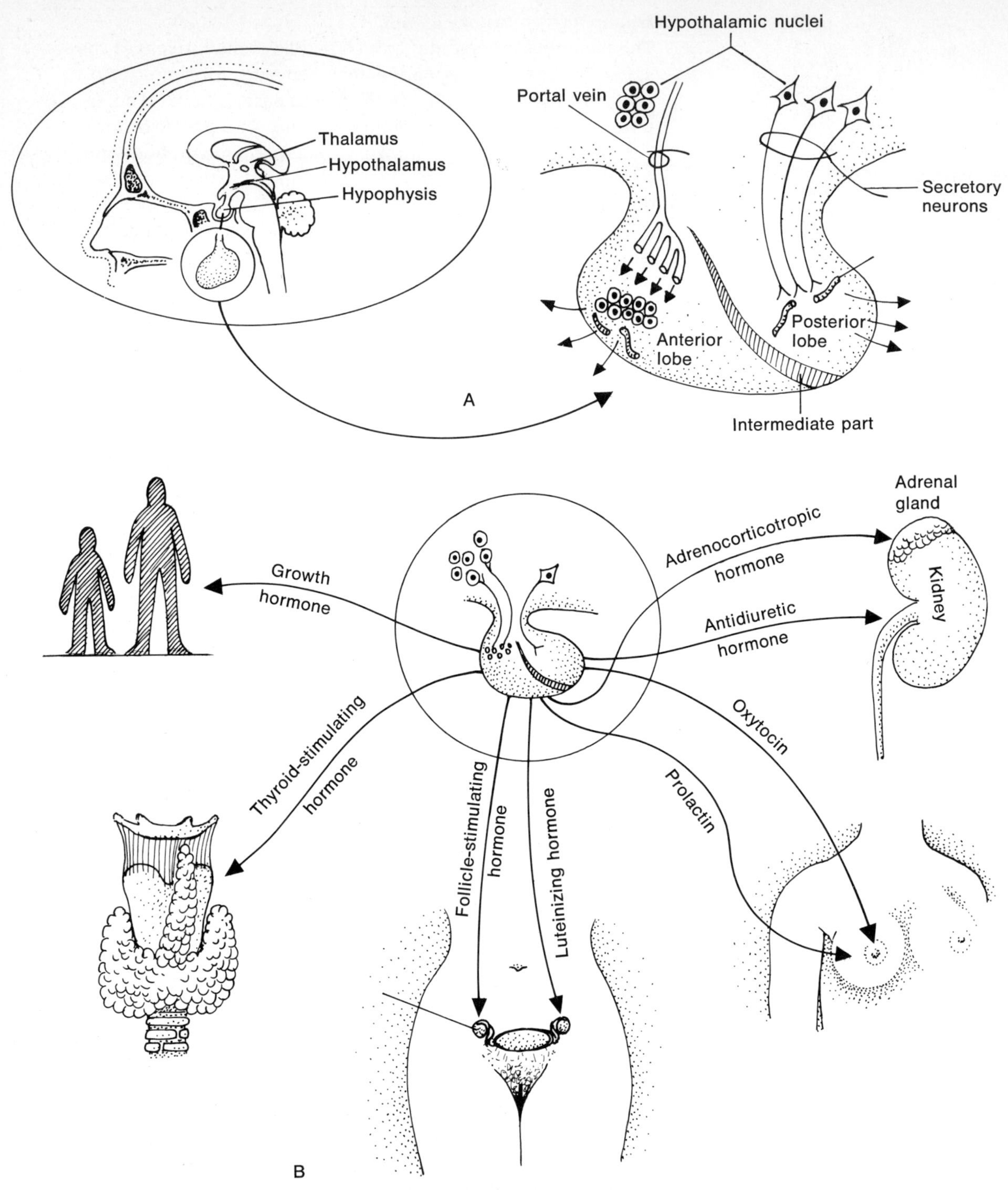

Fig. 6-29 The hypophysis. A. Portal circulation and secretory neurons of the hypophysis. B. Secretory products of the hypophysis and their target organs (diagrammatic).

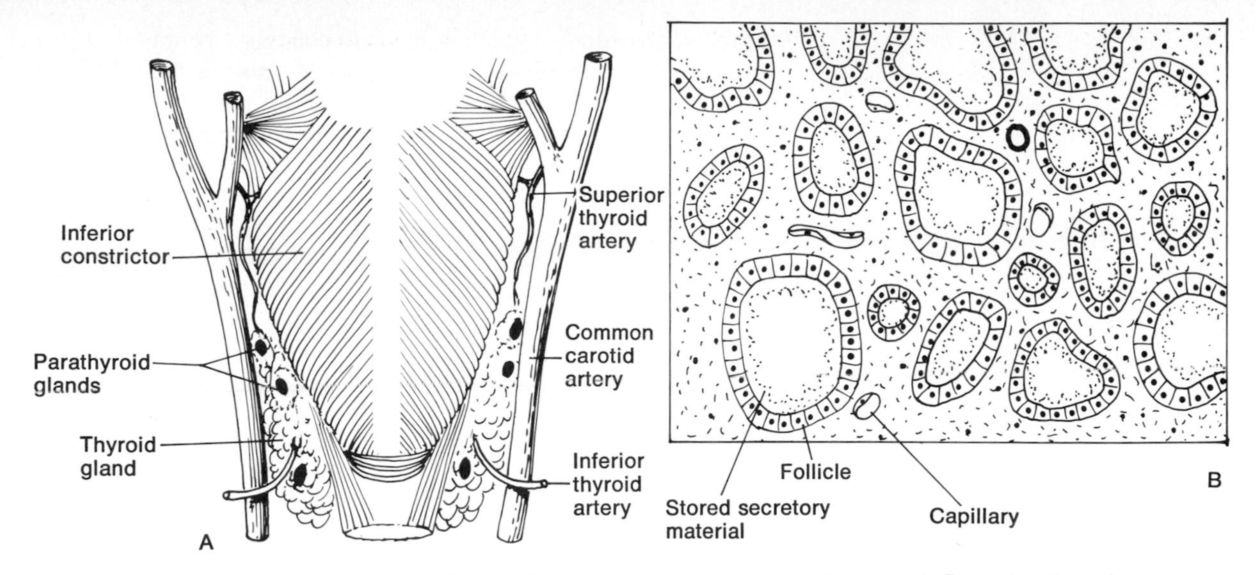

Fig. 6-30 The thyroid and parathyroid glands. A. Posterior view of thyroid gland embracing pharynx. Note the parathyroid glands. B. Cross section of the thyroid gland drawn from a photomicrograph.

(decreased hypophyseal activity) often generates rather dramatic manifestations, e.g., atrophy of the sex organs, supersensitivity to cold and stress, etc. Hypopituitarism in children results in dwarfism. Hyperpituitarism leads to gigantism in children and acromegaly in adults.

Thyroid gland

The thyroid gland is a bilobed, highly vascular structure embracing the anterolateral aspect of the upper trachea and part of the cricoid cartilage (Fig. 6-29). The gland consists of an encapsulated mass of grapelike follicles joined by connective tissue. The glandular, follicular nature of this endocrine structure is best visualized in a cross-sectional view as seen in Fig. 6-30.

The follicles are filled with a fluid (a colloid, thyroglobulin) and lined by simple cuboidal epithelium. These are secretory cells which take up iodine from the plasma, transport it to the colloid where it is used in synthesis of thyroxine. These same cells transfer the thyroxine across the cytoplasm to be discharged into neighboring capillaries. The activity of these secretory cells is a function of TSH secretion by the hypophysis. The principal effect of **thyroxine** is to maintain the proper rate of metabolic activity by stimulating increased oxygen consumption, and it is absolutely essential for proper mental and physical development in the developing fetus and newborn infant. Hypothyroidism in children **(cretinism)** is produced by a maternal lack of thyroxine during fetal development. Before the advent of iodized salt, cretinism was seen more frequently than it is today. Hypothyroidism in adults produces myxedema, combining dry, yellowish skin, decreased mental ability, and a deep, husky voice.

An excessive increase in thyroid activity, **hyperthyroidism,** is characterized by abnormally high metabolic rates, extreme nervousness and, in some cases, protrusion of the eyeballs (exophthalmos).

In those people who lack proper levels of iodine in their diet, the level of thyroxine in the plasma is decreased. Consequently, the levels of TSH are increased and the thyroid is stimulated to try to synthesize thyroxine; but without iodine it cannot—all it can do is become hypertrophied and hyperplastic (like trying to go up a down-moving escalator—you put out a lot of energy but you don't go anywhere!). Such is an iodine-deficiency **goiter.**

The thyroid also secretes a factor which acts to lower the calcium concentration in the plasma. This is **thyrocalcitonin.**

Parathyroid glands

These are four, small, peanut-sized structures which are embedded in the capsule of the thyroid gland on its posterior aspect (Fig. 6-30). Their significance was grasped when it was known that the thyroid itself was not essential to life and yet patients died following total thyroidectomy. The secretion of these bodies is called **parathormone** or simply parathyroid hormone.

Parathormone is known to regulate the blood level of ionized calcium, drawing upon bone for a source if the calcium level drops. Decreased parathyroid activity yields the condition called **tetany** or muscle spasm, resulting in death due to laryngeal constriction if calcium chloride injections are not instituted quickly. Increased parathyroid activity draws excessive calcium out of bone, leaving a highly porous (cystic) bone which is easily fractured (osteitis fibrosa cystica).

regional organization according to structure

We have now considered much of the viscera of the head and neck, in the functional sense, leaving us with vascular and neural structures, some muscles, and a need for general orientation of the region. First, an orientation (refer to Fig. 6-31):

In the classical tradition, the **region of the head** is divided up into subregions of the **cranium** and the **face.** We have completed our study of the cranial region with the exception of the **scalp,** the **temporal region,** and the **infratemporal region.** In the facial region, we have yet to consider the **muscles of facial expression, neurovascular structures,** and the **submandibular** and **parotid regions.** The neck is nicely divided into **anterior, lateral,** and **posterior regions.**

Fig. 6-31 Some regions of the head and neck.

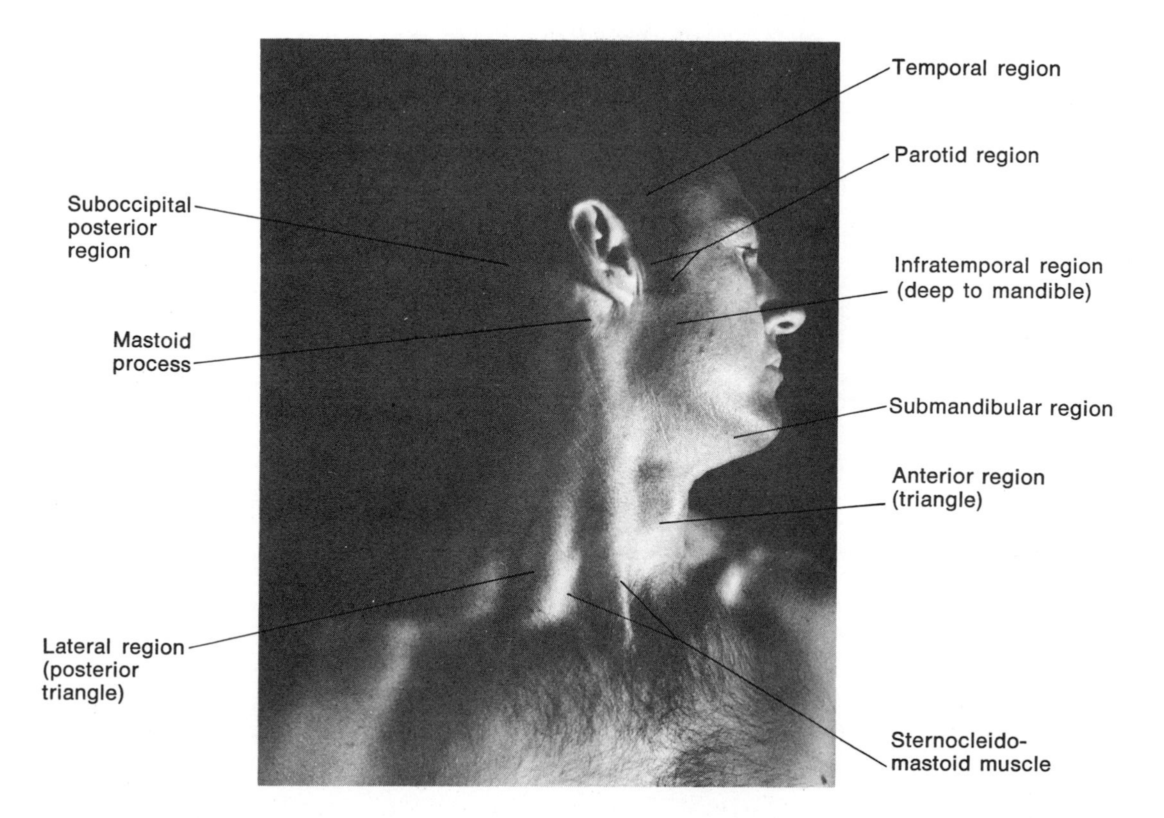

SCALP

The multiple-layered tissue covering the calvaria—often (but sometimes not) replete with hair—is called the scalp. It is quite characteristic of the scalp to bleed profusely when lacerated. Further, infections tend to spread rapidly throughout the scalp and into the cranium as well. The basis for these phenomena are—like most clinical situations—anatomical.

The scalp consists of five layers (Fig. 6-32); from superficial to deep, these are:

1. Skin.
2. Subcutaneous fascia.
3. Aponeurotic layer.
4. Loose connective tissue.
5. Pericranium.

By remembering how to spell SCALP (with subcutaneous spelled subCutaneous), you will recall these layers.

The skin is thin and incorportates many hair follicles. The subcutaneous fascia is a vascular tissue compartmentalized by strong, fibrous strands. Thus when the scalp is lacerated, the blood vessels—strongly enveloped and bound by these strands—cannot contract when cut, and tend to bleed until pressure is applied. The aponeurosis of the scalp includes the

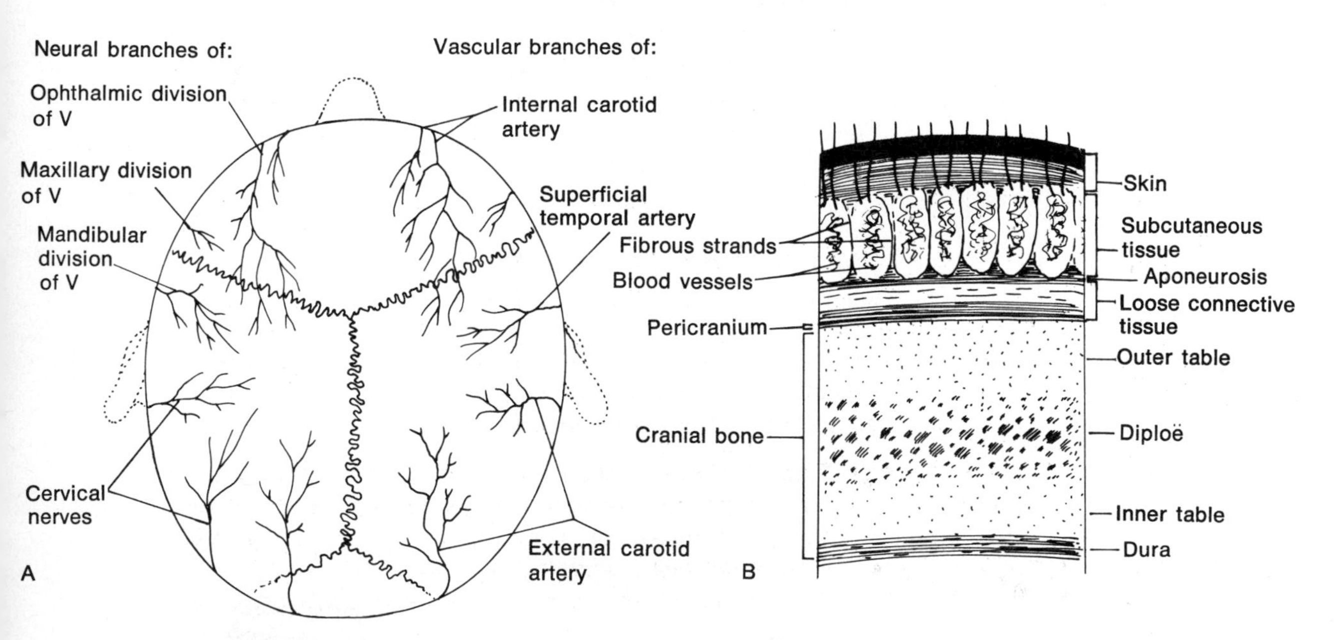

Fig. 6-32 Scalp. A. Vascular and nerve supply of the scalp. B. Sagittal section (diagrammatic).

occipitofrontalis muscle, which may be seen in Fig. 6-34. This muscle comes into play whenever the eyebrows are to be raised as in an expression of surprise or horror. Try these expressions yourself and concentrate on or feel the forehead skin contract. The loose subaponeurotic layer is responsible for the mobility of the scalp over the cranium and was probably the cleavage point for the scalper's knife in the days of yore. Move your own scalp with the palm of your hand and then contract the occipitofrontalis by raising your eyebrows and note how the mobility ceases. Why so? It is in this loose layer that infections introduced into the scalp spread. The pericranium is simply the periosteum of the cranial bones.

The scalp has a generous supply of blood and nerves (Fig. 6-32). The sources of blood all radiate from the internal and external carotid arteries. The veins drain generally into the jugular veins. The superficial temporal artery can be felt pulsating just in front of the ears. The nerves to the scalp are sensory with sympathetic fibers to the blood vessels, sweat glands, and arrector pili (hair) muscles. The motor supply to the occipitofrontalis comes from the facial nerve (VII).

THE TEMPORAL REGION

The temporal region is an area at the side of the skull anterior to and slightly above the ear flap covering the squamous part of the temporal bone (Fig. 6-33). Its lower border is at the level of the zygomatic arch. The principal occupant of this region is the **temporalis muscle** (of mastication) its fascia, and its vessels and nerves, which reach it from the infratemporal fossa. Place your fingers against the temporal region ("temples") and clench and unclench your jaws—that's the temporalis you feel contracting.

THE INFRATEMPORAL REGION

This is a well-filled fossa just below the temporal region, bordered externally by the ramus of the mandible and medially by the lateral pterygoid plate (Fig. 6-33). The infratemporal fossa owes its depth in part to the convex zygomatic arch. Some of the contents of this region include:

1. Three muscles of mastication: part of the **temporalis** and the lateral and medial **pterygoids** (Fig. 6-22). (The remaining muscle of mastication, the masseter, will be considered with the parotid region.)
2. The **maxillary** artery and its branches, which are the source of blood to the dura, the four masticatory

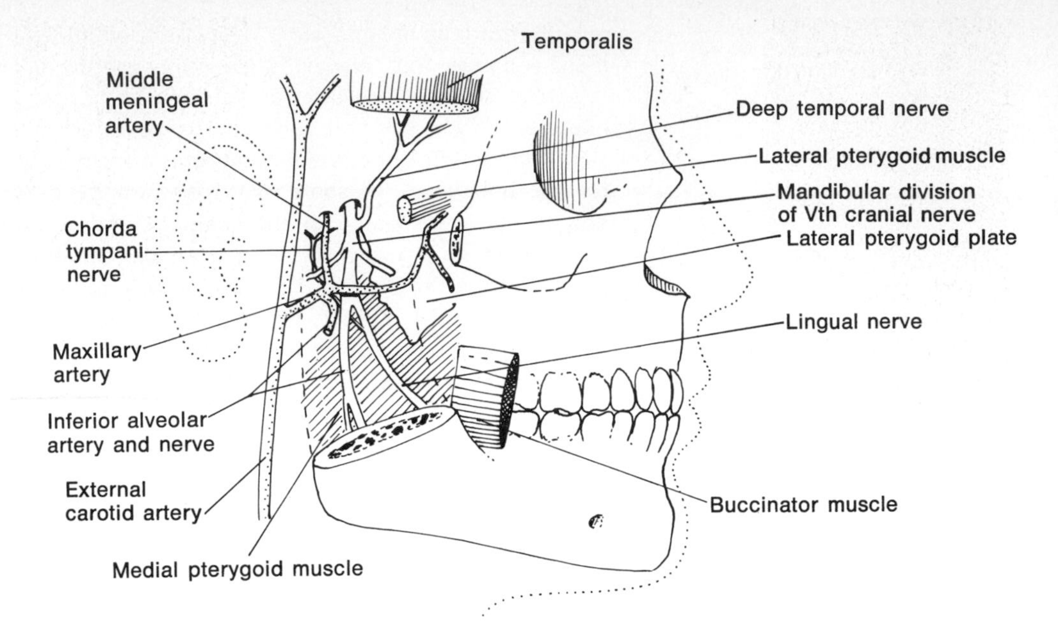

Fig. 6-33 The infratemporal fossa and some contents.

muscles, the gums and teeth of both jaws, the palate, the nasopharynx, and walls of the nasal cavity.

3. The **mandibular** or third part of the trigeminal (Vth) cranial nerve, serving all the masticatory muscles, and a conductor of sensory information from the tongue, lower teeth and gums, and skin of lower face.
4. The **chorda tympani** nerve (branch of cranial nerve VII), innervating the submandibular and sublingual salivary glands and transmitting taste information to the brain from the anterior two-thirds of the tongue.
5. The **pterygoid venous plexus,** one of two important collateral routes of venous return from the brain.

MUSCLES OF FACIAL EXPRESSION

For this section you will need a mirror. As you make facial expressions note that most of the movements of the face occur with the lips, the nose, and the skin around the eye. Thus you may suspect the majority of facial muscles insert at these places, and you are right.

First, let us consider the muscle about the orbit. Close and open your eyes with some force, while concentrating on what you feel. Now look at Fig. 6-34C. The muscle is appropriately titled **orbicularis oculi,** and arises from the surrounding bone to insert on the skin.

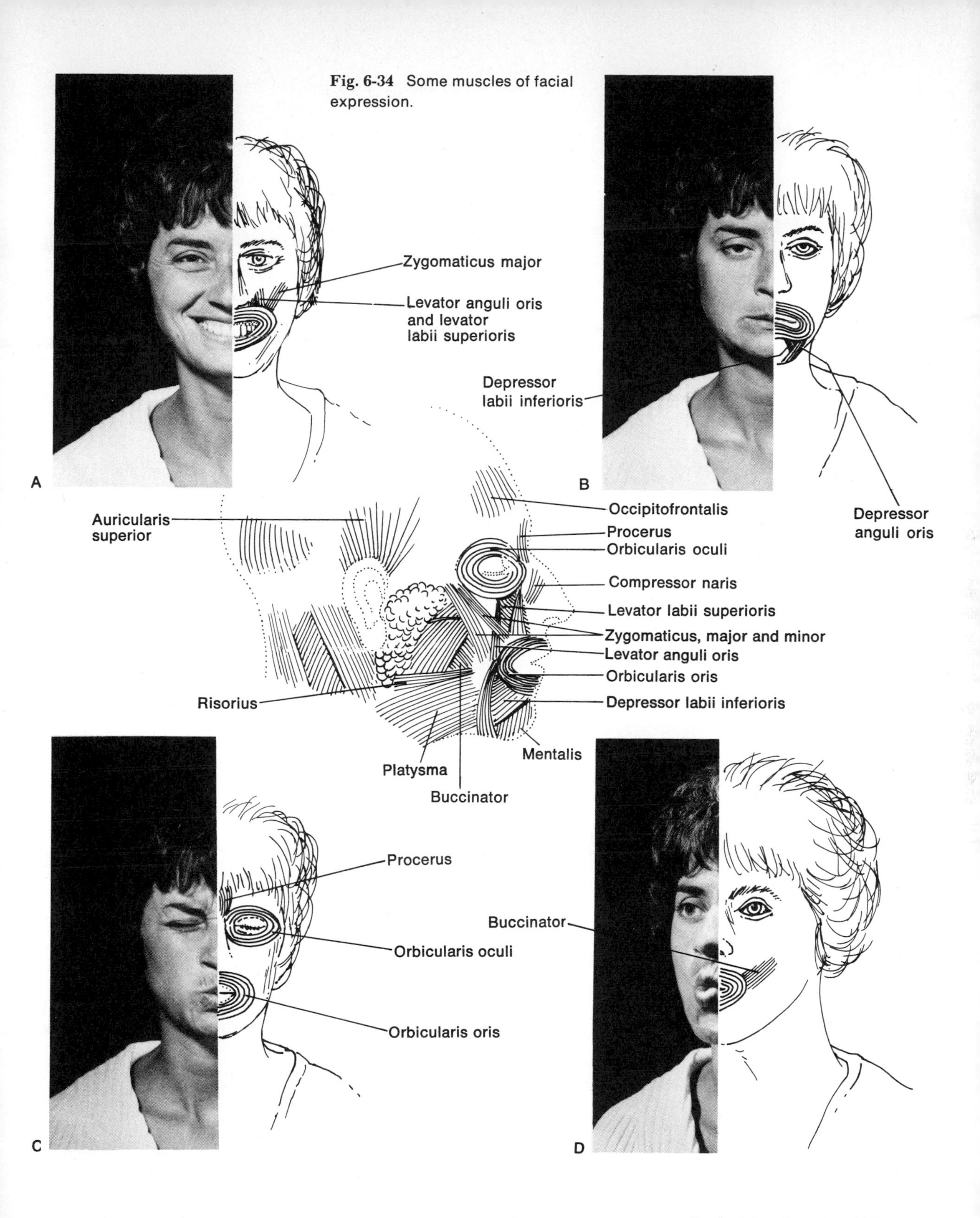

Fig. 6-34 Some muscles of facial expression.

Now the muscles of the nose are smaller but quite effective. Notice you can wrinkle your nose and dilate the nasal apertures (you can compress the nares too). These muscles are called **procerus, dilator,** and **compressor nares.**

Most of the muscles of the face are related to the mouth. Note that you can elevate the lips (laughing), depress them (pouting, grief), pucker them (kissing, whistling), retract the angles of your mouth (chagrin or disdain), and just generally create a myriad of expressions through the subtle employment of discrete muscle groups in the face. See now Fig. 6-34 and note how appropriate are the names for the muscles about the mouth (oris) and lips (labia):

1. Most muscles of the lips insert into the orbicularis sphincter of the mouth **(orbicularis oris).**
2. The elevators of the lips: **levator anguli oris, levator labii superioris, zygomaticus** (named after its origin).
3. The depressors of the lips: **depressor anguli oris, depressor labii inferioris, mentalis.**

One other muscle of the face—deeper than its fellows—is the muscle of the cheek, which you palpated earlier: the **buccinator**—the trumpeter's muscle. It arises near the lateral border of the superior constrictor of the pharynx and traveling just outside the maxilla and mandible enveloped in fascia and mucous membrane, inserts at the angle of the lips. It is employed in whistling, sucking, and playing wind instruments.

The muscles of facial expression have at least two characteristics in common:

1. They are all innervated by the **facial** (VIIth) **nerve.**
2. They are all superficially located and generally insert into the skin. They usually arise from the facial bones.

THE PAROTID REGION

The parotid region is an area of the face just anterior and inferior to the ear flap, below the zygomatic arch, and external to the ramus of the mandible. The chief constituents of this region (Fig. 6-35) are:

1. The **parotid gland** (the largest of the saliva-producing glands) and its **duct** (the occlusion of which—caused by viral infection—is termed mumps).
2. The **facial** (VIIth) **nerve** ramifying throughout the substance of the parotid and supplying all the muscles of facial expression.
3. The massive muscle of mastication, the **masseter,** easily palpated when the jaws are clenched.

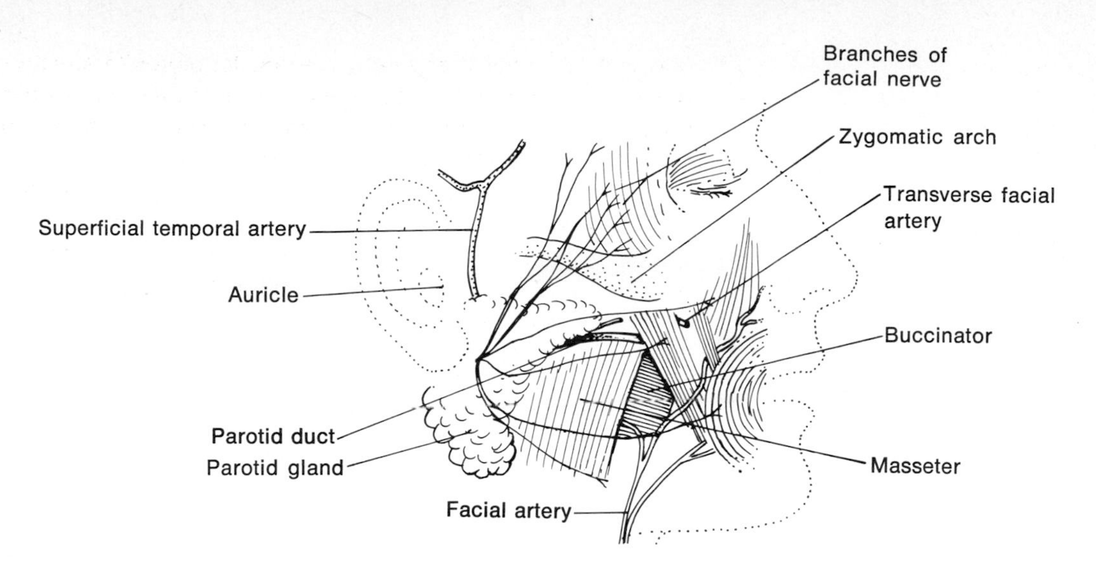

Fig. 6-35 The parotid region.

Palpate this area on yourself, referring to Fig. 6-35, and outline the gland and the probable distribution of the facial nerve on yourself or a fellow student.

THE SUBMANDIBULAR REGION

A region surrounded by the bodies of the mandible bordered below by the hyoid bone and above by the tongue, the submandibular region houses the **submandibular** and **sublingual glands,** associated ducts, vessels, and nerves, and the **suprahyoid muscle group** (Fig. 6-36).

ANTERIOR REGION OF THE NECK

Anteriorly, the midline of the neck and the two anterior borders of the sternocleidomastoid muscles create a pair of triangles, the **anterior triangles** (Fig. 6-37).

The **sternocleidomastoid** (cleido = clavicle) rotates the head to the opposite side while tilting it downward. Both (left and right) muscles flex the head and neck. Place your hand about the front of your neck and press your forehead against the palm of your hand and flex the head against resistance and note how this muscle contracts. The sternocleidomastoid is innervated by the accessory (XIth) nerve. The attachments of the muscle can be observed in Fig. 6-38.

The floor of the anterior triangle is made up of the **supra-** and **infrahyoid muscle groups.** These work together in mooring down the hyoid bone and allowing the latter to function

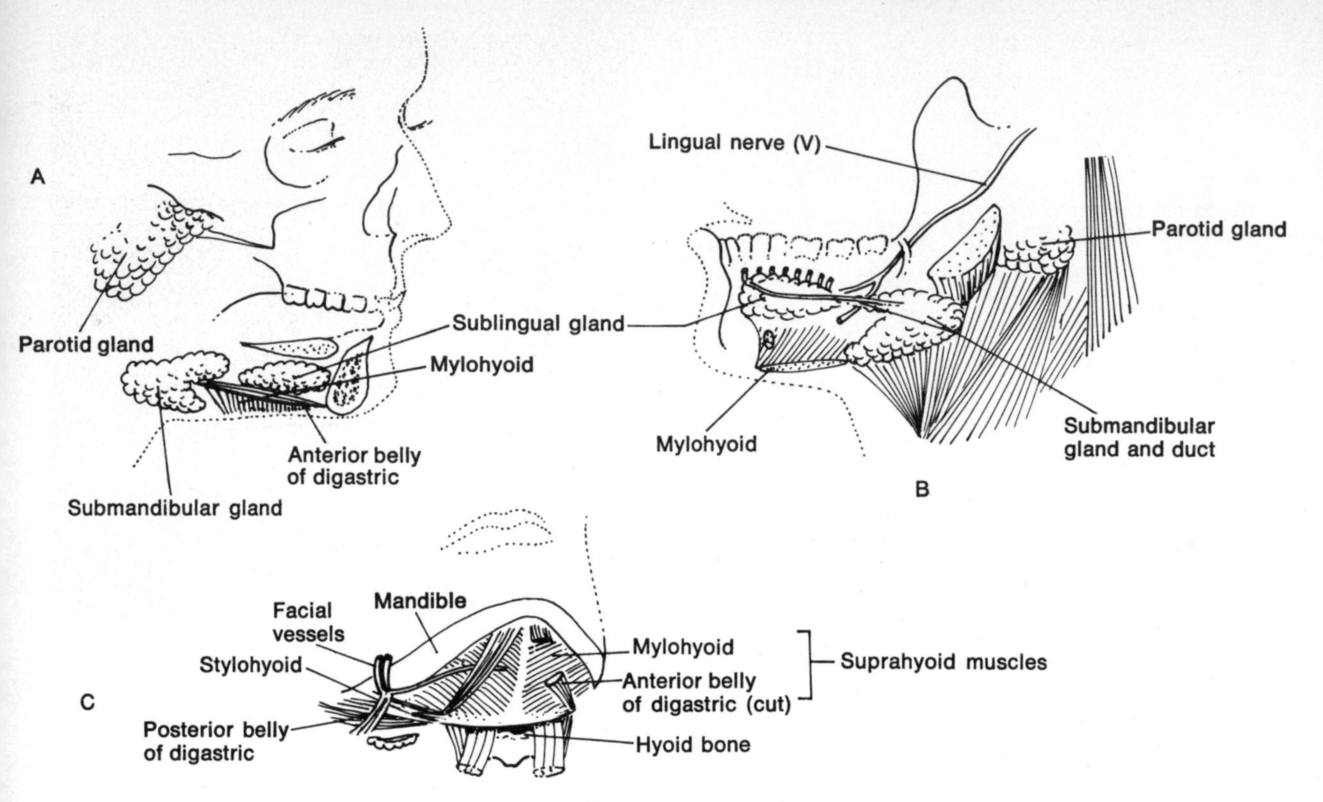

Fig. 6-36 Submandibular region. A. Lateral aspect. B. Medial aspect from inside the mouth, with tongue removed. C. View from below.

as a base for tongue movements. The suprahyoid group are innervated by both cranial and cervical nerves; the infrahyoid group by cervical nerves exclusively.

The roof of the triangles is the most superficial muscle of the neck, the thin **platysma,** used by men to tighten the upper neck during shaving. It attaches to the mandible above and skin about the clavicle below. Feel it on yourself when contracted.

The contents of the anterior cervical region may be studied in Fig. 6-37 and include:

1. The **common, internal,** and **external carotid arteries.**
2. The **internal jugular vein.**
3. The **vagus nerve** (X). These three occupy a common sheath, bordered anteriorly by the sternocleidomastoid muscle, for the most part.

Fig. 6-37 (opposite) Anterior triangle. A. Surface view of platysma overlying the anterior triangle. B. Lateral view partly dissected.

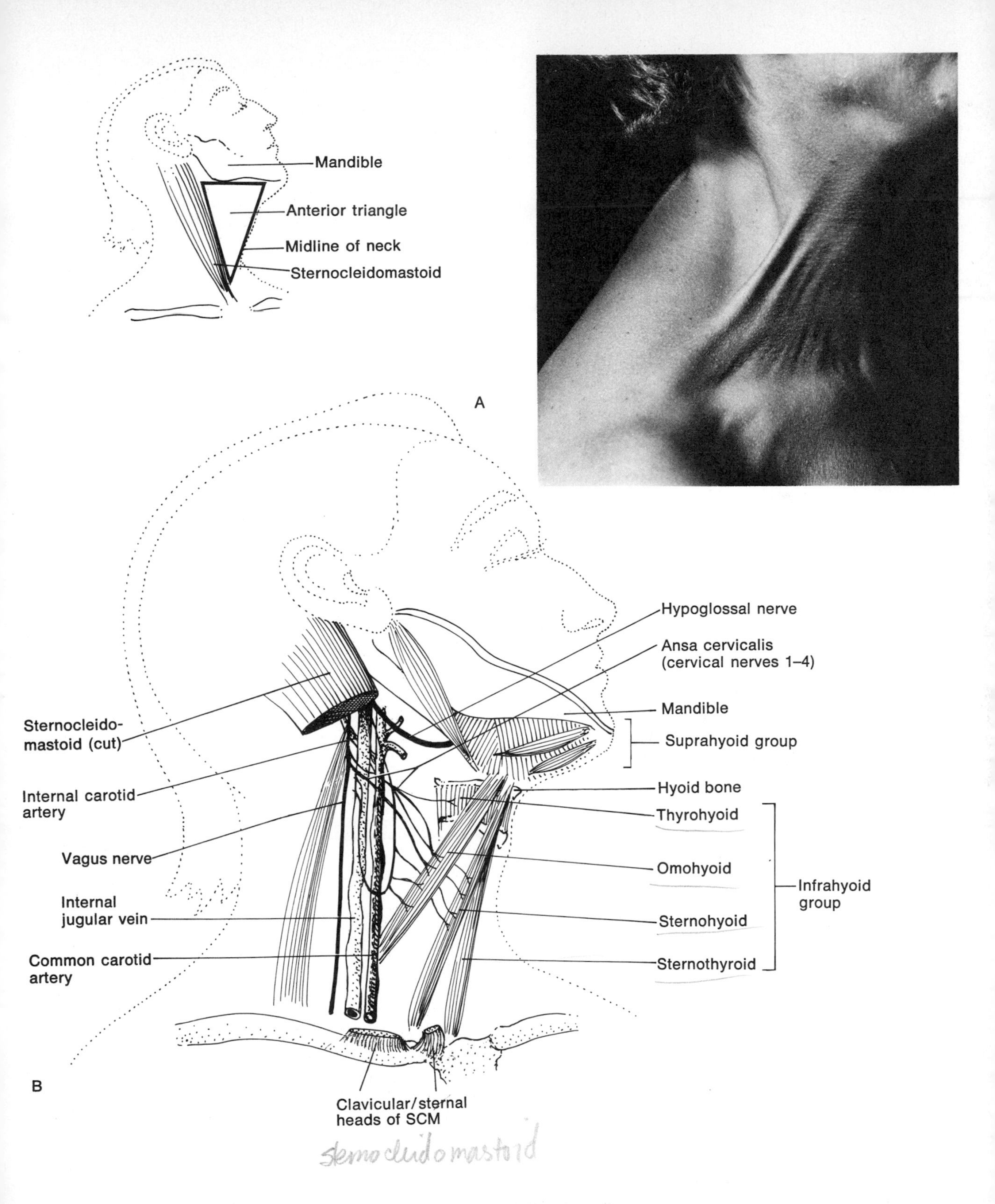

Mandible
Anterior triangle
Midline of neck
Sternocleidomastoid
A
Hypoglossal nerve
Ansa cervicalis
(cervical nerves 1–4)
Mandible
Suprahyoid group
Sternocleido-
mastoid (cut)
Hyoid bone
Internal carotid
artery
Thyrohyoid
Omohyoid
Vagus nerve
Infrahyoid
group
Internal
jugular vein
Sternohyoid
Common carotid
artery
Sternothyroid
B
Clavicular/sternal
heads of SCM
sternocleidomastoid

4. The many branches of the external carotid arteries and tributaries of the jugular veins, and the branches of the **hypoglossal** and **upper cervical nerves** splay out within the triangle (look ahead to Figs. 6-40 to 6-42).
5. The **thyroid gland** and the embedded **parathyroid** glands.
6. The **larynx,** clothed in infrahyoid muscles. The arteries and nerves are difficult if not impossible to palpate because of the intervening fascia and muscle.
7. The **laryngopharynx** immediately posterior to the larynx.
8. Deep muscles of the neck including those which stabilize the first and second ribs during respiration **(scalenes)** and those that flex the head and neck, lying in front of the vertebrae **(prevertebral muscles).**

LATERAL REGION OF THE NECK

Refer now to Fig. 6-38 and note the relationship of the superficial muscle of the back, the trapezius, with the clavicle and the sternocleidomastoid muscle. The triangle formed (often called the **posterior triangle**) is easily palpated on yourself: Turn your head to the right and flex your head sideways to the left while feeling for the outline of the left sternocleidomastoid—it can be seen easily this way in slender people. Feel along the posterior border of sternocleidomastoid from mastoid to clavicle. Now run your fingers posteriorly along the clavicle from the sternocleidomastoid to the next muscle you feel—the anterior border of the trapezius. Follow it up to the head. Note that there is a definite hollow between the two muscles, easily seen on the side from which the head is turned. Palpate it freely. This supraclavicular hollow (triangle) incorporates several vascular and neural structures traversing this area, some of which may be seen in Fig. 6-38 and possibly palpated on yourself:

1. The **external jugular** vein (frequently visible, with visible valves).
2. The **accessory** (XIth) nerve to trapezius.
3. The deeper **brachial plexus** en route to the axilla from the spinal cord.
4. Cutaneous nerves of the neck.
5. Motor nerves to surrounding muscle.
6. Arteries and veins en route to the superficial back from the subclavian artery.
7. **Cervical lymph nodes.**

Fig. 6-38 (opposite) Posterior triangle. A. Surface features. B. Contents.

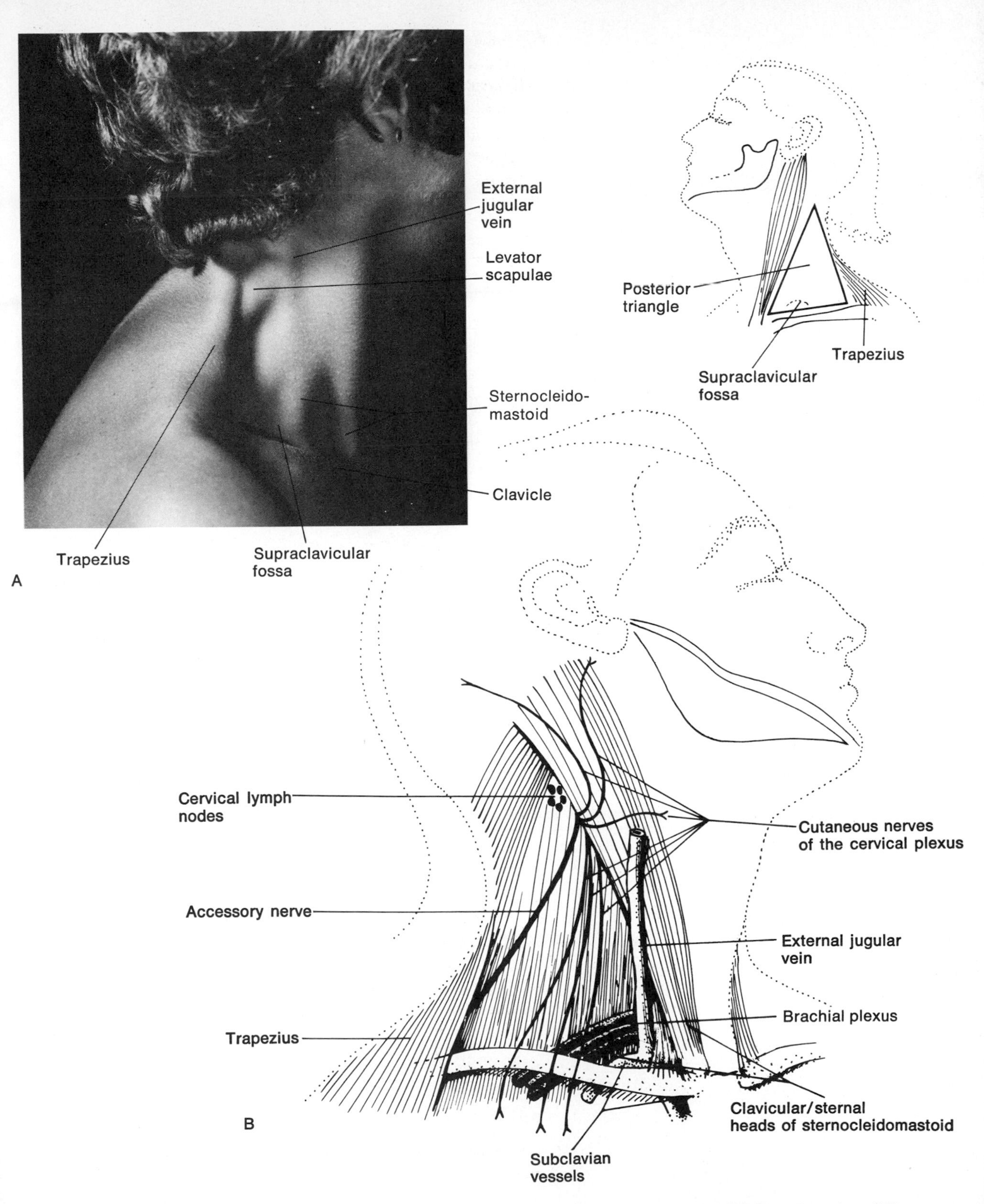
External jugular vein
Levator scapulae
Sternocleido-mastoid
Clavicle
Trapezius
Supraclavicular fossa
A
Posterior triangle
Trapezius
Supraclavicular fossa
Cervical lymph nodes
Cutaneous nerves of the cervical plexus
Accessory nerve
External jugular vein
Brachial plexus
Trapezius
Clavicular/sternal heads of sternocleidomastoid
B
Subclavian vessels

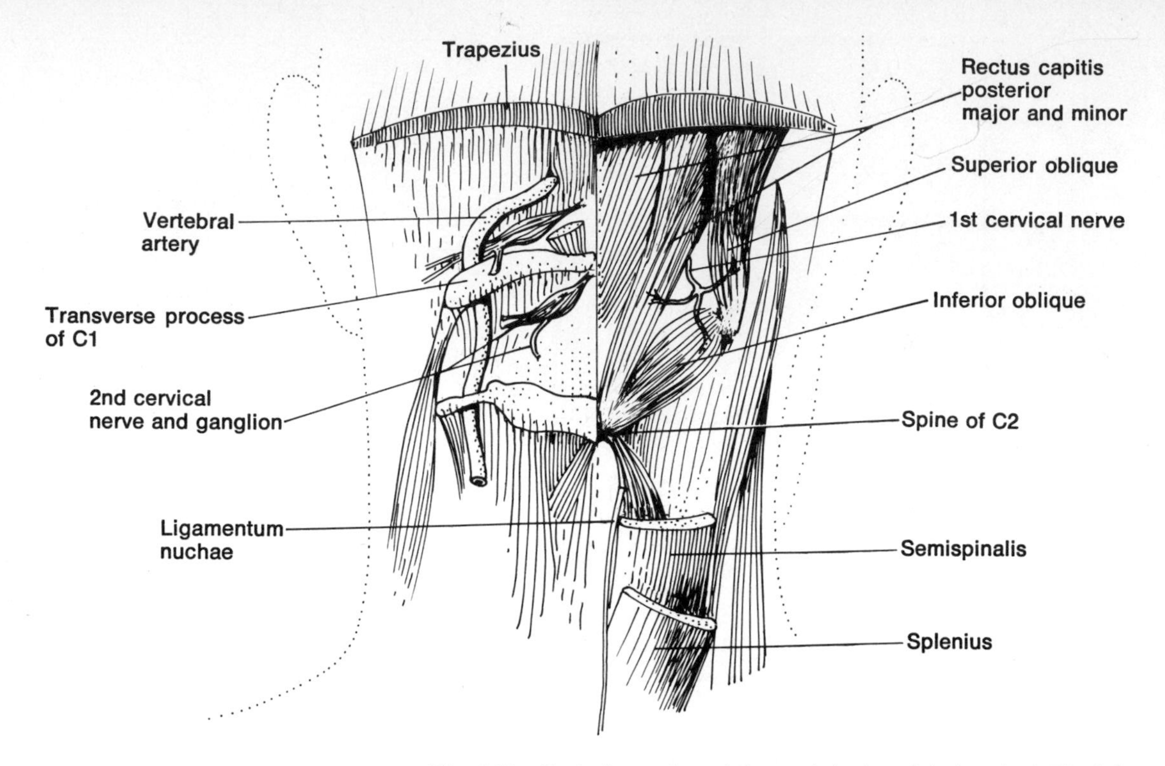

Fig. 6-39 Posterior region of the neck (suboccipital region). The left side illustrates a deeper dissection than the right.

POSTERIOR REGION OF THE NECK

This region is stuffed with layer after layer of muscle and lies within a boundary drawn from C7 below to the occiput above, and from the lateral border of the trapezius on one side to the lateral border on the other side (Fig. 6-39). The deep portion of this region just under the occipital bone (suboccipital region) includes a rather fascinating triangular arrangement of muscle bundles involved in rotating, flexing, and extending the head on the cervical vertebrae such that "yes" and "no" motions are possible. All are supplied by dorsal rami of upper cervical nerves. A probe pushed through the center of the muscular triangle on each side will pass through the **vertebral artery** en route to the brain from the subclavian artery.

Now for a little surface anatomy drill: can you outline the anterior and lateral regions on yourself? Can you palpate the:

- Platysma
- Larynx
- Hyoid bone
- Suprahyoid muscle group (submandibular glands?)

- Infrahyoid muscle group
- Sternocleidomastoid muscle
- Course of carotid sheath
- Cervical lymph nodes
- Approximate course of accessory nerve
- Approximate course of brachial plexus
- Approximate course of subclavian vessels
- Approximate course of exterior jugular vein

ARTERIES AND VEINS SERVING THE HEAD AND NECK

Now refer to Figure 6-40 as you read.

The source of blood to the neck and head are the **common carotid arteries** arising directly (left) and indirectly (right) off the arch of the aorta. These arteries bifurcate into **internal** and **external** branches at about the angle of the mandible (feel for this).

Except for those to the orbit and brain, all major branches to the head and neck arise from the external carotid artery.

The internal carotid goes directly to the underside of the cerebrum and supplies the brain and the orbit.

The various target organs of the arteries and the major arteries themselves can be studied using the diagram provided. The common carotid, facial, and superficial temporal arteries can be palpated on yourself—try it.

Next refer to Fig. 6-41 as you read.

The brain, face, and neck are all ultimately drained by the internal and external jugular veins. These veins drain into the right side of the heart via the brachiocephalic veins and superior vena cava.

With the above exceptions and the one noted below, the veins generally parallel the arteries and are similarly named.

The internal jugular vein is assisted by the **pterygoid** and **ophthalmic venous plexuses** in draining the brain and face.

The location of the tributaries of the jugular veins is extremely variable.

INNERVATION OF THE HEAD AND NECK

The scheme of head and neck innervation may be seen in Fig. 6-42. **Sympathetic innervation** is derived from upper thoracic preganglionic neurons which synapse in the **superior, middle,** and **inferior cervical ganglia** (Fig. 3-51). The postganglionic fibers reach their destinations by "hitching a ride" on one of the local arteries, e.g., the internal carotid artery to the orbit. These fibers are principally vasoconstrictors in effect and also dilate the pupil of the eye (see ahead).

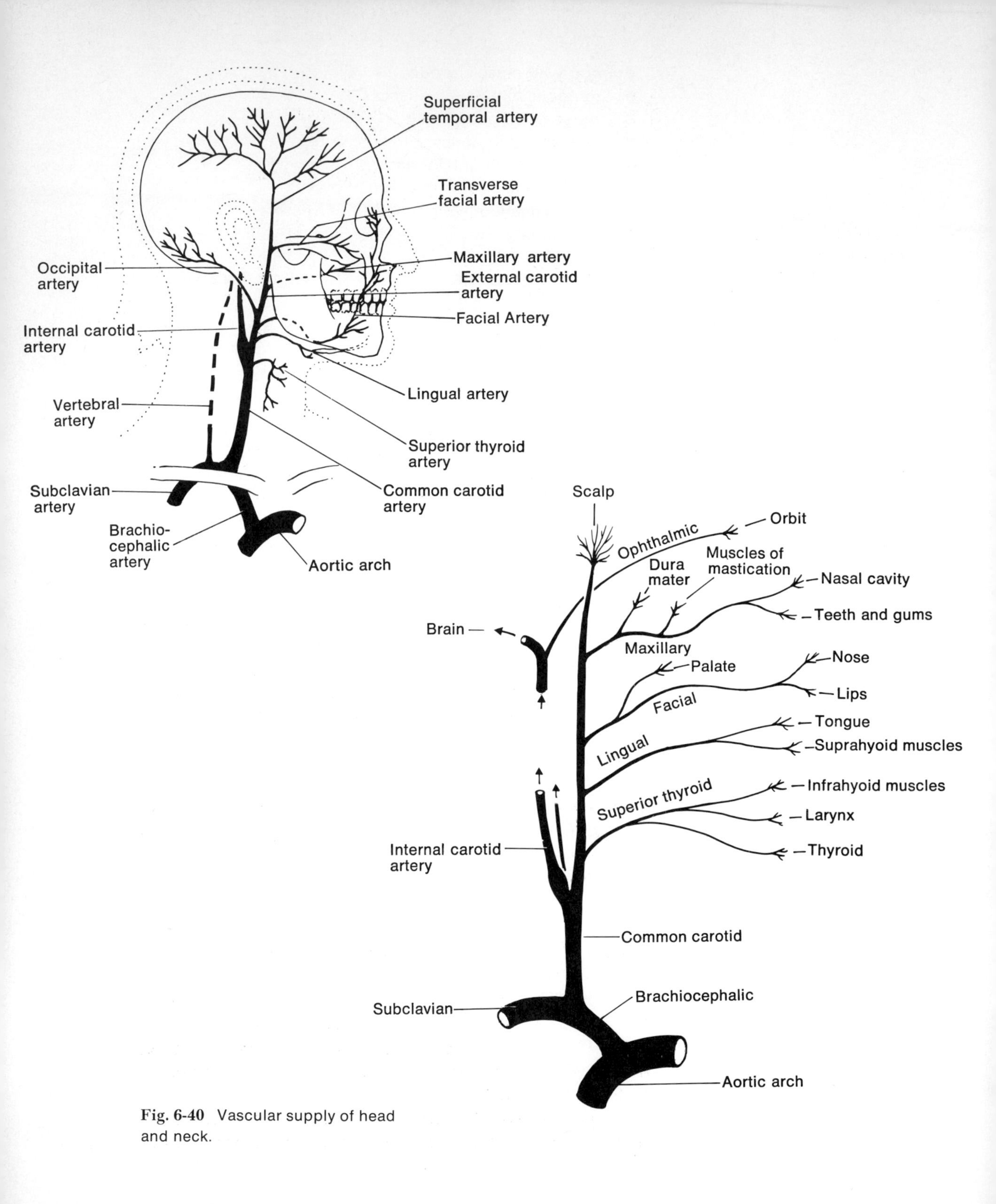

Fig. 6-40 Vascular supply of head and neck.

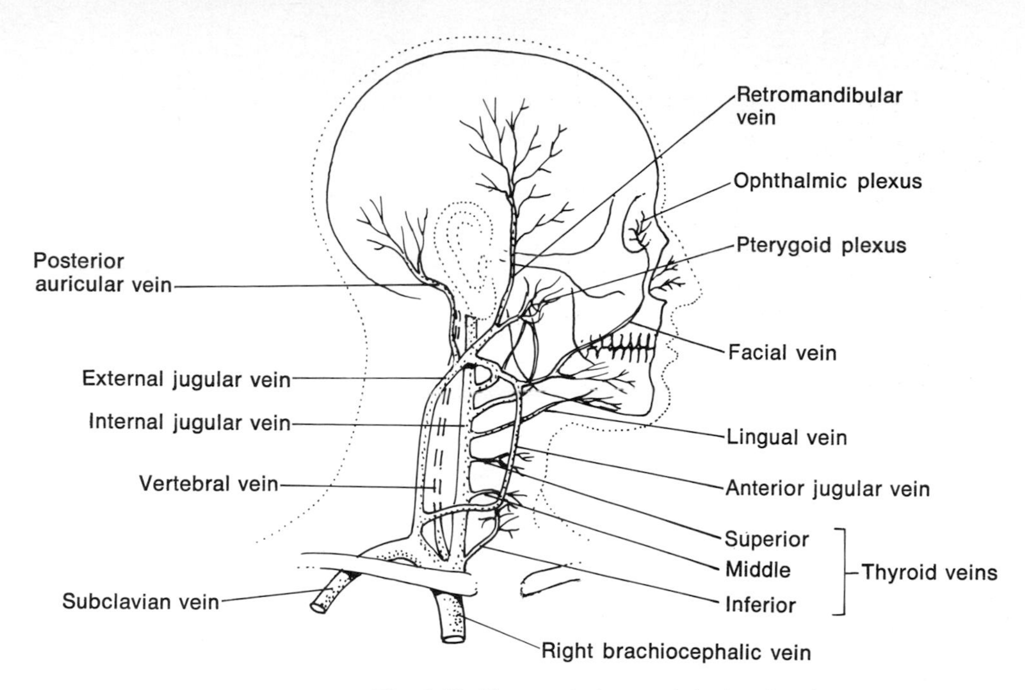

Fig. 6-41 Venous drainage of the head and neck.

Parasympathetic innervation of the head and neck is derived from cranial nuclei associated with the third, seventh, ninth, and tenth cranial nerves. Preganglionic fibers from these nuclei synapse at one of four cranial parasympathetic ganglia (Fig. 3-53). These fibers innervate the glands of the head and muscles of the eye (see ahead).

The general **somatic innervation** of the head and neck may be seen in Figs. 6-42 and 6-43. The cutaneous fibers serving the head and the fibers supplying the muscles of mastication are largely branches of the trigeminal (Vth) cranial nerve. The back of the head and the entire neck are supplied by branches of cervical nerves 1 to 4.

LYMPHATIC DRAINAGE OF THE HEAD AND NECK

Now refer to Fig. 6-44 as you read the following:

All lymphatic vessels of the head and neck ultimately drain into a chain of lymph nodes under cover of the sternocleidomastoid muscle. These are the **deep cervical lymph nodes** and receive the flow of such nodes as the parotid, facial, and submaxillary.

A superficial set of cervical nodes can be felt on yourself in both posterior and anterior triangles along the border of the

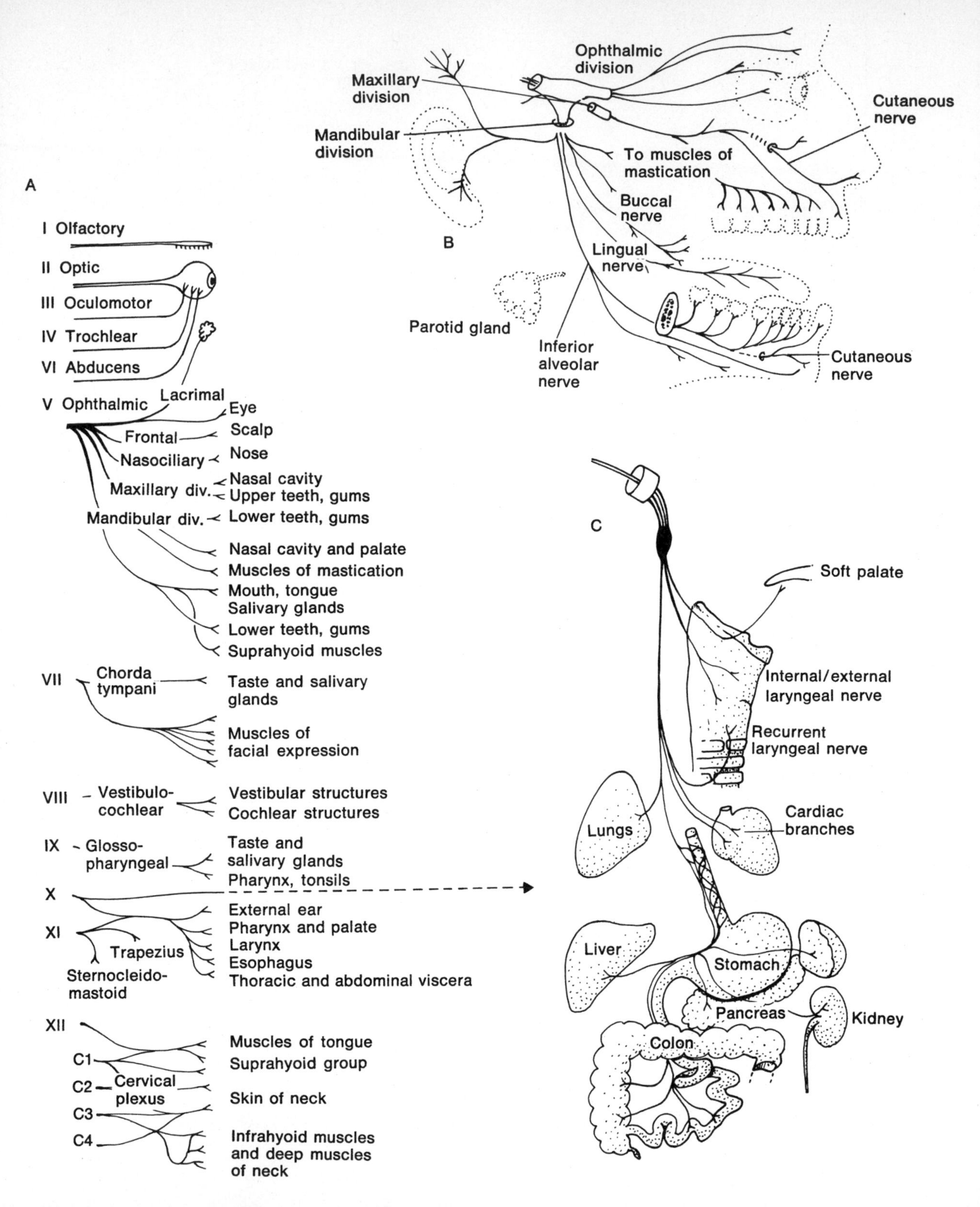

Ophthalmic division
Maxillary division
Mandibular division
Cutaneous nerve
To muscles of mastication
Buccal nerve
B
Lingual nerve
Parotid gland
Inferior alveolar nerve
Cutaneous nerve
A
I Olfactory
II Optic
III Oculomotor
IV Trochlear
VI Abducens
V Ophthalmic
Lacrimal
Eye
Frontal
Scalp
Nasociliary
Nose
Maxillary div.
Nasal cavity
Upper teeth, gums
Mandibular div.
Lower teeth, gums
Nasal cavity and palate
Muscles of mastication
Mouth, tongue
Salivary glands
Lower teeth, gums
Suprahyoid muscles
VII
Chorda tympani
Taste and salivary glands
Muscles of facial expression
VIII
Vestibulo-cochlear
Vestibular structures
Cochlear structures
IX
Glosso-pharyngeal
Taste and salivary glands
Pharynx, tonsils
X
External ear
Pharynx and palate
XI
Larynx
Trapezius
Esophagus
Sternocleido-mastoid
Thoracic and abdominal viscera
XII
Muscles of tongue
C1
Suprahyoid group
C2
Cervical plexus
Skin of neck
C3
C4
Infrahyoid muscles and deep muscles of neck
C
Soft palate
Internal/external laryngeal nerve
Recurrent laryngeal nerve
Lungs
Cardiac branches
Liver
Stomach
Pancreas
Kidney
Colon

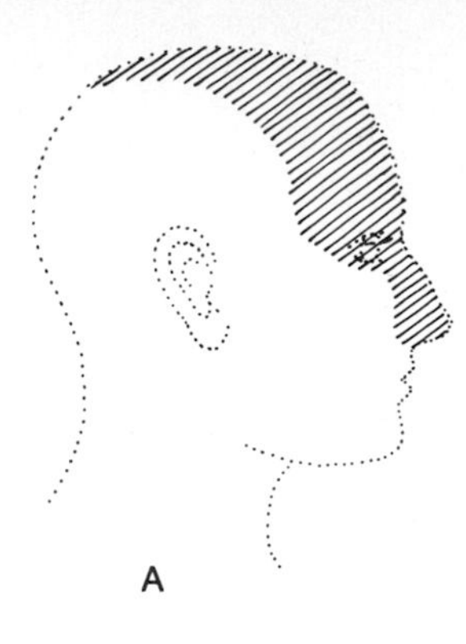

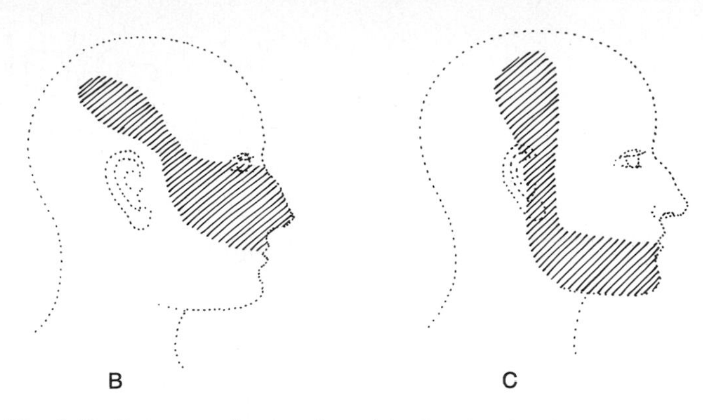

Fig. 6-43 Cutaneous innervation of the face by divisions of the trigeminal (Vth) nerve. A. Area supplied by ophthalmic division. B. Area supplied by the maxillary division. C. Area supplied by the mandibular division.

Fig. 6-42 (opposite) Distribution of cranial and spinal nerves to the head and neck. A. General scheme of distribution. B. Distribution of trigeminal (Vth) nerve. C. Distribution of vagus (Xth) nerve.

sternocleidomastoid muscle, particularly when hypertrophied due to a tonsillar infection, common cold, or other cause.

The efferent vessels of the deep cervical nodes form the jugular trunks (on each side), which terminate in the thoracic duct on the left and the right lymph duct on the right. These ducts empty into venous channels at the junction of the internal jugular and subclavian veins.

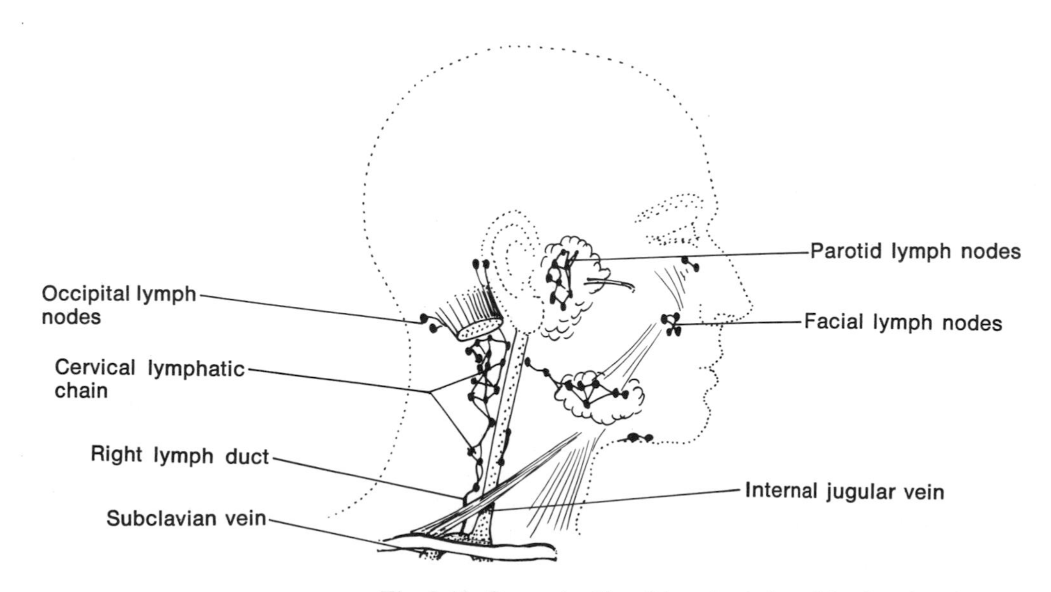

Fig. 6-44 Some significant lymph nodes of the head and neck.

the central nervous system

The central nervous system consists of the brain and spinal cord. It is the headquarters of the entire nervous system, the peripheral part of which you studied in Unit 3.

You will perhaps recall that the fundamental unit of the nervous system is the **neuron,** which consists of a cell body and processes. In the central nervous system collections of cell bodies are called **nuclei;** collections of nerve cell processes (axons) are called **tracts.** Tracts consist of axons which are largely myelinated, hence whitish, and are referred to as **white matter;** nuclei and their related processes are generally not myelinated and so are grayish and known as **gray matter** (Fig. 6-45). White and gray matter, mechanically fastened and metabolically supported by **neuroglia**—non-neural connective tissue found only in the brain and spinal cord (Fig. 6-46)—make up the entire central nervous system (CNS).

The human nervous system is a vast communications network. Incoming "data" from receptors and sensory neurons

Fig. 6-45 General internal structure of the head and neck. A. Level of cerebrum. B. Level of spinal cord.

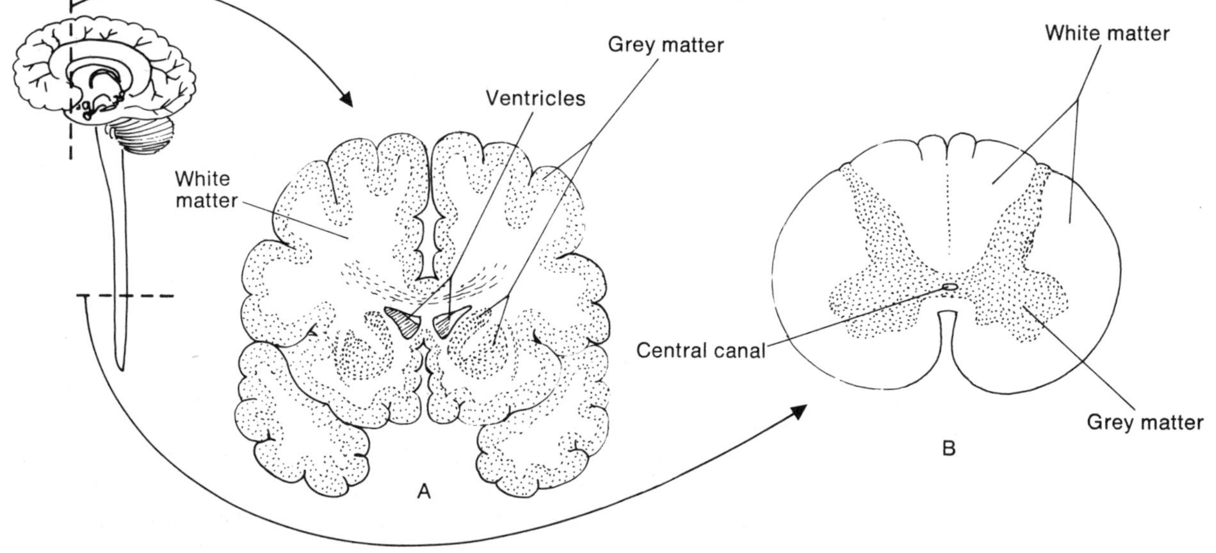

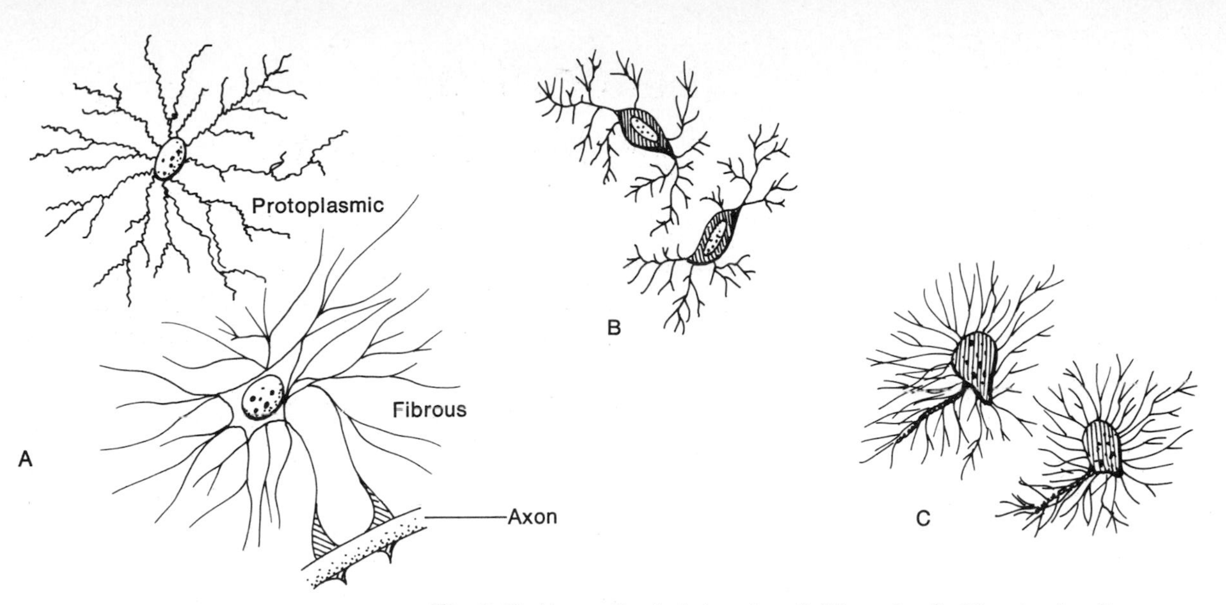

Fig. 6-46 Neuroglia. A. Astrocytes. B. Microglia. C. Oligodendroglia.

of the PNS "percolate" through the myriad neurons of the CNS and, by the coordinated efforts of its nuclei and centers, appropriate responses are issued.

A good portion of the CNS consists of ascending and descending pathways. The ascending tracts conduct sensory-related impulses from sensory neurons (either spinal or cranial) to nuclei at various levels in the brain. The descending tracts carry motor-related impulses to motor neurons of the PNS. These tracts will be considered shortly.

The study of the nervous system has usually been a stumbling block in the learning of anatomy. Therefore, in this text the central nervous system begins with a structure familiar to everybody—a hollow tube. In building block fashion we shall then construct a general picture of CNS structure and function.

A DEVELOPMENTAL CONCEPT

The key to understanding the basic structure of the central nervous system lies in appreciating its humble beginnings. At 25 days of age, the embryo is in the midst of extensive external and internal development. In the posterior midline region just deep to the skin, a hollow tube of embryonic nerve tissue has taken form (Fig. 6-47). Over several months this tube undergoes considerable growth and alteration, es-

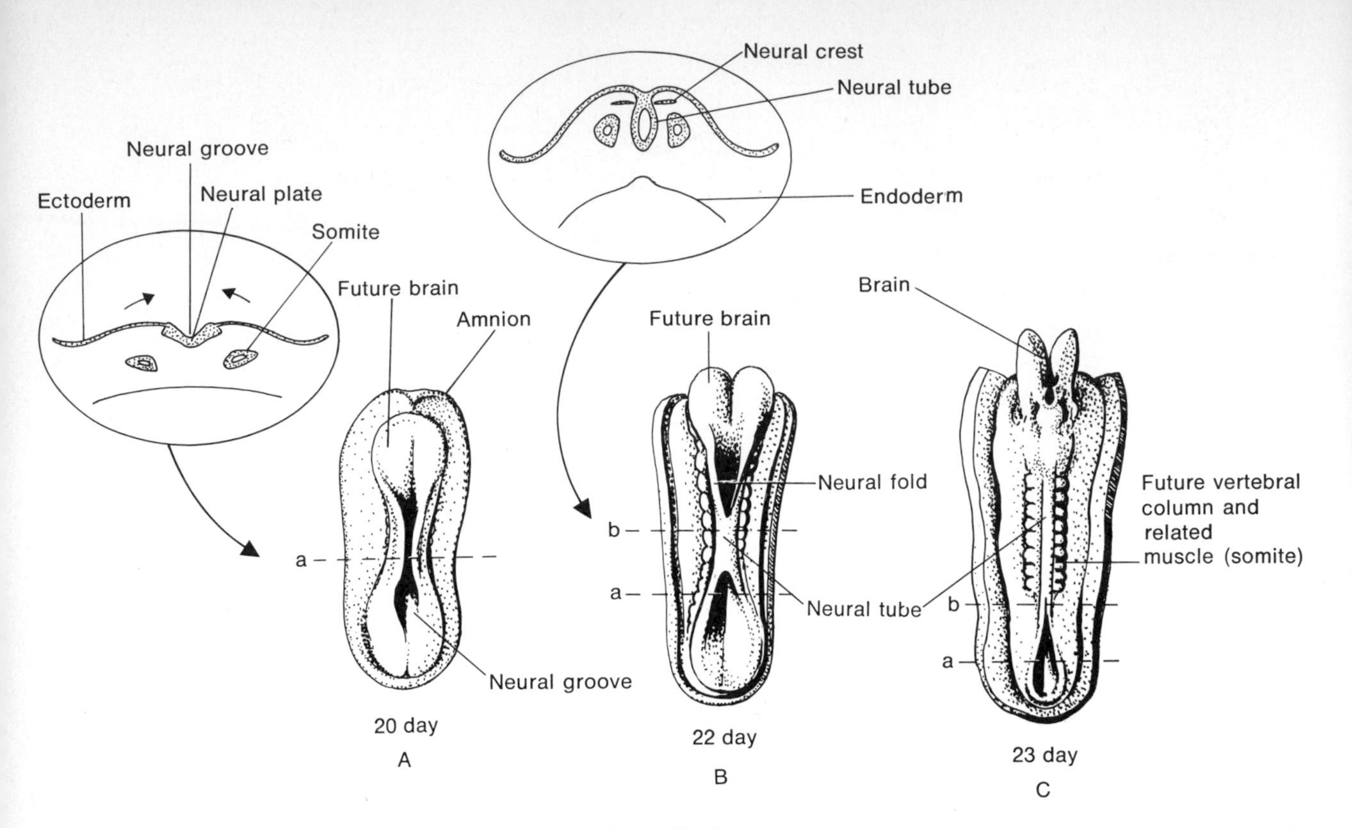

Fig. 6-47 Development of the neural tube, posterior surface. A. At 20 days of age. B. At 22 days of age. C. At 23 days of age. Inserts show cross sections of developing tube at ages indicated.

pecially in the head end, where the brain is to develop. The caudal two-thirds of the hollow neural tube will become the spinal cord. The interior of the tube will be formed into a series of internal cavities (ventricles and central canal of the spinal cord).

Soon after the neural tube has taken shape, three regions of the developing brain can be distinguished (Fig. 6-48):

1. Forebrain.
2. Midbrain.
3. Hindbrain.

As you can see in Fig. 6-49, a number of bends (flexures) appear in the brain region. These result from cells in certain regions growing at a faster rate than cells elsewhere. By this *differential growth* five regions are created out of the original three.

Refer now to Figs. 6-49 and 6-50 and note the following:

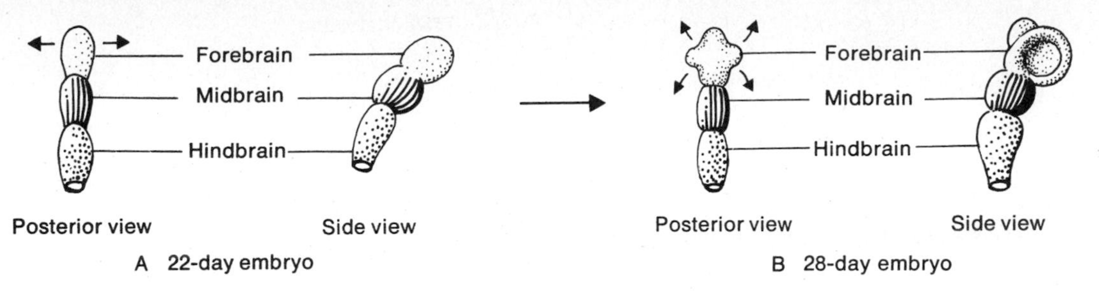

Fig. 6-48 Development of the three primordial regions of the brain. A. At 22 days of age. B. At 28 days of age. (Adapted from E. L. House and B. Pansky, *A Functional Approach to Neuroanatomy*, 2d ed., McGraw–Hill, New York, 1967.)

The forebrain changes significantly; two somewhat hollow hemispheric vesicles grow out, one on each side, creating a new region, the **telencephalon** (end brain).

The remaining portion of the forebrain changes only slightly, being squeezed between the hemispheres of the telencephalon, and is appropriately renamed the **diencephalon** (between brain).

The midbrain largely retains its external shape and **mesencephalon** (middle brain) remains its name.

The hindbrain, as you can see, undergoes considerable rearrangement. The upper part develops characteristic outpocketings anteriorly and especially posteriorly. The region is named the **metencephalon** (after brain).

The lower hindbrain, as one moves inferiorly, looks increasingly like the developing spinal cord and is suitably named the **myelencephalon** (spinal brain).

Fig. 6-49 Development of the five basic regions of the brain from the primordial three. (Adapted from E. L. House and B. Pansky, *A Functional Approach to Neuroanatomy*, 2d ed., McGraw–Hill, New York, 1967.)

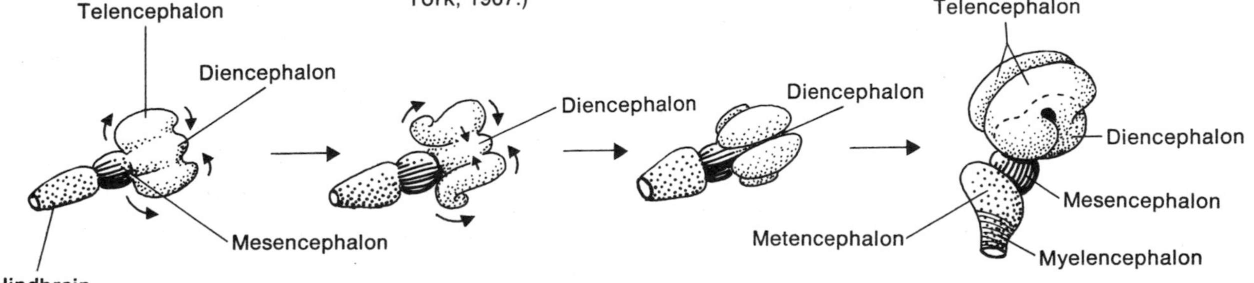

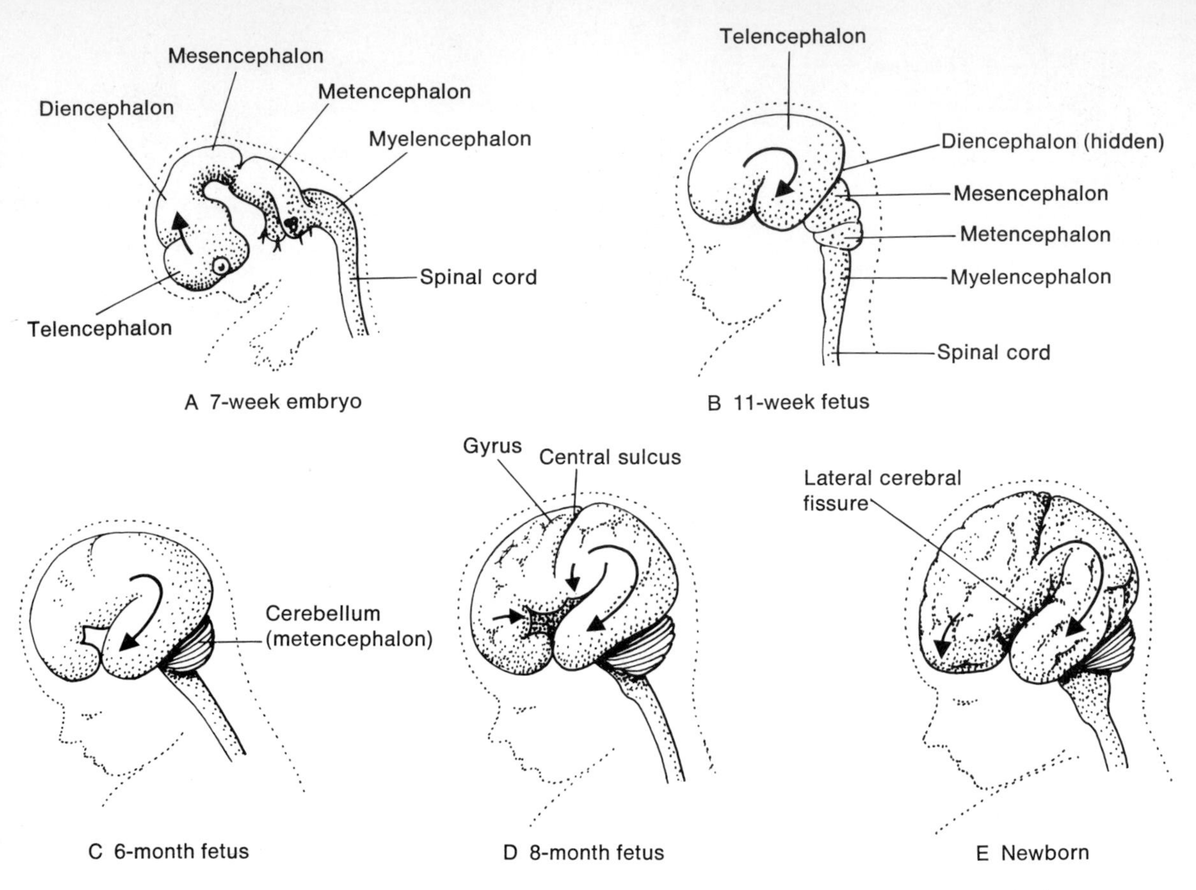

Fig. 6-50 External development of the adult human brain.

In subsequent months of development, the most spectacular growth continues to be in the telencephalon (Figs. 6-50 and 6-51). The diencephalon, hidden between the hemispheres, enlarges. The mesencephalon develops in the shadow of the hemispheres. In the metencephalon, the anterior swellings project posteriorly to join with the highly convoluted posterior prominence. The myelencephalon undergoes relatively little external change except to increase in mass, and the same may be said for the spinal cord. It should be understood that highly complex internal development goes on despite minor external changes.

The important derivatives of these five basic regions are as follows:

1. Telencephalon
   a. Cerebral cortex and related white matter
   b. Basal (subcortical) nuclei

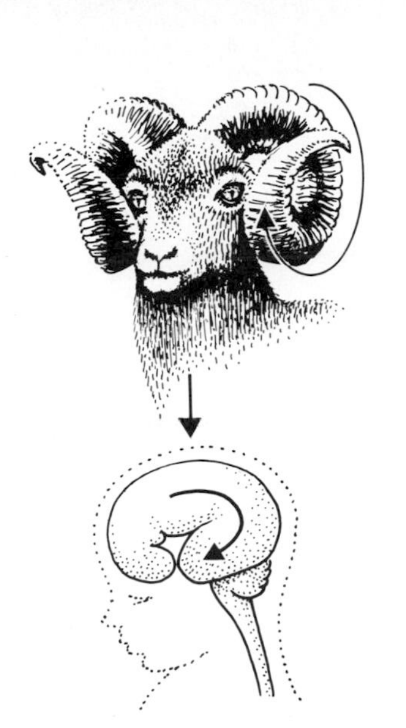

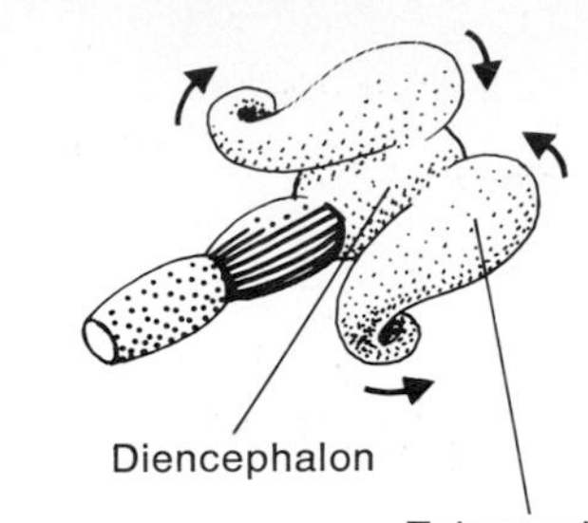

Fig. 6-51 The directions of cerebral growth resemble the horns of a bighorn sheep.

2. Diencephalon
   a. Thalamus
   b. Hypothalamus
3. Mesencephalon
   a. Cerebral peduncles
   b. Superior colliculi
   c. Inferior colliculi
4. Metencephalon
   a. Pons
   b. Cerebellum
5. Myelencephalon
   a. Medulla

The cavity within the developing neural tube changes along with each brain region. Refer to Figs. 6-52 and 6-53 and note:

The cavity of the forebrain undergoes considerable dilation particularly in the telencephalon where it develops into bilateral **lateral ventricles** (I, II). It thins into a midline, flat **third ventricle** (III) in the diencephalon, retaining communications with the lateral ventricles via two interventricular foramina.

Continuing caudally from the third ventricle, the cavity narrows into a tubelike **cerebral aqueduct** within the substance of the midbrain.

Going into the metencephalon, the cavity flares out to become the broad **fourth ventricle.** It narrows within the myel-

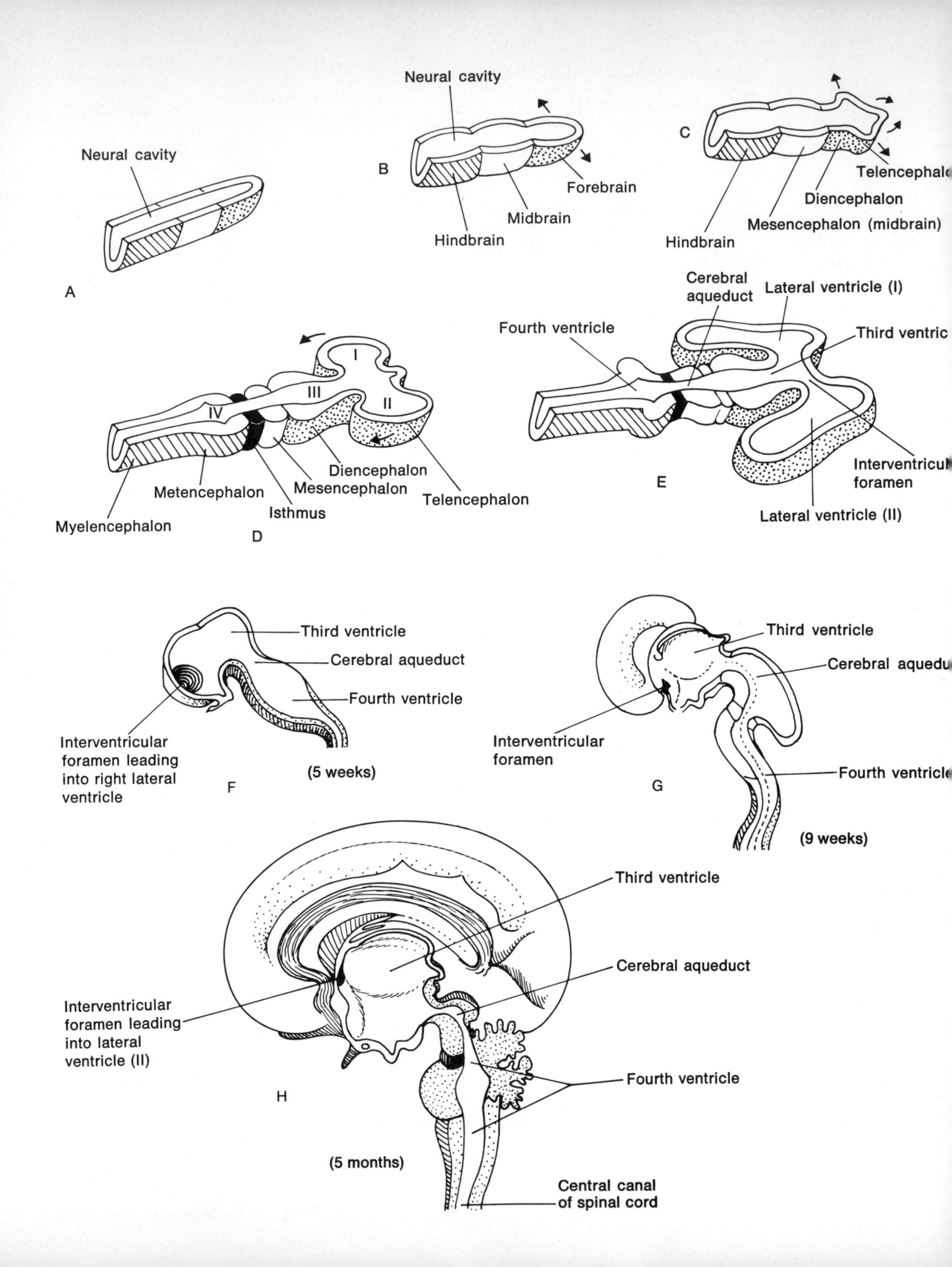
Neural cavity
A
Neural cavity
B
Hindbrain
Midbrain
Forebrain
C
Hindbrain
Mesencephalon (midbrain)
Diencephalon
Telencephalo
I
II
III
IV
Metencephalon
Myelencephalon
Isthmus
Mesencephalon
Diencephalon
Telencephalon
D
Cerebral aqueduct
Lateral ventricle (I)
Fourth ventricle
Third ventric
Interventricul foramen
Lateral ventricle (II)
E
Third ventricle
Cerebral aqueduct
Fourth ventricle
Interventricular foramen leading into right lateral ventricle
F
(5 weeks)
Third ventricle
Cerebral aquedu
Interventricular foramen
Fourth ventricle
G
(9 weeks)
Third ventricle
Cerebral aqueduct
Interventricular foramen leading into lateral ventricle (II)
Fourth ventricle
H
(5 months)
Central canal of spinal cord

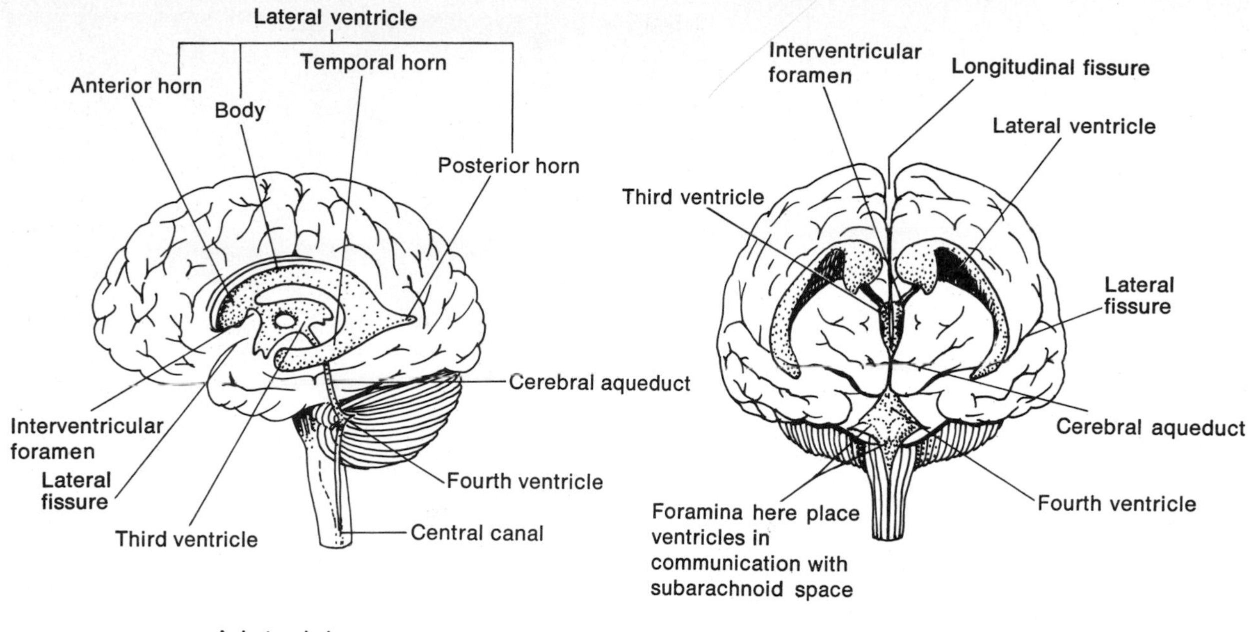

Fig. 6-53 The ventricles shown in situ.

encephalon to merge into the **central canal of the spinal cord.**

The fluid of the ventricles is called cerebrospinal fluid **(CSF)**—a plasma-like material which is secreted from the **choroid plexus**—an epithelial tissue heavily infiltrated with capillaries on the roof of the lateral, third, and fourth ventricles. The routes of CSF flow will be discussed shortly. In about one of every 700 babies born, there is an obstruction in the ventricles—often in the cerebral aqueduct or the fourth ventricle—and the trapped CSF volume dilates the ventricles. These fluid-filled enlarged cavities press against neighboring brain tissue, often cutting off the blood supply (ischemia) and causing atrophy of brain tissue. The soft, distensible head of the newborn is enlarged from the pressure of the swollen ventricles. This condition is called **hydrocephalus** ("water on the brain") (Fig. 6-54).

Fig. 6-52 (opposite) Stages in the development of ventricles of the brain. A–E, frontal sections. F–H, sagittal sections. (Adapted from E. L. House and B. Pansky, *A Functional Approach to Neuroanatomy,* 2d ed., McGraw-Hill, New York, 1967.)

GROSS ANATOMY OF THE BRAIN

At this point you should understand, generally, how the brain is oriented into the five regions just explained. We are now going to take a hard look at the residents of these regions, emphasizing the *continuity of structure* in the central nervous

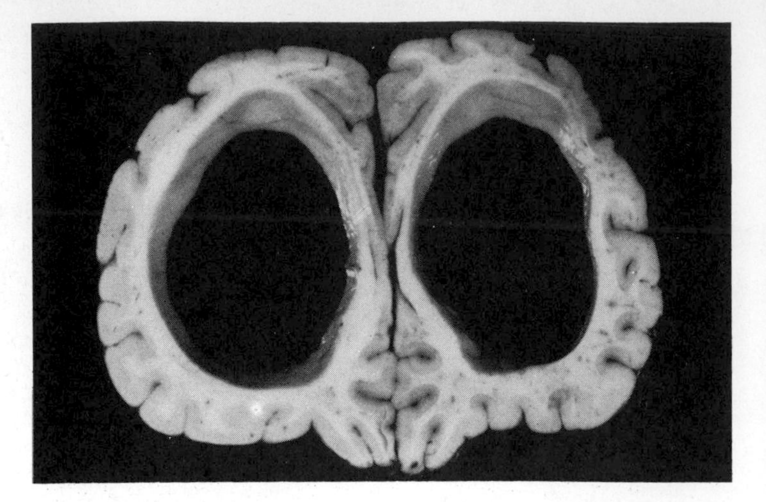

Fig. 6-54 Hydrocephalus. Section of brain illustrating the dilatation of the lateral ventricles and compression of adjacent neural tissues. (Courtesy of R. P. Morehead, *Human Pathology*, McGraw-Hill, New York, 1965.)

system. It is important to understand this interregional continuity when considering tracts and pathways of the CNS and the nuclei with which they relate. With this in mind, we can organize and will consider the CNS as follows (Fig. 6-55):

1. **Cerebral hemispheres.**
2. **Brainstem.**
3. **Cerebellum.**
4. **Spinal cord.**

Cerebral hemispheres

The cerebral hemispheres take up the largest share of the brain mass—those great cauliflower-like structures bulging out above and over the brainstem like a great flower projecting from its stem. The hemispheres are separated by a deep longitudinal fissure (Fig. 6-53) and attached by a flat white band of interconnecting tracts. They are further characterized by a lateral fissure separating the temporal lobe from the more central part of the cerebral hemispheres.

The cerebral hemispheres consist of the following structures:

1. Cerebral cortex.
2. White matter.
3. Subcortical (basal) nuclei.

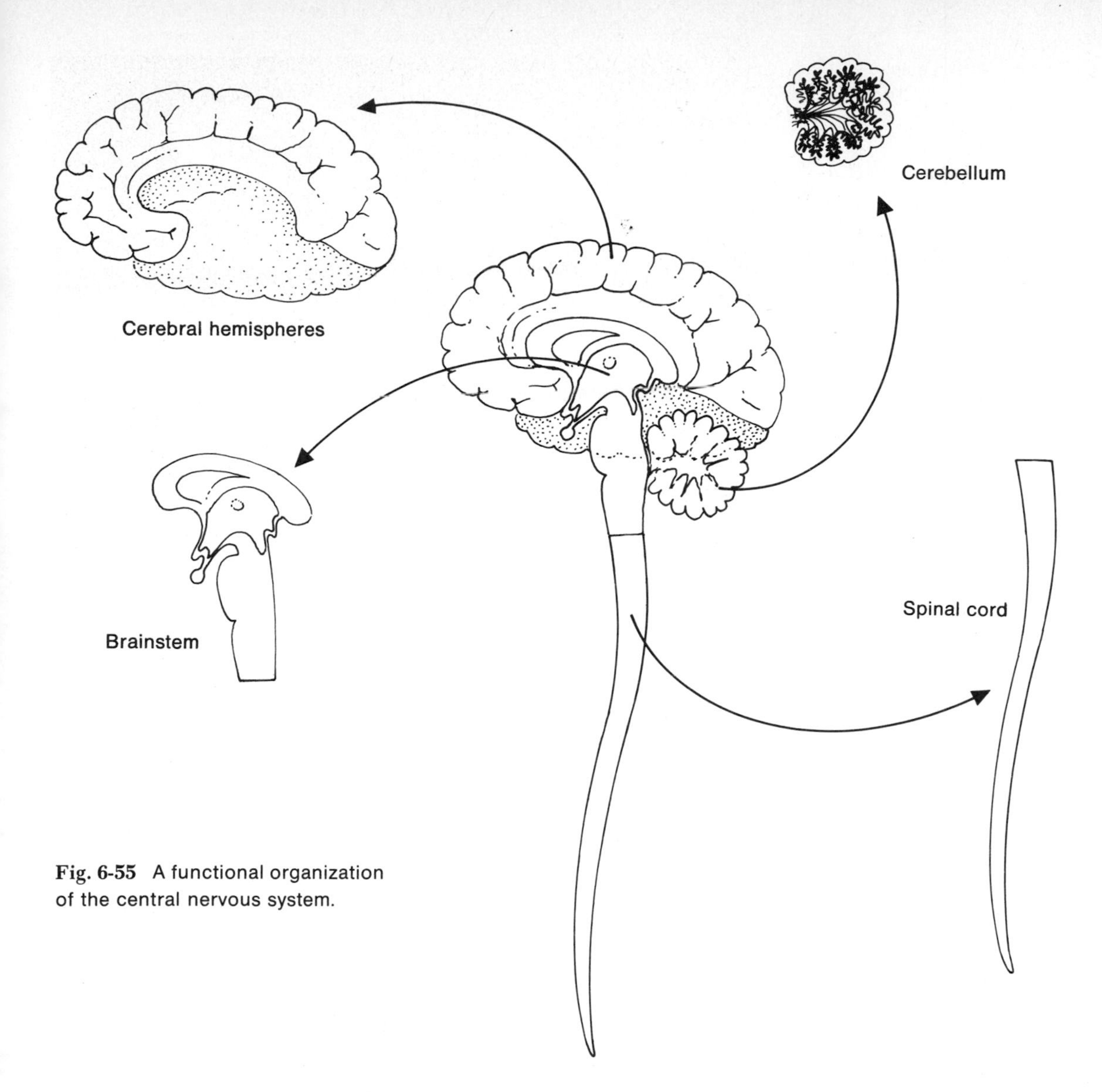

Fig. 6-55 A functional organization of the central nervous system.

Cerebral cortex. The cerebral cortex (L., bark) is that external layer of *gray matter* covering the brain (Fig. 6-45). It has deep furrows and round ridges. In anatomical parlance, such a furrow is called a **sulcus** (pl., -i); a ridge is called a **gyrus** (pl., -i). These sulci and gyri increase the surface area of the brain considerably.

The neurons of the cerebral cortex generally occupy one of the following areas: sensory areas of the cortex, motor areas of the cortex, or association areas of the cortex.

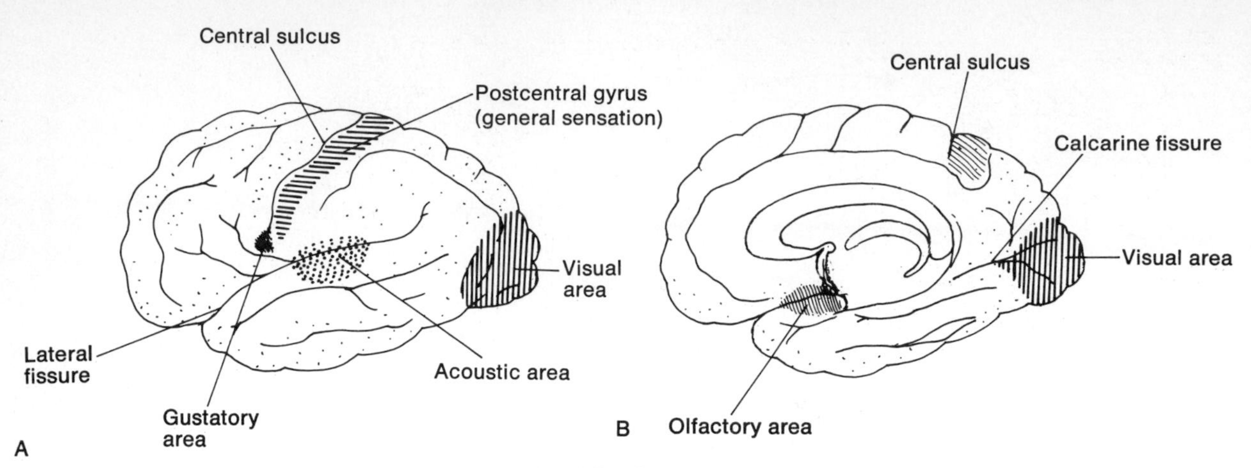

Fig. 6-56 Primary sensory areas of the cerebral cortex. A. Lateral view. B. Medial view.

Sensory areas of the cortex. Neurons in the various sensory areas of the cortex are concerned with the final receipt of ascending impulses from lower centers in the brain and spinal cord. It is in these areas that one is made *aware* of a sensory experience. More specifically, in these areas one is able to *discriminate* among the various sensations the body is equipped to perceive. You will note in Fig. 6-56 the following principal sensory areas of the cortex:

- Impulses related to conscious perception of pain, temperature, touch, pressure, and muscle sense reach for the postcentral gyrus of the parietal lobe. (Lobes are identified in Fig. 6-58.)
- Impulses concerned with vision arrive at the calcarine fissure of the occipital lobe.
- Impulses related to the conscious perception of sound terminate in a portion of the superior temporal gyrus of the temporal lobe.
- Sensations of taste (gustation) are believed to be perceived at the base of the central sulcus just above the lateral fissure.
- The sense of smell (olfaction) is believed to make contact with a complex area relating to the limbic system at the medial and inferior aspects of the temporal lobe.

Motor areas of the cortex. The principal motor area is located largely in the precentral gyrus of the frontal lobe (Fig. 6-57). The neurons in this area initiate willful, skilled movement on the *opposite* side of the body. These impulses are largely

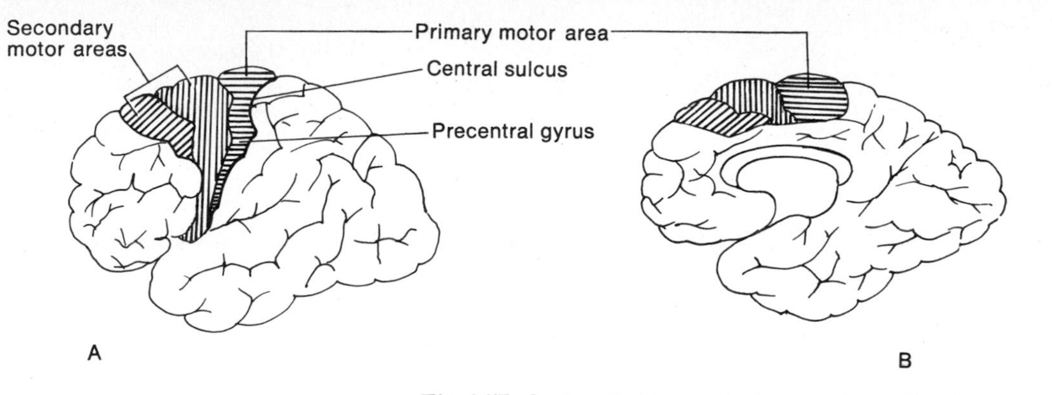

Fig. 6-57 Some principal motor areas of the cortex. A. Lateral view. B. Medial view.

transmitted down a specific tract called the **corticospinal tract*** or **pyramidal pathway** to the motor neurons of the spinal cord. There are also parts of the frontal lobe whose cortical neurons initiate unskilled, reflex postural movements. The axons of these neurons are not part of the corticospinal tract.

Association areas of the cortex. All areas of the cortex not specified as "motor" or "sensory" are generally labeled association areas. Such areas take up the largest share of cortex. They integrate sensory input of different kinds with memory (stored in association areas), pemitting complex perceptions and emotional expressions. For instance, you see a picture of a car and recall the *sound* of a powerful engine; your friend sees the same picture and recalls the *smell* of a new car's interior; another friend sees the picture and immediately *feels anxious* because not long ago he was in an automobile accident. Association areas also provide the mechanism for complex mental activities such as reasoning, thinking in abstract terms, etc. In addition, association areas enable complex motor activities such as talking or walking. *In general, then, association areas integrate sensory input to provide motor output appropriate for the occasion.*

Brain function according to lobes. The sulci and fissures of the cortex are employed by the neuroanatomist to divide the cortex and underlying white matter into lobes (Fig. 6-58) of which:

* Note that many tracts are named to suggest their origin and destination, the corticospinal tract, for instance, which goes from cortex to spinal cord.

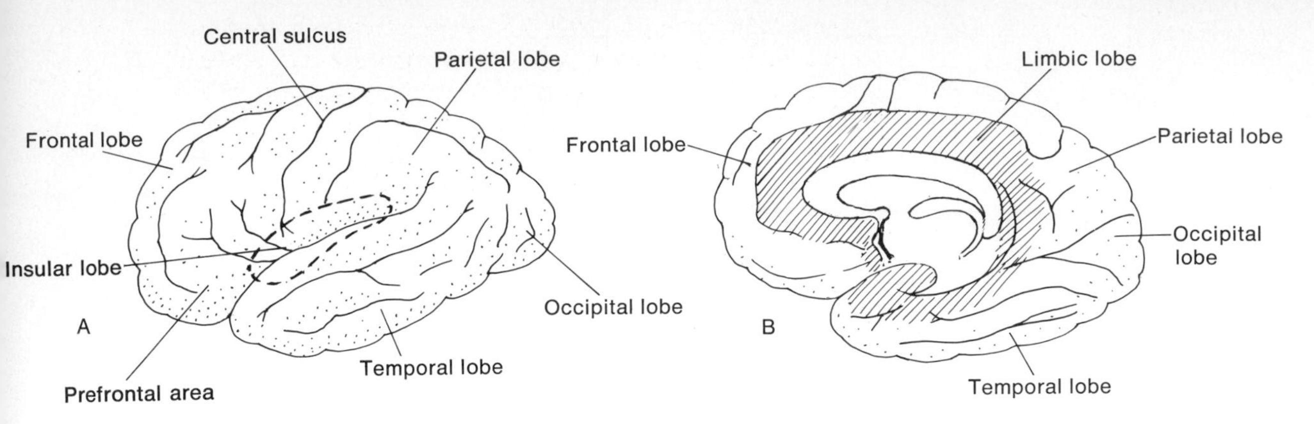

Fig. 6-58 Lobes of the cerebral hemispheres. A. Lateral view. B. Medial view. The shaded area represents the limbic lobe.

- Four take the names of the bones shielding them: **frontal, parietal, temporal,** and **occipital.**
- One is set deep into the lateral fissure and is not visible at the surface: the **insula.**
- One is not a distinct anatomical entity but a composite of several structures from different areas: the **limbic lobe.**

Cortical activities have been assigned to lobes, as discussed below, after observation and testing of animals and patients who suffer lesions in specific cortical areas. Within a lobe, precise areas for specific function do not exist. Functions overlap from area to area and from lobe to lobe.

The **frontal lobe** (Fig. 6-58) is generally concerned with voluntary and reflex motor activity—that is, these activities are largely initiated here. It is believed that the "higher functions" of humans, such as abstract reasoning, learning, and intelligence are a product of frontal activity. Memory takes place here as it does in most areas of the cortex. Expression through speech may be related to this region. Analysis of results of prefrontal lobotomy (a procedure involving cutting tracts coming into or leaving the anteriormost part of the cortex) indicate that certain expressions of emotion (anxiety, fear) are related to the frontal cortex. The frontal lobe maintains dense neural connections to brainstem nuclei, the cerebellum, and other cortical areas.

The **parietal lobe** (Fig. 6-58) incorporates the receiving areas for such somesthetic sensations as pain, temperature, pressure, and touch. Only when these impulses reach the parietal lobe does one become *aware* of these sensations. Recognition of your body's position in space (proprioception) is also registered here. Try this: close your eyes, swing one upper limb around in space, then bring that hand to your face and touch the tip of your nose with one finger. The fact that you could sense where your limb was at all times *with your eyes closed* was due to cortical activity in the postcentral gyrus. The parietal lobe—in conjunction with temporal activity—is also related to speech and the interpretation of language. Like all other cerebral lobes, the parietal lobe has abundant connections to other parts of the cortex, the brainstem, and the cerebellum.

The **temporal lobe** (Fig. 6-58) receives impulses related to hearing and integrates them with other sensory input and memory as well. Localization and awareness of sound is produced here. Bilateral removal of the auditory cortex causes deafness. Because the medial and inferior aspects of the temporal lobe are anatomically and functionally related to the limbic and frontal lobes, certain aspects of behavior and emotional expression as well as memory patterns are associated with the temporal lobe. Comprehension of language—both spoken and written—is also made possible here.

The **occipital lobe** (Fig. 6-58) is almost entirely devoted to reception of visual input and associating this input with memory and the data of other areas of the cortex. Destruction of this area results in a spectrum of defects, the worst being complete blindness.

The **limbic lobe** (Fig. 6-58) and functionally related areas (collectively called the limbic system) have complex tasks not totally understood. Generally the limbic lobe is associated with activities relating to self-preservation and preservation of the species, such as eating, fighting, fear, flight to safety, sexual behavior, and parental care. The limbic lobe plays an important role in making our responses to stimuli subjective (based on feeling) rather than objective (based on intellect). Bias and prejudice may well be related to limbic activity. The limbic lobe has extensive connections with the hypothalamus of the brainstem.

In summary, it is in the cerebral cortex that sensory awareness takes place and that voluntary and reflex motor activity is initiated. In addition, it is here that the higher faculties of humans are found. Our increasing ingenuity throughout evolutionary history is partly a reflection of increasing cortical complexity—a complexity based on expanding numbers of neurons and, as a consequence, their interconnections.

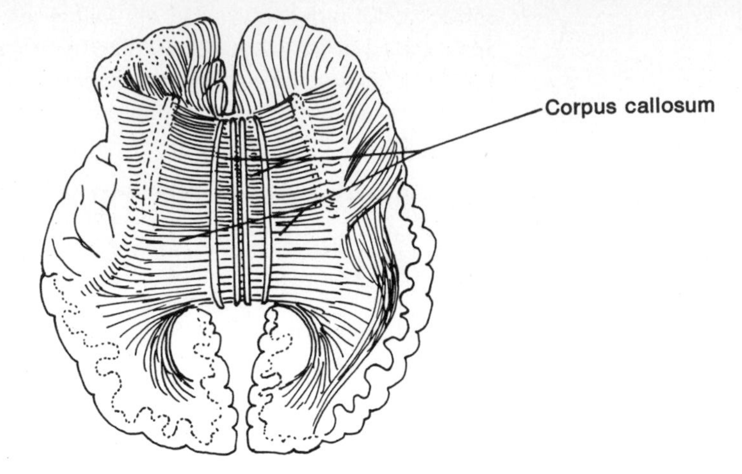

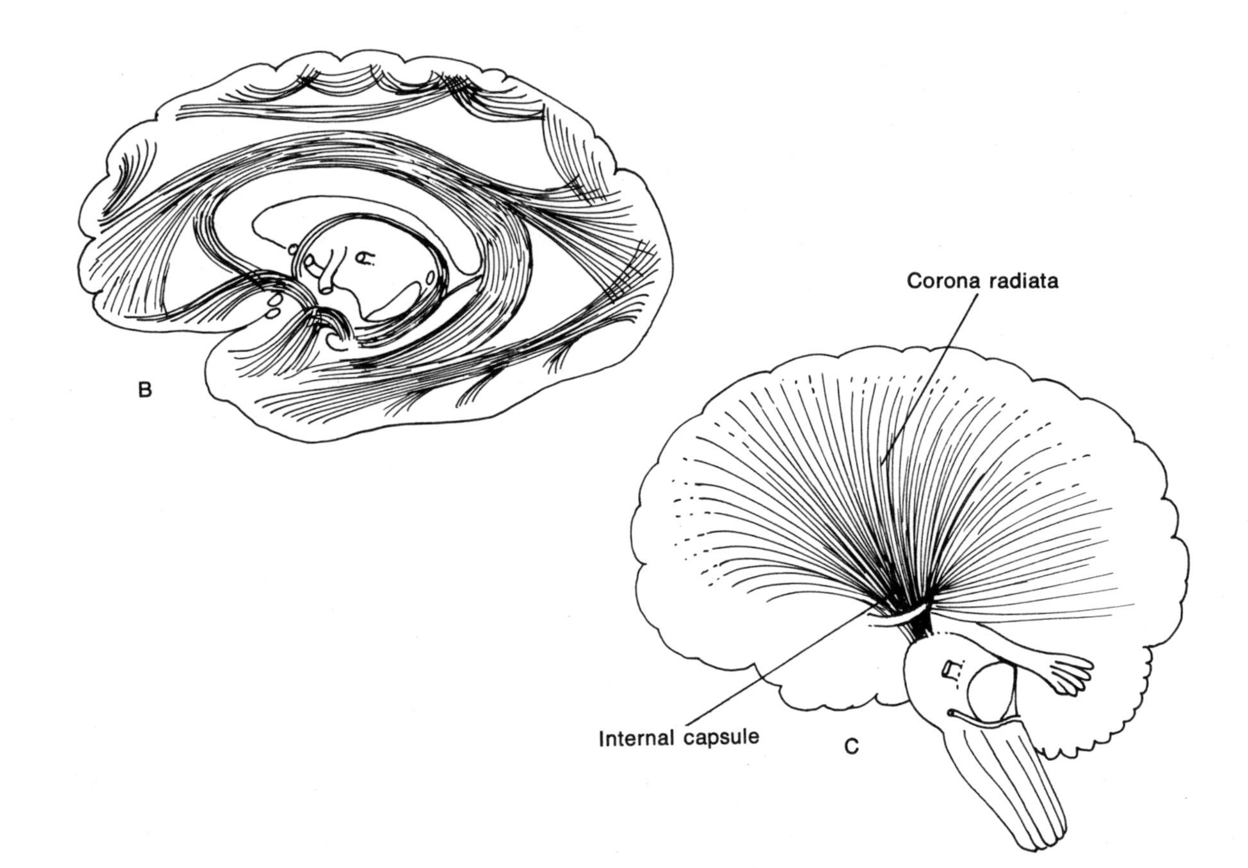

Fig. 6-59 White matter of the cerebral hemispheres. A. The corpus callosum, a type of commissure. B. Association tracts. C. Projection tracts.

White matter of the cerebral hemispheres

The major mass of the cerebral hemispheres consists of collections of myelinated nerve processes (tracts). In general these tracts take one of three courses (Fig. 6-59):

1. Across from one hemisphere to another **(commissures).** The largest of these is the **corpus callosum.**
2. Longitudinally from anterior to posterior, and vice versa **(association tracts).** These lie deep in the cortex, connecting one lobe or region with another in the same hemisphere.
3. Vertically from higher centers to lower centers and vice versa **(projection fibers).** These are probably the most spectacular of all fiber systems in the brain. The chief system is a band of fibers projecting up through the substance of each hemisphere (like the stalks of a flower bouquet) called the **internal capsule.*** As it fans out to all regions of the cortex (like the flowers themselves) it gives one the impression of a "radiating crown" and is so named: **corona radiata.** The internal capsule and corona is the primary avenue of fiber communication between the cerebral cortex and lower centers.

Basal (subcortical) nuclei

These are groups of neuron cell bodies (nuclei) making up, in part, the basal and medial walls of *each* hemisphere (Fig. 6-60). They include:

1. The **caudate nucleus,** lying between the lateral ventricle and the internal capsule. It is a tail-shaped structure following the contours of the lateral ventricle into the medial portion of the temporal lobe where it terminates as the almond-shaped amygdala (L., almond).
2. The **lenticular nucleus,** partly encircled by the caudate and separated from it by the internal capsule, is so named because of its "lens" shape. It consists of two nuclei called the putamen and globus pallidus.

The basal nuclei in conjunction with lower brainstem nuclei are principally concerned with reflex movements such as postural adjustments. Postural adjustments, e.g., positioning the body for a golf swing, a tennis serve, or batting a baseball, form a background for the more skilled, voluntary move-

* So called because the fibers form a white inner wall (capsule) for a group of basal nuclei. There is also a smaller, thinner *external capsule* of fibers.

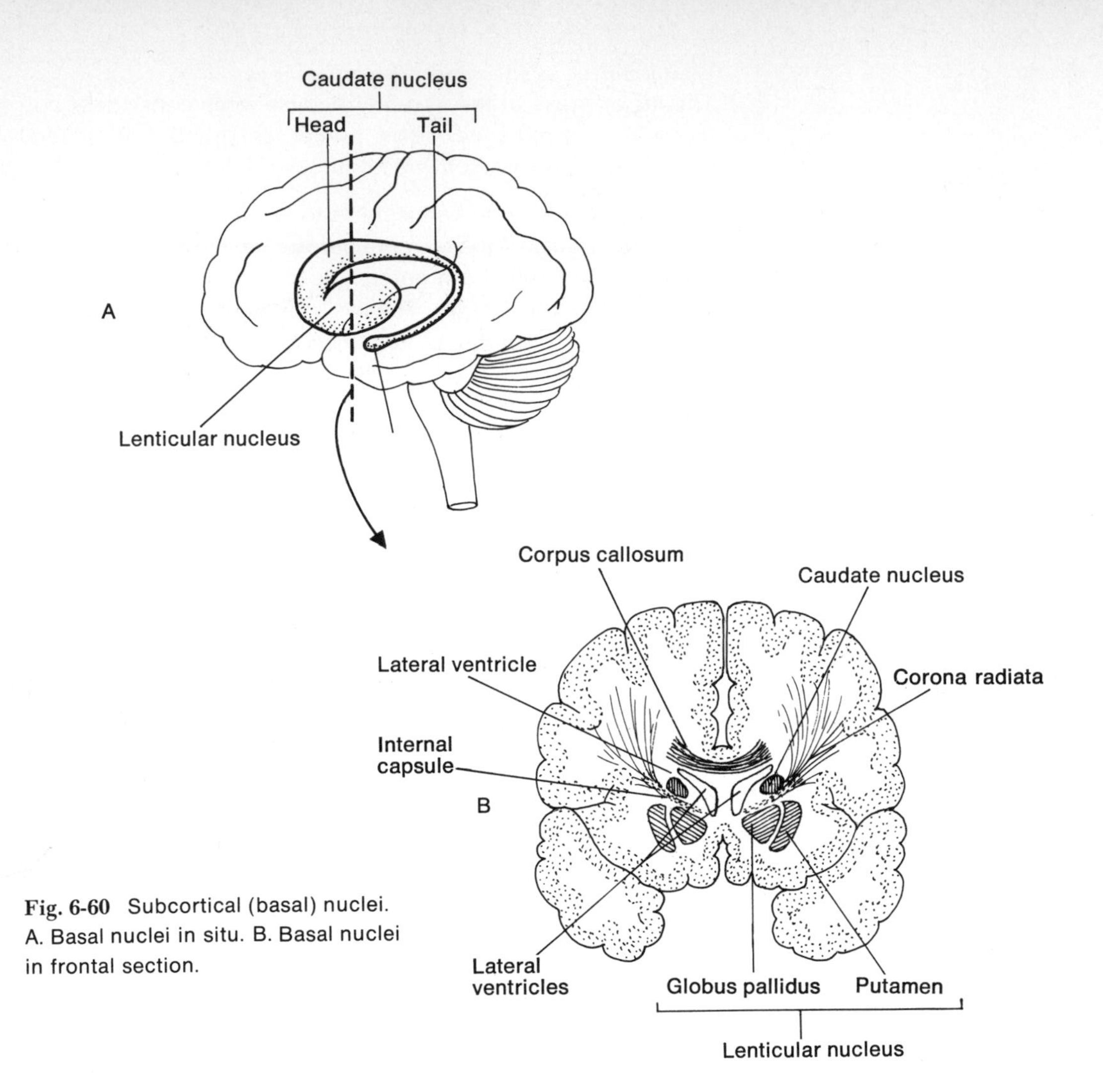

Fig. 6-60 Subcortical (basal) nuclei. A. Basal nuclei in situ. B. Basal nuclei in frontal section.

ments, such as hitting the ball. Impulses from the basal nuclei probably do not initiate movement—they modify cortex-initiated impulses. In this respect, basal nuclei are believed to play an error control function preventing any movement other than that desired. Destruction of certain basal nuclei result in diseases associated with abnormal movements (dyskinesia).

STRUCTURES OF THE BRAINSTEM

The brainstem consists of the brain less the cerebral and cerebellar hemispheres. It includes the thalamus and hypothalamus, midbrain, pons, and medulla. The brainstem makes connections with the cerebral hemispheres by way of the internal capsule (Fig. 6-61). It makes connections with the cere-

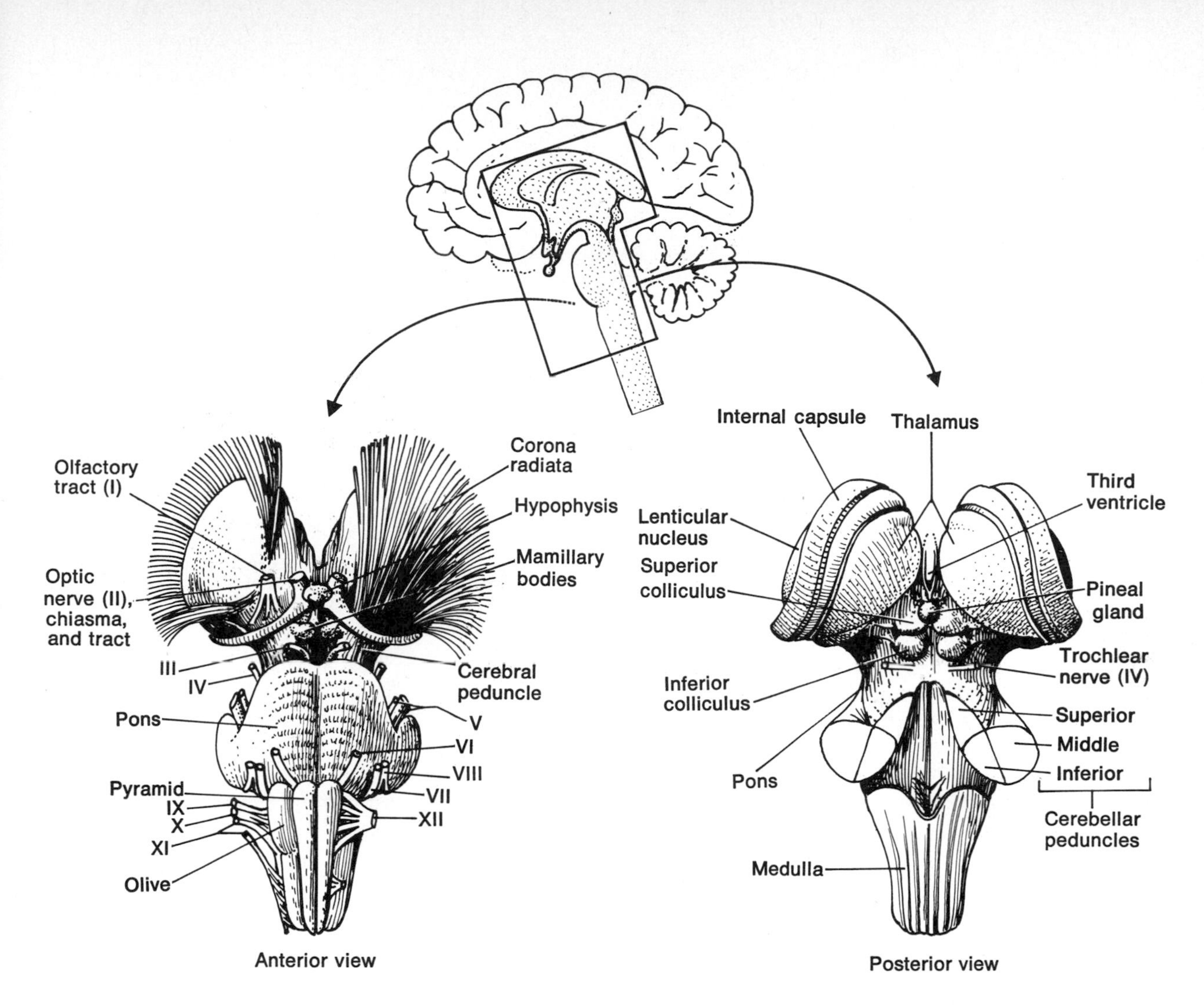

Fig. 6-61 Structures of the brainstem. A. Anterior view. B. Posterior view. (Adapted from C. R. Noback and R. Demarest *The Human Nervous System,* McGraw-Hill, 1967.)

bellum by way of the three paired cerebellar fiber bundles known as peduncles (superior, middle, inferior).

Except in the diencephalon, three regions can be identified in most cross sections taken through the brainstem (Fig. 6-62):

1. The **tectum:** the posterior roof of the cerebral aqueduct and fourth ventricle. Thick in the midbrain, the tectum incorporates the superior and inferior colliculi of that region. It is thin and without much significance in the rest of the brainstem.

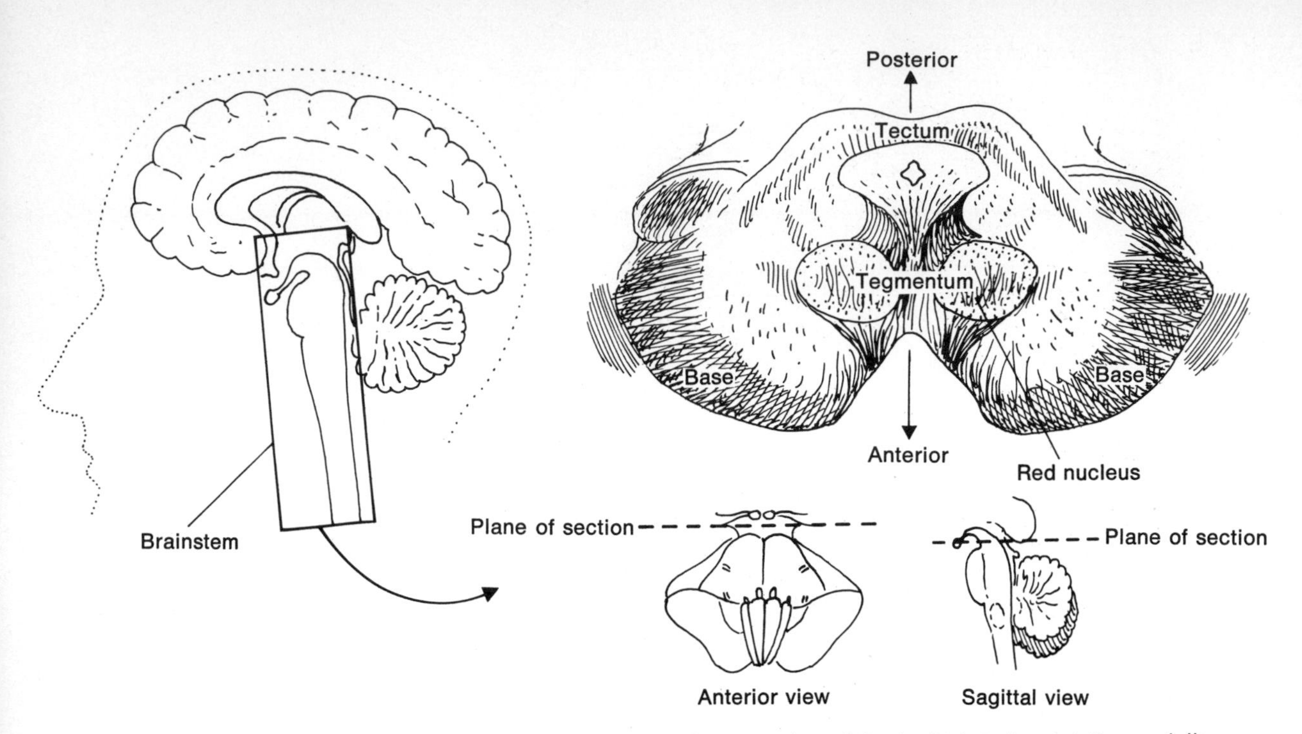

Fig. 6-62 Cross section of the brainstem through the medulla illustrating basic organization.

2. The **base:** the anterior aspect of the midbrain, pons, and medulla. The base includes most of the cerebral peduncles, the superficial pons, and the pyramids of the medulla.
3. The **tegmentum:** the region between the tectum and the base. It is separated from the tectum by the cerebral aqueduct and the fourth ventricle. It contains the great majority of the brainstem nuclei. It is continuous with the reticular formation of the medulla.

Thalamus

The thalamus is a paired, football-shaped mass of many nuclei immediately lateral to the third ventricle (Fig. 6-63). It is connected to its fellow on the opposite side by an intermediate mass of gray matter. In very general terms, the thalamus functions as a center: receiving impulses from ascending (sensory), association, and some descending fibers; correlating, integrating, and distributing impulses to appropriate areas of the cerebral cortex, basal nuclei, hypothalamus, cerebellum, and other brainstem nuclei. With respect to connections with the cerebral cortex, the thalamus contributes great

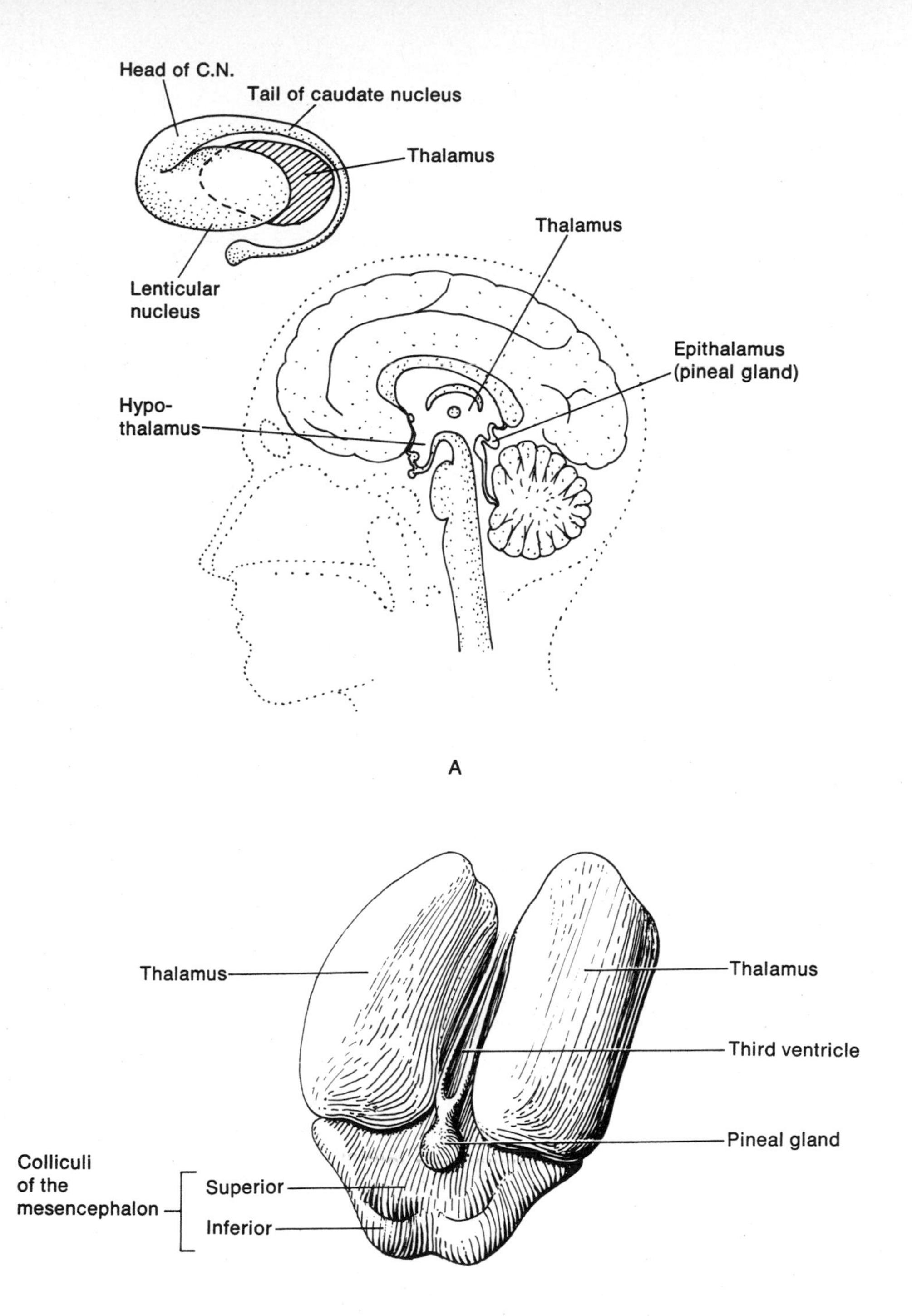

Fig. 6-63 The thalamus. A. Sagittal view. B. Oblique, overhead view. Insert: Thalamus in relation to caudate and lenticular nuclei.

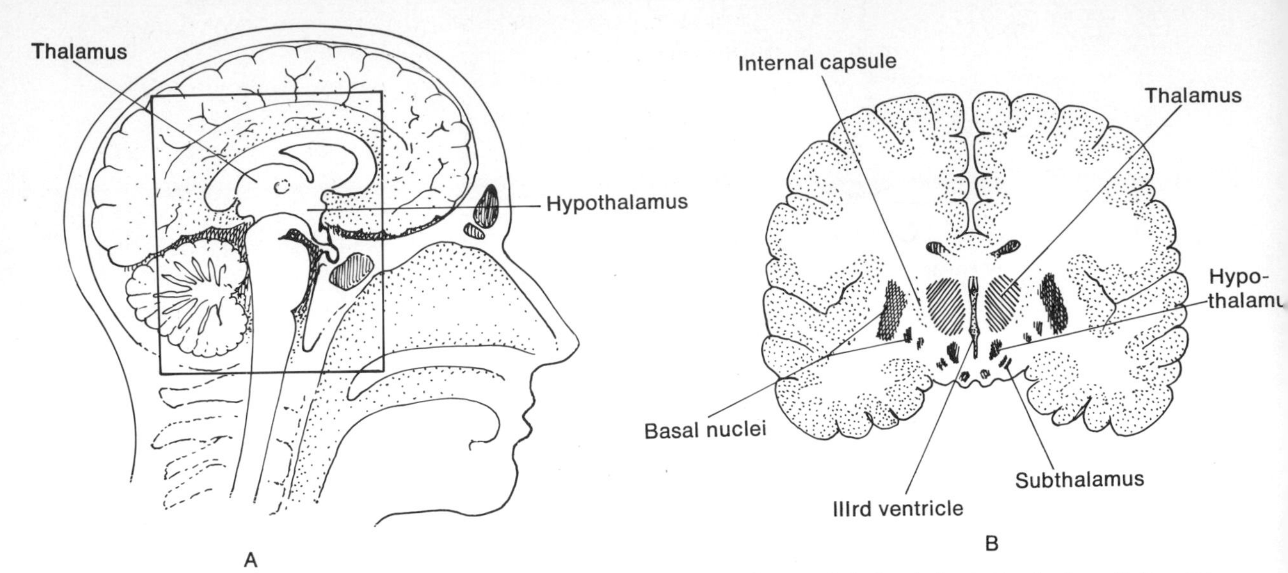

Fig. 6-64 The hypothalamus. A. Lateral view of brainstem. B. Frontal section through hypothalamus.

masses of fibers to the internal capsule. It also sends a mass of fibers to the occipital cortex. Carrying visual impulses, these fibers are known as the optic radiations. The thalamus is believed to figure significantly in awareness and the conscious state. It is much more than a simple sensory relay center. Clinical evidence relating to Parkinson's disease (a disease of dyskinesia) suggests that some parts of the thalamus may influence voluntary movements. The large number of cortical fibers descending to the thalamus suggest that the thalamus may function in part at the command of the cerebral cortex.

Hypothalamus

The single, unpaired hypothalamus (Fig. 6-64) is a most compact and complex aggregation of nuclei and related processes lying below and slightly anterior to the thalamus, hugging the lower part of the third ventricle.

The circuitry to and from the hypothalamus is extensive—connections being made to nuclei throughout the cerebral hemispheres and brainstem. The functions generally attributed to this area are a memorizer's dream:

1. Regulation of sympathetic and parasympathetic activity, working, in part, through the cardiovascular and respiratory centers in the medulla. Any expression of emotion incorporating visceral changes (changes in heart rate,

blood pressure, respiratory rate, etc.) involve hypothalamic nuclei.

2. In association with the hypophysis (pituitary), it influences certain endocrine activity, e.g., lactation, ovulation, onset of puberty, the menstrual cycle, water balance, and general growth and development.
3. It regulates body temperature by influencing mechanisms of heat production and conservation. Such mechanisms include vasodilation (enlargement of vessels' lumina), vasoconstriction (narrowing of the lumina), etc.
4. It regulates appetite, possibly being signalled by subtle changes in sugar concentration in the plasma and distention of the stomach.
5. In association with the reticular system (to be discussed shortly), the hypothalamus influences the sleep-wakefulness mechanism.

Epithalamus

Perhaps better known as the pineal gland, this region of the diencephalon seems to play an inhibitory role in development of the testes, the significance of which is not clear. The pineal overlies the superior colliculi on the posterior aspect of the brainstem (Fig. 6-61).

Midbrain

The midbrain consists of the tectum incorporating the superior and inferior colliculi, the tegmentum, in which may be seen a number of nuclei and important tracts, and the cerebral peduncles, consisting largely of tracts descending from the cerebral cortex.

The **colliculi** (kolick′ ·ulye) are four knobs of gray matter easily observed on the posterior aspect of the brainstem when the cerebellum is bent away from the cerebral hemispheres (Fig. 6-61). The superior pair communicate with the optic tracts and play an important role in the visual reflex phenomenon (the "double-take" or "second-look" reflex) illustrated in Fig. 6-65. This is accomplished by integrating reflex movements of the head with visual input. The inferior pair of colliculi are believed to integrate reflex movements of the head to auditory input, such as seen in one's own response to alarming sounds (firecrackers, horns, etc.).

Fig. 6-65

The tegmental region includes the motor nuclei of the oculomotor (IIIrd) cranial nerve and trochlear (IVth) cranial nerve, the midbrain reticular formation, and certain other nuclei and short-range tracts. The IIIrd and IVth cranial nerves supply some of the extrinsic muscles of the eye.

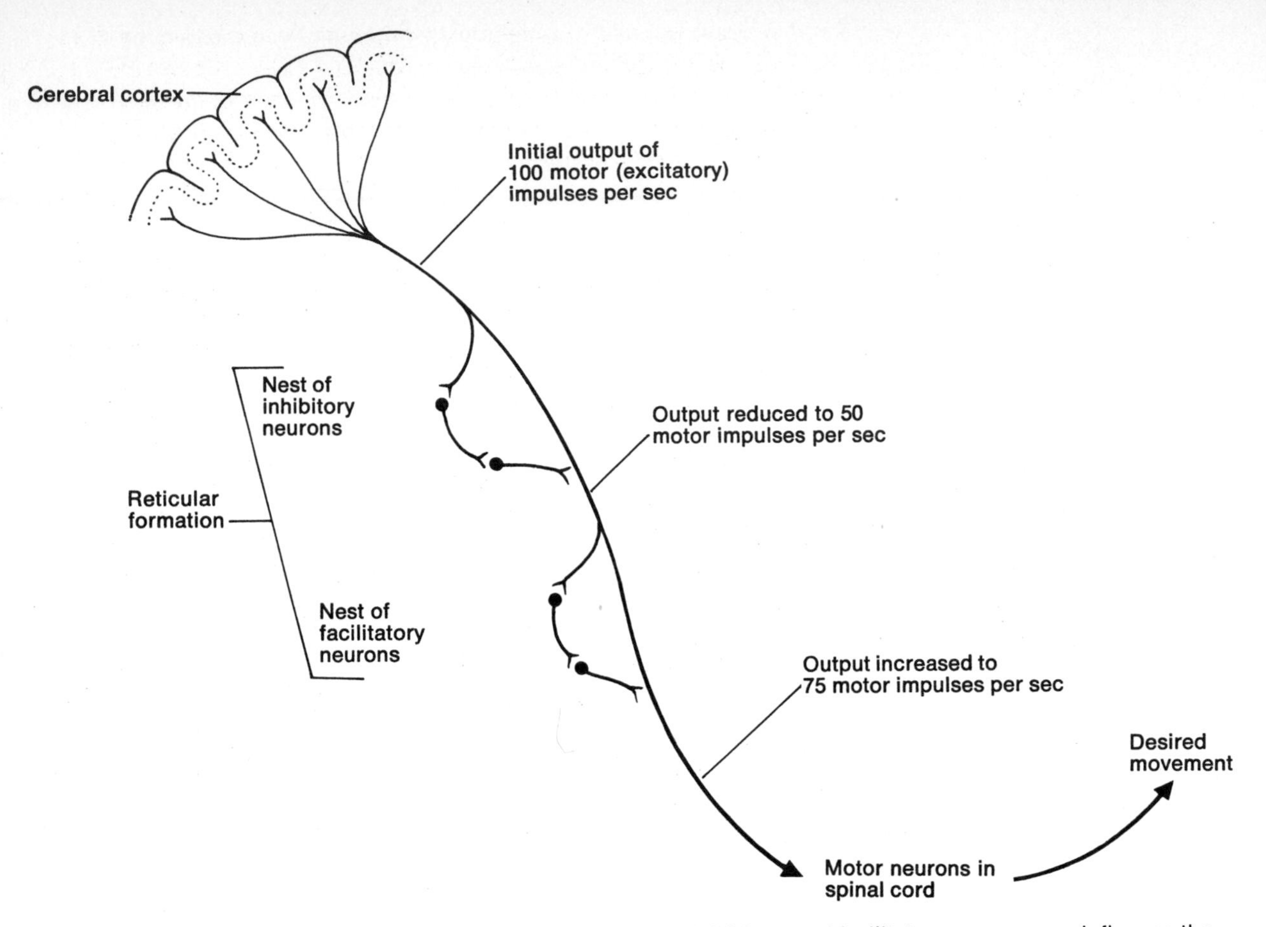

Fig. 6-66 How inhibitory and facilitatory neurons can influence the transmission of motor impulses (schematic). The initial output of excitatory impulses is greater than is necessary to do the desired job. The actual spatial arrangement of the neurons is much more complex.

The **reticular formation** of the brainstem consists of a diffuse network of neurons which percolate ascending and descending impulses through myriads of synapses. These impulses have facilitatory or inhibitory influences (Fig. 6-66) on almost any aspect of sensory input or motor output, e.g., sleep, wakefulness, awareness, reflexive movements, muscle tone, etc. The distinctive **red nucleus** (Fig. 6-62) of the midbrain is a part of this midbrain reticular formation.

The **cerebral peduncles** (Fig. 6-61) consist largely of tracts descending from the cerebral cortex (via the internal capsule) to the spinal cord (corticospinal tract) and to intermediate nuclei (corticopontine and corticobulbar tracts).

The midbrain also communicates with the cerebellum by way of the superior cerebellar peduncles. Most fibers in these peduncles conduct impulses from the deep cerebellar nuclei to the midbrain reticular formation (in particular, to the red nucleus) and thalamus.

Pons

The pons (L., bridge) (Fig. 6-61) consists of a deeper tegmentum and a more superficial basilar part. The latter contains several nuclei and masses of fibers oriented transversely to the brainstem. These transverse fibers wrap around ("bridging") the pontine tegmentum to become the **middle cerebellar peduncles** (Fig. 6-67), conducting impulses from the cerebral cortex (via pontine nuclei) to cerebellar nuclei. Projection pathways in this basilar part include the corticospinal and corticobulbar (cortex-to-brainstem nuclei) tracts.

The tegmentum contains the pons' portion of the reticular formation. The great ascending pathways are found here as well as nuclei associated with the fifth (trigeminal), sixth (abducens), seventh (facial) and eighth (vestibular and cochlear) cranial nerves. The tectum here does not exist; it is "replaced" by the cerebellum. The fourth ventricle, then, intervenes between the pontine tegmentum and the cerebellum.

Medulla

The medulla (Fig. 6-61), clearly separated from the pons by a transverse sulcus, merges indistinctly into the spinal cord at the level of the foramen magnum. The anterior aspect or "base" of the medulla is almost entirely taken up by the corticospinal tract. The brainstem is narrowed in this area and the tract produces paired swellings known as the **pyramids.** Some 80 to 90 percent of corticospinal fibers cross to opposite sides here, creating the visible **decussation** (crossing) **of the pyramids.** Thus you can readily understand that injuries to the *left* precentral gyrus of the cerebral cortex may be manifested as a paralysis of the *right* side of the body.

The core of the medulla (tegmentum) consists largely of the reticular formation and related tracts, the great ascending pathways, and nuclei relating to the eighth, ninth (glossopharyngeal), tenth (vagus), eleventh (accessory), and twelfth (hypoglossal) nerves.

The reticular formation here is believed to be an influence in neural regulation of respiration, heart rate, blood pressure, and other visceral activities.

The two large inferior olivary nuclei take up a significant

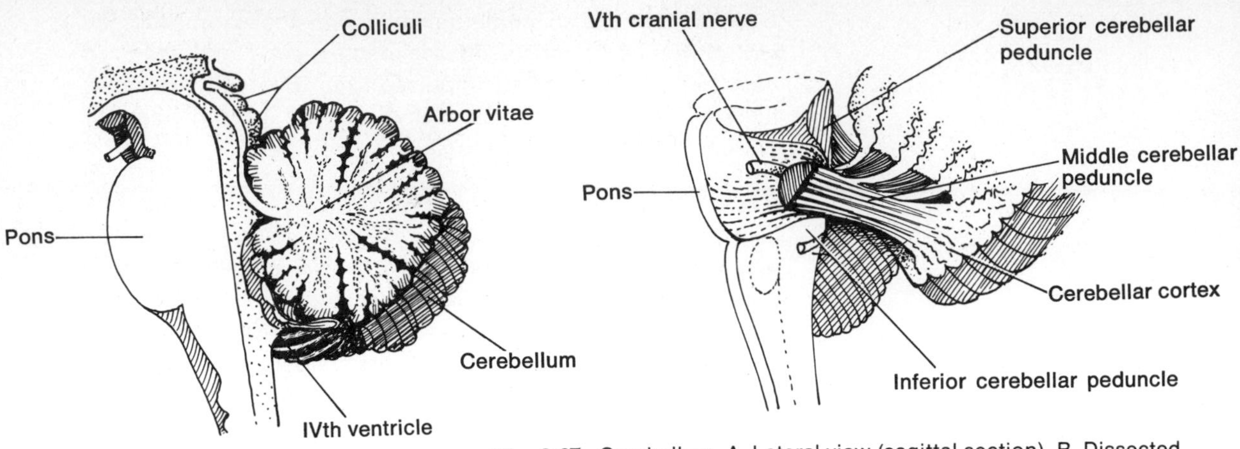

Fig. 6-67 Cerebellum. A. Lateral view (sagittal section). B. Dissected view showing cerebellar tracts.

share of the medulla. Each communicates with the cerebellum (olivocerebellar tract) by way of the inferior cerebellar peduncle on its side.

On the posterior aspect of the medulla (there is no tectum here), two large nuclear masses appear as the fourth ventricle dives toward the central canal of the spinal cord. These, the nuclei cuneatus and gracilis, are related to certain ascending pathways conducting sensations of proprioception, touch, and deep pressure.

As the medulla nears the spinal cord, the various nuclei (gray matter) centralize about the fourth ventricle. The tracts (white matter) become oriented around the gray matter in huge bundles called *funiculi.*

CEREBELLUM

The cerebellum, although derived from the metencephalon along with the pons, is not part of the brainstem. Its Latin name, meaning "little brain," refers to its structural similarity to the cerebral hemispheres. A look at Fig. 6-67 will show that the cerebellar cortex is highly convoluted and consists of cell bodies and small fibers (like the cerebral cortex). Projections of white matter radiate from the cortex toward the three peduncles as branches of a tree reach for the trunk, hence their name *arbor vitae* (L., tree of life). Several important nuclei occupy the central region of the cerebellum.

The cerebellum is principally responsible for muscular coordination. Based on input from proprioceptors in the muscles and tendons (via the inferior and middle cerebellar peduncles), the cerebellum assures proper muscle tension and the coordinated contraction of several muscle groups to effect a precise movement, such as picking up a marble. It does this by conducting motor-related impulses from the deep cerebellar nuclei to the tegmentum of the pons and midbrain and to the thalamus via the superior cerebellar peduncle. The The cerebellum has extensive communications with the vestibular nuclei of the pons and medulla and plays a significant role in maintenance of equilibrium. For example, the cerebellum strongly influences the success of coordinated actions such as this one: Close your eyes and touch the tip of your nose, then touch the tips of fingers of opposite hands.

SPINAL CORD

The spinal cord takes up the lower two-thirds of the central nervous system, beginning at the level of the foramen magnum as the distal continuation of the medulla. Locked within the framework of the vertebral column, the spinal cord ends at the level of the second lumbar vertebra.

Here it might be wise to invest in a brief conceptual analysis of the spinal cord. The brain has been described as the grand integrator, coordinator, and modifier of sensory information and motor response. The spinal cord functions to extend the authority of the brain to the body proper and provides a means by which the body's peripheral structures can get quick access to the brain for awareness and decisions. (Stub your toe and see how long it takes for cortical awareness!) The spinal cord is generally responsible for:

1. Providing the connections between incoming and outgoing limbs of reflexes (see Unit 3).
2. Routing input from sensory (peripheral) nerves to appropriate motor cells in the cord (for the simplest reflexes), to cells of the brainstem (for all but the simplest reflex acts), and to the cerebral cortex (for awareness).
3. Routing all impulses descending below the medulla on to the motor neurons which generate the actual motor impulses to the muscles and glands of the body.

External features

Refer now to Fig. 6-68 as you read the following description.

The spinal cord is wrapped in three fibrous coverings, or meninges, the outermost of which is the dura mater—a structure to be discussed later.

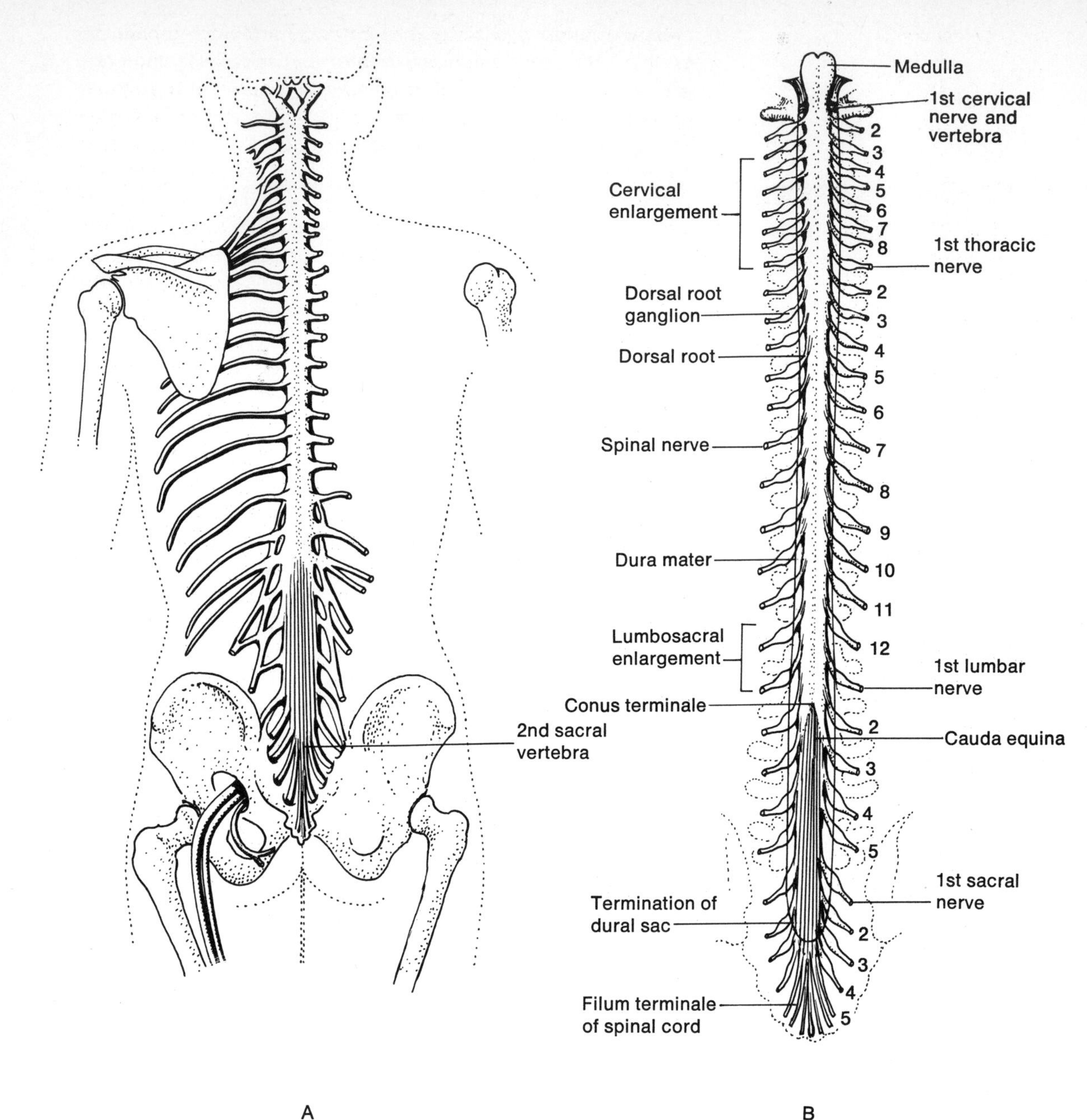

Fig. 6-68 The spinal cord. A. In situ. B. With neural arches removed. Note how the spinal nerves descend within the vertebral canal before exiting in the lower half of the column.

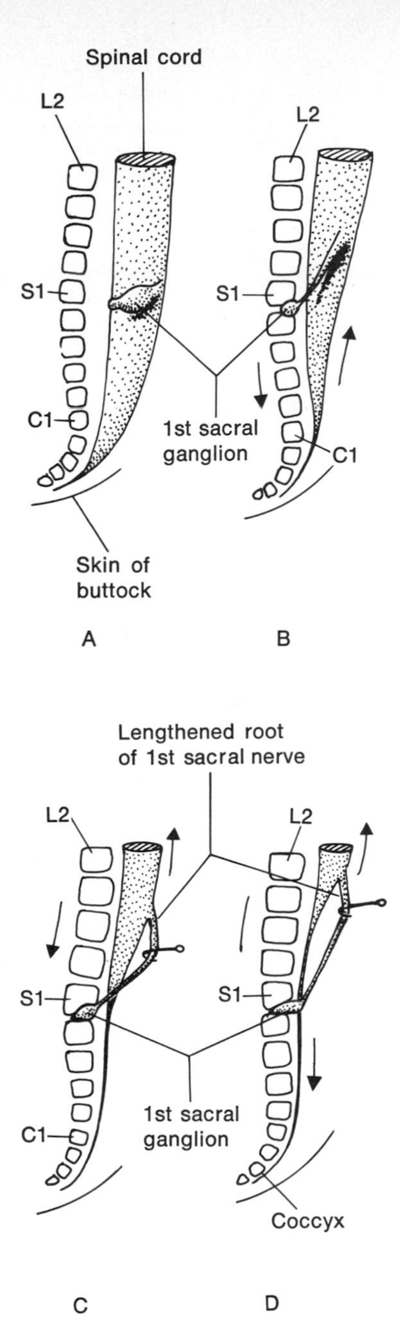

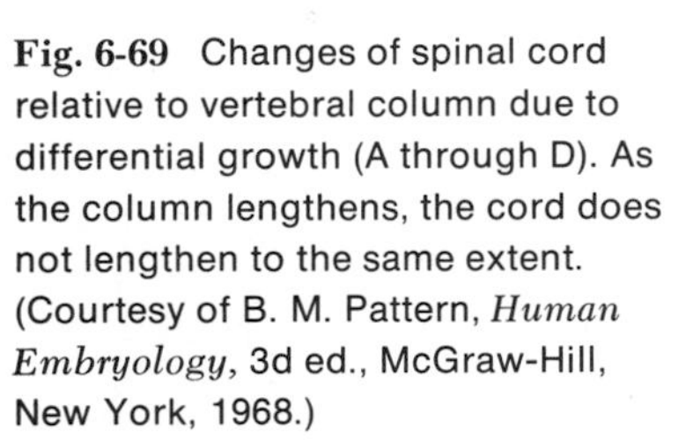

Fig. 6-69 Changes of spinal cord relative to vertebral column due to differential growth (A through D). As the column lengthens, the cord does not lengthen to the same extent. (Courtesy of B. M. Pattern, *Human Embryology*, 3d ed., McGraw-Hill, New York, 1968.)

There are five regions of the spinal cord, which correspond to the vertebrae enclosing them, i.e., cervical, thoracic, lumbar, sacral, etc. The cord gives off the **roots of the spinal nerves.** The dorsal and ventral roots merge to form a single unit (the **spinal nerve**) as they pass out of the vertebral canal through the intervertebral foramina (spaces between adjacent vertebrae). In so doing, they take with them their covering of dura mater, which is continuous with the neurilemma of the peripheral nerves. Each bilateral pair of spinal nerves is numbered according to the region of the spinal cord from which it arises. The first seven cervical nerves emerge above their respective vertebrae, that is, C1 nerve emerges above C1 vertebra. The eighth cervical nerve and all subsequent spinal nerves emerge below their respective vertebrae, that is, C8 nerve emerges below C8 vertebra.

Enlargements of the cord are visible at the lower cervical region and upper lumbar regions of the vertebral column. These cervical and lumbosacral enlargements represent those portions of the cord contributing to the brachial and lumbosacral plexuses (see Fig. 3-41).

The roots coming off the cord are directed progressively downward as one moves along the cord caudally. The cord does not fill the full extent of the vertebral canal. At one period during development, the spinal cord did occupy the whole canal, but because the vertebral column continued to develop after the central nervous system had ceased to grow, there resulted a situation in which the spinal cord and its spinal nerve roots appear to have been dragged upward through the vertebral canal (Fig. 6-69). Thus the lumbar and sacral regions of the *cord* are at the lower thoracic and upper lumbar regions of the *vertebral column.*

The cord comes to a cone-shaped end **(conus terminale),** surrounded by streams of spinal nerve roots [**cauda equina** (''horse's tail'')] at the level of the L1–2. All of these structures are within the **dural sac,** which ends at S2.

The cord is tied down to the coccygeal vertebrae by a thin thread (filum terminale) of meningeal tissue extending from the conus terminale.

Internal features

Refer to Fig. 6-70 and note that the gray matter of the cord is arranged in the shape of an H. The thinner uprights of the gray matter are the **posterior horns** (columns*) receiving the

* Column is the preferred term because it refers to a three-dimensional structure, which the spinal cord is.

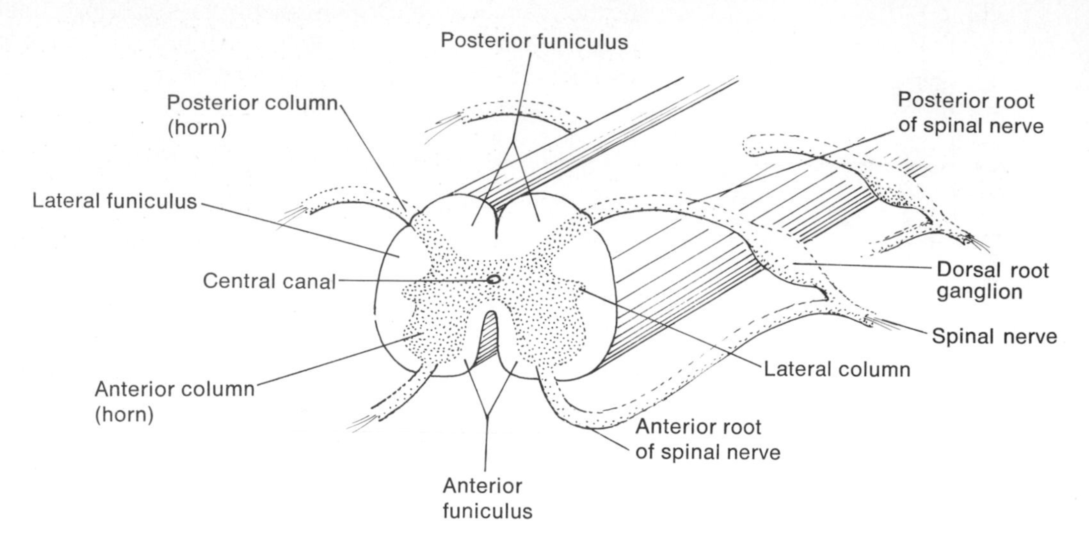

Fig. 6-70 Internal arrangement of spinal cord.

posterior roots of spinal nerves. The thicker uprights are the **anterior horns** (columns). The anterior horn motor neurons reside here, dispatching the anterior (motor) roots of the spinal nerves. The bar connecting the two sides of the gray columns is the gray commissure pierced by the midline **central canal.** In the thoracic, upper lumbar, and sacral segments of the cord, there are lateral projections (columns) of gray matter between the posterior and anterior gray columns; these incorporate the motor nuclei of autonomic nerves.

White matter surrounds the gray matter. The mass of myelinated fibers between the posterior gray columns constitutes the **posterior funiculi.** The mass of fibers between the anterior gray columns are called, as you might suspect, the **anterior funiculi.** The regions of white matter between adjacent pairs of posterior and anterior gray columns are **lateral funiculi.** At the cervical level more area is occupied by tracts than by gray matter because the motor neurons on which the tracts terminate have not been reached in any great number. Toward the caudal end of the cord more and more of the tracts have ended in synapsis with motor neurons, so the area occupied by tracts becomes smaller relative to the gray area. In the sacral and coccygeal regions the entire spinal cord diminishes in cross-sectional area quite rapidly; because the white areas do so more quickly than the gray, the gray matter occupies the greater part of the cord.

FUNCTIONAL-STRUCTURAL INTEGRATION: PATHWAYS

You have labored over the brain and spinal cord and learned about numerous nuclei and other structures. It was not in vain, for in this section we shall tie it all together.

The ultimate expression of central nervous system activity is muscular contraction, which usually follows some kind of sensory input, often from receptors in skin or muscle. Thus the common sequence is sensory input and motor output, each of which has separate pathways through the nervous system.

Sensory, or **ascending, pathways** carry impulses from the receptors into the spinal cord and up the cord through the brainstem and cerebral hemispheres to appropriate nuclei and centers for interpretation and reaction. **Motor,** or **descending, pathways** are the routes for motor-related impulses from the cerebral cortex and subcortical nuclei through the brainstem (where they come under the influence of the cerebellum), down the spinal cord and out to the periphery to the muscles and glands of the body.

There are other significant pathways for the cranial nerves; these, of course, bypass the spinal cord.

Ascending pathways

The first-order (1°) neuron is the term for the first link in an ascending pathway. With its receptor the 1° neuron lies outside the central nervous system, its cell body in the dorsal root ganglion. Its central process extends into the spinal cord. The cell body of the second-order (2°) neuron lies in the gray matter or, for some pathways, in the brainstem. The axon of the 2° neuron ascends to the thalamus. The third-order (3°) neuron lies in the thalamus and sends its axon to the cerebral cortex. Each link in the chain makes synaptic contact with the succeeding one.

The various sensory modalities are channeled into one of several specific tracts in the white matter of the cord. The principal ones are as follows:

1. Posterior columns (position sense, movement sense, discriminative touch, pressure).
2. Anterior spinothalamic tract (light touch).
3. Lateral spinothalamic tract (pain, temperature).

Other ascending tracts worthy of note but not to be examined in detail are the spinocerebellar tracts (conveying position and movement sense to the cerebellum) and the spinoreticular tracts (conducting position and movement sense to the reticular formation).

Posterior columns. Within the posterior columns on either side of the midline are two tracts or fasciculi. The medial one is the **fasciculus gracilis,** the lateral one the **fasciculus cuneatus.**

The 1° neuron enters the spinal cord via the dorsal root and immediately ascends in one of these fasciculi on its side of the midline. If it enters at lower sacral and lumbar regions it ascends in the fasciculus gracilis. If it enters higher on the cord it ascends in the fasciculus cuneatus. The 1° neuron terminates in the nuclei gracilis or cuneatus, respectively, located within the posterior medulla.

The 2° neuron sends its axon across the medulla to the other side, where it ascends as the **medial lemniscus** up through the brainstem tegmentum to the thalamus.

The 3° neuron in the thalamus projects its axon through the internal capsule to the postcentral gyrus of the cerebral cortex. It is only at this point that an impulse evokes conscious awareness of the sensations.

Anterior spinothalamic tract. The 1° neuron makes synaptic contact with the 2° neuron in the posterior horn of the central gray matter. The 2° neuron sends its axon across the anterior white commissure to the anterior funiculus on the other side of the midline. Here the fiber joins the anterior spinothalamic tract and ascends to the thalamus.

The 3° neuron, in the thalamus, directs its axon through the internal capsule and corona radiata to the postcentral gyrus of the cerebral cortex.

Lateral spinothalamic tract. This tract (Fig. 6-71), conducting pain and temperature sensations, takes the same route as the anterior spinothalamic tract but is located in the lateral funiculus.

Descending pathways

Descending tracts may be long or short; if the former, they arise in the cerebral cortex and pass through the corona radiata and internal capsule, the brainstem, and into the spinal cord. Intermediately long pathways may arise somewhere in the brainstem and pass into the spinal cord or arise in the cortex and end in the brainstem. Short fiber systems may descend only two or three cord segments; they are largely involved in spinal reflexes, giving some flexibility to the cord's response to incoming stimuli.

Since, as was said before, the ultimate expression of CNS

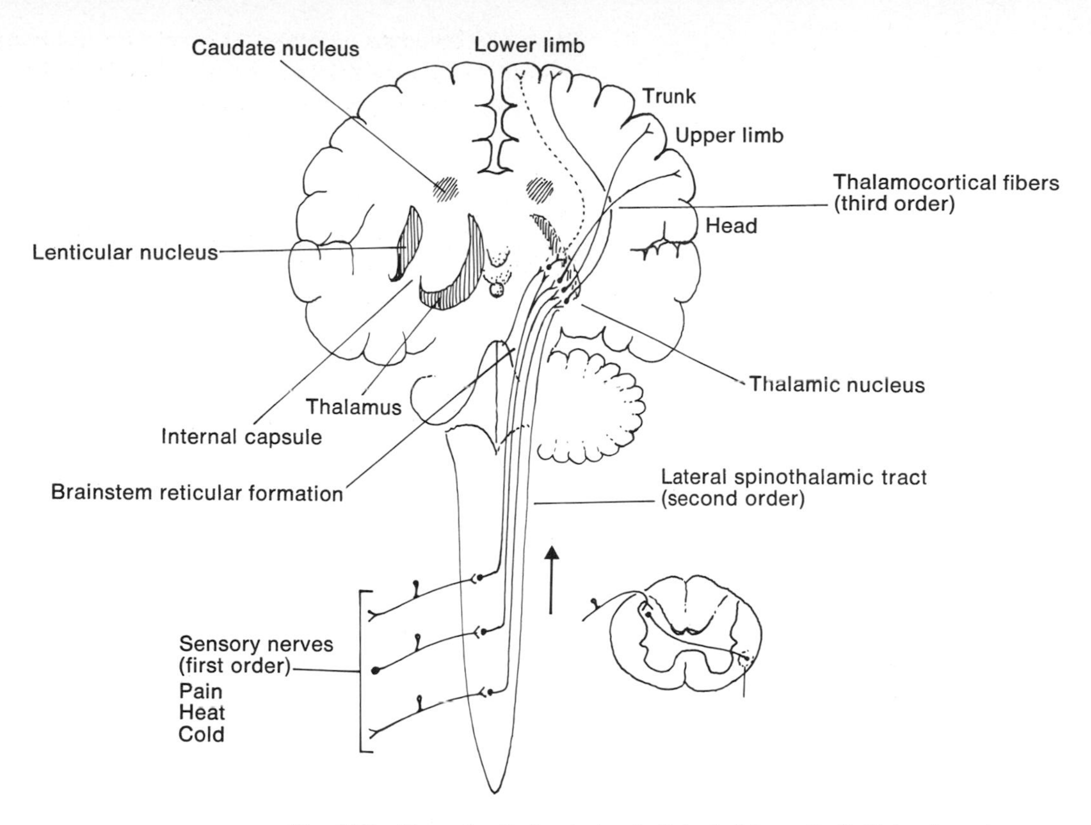

Fig. 6-71 The spinothalamic tract. (Adapted from C. R. Noback and R. Demarest, *The Human Nervous System*, McGraw-Hill, New York, 1967.)

activity is muscular contraction, you can understand that descending influences must eventually come to bear upon the motor neurons of the PNS in the brainstem or spinal cord (Fig. 6-73). The axons of these neurons, the motor component of spinal nerves, represent the **final common pathway** to the muscles and glands. The impulses in those axons represent the algebraic summation of all cortical, brainstem, cerebellar, and spinal cord activity brought together to generate a spectrum of coordinated body movement from precise movement to rough postural changes to slight changes in the background muscular tension (tone).

The principal descending pathways are:

1. **Corticospinal** or **pyramidal tract** (voluntary, skilled movement).
2. Nonpyramidal pathways (postural adjustments, maintenance of muscle tone, facilitation or inhibition of voluntary movement).

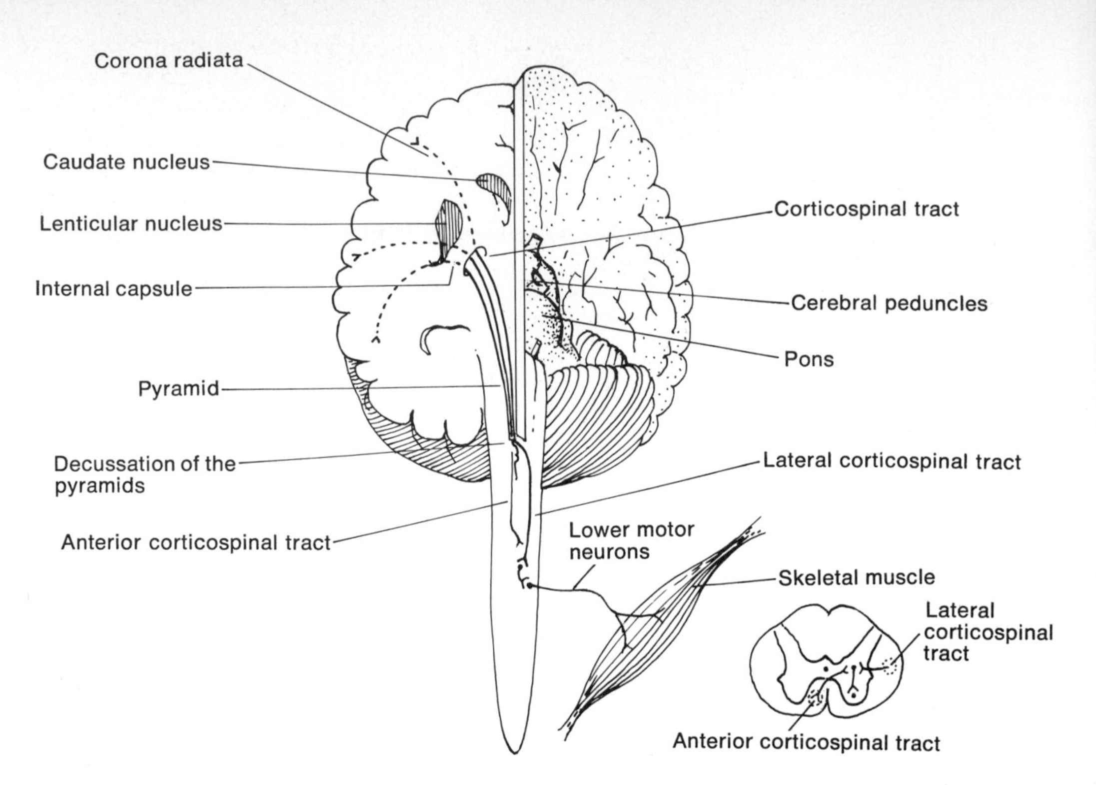

Fig. 6-72 The corticospinal tract. (Adapted from C. R. Noback and R. Demarest, *The Human Nervous System,* McGraw-Hill, New York, 1967).

Corticospinal tract. This tract finds its origin primarily in the precentral gyrus of the cerebral cortex. Large multipolar neurons there give off axons which take the following course (Fig. 6-72): Cerebral cortex (via corona radiata) → internal capsule (between the lenticular nucleus and thalamus) → cerebral peduncles of midbrain → pons → anterior medulla. In the medulla they create the pyramids. Also, some 80 to 90 percent of the axons cross there (decussation of the pyramids) to opposite sides of the medulla and descend in the lateral funiculus as the **lateral corticospinal tract.** The uncrossed fibers pass into the anterior funiculus as the **anterior corticospinal tract.**

The corticospinal tracts get progressively smaller as they descend the cord because increasing numbers of fibers have turned off into the anterior horn where they synapse with the anterior horn motor neurons of the PNS.

The axons of the anterior horn motor cells depart the CNS via the ventral root, join a spinal nerve, and head off to terminate in the effector organ: skeletal muscle.

Fibers of the corticospinal tract are believed not to synapse with intervening nuclei between their origin and termination. The corticospinal tract conveys impulses related to the voluntary contraction of muscles employed in skilled movements, e.g., playing a piano. The background postural movements and maintenance of muscle tone (the "fine tuning") are the tasks of descending tracts which do *not* pass through the pyramids of the medulla.

Nonpyramidal (descending) pathways. These are relatively short pathways (with respect to the long corticospinal tract) from cortex to spinal cord with numerous intermediate synapses (Fig. 6-73). As you can see, these short fiber systems go from cortex to basal nuclei to midbrain nuclei, from thalamus to basal nuclei, from basal nuclei to hypothalamus, and so on. These pathways are primarily related to the reflex, unskilled postural movements. They probably influence the anterior horn motor neurons by way of unnamed short-chain pathways, possibly the reticular formation.

The reticular formation through the reticulospinal tract is believed to have primarily facilitatory and secondarily inhibitory influences on reflex movements. You will remember that the reticular formation consists of numerous interconnected neurons and is seen throughout the tegmentum of the brainstem.

The cerebellum, mediator of fine, precise movement and equilibrium, is kept informed of descending neural activity by brainstem nuclei and of muscle tension (tone) by spinocerebellar fibers carrying impulses from stretch receptors in muscles. These fiber systems employ both inferior and middle cerebellar peduncles. Fiber tracts leave the cerebellum primarily by the superior cerebellar peduncle for the midbrain nuclei and the thalamus. Via these nuclei, cerebellar influences are injected into the corticospinal tract and other descending pathways.

Facilitation of muscle tone and reflex muscular activity is also influenced by the vestibular nuclei of the medulla, which receive input regarding the resting tension of the body musculature. Anterior horn motor neurons receive these influences via the vestibulospinal tract.

In summary then, skilled movement is effected principally by the corticospinal tract; postural adjustments and other reflex movements by pathways involving the basal nuclei, thalamus, and the reticular formation; control of muscle tone and maintenance of equilibrium is regulated by cerebellar influences on midbrain and thalamic nuclei and by the ves-

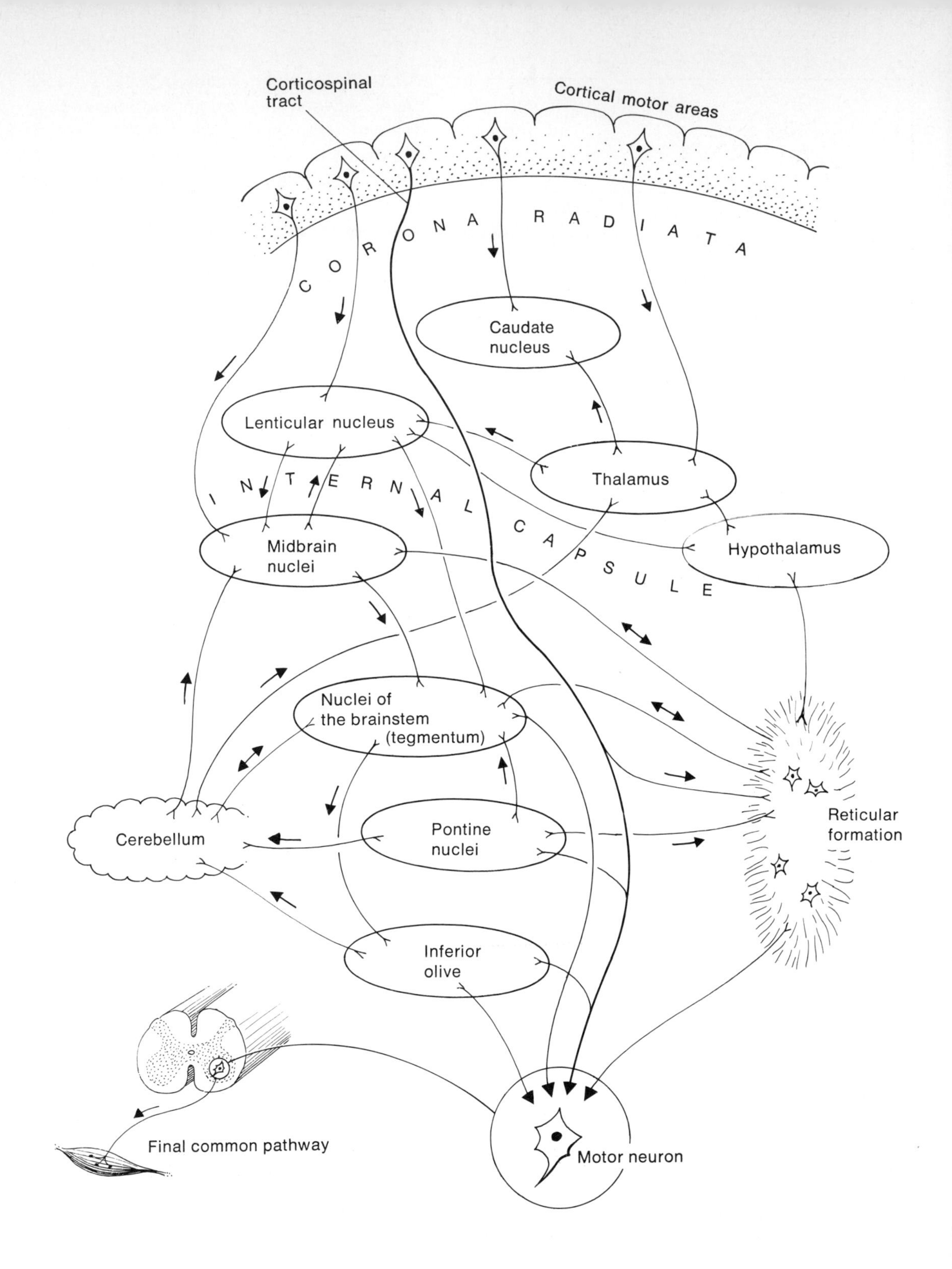
Corticospinal tract
Cortical motor areas
CORONA RADIATA
Caudate nucleus
Lenticular nucleus
Thalamus
INTERNAL CAPSULE
Midbrain nuclei
Hypothalamus
Nuclei of the brainstem (tegmentum)
Cerebellum
Pontine nuclei
Reticular formation
Inferior olive
Final common pathway
Motor neuron

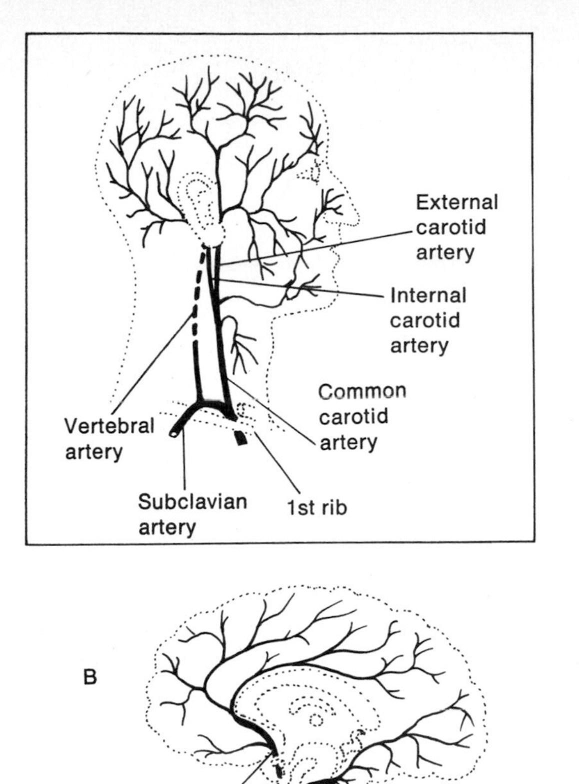

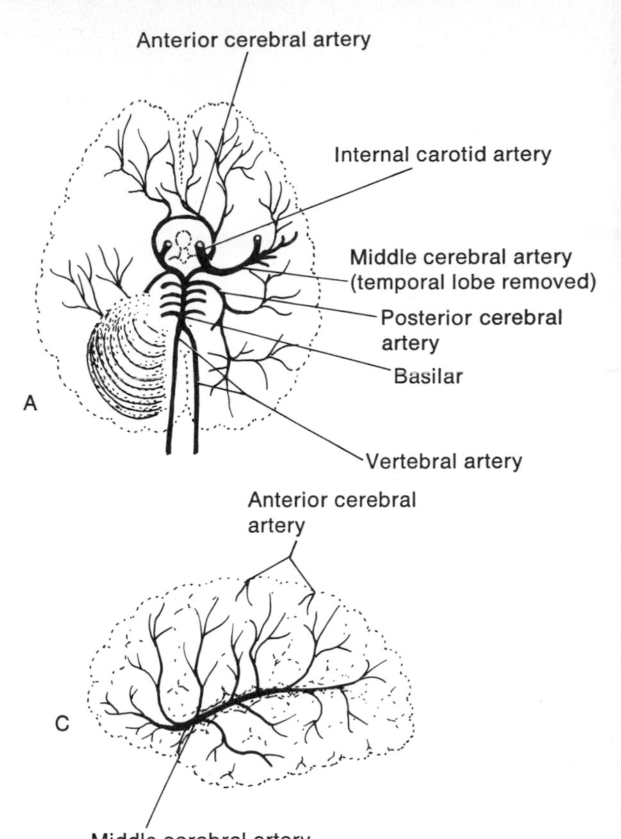

Fig. 6-74 Arterial supply to the brain. Insert: sources of arterial blood to the brain. A. Vascular pattern on anterior brainstem. B. Distribution of arteries to medial and lateral surfaces of the brain.

tibular nuclei of the medulla via the vestibulospinal tract. All of these influences filter down to the motor neuron which represents the final common pathway to the muscles of the body.

Fig. 6-73 (opposite) A generalized scheme of the source and distribution of descending influences raining upon an anterior horn motor neuron. The multiple short neuron chain collectively constitutes the non- or extrapyramidal system.

ARTERIES AND VEINS SERVING THE BRAIN

The supply of blood to all parts of the CNS is critical. Interruptions in blood flow and hence oxygen supply for more than 5 to 10 sec cause irreparable brain damage. The vascular circuits to the brain may be studied in Fig. 6-74. Note that the blood supply to the brain is derived from two sources:

1. The **vertebral arteries**—arising from the subclavian arteries.

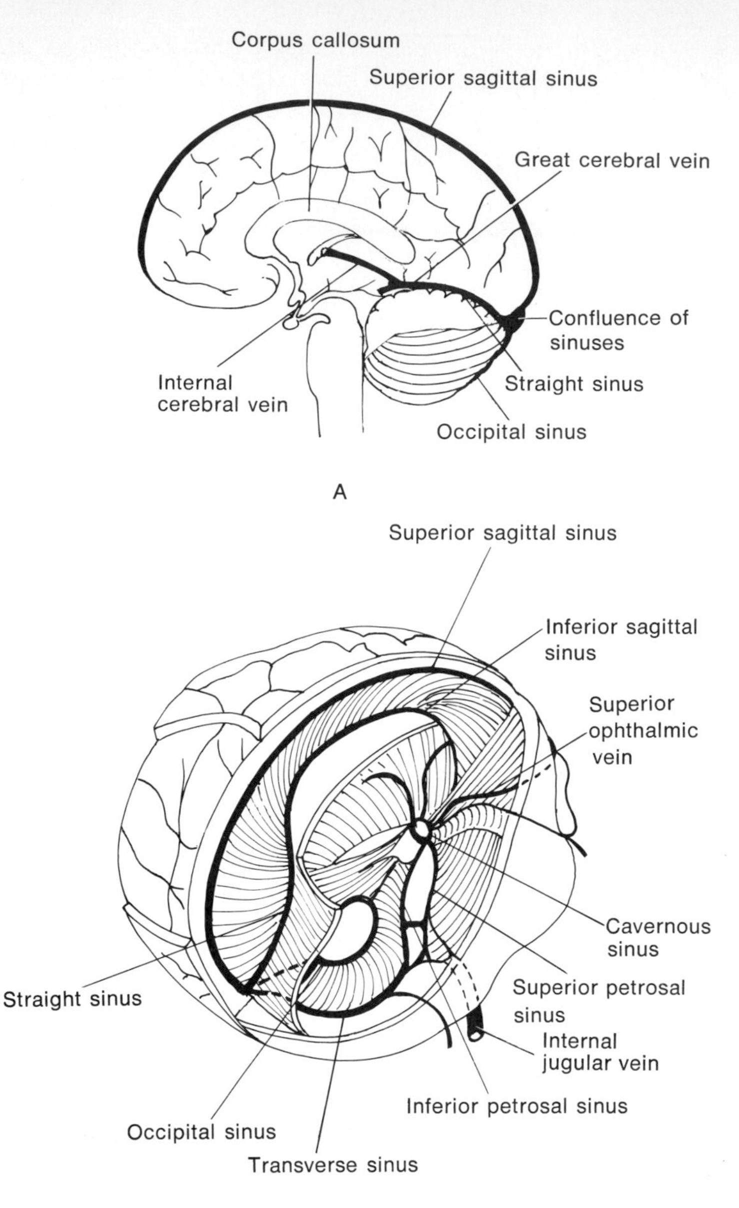

Fig. 6-75 Venous drainage of the brain. A. Medial surface drainage. B. Dural sinuses.

2. The **internal carotid arteries**—arising from the arch of the aorta via the common carotid arteries.

Note further that the internal carotids terminate by giving off the middle and anterior cerebral arteries supplying the major part of the cerebral hemispheres.

An irregular ring of vessels may be seen around the optic

chiasma. This is known as the circle of Willis, and in it the blood of the vertebrals mixes with that of the internal carotid arteries, possibly creating a "safety-valve" mechanism during periods of differential blood pressure in the two pairs of arteries. The circle of Willis does not provide effective collateral circulation, however, and interruption of one of the four main arteries will seriously diminish cerebral circulation and could bring about coma and subsequent death.

Venous drainage of the brain (Fig. 6-75) is accomplished by two sets of veins, superficial and deep, with generous anastomoses between them. The veins all drain into **venous sinuses,** and some also drain into venous plexuses outside the skull.

The venous sinuses are generally bound to the periosteum of the cranial bones by the dura mater or an extension of it. They are large channels receiving venous blood from (1) tributaries of the superficial cerebral veins, and (2) the internal cerebral vein, which in turn drains smaller veins of the deep cerebral hemispheres and upper brainstem.

ARTERIES AND VEINS SERVING THE SPINAL CORD

Blood supply to the spinal cord is accomplished by anterior and posterior spinal arteries from the vertebral arteries. There are abundant anastomoses among these vessels and arteries along the trunk adjacent to the vertebral column.

The veins draining the spinal cord make up complicated plexuses on the anterior and posterior aspect of the spinal cord. These plexuses generally drain into veins adjacent to the vertebral column which themselves drain into the superior or inferior vena cava.

MENINGES AND CEREBROSPINAL FLUID

The meninges consist of three connective tissue envelopes of the central nervous system. These coverings serve to do three things:

1. Protect the brain and spinal cord in association with cerebrospinal fluid.
2. Support the brain and spinal cord within their bony housings.
3. Serve as a vehicle for vessels supplying and draining the brain and spinal cord.

The innermost covering is closely applied to the surface of

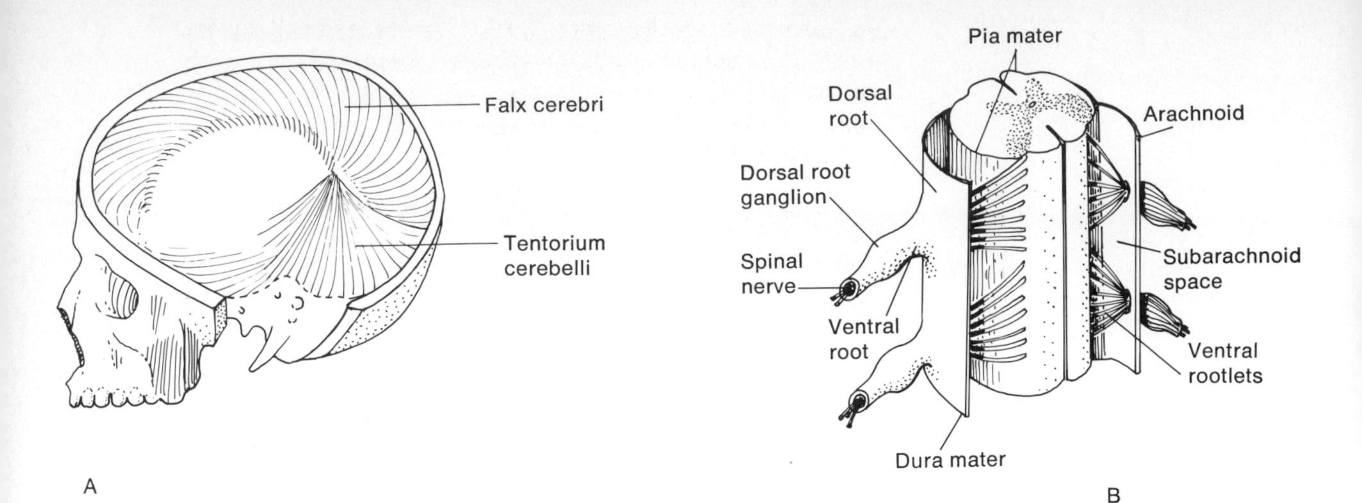

Fig. 6-76 Coverings of the CNS. A. Dural reflections in cranial cavity. B. Meningeal relations of the spinal cord.

the brain and spinal cord. It is a delicate, soft layer, hence its name: **pia mater** (Fig. 6-76). It cannot be easily separated from the underlying nervous tissue, as it is believed to send projections into it.

The pia embraces the CNS in a continuous sheet except where three small foramina penetrate it in the roof of the fourth ventricle and where the many blood vessels supplying the brain and spinal cord pass through. The pia terminates caudal to the conus terminale as the filum terminale.

The middle layer of the meninges is a filmy, weblike membrane, termed the **arachnoid.** It is separated from the underlying pia by a space which is crossed by small beams of connective tissue (trabeculae) (Fig. 6-77). This **subarachnoid space** conducts the cerebrospinal fluid. The arachnoid ends caudal to the conus terminale as the internal lining of the dural sac.

The **dura mater** is the tough, fibrous outer covering of the CNS. It consists of two layers, an outer layer serving as the periosteum of the cranial bones and terminating at the foramen magnum, and an inner layer, separated from the arachnoid by a potential **subdural space.** Lying within the two layers are the previously discussed dural sinuses. The inner layer of dura dips into great crevices between certain parts of the brain, forming dividers or septa.

The dura is a vascularized structure, being supplied by the **middle meningeal artery** and its branches, which leave an impression on the inner surface of the calvaria. This important artery, a second-order branch of the external carotid artery, is subject to hemorrhage with fractures of the skull in the parietal region—hemorrhages which are frequently fatal (epidural hemorrhage). The dura terminates caudally as a sac at the S2 vertebral level.

CSF SECRETION, SPACES, AND FLOW

Cerebrospinal fluid (CSF) is a part of the extracellular fluid volume, being of similar composition. It functions as a shock absorber for the CNS. Other functions, if any, are not known. It is believed to be secreted by a capillary-epithelial complex **(choroid plexus)** found in the roofs of the third and fourth ventricle as well as the medial walls of the lateral ventricles.

The CSF flows within the ventricles, through the foramina in the roof of the fourth ventricle, and into the subarachnoid spaces (Fig. 6-77). Flow through the central canal of the spinal cord is doubtful, since the canal is small and usually occluded. In some places the subarachnoid space is expanded, and there it is called a cistern; for example, where the cerebellum is some distance from the overlying dura. Cisterns are found primarily around the brainstem. The CSF flows through

Fig. 6-77 Diagram of the meninges, ventricles, and direction of cerebrospinal fluid flow.

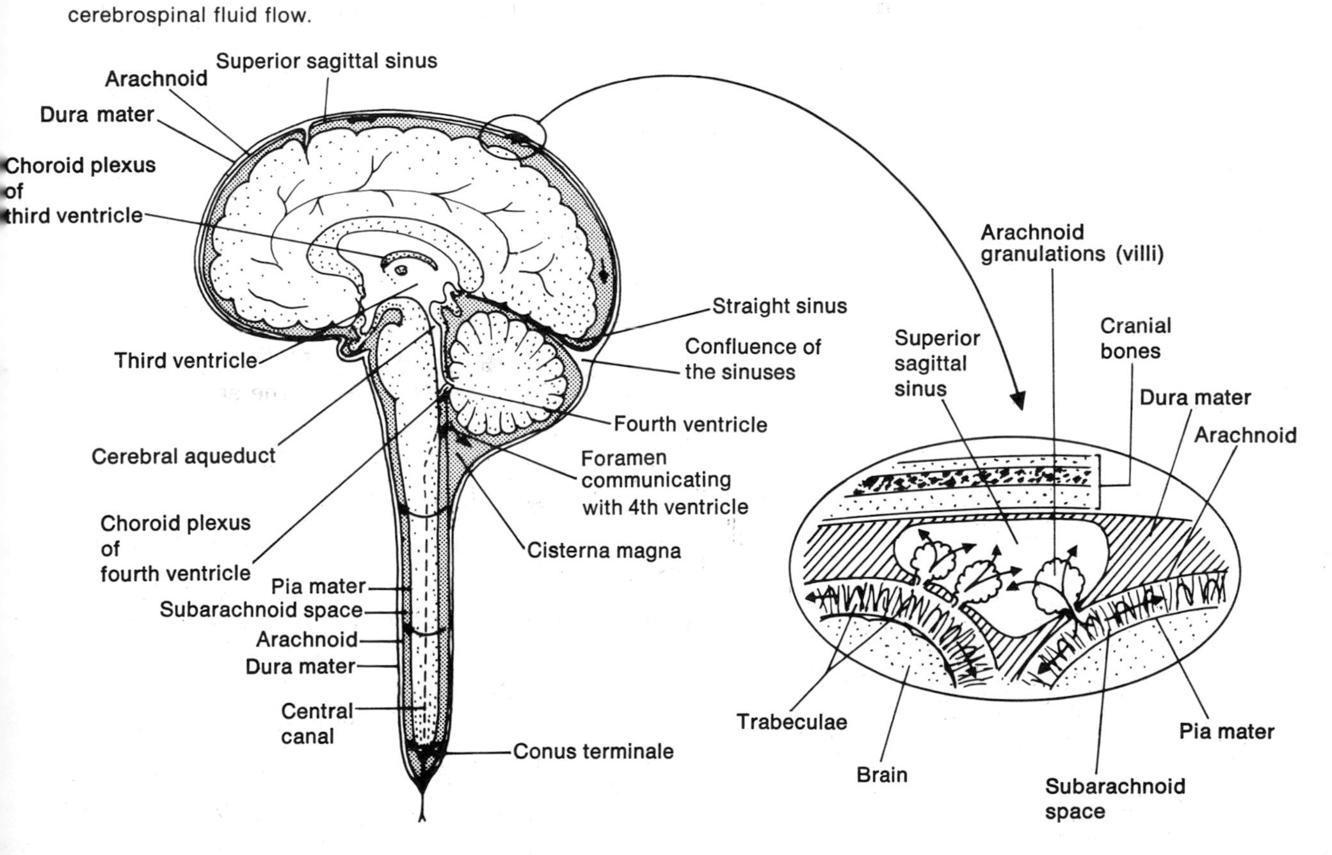

the subarachnoid spaces, through the cisterns, and ultimately passes into the dural venous sinuses via the arachnoid granulations, which are villous projections of arachnoid that poke up through the dura into the sinuses of the skull. These granulations are most easily seen where they pierce the superior sagittal sinus (Fig. 6-77).

Study of CSF by the physician is a valuable diagnostic tool. The pressure of the fluid in the subarachnoid space (measured usually at the lumbar region of the vertebral column) as well as its chemical composition can disclose venous obstruction, ventricular obstruction, tumor formation, subdural hemorrhage, and a variety of other pathologies.

QUIZ

1. The central nervous system is composed of the ______ and ______.
2. At 3 weeks of age the following regions of the brain can be distinguished: ______, ______, and ______.
3. Subsequently, the forebrain changes sufficiently to form two regions: the ______ and ______.
4. The three principal derivatives of the telencephalon are: ______ ______, ______ ______, and the lateral ______.
5. The occipital lobe is associated principally with the receipt and interpretation of ______ impulses.
6. ______ tracts conduct impulses from one side of the cerebral hemispheres to the other.
7. The brainstem does not include in its composition the ______ and the ______ ______.
8. The principal role of the thalamus is/is not to serve as a sensory relay center. (Cross out wrong answer.)
9. The reticular formation and major nuclei of the midbrain can be found in the ______
10. The motor neuron of the PNS whose cell body resides in the anterior horn of the spinal cord is often referred to as the ______ ______ pathway.
11. The tract conducting impulses relating to pain and temperature involves ______ orders of neurons, the second of which terminates in the ______
12. The tract conducting impulses relating to skilled, voluntary movement toward the spinal cord is the ______ tract.

special sensory receptors and associated organs

Throughout their evolutionary history, animals have consistently greeted their environment head first. This is not by accident, for the head is gifted with a set of exteroceptors unlike any others in the body. Consider: a wildcat apprehensively checks out a cave as potential quarters for his mate. Approaching the entrance, his nose wrinkles, nostrils flared for suspicious scents; his ears prick up smartly and turn toward the cave, surveying every sound; his dilated eyes dart from point to point, seeking movement of the slightest sort. Preservation at stake, do we not emulate that very pattern? When investigating new ground, as in war, do we not crouch forward—head first—using our receptors of sight, sound, and smell to inform ourselves of the state of things?

These special sense receptors are:

1. Visual sensors located in the globe of the eye.
2. Auditory and head position sensors in the inner ear cavity of the temporal bone.
3. Olfactory sensors in the roof of the nasal cavity.
4. Gustatory sensors in the surface lining of the tongue.

VISUAL RECEPTORS AND RELATED STRUCTURES: THE EYE

Those receptor elements of the eye sensitive to light stimuli (photoreceptors) make up a layer (the retina) lining the inner surface of the posterior two-thirds of the eyeball. When stimulated by light these photoreceptors generate and transmit impulses to the occipital lobe (visual cortex) via the circuitry illustrated in Fig. 6-78.

It is at the occipital lobe that the visual impulses are interpreted and an image created which is meaningful to us.

However, the retina is not all there is to the eye—not by a long shot! If one were to design a functioning visual receptor system, one would have to include in the plans a number of structures besides the retina. The full list would be as follows:

1. Protective housing.
2. Lightproof shroud.
3. Refractive media.

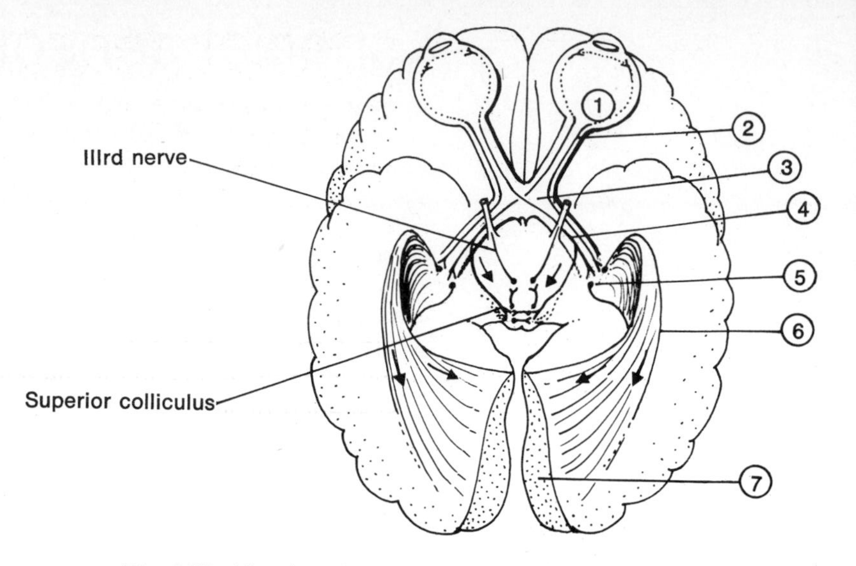

Fig. 6-78 Visual pathways. Impulses are generated at retina (1) in response to light stimuli. Impulses pass through: optic nerve (2), optic chiasma (3), optic tract (4), lateral geniculate body of the thalamus (5), optic radiations (6) to calcarine fissure of the occipital cortex (7). Fibers to superior colliculus are related to visual reflexes. (After C. R. Noback, *The Human Nervous System,* McGraw-Hill, New York, 1967.)

4. Photoreceptors.
5. Light-regulating and distance accommodation mechanism.
6. Source of nutrition and innervation.
7. Mechanism for eye movement.
8. Mechanism for maintaining the external surface of the eye.

And we have all of these.

Protective housing

The eyeball and related structures are firmly entrenched within the deep recess of the bony skull, the **orbit,** protected circumferentially by packings of fat and held firmly in place by extrinsic muscular and ligamentous structures. Nerves, arteries, and veins reach the eyeball through the optic canal (optic nerve, ophthalmic artery) and the superior orbital fissure (Fig. 6-3). Anteriorly, the eyeballs are protected by two mobile eyelids (palpebrae).

The eyeball (or globe of the eye), consisting of three incompletely uniform layers investing a lens and other refractive media, has as its outer layer a dense, fibrous, rubberlike housing, the **sclera** (Fig. 6-79). It is white and quite resilient, as may be demonstrated on yourself: Look into a mirror and note the white part of your eye; this is the sclera. Over your upper lid, press gently on the sclera and note its quality. The structure is penetrated posteriorly by the optic tract and assorted vessels and nerves. The muscles which operate the eyeball insert on the sclera.

Lightproof shroud

Light entering the eyeball must be directed (by refractive media) to the retina and to the retina alone. No reflections or scattering of light can be tolerated. Thus a darkly colored shroud is provided to coat the inner surface of the sclera and the outer surface of the retina. This shroud, the middle layer of the eyeball, highly pigmented and extremely vascular, is

Fig. 6-79 Structure of the eye, horizontal section.

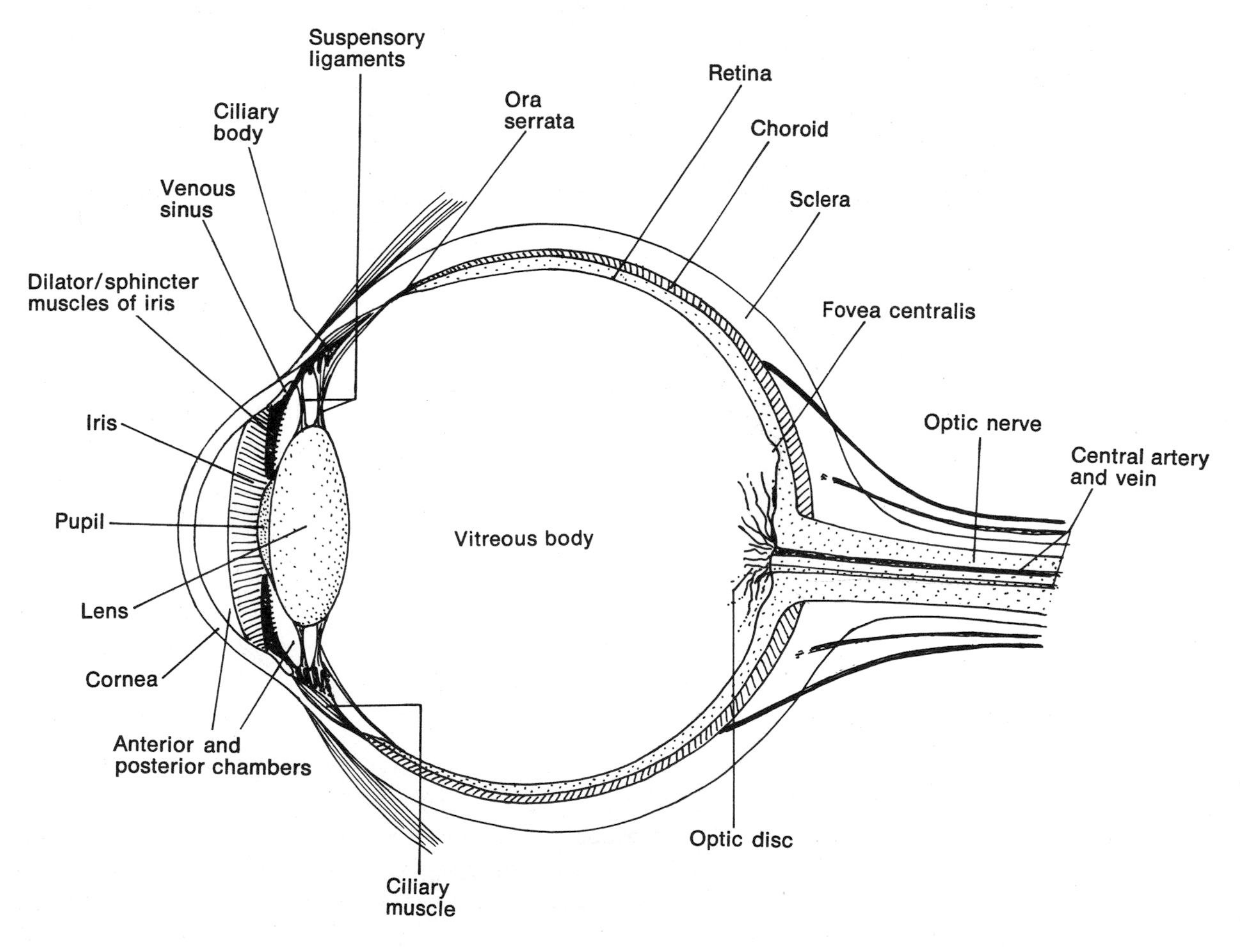

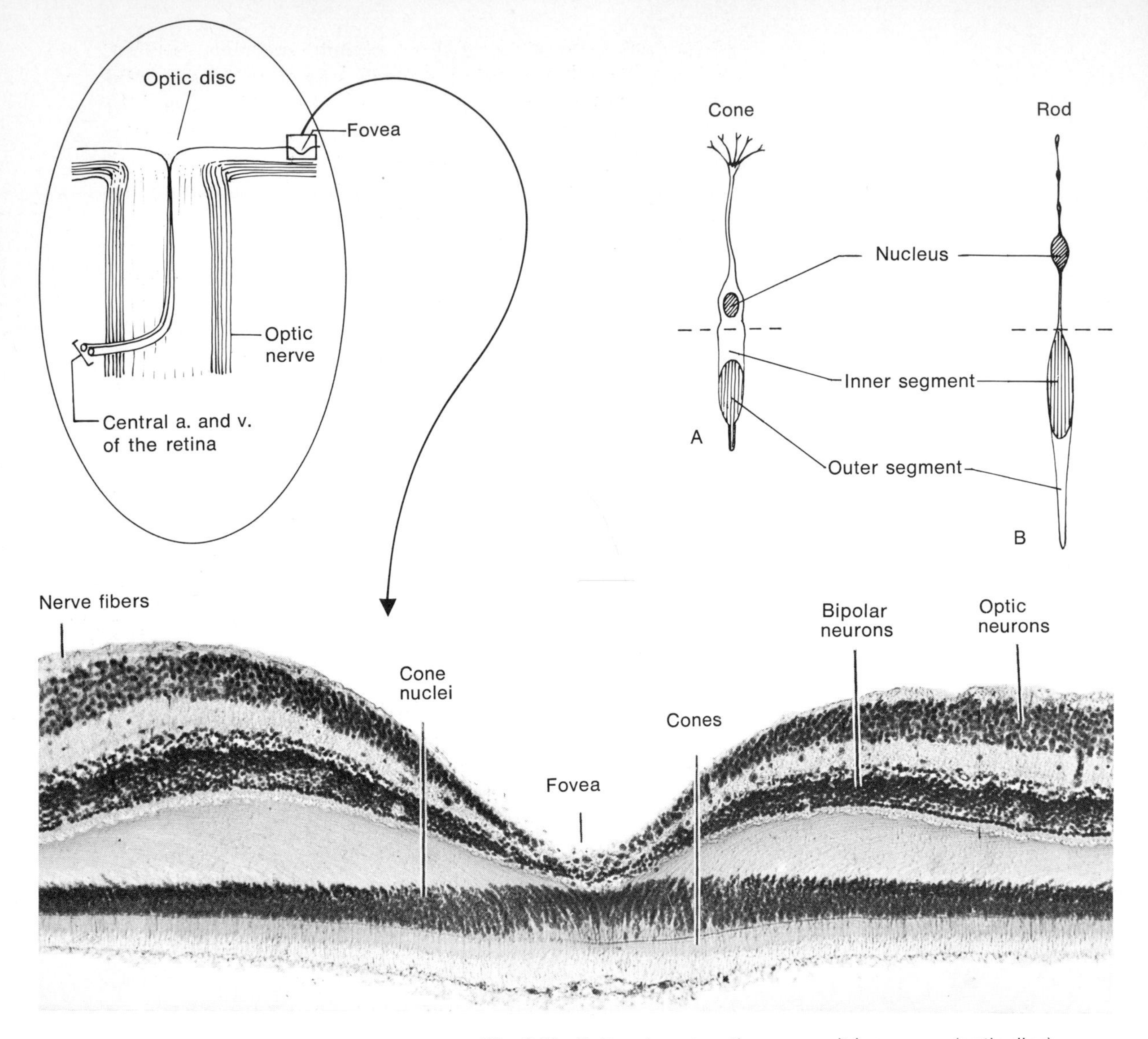

Fig. 6-80 Retina. Insert: optic nerve as it leaves eye (optic disc). Center, photomicrograph of retina at fovea centralis. A. Cone cell. B. Rod cell. (Photomicrograph courtesy of W. F. Windle, *Textbook of Histology*, 4th ed., McGraw-Hill, New York, 1969.)

named the **choroid** (Fig. 6-79). The critical blood supply of the retina is derived from vessels of the choroid.

Photoreceptors

The third, inner layer of the posterior two-thirds of the eyeball is the **retina.** The retina is a complex structure—some 10 layers have been demonstrated (Fig. 6-80). One of the deeper layers is occupied by the photoreceptor cells—**rod** and **cone** cells. Generally the organization of the retina is uniform throughout, with rods outnumbering cones 20 to 1. However, there is an indented, thinned area densely populated with cones and entirely without rods at the center of the retina, hence its name, **fovea centralis** (Figs. 6-79, 6-80). The fovea is the region of sharpest visual acuity.

The sensory portion of the retina terminates at the peripheral, serrated margin (ora serrata) of the ciliary body.

Rod cells are very sensitive but only to the black and white components of light, and function in dim light. Cones are color sensors and are not as sensitive to light. In this way the following phenomena can be explained.

1. One needs bright light for the greatest visual acuity.
2. Colors cannot be easily differentiated in shadows or dim light (as in trying to distinguish a blue suit from a brown one in an unlighted closet).
3. In bright light, one sees best by looking directly at the object in view (bringing the image onto the fovea), but in moonlight, one sees best by looking just to the side of the desired object (employing the cones of the periphery).

One other region of the retina is nonuniform, and this is the **optic disc** (Fig. 6-80), where the optic tract fibers converge to form a bundle and leave the globe of the eye posteriorly. There are no photoreceptors here, so a "blind spot" is created.

Refractive media

Light is a phenomenon which travels in wave form along straight lines. In order to collect these waves and concentrate them at a focal point on the retina, they must be bent or *refracted* as they pass through the eyeball. This is accomplished by a number of structures constituting the refractive* media of the eye; these are listed below in order of contact with incoming light waves:

1. **Cornea.**
2. **Aqueous humor.**

* Refraction can be demonstrated easily in the home. Fill a glass half-full with water and place a straw or similar structure in the glass at an angle. Note how the straw appears bent as it leaves the air medium and enters the watery medium. This phenomenon of bending light waves is called refraction.

3. **Lens.**
4. **Vitreous body.**

The **cornea** (Fig. 6-81) is the anterior continuation of the sclera but is transparent. Composed of lamellae of connective tissue bounded by epithelium, the cornea is extremely sensitive and a good dose of polluted air, not to mention touching it, will set off the familiar blink reflex. The cornea is avascular and receives its nutrition by diffusion.* The cornea is the principal refractive medium of the eye.

Immediately deep to the cornea is a space, the **anterior chamber,** filled with a watery—hence aqueous—fluid (humor), behind which is the lens. The anterior chamber is in communication with the posterior chamber via a small channel between iris and lens (Fig. 6-79).

Aside from acting as a refractive medium, the aqueous humor places a uniform pressure on the eyeball interior,

* Lack of blood vessels may be why transplanted corneas run a low risk of rejection. Lymphocytes and other cells related to immunity don't set in motion the process of tissue rejection (which they normally do in more vascular areas).

Fig. 6-81 Surface features of the eye. The cornea lies over the pupil and the iris.

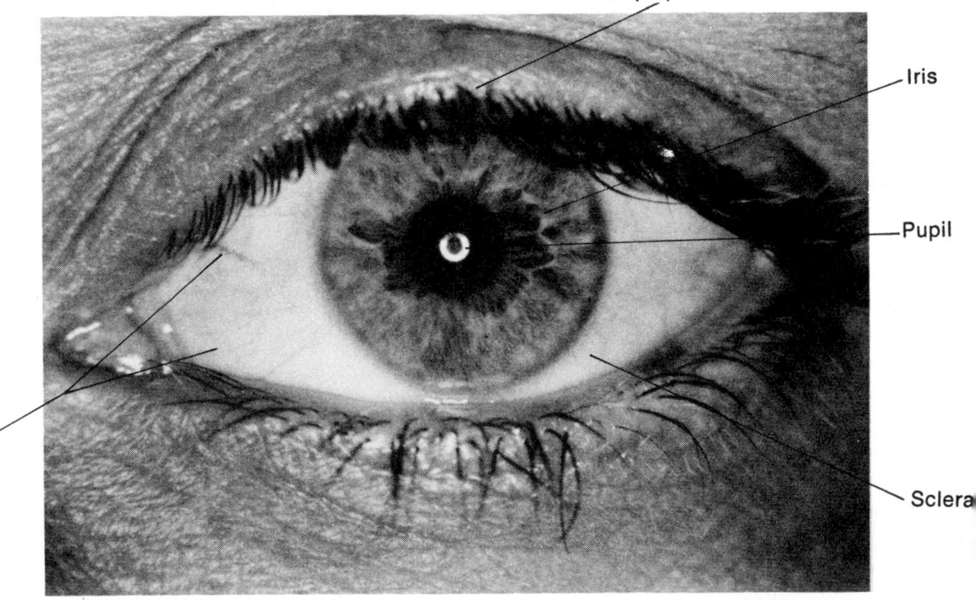

helping to maintain its roundness. Aqueous humor is apparently derived from epithelial cells at the junction of iris and ciliary body (see ahead), and it is absorbed by venous sinuses at the junction of cornea, sclera, and iris.

Situated just deep to the anterior chamber, suspended by fine ligaments extending to the ciliary body, is the **lens** (Fig. 6-79). This encapsulated epithelial body is a flexible structure, unlike the lens in eyeglasses. Functioning to refract light waves to a focal point on the retina, the lens can alter its shape in accommodation of near or far vision. This mechanism will be discussed.

Immediately deep to the lens and filling the posterior two-thirds of the eyeball interior is the gelatinous **vitreous body** (Fig. 6-79). A refractive structure, its anterior border supports the lens and contributes fibers toward the suspensory ligament of the lens.

Light-regulating and distance accommodation structures

The anterior extension of the choroid is the **ciliary body,** which resembles the petals of a daisy when viewed from behind. It contains muscles which regulate the lens shape.

The lens is suspended by suspensory ligaments arising largely from the ciliary body.

The anterior and medial radial projections from the ciliary body are part of the disclike **iris.** The iris has a central aperture, the **pupil,** which transmits light from the outside world to the retina. Note the relations of the iris.

The ciliary body and iris have a basic matrix of connective tissue. The inner border of the ciliary body is thrown into a series of folds—the **ciliary processes,** from which the **suspensory ligaments** arise. Within the substance of the ciliary body is a mass of variously arranged smooth muscle fibers, **ciliary muscles.** These muscles are concerned with altering the shape of the lens. The iris has two sets of muscle fibers within its matrix, the radial **dilator pupillae** muscles and the more circular **sphincter pupillae** muscles (Fig. 6-79). These muscles function to dilate or constrict the diameter of the pupil, regulating the amount of light impinging on the retina. The posterior surface of the iris is lined with a richly pigmented layer. The color of the iris is dependent upon:

1. The thickness of the iris.
2. The number and arrangement of pigment cells in the stroma.

Increases in these factors darken the iris toward the color brown. A reduced concentration of pigment generates lighter colors such as green or blue. This pigment reduces considerably the amount of light coming into the pupil, thus blue-eyed people are generally more sensitive to bright light than the brown-eyed.

Mechanism for regulation of light input and accommodation of vision. The amount of light impinging on the lens is largely regulated by increasing or decreasing the *diameter* of the *pupil.* This is an involuntary (*reflex*) phenomenon. In response to increased brightness, the *sphincter* muscles of the iris, situated around the pupil like a drawstring, contract and the iris closes around the pupil, *decreasing* the latter's diameter. Conversely, in reduced light the *dilator* muscles of the iris are stimulated to contract and the iris, like radial venetian blinds, opens about the pupil, *increasing* the latter's diameter. The neural pathways for pupillary light reflexes involve optic tract (*sensory*) impulses passing to the superior colliculus, *motor* impulses from sympathetic neurons to the *dilator* muscles, and parasympathetic impulses to the *constrictor* muscles of the iris. Thus, in states of fear or anger, the pupils are *dilated.*

The mechanism by which the eye adjusts itself to viewing things close up is termed **accommodation.** Accommodation involves:

1. Altering the shape of the *lens.*
2. Altering the diameter of the *pupil.*
3. Converging the eyes.

Light rays from distant objects approach the lens of the pupil in parallel.

Light rays from close-in objects approach the lens of the eye divergently.

In the latter case, the lens must be more curved or rounder in order to focus the light onto the same retinal area as in the former case.

For distant viewing, the curvature of the lens is reduced. The lens, being naturally elastic, will tend to round up if left to its own devices—bad for distant viewing! But, the lens is normally held under tension by the suspensory ligaments—no muscular effort is required. Thus, no effort is needed for far vision.

For in-close viewing, or **accommodating,** the curvature of the lens must be increased—the lens must be allowed to round up. In other words, tension must be taken off the suspensory ligaments. This is accomplished by the ciliary muscles, which contract and so relax the ciliary body sufficiently to loosen the tension on the ligaments and allow the lens to round up in accordance with its natural elasticity. Reasonably divergent incoming light waves can now be focused on the retina.

Source of nutrition

The primary source of blood to the orbital contents is the **ophthalmic artery** (Fig. 6-82), a branch of the internal carotid artery. It enters the orbit through the optic canal and branches extensively. The artery to the retina is the **central artery,** traveling in the substance of the optic nerve. Gener-

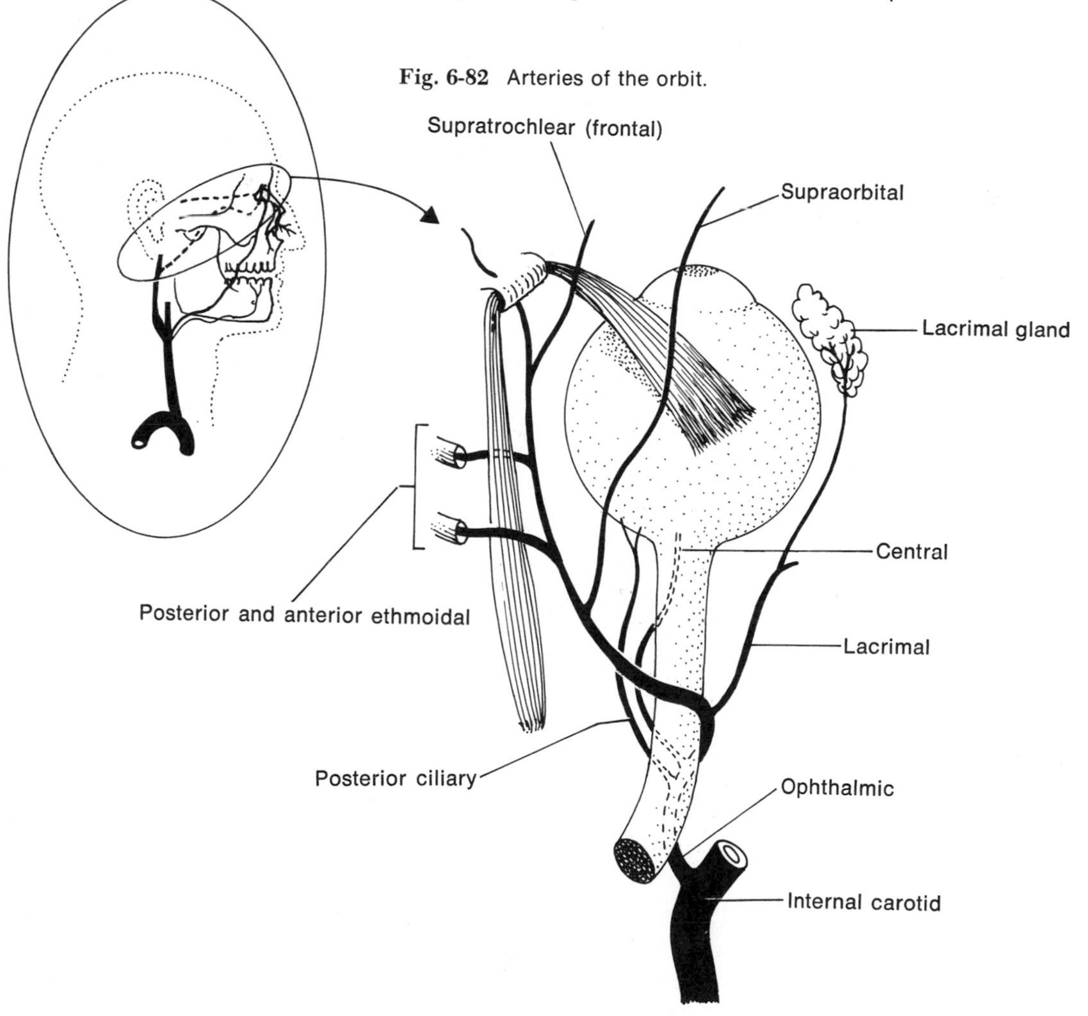

Fig. 6-82 Arteries of the orbit.

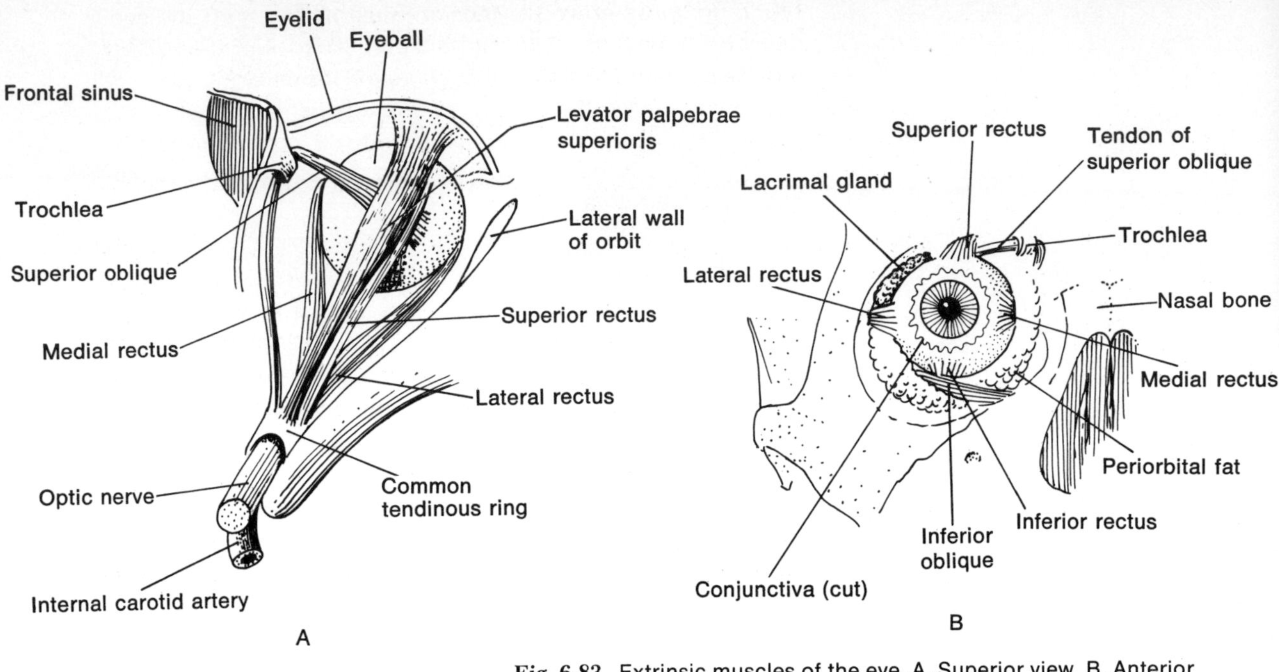

Fig. 6-83 Extrinsic muscles of the eye. A. Superior view. B. Anterior view.

ally, the eyeball structures are serviced by the ciliary arteries, all of them branches of the ophthalmic artery.

Venous drainage of the iris, retina, and related structures is accomplished by tributaries of the **vorticose veins.** These latter veins drain into the cavernous sinus via ophthalmic and ciliary veins (Fig. 6-75).

Mechanism for eye movement

Movement of the orbit is a function of six slender skeletal muscles inserting on the sclera which (with two exceptions) arise from a common circular tendon ringing the optic canal. These **extrinsic muscles** may be seen in Fig. 6-83. Note that

1. Four of the muscles are straight and are so named: rectus. Two are angled with reference to the straight muscles and are also so named: oblique.
2. The names of the muscles are prefixed with a term of orientation, e.g., superior, inferior.

The function of the extrinsic muscles is often complex—apparently no one muscle acts alone. Experiment yourself with eye movements, referring to Fig. 6-84.

Innervation of orbital structures

The nerves relating to the orbit and its occupants are many and may be divided into three functional categories:

1. *Sensory:* e.g., the optic nerve conducting visual impulses, and the other nerves (cranial nerves III, IV, V, and VI) conducting impulses of pain, pressure, and proprioceptive stimuli.
2. *Autonomic:* e.g., the parasympathetic preganglionic nerves riding with the oculomotor (IIIrd) nerve to the **ciliary ganglion** (Fig. 6-85); parasympathetic postganglionic nerves to the ciliary (accommodating) muscles and pupillary sphincter muscle; sympathetic postganglionic nerves to the blood vessels and dilator muscle of the pupil.
3. *Somatic motor:* the cranial nerves innervating the extrinsic muscles of the eye, e.g., oculomotor nerve to all but the lateral rectus muscle, which is innervated by the abducens (VIth) nerve, and the superior oblique muscle, which is supplied by the trochlear (IVth) nerve.

Maintenance of external surface of eye

The cornea and anterior sclera are in constant contact with airborne pollutants of the external environment. More and

Fig. 6-84 Diagram of eye movements. (From Eugene Wolff, *Anatomy of the Eye and Orbit,* H. K. Lewis and Co., Ltd.)

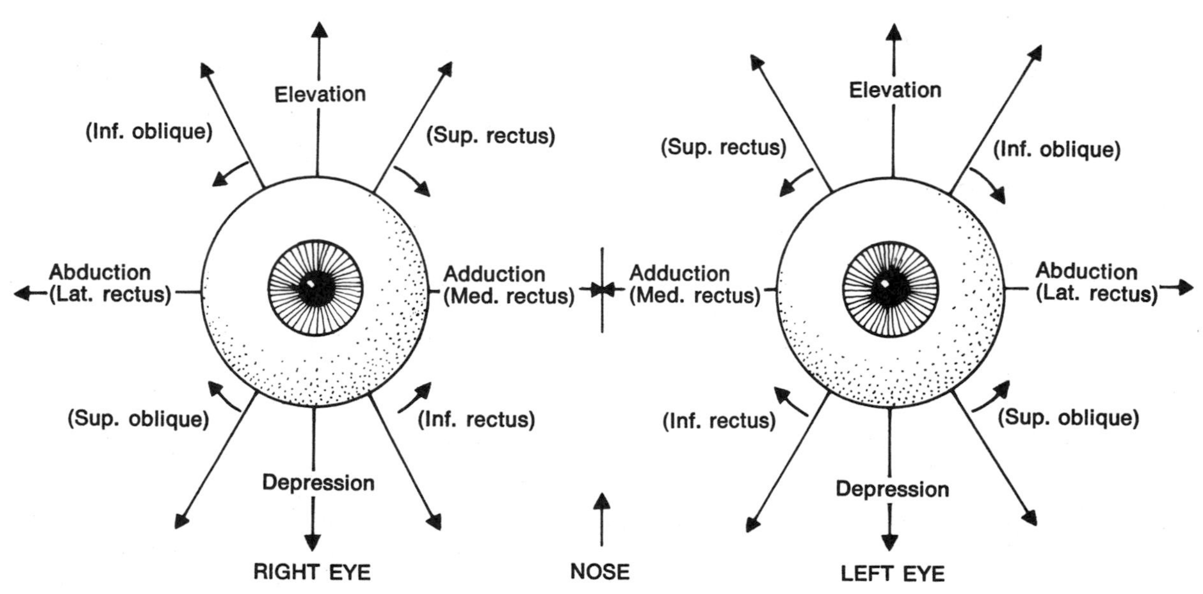

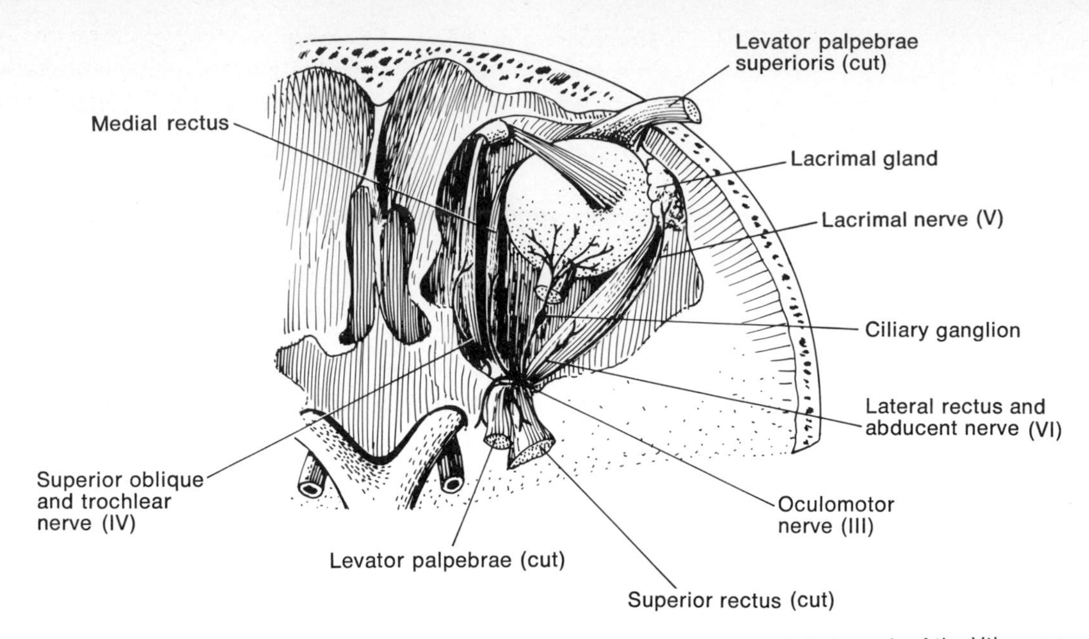

Fig. 6-85 Nerves of the orbit. The ophthalmic branch of the Vth nerve (sensory) is not shown except for one of its branches (lacrimal).

more, demands are put upon the protective elements* of these surfaces:

1. The **eyelids.**
2. The **conjunctiva.**
3. The **lacrimal apparatus.**

Eyelids. The eyelids (palpebrae) are fibrous connective tissue structures incorporating a very thin layer of skin externally, a mucous membrane (conjunctiva) on its internal surface and groups of smooth and skeletal muscle and sebaceous-like (tarsal) glands within. The free margin of the upper and lower lids, as you can see for yourself, is characterized by cilia (eyelashes) and associated glands. The upper lid is retracted by muscular effort (levator palpebrae superioris); the lower by gravity. The lids are closed through relaxation of the above muscle and the influence of the cutaneous circular muscle of the orbit.

Inspection of the free surface (edge) of your own lower or

* If evolutionary history is any guide to the future, our descendants many times removed should have elephant-sized eyelids and tear glands, if air pollution continues to worsen.

upper lid will demonstrate several small perforations—the openings of the tarsal glands (Fig. 6-86). At the medial aspect of each lid is a small lacrimal papilla with its opening (lacrimal punctum).

The eyelids protect the eye against such missiles as splintering wood, metal shavings, or just plain dust. With each closing of the lids, the cornea get a "wash job" much as a car window is brushed by a window wiper blade.

Conjunctiva. The conjunctiva is the thin, vascular mucous membrane lining the internal surfaces of the eyelids and reflecting onto the cornea and anterior sclera (Fig. 6-86). The upper, wrinkled reflection of conjunctiva contains the several ducts of the lacrimal gland.

Evert your own eyelids while peering into a mirror and note:

- The very vascular conjunctiva on the inner surface of each eyelid.
- The layer of conjunctiva over the cornea and sclera, which is less vascular but which can become blood red with irritation.
- At the lowest point of the internal surface of the eyelid, the reflection of conjunctiva from eyelid to cornea.

The conjunctiva plays an important role in protecting the cornea from physical contact with damaging particles and, perhaps most importantly, it keeps the cornea moist—a prerequisite to corneal transparency.

Lacrimal apparatus. The lacrimal apparatus consists of:

1. A secretory gland, the lacrimal gland.
2. Drainage sac and associated tubes/ducts.

Look at Fig. 6-86 as you read the following:
The **lacrimal gland** occupies the anterolateral corner of the orbit just above and deep to the upper eyelid, reposing on the lateral and superior rectus muscles. Some one dozen ducts drain this autonomically innervated tear gland.

Flowing through the conjunctival sac, the tears drain medially along the edge of the eyelids to the lacrimal puncta, the orifices of the lacrimal papillae.

The papillae are drained by a pair of small canals reaching toward the lacrimal sac. The lacrimal sac drains into the nasal cavity (inferior meatus) via the nasolacrimal duct. Thus when you cry you blow your nose.

The lacrimal gland performs the obvious function of mois-

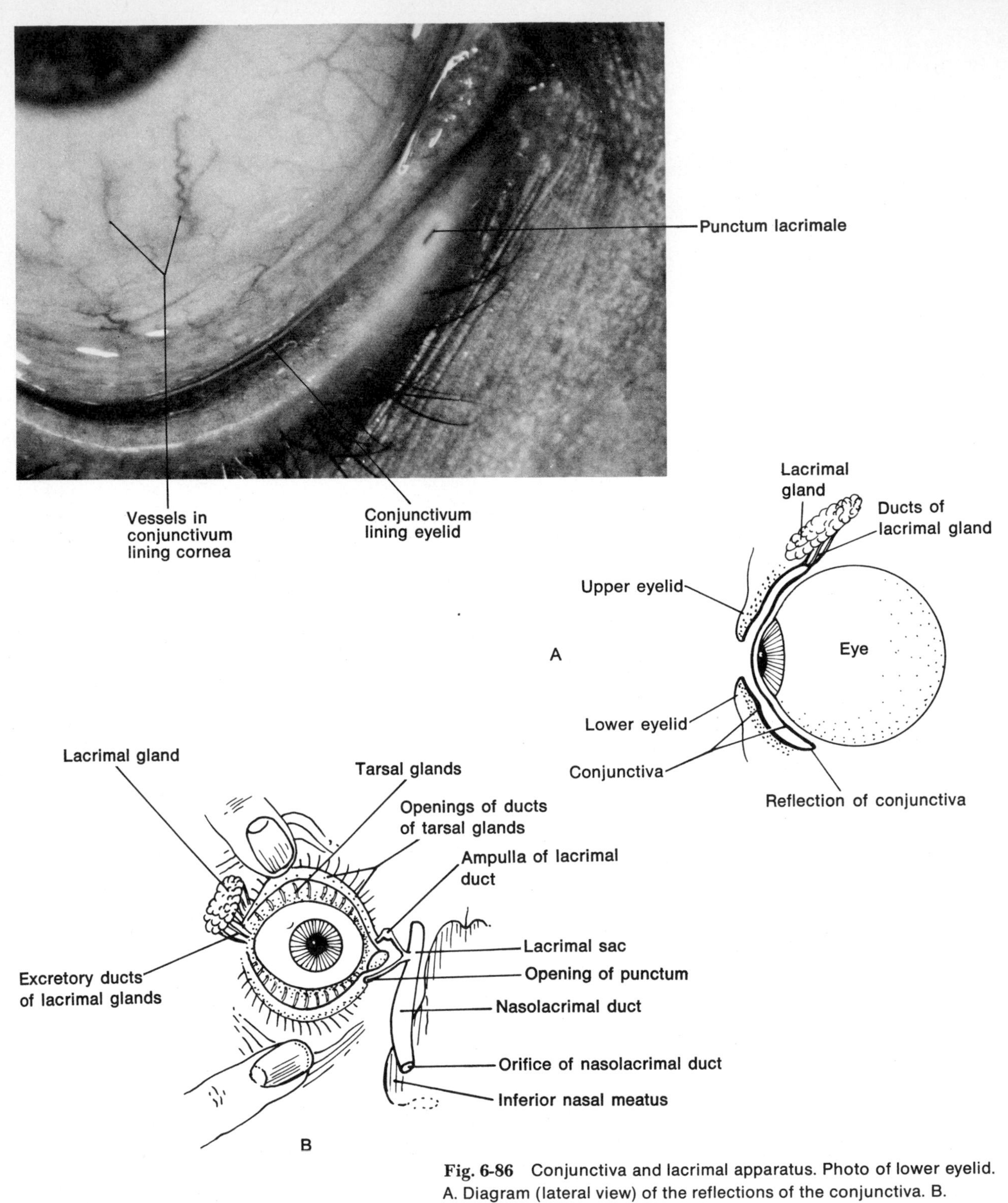

Fig. 6-86 Conjunctiva and lacrimal apparatus. Photo of lower eyelid. A. Diagram (lateral view) of the reflections of the conjunctiva. B. Nasolacrimal apparatus in situ.

tening and cleansing. In certain emotional situations, the lacrimal gland is stimulated to secrete and the excess tears overflow the lids.

AUDIOVESTIBULAR RECEPTORS AND RELATED STRUCTURES: THE EAR

There are no sensory receptors in the nervous system specifically for sound energy, and therein lies the basis for the ear's complexity. Sound waves collected by those often elaborate appendages we call ears are funneled into the cavernous interior of the temporal bone and converted into mechanical energy. Then, still deeper, this energy is translated into nervous impulses, the currency which the CNS can handle. Also in the internal ear are vestibular cells for sensing head movements.

The entire audiovestibular apparatus is called the **ear.** Anatomically it is divided into three areas (Fig. 6-87):

Fig. 6-87 The ear.

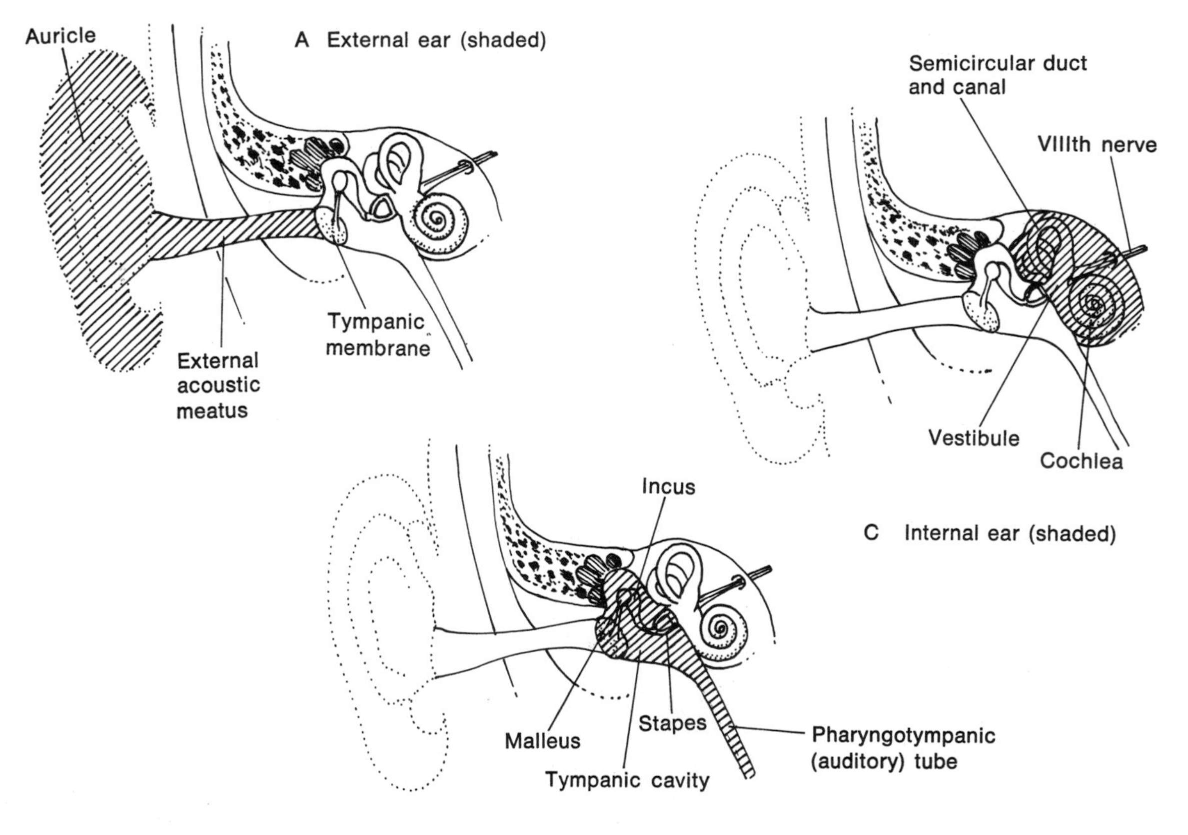

1. External ear: a funnel-like apparatus which ducts air back to the tympanic membrane, which bars the way to the . . .
2. Middle ear: a cavity housing bony and muscular structures which convert sound energy into mechanical energy and transmit it to the chambers of the . . .
3. Internal ear: the home of sensory organs of audition and equilibrium. From here the vestibulocochlear nerve (VIIIth cranial nerve) conducts auditory and vestibular-related impulses to the lower brainstem for interpretation.

External ear

That flap of skin protruding from each side of your head is more precisely termed the **auricle.** On yourself you can feel that it has an elastic cartilaginous framework. The lower free portion is the lobule—a fat-filled bag of skin, quite vascular but not well innervated (thus, a frequent site for sampling blood). The orifice at the auricle projects medially via a crooked path, the **external acoustic meatus.**

The external acoustic meatus is a tube, shaped somewhat like an S. The lateral part can be palpated on yourself; it is supported by cartilage, lined with stratified squamous epithelium characterized by stout hairs and glands—all of which have a protective function. Medially, the meatus is supported by the temporal bone, and it terminates at the entrance to the middle ear cavity, which is sealed by the **tympanic membrane** or "eardrum." This structure fits obliquely in a bony groove of the meatus. Medially it is carpeted with a layer of mucous membrane; laterally it is covered with skin.

To appreciate the effectiveness of the auricle as a sound collector, cup your hand about your ear while listening to the radio or other sound source and note how the volume is enhanced. Many animals have enough muscular control over their auricles to be able to rotate them toward a sound source. Although humans have such muscles, they are generally ineffective for more than "ear wiggling."

Middle ear

The middle ear, or **tympanic cavity** (Fig. 6-88), is a region which may be likened to a room. It has a roof, a floor, four walls, two windows, and a carpet of mucous membrane. It even has a front and a back "door"—and it is furnished with bony, ligamentous, muscular, and neural "furniture." Consult Fig. 6-88 as you read the following:

The **roof** (tegmen tympani) constitutes part of the petrous portion of the temporal bone.

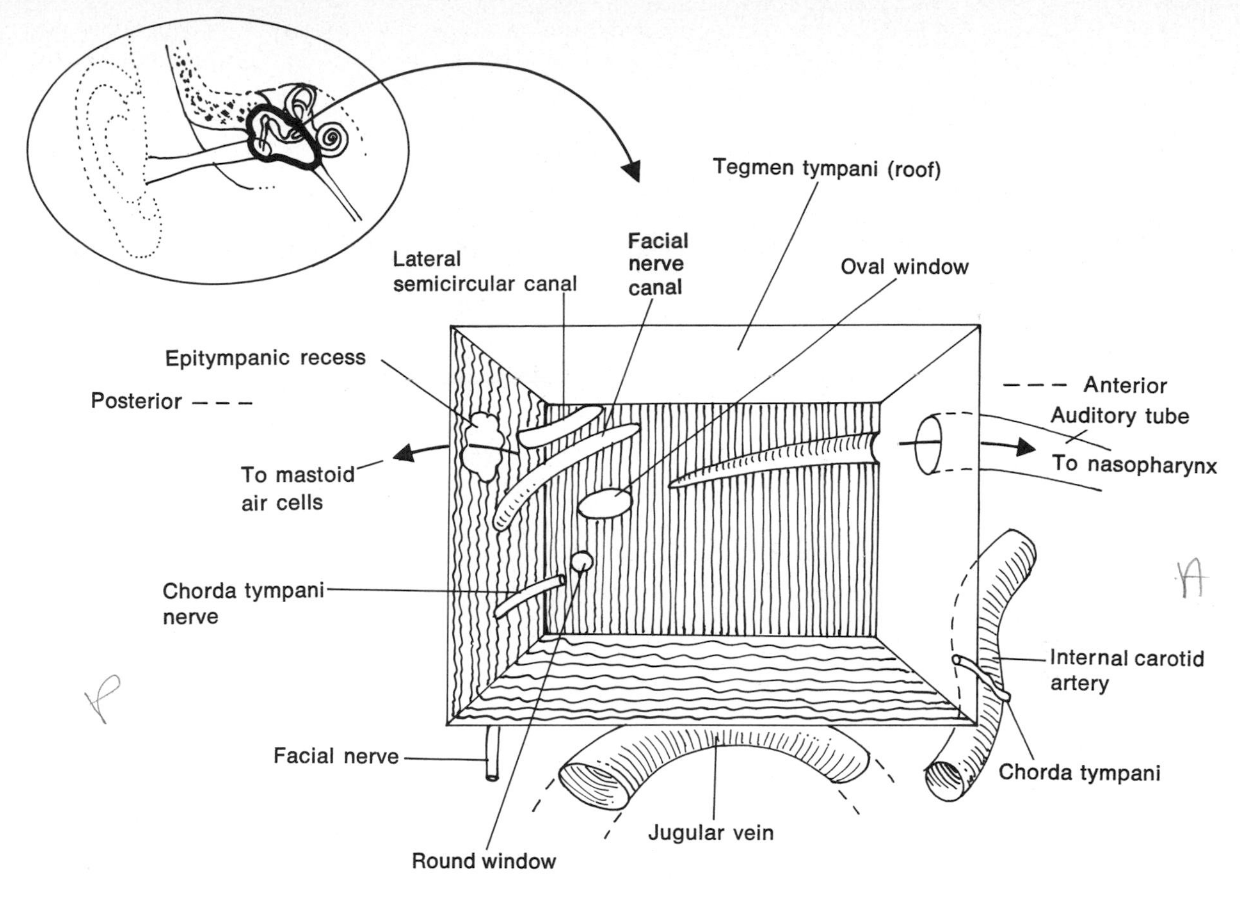

Fig. 6-88 Schematic of the middle ear cavity and related structures. View looks medially. (Adapted from J. Maisonnet and R. Coudane, *Anatomie Clinique et Operatoire,* Doin, Paris, 1950.)

The **floor** (jugular wall) consists of the lower plate of the petrous portion of the temporal bone. It overlies the internal jugular vein.

The **lateral wall** is the tympanic membrane.

The **medial wall** of the tympanic cavity is the lateral wall of the internal ear; penetrating it are an **oval window,** with a membranous "windowpane," and a **round window,** also sealed by a membrane.

The **posterior wall** contains the epitympanic recess (the back "door"), which communicates posteriorly and inferiorly with the mastoid air cells. You can palpate the mastoid process (which encloses the air cells) just behind your ear.

The **anterior wall** (carotid wall) carries the impression of the internal carotid artery, which passes alongside the middle ear as it ascends to the brain. In the anteromedial corner of this

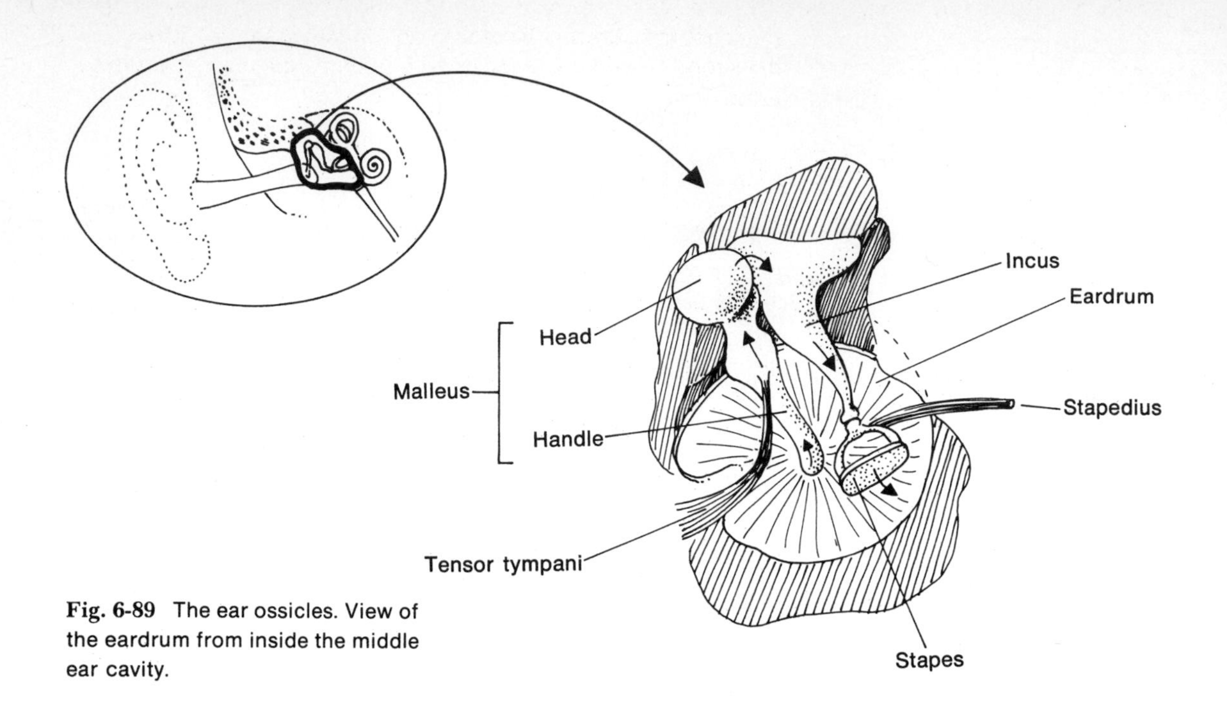

Fig. 6-89 The ear ossicles. View of the eardrum from inside the middle ear cavity.

wall is an orifice, the termination of the pharyngotympanic (auditory) tube (the front "door").

Ossicles. Three small articulated bones, or **ossicles,** bridge the middle-ear cavity between the tympanic membrane and the oval window (Fig. 6-89). They articulate with each other by means of freely movable synovial joints. The **malleus** attaches by its handle to the tympanic membrane. Its neighbor is the **incus,** which attaches to the head of the stapes. The **stapes** is joined to the oval window by its firmly embedded footplate. Thus vibrations received by the stapes from its fellow ossicles produce pistonlike movements in the oval window and so set the fluid inside in motion.

Slender skeletal muscles attach to two of the ossicles from origins on the wall of the middle ear. The **tensor tympani** attachs to the handle of the malleus, and the **stapedius** inserts on the head of the stapes. By contracting these muscles, one diminishes the response of the ossicles to the sound vibrations reaching the tympanic membrane and so protect the ear against dangerous overload.

Pharyngotympanic tube. This tube extends from the middle-ear cavity to the nasopharynx. It is carpeted with mucous membrane. It serves to equalize the air pressure between the two cavities. It is sometimes called the auditory tube.

Internal ear

The internal ear is a series of fluid-filled bony chambers **(bony labyrinth)** housing a series of fluid-filled membranous chambers **(membranous labyrinth).** Read that again carefully, because if you haven't got those basic facts down, you are headed for frustation in attempting to learn about the internal, or inner, ear. Refer now to Fig. 6-90 and note that the bony labyrinth consists of:

1. A **vestibule** opening into both the semicircular canals and the scala vestibuli of the cochlea. It is also in communication with the oval window.
2. Three **semicircular canals.**
3. The **cochlea,** consisting of the scala vestibuli, scala tympani, and housing the membranous cochlear duct.

The bony labyrinth is actually a series of passageways in the temporal bone. It is a closed set of channels and contains **perilymph,** an extracellular fluid.

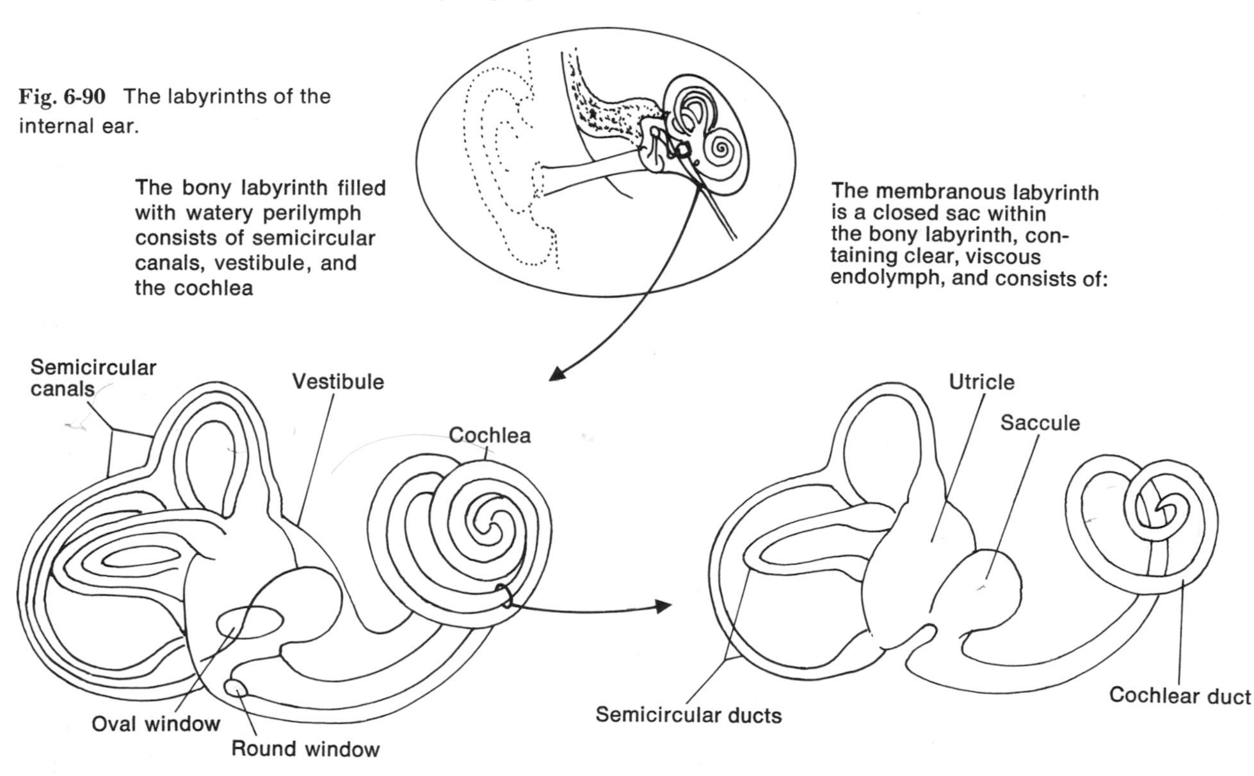

Fig. 6-90 The labyrinths of the internal ear.

The membranous labyrinth is an epithelium-lined series of structures within the bony labyrinth including:

1. The large **utricle** and the smaller **saccule,** both within the vestibule.
2. The semicircular **ducts** within the bony semicircular canals.
3. The **cochlear duct,** more or less centered within the cochlea.

The membranous labyrinth is also fluid-filled—with **endolymph,** a fluid similar to intracellular fluid. The membranous labyrinth is *not* in communication with the bony labyrinth.

Cochlea. Refer to Fig. 6-91 as you read. Deep to the oval window is the **vestibule.** This chamber is open to the **scala vestibuli,** a canal within the bony cochlea. Thus the oval window is in direct communication with the scala vestibuli. The scala vestibuli spirals 2½ times around the bony core, or **modiolus,** of the cochlea. Just below it, but following the same spiral course, is another canal in the bone of the cochlea, the **scala tympani.** The two canals communicate only at the top of the spiral, through a hole, the **helicotrema.** Thus the one connection between the oval window and the scala tympani is through the scala vestibuli. Lying between the two canals, but along the edge of the scala vestibuli, is a third, smaller canal, the **cochlear duct.** It is part of the membranous labyrinth, hence surrounded by membrane.

Put another way, if you were in a submarine within the vestibule, someone in the middle ear who pushed against the oval window (as a movement of the stapes does) could send you through the perilymph in one of two directions: (1) into the scala vestibuli or (2) into the semicircular canals (not ducts). If you took the first route you would go up around the modiolus, completing a circle 2½ times before turning abruptly into the scala tympani at the helicotrema. Continuing this odyssey, you would drift down the scala tympani, again circling the modiolus 2½ times, and finally come up against a membrane covering the round window. This secondary tympanic membrane marks the end of the scala tympani.

The cochlear duct, triangular in cross section, may be likened to a long curved tent placed at the modiolar edge of the scala vestibuli and following its course all the way to the helicotrema.

Now the cochlear duct

- Is closed at both ends.
- Has a delicate membranous (vestibular) roof.

Fig. 6-91 Anatomy of the internal ear greatly magnified.

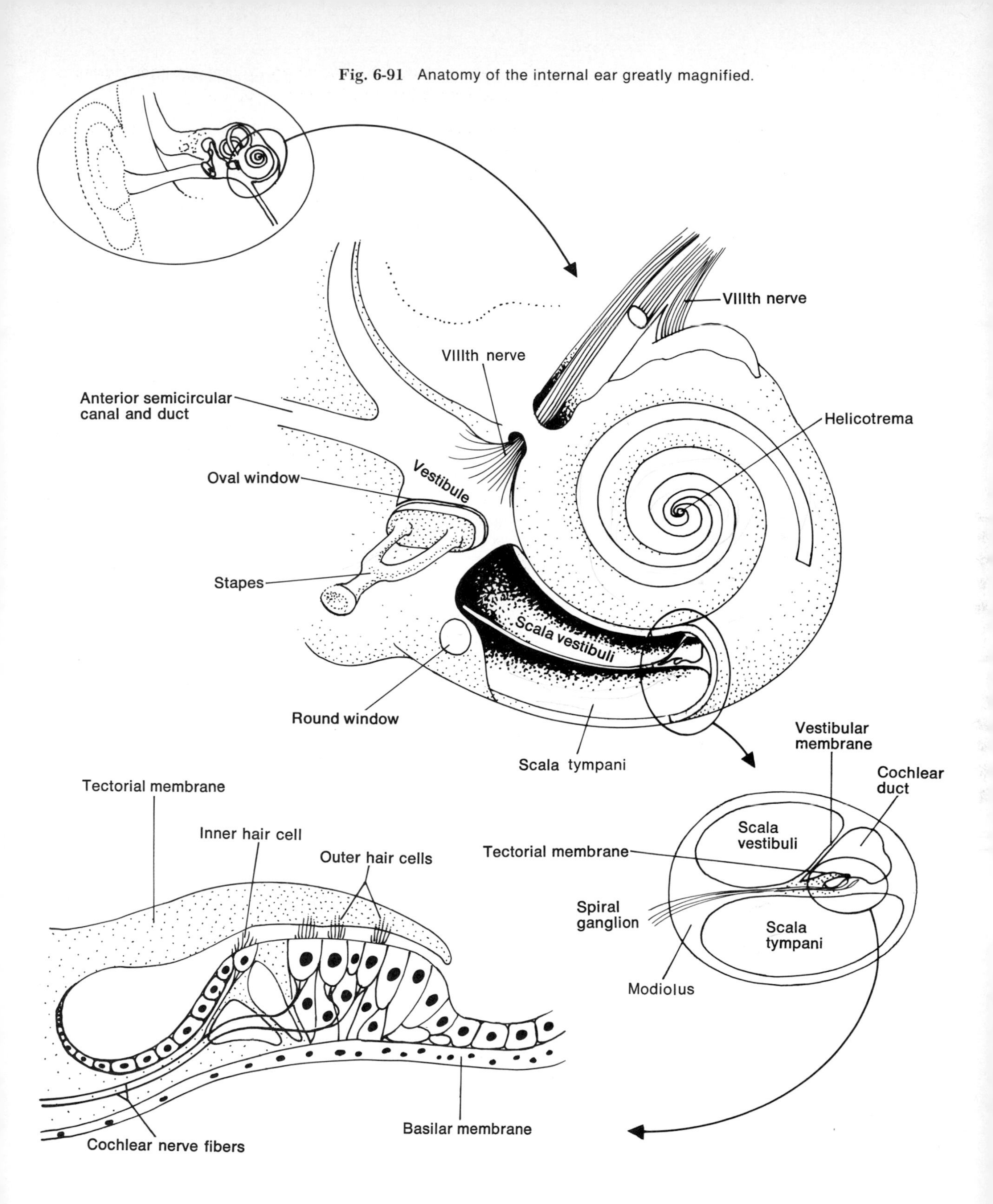

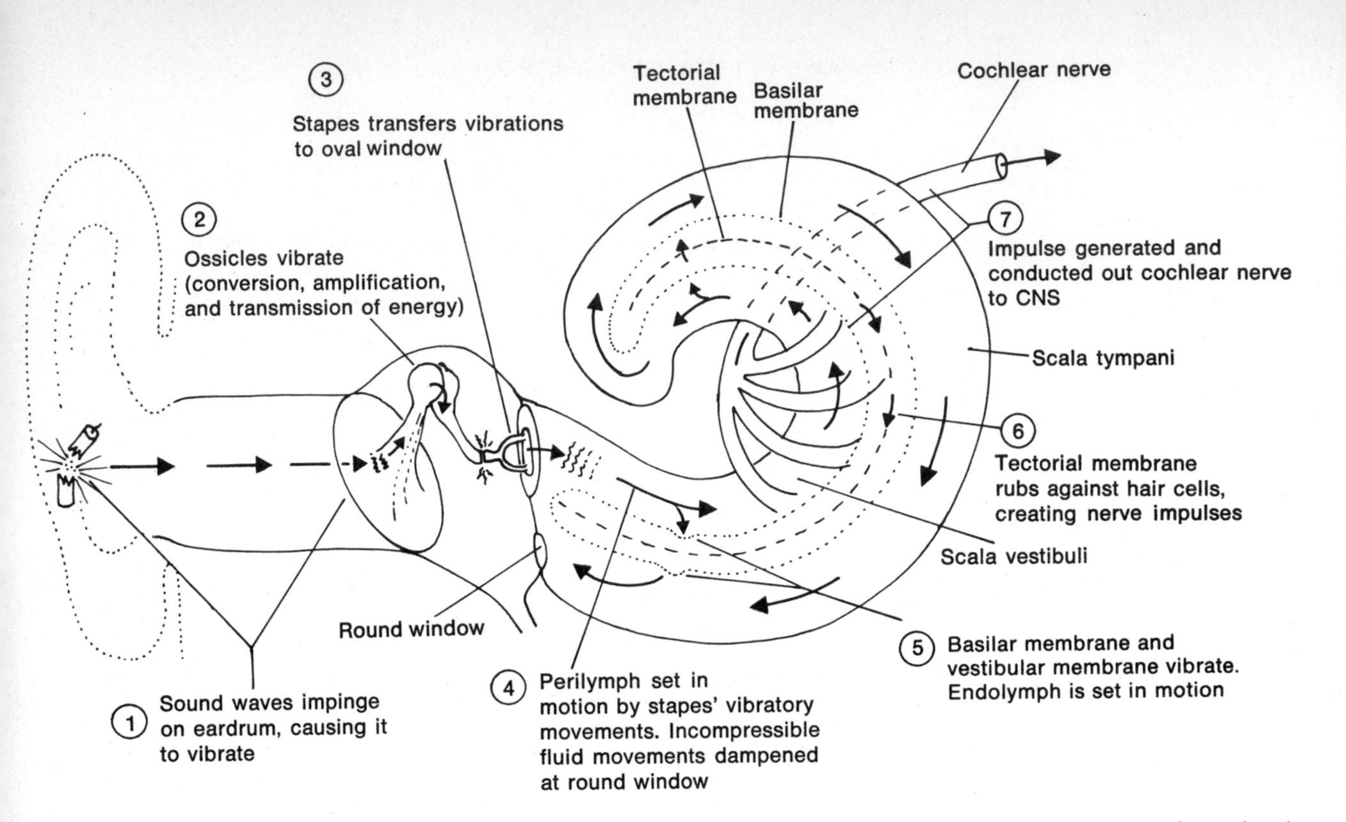

Fig. 6-92 Events that lead up to generation of nerve impulses related to sound.

- Has a vascular and ligamentous lateral wall.
- Has a partly bony, partly membranous floor projecting from the core of the cochlea: the spiral lamina and the basilar membrane.
- Has a complicated and specialized structure arising from the floor of the duct—the **organ of Corti**—composed of sensitive hair cells and supporting cells, and an overlying **tectorial membrane** arising indirectly from the osseous (spiral) lamina.

Now reference to Fig. 6-92 will demonstrate how the structures just explained take part in the modification of sound energy for subsequent interpretation by the CNS.

The cochlear portion of the VIIIth nerve transmits the impulses to the cochlear nuclei in the pons, where they are sent on to the inferior colliculi, the thalamus, and the temporal lobe.

Vestibular structures. The structures of the internal ear associated with equilibrium (Fig. 6-93) are:

1. The three bony semicircular canals housing the semicircular ducts.
2. The larger utricle and the smaller saccule—both membranous structures within the bony vestibule.
3. The pressure-relief valve of the internal ear, the endolymphatic duct, which may be involved in fluid reabsorption.

Hair cells embedded in a gelatinous material (Fig. 6-93) lie in the ampullae (dilated ends) of the **semicircular ducts.** The hair cells are sensory receptors of the vestibular portion of nerve VIII. When movement of the head causes the endolymph in the appropriate duct to move, the hair cells are stimulated to fire.

In the **utricle** and **saccule** too hair cells are embedded in a gelatinous material (Fig. 6-93). Moving the head also forces these top-heavy sensory receptor cells to bend and fire an impulse via the vestibular portion of VIII. Thus any movements of the head stimulate one or more of these sensory cell patches to fire afferent impulses. The architecture of the canals—oriented in three different planes—is such that all movements of the head can be detected. The impulses are distributed to the brainstem, cerebellum, and spinal cord.

Fig. 6-93 Sensors of head movement.

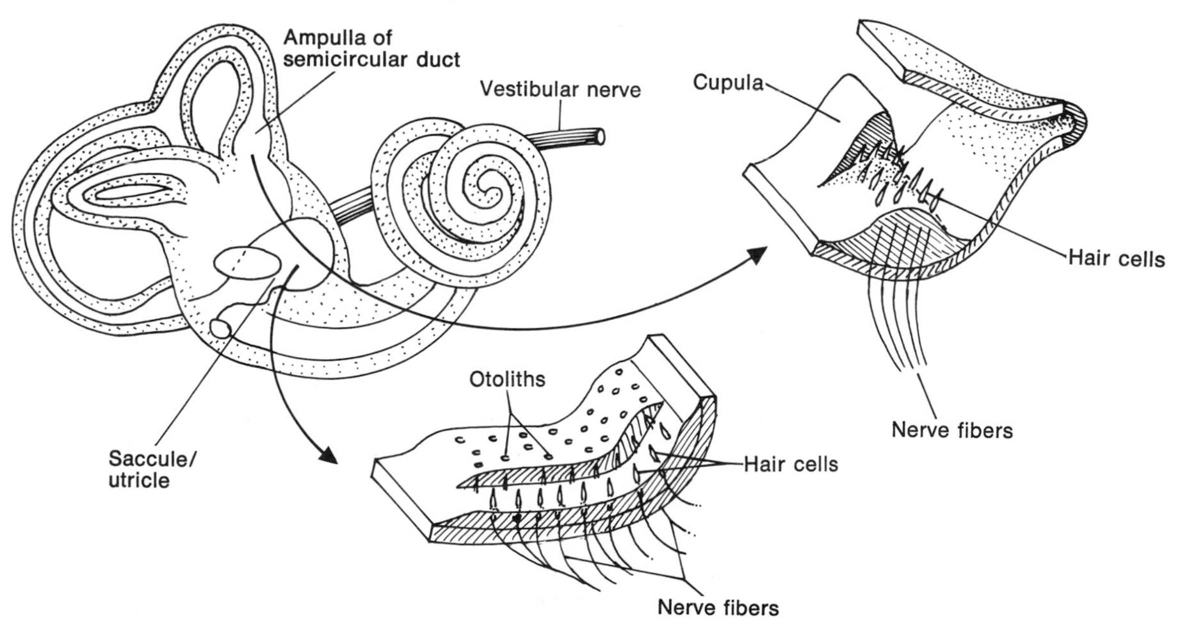

RECEPTORS OF TASTE

Since the things we require for maintenance of our bodies normally enter by way of the mouth, it is no surprise that there is a taste receptor just inside. Some 9,000 chemoreceptors are located on the tongue and pharynx. The receptors look like the buds of a flowering plant and, therefore, are captioned **taste buds,** or gustatory organs.

Refer to Fig. 6-94 as you read the following:

Taste buds are epithelial structures (less than 1 mm in height) located on the walls of the pharynx and palate to a lesser degree and on the **circumvallate papillae** of the tongue to a greater degree.

Fig. 6-94 Receptors of taste. A. Dorsum of tongue. B. Circumvallate papilla in section and highly magnified, showing taste buds along the crypts. C. Diagram of a taste bud.

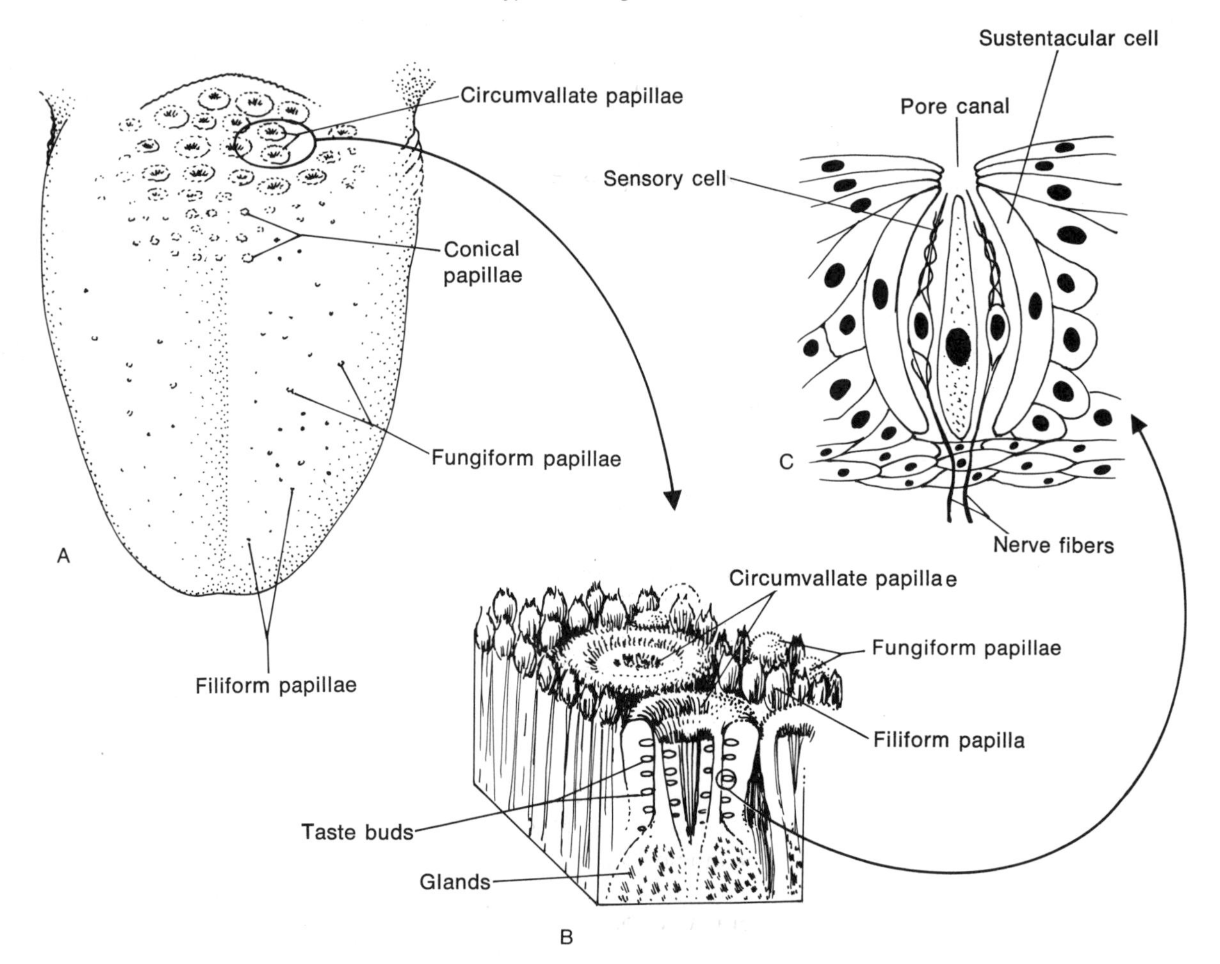

Microscopically, taste buds are characterized by receptor cells surrounded by supporting cells. Embedded into the papilla, taste buds open onto the surface of the tongue through pores. At the basement membrane, nerve fibers from the receptor cells conduct impulses of taste sensation to the cerebral cortex via the cranial nerves VII and IX, the medial lemniscus, and the thalamus.

Taste receptor cells can discriminate among four chemical modalities: sweet, sour, bitter, and salt. Some receptors are sensitive to more than one mode but probably never to all four. Substances placed in the mouth dissolve sufficiently to bathe the crypts of the papillae. These solutions stimulate certain receptors to fire sensory impulses to the brain for interpretation. There is no specific cortical area for taste reception. Interpretation is made at the sensory cortex subserving the facial region.

The phenomenon of "discriminating taste" displayed by the gourmet is really based on a differential mixture of (1) the four basic tastes, (2) the temperature and texture of the food, and (3) the smell of the food.

RECEPTORS OF SMELL

In evolutionary terms, the sense of smell—**olfaction**—has great significance, particularly as it relates to the development of instincts and emotions. In less complex animals, the part of the brain (rhinencephalon) devoted to interpretation of smell is quite large. In man, no part of the rhinencephalon is devoted to olfaction, as all of it is taken up by neuronal circuits related to emotional behavior.

Olfactory receptors are located in the nasal mucosa at the roof of the nasal cavity (Fig. 6-95). These neurons are wedged in between supporting cells, and their cilia project into the mucus covering the nasal mucous membrane. Actually, olfactory receptors are sensitive to certain chemicals dissolved in the mucus. The basis for olfactory discrimination is unknown. When odors pass into the nose, they generally pass straight through into the nasopharynx. When one wants to partake of the odor in greater magnitude, one dilates the nares, tilts the head back to expose the sensory cells directly, and inhales. This activity is commonly called sniffing.

Once stimulated, the olfactory neurons transmit the impulses to the olfactory bulbs lying atop the cribriform plate of the ethmoid in the anterior cranial fossae. From these, the impulses pass via the olfactory stria to the underside of the temporal and frontal lobes.

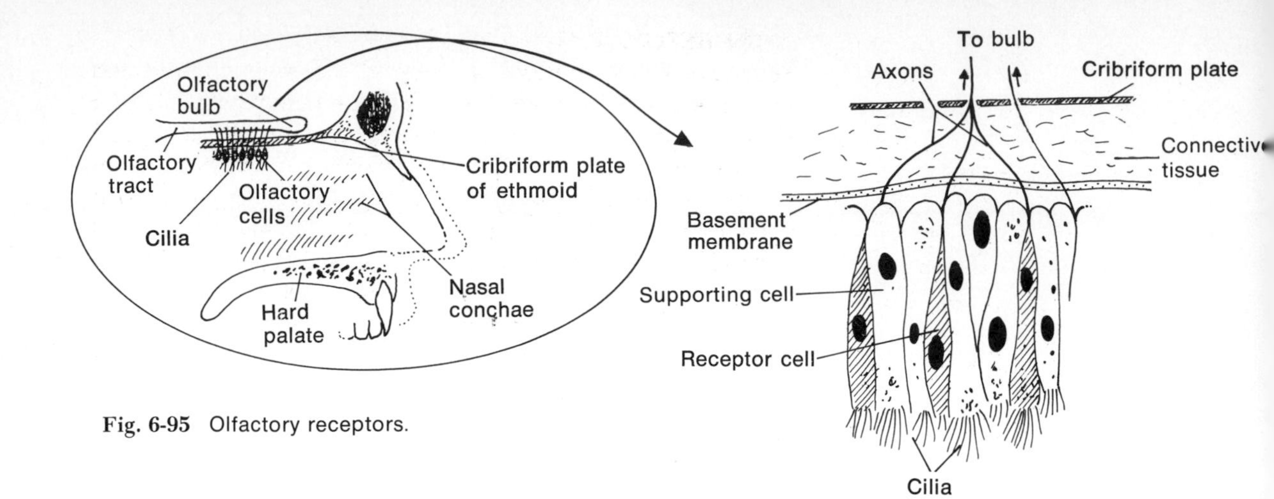

Fig. 6-95 Olfactory receptors.

QUIZ

1. The sensory receptors of visual stimuli form a layer called the ____________.
2. Visual interpretation occurs in the cerebral cortex of the ____________ lobe.
3. The globe of the eye is constructed of three layers: an outer protective housing (____________), an intermediate light proof shroud (____________), and a layer of photoreceptors (____________).
4. The refractive media of the eye include the ____________, aqueous humor, ____________, and ____________ ____________.
5. The color of the iris is dependent upon ____________ as well as number and arrangement of ____________ ____________.
6. The mechanism by which the eye adjusts to vision is called ____________.
7. The bony structures of the middle ear convert ____________ energy into ____________ energy.
8. The pharyngotympanic tube serves to ____________ air pressure between the middle ear cavity and the ____________.
9. The canals of the cochlea consist of the ____________ ____________, ____________ ____________, and the ____________ ____________.
10. The cranial nerve associated with vestibular sensations is the ____________ cranial nerve.

UPON REFLECTION

You have just completed a study of the most complicated region of the body. Your objective upon completing this unit should be to try to understand the regional breakdown based on structure within the head and neck, e.g., cranial region housing the brain, regions of the face and of the neck. Then consider the route of air passing through the nasal cavity and pharynx into the larynx and the various structures it would contact and how they affect the air. Next think about the effect of the various structures in the oral cavity on a piece of meat introduced into the mouth. Mentally organize the CNS into sensory and motor components and review how these functional systems are manifested in your daily activities and how they are influenced by higher centers in the brain.

REFERENCES

1. Anson, B. J. (ed.): *Morris's Human Anatomy,* 12th ed., McGraw Hill Book Co., New York, 1966.
2. Grant, J. C. B., and J. V. Basmajian: *A Method of Anatomy,* 7th ed. Williams & Wilkins Co., Baltimore, 1965. (Excellent for concepts of head and neck.)
3. Noback, C. R., and R. Demarest: *The Human Nervous System,* McGraw-Hill Book Co., New York, 1967. (Good illustrations. Text is difficult.)
4. Goss, C. M.: *Gray's Anatomy of the Human Body,* 2d ed, Lea and Febiger, Philadelphia, 1966. (Excellent three-dimensional illustrations of the CNS.)
5. Peele, T. L.: *The Neuroanatomic Basis for Clinical Neurology,* 2nd ed., McGraw-Hill Book Co., New York, 1961. (A treatise.)
6. Magoun, H. W.: *The Waking Brain,* 2d ed., Charles C. Thomas, Springfield, Ill., 1963. (On the reticular system. Sets forth some digestible concepts.)
7. *Scientific American* offprints:
   - 11 What Is Memory (Gerard, 1953).
   - 13 The Great Ravelled Knot (Gray, 1948).
   - 38 The Cerebellum (Snider, 1958).
   - 66 The Reticular Formation (French, 1957).
   - 72 Growth of Nerve Circuits (Sperry, 1959).
   - 73 Electrical Activity of the Brain (Walter, 1954).
   - 129 The Human Thermostat (Benzinger, 1961).
   - 134 Satellite Cells in the Nervous System (Hyden, 1961).
   - 174 The Great Cerebral Commissure (Sperry, 1964).
   - 1077 Memory and Protein Synthesis (Agranoff, 1967).

clinical considerations

PILONIDAL SINUS

During fetal development, the length of the body begins to grow at a faster rate (or for a longer time) than the vertebral column or spinal cord. In the embryo, the vertebral column and spinal cord occupied the full length of the posterior aspect of the body. In later fetal development with accelerated body elongation the column appears to have been pulled up so that the lower tip of the spinal cord no longer attaches to the skin. In the newborn child, a pit or fovea often marks the site where the spinal cord once attached to the skin. (Remember that the brain and spinal cord formed from an invagination of the future skin in the embryo.)

When this fovea becomes infected, as it frequently does, the condition is known as a pilonidal sinus (L. *pilus,* hair; *nidus,* nest). If a pathway between this sinus and the lower spinal cord or dural sac exists, bacteria have access to the CNS and abcess and inflammation of the meninges (meningitis) can follow. In such cases, the pathway between fovea and CNS must be closed surgically.

SPINA BIFIDA ("split spine")

This is a developmental defect in which the arch of one or more vertebrae fails to form properly and the spinal cord is left unprotected posteriorly (Fig. 6-96). The underlying dura

Fig. 6-96 Spina bifida. A. Normal lumbar vertebra encasing a section of spinal cord. B. Lumbar vertebra with underdeveloped neural arch—an uncomplicated spina bifida.

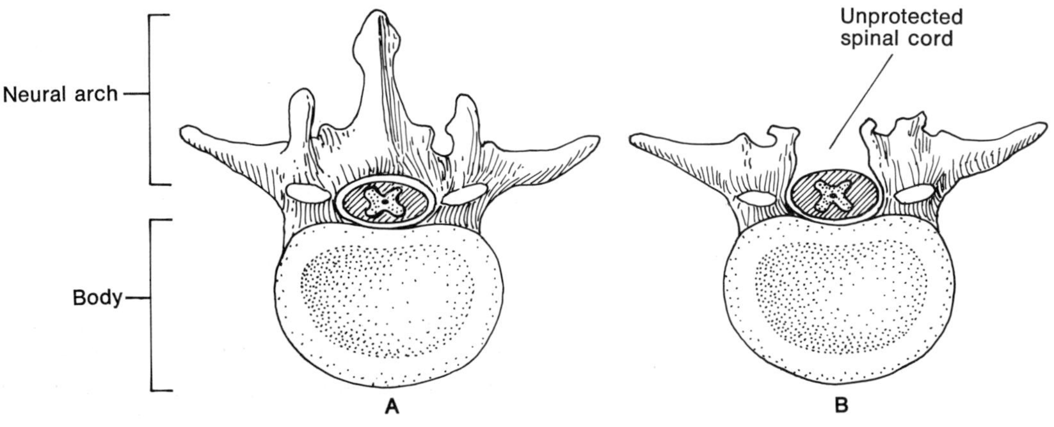

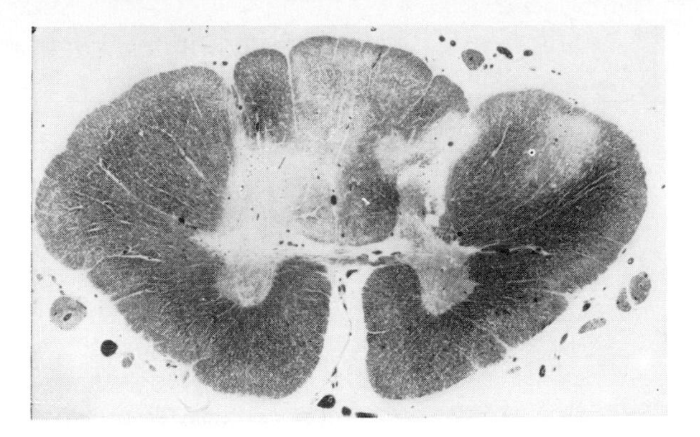

Fig. 6-97 Multiple sclerosis. Areas of demyelinization on the spinal cord appear bleached. (Courtesy of R. P. Morehead, *Human Pathology*, McGraw-Hill, New York, 1965.)

mater and arachnoid may balloon through the defect, often creating a rather large cystlike structure. Such a sac, devoid of nervous tissue, is termed a meningocele (Gk., *meninges*, membrane; *cele*, hernia or rupture). If the herniated sac contains nerve roots or part of the spinal cord it is termed a meningomyelocele. If the cord itself occupies the sac, it is a myelocele. In the latter two cases, weakness (paresis) or paralysis may accompany the defect.

MULTIPLE SCLEROSIS

Multiple sclerosis (sclerosis, a condition of hardening) is a progressive disease of unknown cause characterized by a proliferation of neuroglia and a loss of myelin in the white matter of the brain and spinal cord (Fig. 6-97). These multiple lesions can produce a variety of symptoms including sensory loss and motor defects, e.g., loss of vision, position sense, muscular weakness, or paralysis of the limbs. The disease progresses rapidly in a few cases but more often the course is marked by remissions and relapses.

PARKINSON'S DISEASE

A disease of the nervous system, paralysis agitans, or Parkinson's disease, has no effect on the sensory systems or the intellect. The cause of the disease is obscure but the regions of the brain affected (as indicated by reduction in number of neurons, increase in number of glial cells, and increase in concentrations of pigment in nerve cells) are the lenticular and caudate nuclei and related areas. Clinically the disease is characterized by tremor of the limbs, slow and stiff movements, fixed facial expressions, and a "bent-over" posture. The tremor is most often in the hands but is seen elsewhere. Interestingly, when a Parkinson's patient begins voluntary movements the tremors often subside during that moment. Complete relaxation also reduces tremor.

TONSILLITIS

It is a well known fact among streptococcal bacteria that masses of lymphatic tissue in the oral cavity are great convention centers. This is possibly so because of their low resistance capability and their proximity to the outdoors; whatever the reason, the tonsils are subject to hypertrophy and infection, particularly in children. Normally tucked away between the pillars of the passageway connecting oral cavity and pharynx, the palatine tonsils when inflamed and infected often droop down onto the base of the tongue and set off the gag reflex. Infections of the pharyngeal and tubal tonsils often spread to the pharyngotympanic tube and then to the middle ear cavity. Frequently, inflammation of the middle ear with pus drainage and considerable pain follows. In point of fact, tonsillectomy will often eliminate chronic middle ear infections of the young.

Removal of the palatine tonsils can be managed easily with rare complications. This is so because the tonsil is attached to its bed (fossa) by only a thin pedicle of vascular tissue, and covered by a loose fibrous capsule and oral mucosa. In one common surgical procedure, an incision is made in the anterior pillar, and the tonsil is shelled out of its fossa (Fig. 6-98). When sufficient tonsil is freed from its bed, a snare is placed around it and it is guillotined from its vascular attachment. Bleeding is controlled by impaction of cotton sponges.

In children, bleeding generally ceases without complication. This is often so because of the plasticity of young vessels, which easily constrict when cut. In adults, complications of tonsillectomy are more frequent. This is so because the tonsils have probably been inflamed and infected on and off for several years. As a consequence, adhesions of tonsillar tissue often develop with the lateral pharyngeal wall. The tonsil, under these circumstances, does not "shell out" as easily and one or more respectable arteries/veins are cut in the process. These may not contract as well as those of the young vessels and bleeding may persist. Thus, following the

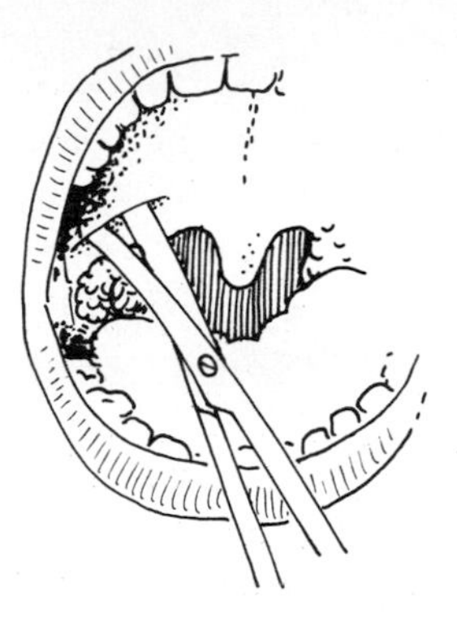

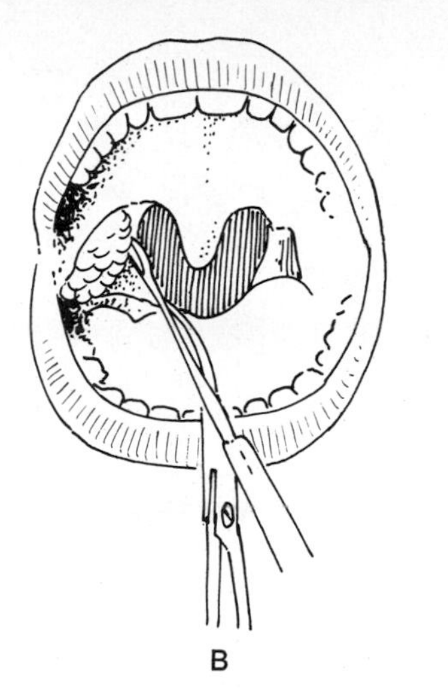

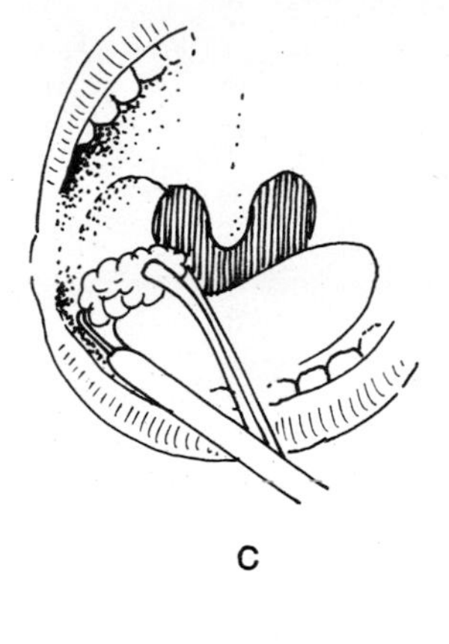

Fig. 6-98 Tonsillectomy. A. Incision is made into anterior pillar of a tonsil (palatoglossal arch). B. The tonsil is separated from its fascial bed. C. The tonsil is relieved of its vascular pedicle by a snare. (Courtesy of Boies, Hilger, and Priest, *Fundamentals of Otolaryngology*, W. B. Saunders Co., Philadelphia, 1964.)

operative procedure, one may have to stay under close observation for 24 hours or so. If bleeding continues absorbable stitches are taken and the problem is solved.

unit 7 the body wall

The body wall consists of the skin, fascia, and musculoskeletal structures serving as boundaries to the trunk. It includes

1. Vertebral column.
2. Ribs and sternum.
3. Intercostal musculature.
4. The diaphragm.
5. Muscles of the abdominal wall.
6. Muscles of the deep back.

Interrelations among these structures permit us to demonstrate the following regions or cavities (Fig. 7-1):

Thorax: bounded by the sternum and costal cartilages anteriorly, the ribs antero- and posterolaterally, the vertebral column posteriorly, and the diaphragm below. The intercostal (between-the-ribs) musculature and muscles of the deep back contribute to these bony boundaries. The space created within constitutes the thoracic cavity and incorporates the lungs, heart, and related structures. The thoracic cavity is discussed in Unit 8.

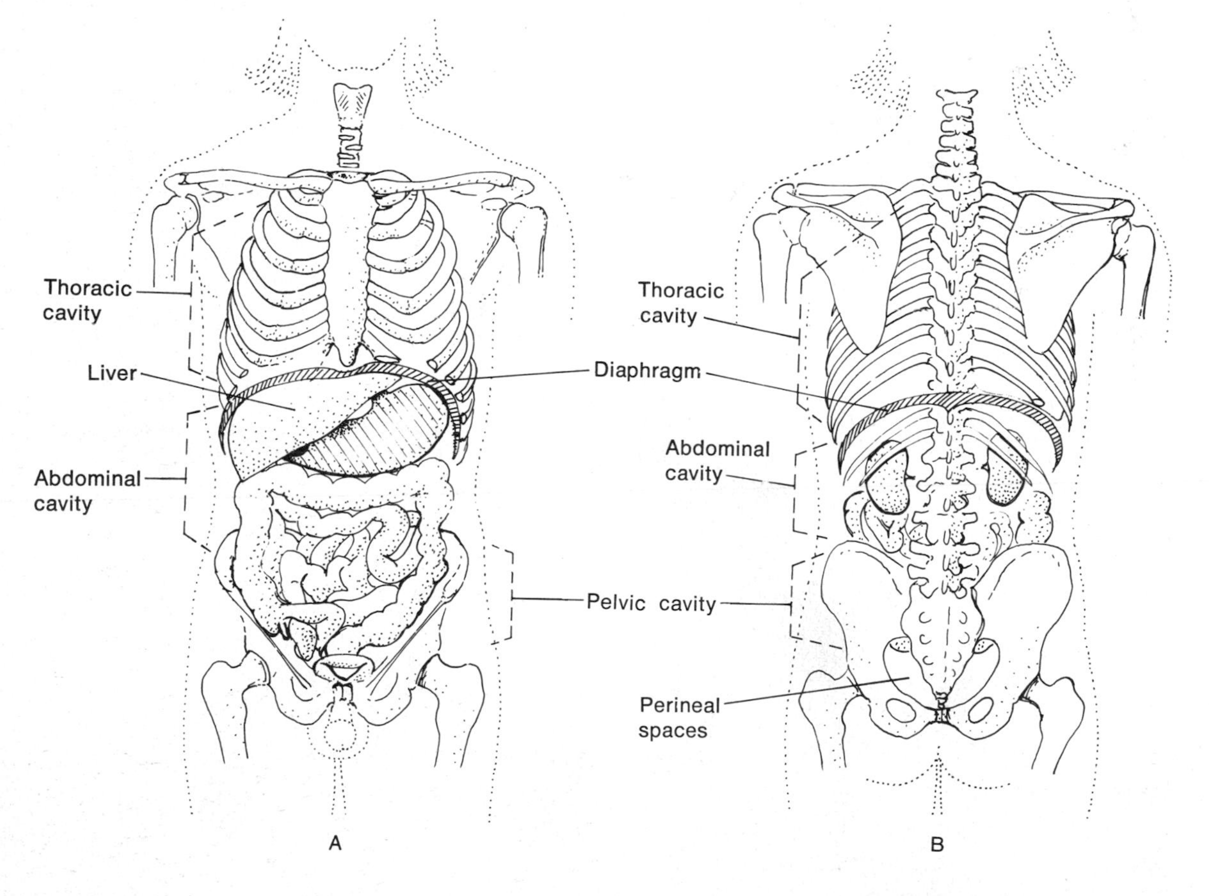

Fig. 7-1 The great cavities of the body and their boundaries. A. Anterior view. B. Posterior view.

Abdomen: bounded by the abdominal musculature anteriorly and laterally, by the thoracic diaphragm and lower ribs above, by the posterior abdominal and deep back musculature posteriorly, and continuous with the pelvis below. The abdominal cavity created within contains the alimentary canal and related viscera, the spleen, the kidneys, and assorted endocrine glands, ducts, vessels, and nerves—all to be topics of interest in Unit 9.

The anterior abdominal wall contributes to the boundaries of the **pelvis,** a cavity continuous above with the abdominal cavity, surrounded largely by bone and containing certain structures of the digestive, urinary, and reproductive systems. The pelvis is tackled in Unit 10.

the vertebral column

GENERAL

The column of blocklike bones in the back constitutes the basic pillar which our body depends upon for support. If you train yourself to walk and sit truly erect, and do not apply unduly stressful, abnormal forces to your back, you will be rewarded with the ability to make painless postural adjustments. If you carelessly carry your weight, slouch, and use your backbone as a crane for lifting heavy objects, you are destined for *big trouble.*

The vertebral column articulates with the skull above and terminates below in a series of tail bones. Just cephalic to the tail bones, a wedge-shaped mass of five fused vertebrae, the sacrum, forms a slightly movable articulation with the hip bones. In the thorax, the column gives attachment to the ribs. You will perhaps remember that the vertebral column has no direct articulation with the pectoral girdle. The vertebral column is moored securely throughout by the deep muscles of the back.

Reference to Fig. 7-2 will show that the column consists of a collection of some 26 separate bones. Note that the bodies of the individual vertebrae are not square; they differ in height between front and back sides, resulting in a series of curves along the length of the column. The group of vertebrae associated with a particular curve look reasonably alike and differ intelligibly from groups of vertebrae in other curves. Thus a basis is provided for grouping the vertebrae. In general, each vertebra is progressively larger than its fellow above, culminating in the sacrum, suggesting that more and more weight is borne by the column caudally. The bodies of the vertebrae are separated from their adjacent fellows by a fibrocartilaginous **intervertebral disc.** The presence of intervertebral discs implies a certain flexibility in the column as a whole. Happily, stability is not sacrificed for this flexibility because, in part, one curve compensates somewhat for the other.

The length of the column in normal adults is amazingly constant, "a constancy upon which the ready-made clothing industry depends."* The height of an individual is more

* From Lockhart, Hamilton, and Fyfe, p. 63.

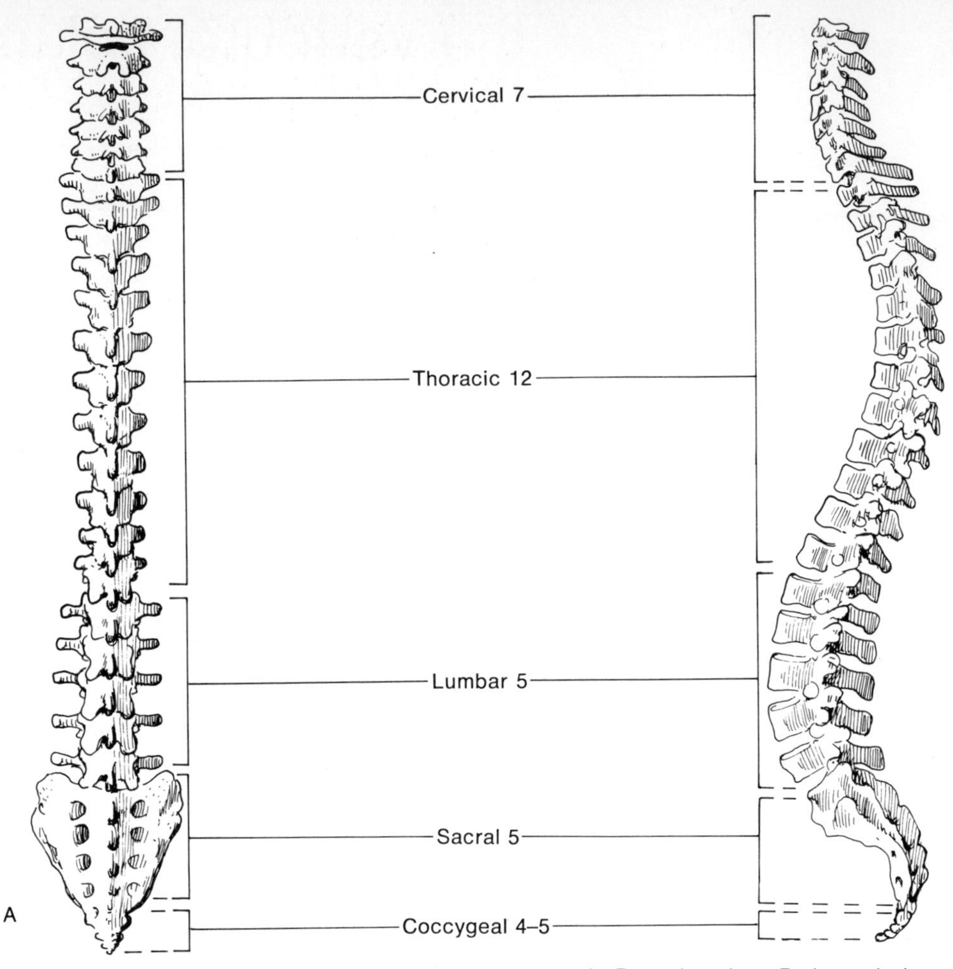

Fig. 7-2 The vertebral column. A. Posterior view. B. Lateral view.

function of the length of the lower limbs, although the column of females is 10 to 15 percent shorter than that of males.

CURVATURES OF THE COLUMN

An infant *in utero* or newly born has a vertebral column with two curvatures, the concave surfaces of both facing anteriorly. The upper curve involves the thoracic region of the column, while the lower is formed in the sacral region. Since one is born with these curvatures, they are said to be **primary curvatures** (Fig. 7-3A).

By about 3 months of age, baby has learned to hold its head up without support—a significant neuromuscular accomplishment. As a consequence the vertebrae of the neck (cervical region) compensatorily begin to curve backward to sustain the weight of the skull. The concavity of this new curve faces posteriorly. Because it is created secondary to the baby's holding up its head this development is called a **secondary curvature** (Fig. 7-3B).

By about 1 year of age, baby has left creeping and crawling to younger folk and has added walking to its list of accomplishments. During the preliminary stages of the feat, baby leans back to prevent falling forward. In time, the lumbar vertebrae compensate for this new stress by forming a new curve, with concavity facing posteriorly. Because it is created as a consequence of walking, this lumbar development is said to be a **secondary curvature** (Fig. 7-3C).

These then are the curvatures seen in the vertebral column: two primary and two secondary, the curve of each region of the column facing oppositely the one above (Fig. 7-3D), except that the coccyx does not curve reciprocally with the sacrum.

Under certain postural conditions and in certain disease processes, the vertebral curvatures can become exaggerated, or a new curvature can be found (see page 474).

Fig. 7-3 Development of the vertebral curvatures.

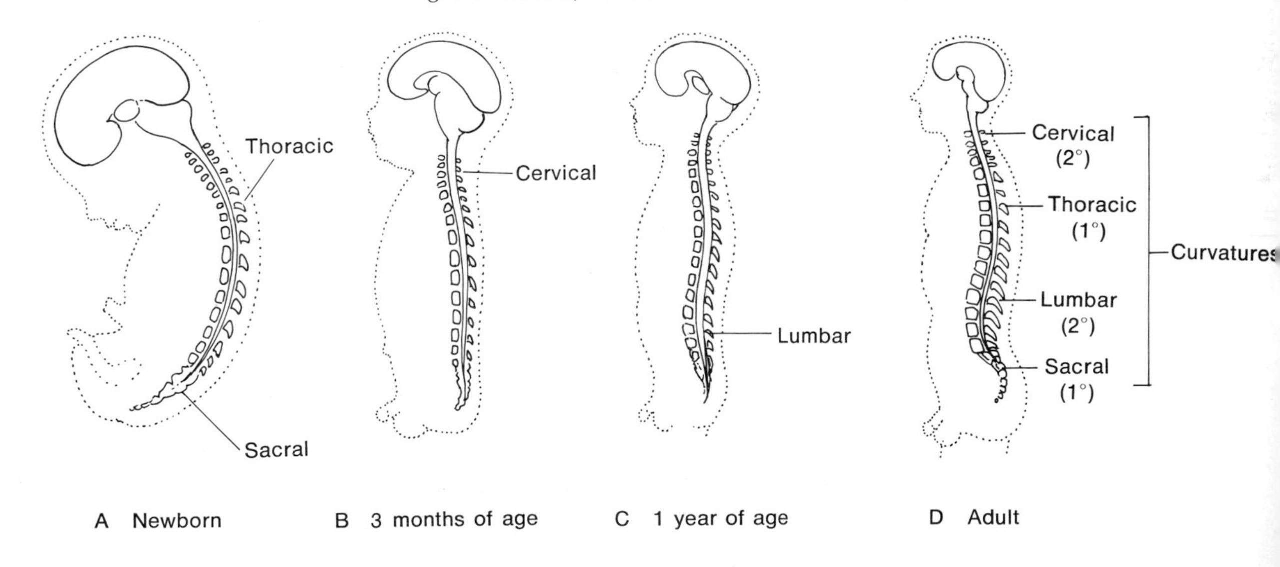

ANATOMY OF A VERTEBRA

From above a classic, typical vertebra looks like a pagoda. A typical vertebra (Fig. 7-4) basically consists of a **body** and **vertebral arch,** the latter so-called because it arches over the spinal cord. The vertebral arch has two sides—the **pedicles**—and a roof consisting of two plates, the **laminae.** Projecting from the junction of the pedicle and lamina on each side is a process, the **transverse process.** These are simply "muscle hangers"—as many as 16 different muscles may insert on or arise from one lumbar transverse process. In the thoracic region, each transverse process gives attachment to a rib. Also projecting from the junction of pedicle and lamina on each side but directed inferiorly and superiorly are **articular processes,** which articulate with those of adjacent

Fig. 7-4 Anatomy of a vertebra. A. Classical vertebra. B. Chinese pagoda on a rock.

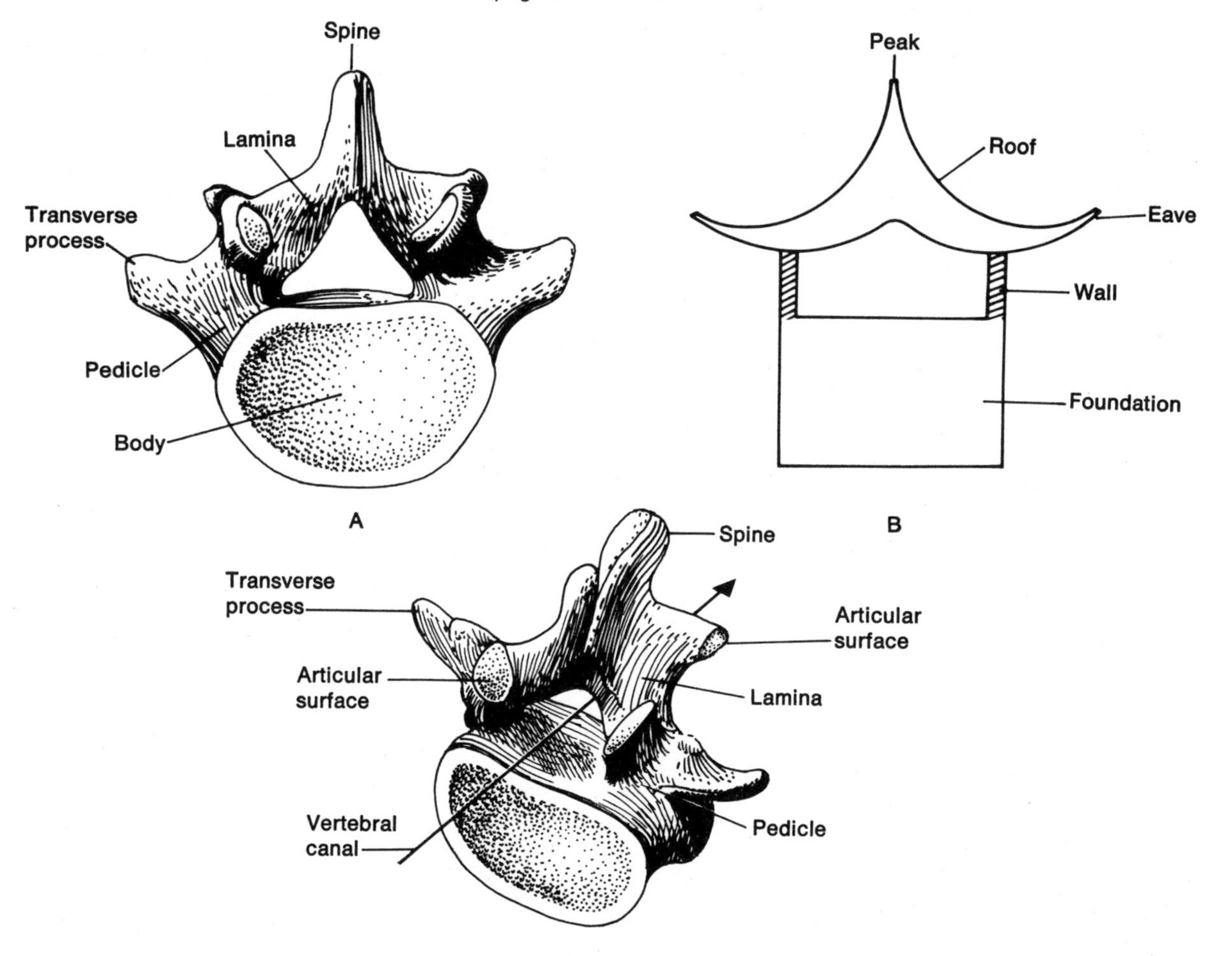

vertebrae. These are synovial joints. A **spinous process** or spine projects posteriorly from the fused lamina. These can be seen in the slender individual and felt on yourself at the back of the lower neck (Fig. 7-5). These too are grand "muscle hangers."

When the vertebrae are in articulation, two holes become visible (Fig. 7-7):

1. The **vertebral canal,** a passageway for the spinal cord created by the vertebral arches in series.
2. **Intervertebral foramina** along both sides for the passage of spinal nerves exiting the cord for the periphery. The foramina are the spaces between the opposed pedicles.

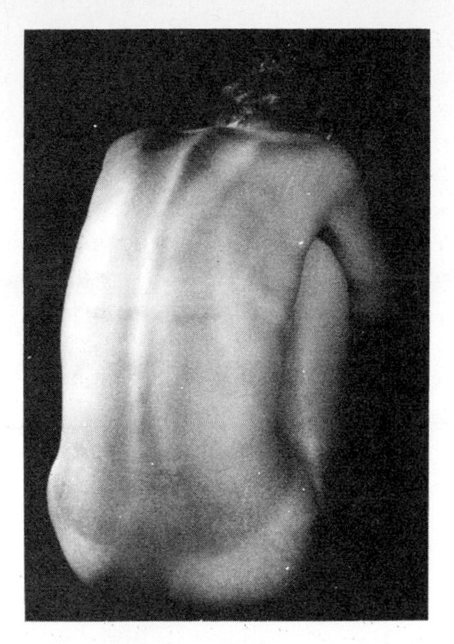

Fig. 7-5 Surface view of the vertebral spines with the back flexed. The majority of the spines seen belong to thoracic vertebrae.

Fig. 7-6 A demonstration of the flexibility of the vertebral column. The x-ray shows the column in extreme extension. The space between the vertebrae is occupied by intervertebral discs (not dense enough to be picked up by the x-ray).

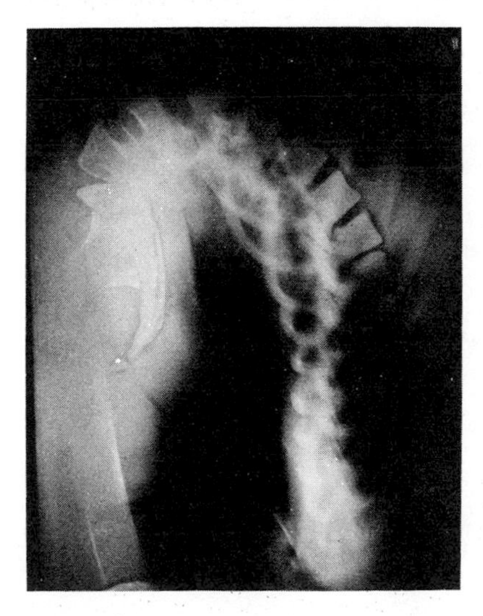

INTERVERTEBRAL DISC

An intervertebral disc may be found between each pair of vertebrae. Discs consist of concentric layers of fibrocartilaginous tissue, termed the **annulus fibrosus,** enclosing a soft, jellylike core, the **nucleus pulposus** (Fig. 7-7). They function as shock absorbers and give flexibility to the column, besides acting as spacers or shims so that one vertebra can bend over another without touching its processes (Fig. 7-6). In the young person discs are shaped not unlike marshmallows, but they lose moisture and elasticity and become thinner with age, so that an elderly person actually becomes shorter.

These gelatinous cores are the nemesis of people with "slipped discs." When the column is flexed, the portion of the disc on the side being flexed is compressed more than the opposite side (Fig. 7-7). The nucleus pulposus comes under great pressure (equal to 500 lb per square inch or more) and can herniate through the concentric layers of the annulus and enter the vertebral canal, or it can press against a spinal nerve entering the intervertebral foramen. The result often is a stabbing pain in the limb supplied by that nerve, often followed by weakness, loss of reflexes, and paralysis there. If natural healing doesn't occur, the situation can sometimes be remedied by the orthopedic surgeon, who must enter the back and remove the damaged disc. Herniated discs are often generated in people who insist on using their backs rather than their limbs for lifting (Fig. 7-8).

REGIONAL VERTEBRAE

On the basis of shape, size, and other characteristics, the vertebrae of the column may be divided into five regions (Fig. 7-2):

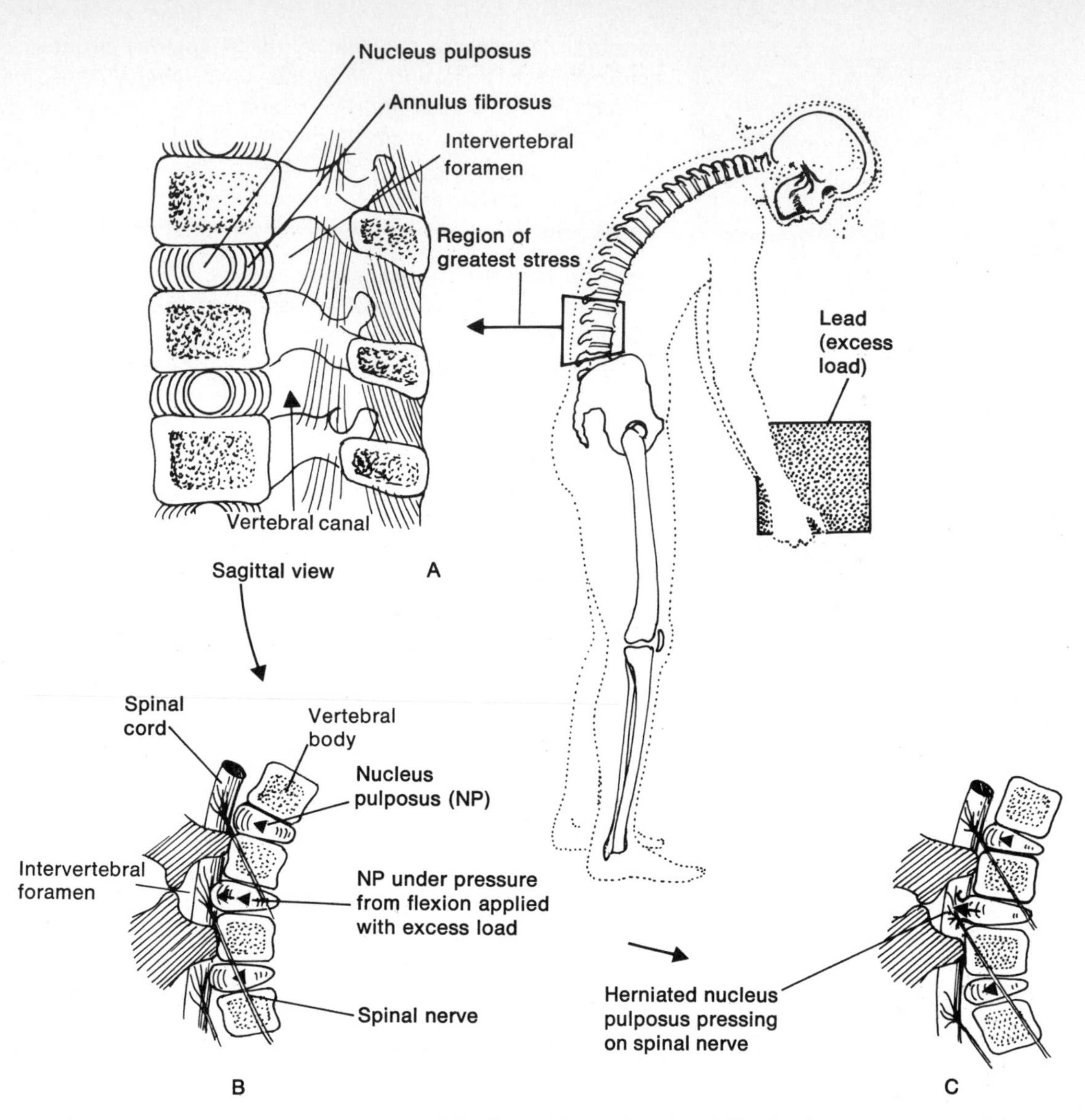

Fig. 7-7 Preamble to a herniated disc (or the wrong way to pick up a heavy object). A. Anatomy of an intervertebral disc in relation to adjacent vertebrae, vertebral canal, and intervertebral foramen. B. Displacement of nucleus pulposus when disc is under side pressure. C. Rupture of the disc (nucleus pulposus) into intervertebral foramen.

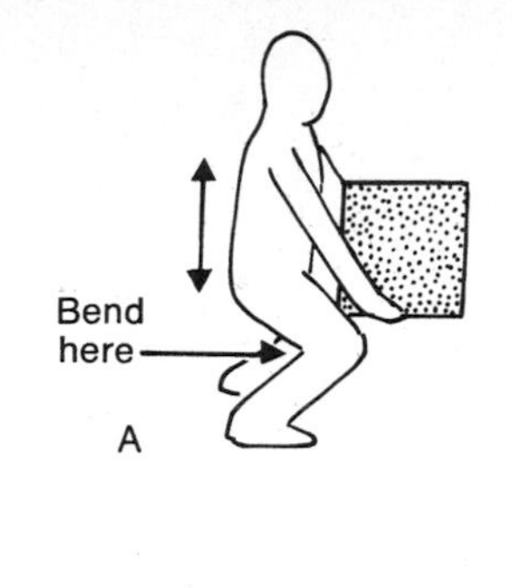

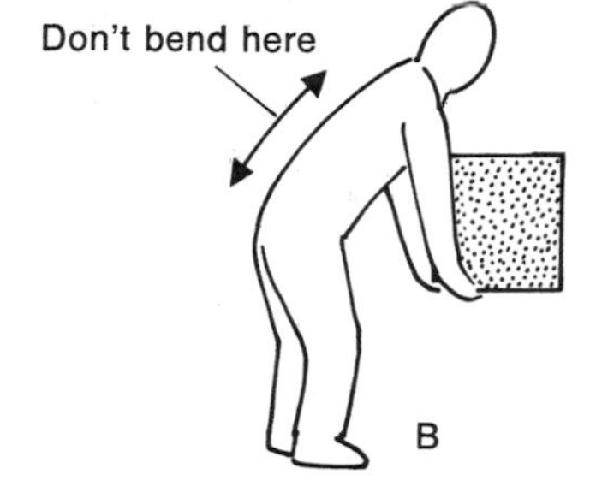

Fig. 7-8 The right (A) and wrong (B) ways of picking up an object.

1. **Cervical** vertebrae, numbering 7.
2. **Thoracic** vertebrae, numbering 12.
3. **Lumbar** vertebrae, numbering 5.
4. **Sacral** vertebrae, numbering 5, but fused into one mass **(sacrum).**
5. **Coccygeal** vertebrae, usually numbering 4, and collectively called the **coccyx.**

Cervical vertebrae

The seven cervical vertebrae are the vertebrae of the neck. The first cervical vertebra (C1) supports the skull much as the god of Greek myth, Atlas, carried the globe on his shoulders, hence its name: **atlas.** Its structure differs from the other cervical vertebrae in that it has no body; in its place there is an oval ring of bone. Although the skull can move anteroposteriorly (flex/extend) on the atlas as in nodding, it cannot rotate or pivot well on it.

The second cervical vertebrae (C2) projects superiorly a tooth-shaped (odontoid) process about which the oval ring of the atlas pivots in side-to-side, "no" movements. Thus C2 performs the task of an axle and its name is **axis** (Fig. 7-9).

Most of the seven cervical vertebrae (Fig. 7-9) have qualities which characterize them as a group:

1. Small bodies, which progressively increase in mass from C2 to C7.
2. Foramina in the transverse processes of C1 to C6 for the passage of the vertebral artery (and vein) to the brain.
3. Articulating surfaces (facets) on C1 to C7 that face superiorly and inferiorly along a horizontal plane, suggesting that rotation of the neck is mechanically plausible—try it and see.
4. Split or bifid spines on C2 to C6, which become progressively longer caudally. In fact, the spine of C7 can be felt on yourself at the base of the neck. Why can't you feel the spines of C3, C4, or C5?* Why not the spine of C1?†

Thoracic vertebrae

The 12 thoracic vertebrae are the vertebrae of the thorax and, therefore, give attachment to 12 ribs on each side. This fact suggests certain characteristic features of these vertebrae, i.e., facets at the sides of the body and at the tip of the transverse process for articulation with the ribs.

* Because the cervical curvature is concave posteriorly, thus vertebrae C1 to C5 are deep in the neck; C6 and C7 lie more to the posterior.

† C1 has no spine.

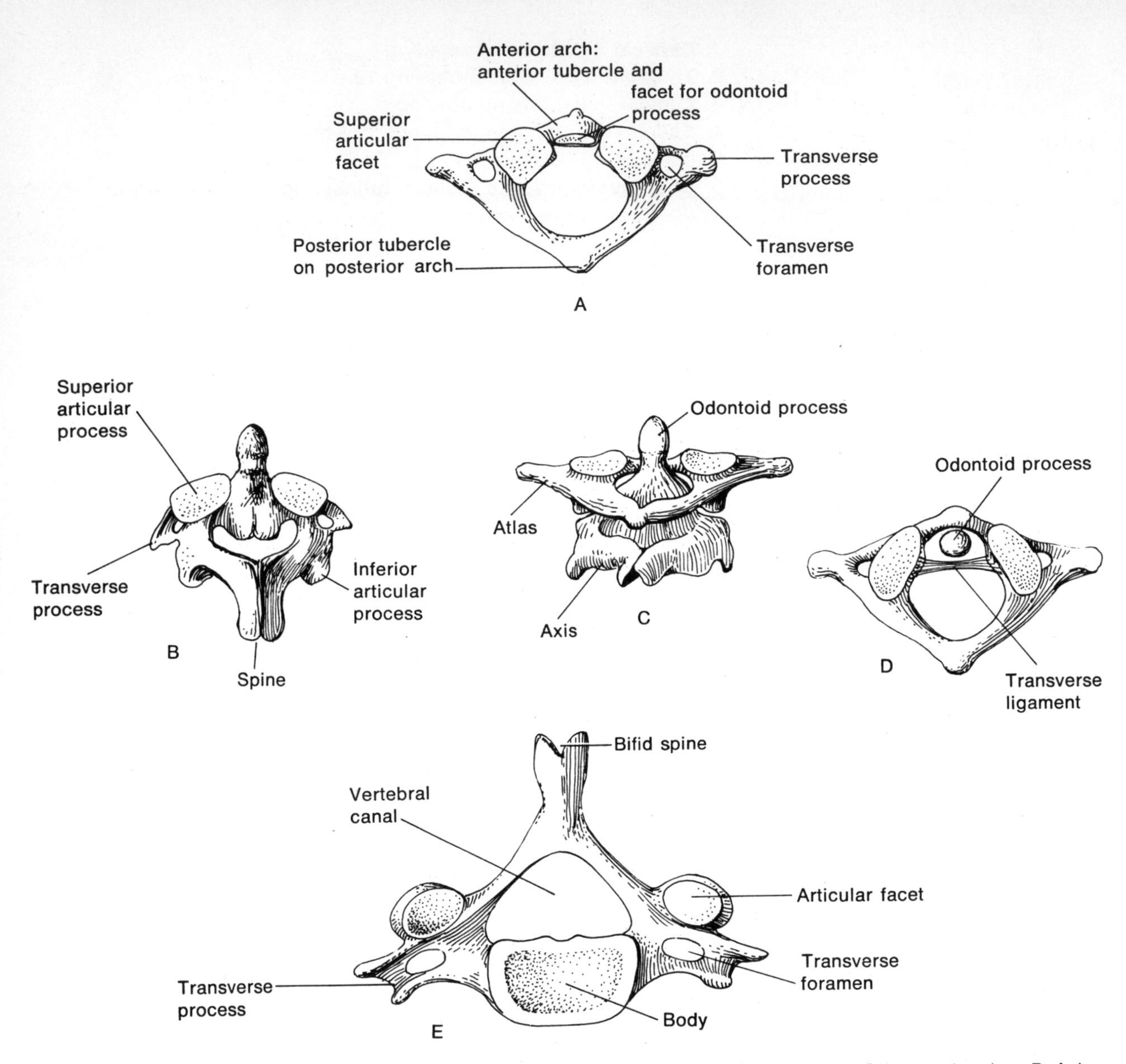

Fig. 7-9 The cervical vertebrae. A. Atlas (C1), superior view. B. Axis (C2), posterior view. C. Atlantoaxial articulation, posterior view. D. Pivot joint between odontoid process of axis and anterior arch of atlas, superior view. E. Typical cervical vertebra, superior view.

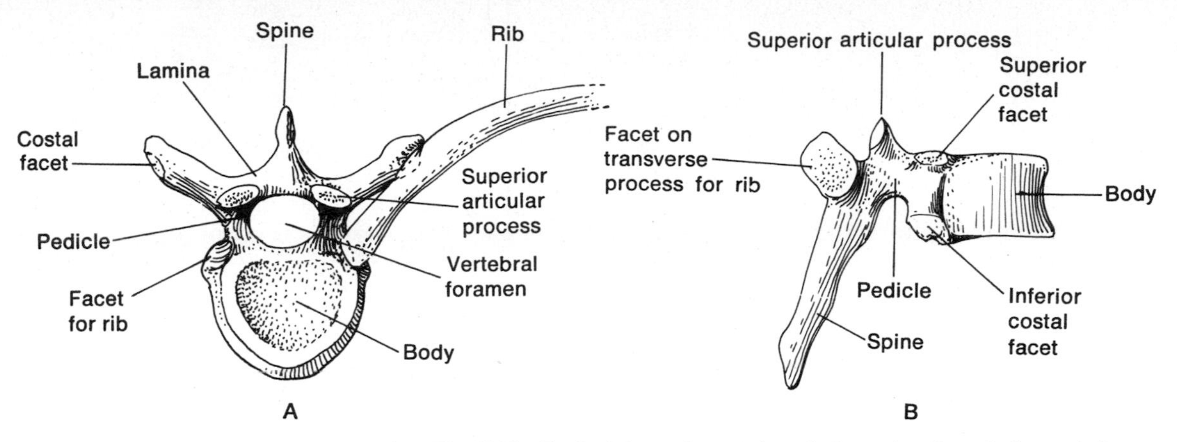

Fig. 7-10 Typical thoracic vertebra. A. Superior view. B. Lateral view.

Other characteristics of thoracic vertebrae (Fig. 7-10) are:

1. **Heart-shaped bodies** (thought association: The heart is in the thorax).
2. Long, bladelike **spines** extending down over the vertebrae below.
3. Articular **facets** lying along a frontal plane and facing in an anteroposterior direction.

Again, rotation is possible, but lateral bending of thoracic vertebrae is impossible because the ribs get in the way.

Lumbar vertebrae

The five lumbar vertebrae support the thorax above and, in effect, function as a pillar, resting solidly on the sacrum below. As one might expect, the lumbar vertebrae (Fig. 7-11) are

Fig. 7-11 Typical lumbar vertebra. A. Superior view. B. Lateral view.

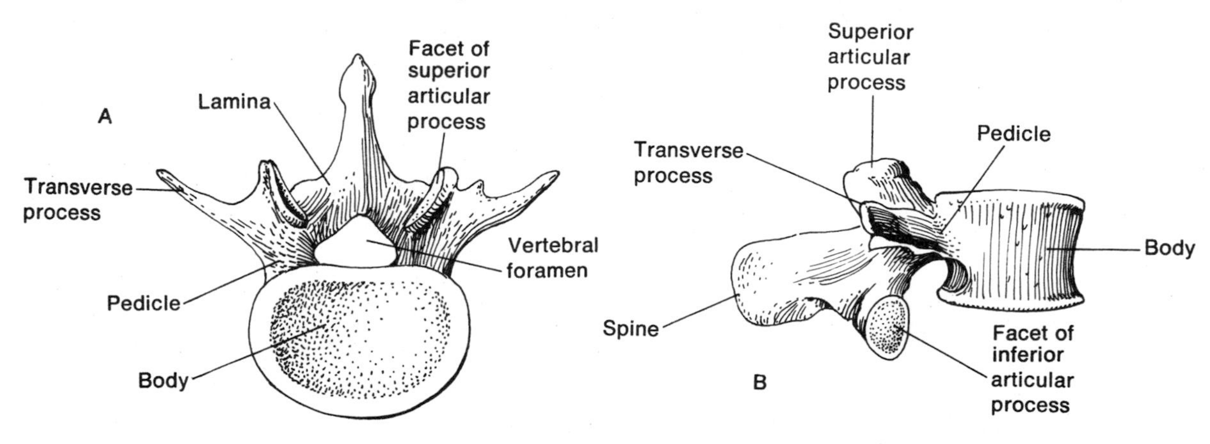

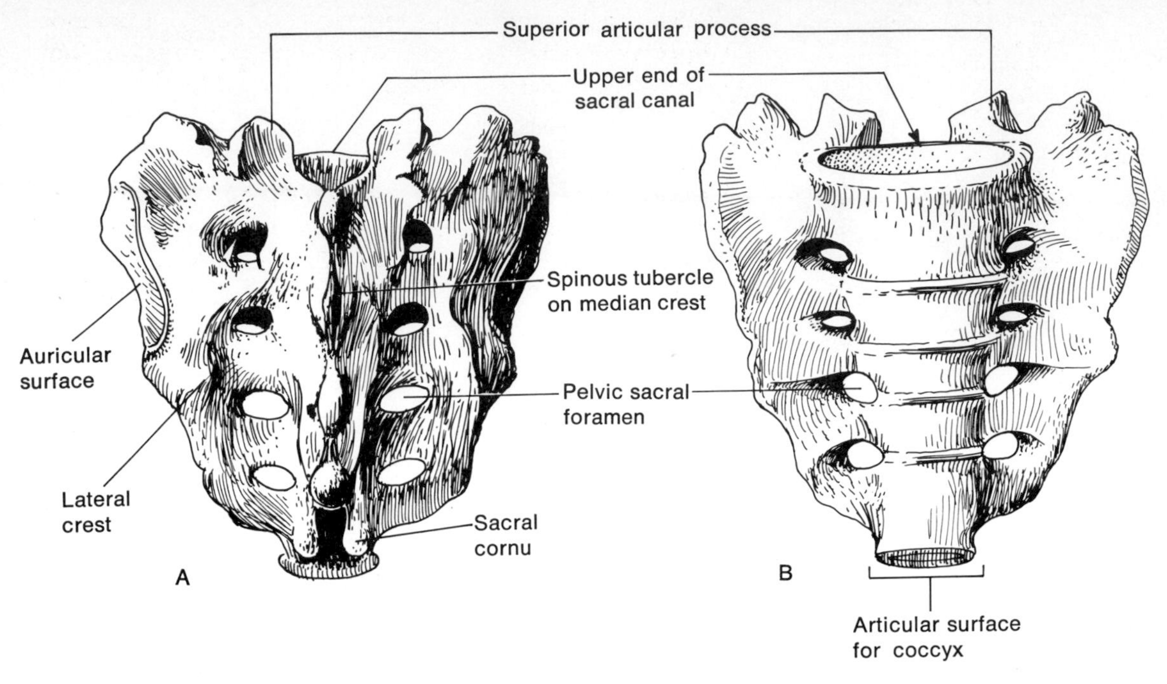

Fig. 7-12 Sacrum. A. Posterior view. B. Anterior view.

stocky and massive, with large processes for muscle attachment. Other characteristic features:

1. The bodies are broad and oval-shaped.
2. The spines are short, quadrilateral stumps.
3. The transverse processes are broad platforms.
4. The articular processes *face one another* on a *sagittal* plane, suggesting that rotation of the lumbar vertebrae is virtually impossible.

Sacral vertebrae

The five sacral vertebrae are fused to form one structure, the **sacrum** (Fig. 7-12). The sacrum is wedged in between the hip bones and articulates with them via auricular (earlike) surfaces. Tightly bound in ligaments to the hip bones, the sacrum bears the weight of the torso. It can be palpated easily on yourself, as the convex portion of the bone is subcutaneous just below the iliac crests and just above the gluteal crease.

The dorsal surface of the sacrum must look to the passing ant as the Rocky Mountains look to us, many of the "peaks" representing spinous processes and articular processes fused together. The intervertebral foramina are open on anterior and posterior sides and transmit anterior and posterior rami of spinal nerves, respectively.

Coccyx

Man's rudimentary tail, the four (variable) coccygeal vertebrae, form the **coccyx,** the upper component of which articulates with the fifth sacral vertebra (Fig. 7-13). The coccygeal vertebrae are not completely fused, allowing some movement, as during parturition.

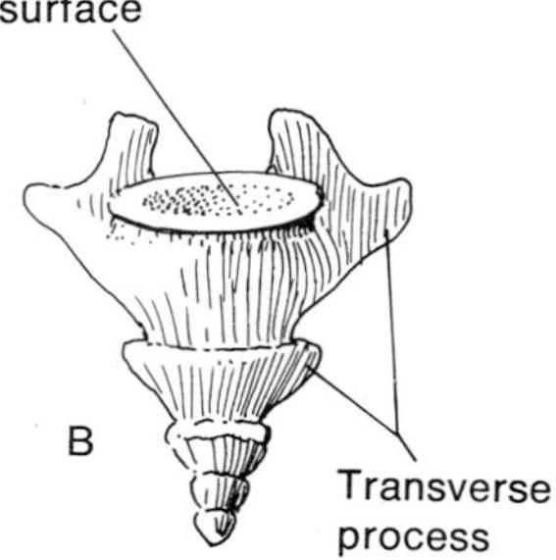

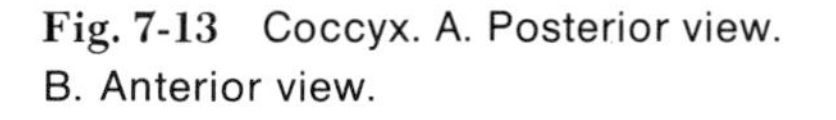

Fig. 7-13 Coccyx. A. Posterior view. B. Anterior view.

JOINTS AND LIGAMENTS OF THE VERTEBRAL COLUMN

Each pair of vertebrae are united by three joints (Fig. 7-14):

1. The partly movable joint between the vertebral body and the intervertebral disc. The joint is reinforced by a pair of longitudinal ligaments on the anterior and posterior faces of the vertebral body and disc.
2. The two joints between articular processes of the arches. These are *synovial* joints of the plane or gliding variety. These joints are reinforced by ligaments joining the vertebral lamina (ligamenta flava), the spines, and the transverse processes.

Fig. 7-14 Ligaments and joints of the vertebral column. A. Anterior view of the neural arches, bodies removed. B. Sagittal view of the vertebral column.

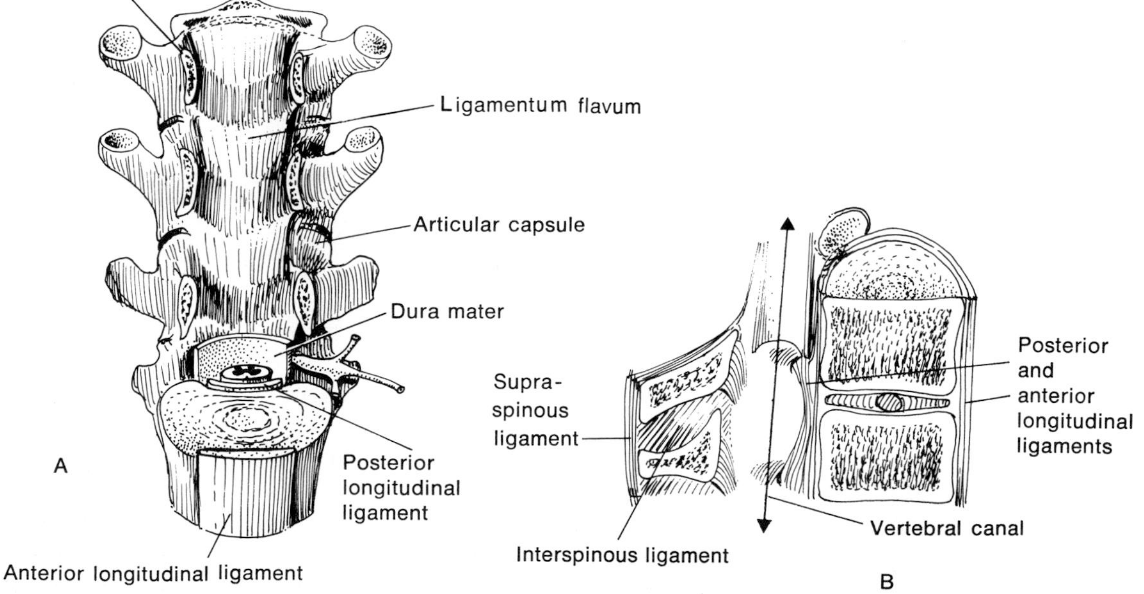

In certain regions of the column, listed below, the ligamentous arrangement is so specialized as to draw individual attention:

1. Atlanto-occipital joint.
2. Atlantoaxial joint.
3. Cervical supraspinous ligament.
4. Sacroiliac and iliolumbar joints.

The first two joints incorporate a number of ligaments which stabilize the pivoting process of the axis and yet allow full movement of the head (Fig. 7-15).

The curvature of the cervical vertebrae creates a fossa which is filled with deep muscles of the neck. Intervening in the midline is a saillike ligament extending from the occiput to C7, and outwardly from spine to skin (Fig. 7-16). A site of numerous muscular insertions, this **ligamentum nuchae** is largely elastic tissue in certain four-footed beasts and serves to sustain the weight of the massive skull. In people, it counteracts the weight of the head during flexion.

The ligaments binding the sacrum to the hips are all-important in maintaining the security of this region. The weight of the body (when standing or sitting) centers on the sacrum, which transmits the forces to the ilia of the hips via the sacroiliac joint. The **sacroiliac** and **iliolumbar ligaments** strengthen these bony connections (Fig. 7-17). The ligaments can become strained if a person insists on utilizing his back as a crane arm.

Fig. 7-15 Atlanto-occipital and atlantoaxial joints, posterior view, with neural arches and spinal cord removed.

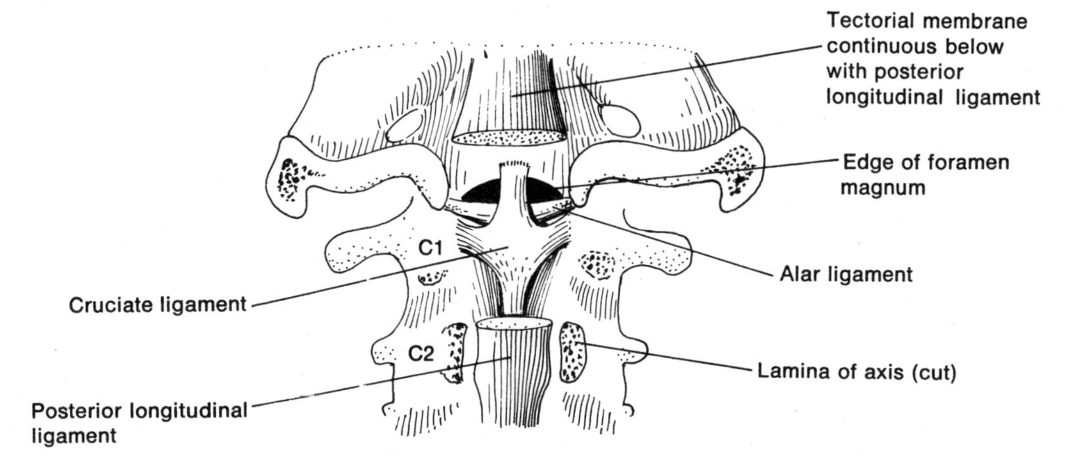

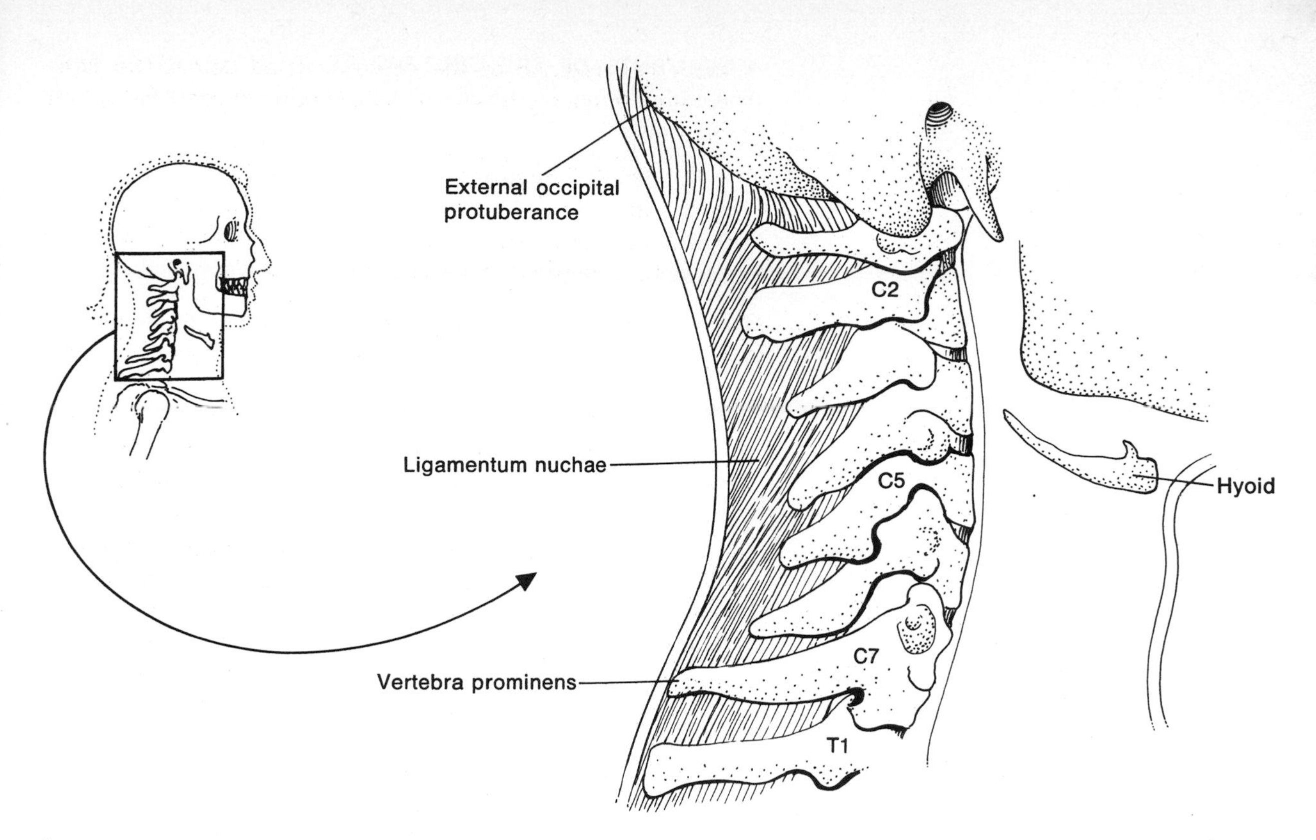

Fig. 7-16 The ligamentum nuchae.

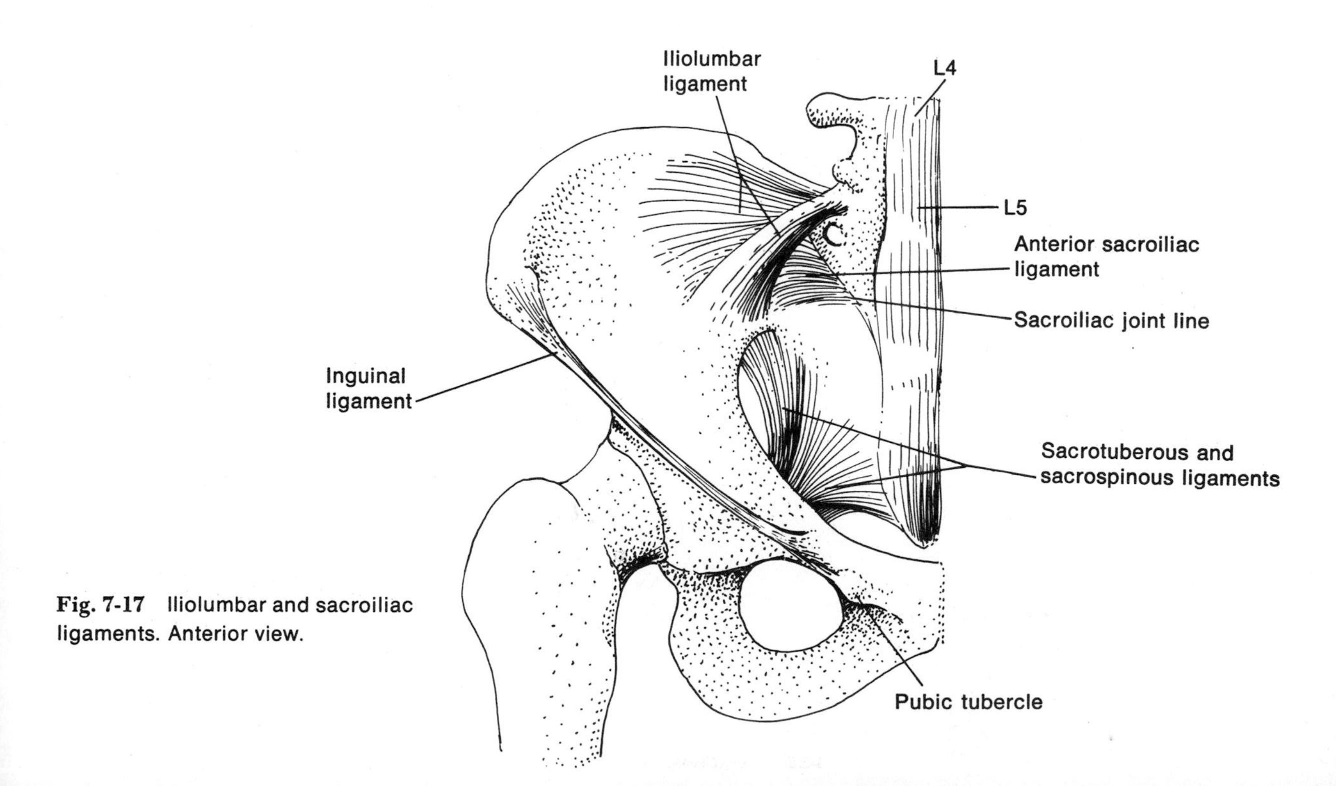

Fig. 7-17 Iliolumbar and sacroiliac ligaments. Anterior view.

MOVEMENTS OF THE VERTEBRAL COLUMN

The movements of the vertebral column are a function of three things:

1. The type of articulation between vertebral arches.
2. The thickness of intervertebral discs.
3. The functional state of related ligaments and muscles, as well as the vertebrae and discs themselves.

The movements allowed in the cervical region are flexion, extension, and a combined lateral flexion and rotation. Movement is more limited in the thoracic region, where extension is inhibited by the spinous processes and lateral flexion by the ribs. The lumbar region is as flexible as the cervical region except that the articular processes of adjacent vertebrae (Fig. 7-11) make rotation almost impossible. The sacrum, of course, is immobile and the coccyx only passively movable.

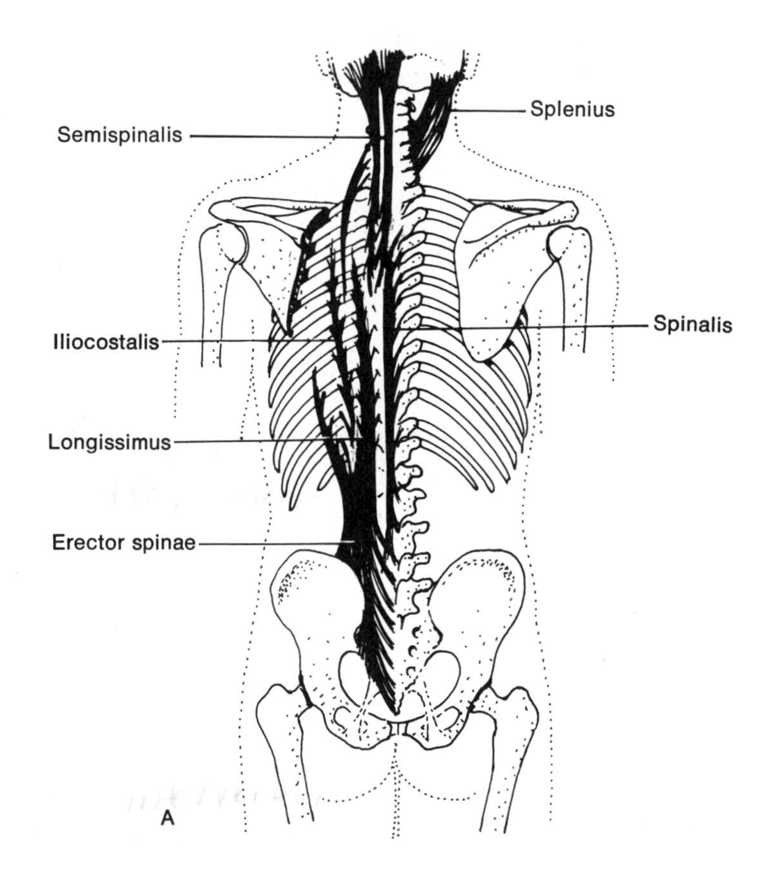

the back

The superficial regions of the back relating to the upper limb have already been covered in Unit 4. This leaves the muscles of the deep back overlying the posterior aspect of the abdomen and thorax. They can be considered in four groups:

1. Small bundles between adjacent vertebrae, e.g., interspinalis.
2. Obliquely oriented muscles spanning two or more vertebrae running from transverse processes to spines, e.g., multifidus, semispinalis.
3. The long **erector spinae** muscle group (Fig. 7-18A), which may be seen on yourself as muscular bands running parallel to the vertebral column in the lower back. These are the most significant of the deep back muscle groups.
4. A strap muscle bandaging the deep muscles in the neck: **splenius.**

These muscles—acting as groups—play an important role in maintenance of erect posture and movements of the head, hip, and trunk, including extension, lateral flexion, and rotation.

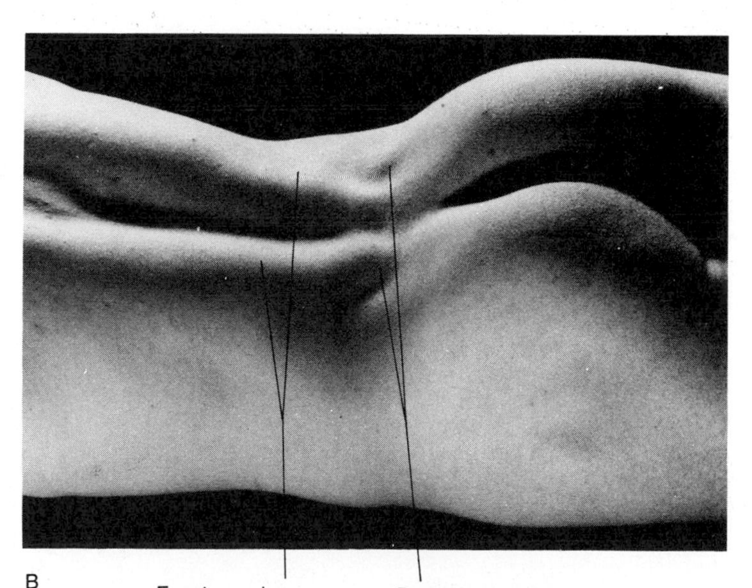

Fig. 7-18 A (opposite). Principal deep muscles of the back. B. Surface features of the lower back. C (turn page). Innervation scheme of the deep back muscles.

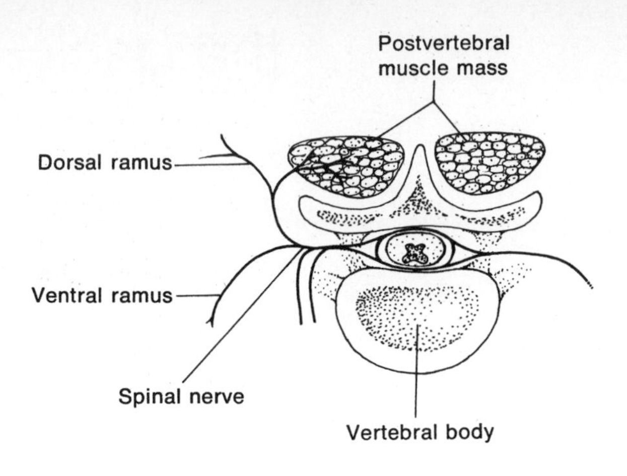

The mass of the muscles of the back (postvertebral) is considerably greater than that of the prevertebral muscles, and this is probably because they must resist the forces of gravity in maintaining the extended trunk.

The muscles of the back are innervated by the dorsal rami of spinal nerves emerging from each vertebral segment from C1 to the sacrum (Fig. 7-18C). The vascular supply is derived from neighboring arteries, i.e., the intercostal arteries (see ahead).

QUIZ

1. The thorax is bounded by the sternum and costal cartilages anteriorly, the diaphragm below, and the vertebral column posteriorly.
2. There are four curvatures in the vertebral column: two primary and two secondary curvatures.
3. A typical vertebra consists of a body and a vertebral arch; the latter has two sides (pedicles) and a roof consisting of two plates (the laminae).
4. An intervertebral disc may be found between each pair of vertebrae. They act as shock absorbers and give flexibility to the column.

5. There are seven cervical vertebra, twelve thoracic vertebrae, Five lumbar vertebrae, one sacrum and one coccyx.
6. Cervical vertebrae are characterized by foramina in the transverse processes for the passage of the vertebral artery to the brain.
7. Rotation of the lumbar vertebrae is virtually impossible because the articular processes face one another on a sagittal plane.
8. The two joints between articular processes of adjacent vertebrae are synovial joints.
9. The movements of the vertebral column are partly a function of the thickness of intervertebral discs.
10. The most significant muscle group of the deep back is erector spinae.

the thoracic wall

GENERAL

The **thorax** is a cage for the lungs, heart, and other significant structures. It is more or less open into the neck* above at the **superior aperture** and closed below at the **inferior aperture** by the domelike **thoracic diaphragm.** The thoracic walls can be easily palpated on yourself. Try this:

Place your hand behind your back at the level of the lower part of the scapula—as if you were scratching your back—to feel the spines of the **thoracic vertebrae.**

Moving your hand laterally, note a longitudinal mass of muscle running from sacrum and iliac crests to the skull. This collection of muscles, the erector spinae, occupies a groove (gutter) between the vertebral spines and the posterior projection of the ribs.

Continuing laterally, the **ribs** become palpable, blanketed only by the latissimus dorsi muscle below the scapula. Note the ribs descend somewhat as they project anteriorly.

Wrapping your arm about yourself, you can feel the ribs become easily palpable as you pass the border of latissimus dorsi, at about the level of the midaxillary line (a line extending vertically along the side of the body and into the middle of the axilla).

At the midaxillary line just below the axilla, you may note the serratus anterior muscle clothing the ribs.

Following the ribs anteriorly, you will note that they terminate at a midline bony structure, the **sternum** or breastbone.

STERNUM

All boundaries of the sternum can be felt. Refer to Fig. 7-19 as you read, and palpate as follows:

Start above, where you feel the concave, crescent-shaped upper border, the **jugular notch** (at about T2). Spreading your fingers laterally from the notch, palpate the clavicles. Immediately deep and inferior to the **sternoclavicular** joint, the first costal cartilage, in a short space becoming the first rib, starts its journey around to the first thoracic vertebra. The circle of bone thus created, which cannot be felt, constitutes the **superior aperture.**

* Actually closed by a fascial membrane.

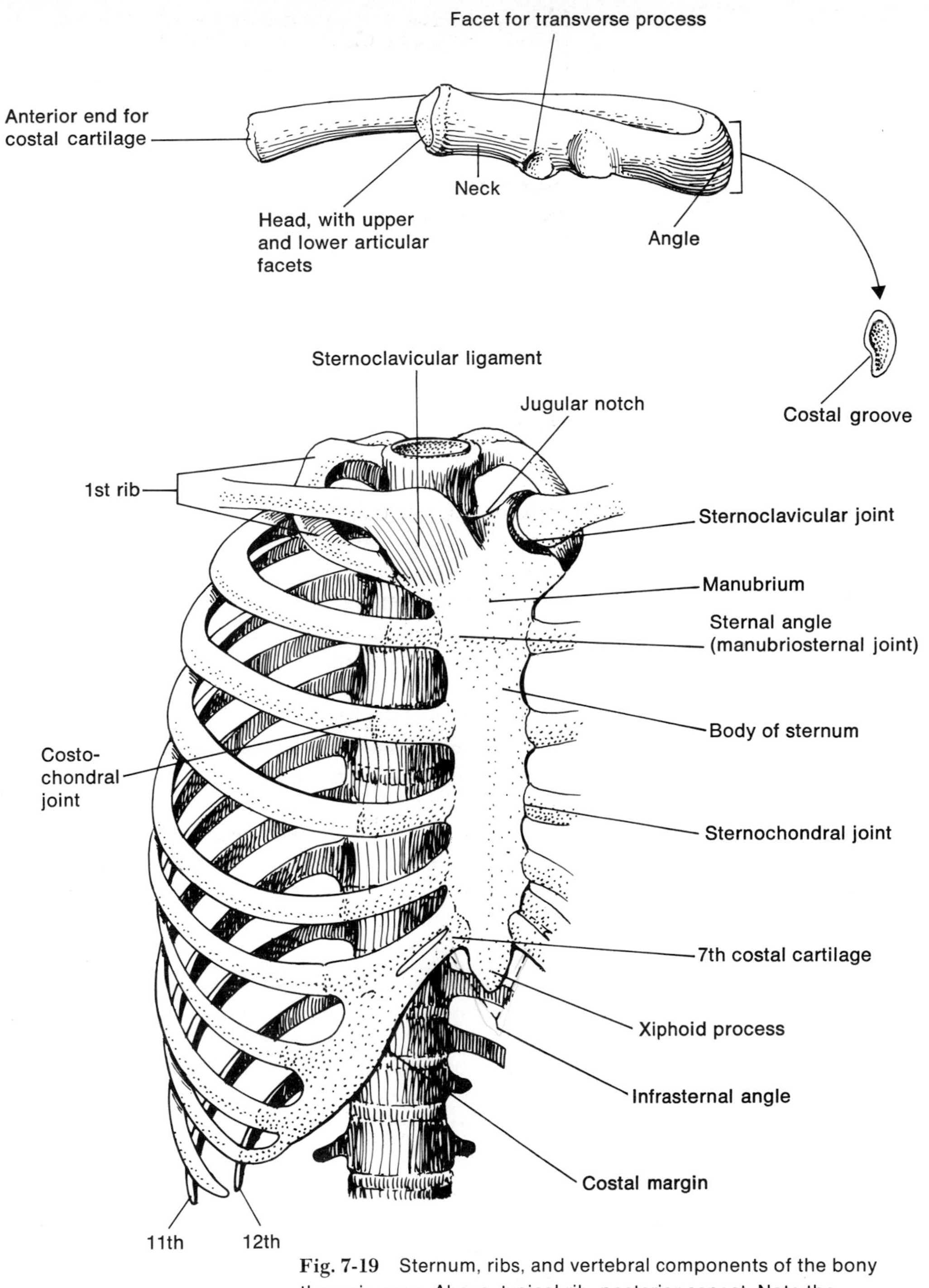

Fig. 7-19 Sternum, ribs, and vertebral components of the bony thoracic cage. Above, typical rib, posterior aspect. Note the cross-sectional view of the rib.

Now feel down two or three fingersbreadth below the jugular notch for a slight bony bump or rise signalling the **sternal angle**—a synovial joint between the handle **(manubrium)** and the **body** of the sternum. The sternal angle represents a landmark for deeper visceral structures in the thorax and is usually on a horizontal plane with T4.

Along the lateral borders of the sternum are the sockets for the sternochondral articulations, best appreciated in the laboratory.

The sternal **body**—at the level of the seventh costal cartilage—articulates with the spear-tip **xiphoid process** (third part of the sternum), palpated only with difficulty.

The sternum, located just deep to the thin superficial fascia, is a frequent site for sampling of bone marrow (in a process called sternal puncture).

RIBS

Although there are 12 ribs on each side, no one is exactly like another. Ribs have both bony (costo-) and cartilaginous (chondro-) components.

Each bony rib is characterized by a **shaft** which begins anteriorly with the costochondral joint and arches backward and laterally to a point short of the vertebra (Fig. 7-19). It then turns abruptly forward at the **angle** of the rib, narrows to form a **neck** and terminates at the vertebral articulation as the **head.** The neck incorporates an articular tubercle for the facet of the transverse process of the respective vertebra. The head incorporates two articular surfaces for articulation with the bodies of two vertebrae. The vertebra below corresponds in number to the rib. The costovertebral articulations are generally all synovial joints and capable of gliding movements.

The **costal cartilages** are of the hyaline variety and join the bony rib to the sternum anteriorly. The 8th, 9th, and 10th cartilages joint with the 7th to form an arch, the **costal margin,** near the infrasternal angle (Fig. 7-19). Both the margins and the angle are easily palpated on yourself and constitute good surface landmarks for deeper structures (liver, stomach, etc.). The 11th and 12th costal cartilages terminate freely in the lateral abdominal wall.

The ribs can be accurately counted: find the sternal angle, move laterally to the adjacent rib (costal cartilage). This is no. 2, as no. 1 is hidden under the clavicle. With your fingers step down to rib 3, etc. Tracing downward, seven costal cartilages should be felt joining with the sternum. These are **true ribs,** because they articulate (via costal cartilages) directly with the sternum. The other five ribs do not, and are called **false ribs.** Of these five, three merge with the seventh costal cartilage, and the lower two "float" freely in the abdominal wall in the sense that there is a vertebral attachment but no anterior attachment.

INTERCOSTAL MUSCLES, NERVES, AND VESSELS

The space between adjacent pairs of ribs is called an **intercostal space.** These spaces are the avenues through which needles are introduced into the pleural cavity to draw fluid or collapse a lung, and into the pericardial cavity to draw fluid that compromises heart action. For these reasons, the architecture of the intercostal space must be well known by the operator.

Refer now to Figure 7-20. The musculature of each of the 11 intercostal spaces is oriented in three incomplete layers.

The **external** layer of **intercostal** musculature arises from the lower border of the rib above and descends downward and medially to insert on the upper border of the rib below. (You might want to read that again!)

The **internal intercostals** have similar attachments except that the fibers are oriented 90° with respect to the external intercostals.

The **innermost** intercostals are composed of discontinuous sheets of muscles spanning one or more intercostal spaces. Again, these fibers are oriented 90° with respect to the internal intercostals.

The nerves and vessels supplying the intercostal structures course between the innermost and internal muscle layers, fitting snugly under the lower costal groove of each rib.

The external intercostal muscles play a role in the inspiratory phase of respiration by collectively acting to increase the anteroposterior dimensions of the thoracic cage via a mechanism to be discussed later. The internal intercostal muscles probably assist in forced expiration via a similar mechanism.

The intercostal muscles are supplied by anterior rami of the thoracic spinal nerves 1 to 11. Branches of these nerves perforate the intercostal musculature to ramify in the superficial fascia as cutaneous nerves (Fig. 7-21). As important as these nerves are, it is interesting to note that respiration does not cease following transection of the nerves. Why not?

The blood supply to the intercostal musculature and overlying skin and fascia is derived from the thoracic aorta—both directly and indirectly (Fig. 7-21); in the latter case via the in-

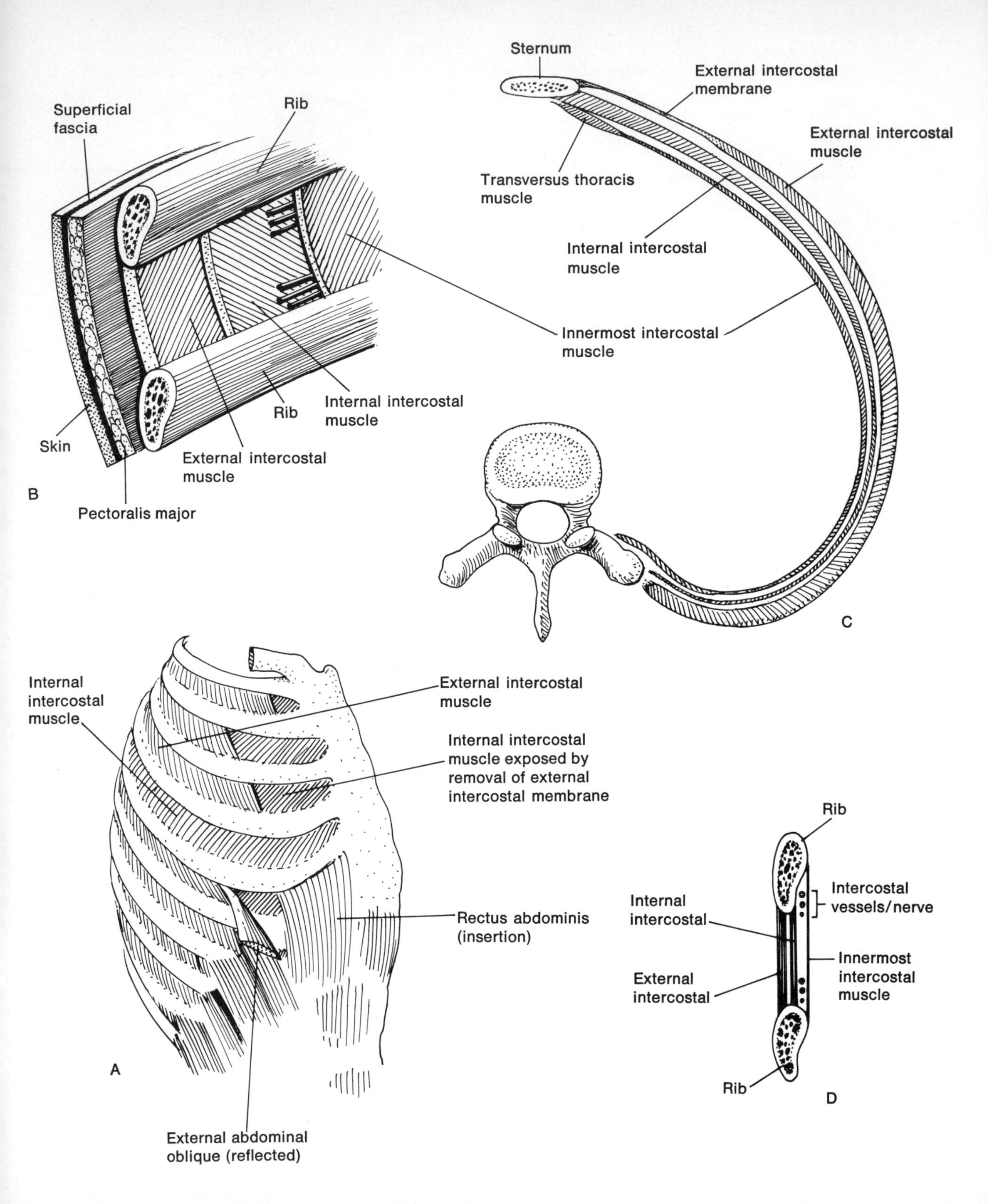

Superficial fascia
Rib
Skin
Rib
Internal intercostal muscle
External intercostal muscle
B
Pectoralis major
Sternum
External intercostal membrane
External intercostal muscle
Transversus thoracis muscle
Internal intercostal muscle
Innermost intercostal muscle
C
Internal intercostal muscle
External intercostal muscle
Internal intercostal muscle exposed by removal of external intercostal membrane
Rectus abdominis (insertion)
A
External abdominal oblique (reflected)
Rib
Intercostal vessels/nerve
Internal intercostal
Innermost intercostal muscle
External intercostal
Rib
D

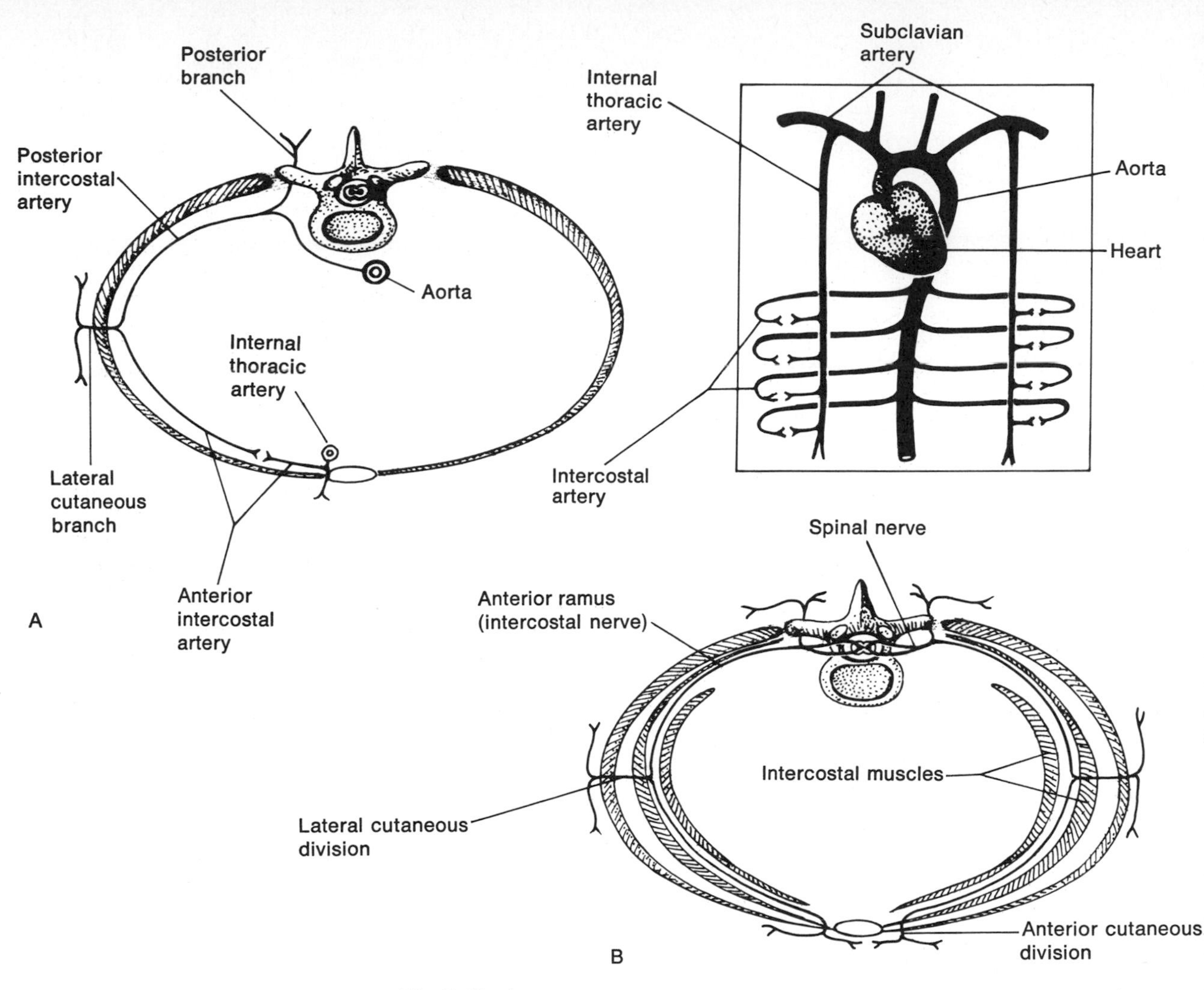

Fig. 7-21 Arrangement of thoracic spinal nerves and intercostal arteries. A. Arterial pattern of an intercostal space. B. Distribution of an intercostal nerve. Insert: anterior and posterior intercostal arterial sources.

Fig. 7-20 (opposite) The intercostal musculature and spaces. A. Intercostal musculature, anterolateral view. B. Interior view of an intercostal space showing contents. C. Cross section of the thoracic cage through an intercostal space. D. Vertical section through an intercostal space.

ternal thoracic artery. The posterior and anterior intercostal arteries anastomose in the intercostal space.

DIAPHRAGM

The diaphragm is a musculotendinous structure intervening between the abdomen and the thorax (Fig. 7-22). The muscular fibers of the diaphragm originate from the ribs, the xiphoid process, and vertebrae T12 to L3, that is, from all sides around the thoracoabdominal opening. The muscle fibers project inward toward the **central tendon,** which is punctured by three openings (hiatuses). On the right side, the inferior vena cava (draining the lower limbs, pelvis, and the body wall) passes through. Centrally, just in front of the vertebral bodies, the aorta passes through. Slightly to the left, the esophagus passes through to become the stomach below. Since the diaphragm arches upward centrally, it follows that the vena cava passes through the middle of the diaphragm at a higher vertebral level (T8) than the more peripheral esophagus (T10) or aorta (T12).

When contracted, the diaphragm flattens, forcing such abdominal viscera as the liver against the anterior abdominal wall. You can see this on yourself: Place your hand over your

Fig. 7-22 The diaphragm. A. Sagittal view of left portion of diaphragm. B. The diaphragm, anteroinferior view.

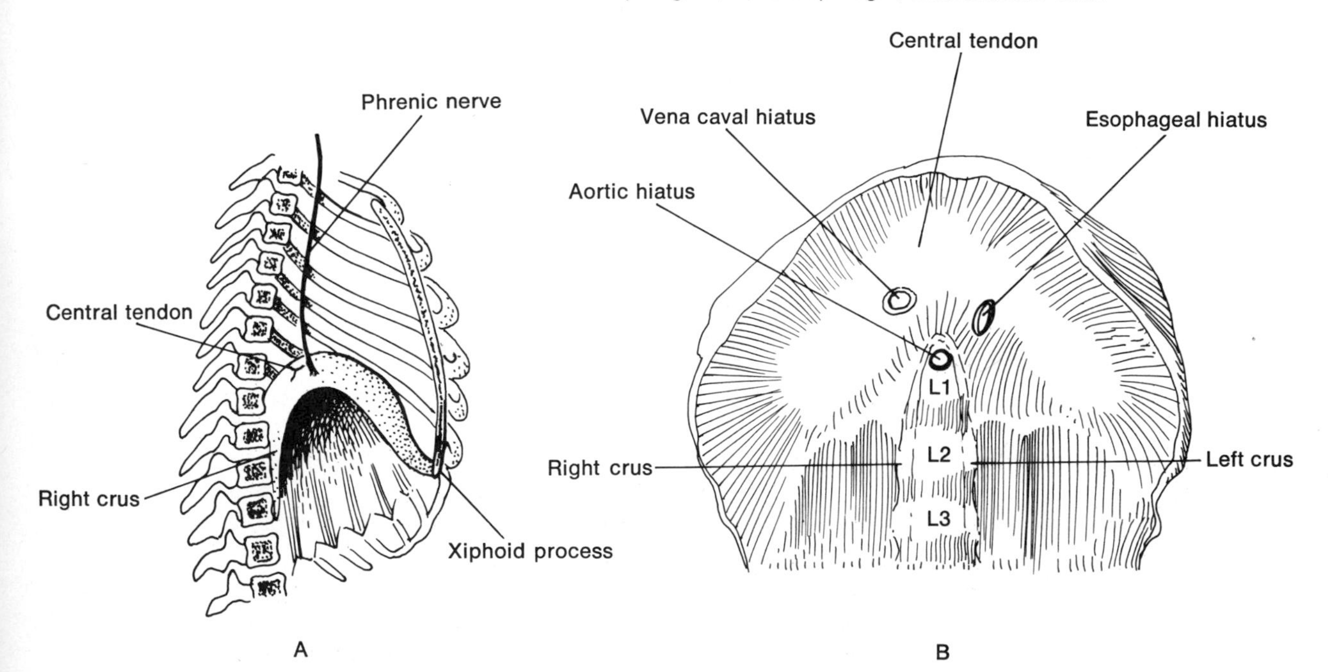

upper abdomen, inhale deeply, and do not tighten your abdomen. Note how the abdominal contents, i.e., the liver, pushes outward. On exhalation, each half of the diaphragm is pushed upward by the visceral structures below. As it is contracted, the diaphragm simultaneously decreases the pressure in the thorax and increases the pressure in the abdomen. Thus the blood returning to the heart in the inferior vena cava gets an extra push from the pressure differential as it passes up to the heart. The diaphragm is innervated by the **phrenic nerves,** which arise from spinal nerves C3 to C5 via the cervical plexus.

The diaphragm may be voluntarily contracted or inhibited—to a point. Hence, you can "hold your breath" (voluntarily suspend diaphragmatic contraction) almost to a point of loss of consciousness before the reflex act of diaphragmatic relaxation occurs. You can contract your diaphragm preparatory to generating a cough which requires an increased volume of air in the lungs (try it). You may also contract your diaphragm to increase abdominal pressure (straining) in tightening your abdominal wall or in defecation.

Respiration is made difficult but not impossible by devastating injuries to the phrenic nerves, thanks to the intercostal musculature. Near full-term pregnancies often limit the action of the diaphragm and force a somewhat labored intercostal breathing. A broken neck above the C3 to C4 level, if severe enough, can result in fast fatality due to loss of respiration, for here both intercostals (T1–T11) and diaphragm (C3 to C5) are affected.

The diaphragm receives its vascular supply from the terminal branches of the internal thoracic artery in the thorax and phrenic branches of the aorta in the abdomen.

RESPIRATION

Respiration, or the act of breathing, consists of two phases: **inspiration** (taking air into the lungs) and **expiration** (taking air out of the lungs).

The thoracic cavity is a closed cavity—not in any way in communication with the outside. The walls of the cavity are movable and therefore can change the volume of the cavity. The elastic lungs within the thorax are open to the outside, of course, via the conducting airways of the trachea, larynx, pharynx, and nasal and oral cavities.

Inspiration

The pressure of air in the atmosphere at sea level is generally stable at about 15 lb per in.^{2} of space, or sufficient to depress

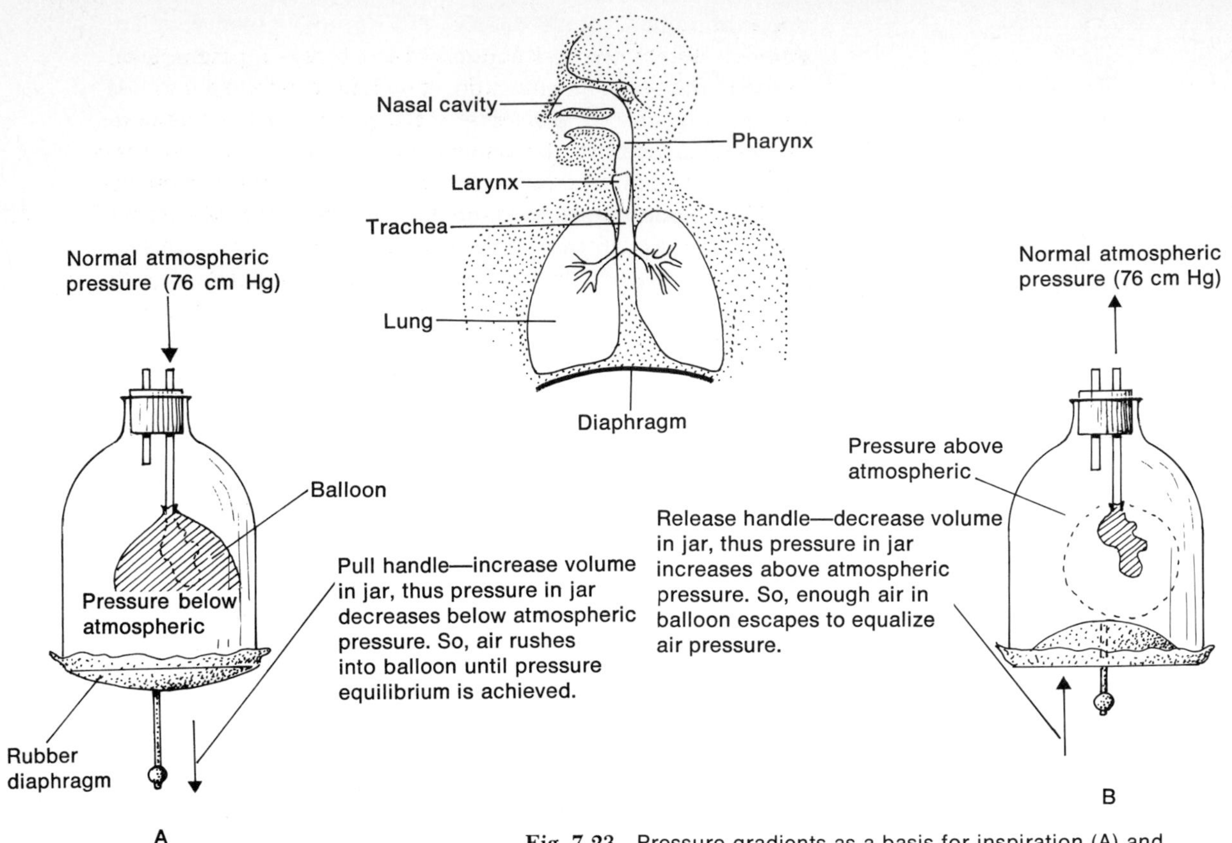

Fig. 7-23 Pressure gradients as a basis for inspiration (A) and expiration (B).

a column of mercury (Hg) in a tube a distance of 76 cm.

The pressure within the thoracic cavity at rest is less than atmospheric pressure.

If the volume of the thoracic cavity is increased by increasing the dimensions of the thoracic cage, the pressure within the cavity will drop further with respect to the outside.

When the thoracic cage expands, the lungs will expand similarly, increasing their own volume and lowering their pressure.

An unequal set of *air pressures* now occurs between the *outside atmosphere* around the nose and mouth and the spaces of the *lung*.

As "nature abhors a vacuum," air will immediately rush from the outside to the lungs via the airways to create equilibrium.

Presto: *inspiration* (Fig. 7-23A).

Expiration

The key to inspiratory success was to create a pressure differential between the outside atmosphere and the lung air space, air thus being forced into the lungs. The key to expiratory success involves the same phenomenon—only in reverse:

As the muscles of the thorax relax, the diaphragm relaxes and, as a consequence, the volume of the thorax decreases, building up pressure. The elastic lungs change in size together with changes in size of the thorax (they recoil). Thus their volume decreases and the pressure within increases.

Momentarily, the pressure in the lungs and airways surpasses that of the atmosphere outside and air rushes out of the lungs to create equilibrium.

Presto: *expiration* (Fig. 7-23B).

Mechanics

The intercostal musculature and the diaphragm act to *increase intrathoracic dimensions* so inspiration can occur (Fig. 7-23). The act of expiration normally requires no muscular assistance since the natural elasticity of the lungs and thoracic wall brings about decreases in lung and thoracic wall dimensions.

Now see Fig. 7-24 as you read:

In inspiration, the diaphragm flattens out, *increasing the vertical dimension* (length) *of the thorax.*

In inspiration, the external intercostals contract. The first rib being secured by the scalene muscles, the ribs are moved upward by the external intercostals. Furthermore, the lower ribs are generally longer than the upper ribs. Thus, with external intercostal muscles contracting, the ribs (hinged at their vertebral attachments) are brought upward and *outward.* Compensatorily, the body of the sternum is pushed forward, for it too is hinged—at the sternal angle. Result: the *anterior-posterior dimensions of the thorax are increased.*

As a consequence of these muscular actions there is an increase in volume of the thoracic cavity and a drop in pressure. The pressure equilbrium is upset and inspiration occurs.

In hiccoughs the diaphragm contracts out of sequence and out of phase with the intercostal musculature. The noise created is air being sucked down through the larynx.

Every muscle which attaches to the ribs or sternum is potentially an accessory muscle of respiration. With training, certain accessory muscles of the abdominal wall and other areas can help generate a sustained, controlled flow of air through the larynx, of value in singing. In exercise, accessory

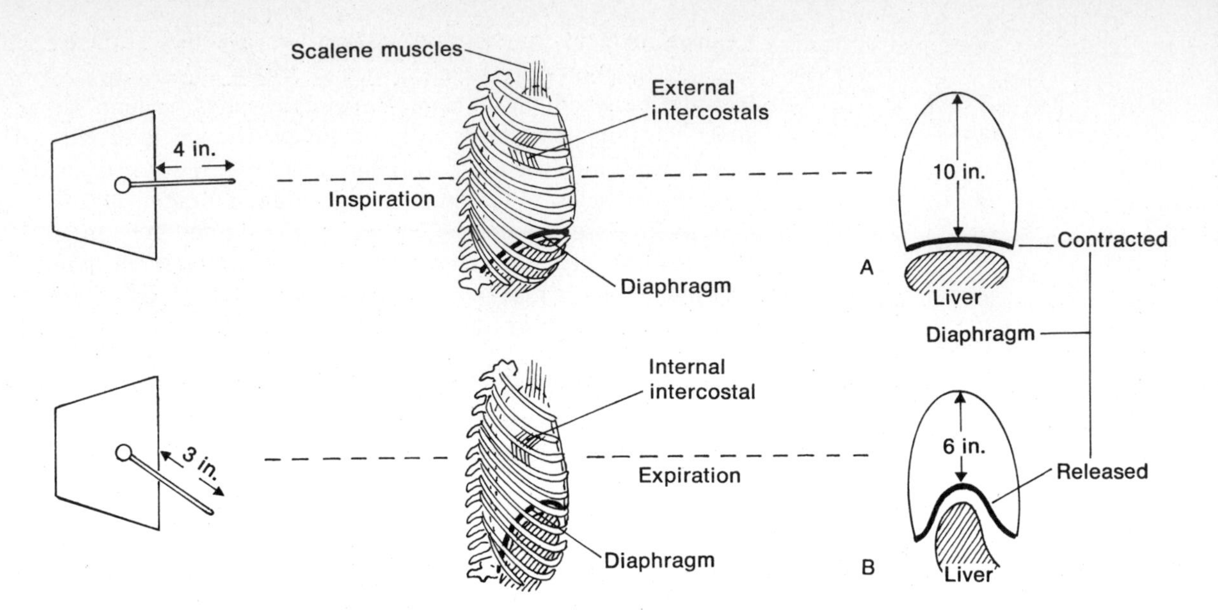

Fig. 7-24 Influence of thoracic musculoskeletal structures in respiration. Increase/decrease in thoracic dimensions as seen in inspiration (A) and expiration (B), respectively. Bucket-handle model represents changes in anteroposterior dimensions as a result of lifting (A) and lowering (B) the rib cage.

muscles of respiration are employed to increase the capacity for movement of air into and out of the lungs, by expanding the thorax in inspiration and increasing intrathoracic pressure during expiration.

QUIZ

1. The upper seven ribs, articulating with the vertebral column and the sternum, are called true ribs. The lower five are false ribs because the upper three of these do not attach to the sternum directly and the lower two (11 and 12) float freely in the abdominal wall.
2. The sternum is a frequent site for sampling bone marrow.

3. The head of the rib incorporates two articular surfaces for articulation with the bodies of two vertebra. The vertebra below corresponds in number to the rib.
4. The space between adjacent ribs is called an inter-costal space.
5. The intercostal muscles are supplied by anterior rami of thoracic spinal nerves 1 to 11.
6. The diaphragm is innervated by the phrenic nerves, from spinal nerves C3, C4, and C5.
7. A severe fracture of neck above the level of C3 to C4 can instantaneously arrest respiration.
8. Respiration consists of two phases: Inspiration and expiration.
9. Inspiration and expiration is a consequence of an unequal set of air pressures existing between the outside atmosphere and the lung.
10. The intercostal musculature and the diaphragm act to increase intra-thoracic dimensions so inspiration can occur.

abdominal wall

Because of their muscular configuration, the walls of the abdominal cavity are described as being anterolateral and posterior. The orientation of the muscles in the anterolateral walls is quite reminiscent of the muscles in the thoracic wall, suggesting some degree of continuity in development.

BOUNDARIES OF THE ANTEROLATERAL ABDOMINAL WALL

Unlike the thoracic wall, there is no bony framework to the major portion of the abdominal wall. You can feel this on yourself: palpate the margin of the ribs from the xiphoid process laterally. Below these bony structures there is only muscle, fascia, and skin protecting the fragile abdominal viscera from the external environment (Fig. 7-25). At the level of the umbilicus feel laterally for the crests of the ilium, and feel for the sharp drop off anteriorly to the pubic bones. These bony landmarks not only indicate to you the boundaries of the anterolateral abdominal wall but they are also the points of origin and insertion of certain important muscles.

EXTERNAL OBLIQUE

The **external oblique** is the most superficial of the three anterolateral muscles and its fiber orientation is like that of the external intercostals. In lean individuals, the fibers of origin of the external oblique may be seen descending obliquely from the ribs below the origin of the serratus anterior (Fig. 7-26). The left and right fibers of the external oblique form a massive aponeurosis anteriorly and interdigitate in the midline as part of the **linea alba** (white line). Inferiorly, the muscle ends as a free border, turned or curled upon itself to form a ligament between anterior superior iliac spine and pubis. Can you palpate this **inguinal ligament** on yourself? Contract your abdominal muscles as in straining and the external oblique may be palpated deep to the skin and the variably fatty superficial fascia. Because of the density of the superficial fascia in the abdominal wall and the flatness of this muscle, the external oblique is rarely seen bulging in the abdominal wall. Its relationship to the rectus abdominis muscle and the inguinal region will be discussed shortly.

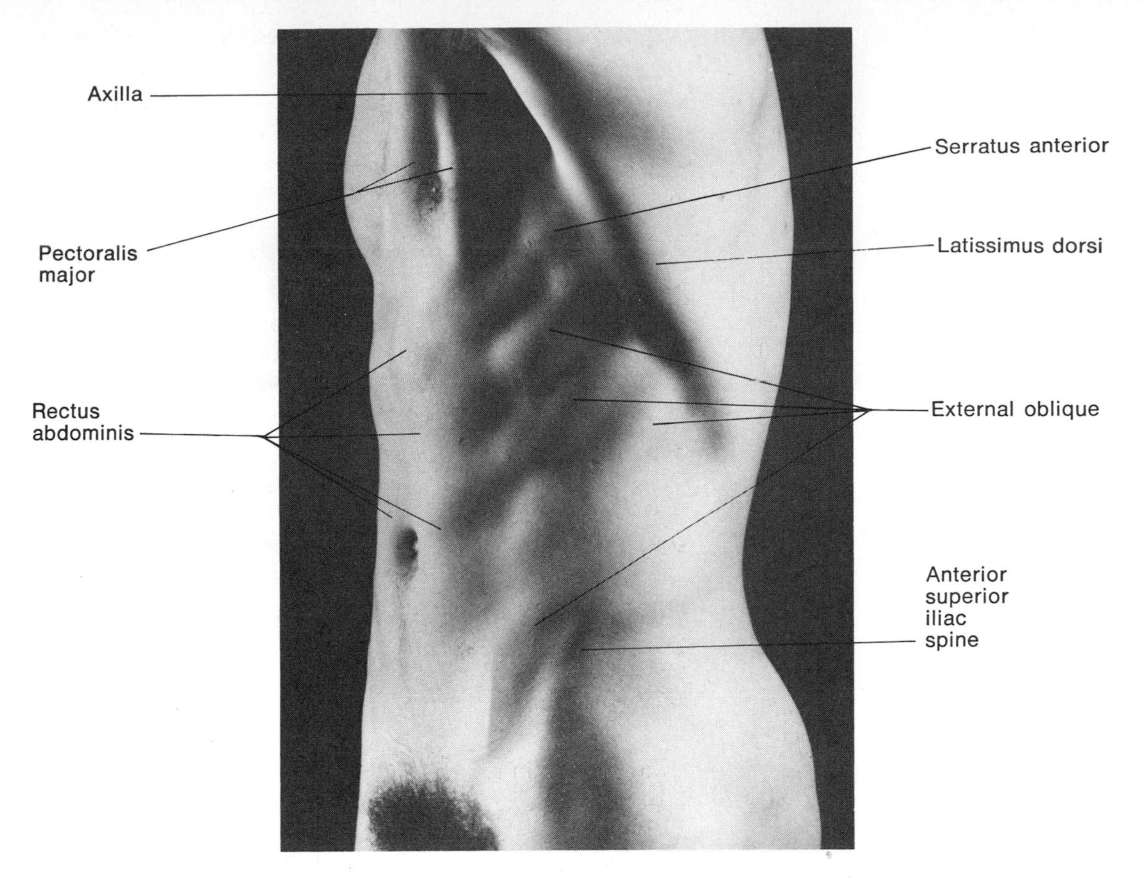

Fig. 7-25 Surface view of anterolateral abdominal and thoracic walls.

INTERNAL OBLIQUE

Deep to external oblique, its fibers oriented somewhat like those of the internal intercostals, one finds the **internal oblique.** It is, of course, not palpable, and its attachments may be noted in Fig. 7-26. Its relationship to the rectus abdominis muscle and the inguinal region will be discussed shortly.

TRANSVERSUS ABDOMINIS

Deep to the internal oblique, the transversus abdominis reinforces the muscular anterolateral wall. Its attachments may be seen in Fig. 7-26. Like its more superficial fellows, the transversus abdominis contributes to the sheath of the rectus abdominis and to the inguinal canal as well.

RECTUS ABDOMINIS AND ITS SHEATH

The straight muscle of the abdomen is easily seen in well-developed, lean males. The lateral margin of the muscle creates a skin shadow on the abdominal wall as well (Fig. 7-26). The borders and extent of the rectus abdominis can probably be ascertained on yourself. The attachments of this muscle may be seen in Fig. 7-26.

The fascial sheath of the rectus is reinforced by the interlacing aponeuroses of the three anterolateral muscles just described. The manner in which these aponeuroses form the anterior and posterior layers of the sheath is of significance to the surgeon. In the midline between the recti, the intertwining aponeuroses form a dense white line (linea alba) from xiphoid to pubis.

The segmental origin of the rectus abdominis is reflected by the segments of muscle interrupted by intersections of tendon (tendinous intersections), frequently seen in the well-developed male (Fig. 7-26). Each of these segments developed embryologically as separate muscle masses joined in series by connective tissue.

FUNCTION

The three anterolateral muscles generally act together in flexing the trunk. The rectus abdominis helps out only when there is considerable resistance to that flexion, as in sit-up exercises. The sheath of the rectus prevents the muscle from bulging forward as though to pop out of the anterior wall (i.e., bowstringing effect). Individual muscles on one side have been shown to be active in rotation of the trunk and lateral flexion.

The anterolateral muscles except the rectus are active in tensing the abdominal wall as in straining and coughing and are of primary importance in forced expiration.

Fig. 7-26 (opposite) Muscles of the anterolateral abdominal wall. A. External oblique. B. Internal oblique. C. Transversus abdominis. D. Rectus abdominis. The sheath of rectus may be seen in A, B, and C. Center, surface view of anterior abdominal wall. The semilunar line represents the junction of the rectus sheath with the aponeuroses of internal oblique and transversus abdominis.

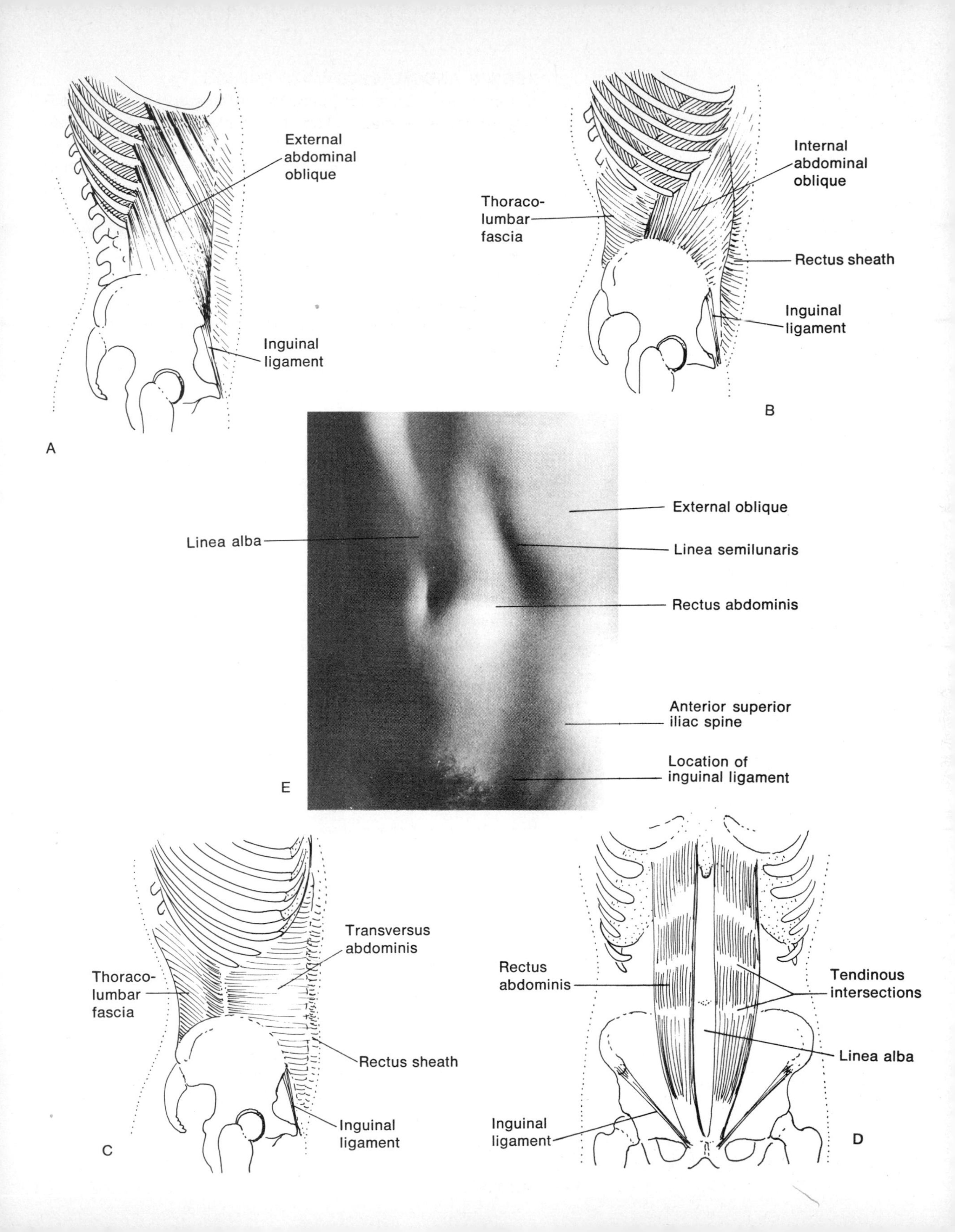
External
abdominal
oblique
Inguinal
ligament
A
Internal
abdominal
oblique
Thoraco-
lumbar
fascia
Rectus sheath
Inguinal
ligament
B
External oblique
Linea alba
Linea semilunaris
Rectus abdominis
Anterior superior
iliac spine
Location of
inguinal ligament
E
Transversus
abdominis
Thoraco-
lumbar
fascia
Rectus sheath
Inguinal
ligament
C
Rectus
abdominis
Tendinous
intersections
Linea alba
Inguinal
ligament
D

NERVES AND VESSELS

The muscles of the anterolateral abdominal wall are innervated by anterior rami of spinal nerves T7 to L1, which wrap around the wall not unlike the intercostal nerves around the thoracic wall. Thus, you might suspect that since the intercostal nerves pass between the second and third layers of intercostal musculature, the nerves to the abdominal wall must pass between internal oblique and transversus abdominis, and indeed they do, giving off branches as they go. The cutaneous distribution to the abdominal wall is oriented in the same fashion as that to the thoracic wall.

The cutaneous distribution of nerves to the abdomen and thorax is important to the physician, for often visceral pain will be referred to cutaneous regions that have the same spinal cord level of innervation. Thus, an itching or painful sensation on the skin of the abdomen at the level of the navel (T10 dermatome) may be a reflection of an inflammation of the lining of the lungs (pleurisy) at the level supplied by the 10th thoracic nerve.

The general cutaneous distribution over the trunk is easily learned (Fig. 7-27). Place a flat hand over the center of your chest with the upper border (thumb side) level with the sternal angle. The area covered by your hand is approximately the cutaneous region covered by T2 (second thoracic nerve). Now place your other hand immediately below the one on the chest; this is the region innervated by T4. A hand placed below this one would represent the T6 dermatome. Note while continuing distally in this fashion that the dermatome about the navel is supplied by T10. The area between adjacent hands is supplied by the odd-numbered spinal nerves. There is a significant degree of sensory overlap from one dermatome to another, however. Thus pain in a specified area of the chest may not arise from the nerve serving that area but from fibers of the nerve immediately above or below it.

Figure 7-28 illustrates the local arteries responsible for supplying the anterolateral abdominal wall. These arteries provide an important anastomotic circuit for blood flow to the lower limb should the aorta become obstructed. Venous drainage parallels the arteries.

Fig. 7-27 (opposite) A. Cutaneous nerve distribution of the anterolateral abdominal wall. B. The lumbar plexus. L1 and L2 nerves contribute to the cutaneous innervation of the lower abdominal wall.

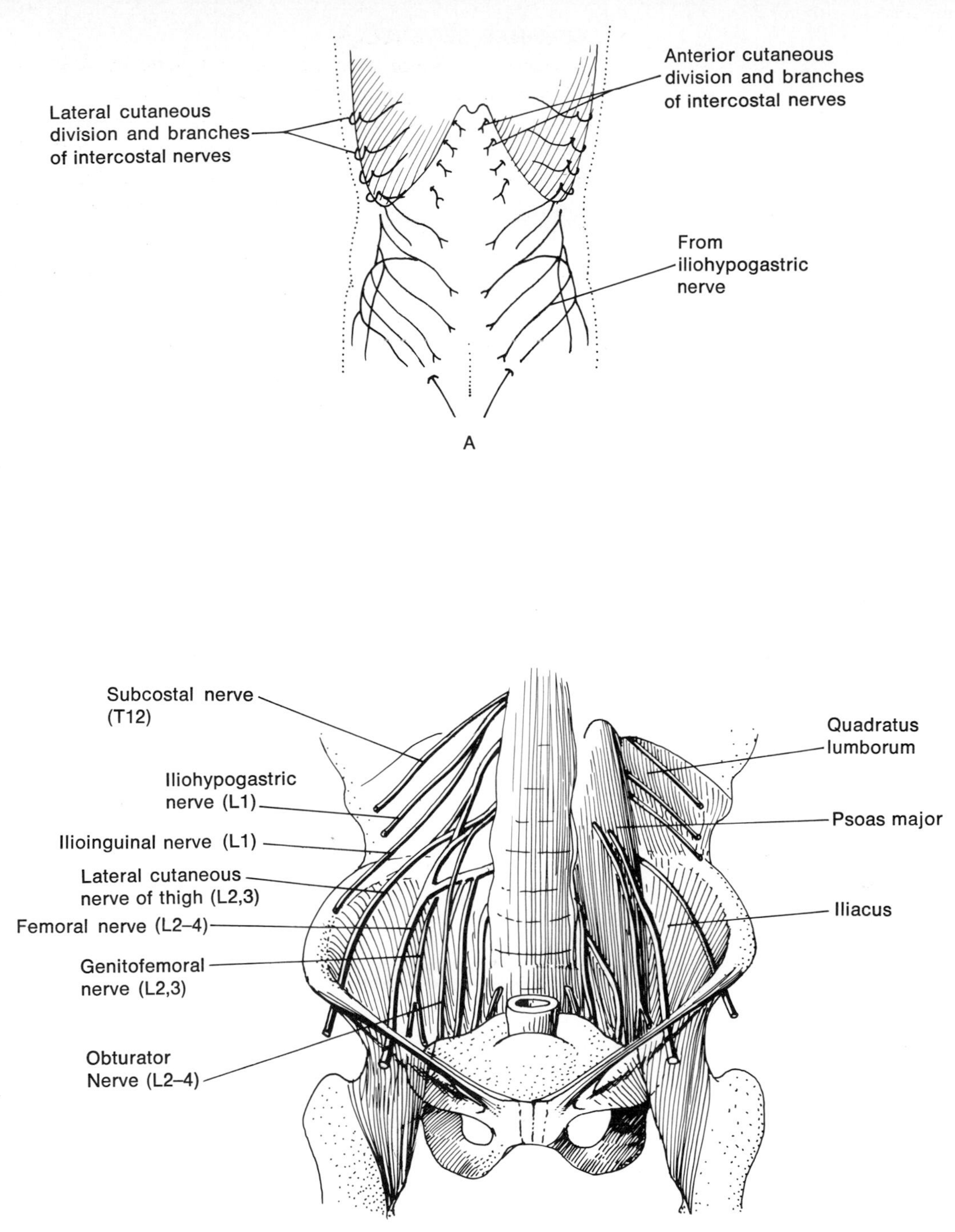
Lateral cutaneous division and branches of intercostal nerves
Anterior cutaneous division and branches of intercostal nerves
From iliohypogastric nerve
A
Subcostal nerve (T12)
Iliohypogastric nerve (L1)
Ilioinguinal nerve (L1)
Lateral cutaneous nerve of thigh (L2,3)
Femoral nerve (L2–4)
Genitofemoral nerve (L2,3)
Obturator Nerve (L2–4)
Quadratus lumborum
Psoas major
Iliacus
B

INGUINAL REGION

The inguinal region is a portion of the lower quadrants of the anterior abdominal wall characterized by a fibromuscular canal transmitting the spermatic cord (in males) from the testes in the scrotum. It is an area of paramount interest to surgeons for repair of ruptures of the abdominal wall (herniorraphy). Thus it is an area of interest to the nonsurgeon as well, for it is we who suffer hernias too.

Ruptures are common here largely because the canal represents an imperfection in the fibromuscular abdominal wall. With excessively high intra-abdominal pressures, as in straining while lifting heavy objects, a loop of intestine can be forced into the canal, forming a variably sized bulge (hernia) in the abdominal wall. If the vascular supply to the herniated tissue is reduced by enough to cause ischemia, necrosis and gangrene may result.

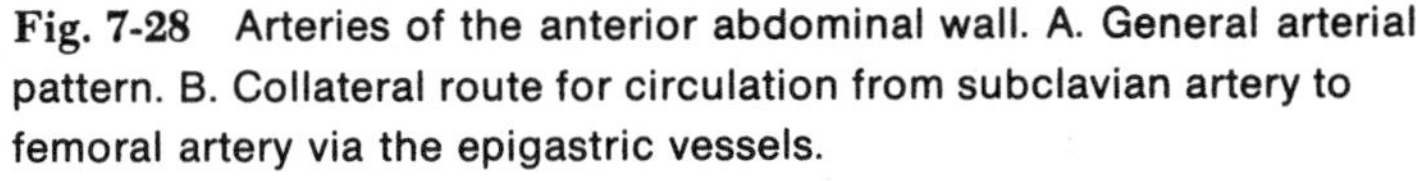

Fig. 7-28 Arteries of the anterior abdominal wall. A. General arterial pattern. B. Collateral route for circulation from subclavian artery to femoral artery via the epigastric vessels.

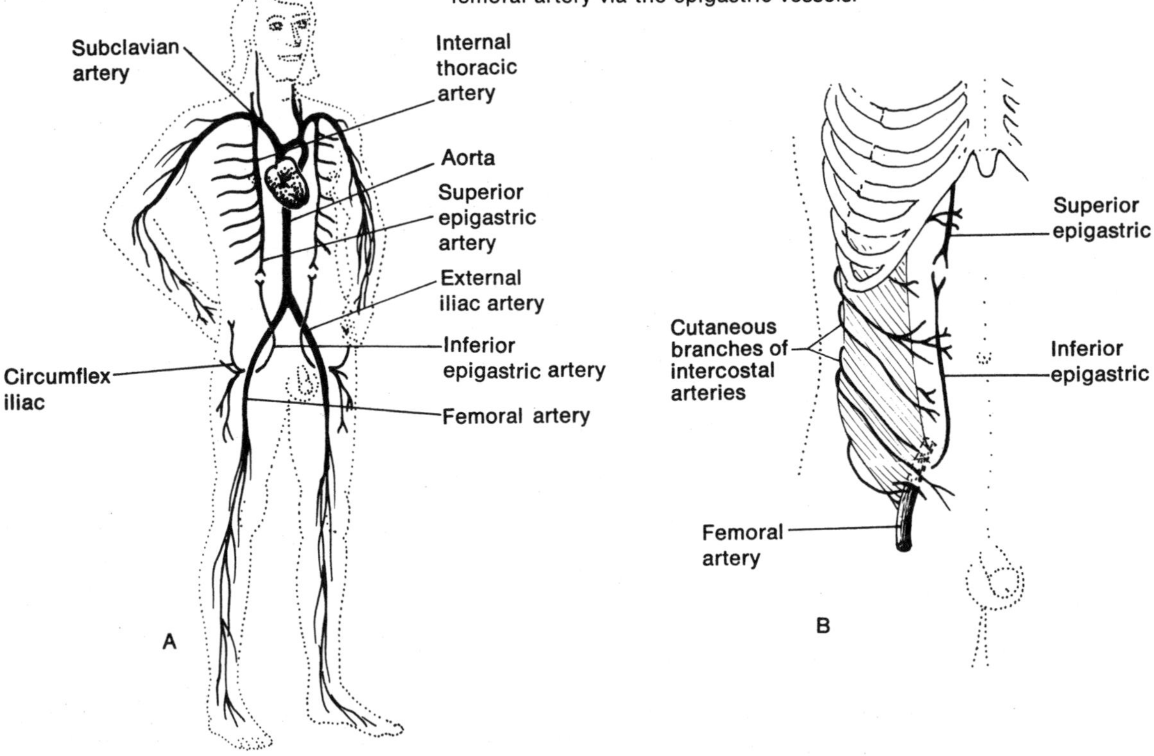

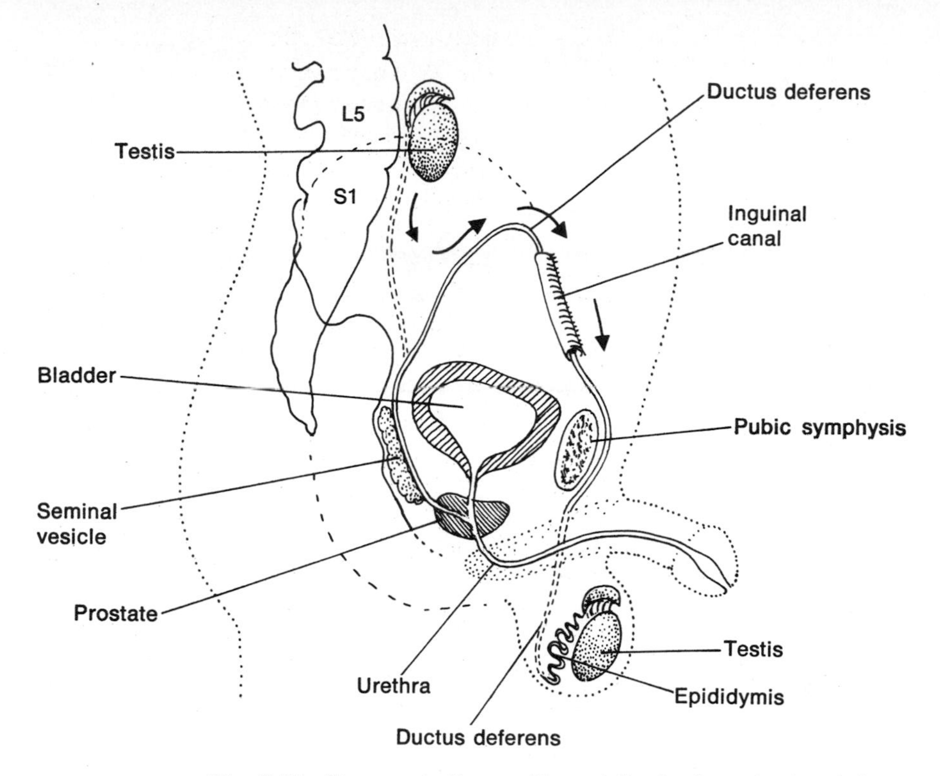

Fig. 7-29 Changes in the position of the testis and related ducts during fetal development. The testis migrates from abdomen to scrotum via the inguinal canal.

Refer now to Fig. 7-29 as you read the following:

The **spermatic cord** consists of blood vessels and a duct. The duct transports sperm cells from the testis (where they are produced) through the abdominal wall, by way of the inguinal canal, to the urethra in the prostate gland. The urethral duct passes through the penis and is, of course, open to the outside. In this way sperm cells leave the body of the male.

Embryologically, the testes developed in the abdomen. Each testis and its cord "descended" through the abdominal wall into the outpocketing of the anterior abdominal wall, which is what the **scrotum** is.

Now look at Fig. 7-30 as you read:

As the testes pushed its way through the abdominal wall it took a prolongation of each of the layers of the wall with it. From deep to superficial these layers consist of the transversalis fascia and the muscles transversus abdominis, internal

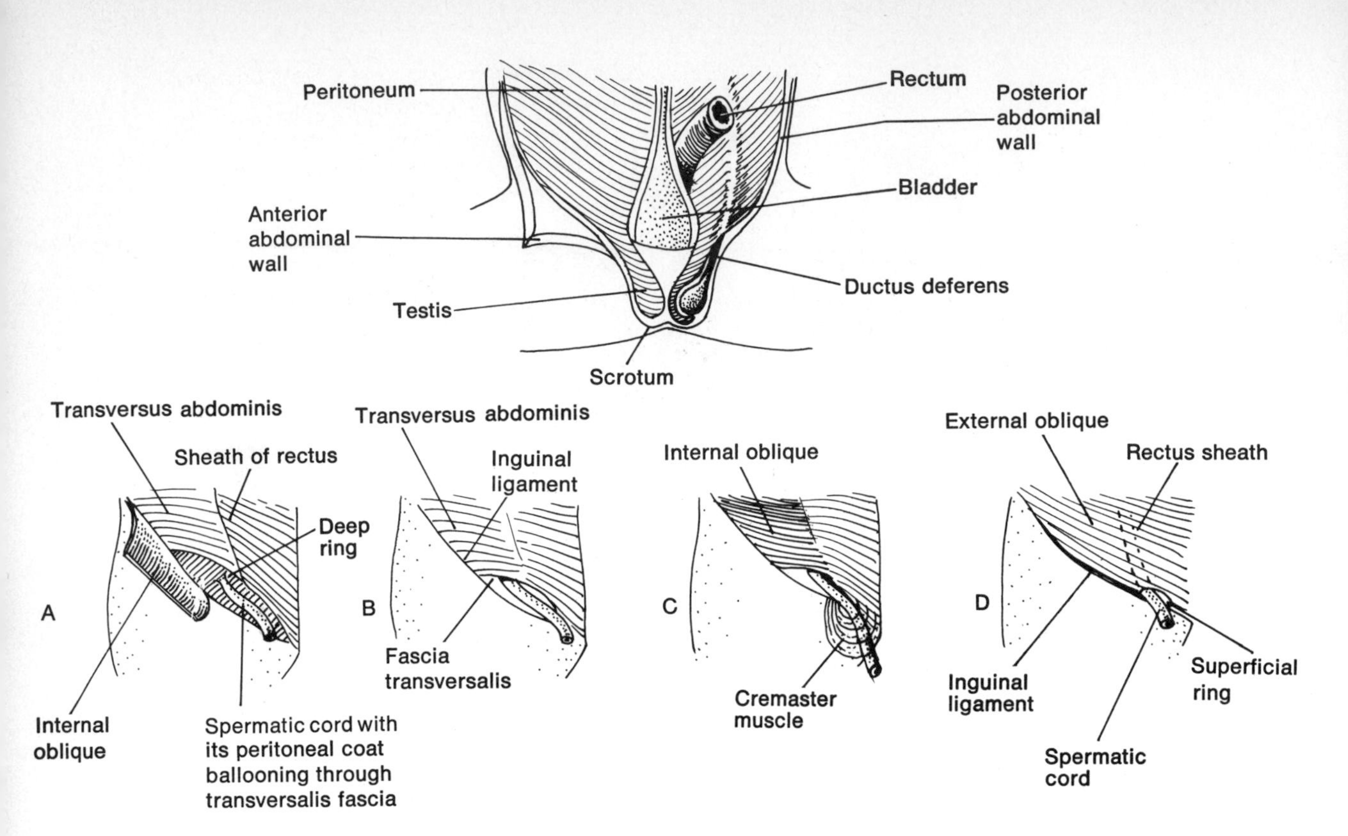

Fig. 7-30 Constitution of the inguinal canal. As spermatic cord passes through the abdominal wall, it takes with it the layers of that wall, from deep to superficial, transversalis fascia (A), transversus abdominus (B), internal oblique (C), and external oblique (D). Above, testis takes peritoneum with it in its descent. Right side, anterior layer of peritoneum is shown. Left side, posterior layer of peritoneum is shown. The testis and ductus deferens are behind the posterior layer of peritoneum.

oblique, and external oblique. Together they are **coverings of the cord.** The orientation of these layers around the cord as it passes through the wall creates the walls of the canal as well as the coats for the cord:

1. The ballooning of the transversalis fascia (Fig. 7-30A) created by the penetrating ductus deferens and related vessels constitutes the **internal (deep) inguinal ring** or internal opening of the inguinal canal. This fascia's contribution to the cord is an internal layer of fascia.
2. The lower fibers of the transversus abdominis and internal oblique join to form a fibromuscular arcade over the spermatic cord structures, thus largely making up

the walls of the canal. Their contribution to the cord consists of several loops of muscle (cremaster).

3. The lower free fibers (aponeurosis) of the external oblique roll under to form the inguinal ligament between the anterior superior spine of the ilium and the tubercle of the pubic bone. Just above this ligament, the aponeurosis balloons outward carried away by the spermatic cord structures. At this point is the **external (superficial) inguinal ring** or external opening of the canal. The aponeurosis covers the canal created by the underlying muscles and completes its form as a tubular canal.

Therefore, because of the manner in which the layers of the abdominal wall contribute to this inguinal canal, the strength and security of the wall at this region is in jeopardy. If there is excessive intra-abdominal pressure, mobile intra-abdominal structures, i.e., a loop of intestine, can herniate directly *through* the wall of the canal (direct inguinal hernia) or enter "indirectly" via the deep ring (indirect inguinal hernia), following the path of the descended testis.

POSTERIOR ABDOMINAL WALL

The posterior wall of the abdomen (Fig. 7-31), internal to the muscles of the back and anterior to the transverse processes of the lumbar vertebrae, consists of five major structures:

1. Five lumbar vertebrae.
2. The quadrate-shaped muscle of the lumbar region (quadratus lumborum) located between the 12th rib and the iliac crest.
3. The psoas muscle (a flexor of the thigh and vertebral column).
4. The iliacus muscle (a flexor of the thigh and vertebral column).
5. The diaphragm (in part).

The significance of this wall lies not so much in what it is composed of but in what relates to it. Here one can generate a long list, for between this wall and the peritoneum lining it (which invests much of the abdominal viscera) are most of the nerves and vessels supplying the abdomen, not to mention the kidneys, adrenal glands, and portions of the large intestine. Some of these can be visualized in Fig. 7-31. Note:

- The **abdominal aorta** which courses downward along the bodies of the vertebra from the thoracic aorta above to its bifurcation into the common iliac arteries below. Its

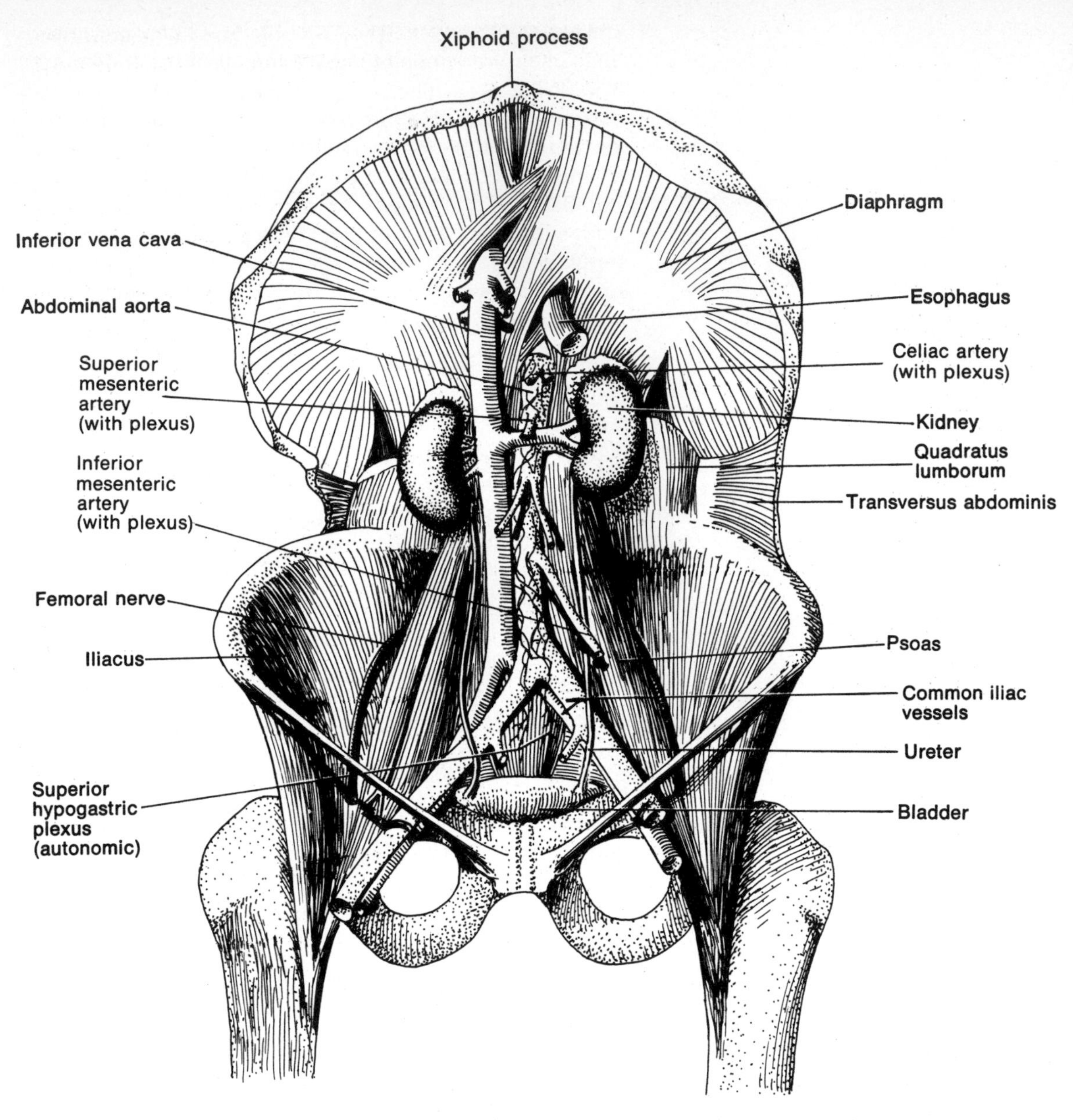

Fig. 7-31 The posterior abdominal wall and some of its residents.

branches supply the abdominal viscera, the pelvis, lower limbs, as well as the posterior abdominal wall.

- The **inferior vena cava** (IVC) adjacent and parallel to the aorta; IVC passes upward from its principal tributaries, the common iliac veins (draining the lower limbs, the pelvis, kidneys, adrenals, testes, ovaries, and posterior wall). In the thorax, the IVC enters the heart.

- The **autonomic plexuses** on the aorta, supplying sympathetic and parasympathetic innervation to the abdominal viscera.
- The retroperitoneal (behind-the-peritoneum) viscera immediately adjacent to the posterior wall, e.g., kidneys.
- The **lumbar plexus** of somatic nerves (Fig. 7-27), whose members not only supply muscles of the posterior wall (T12 to L3) but pass into the anterior thigh compartment as the femoral and obturator nerves to supply the anterior and medial femoral compartments. Nerves L4 and L5 join members of the sacral plexus to form the **lumbosacral trunk,** source of the great sciatic nerve.

UPON REFLECTION

People are constantly concerned about their backs, e.g., "Oh, my aching back!" "Boy, does my side hurt!" Having just completed your study on the body wall, you are now in a position to understand the general structure of the chest wall, the back, and the abdominal wall. No longer need you be in the dark about what's there that can hurt or what a back is and why it can so easily be abused. Perhaps now you know how not to use your back and what will happen if you exceed its limitations.

REFERENCES

1. Anson, B. J. (ed.): *Morris' Human Anatomy,* 12th ed., McGraw-Hill Book Co., New York, 1966.
2. Lockhart, R. D., G. F. Hamilton, and F. W. Fyfe: *Anatomy of the Human Body,* 2d ed., J. B. Lippincott Co., Philadelphia, 1965.

clinical considerations

ABNORMAL VERTEBRAL CURVATURES

Curvatures can become affected by diseases of individual vertebrae in which there is degeneration of vertebral bone. In response, the body increases the rate of osteogenesis and bony growths (spurs) can overlap intervertebral joints, causing fusion and rigidity (ankylosis*). Generally, this collection of disease processes is termed arthritis.

Like a tall stack of blocks, the collection of vertebrae with its curves represent a delicately balanced column. Additions to or subtractions from the degree of curvature have immense implications structurally. Ligaments and muscles reinforcing the stability of the column will contract following an alteration, in an attempt to bring the column back in alignment. Sustained contractions may lead to spasmodic contractions and subsequent inflammations, all quite painful.

An abnormal thoracic curvature—an exaggerated posterior convexity of the thoracic portion of the column—is termed **kyphosis** (Gr., *kypho,* hump). This condition often leads to cardiopulmonary problems because of the unusual position of the organs with respect to the forces of gravity.

An abnormal lateral curvature of the vertebral column is termed **scoliosis** (Gr., *scolios,* crooked). Seen infrequently in children, the cause of juvenile scoliosis is unknown (idiopathic) (Fig. 7-32). Another kind of scoliosis occurs because of a long-standing strain or force acting on one side of the body. This is a nonprogressive curvature which compensates for the added stress put upon the vertebral column. Scoliosis of varying degrees is an occupational hazard for such people as career mailmen who may carry their bags over their favorite shoulder, thus creating the foundation for the abnormal curvature.

A less serious exaggerated curvature, **lordosis,** involves the lumbar vertebrae and is frequently seen transiently in pregnant women who compensate for the muscular strain of the added anterior "baggage" by taking on a "proud stance." Exercises following parturition usually reduce this exaggerated curve.

* Gr., *ankylo,* bend, crooked; Gr., *-osis,* condition of.

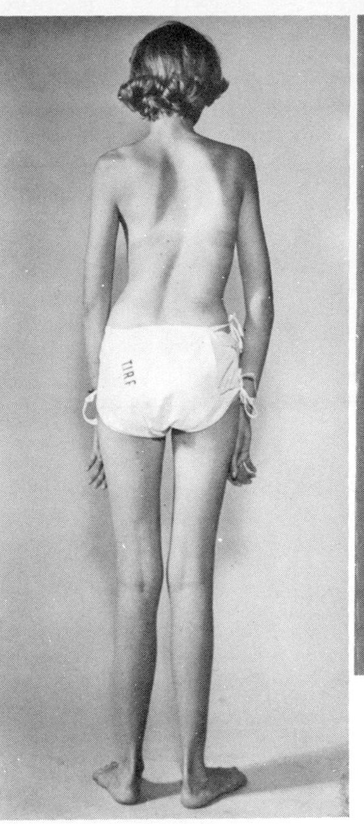

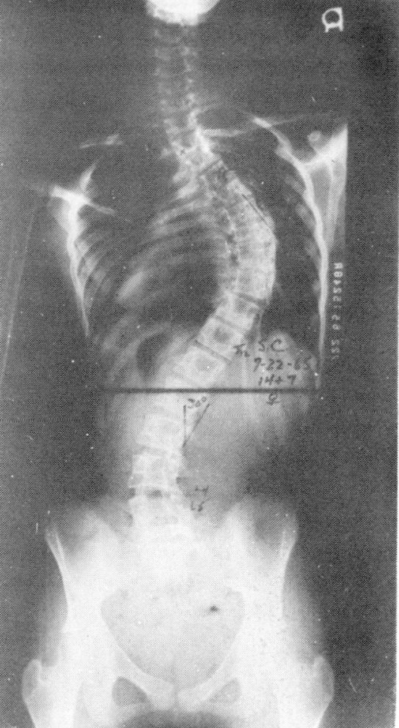

A

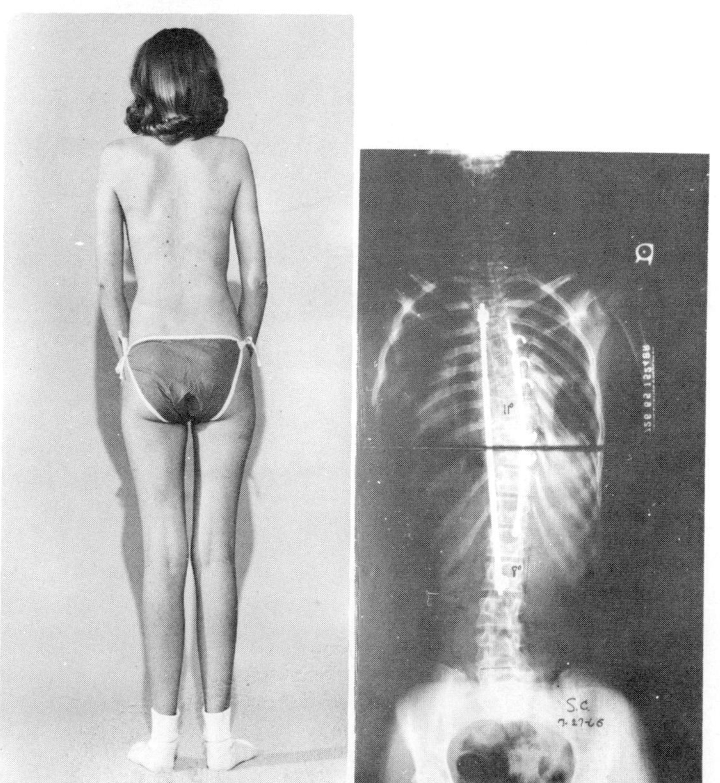

B

Fig. 7-32 Scoliosis. A. A young patient with severe congenital idiopathic scholiosis. B. Same patient after surgical correction. In the x-ray, note the prosthetic devices straightening the vertebral column. (Courtesy of Dr. Paul Harrington, Texas Institute of Rehabilitation and Research, Houston, Texas.)

unit 8 thoracic viscera

The viscera of the chest are harbored within, and thus protected by, a cage created from slender curved bones (ribs) attached behind to the twelve thoracic vertebrae and in front to the sternum. In the space between each pair of ribs is a sheet of muscle, three layers deep (intercostals).

The *heart* and the *lungs* are the principal residents of the thorax. It is a reasonably efficient housing for both; for the heart because it can pump oxygen, within the medium of blood, a short distance to the head and upper limbs, overcoming the force of gravity without undue strain; for the lungs because they rest on the diaphragm, the floor of the thorax, which performs as a muscular bellows, drawing air to and fro through the lungs and their airway system.

Anatomically, the organization of structures within the thorax is quite simple: the lungs, their lining, and their vessels and nerves occupy the outer two-thirds of the cavity; the heart and transient structures, lying deep to the sternum, fill the region between the lungs, called the mediastinum.

ORGANIZATION OF THE MEDIASTINUM

The **mediastinum,** as you can see in Fig. 8-1, is arranged into four compartments: anterior, posterior, middle, and superior. The upper limit of the mediastinum is set by the superior aperture of the thorax, a thoroughfare between neck and chest. The remaining boundaries are, below, the diaphragm, laterally, the medial surface of the lungs, anteriorly, the sternum, and posteriorly, the bodies of thoracic vertebrae.

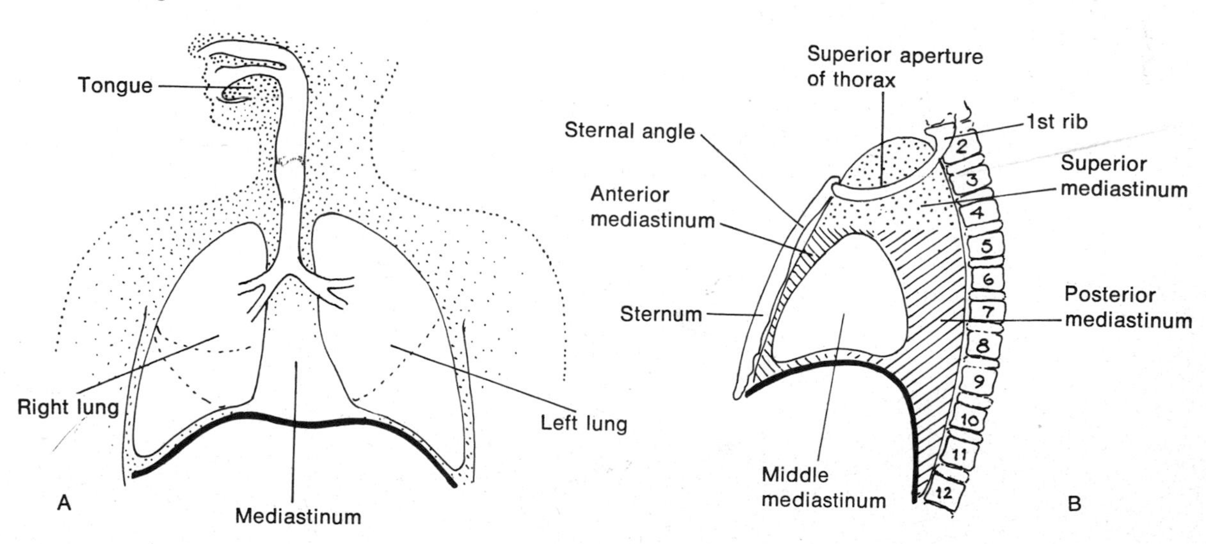

Fig. 8-1 The mediastinum. A. Anterior view. B. Lateral view demonstrating subdivisions.

As you can verify in Fig. 8-2:

1. The principal resident of the mediastinum is the heart, occupying the middle compartment.
2. The anterior mediastinum is virtually free of major structures.
3. The posterior compartment is largely filled with structures en route to the lungs (trachea, bronchi), and abdomen (aorta, esophagus, nerves, etc.).
4. The superior compartment contains structures in transit from the neck to the posterior mediastinum and points below (esophagus, etc.), the great vessels springing from and draining into the heart, as well as local viscera such as the thymus gland.

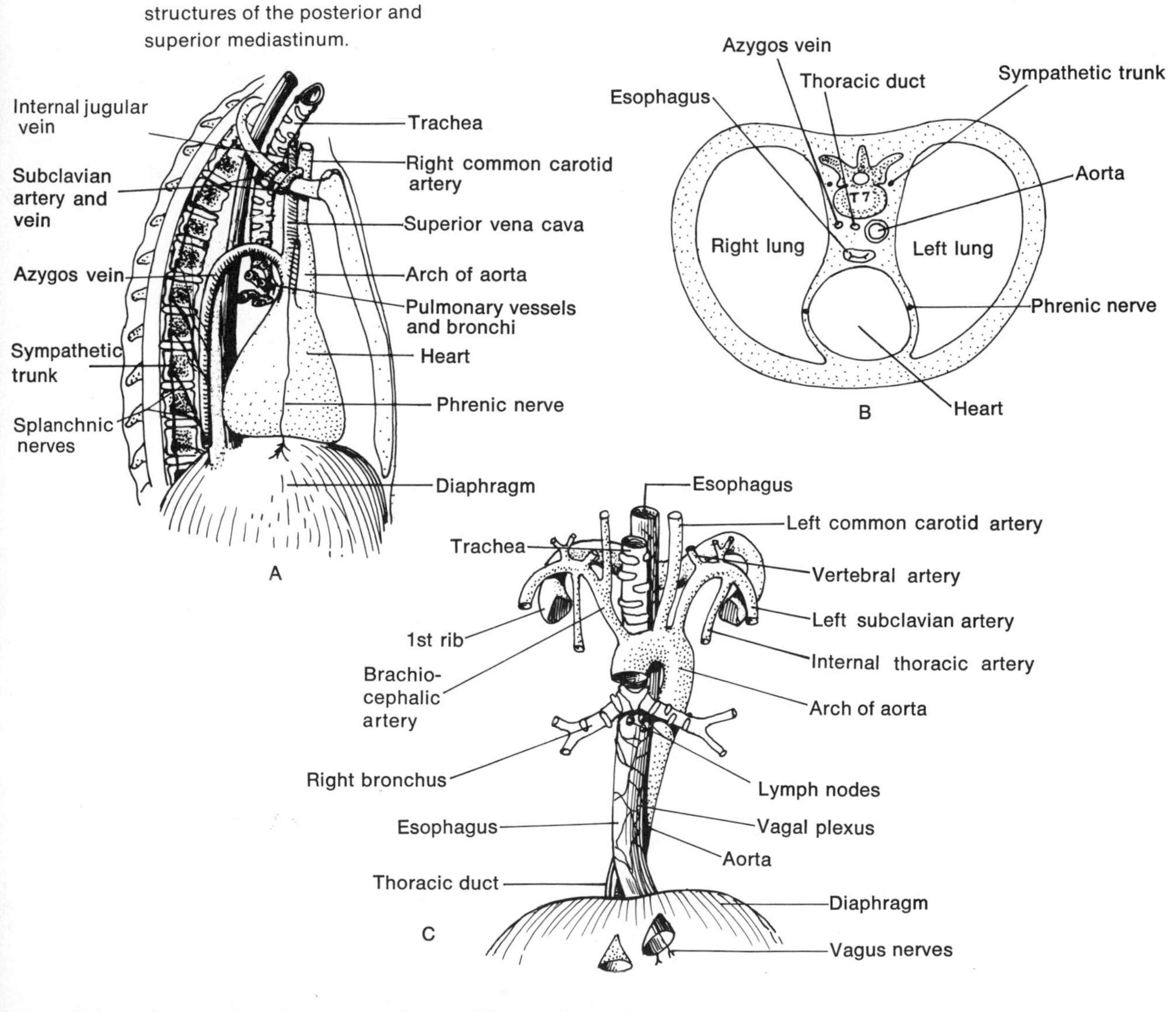

Fig. 8-2 Contents of the mediastinum. A. Lateral view of the posterior and middle mediastinum. B. Cross section of the middle and posterior mediastinum. C. Some structures of the posterior and superior mediastinum.

cardiovascular structures of the mediastinum

THE HEART

A concept

The heart is a muscular pump which provides the force for the movement of blood through all the vessels within the body. Thanks to the heart:

1. The tissues of the body receive oxygen and nutrition.
2. The tissues of the body have a disposal service for discharge of carbon dioxide and other "undesirable" elements.
3. The tissues of the body can release secretory material that within seconds exerts an influence some distance from its source.
4. Medications injected into the body are distributed to all the "nooks and crannies."
5. The body's defensive weaponry, e.g., antibodies and phagocytes, may be mobilized to areas of infection and inflammation.

Conversely, diminished heart action can result in insufficient tissue oxygenation, venous pooling (stasis), filling of tissue spaces (edema) due to venous back-pressure, and other conditions. Arrested heart action causes death because the cells of the brain initially, and all cells subsequently, die when the oxygen supply stops.

As is the case with any pump, there is an input as well as an output side (Fig. 8-3). The input chamber of the heart is called an **atrium;** the output chamber, a **ventricle.** The apparent simplicity of this operation is complicated by the fact that two pumps are necessary (Fig. 8-4):

Pump number 1 collects deoxygenated blood from the body tissues via two great veins (input) and drives it into the lungs for aeration by way of one great arterial* trunk (output).

Fig. 8-3 Schematic of a theoretical one-pump heart. Note the absence of an oxygenation circuit.

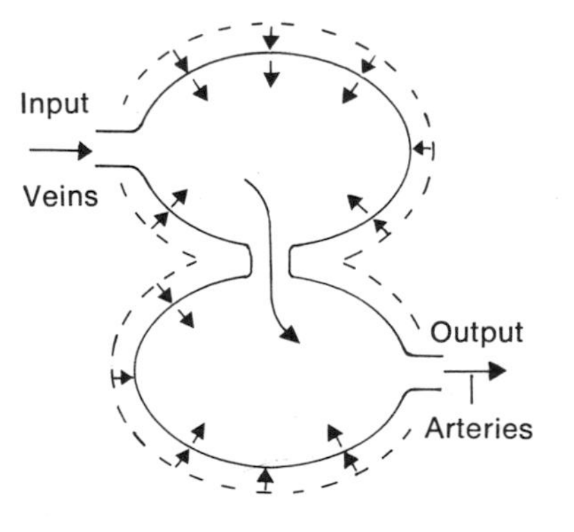

* Arteries conduct blood (regardless of its oxygen state) *away from* the heart, and veins carry it *toward* the heart.

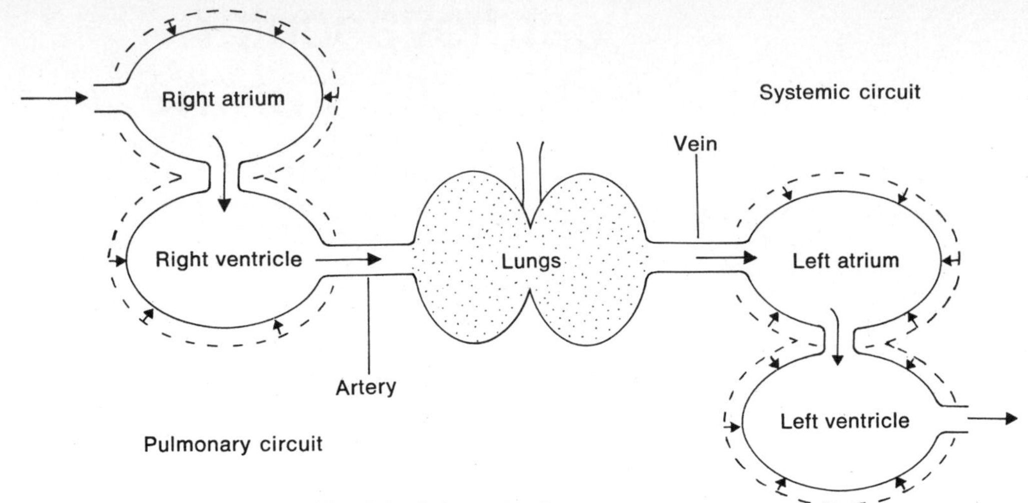

Fig. 8-4 Schematic of a two-pump heart.

Pump number 2 collects freshly oxygenated blood from the lungs by means of four veins (input) and drives it to the body's tissues through one great artery (output).

These two pumps, composed of two atria and two ventricles in a triple-layered bag (pericardium) and punctured by orifices for six large veins and two great arteries, constitute the **heart.**

Orientation of the heart in the chest

The heart is the size of a fist and is shaped like a cone, and tilted slightly forward and to one side. It is within the middle mediastinum. Its apex points downward, forward, and to the left. Its base faces backward, upward, and to the right. This can be verified in Fig. 8-5. As you can see, the base (consisting of the two atria) is thickly populated with vessels. Projected onto the surface, the base is roughly 1 in. below the sternal angle. The apex, part of which is the left ventricle, is represented on the surface in the 5th intercostal space along the midclavicular line. Can you find these landmarks on yourself?

Because of the way the heart developed, the chambers are *not* oriented along vertical and horizontal planes. Instead, as you can see in Fig. 8-6B, the heart is turned and inclined such that:

1. The right atrium, occupies the right quadrant of the heart, from posterior to anterior.

2. The right ventricle occupies the anterior central quadrant.
3. The left ventricle occupies much of the left and posterior quadrant, below the left atrium.
4. The left atrium occupies most of the posterior quadrant, somewhat above and behind the left ventricle.

On the heart's external surface the atria are set off from the ventricles by a groove (sulcus). Since it houses, in part, the coronary sinus, it is named the **coronary sulcus.** The ventricles are separated externally by a pair of sulci—anterior and

Fig. 8-5 Orientation of the heart in the thorax. A. Posterior surface (base of the heart) viewed from behind. B. Anterior surface facing the sternum. C. Diaphragmatic surface and some of the posterior surface, viewed from below and behind.

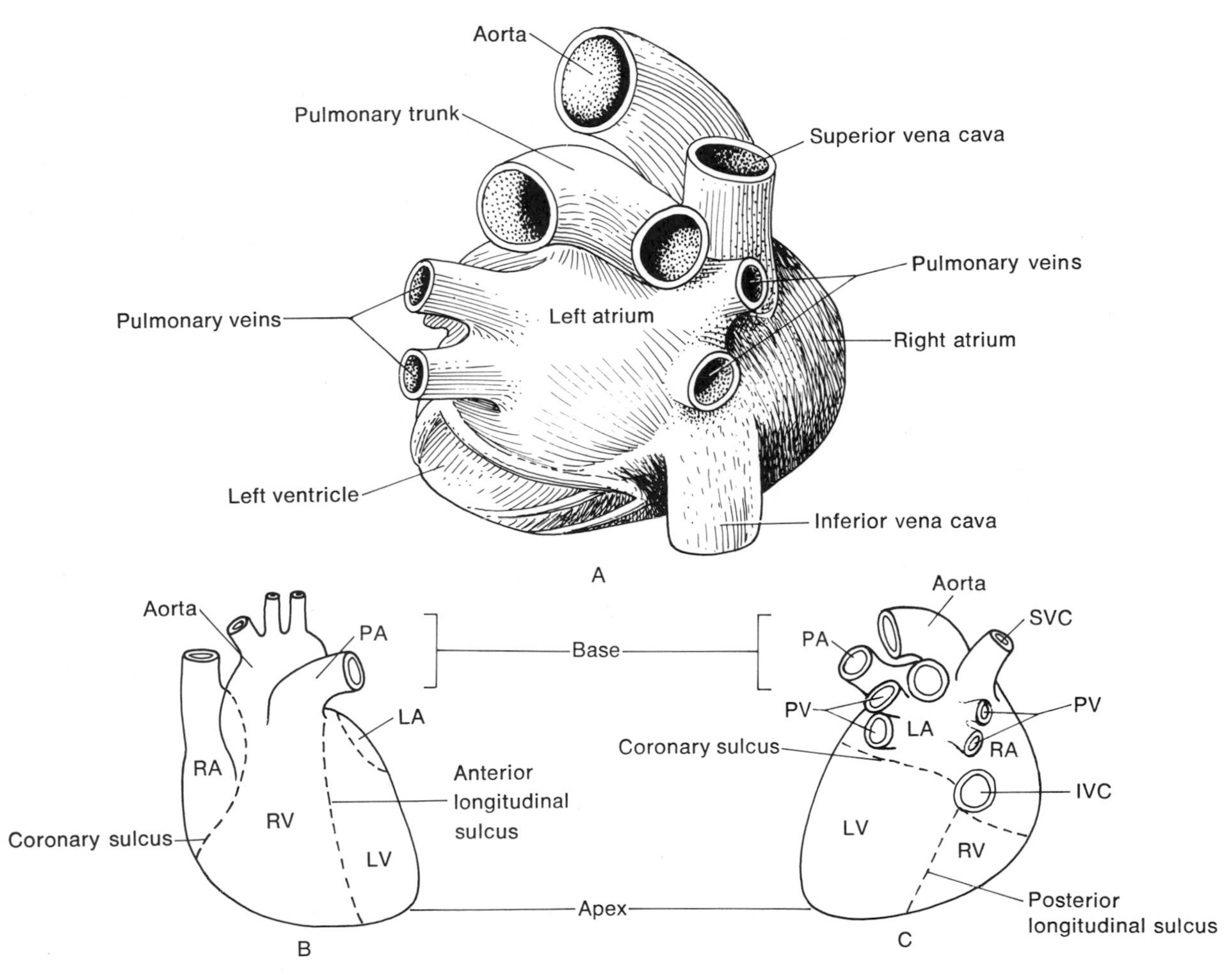

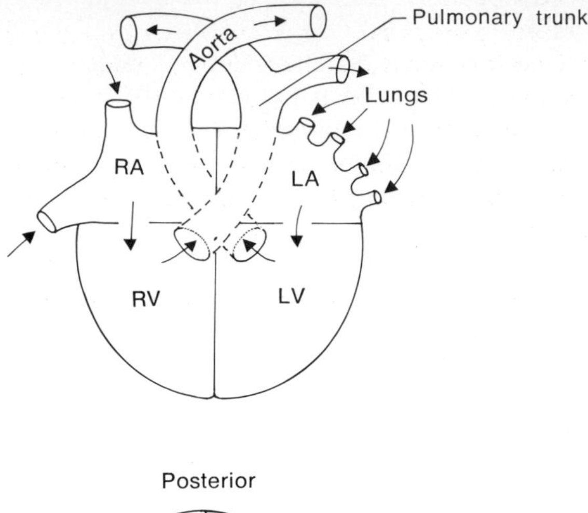

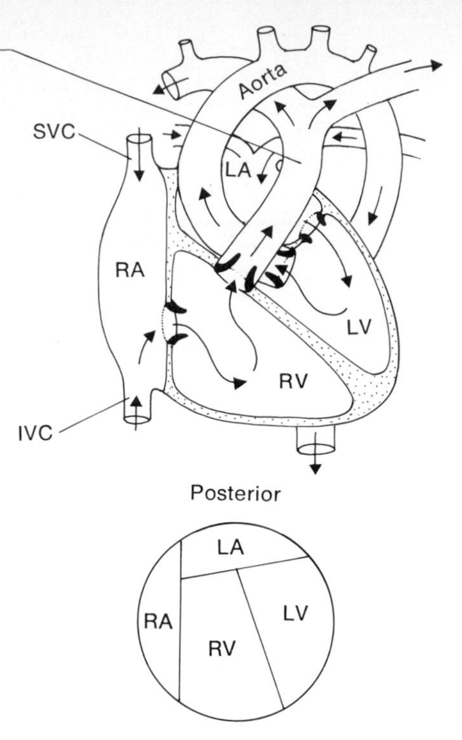

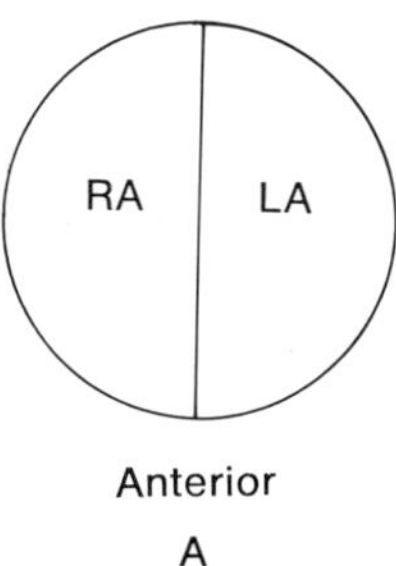

Fig. 8-6 Schematic of compartment orientation of the heart. A. Above, anterior view of idealized vertical-horizontal compartmentalization. Below, view from above of such compartmentalization. B. Above, anterior view of actual compartments. Below, superior view of such compartments.

posterior—oriented along the longitudinal axis of the heart. These are the **longitudinal sulci** (Fig. 8-5). They transmit the coronary arteries. Because the heart, from below upwards, is inclined to the right and backward (Fig. 8-7):

1. The left atrium is related to structures of the posterior mediastinum, e.g., esophagus.
2. The left ventricle rests almost entirely on the diaphragm and is related laterally to the left lung and its membranes.
3. The right ventricle is related anteriorly to the sternum and below to the diaphragm.
4. The right atrium is related to the right lung and its membrane.

Pericardium

The heart is enveloped by pericardium, a double-walled sac. The inner wall, or **visceral pericardium,** adheres closely to the myocardium. At the base of the heart the inner wall turns outward and heads back (is reflected) to become the inner surface of the **parietal pericardium.** The potential space between these two pericardial layers constitutes the **pericardial cavity** (Fig. 8-8).

The layers are lined with mesothelial cells that secrete a serous fluid which allows for frictionless movement of the heart against the pericardium. Loss of serous secretion may lead to the inflammatory condition of pericarditis.

The outer surface of the parietal pericardium is **fibrous** and is tied down to the great vessels at the base of the heart, to the sternum in front, to the vertebral column in back, and to the diaphragm below.

Flow circuits

In the body there are two circuits of blood flow (Fig. 8-9). One involves the "right heart" (right atrium and ventricle), the other, the "left heart" (left atrium and ventricle).

Pulmonary circuit: right ventricle (pump) → lungs → left atrium (reservoir)

Systemic circuit: left ventricle (pump) → body tissues → right atrium (reservoir)

Fig. 8-7 Relations of the heart. A. Anterior view. B. Posterior view.

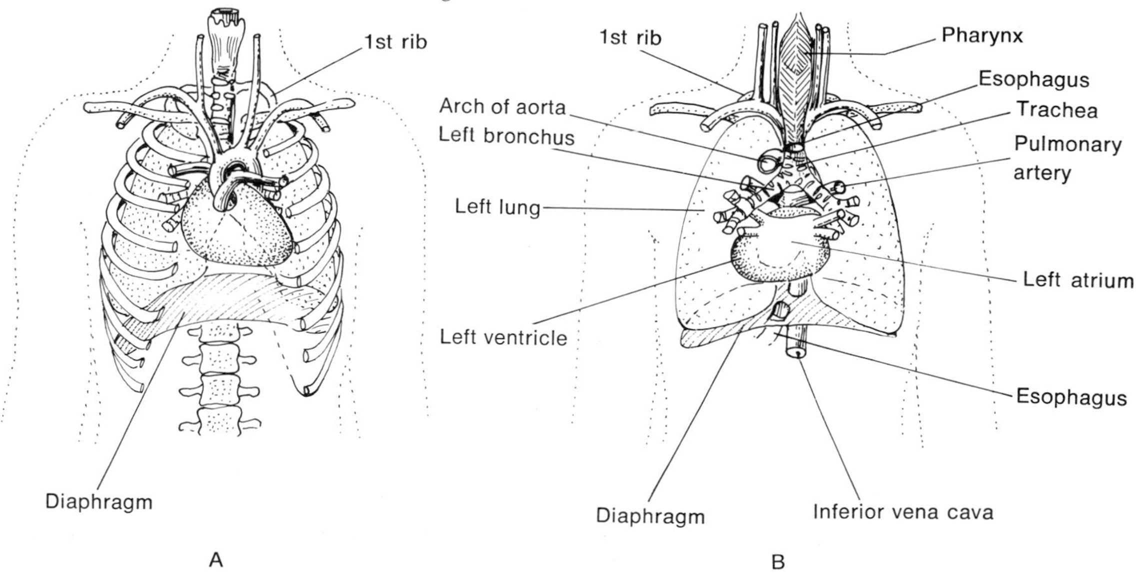

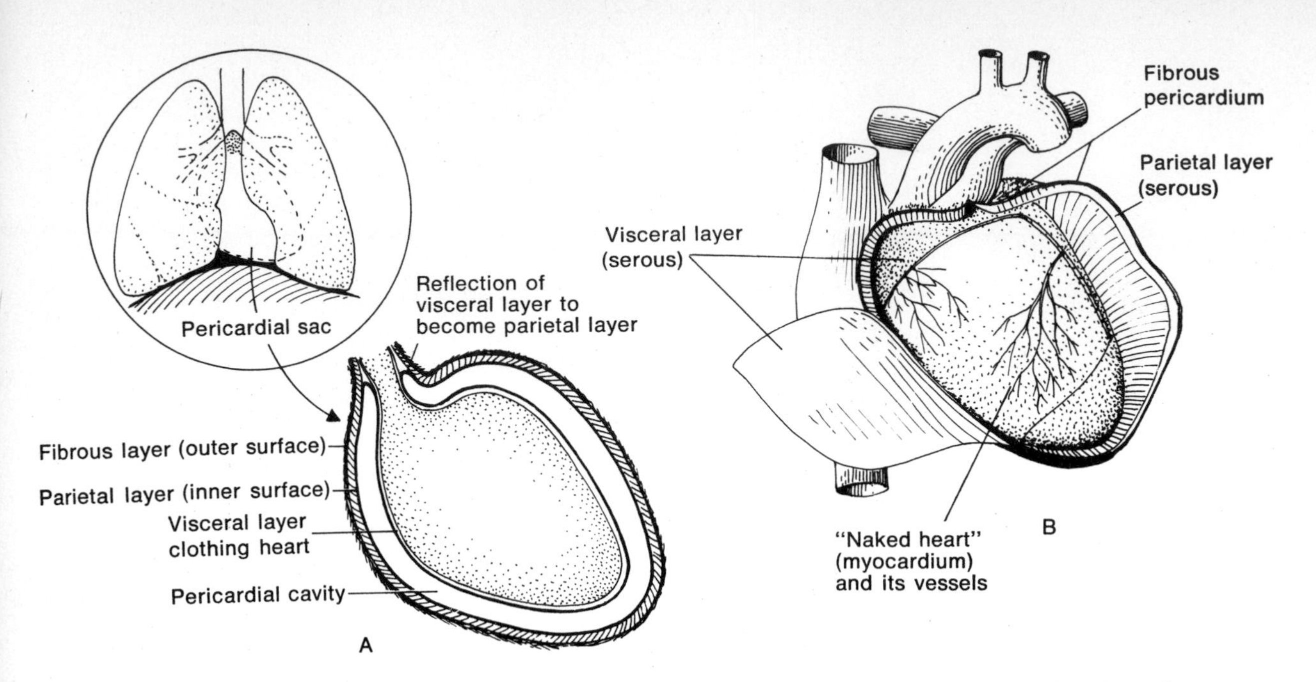

Fig. 8-8 The pericardium and its cavity (slightly enlarged).
A. Scheme of pericardial layers. B. Pericardial layers as seen about the heart in situ. Note that the heart lies in the pericardial sac but not the pericardial cavity.

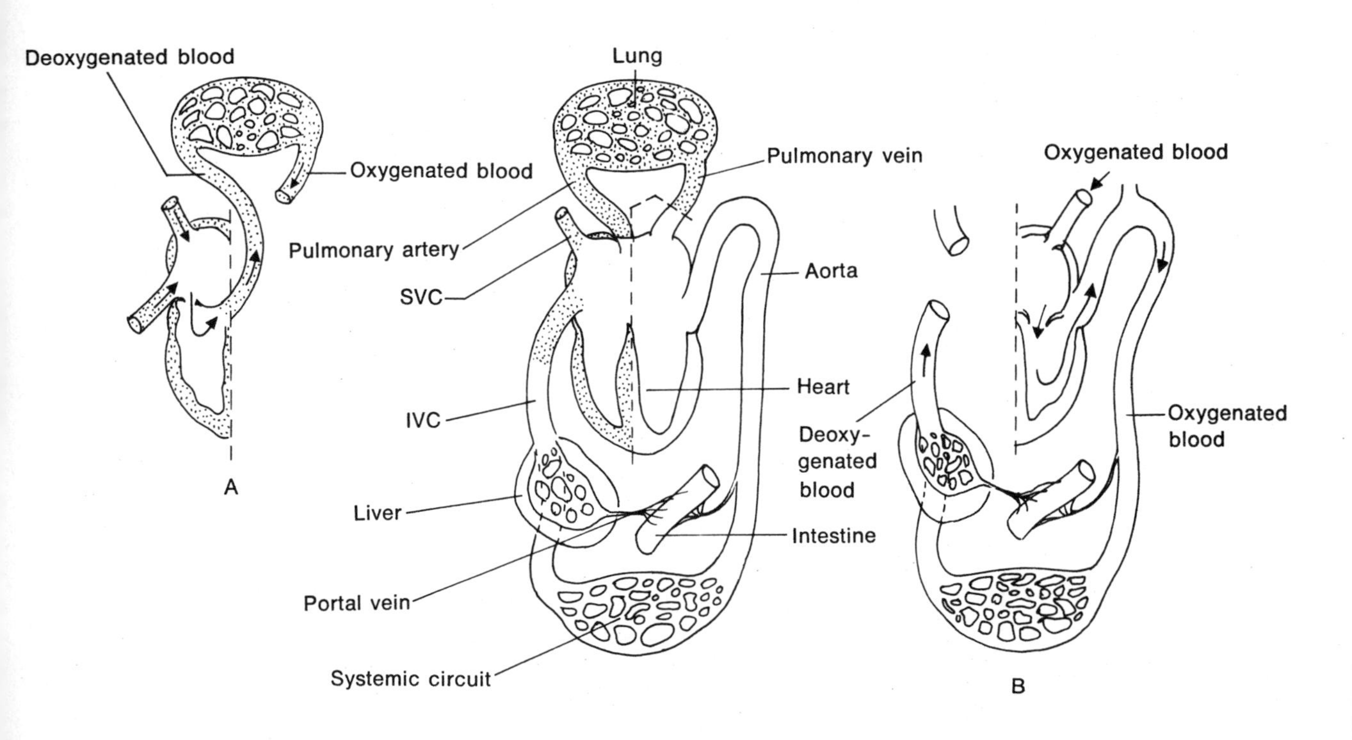

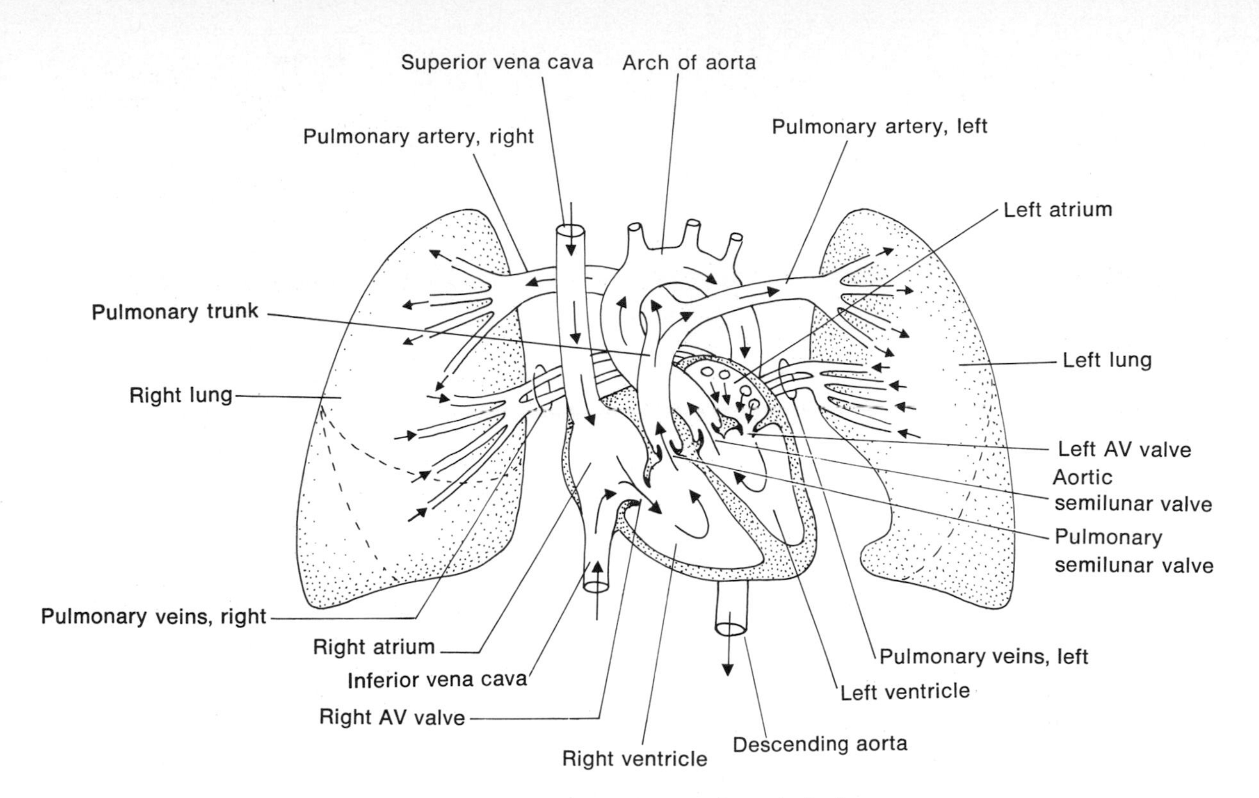

Fig. 8-10 Flow diagram through the heart.

The pulmonary circuit functions to oxygenate blood returning to the heart from the body tissues. The systemic circuit functions to pump oxygenated blood to the tissues of the body and return it—deoxygenated—to the heart. The difference in pressure required to drive the blood through each circuit is considerable and is reflected in the structure of the ventricles.

Refer now to Fig. 8-10 as you read the following description:

Deoxygenated blood enters the **right atrium** via the **inferior** and **superior venae cavae** (IVC, SVC). The IVC generally drains the lower extremities and the abdomen; while the SVC generally drains the upper extremities, the head, the neck and the thorax.

Fig. 8-9 (opposite) Flow circuits of blood. A. Pulmonary circuit (related vessels are stippled). One pulmonary artery and vein are shown. B. Systemic circuit. Center, general circulation plan of the body (pulmonary circuit is stippled).

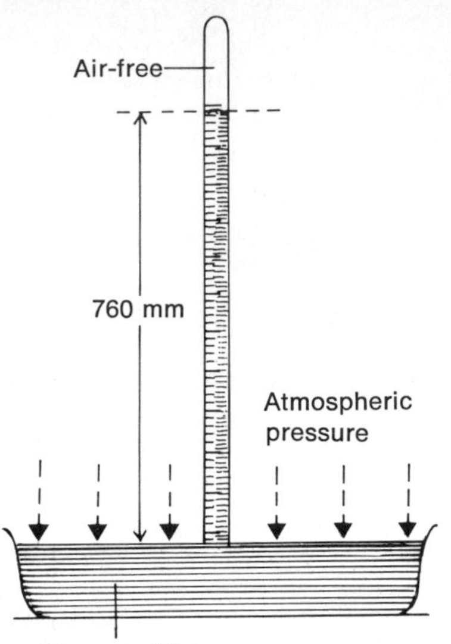

Fig. 8-11 The atmosphere is a collection of gases exerting a pressure. Conventionally, this pressure is found by measuring the distance (in inches or millimeters) it will force mercury (Hg) upward into an evacuated tube. Blood pressures are measured from a starting point of 1 atmosphere.

The blood in the right atrium is passed into the **right ventricle** at a pressure of about 5 mm Hg (see Fig. 8-11) by way of the right atrioventricular orifice, guarded by a tricuspid (three-cusped) valve.

The still-deoxygenated blood is pumped out of the right ventricle at a pressure of about 26 mm Hg into the **pulmonary trunk,** whose branches, the **right** and **left pulmonary arteries,** deliver the blood to the capillary plexuses of the lung.

Differences in pressure (pressure *gradients*) cause oxygen to diffuse into the capillaries of the lung from the lung's air cells (alveoli). Conversely, carbon dioxide diffuses into the air cells from the capillaries. Consequently, as the blood leaves the lungs for the heart, it is very nearly saturated with oxygen.

The oxygenated blood is conducted to the **left atrium** via four **pulmonary veins**—two from each lung.

The blood in the left atrium passes into the **left ventricle** at a pressure of about 5 mm Hg via the left atrioventricular orifice, guarded by a bicuspid (two-cusped) valve.

The highly oxygenated blood is pumped out of the left ventricle into the **aorta** (and so on to the tissues of the body) via the aortic valve at a peak pressure of about 120 mm Hg.

Right atrium

The right atrium receives the venous blood from all areas of the body by way of the superior and inferior venae cavae. In effect, it is a reservoir. Thin-walled, its contraction drives deoxygenated blood forward into the right ventricle. This action takes about 0.1 sec.

Much of the structure characteristic of the right atrium is best seen in the laboratory. Functionally the following intra-atrial structures (Fig. 8-12) are significant:

1. The orifices of the superior vena cava, the inferior vena cava, and the coronary sinus (draining the vessels that supply blood to the heart)—all of which discharge venous blood into the atrium.
2. The orifice of the atrioventricular channel, through which the venous blood is forced into the right ventricle.
3. The wall between the right and left atria **(interatrial septum).** In the developing fetus, there is a communication (foramen ovale) between the atria which allows oxygenated blood to bypass the pulmonary circuit, since the lungs are not inflated and so not involved in aeration of fetal blood (Fig. 8-12). After birth, this communication in the septum closes, leaving an oval fossa and its rounded margin (limbus).

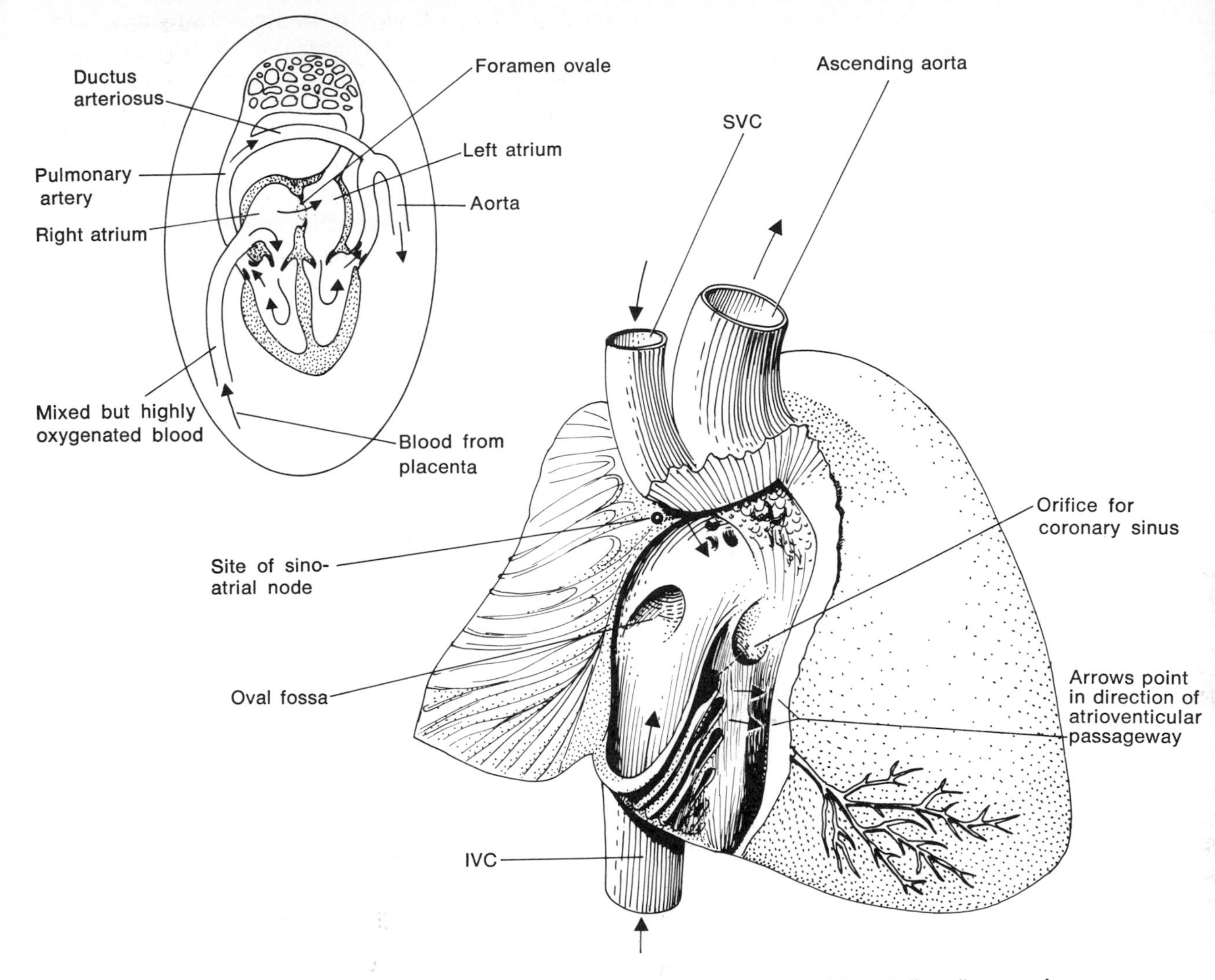

Fig. 8-12 Right atrium, anterior flap opened. Insert: flow diagram of a fetal heart. As lungs do not function in fetal life (when the fetus is in an aqueous environment), routing of placental (oxygenated) blood through the patent (open) foramen ovale and ductus arteriosus ensures a supply of oxygenated blood.

Right ventricle

The right ventricle receives deoxygenated blood from the right atrium via the right atrioventricular orifice. Whereas the atrium is a reservoir, the ventricle is a pump. The thickness of the ventricular wall is a testament to this increased pressure. Blood is pumped from the right ventricle into the pulmonary trunk, an action taking about 0.3 sec.

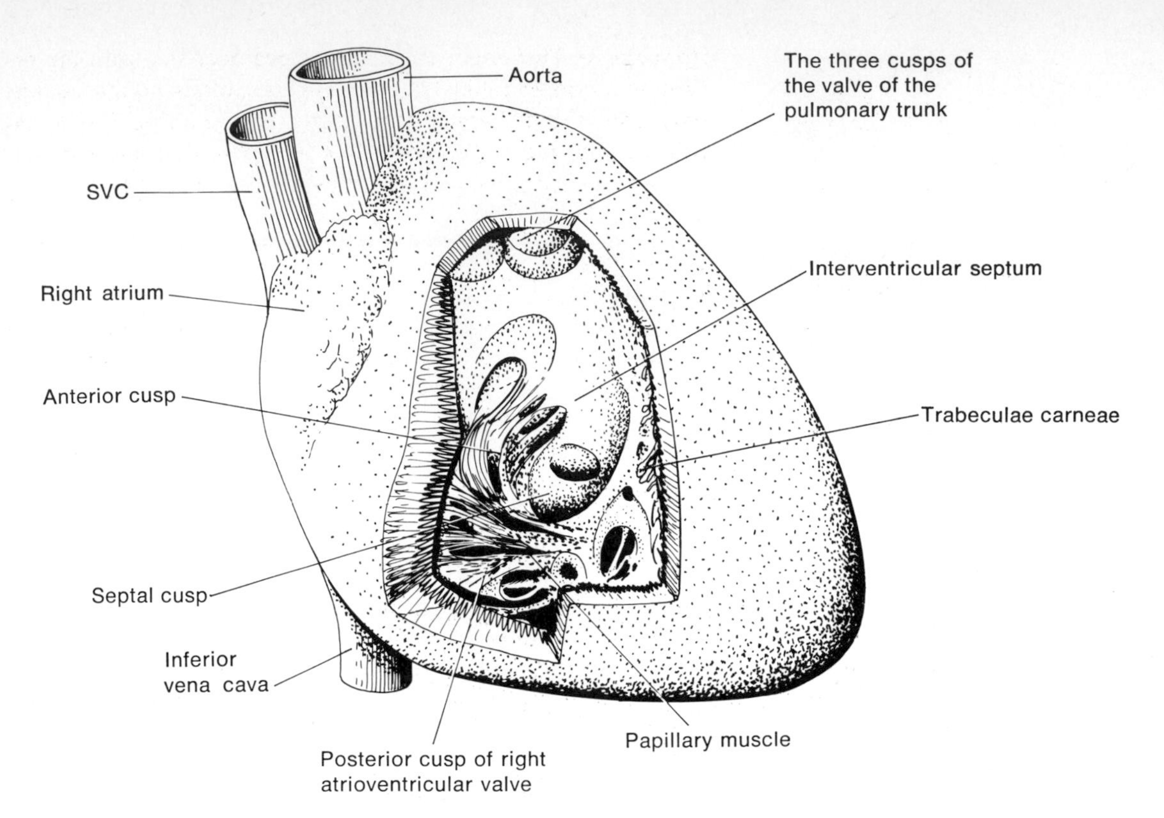

Fig. 8-13 Right ventricle, anterior flap opened. Relate thickness of the right ventricular wall to that of the left ventricular wall (Fig. 8-14).

The right ventricle may be seen internally in Fig. 8-13. The atrioventricular orifice is guarded, on the ventricular side, by a valve consisting of three parachute-shaped cusps, hence its name, **tricuspid valve.** These cusps are tied down by tendinous cords to specialized (papillary) muscle masses. When the blood in the ventricle is forced upward by the pumping of the myocardium, it also forces the cusps upward, toward the atrioventricular orifice, closing the pathway back to the atrium and thereby preventing regurgitation. Overflap is inhibited by the tendinous cords, which are tensed by the contraction of the papillary muscles at the moment of ventricular contraction.

The pulmonary arterial trunk is at the upper extreme of the right ventricle. The **pulmonary valve** is decidedly different in structure from the atrioventricular valve. The three saclike cusps of the pulmonic valves are forced flush against the sides of the vessel as the blood rushes upward during ventricular contraction. As the contraction ceases and the pressure diminishes gravity tends to return blood to the ventricle from above. The upper surface of these sacs fills with the returning blood and they fall back in place, blocking the pulmonary orifice and thereby preventing regurgitation. The blood passing into the trunk is diverted into the right and left pulmonary arteries, which conduct the deoxygenated blood laterally toward the lungs.

Left atrium

The left atrium receives the freshly oxygenated blood from the lungs via four pulmonary veins. As with the right atrium, the left is essentially a thin-walled reservoir, driving oxygenated blood into the left ventricle via the left atrioventricular orifice at a pressure of about 5 mm Hg. The interior of the left atrium is similar to that of the right. The left and right atria contract simultaneously.

Left ventricle

The left ventricle receives the oxygenated blood from the left atrium and drives it into the ascending aorta at a peak pressure of 120 mm Hg. As could be predicted, the wall of the left ventricle is significantly thicker than the wall of the right ventricle.

The interior of the left ventricle can be seen in Fig. 8-14. The left atrioventricular orifice is guarded by a valve of two flaplike cusps, hence the name **bicuspid valve.** The left AV valves are similar in anatomy and function to the right AV valves.

The cusps of the aortic valve function like their pulmonary counterparts. At the anterior and lateral cusps there are orifices **(sinuses)** through which blood enters the coronary arteries and is distributed to the heart muscle itself. The ventricles of the heart contract simultaneously.

Skeleton of the heart

When viewed from above, dense connective tissue, often infiltrated with cartilaginous tissue (fibrocartilage), may be seen surrounding and giving support to the four valve-guarded orifices of the heart (Fig. 8-15). In addition, the cardiac muscle arises from and inserts into this fibrous "skeleton" of the heart. The myocardium generally arranges itself in spiral sheets, the ventricles twist during contraction and literally squeeze out the blood they contain. The spirally oriented myocardium of the ventricles is separated from the musculature of the atria by the fibrous skeleton.

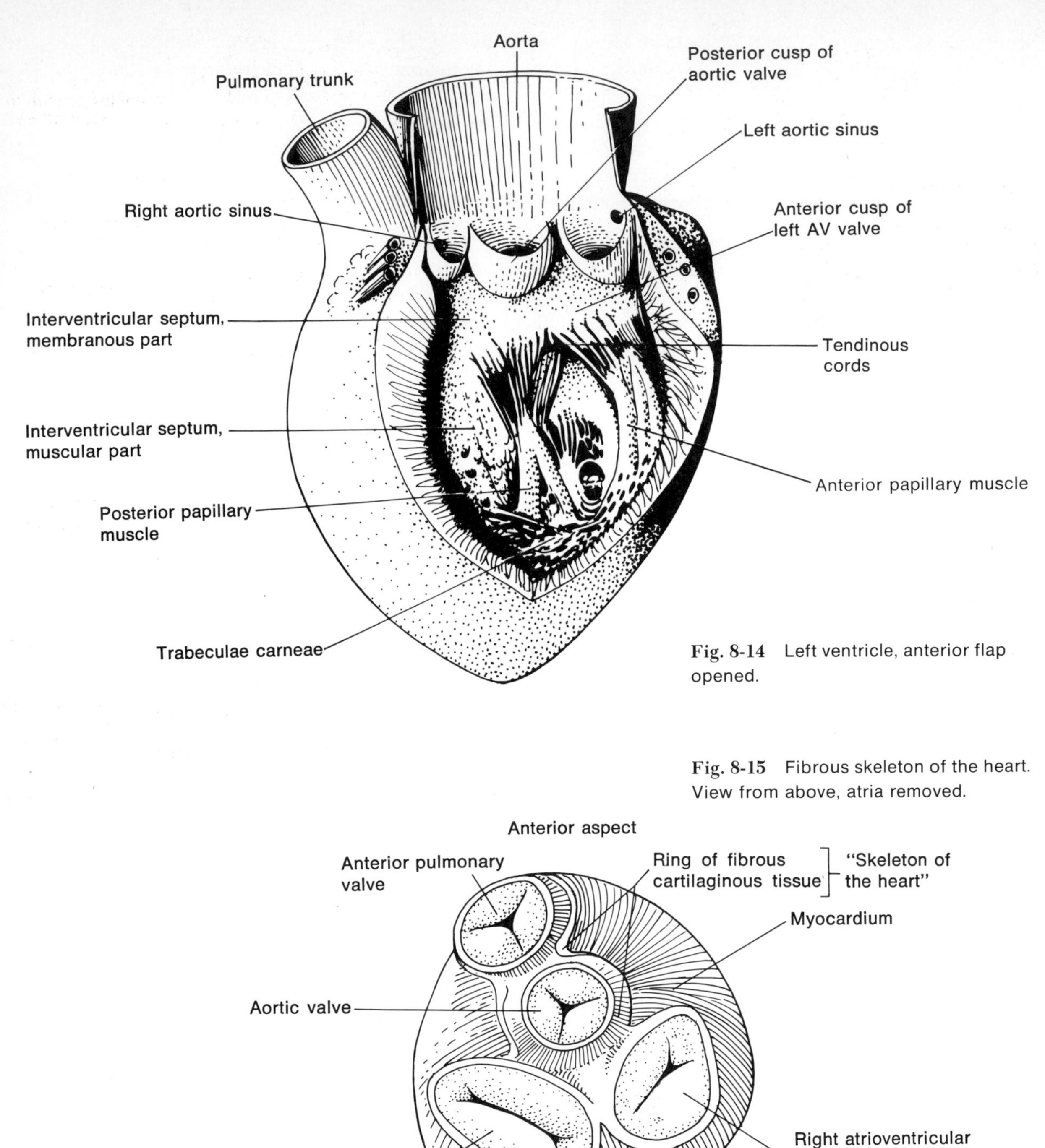

Fig. 8-14 Left ventricle, anterior flap opened.

Fig. 8-15 Fibrous skeleton of the heart. View from above, atria removed.

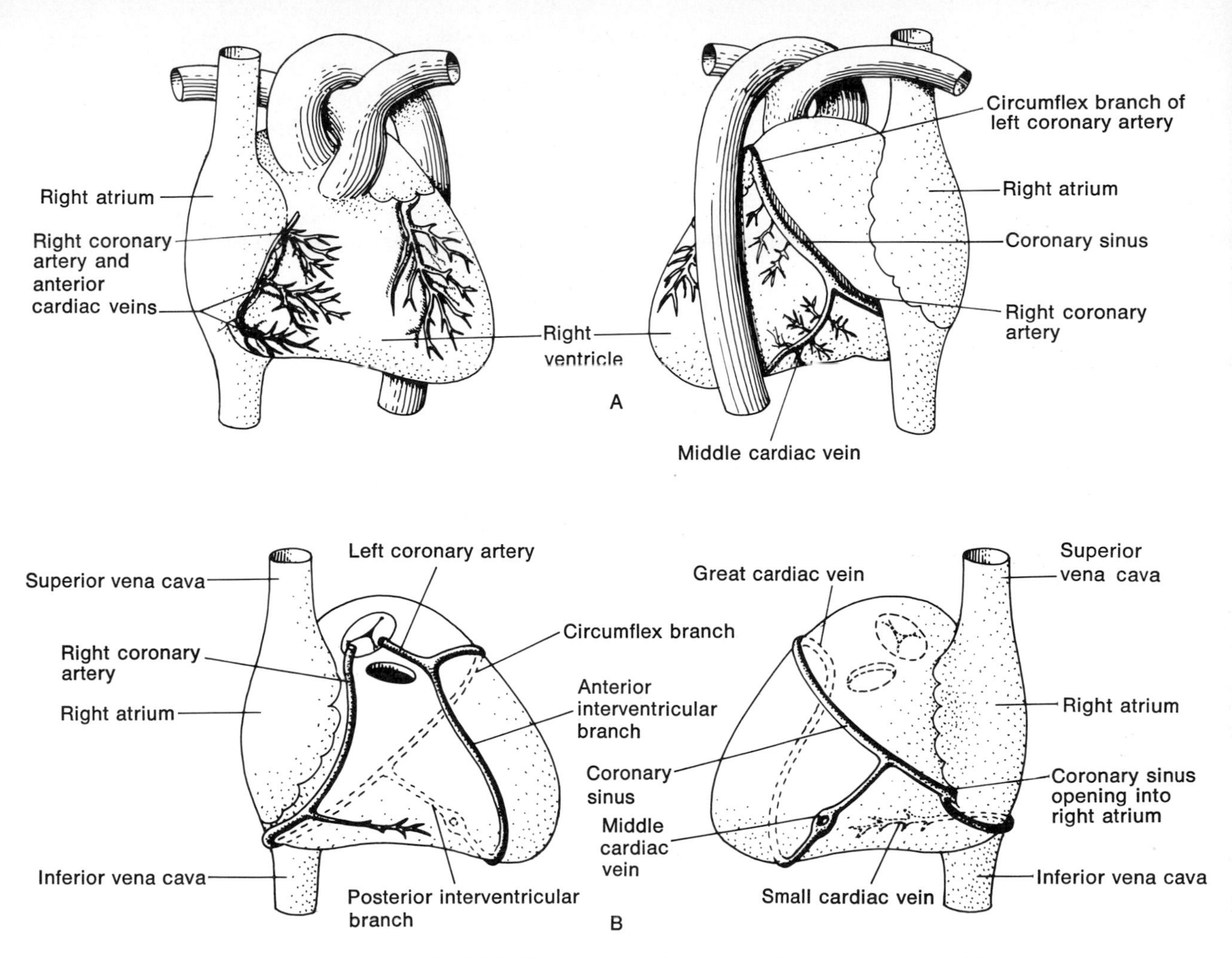

Fig. 8-16 Coronary circulation. A. Left, anterior surface; right, posteroinferior view. Vessels are drawn in situ. B. Same as above only shown diagrammatically.

Coronary circulation

The arteries supplying the heart form a crown, or corona, around it and so collectively have come to be called the **coronary arteries** (hence the term coronary circulation). Study Fig. 8-16 as you read the following:

There are two principal arteries, right and left coronary arteries, arising from the stem of the aorta.

The right coronary artery runs in the groove separating the atria from the ventricles (coronary sulcus) and there anastomoses with the circumflex branch of the left coronary artery.

The left coronary artery divides such that the anterior descending branch anastomoses with the posterior descending branch of the right coronary artery. Both arteries ride the interventricular grooves (longitudinal sulci) and so are named interventricular branches of the corresponding arteries.

The veins of the heart, which do not form a corona about the heart, are the **cardiac veins.** Most drain into the coronary sinus, which empties into the right atrium. Others drain directly into the right atrium.

Since the myocardium contracts 0.4 sec out of every 1 sec, it is imperative that the tissue be oxygenated at all times. Because the coronary arteries draw their blood from the ascending aorta this is insured unless foreign material (an embolus) blocks one of the arteries or its branches. Should this occur, the portion of the heart muscle served by the occluded vessel will die from lack of oxygen and, if the portion is large, the heart will be unable to function. Such a phenomenon is called a heart attack, a coronary occlusion, or a myocardial infarction.

Cardiac conduction system

The muscle fibers of the heart are unique in that they have an inherent capacity for rhythmic contraction; that is, they contract in the absence of neural stimulation. However, if each of the muscle cells contracted independently, the pump would not be very efficient. For better results all the fibers must contract in the appropriate sequence so the cavities can wring out the blood under a good head of pressure. Enter the **cardiac conduction system.**

The cardiac conduction system consists of a pair of specialized muscle cell masses (nodes) which generate a coordinated contraction (beat) of heart muscle, a set of specialized muscle tracts between the nodes, and a pair of specialized muscle bundles conducting the nodal-originated impulses to the myocardium of the ventricles. The rate of firing from the nodes sets the pace for the rate of heartbeat, hence the name for the principal node: pacemaker. The electrical activity of the conduction system and myocardial activity can be measured in an **electrocardiogram.**

Now refer to Fig. 8-17 as you read the following:

The node of special muscle cells **(SA, or sinoatrial, node)** in the right atrium "fires" a set of impulses which in about 0.1 sec races through the **internodal tracts,** causing the atrial musculature to contract (atrial systole). In the electrocardiogram (EKG) this event is represented by the P wave.

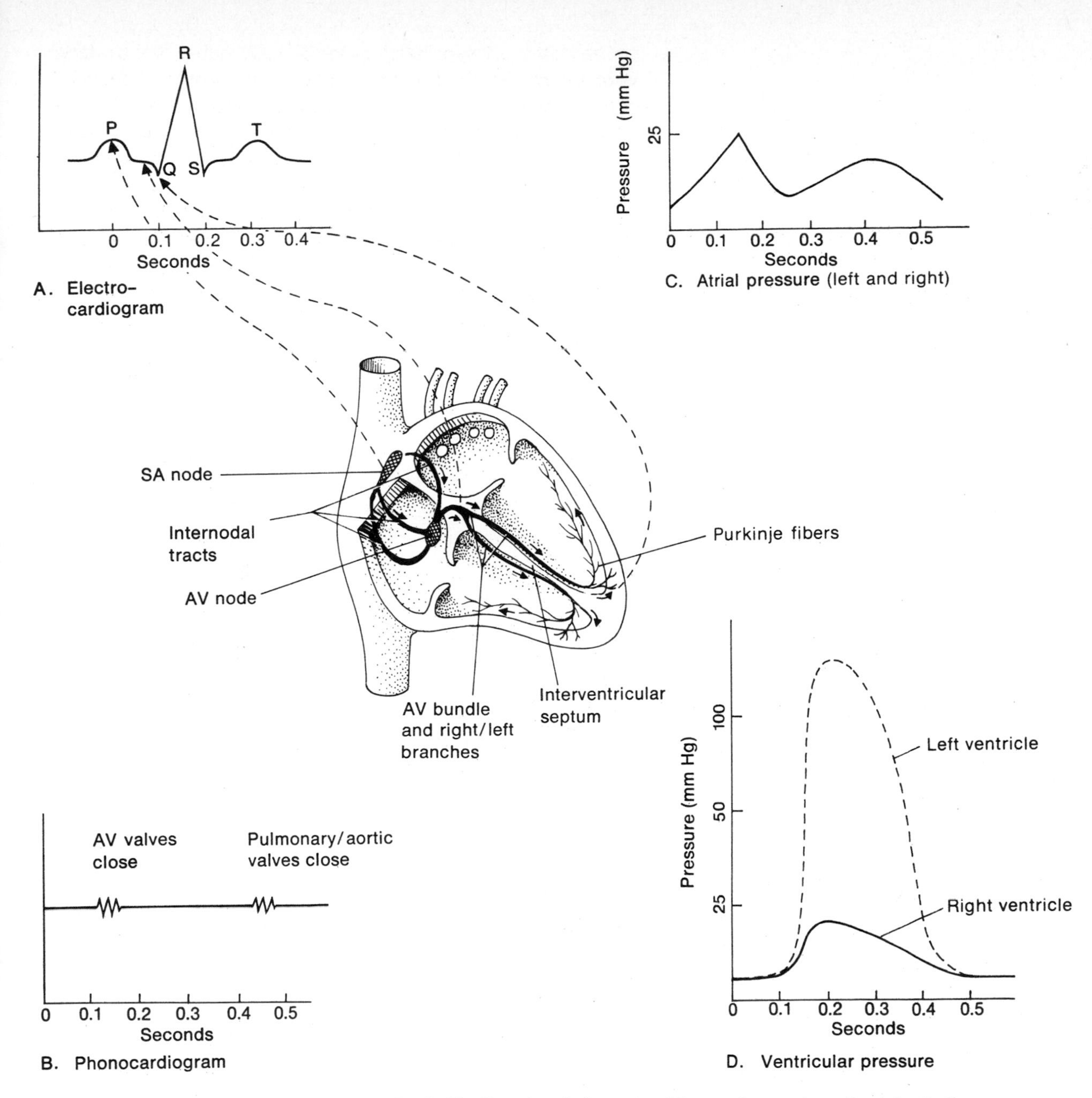

Fig. 8-17 Functional elements of the cardiac cycle as it relates to the cardiac conduction system. As you read the text, refer to Parts A to D.

There is a 0.1-sec delay as the second node (AV, or **atrioventricular, node)** in the medial wall of the right atrium becomes activated by the impulses reaching it from the SA node. The AV node then fires off impulses down two specialized muscle bundles lying superficially on the intermuscular septum. These bundles branch to become small (Purkinje) fibers perforating the ventricular myocardium. This movement of impulses takes about 0.1 sec and the ventricular muscle, in response, contracts from above downward. The electrical activity representing ventricular contraction is indicated on the EKG as the QRS complex. The subsequent T wave indicates the ventricular myocardium is returning to a resting state. Changes in these waveforms or the intervening segments may indicate abnormalities in heart function.

Extrinsic innervation of the heart

Although heart muscle is inherently contractile there are normally nerves, which cluster in the **cardiac plexus,** serving the cardiac conduction system and the coronary arteries (Fig. 8-18). These nerves regulate heart rate and the diameter of the coronary vessels. Since you cannot voluntarily regulate

Fig. 8-18 The cardiac plexus. These extrinsic nerves supply nodal tissue and the coronary arteries.

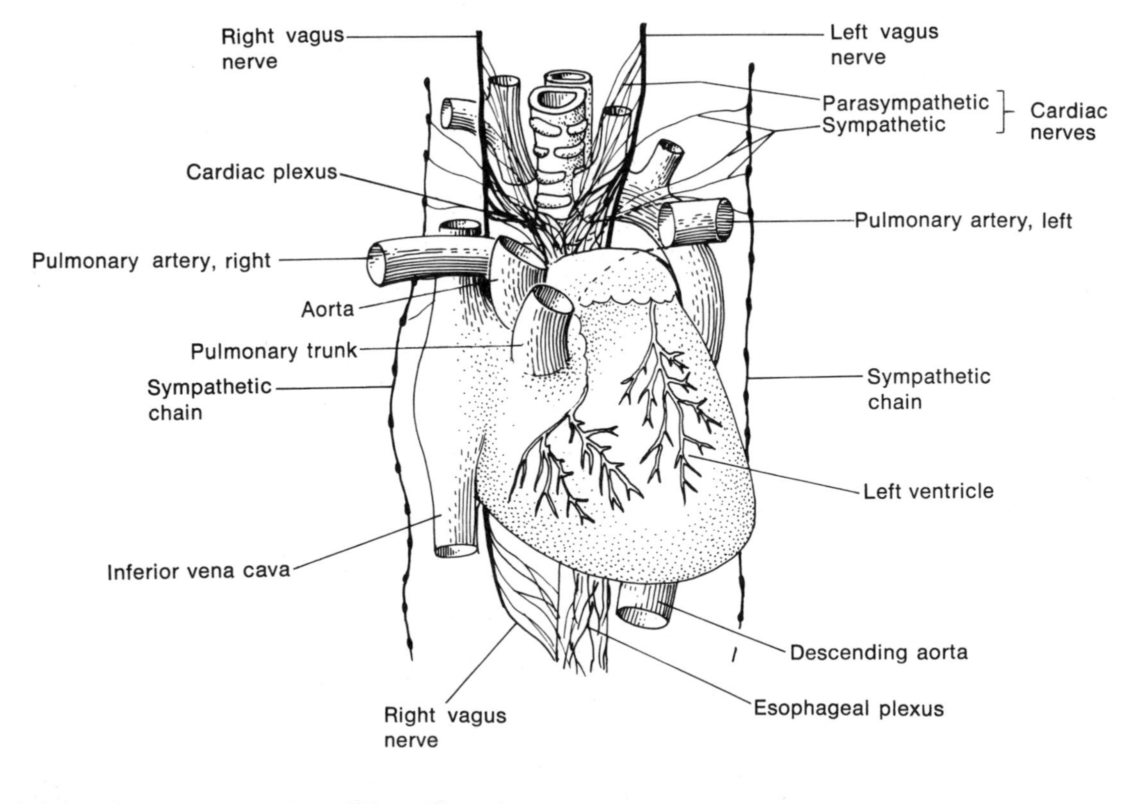

heart rate, you will not be surprised to learn that these nerves belong to the autonomic nervous system.

Nerves from the sympathetic chain stimulate the SA and AV nodes to increase the heart rate; they also cause relaxation of the musculature of the coronary arteries, the result being vasodilatation. These cardiac nerves—postganglionic fibers from the upper three ganglia of the sympathetic chain—can be seen in the posterior and superior mediastinum.

Nerves of the parasympathetic division (cardiac branches of the vagus nerves in the superior and posterior mediastinum) act on the SA and AV nodes to slow down the rate of heart beat.

The great vessels

The eight large veins and arteries at the base of the heart constitute the **great vessels** (Fig. 8-19). They populate the regions of the superior and posterior mediastinum. The eight are as follows:

1. Superior vena cava (SVC).
2. Aorta.
3. Pulmonary trunk.
4 to 7. Pulmonary veins.
8. Inferior vena cava (IVC).

Fig. 8-19 The great vessels. Turn the page for general views of the great veins and tributaries (A) and great arteries and branches (B).

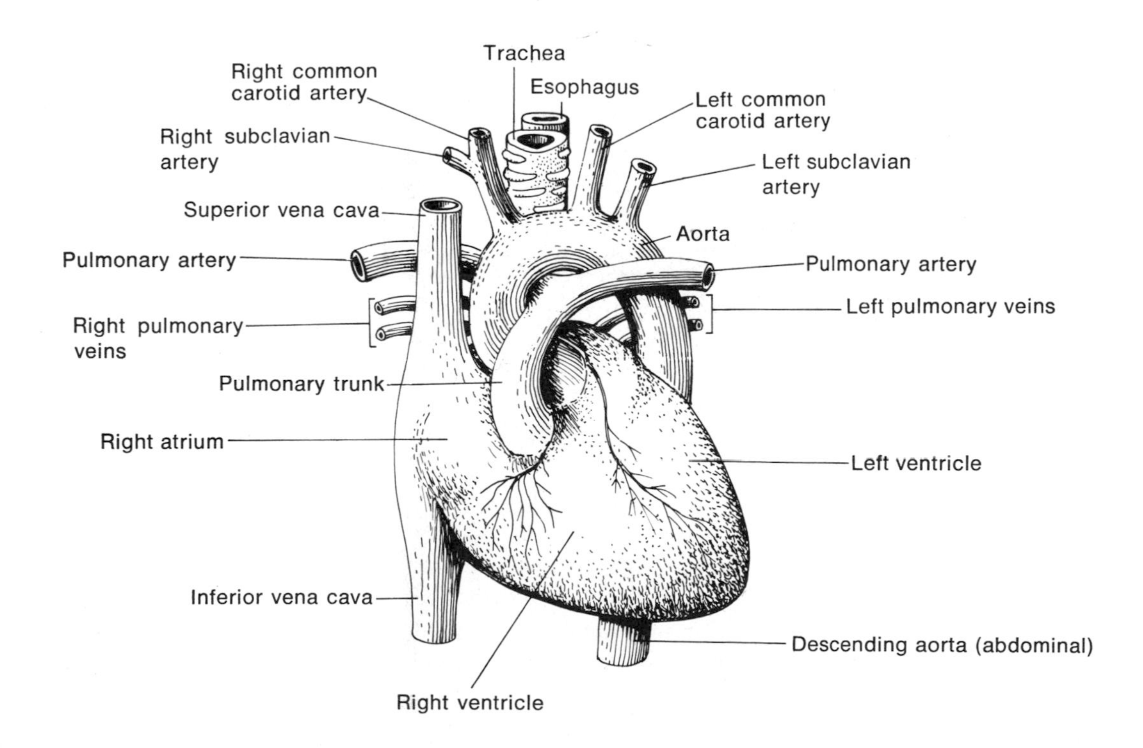

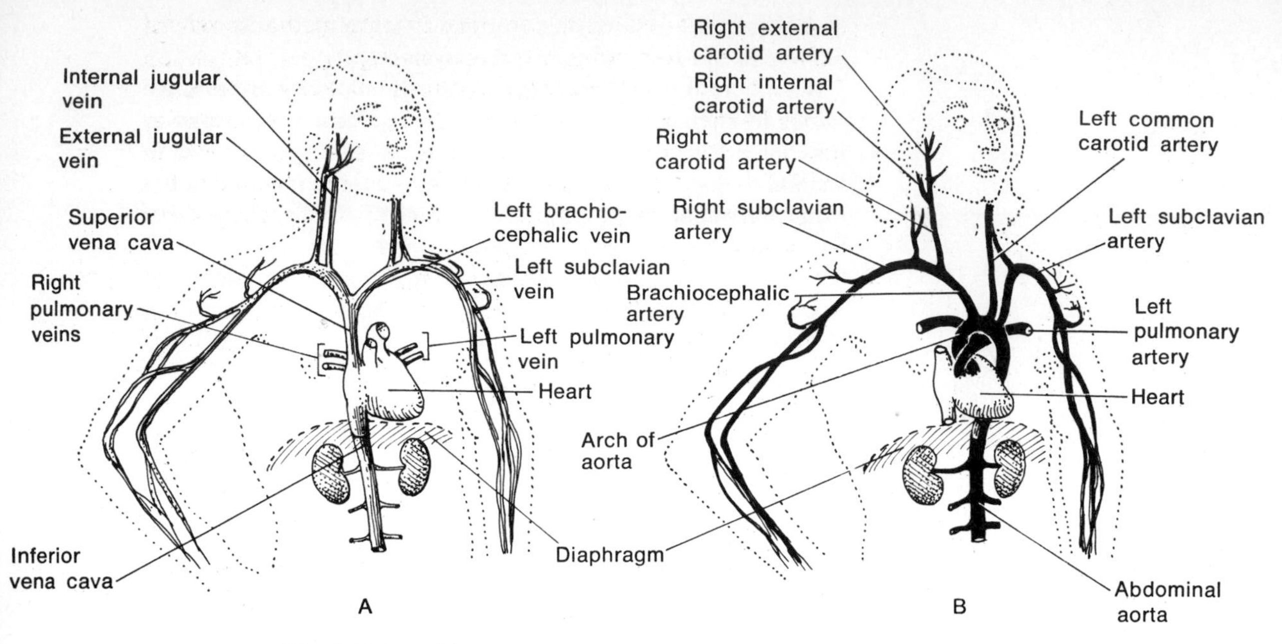

Fig. 8-19 (contd.)

The **superior vena cava,** entering the right atrium, receives venous blood from the brachiocephalic veins, which drain the head and neck (via internal jugular veins) and the upper limb (via subclavian veins); and from the azygos vein (thorax).

The **aorta,** the great artery arising from the left ventricle, is the immediate left neighbor of the SVC and forms an arch over the pulmonary trunk and arteries. The arch has an ascending portion and a descending one, the latter being continuous with the thoracic aorta. The arch gives off the brachiocephalic artery to the head and neck by way of the right common carotid, and to the upper limb on the right side via the subclavian. The arch also gives off the left common carotid and the left subclavian arteries. In the posterior mediastinum the thoracic aorta comes to lie between the esophagus and the vertebral column, slightly to the left of the midline. Also in the posterior mediastinum the thoracic aorta gives off bilateral segmental arteries to the posterior intercostal spaces, as well as a pair of arteries to the lung.

The **pulmonary trunk** is a great artery that rises from the right ventricle. It crosses under the arch of the aorta at the base of the heart and there bifurcates into right and left pulmonary arteries. Immediately deep to the atria the pulmonary arteries pass laterally in front of the bronchial tubes, leave the mediastinum, and disappear within the substance of the lung.

Four **pulmonary veins** emerge from the lungs and project medially to the left atrium of the heart, below and somewhat in front of the pulmonary arteries, indicating the frontward slope of the heart. Why is the relationship of the left atrium and pulmonary veins to the esophagus clinically important?

The **inferior vena cava** (IVC) rises up from a passageway (hiatus) within the diaphragm and joins the right atrium, in front of and to the right of the esophagus. The IVC drains the lower limbs, pelvis and perineum, gonads, kidneys, adrenals, and liver.

QUIZ

1. The heart and the lungs are the principal residents of the thorax.
2. The mediastinum is the central region of the chest cavity.
3. The middle compartment of the mediastinum is taken up by the heart; the posterior compartment is largely filled with structures en route to the lungs, diaphragm, and abdomen.
4. The heart consists of two atria and two ventricles, enclosed within a tripple layered sac, and punctured by the orifices of six large veins and two great arteries.

pericardium

5. Within the thoracic cavity, the heart is turned and twisted to the left, inclined upward, backward and to the right.
6. There are two circuits of blood flow: pulmonary and systemic.
7. Blood flow through the heart is as follows: IVC/SVC to right atrium to right AV orifice to right ventricle to pulmonary valve to pulmonary arteries to lungs to pulmonary veins to left atrium to left A/V orifice to left ventricle to aortic valve to aorta.
8. In effect, the atria function as reservoirs, the ventricles as pumps.
9. The skeleton of the heart consists of thick connective tissue surrounding the four main valve guarded orifices of the heart, giving support to the orifices and the myocardium as well.

respiratory structures of the mediastinum

TRACHEA AND BRONCHI

The trachea, popularly known as the "windpipe," is the distal continuation of the larynx. It passes through the superior mediastinum and into the posterior mediastinum, where it terminates by bifurcating into the right and left primary bronchi, each of which supply a lung. Its upper border is at the level of the C6 vertebra; the bifurcation occurs at about T4.

The relations of the trachea (Fig. 8-20) may be summarized:

Just above the sternal angle, the first few rings of the trachea are largely encircled by the thyroid gland.

Anteriorly, the trachea is related to great vessels stemming from and entering the base of the heart.

Laterally the trachea is related to the nerves en route to the larynx, recurrent branches of the vagus nerve.

Posteriorly, it is related to the esophagus.

The bronchi, made of the same stuff as the trachea, project laterally and downward from the bifurcation, destined for the root of the lungs where they are embraced and surrounded by branches and tributaries of the pulmonary vessels. As you can see (Fig. 8-2), the esophagus and thoracic aorta, largely flush against the vertebral column, pass behind the bronchi. When studying x-rays, these relationships are important.

Figure 8-20 shows that there is a significant difference in the length, angle, and diameter of the primary bronchi as they project from the trachea. The right, being shorter, more vertical, and wider, is more likely to accept aspirated foreign bodies, e.g., buttons, chicken bones, pins.

Structure

The supporting framework of the trachea and bronchi is hyaline cartilage in the form of horseshoe-shaped rings with fibrous interspaces (Fig. 8-20). As one proceeds along the bronchi distally from the bifurcation, the rather regular

Fig. 8-20 (opposite) Trachea and bronchi. A. Anterior view. B. Cross section of the trachea and its neighbors at the level of T1. C. Cross section of the trachea, ×6. D. Drawing of the tracheal mucosa, about ×150. Center, Some relations of the trachea, anterior view.

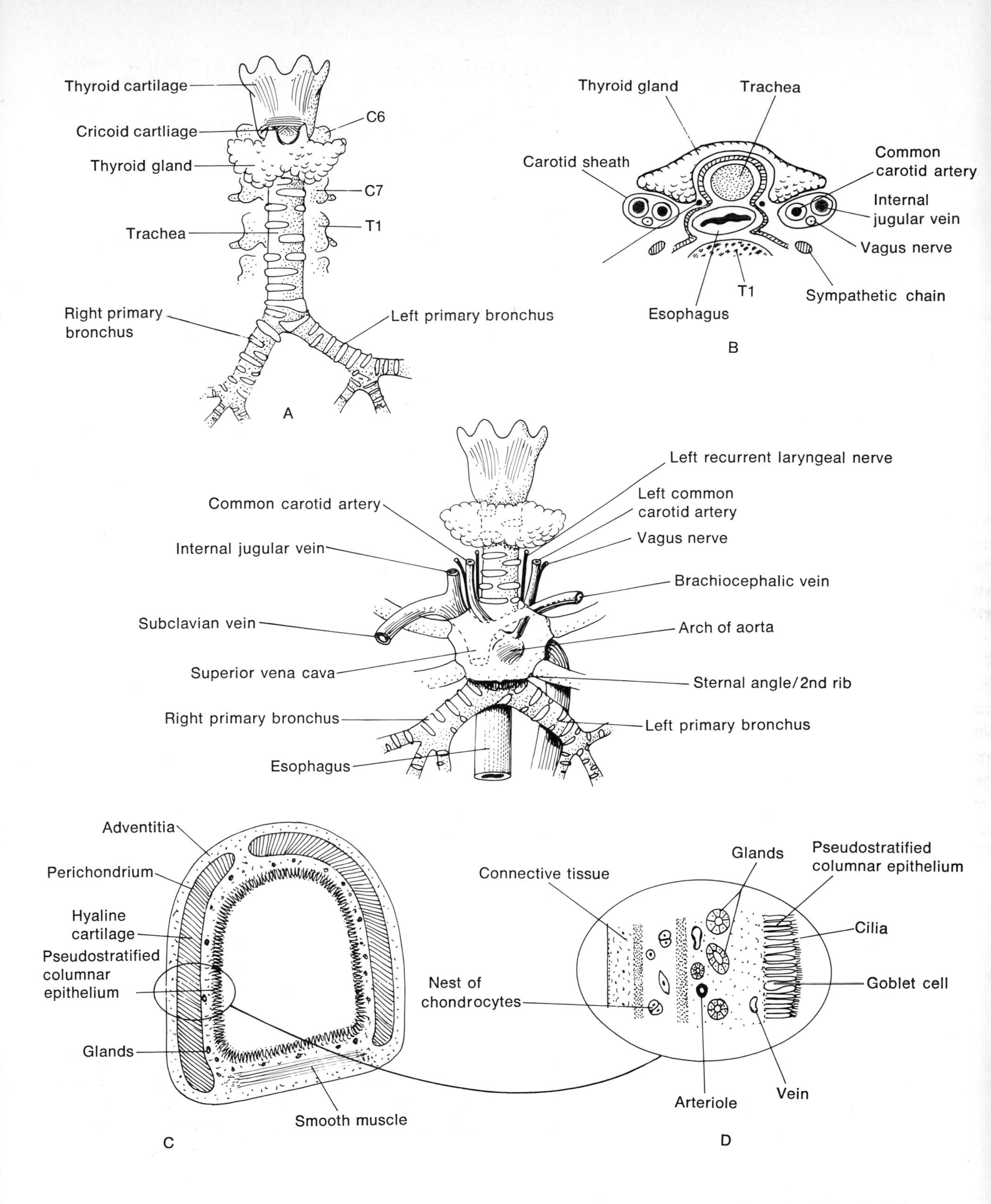
Thyroid cartilage
Cricoid cartilage
C6
Thyroid gland
C7
Trachea
T1
Right primary bronchus
Left primary bronchus
A
Thyroid gland
Trachea
Carotid sheath
Common carotid artery
Internal jugular vein
Vagus nerve
Sympathetic chain
T1
Esophagus
B
Left recurrent laryngeal nerve
Common carotid artery
Left common carotid artery
Internal jugular vein
Vagus nerve
Brachiocephalic vein
Subclavian vein
Arch of aorta
Superior vena cava
Sternal angle/2nd rib
Right primary bronchus
Left primary bronchus
Esophagus
Adventitia
Perichondrium
Hyaline cartilage
Pseudostratified columnar epithelium
Glands
Smooth muscle
C
Glands
Pseudostratified columnar epithelium
Connective tissue
Cilia
Nest of chondrocytes
Goblet cell
Arteriole
Vein
D

arrangement of rings can be seen to fragment into irregular plates. Posteriorly, the two free ends of the cartilaginous bars in the trachea are connected by a band of smooth muscle.

Within the trachea and primary bronchi is a mucosa lined with pseudostratified columnar epithelium, including goblet (mucous) cells and cilia. Mucous glands heavily populate the underlying connective tissue.

Function

The smooth muscle and glands of the air conduction tubes (extending from larynx to bronchi) are innervated by the autonomic nervous system. Thus under "panic" conditions, the sympathetic nerves initiate relaxation of the muscles. The result is dilated tracheal and bronchial passageways and a greater capacity for aerating blood. In parasympathetic stimulation (as might be caused by foreign particles), contraction of the smooth muscle constricts the tubes to prevent ingestion. The glands are stimulated to secrete copious amounts of mucus, as if to trap the unwanted particles.

The cilia of the respiratory epithelium wave rhythmically and in unison, sweeping particulate matter engulfed in mucous secretions upward and out of the tubes into the oral cavity, to be expelled from the body. Interestingly, the cilia of epithelium found in the trachea and bronchi of chronic cigarette smokers are often severely damaged or nonexistent, the tall columnar cells having been crushed by the smoke particles inhaled. Under these conditions, mucus can be extruded from the airways only by violent coughing, and chronic inflammation of the bronchi is not uncommon.

VAGUS NERVES AND BRANCHES

The vagi, you will remember, are cranial parasympathetic nerves originating in the brainstem which are distributed to the regions of the pharynx and larynx, and to the heart, lungs, and abdominal viscera as well.

The vagi pass into the superior mediastinum in company with the common carotid arteries and internal jugular veins (see Fig. 8-20B). While in the neck they give off the cardiac branches which form a plexus at the base of the heart with cardiac branches from the sympathetic chain.

In the posterior mediastinum (Fig. 8-21) the vagi may be seen to give off branches to:

1. The bronchi and lungs (pulmonary nerves).
2. The larynx and trachea (recurrent laryngeal nerves).
3. The esophageal plexus.

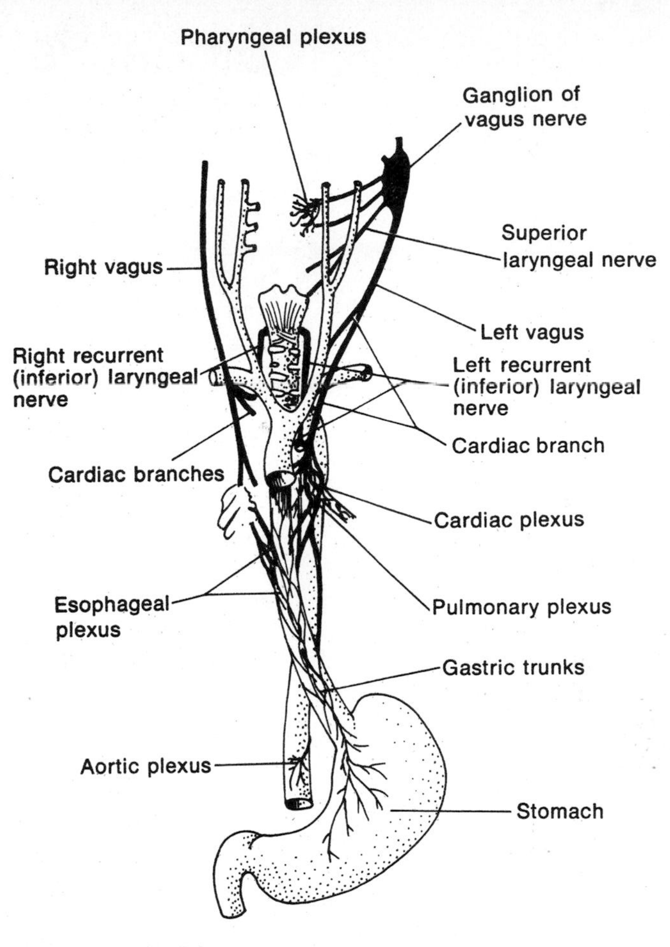

Fig. 8-21 The vagus nerves and their distribution in the thorax. They continue into the abdomen as the gastric trunks.

PHRENIC NERVES

These nerves, branches of cervical spinal nerves 3, 4, and 5, pass downward in the anterior cervical regions, enter the superior mediastinum, and, lying snugly between pleura and pericardium, reach for the diaphragm (Fig. 8-2). These are crucial nerves, responsible for the motor supply to the thoracic diaphragm; cut them and one is left with intercostal breathing (see page 457). Since the intercostal nerves arise from thoracic spinal nerves 1 to 12, damage to the spinal cord above C3 is usually a cause of "instant" death—a lethal break in the circuit between the respiratory center in the medulla and the origins of phrenic and intercostal nerves. As an aside, it might be mentioned that the phrenic nerves' roots (cervical spinal nerves 3, 4, and 5) are also the roots for the sensory nerves to the neck and shoulder. Thus irritation of the membranes of the lung adjacent to the diaphragm may show up in pain of the anterior neck and shoulder (another example of referred pain).

digestive structures of the mediastinum

ESOPHAGUS

The **esophagus** or gullet is the distal extension of the pharynx and proceeds through the superior and posterior mediastinum, posterior to the trachea and anterior to the aorta, to pass through the diaphragm, dilate, and become the stomach (Fig. 8-22). The esophagus extends from C6 to the diaphragm at T10. Can you visualize these levels on yourself?

The esophagus is a collapsed fibromuscular tube that incorporates a type of mucosa consistent with its function, i.e., stratified squamous epithelium. The underlying connective tissue contains glands whose secretions, as well as those of the oral cavity and pharynx above, facilitate passage of the food material, or bolus. The undulating action of the smooth muscle (peristalsis) moves the bolus swiftly through the esophagus. The two muscular layers, oriented 90° to each other, are of the skeletal type in the upper third of the esophagus, of the smooth variety in the lower third, and mixed in between. These transitions mirror the fact that food ingested into the esophagus passes from a stage of processing which is voluntarily initiated, i.e., chewing and swallowing, to a stage which is involuntary, i.e., passage through the esophagus. In terms of digestion the esophagus plays no significant role.

The esophagus is supplied by fibers of the esophageal plexus, combining both vagal and sympathetic (vagal-inhibiting) nerves (Fig. 8-21). The fibers orient themselves into two trunks on anterior and posterior sides of the esophagus and enter the abdominal cavity as the continuation of the vagus nerves.

The blood supply of the esophagus, like that of the trachea and bronchi, derives from many small local arteries springing from the thyroid, bronchial, and intercostal arteries as well as the thoracic aorta (Fig. 8-23).

The venous drainage of the esophagus involves communications with the inferior and superior vena cava, the azygos veins, and the hepatic portal system (Figs. 8-24 and 8-25). Thus it sometimes happens that occlusion of hepatic veins reveals itself in enlargement of esophageal veins due to back-pressure.

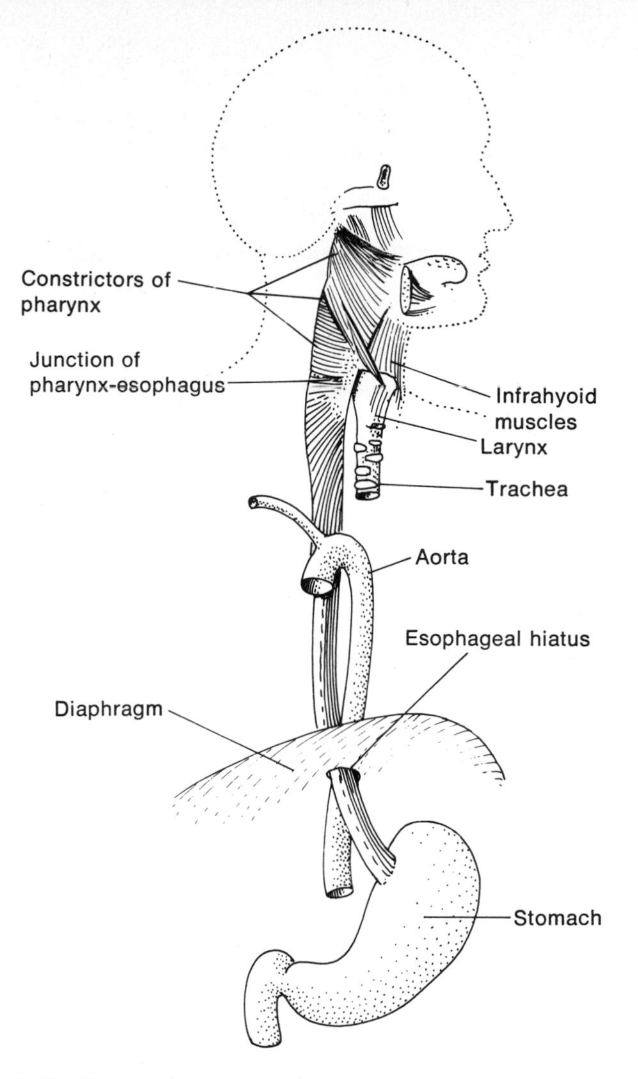

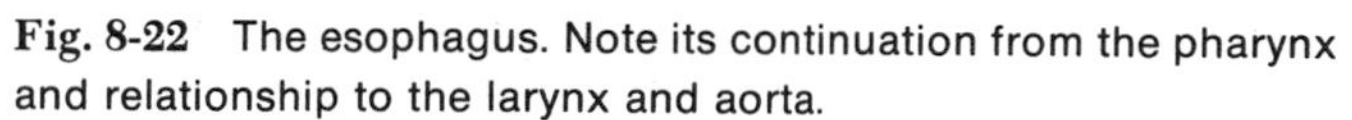
Fig. 8-22 The esophagus. Note its continuation from the pharynx and relationship to the larynx and aorta.

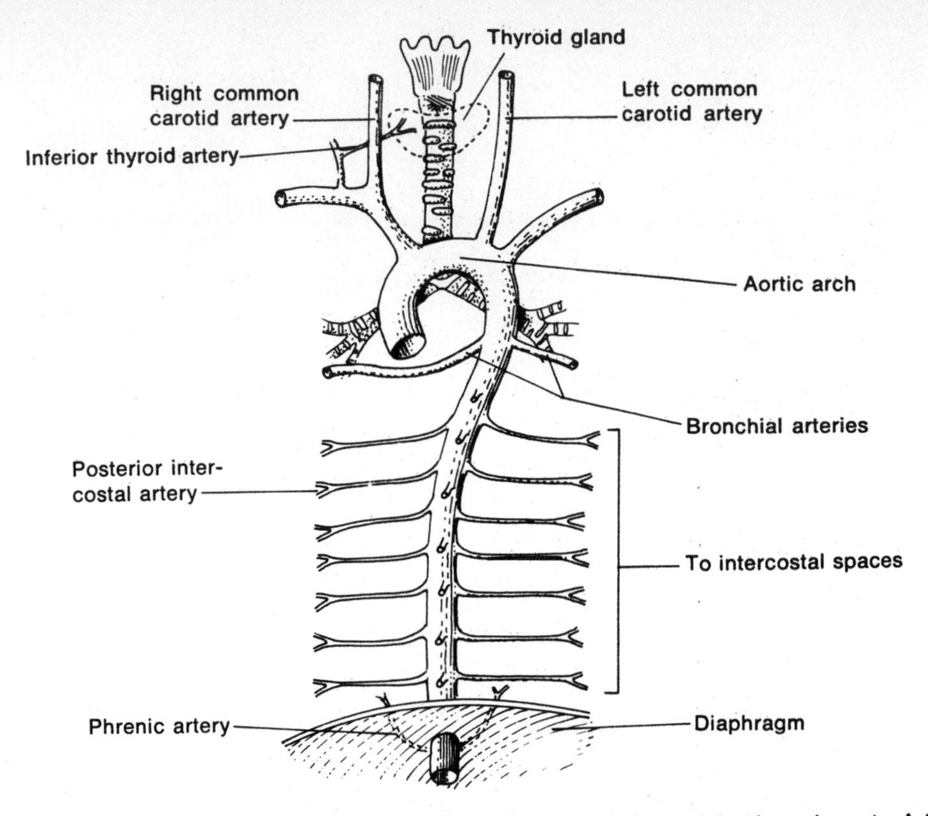

Fig. 8-23 Distribution of the branches of the thoracic aorta. Arterial stubs on the anterior aspect of the thoracic aorta supply the esophagus.

Fig. 8-24 (opposite) Scheme of venous collateral circulation of the thorax and abdomen. Several routes of venous return to the heart are shown. In the event of occlusion, such collateral routes can take on the added burden.

Fig. 8-25 (opposite) The azygos vein and its tributaries, shown in situ.

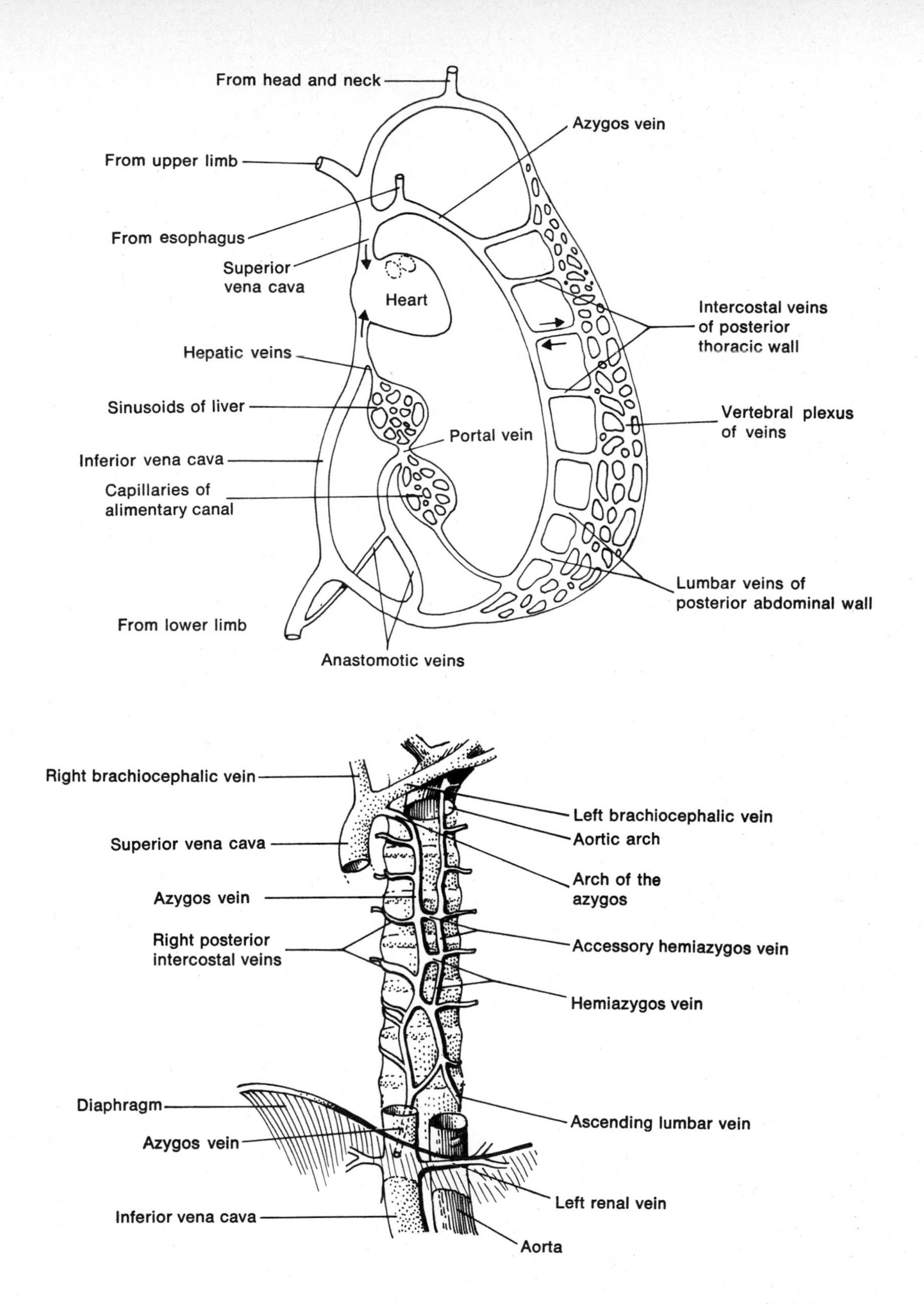
From head and neck
Azygos vein
From upper limb
From esophagus
Superior
vena cava
Heart
Intercostal veins
of posterior
thoracic wall
Hepatic veins
Sinusoids of liver
Vertebral plexus
of veins
Portal vein
Inferior vena cava
Capillaries of
alimentary canal
Lumbar veins of
posterior abdominal wall
From lower limb
Anastomotic veins
Right brachiocephalic vein
Left brachiocephalic vein
Superior vena cava
Aortic arch
Arch of the
azygos
Azygos vein
Right posterior
intercostal veins
Accessory hemiazygos vein
Hemiazygos vein
Diaphragm
Ascending lumbar vein
Azygos vein
Left renal vein
Inferior vena cava
Aorta

miscellaneous structures of the mediastinum

AZYGOS VEIN

Lying on the vertebral bodies along the posterior thoracic wall, the azygos vein and its tributaries drain the posterior intercostal spaces and the posterior abdominal wall (Fig. 8-25). Remember that the inferior vena cava enters the right atrium of the heart immediately on passing through the diaphragm and so is not available for venous drainage in the chest; the azygos system of veins has that chore.

The azygos vein empties into the superior vena cava, and therein lies its special significance: Should the IVC or one of its major tributaries become blocked, or should the hepatic portal vein draining the intestinal tract become obstructed, venous blood can find its way to the heart by means of the azygos system. These collateral routes of venous blood flow are visualized in Fig. 8-24. A thorough understanding of them is of great importance for the physician and surgeon.

THORACIC DUCT

The thoracic duct (Fig. 8-26) is the principal vessel of the lymphatic system. It lies on the bodies of thoracic vertebrae and is deep to the thoracic aorta. Furthermore:

The thoracic duct drains the lymphatic trunks of the lower limbs, pelvis, and abdomen.

It drains the lymph nodes of the right chest, neck, and upper limb.

It empties into the junction of the left subclavian and internal jugular veins.

Upward of 3 quarts of lymph fluid pass through the duct each day. Occlusion of the thoracic duct allows excess fluids to accumulate in the body spaces and tissues (edema). It is apparent, then, that the vascular system cannot handle the draining of body tissues without the lymphatic vessels.

THYMUS GLAND

The thymus gland lies in the anterior mediastinum atop the pericardium at the base of the heart. It is just deep to the manubrium of the sternum (Fig. 8-27). The thymus is a highly

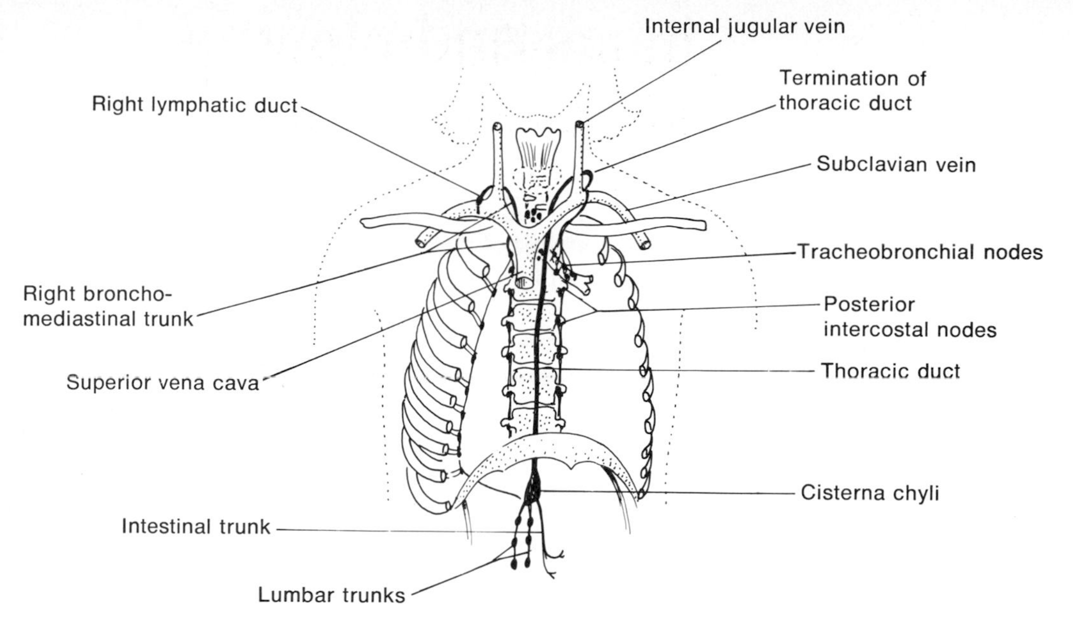

Fig. 8-26 Principal lymphatic trunks and ducts of the thorax and abdomen. Lymph enters the venous system via the right lymphatic and thoracic ducts.

complex gland which is most active in the child. It tends to decrease in size and activity afterward (Fig. 8-27B).

The thymus produces lymphocytes. It is thought by some investigators to "seed" the body with lymphatic tissue cells (lymphocytes, plasma cells, etc.) which in some unexplained manner provide the body with the competence to resist the invasion of foreign bodies, such as bacteria and viruses. It is thus part of the body's immunological apparatus.

Fig. 8-27 A. Thymus gland of a child. B. In the adult most of the gland has become connective tissue.

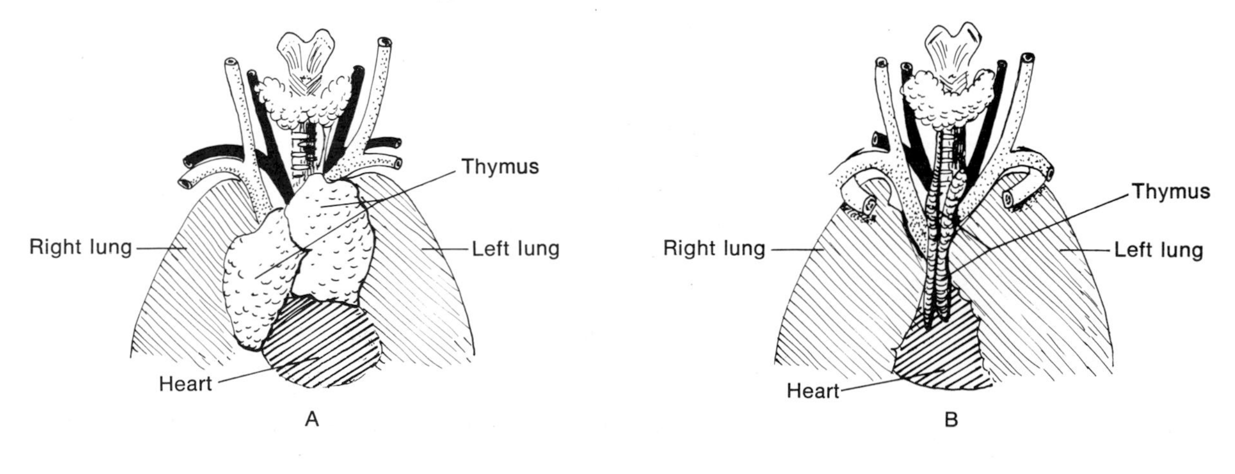

lungs and pleurae

The lungs occupy the greatest portion of the thorax and represent that part of the body in which the internal milieu (i.e., the vascular system) comes closest to contacting the external environment (i.e., the atmosphere). Indeed, the blood–air cell interfaces of the lung are so thin that during exercise almost a quart of oxygen can diffuse into the blood per minute. In terms of volume, 60 percent of the lung is blood. Every minute blood flows through the vessels of the lungs, over a gallon of air is brought into contact with it. The way the bellows (diaphragm) provides the force for air intake and exhaust has been outlined in Unit 7. Now on to the structure of the lungs.

GROSS STRUCTURE

The lungs, resembling light and frothy sponges, fill up the lateral two thirds of the chest cavity and are separated from one another by the mediastinum. Study Fig. 8-28 as you read the following:

Each lung is pyramid-shaped, with a base resting on the diaphragm and the apex projecting up through the superior thoracic aperture into the root of the neck.

Each lung is anchored firmly to the mediastinum at the root (hilus) by the bronchi and pulmonary vessels.

Each lung consists of lobes—three in the right (upper, middle, lower) and two in the left (upper and lower). Oblique and horizontal fissures demarcate the lobes.

Each lung is characterized by a sharp anterior border, a rounded posterior border, a convex lateral surface, and a concave medial surface.

The lungs, being air-filled and malleable, contain notches and grooves created by the heart, aorta, ribs, azygos vein, and the other neighboring structures.

PLEURAE

The pleurae are the membranes of the lung and, somewhat like those of the heart, consist of visceral and parietal layers. These layers are continuous with one another, creating a sleeve-like pulmonary ligament at the root of the lung. The **visceral pleura** clothes the lung tissue while the **parietal**

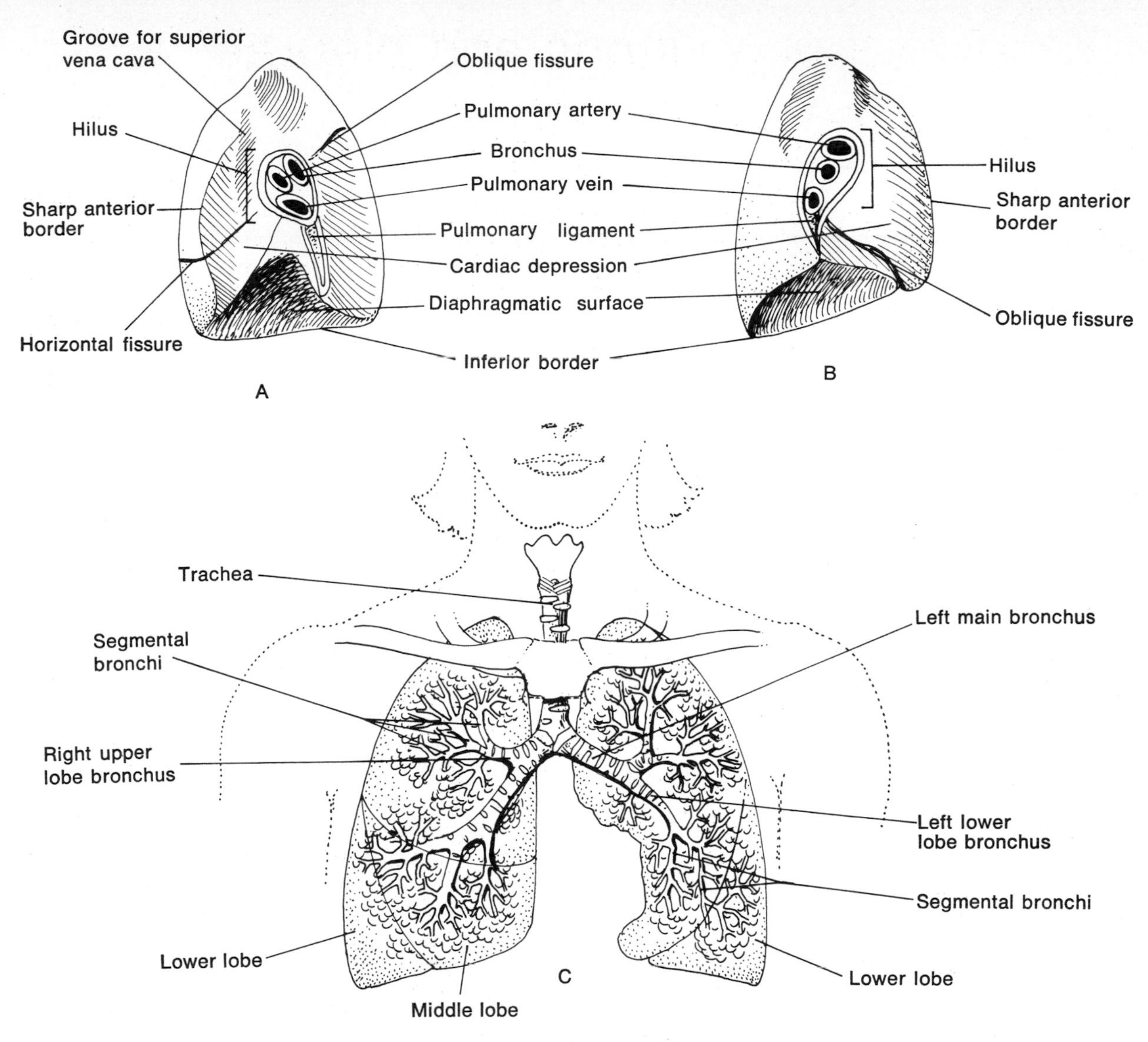

Fig. 8-28 The lungs. A. Mediastinal surface (medial aspect) of the right lung. B. Mediastinal surface of the left lung. C. Anterior view of bronchial distribution. Note the fissures marking lobes of the lung.

pleura is plastered against the inner thoracic wall and forms the sides of the mediastinum (Fig. 8-29). The potential space between the two layers constitutes the **pleural cavity.** It is important to stress the word "potential" here, for the lungs generally go where the chest wall goes throughout the respiratory cycle—as the chest heaves and the thoracic cavity's dimensions shift, the lungs expand and recoil. During quiet respira-

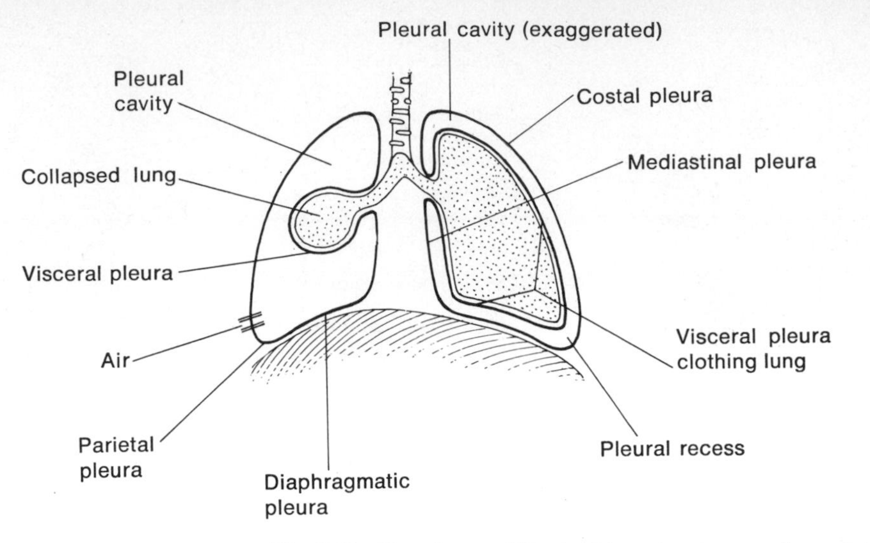

Fig. 8-29 The pleurae. The right lung has been collapsed. Note how the parietal pleura remains attached to the thoracic wall. The pleural cavity has been exaggerated to demonstrate pleural layers.

tion, however, there are "corners" (recesses) where there is space between adjoining pleural layers, and here a needle can be introduced into the pleural cavity through the chest wall without piercing the visceral layer. Excess fluid can be withdrawn or air introduced to cause the lung to deflate (pneumothorax).

The pleural membranes are mesothelial in character, secreting a film of serous fluid which eliminates friction between them and provides a degree of surface tension (adherence) between them. When the pleurae become inflamed and their serous secretions are interrupted a painful condition known as **pleurisy** develops.

THE BRONCHIAL TREE AND COMPARTMENTS OF THE LUNG

There is a hierarchy of structure within the lung ranging from the lobe to the tiny air sac (Fig. 8-30). This hierarchy is organized into two functional components: a **conduction system,** i.e., a series of tubes which conduct air to the air sacs and in which no air exchange (respiration) occurs; and a set of **respiratory units,** i.e., air sacs and their ducts, in which respiration actually occurs. The lung's air conduction tubes (bronchi) branch continuously, becoming progressively smaller from the trachea to the smallest bronchiole, and

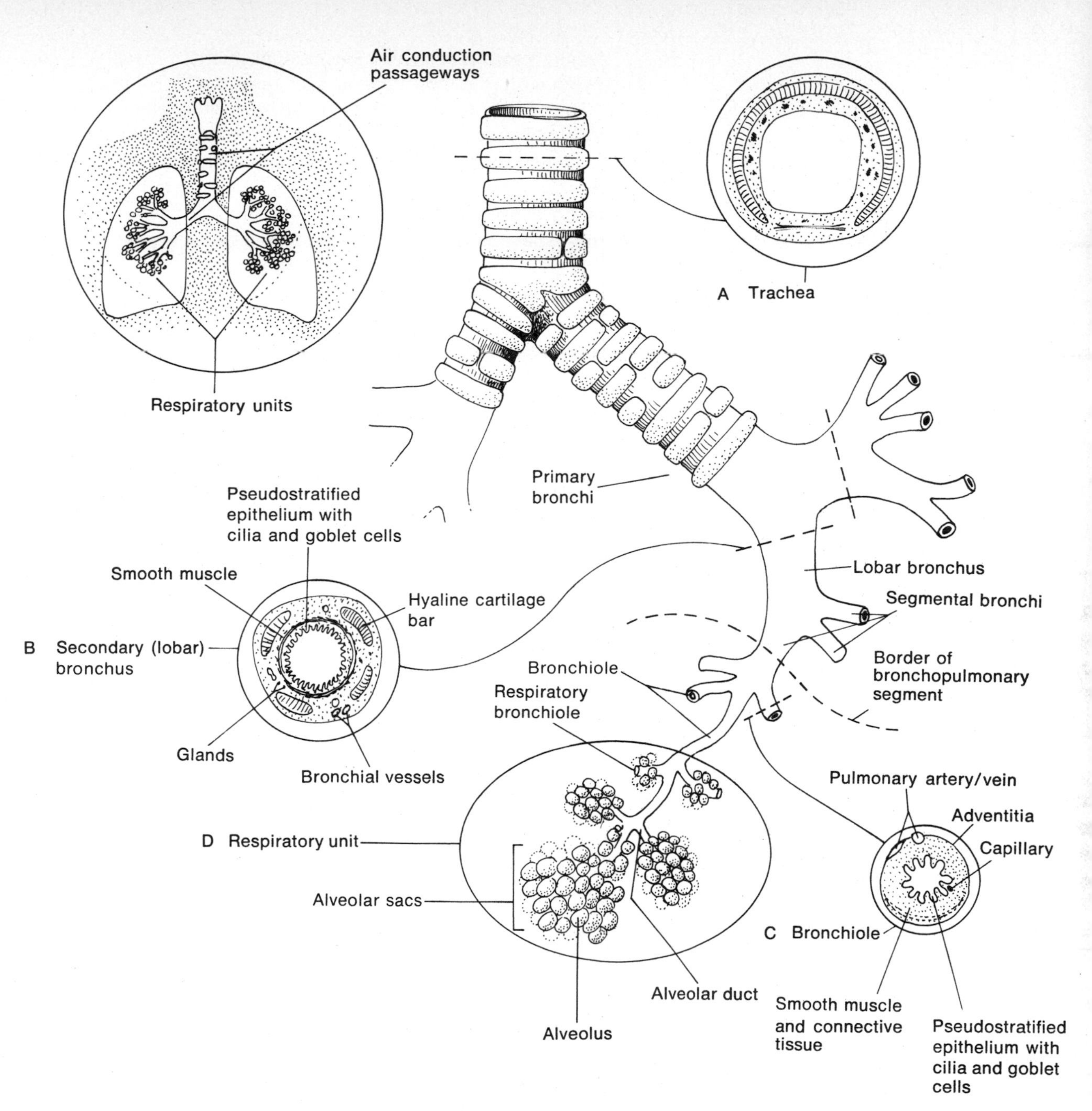

Fig. 8-30 The respiratory tree. Read the text as you study the branchings. Parts A to C were drawn from photomicrographs. A cross section of D may be seen in Fig. 8-31.

serving progressively smaller compartments, ultimately supplying air to the individual air sacs (alveoli).

The microscopic character of the respiratory elements from nasal cavity to trachea is similar: a mucosa characterized by pseudostratified columnar epithelium with goblet (mucous) cells and cilia. The underlying framework for this mucosa changes from bone to cartilage at the level of the larynx. From the trachea to the respiratory unit there is a gradual change in the structure of the bronchial tree—a change in which cartilage is reduced and replaced by increasing amounts of smooth muscle. Study Fig. 8-30 as you read the following paragraphs:

The trachea bifurcates into **primary bronchi** at the level of T4 or T5. These bronchi plunge into the substance of the lung a couple of inches beyond. Primary bronchi have irregular plates of hyaline cartilage and increased amounts of smooth muscle, as well as the "standard" respiratory mucosa.

The primary bronchi divide into smaller **secondary** or **lobar bronchi,** each of which supplies a lobe. Since there are three lobes in the right lung and two lobes in the left, there are, respectively, three and two lobar bronchi. Their patency maintained by cartilage plates, lobar bronchi have the standard respiratory mucosa with increased amounts of smooth muscle.

Within each lobe, the lobar bronchi sprout **tertiary** or **segmental bronchi** which supply the **bronchopulmonary segments** of the lung. These segments are defined by connective tissue partitions (septa). There are 10 in the right lung and 8 in the left. The surgeon can remove a segment as one might remove a section from a grapefruit, without undue hemorrhage or trauma to the neighboring tissue. A segment consists of one segmental bronchus, a number of branches of the bronchus (bronchioles), and many respiratory units. A segmental bronchus has still more smooth muscle and less cartilage than a secondary bronchus.

Within each segment, the bronchus breaks up into nonrigid, muscular **bronchioles** having an inside diameter of 1.0 mm or less. Here the cartilage has been completely replaced by smooth muscle. The glands—so typical of the bronchial tree above this level—are diminished in number. The amount of elastic tissue is significant.

The bronchioles decrease in aperture to about 0.5 mm and alveolar sacs pop out between bands of smooth muscle, hailing the termination of the air conduction system and the beginning of the respiratory unit. The bronchiole here is termed the **respiratory bronchiole.** It supplies—and indeed is

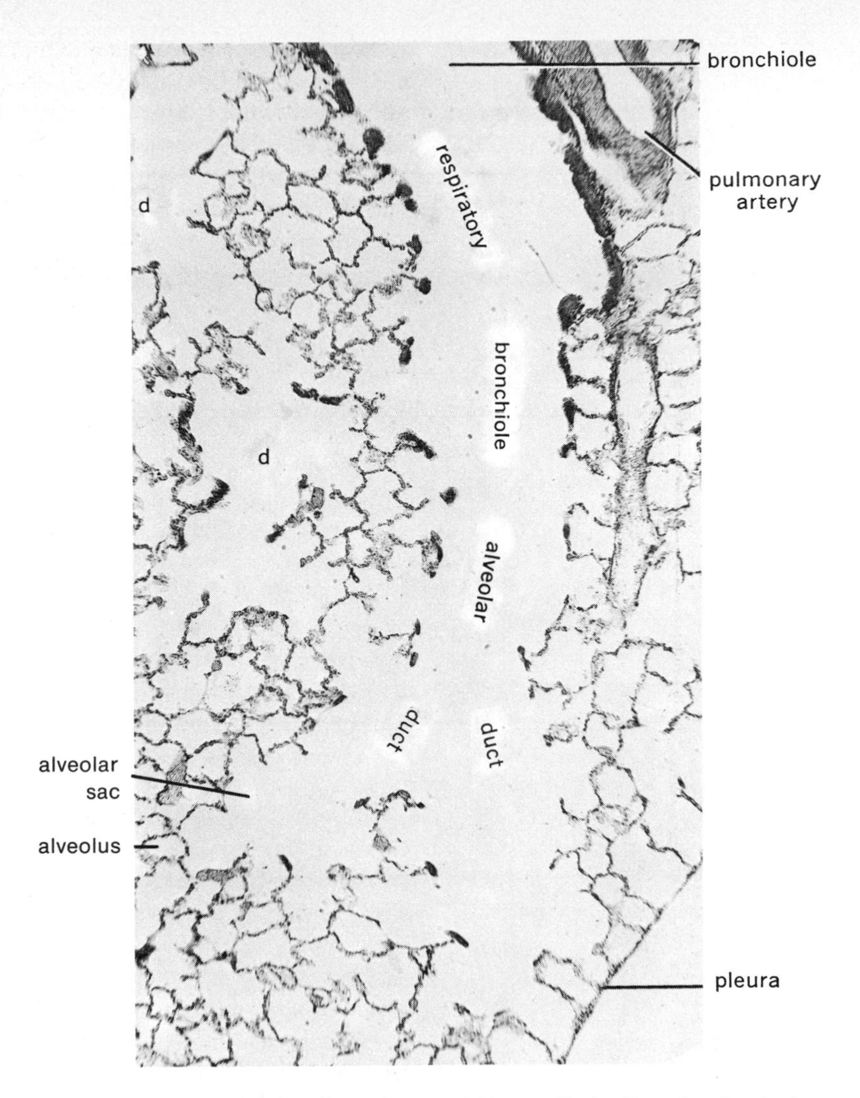

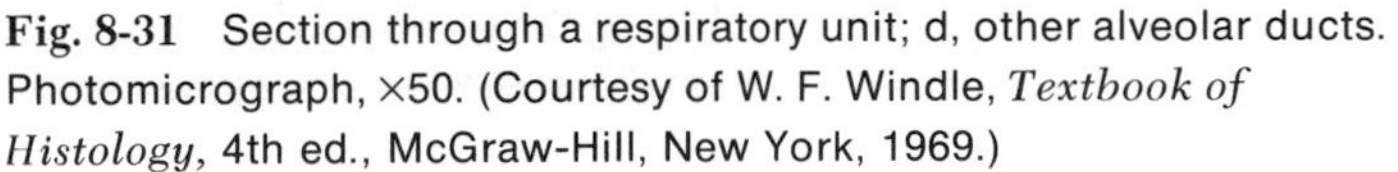
Fig. 8-31 Section through a respiratory unit; d, other alveolar ducts. Photomicrograph, ×50. (Courtesy of W. F. Windle, *Textbook of Histology*, 4th ed., McGraw-Hill, New York, 1969.)

a part of—the respiratory unit, and is characterized by cuboidal epithelium, disappearance of glands, and a mixture of smooth muscle and elastic fibers. The cilia remain, however, and thus glandular secretions which drain distally can be swept upward away from the alveoli.

Respiratory bronchioles become increasingly small, each bronchiole branching out into several **alveolar ducts.** These passageways are surrounded by clusters (sacs) of grapelike alveoli and ultimately end as alveolar sacs (Fig. 8-31).

The wall of an alveolus, as demonstrated by electron microscopy consists of (1) a layer of nonciliated, simple squamous respiratory epithelium (diminished considerably in size from the cuboidal and columnar respiratory epithelium above), (2) its underlying strand of connective tissue, (3) a layer of simple squamous capillary endothelium, and (4) its underlying strand of connective tissue. This tissue—about the thickness of one red blood cell—is all that separates the blood from the air of the alveolus.

The microscopic structure of the lungs provides the clue for understanding how the lung so well serves its function of allowing gaseous exchange. The ducts of the lungs as they fan out radially from the stem bronchus at the hilus become progressively more elastic and less cartilaginous. Thus one finds a good deal of rigidity at the hilus and more elasticity peripherally. The lung is therefore light and capable of considerable distention. The inflated lung is roughly 60 percent blood, 38 percent air, and 2 percent solid tissue. The capillaries of the lung are exceedingly thin-walled, being constructed of endothelium lying over connective tissue. Each alveolar sac—indeed each alveolus—has simple squamous epithelial walls and a reinforcement of elastic and reticular connective tissue. As the lungs fill with air, the elastic walls and their capillaries are stretched, increasing the amount of capillary area exposed to alveolar epithelium while at the same time reducing the thickness of the capillary-epithelium barrier, all of which enhances gaseous exchange.

BLOOD AND NERVE SUPPLY TO THE LUNGS

The principal source of oxygenated blood to lung tissue are the bronchial arteries, which spring from the aorta. These pass into the hilus of the lung with the bronchi and pulmonary vessels. Venous blood returns to the heart by way of bronchial veins which feed into the venae cavae or azygos veins.

Postganglionic sympathetic nerves supply the bronchi and stimulate a relaxation of smooth muscle (bronchodilatation). Stimulation of vagal fibers, on the other hand, induces bronchoconstriction and increased secretion of mucus. Fine vagal filaments may also be found in the alveolar walls. These are related to stretch receptors which sense the alveolar expansion during inspiration. In response to stimulation by these receptors, vagal fibers transmit the information to the respiratory center in the medulla, where inspiration is halted and expiration initiated. This center influences the firing of the phrenic and intercostal nerves and so maintains involuntary control over respiration.

UPON REFLECTION

People are always getting chest pains. When they do, most of them are sure they are having a "heart attack"—and most of them are horrible diagnosticians, simply because they don't know their anatomy. But you do—and because you do, you know that pains in the chest can come from a variety of sources, e.g., pleurisy, esophagitis, bronchitis. Further, you can understand how important a knowledge of relationships within the thorax is to the physician as well as the surgeon.

An appreciation of structural relationships and organ function is important to all of us. Consider the heart and lungs—an appreciation of their structure and function is fundamental, for example, in understanding the effects of exercise on them as well as the rest of the body.

A knowledge of the structure of the respiratory airways will help you understand the ravaging effects of smoking and other pollutants on the rather delicate surface tissues.

We have a responsibility to ourselves and to society to maintain good health and to be able to predict what probably is damaging to the functioning of our anatomical "selves." This responsibility can be realized through an understanding of one's anatomy. Most of us understand a car well enough to know how to operate it without damaging it and we don't want to damage it because it costs us money to fix or replace it. The same rationale should apply to the care of our own "physical plant"—because it costs us infinitely more in money and mobility to fix it and it most assuredly cannot be replaced!

REFERENCES

1. Anson, B. J. (ed.): *Morris' Human Anatomy,* 12th ed., McGraw-Hill Book Co., New York, 1966.
2. Guyton, A. C.: *A Textbook of Medical Physiology,* 3d ed., W. B. Saunders Co., Philadelphia, 1970.
3. Gardner, E., D. J. Gray, and R. O'Rahilly: *Anatomy,* 3d ed., W. B. Saunders Co., Philadelphia, 1969.

clinical considerations

SMOKER'S COUGH, CHRONIC BRONCHITIS, EMPHYSEMA: POLLUTION'S HARVEST

The average human adult, in a normal nonexertive stage, respires about 12 to 16 times per minute, or about 20,000 times per 24 hour day. With each breath, you inspire about 500 ml or one-half quart of air. In a 24-hour day, you pour about 2,500 gallons of air down your respiratory airway, of which 1,500 gallons get to the alveoli—the rest get hung up in the tubing. In our affluent, modern, civilized world we require a great deal of industrialization to satisfy our creature comforts. Spewing forth from our industrial and automotive smokestacks, particles of sulfur oxides, nitrogen oxides, and carbon monoxides contaminate the air and compete with oxygen molecules for space. To soothe our anxiety about pollution, some of us choose (unwisely) to burn finely chopped dried tobacco leaves wrapped in tissue paper and inhale the particulate matter therefrom. Thus our respiratory mucosa is about as busy 24 hours a day as our stomach lining is at suppertime.

For some of us, this creates no great problem. Our nasal, pharyngeal, tracheal, and bronchial mucosa can handle most of the larger particles by trapping them in mucus (from goblet cells) and sweeping them (by means of cilia) out to the oral pharynx. Many particles are absorbed into the lymphatics and phagocytosed in the lymph nodes.

For the chronic smoker and others genetically predisposed, things often get a little sticky—literally. The constant bombardment of ciliated columnar epithelium by pollutants at each breath may ultimately bring about a situation in which the columnar cells become flattened, lose their cilia, and pour out mucus. The mucus, unable to be moved out by cilia, either sits there and occludes the airway, bringing on shortness of breath (dyspnea) and wheezing, or generates repetitive cough reflexes (smoker's cough). Return to a relatively pollutant-free atmosphere will often bring the respiratory mucosa back to happier times—columnar cells with cilia and normal mucosecretion.

For nonbelievers and those of ill luck, continuously traumatized respiratory epithelium becomes inflamed. Chronic inflammation of the bronchial mucosa with hypersecretion of mucus and increased growth of epithelium constitutes **chronic bronchitis.** The lumen of the airway is shrouded in

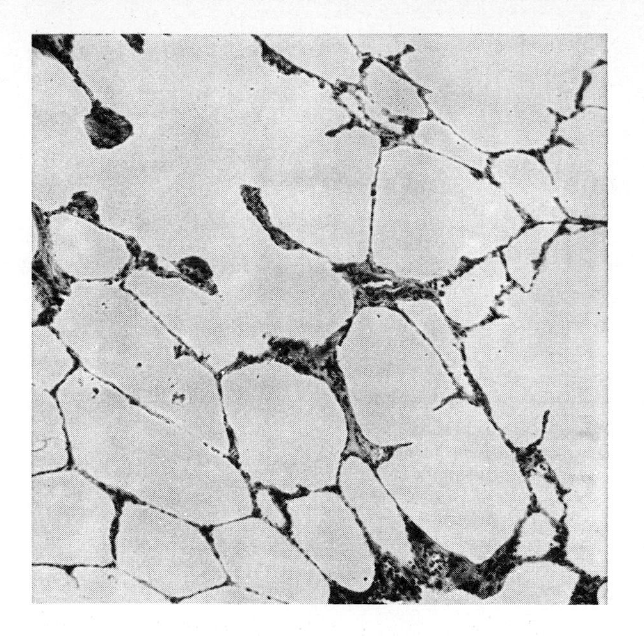

Fig. 8-32 Photomicrograph of emphysematous lung, showing distended and deformed alveoli. Compare with Fig. 8-31. (Courtesy of R. P. Morehead, *Human Pathology*, McGraw-Hill, New York, 1964.)

mucus which may become laden with pus if bacterial invasion overcomes the body's defense forces. Irritating coughs, expectoration (spitting) of yellow-green sputum, and difficult breathing are the trademarks of bronchitis.

One may, and many do, live with chronic bronchitis for years and, except for the signs just described, suffer no serious complications. In other people over a period of years the infection and inflammation extend to the bronchioles and alveoli. The elastic tissue in their walls is damaged and degenerates, making expiration due to elastic recoil ineffective. The combination of smooth muscle enlargement and stuck mucus (plugs) make respiration difficult and exertive. Consequently, the alveoli swell with trapped air; their thin walls break down and fuse with adjacent alveoli (Fig. 8-32). In this manner, great bullae (enlarged air spaces) are created. These symptoms are collectively termed **emphysema,** specifically, obstructive pulmonary emphysema (Gk., *emphysema,* an inflation). As the disease progresses, effective ventilation of the alveoli is impaired as air comes in with ease but goes out with difficulty. The distention of the alveoli compresses the peripheral pulmonary capillaries and increases the resistance to blood flow through them. The vascular resistance causes a backward pressure all the way to the right ventricle, ultimately resulting in right heart failure (cor pulmonale).

unit 9 abdominal viscera

The abdomen consists of a great cavity filled with viscera and membranes bound by musculoskeletal walls. It is not, as is often said, "the stomach"; indeed, the latter is an occupant of the abdomen. Above, the abdomen is roofed by the thoracic diaphragm; below, it is continuous with the cavity of the pelvis. Anterolaterally, the abdomen is bound by the lower extent of the thoracic cage and three musculotendinous sheets interrupted in the midline by the straight muscle of the abdomen, known as the rectus abdominis. Posteriorly, the abdomen is supported by the stanchions of the lower thoracic and lumbar vertebrae and reinforced by the muscles of the posterior wall and back.

Interiorly, the cavity is lined with a layer of serous membrane (peritoneum) which is reflected onto and fitted tightly around most of the abdominal viscera—much as upholstery clothes a sofa.

The viscera found in the abdomen consist of:

1. A string of structures, the **alimentary tract,** continuing distally from the esophagus to the anus.
2. **Secretory organs** associated with and embryologically derived from the alimentary tract.
3. Certain **endocrine organs.**
4. Structures associated with the **urinary system.**
5. The **spleen,** part of the lymphatic system.

Include among these appropriate nerves, vessels, lymphatics, and ligaments and the contents of the abdomen are accounted for.

SURFACE RELATIONSHIPS AND QUADRANTS

There is a variety of structures in the abdominal cavity and the arrangement of structures can be visualized through the medium of surface landmarks. Refer to Fig. 9-1 as you read below.

Two imaginary intersecting lines represent planes of reference commonly used in locating abdominal viscera. The vertical or **median plane** divides the abdomen into right and left halves; the horizontal or **transtubercular** (umbilical) **plane** divides each half into upper and lower **quadrants.** Specific abdominal viscera generally fit into one of these four quadrants. For example, the appendix is located in the lower right quadrant.

Several bony prominences serve as landmarks for placing abdominal viscera. These include the iliac crests (whose tubercles are employed in the transtubercular reference line),

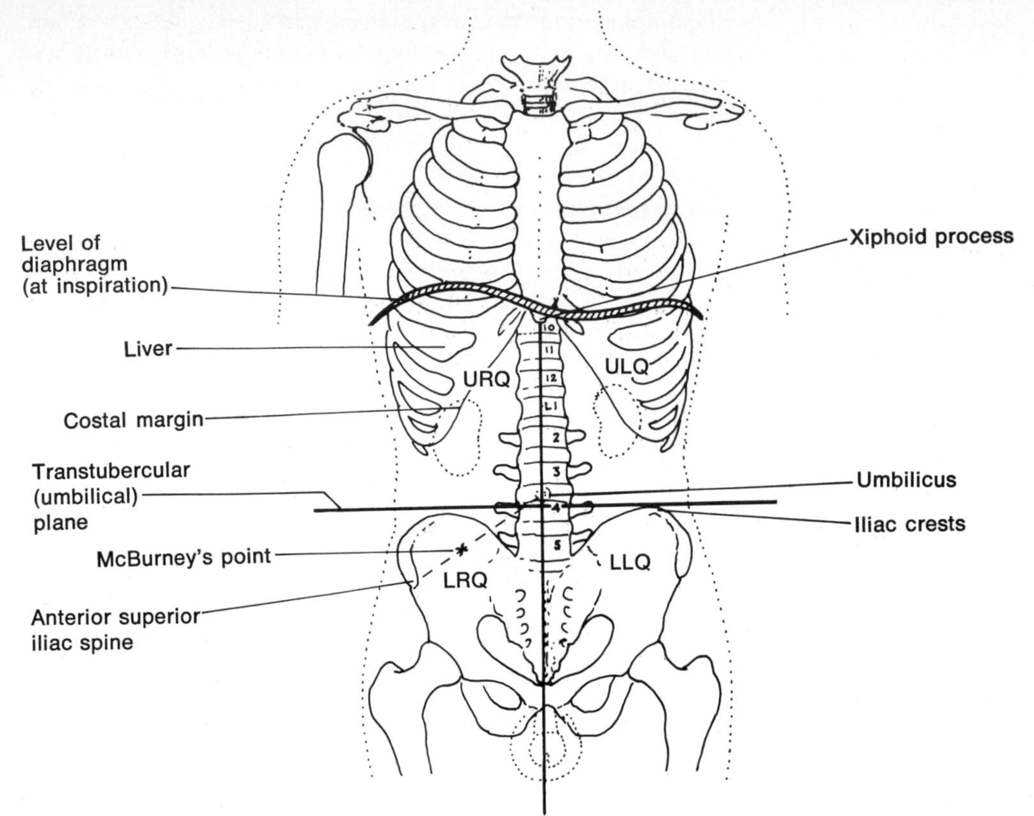

Fig. 9-1 Surface landmarks and reference points of the abdomen. URQ, upper right quadrant. LRQ, lower right quadrant.

the xiphoid process, the costal margins, and the lower thoracic and lumbar vertebrae.

Except for certain fixed viscera the location of abdominal organs is subject to great variation and can only be approximated. The shape and position of the stomach, for example, is dependent upon the amount of food in it and the posture of the individual. The small intestines, suspended from the body wall by elastic membranes, can only be roughly placed within the abdomen as they move during digestion. Furthermore, movement of the diaphragm in breathing rhythmically alters the position of the liver, the kidneys, and the spleen. Thus in this text the vertebrae serve as the primary points of reference. Generally, the xiphoid process is at the level of the tenth thoracic vertebra (T10), the lowest point of the costal margin is at L2, and the umbilicus and the crests of the ilia are at L4. As you proceed through this unit, relate the approximate positions of the various viscera to surface landmarks on yourself. The ability to estimate by palpation the shape, state, and position of various abdominal organs is a particularly critical one for the clinician.

ORGANIZATION OF ABDOMINAL VISCERA

The organs of the abdomen will be described by body system, but before going into a specific analysis, it is appropriate to get an overview of the arrangement of things. Refer to Fig. 9-2 as you read the following:

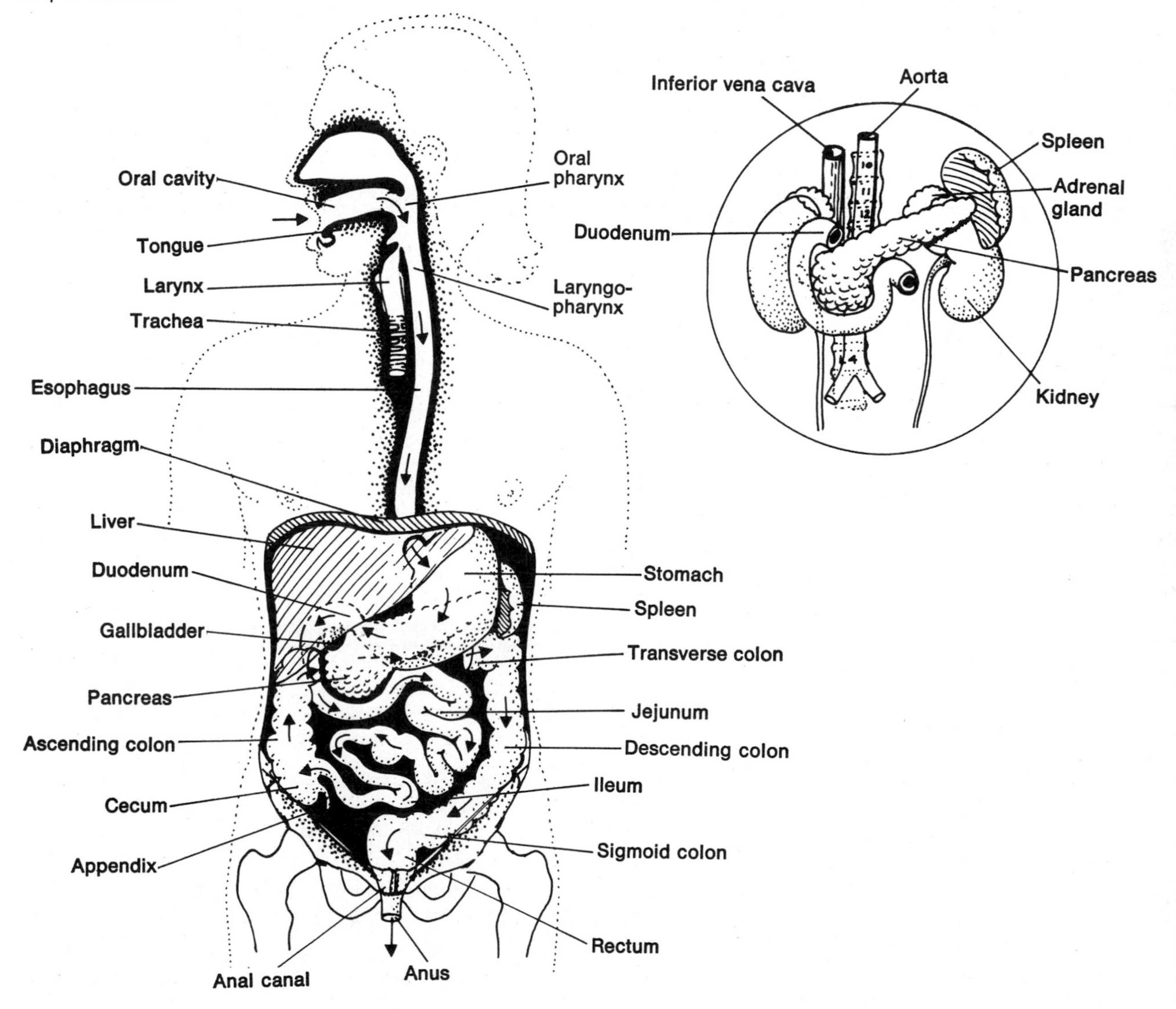

Fig. 9-2 Organization of abdominal viscera. Insert: The pancreas and related organs. The transverse colon has been removed to expose deeper structures.

In the upper left quadrant, just below the diaphragm, is the **stomach**—the swollen continuation of the esophagus—suspended by a peritoneal membrane from the **liver.** The stomach bends around toward the liver to become the **duodenum**—the first part of the small intestine.

The duodenum—C-shaped—hooks around the head of the **pancreas** (a gland), but first bumps into the liver. Both the liver and the pancreas pour their secretions into the duodenum. The gallbladder—a bile reservoir attached to the underside of the liver—has a duct that conducts bile to the duodenum.

Slightly to the left of the median plane at the level of about L2 (or roughly one handsbreadth above the navel) the duodenum distally becomes the **jejunum**—the second part of the small intestine. Highly coiled, mobile, and loosely suspended, the jejunum circuitously continues on to become the **ileum**—the third part of the small intestine. The ileum merges with the **large intestine** at the **cecum** in the lower right quadrant at about the level of L5.

At the cecum, the large intestine ascends toward the liver and abruptly turns to cross the cavity like a watch chain suspended between pockets of a vest. On the left side, the large intestine makes contact with the **spleen** (T12 to L2) and turns downward, ultimately disappearing into the pelvis as the **rectum.** Note how the large intestine "frames" the small intestine.

Posterior to the viscera just discussed, plastered against the body wall by peritoneal membranes, are the **kidneys** (T11 to L2) and their ducts **(ureters),** the latter emptying into the urinary bladder—a resident of the pelvis.

Now let's discuss the membranes (peritoneum) of the cavity and how they support and separate the abdominal viscera.

PERITONEUM

The interior of the abdominal cavity is lined with a serous membrane, peritoneum, into which many of the abdominal viscera grew during development. Thus these viscera are enveloped by peritoneum. These facts concerning invagination of viscera into serous membranes—creating parietal and visceral layers—should be fairly clear by now, for it has been described in the case of the lungs and the heart in Unit 8.

Now the arrangement of peritoneum in the abdomen is difficult to describe and even more difficult to understand. This is so because what begins simply as sheets of membrane in the 6-week-old embryo develops quickly into complex ligaments and sacs by means of various organ rotations and par-

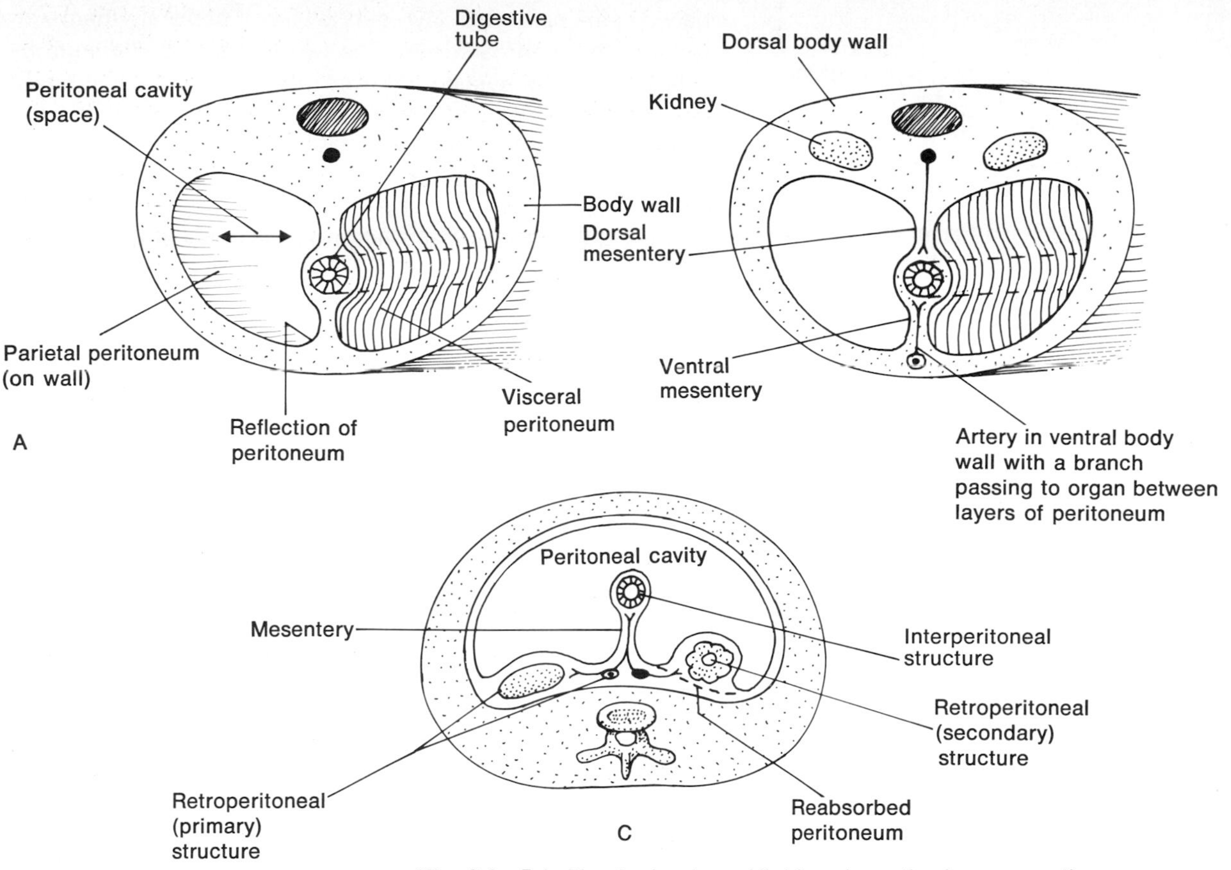

Fig. 9-3 Primitive body plans. Highly schematized cross sections. A–C. Arrangement of peritoneal membranes.

tial obliterations of peritoneum. Consider Figs. 9-3 to 9-5 as you read the following:

The primitive embryonic gut (digestive tube) may be likened to a tube passing through the abdominal cavity. It is largely covered by two layers of peritoneum suspended between anterior and posterior walls of the cavity (Fig. 9-3A).

These layers are reflected from the **parietal peritoneum,** i.e., the peritoneum lining the walls of the abdominal cavity. The reflected peritoneum is termed **visceral peritoneum.** The space between visceral and parietal layers is the **peritoneal cavity.** It is an empty space and, except in females, a closed cavity. That is to say, it is sealed, with nothing entering or leaving. The space of the peritoneal cavity becomes largely "potential" in the developing fetus as organs grow and fill the abdominal cavity. Blood vessels and nerves supplying the gut approach from behind the parietal peritoneum and must pass

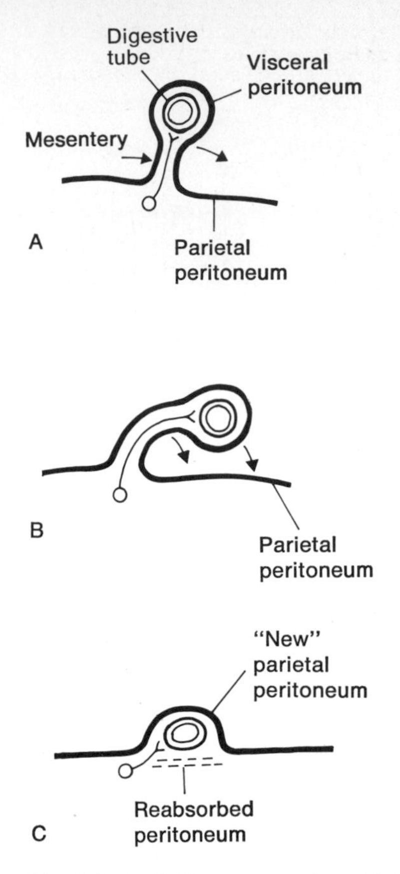

Fig. 9-4 A–C. The process by which an organ becomes secondarily retroperitoneal.

between the peritoneal layers—they never perforate them—en route to the terminal organ. The two layers of peritoneum by which the gut is suspended from the body wall constitute a **mesentery.** Nerves and vessels pass between the layers.

Structures which develop on the inner surface of the body wall behind the parietal peritoneum are covered on one surface by that peritoneum. They are thus plastered to the body wall the way a light switch might be covered with wallpaper by an absent-minded paperhanger. These **primarily retroperitoneal** structures include all blood vessels, nerves, and lymphatics on the body wall, and the kidneys and ureters.

Some structures suspended by a mesentery in the embryonic stage, such as the ascending and descending colon, pancreas, and duodenum, change position during development and eventually become fixed against the body wall. Their intervening layers of visceral and parietal peritoneum are reabsorbed, leaving the structure covered by the visceral peritoneum of its free surface (which in effect becomes parietal peritoneum). Such structures are said to be **secondarily retroperitoneal** (Fig. 9-4).

The peritoneum that links the stomach with other structures of the abdomen is called an **omentum.** The peritoneum that links abdominal structures other than the stomach to one another is a **ligament** (Fig. 9-5).

So much for concepts. After all developmental processes have taken place, the following is the situation (Fig. 9-5):

The liver is largely but not entirely enveloped in peritoneum. It is attached to the stomach and first part of the duodenum by a two-layered sheet of peritoneum, the **lesser omentum.** Between the two layers are the vessels and ducts serving the liver.

The stomach is attached to the transverse colon, diaphragm, and spleen by a great peritoneal apron called the **greater omentum.**

The duodenum is mostly retroperitoneal—that part coming off the stomach is not. The pancreas is secondarily retroperitoneal.

The small intestine is hung to the posterior body wall by an oblique line of mesentery called the **mesentery proper.**

The ascending and descending colon is secondarily retroperitoneal, but the transverse colon is not, being attached to the posterior body wall by a mesentery (which is given its own name, **transverse mesocolon**) and to the stomach by the greater omentum.

The sigmoid colon is attached posteriorly by a mesentery. The rectum and other pelvic structures are draped by a blanket of parietal peritoneum.

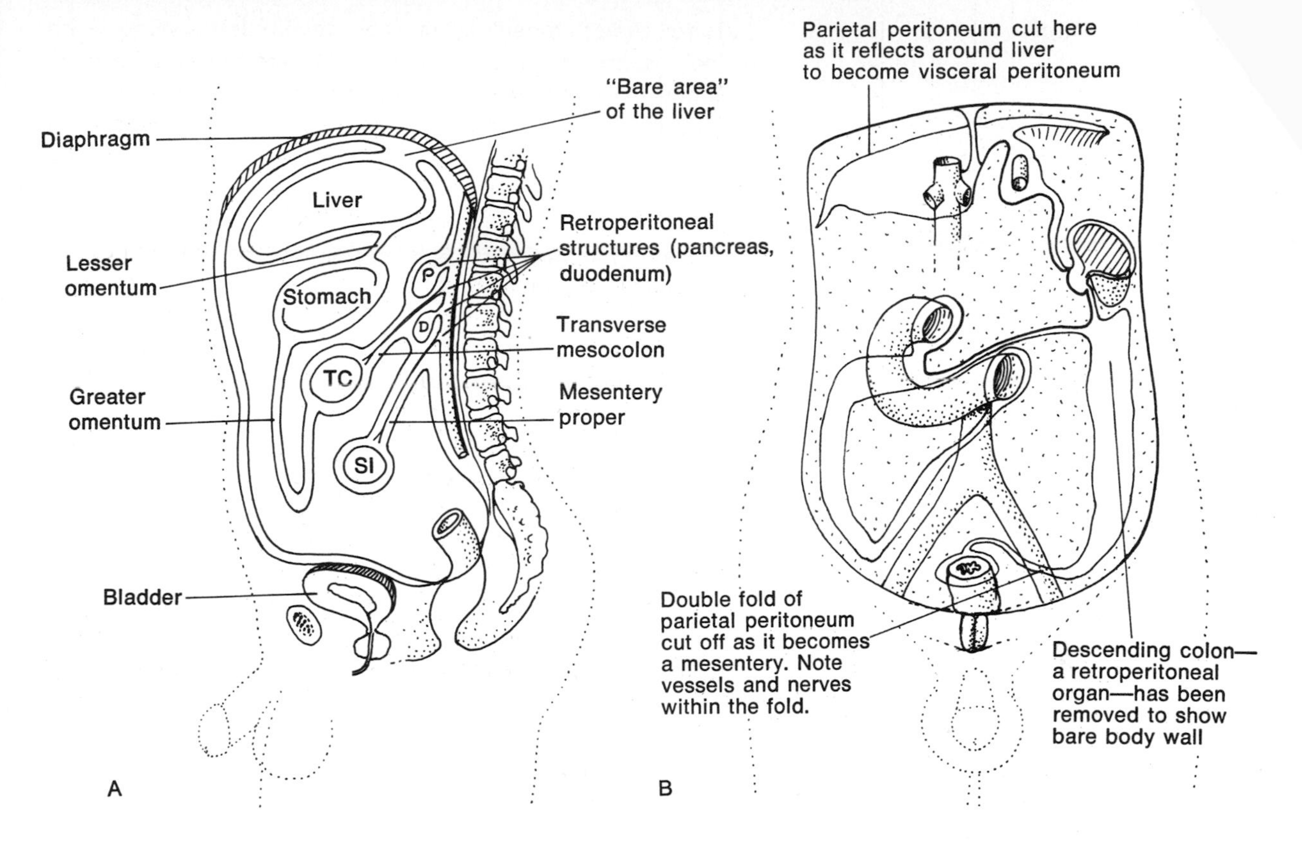

Fig. 9-5 Arrangement of peritoneum as omenta and mesenteries (highly diagrammatic). A. Sagittal section. B. Anterior view of posterior body wall illustrating peritoneal reflections. Most viscera have been removed. The dotted areas indicate body wall covered by parietal peritoneum. The bare areas are filled by retroperitoneal structures or represent cut off folds of peritoneum (mesenteries).

Consequently one can expect considerable mobility from those structures swinging from a mesentery (stomach, small intestine, transverse colon) but not much from the retroperitoneal structures (pancreas, duodenum, ascending and descending colon, kidneys), which are fitted tightly against the posterior abdominal wall.

Clinically, knowing the orientation of peritoneum is important. Segments of intestine can become intertwined in peritoneum and strangulated. The surgeon needs to know peritoneal attachments so as to provide a bloodless field when he operates to free the colon or kidneys. The inner surface of the peritoneum (facing the peritoneal cavity) is lined with mesothelial cells that secrete a serous fluid which enables friction-free movement between adjacent viscera. With handling, as in surgery, or with inflammation (peritonitis), peritoneum of adjacent viscera can adhere to one another (adhesions), creating undue pressures, pain, and possibly obstruction of the bowel.

QUIZ

1. The abdominal cavity is roofed by the ______ ______ and is continuous with the ______ cavity below.
2. Interiorly the abdominal cavity is lined with ______ which is ______ onto the abdominal viscera.
3. The abdominal cavity and its anterior wall is broken up into ______ ______ by two imaginary intersecting lines.
4. The stomach bends around the liver to become the ______. The ______, in turn, hooks around the ______ of the ______.
5. Both the ______ and the ______ pour their secretions into the duodenum.
6. The peritoneum reflected onto the abdominal viscera is termed ______ ______.
7. Structures which develop behind the parietal peritoneum are said to be ______.
8. Structures attached to one another by peritoneum are said to be connected by a ______.
9. One cannot expect much movement from such retroperitoneal structures as the ______, ______, ______, ______ and ______ colon.
10. Adjacent viscera whose layers of peritoneum are stuck together are said to have ______.

abdominal viscera of the digestive system

Stomach

The stomach (L., *gaster*) is the tubular receptacle of the gastrointestinal tract (Figs. 9-1 and 9-2). It is the dilated continuation of the esophagus, yet differs significantly in its microscopic make-up. Elastically hinged by the greater and lesser omenta, the stomach has a great deal of latitude in its movement. The shape of the stomach is subject to a great deal of variation dependent upon gastric contents, position of the body, position of neighboring viscera, etc. When full, the stomach may droop into the pelvis. The two extremities of the stomach are fairly well fixed—above by the diaphragm at about the tenth thoracic vertebral level—below by the duodenum, which is for the most part retroperitioneal.

Refer to Fig. 9-6 and you will see that:

1. The stomach consists of three parts: the **fundus** above, the **pyloric portion** below, and the **body** in between.
2. The stomach has two curvatures: the convex **greater curvature** and the concave **lesser curvature.**
3. The stomach wall consists of four layers: an inner

Fig. 9-6 The stomach. A. External view showing the arrangement of muscle layers. B. Interior (sagittal) view.

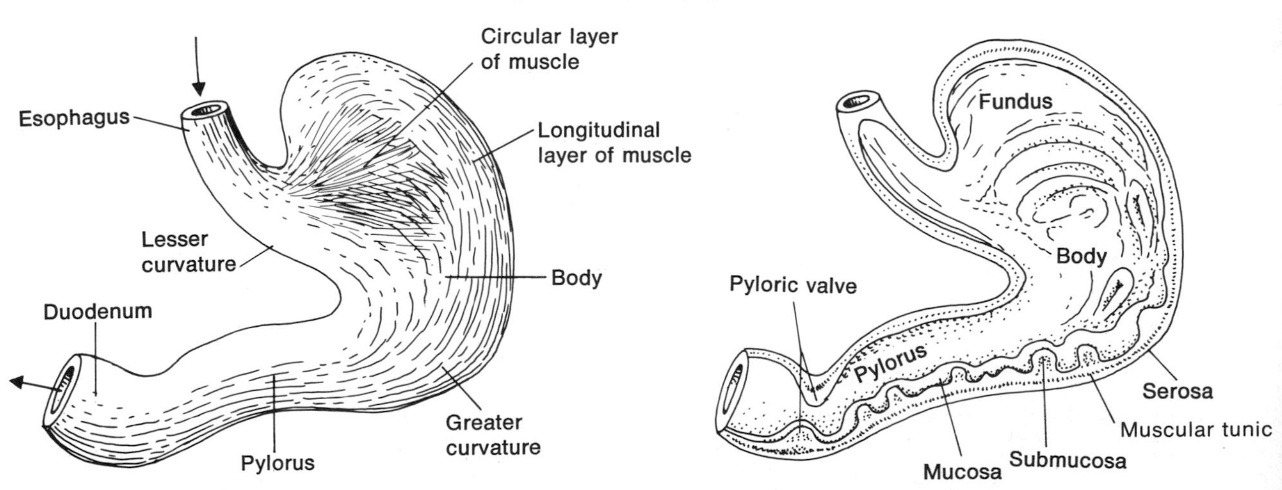

mucosa, an underlying **submucosa,** a **muscular tunic,** and an outer coat of peritoneum **(serosa).**

4. The interior surface of the stomach is characterized by longitudinal folds (rugae) which serve to increase the secretory and absorptive capacity of the stomach. In life, the surface is coated with mucus.

Now the function of the stomach is to receive relatively undigested, bulk foodstuffs from the mouth via the esophagus, mechanically mix the material, and treat it with chemicals in one of the initial stages in digestion. Its target is primarily protein, but it generally prepares all ingested material for intestinal absorption by putting it into solution.

The microscopic structure of the stomach alludes to these functions (refer to Figs. 9-7 and 9-8):

Fig. 9-7 The gastric wall. A. Sagittal view. B. Longitudinal section, x5. C. Cross section of the mucosa of the fundus, about x150.

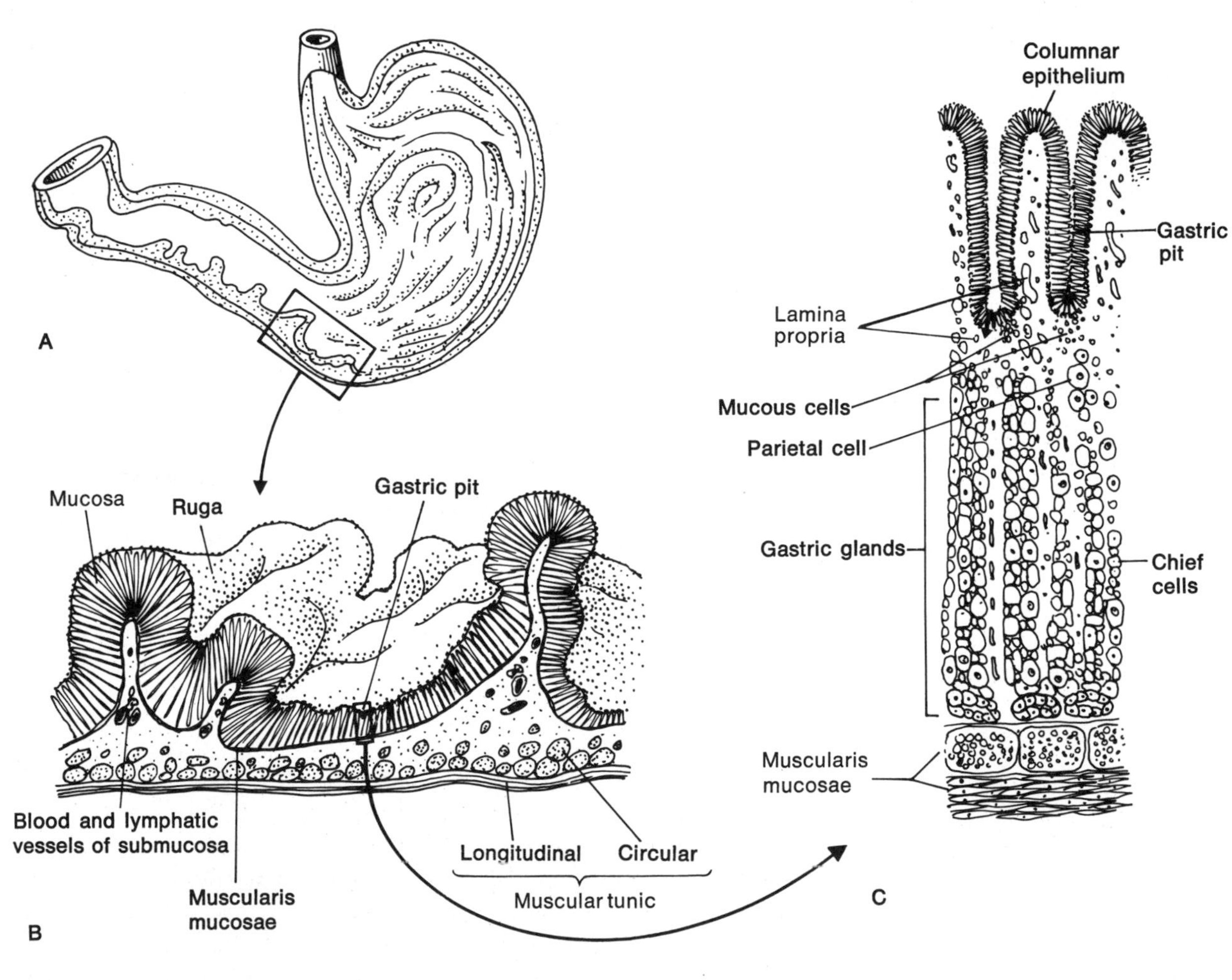

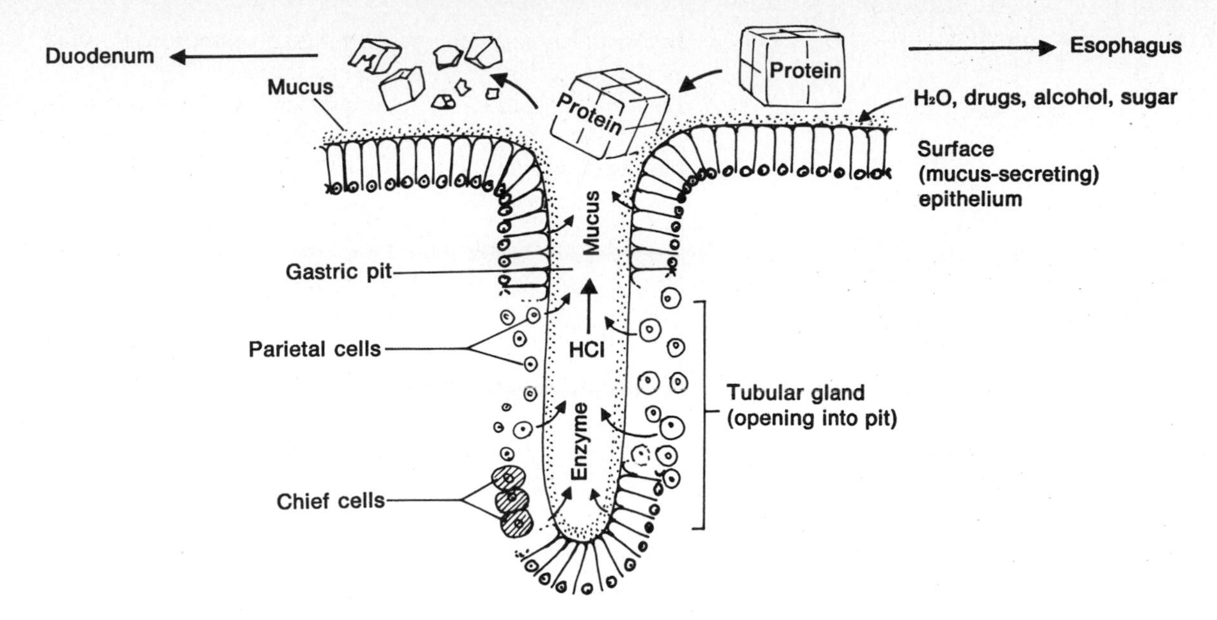

Fig. 9-8 The function of the gastric mucosa, schematic.

The mucosa consists of a single layer of columnar epithelial cells supported by irregular layers of connective tissue (lamina propria). The epithelial layer is arranged roller-coaster fashion into gastric pits, into which glands open from below (Fig. 9-8). The cells lining the pits secrete mucus, which covers the epithelia, protecting them from the acid and enzymatic action of the gastric secretions. The glands of the gastric mucosa generally differ in the three parts of the stomach. Three kinds of cells generally may be found in one or more of these glands:

1. **Mucous cells.**
2. **Chief cells,** which secrete an enzyme (pepsin) activated by hydrochloric acid. Pepsin acts on the linkages binding polypeptides and breaks the proteins into more digestible bits.
3. **Parietal cells,** which secrete a material that becomes hydrochloric acid upon entering the lumen of the gland.

Just deep to the glands is a thin layer of smooth muscle in the mucosa, the **muscularis mucosae,** which provides for a degree of motility within the mucosa, enhancing digestion.

The **submucosa** is a layer of connective tissue thick with blood vessels and nerves. Preganglionic fibers of the parasympathetic division of the autonomic nervous system terminate here by synapsing with postganglionic neurons whose axons innervate the muscularis mucosa and gastric glands.

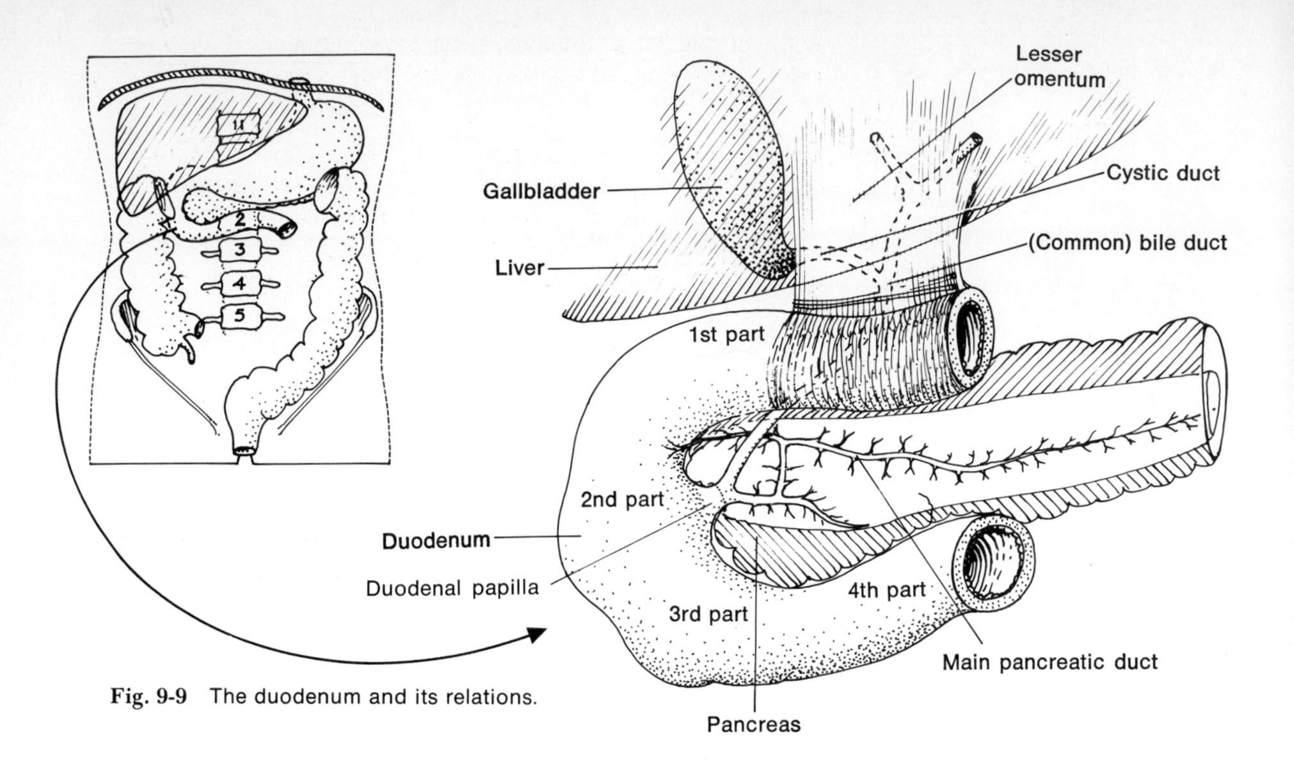

Fig. 9-9 The duodenum and its relations.

The **muscular tunic** consists of three layers of smooth muscle, arranged in a criss-cross pattern, providing the stomach's churning capability for mechanical digestion of food. It is this layer of the GI tract which often cramps and gives the well-known stomach ache. At the duodenal end of the stomach, the circular layer of muscle thickens to form a sphincter, guarding and controlling the flow of the stomach contents into the duodenum.

The **serosa** is a layer of visceral peritoneum with underlying supportive connective tissue enveloping the stomach and continuous with the lesser omentum above and the greater omentum below. These elastic membranes allow the stomach a great deal of mobility.

The vascular, nerve, and lymphatic network about the stomach will be discussed shortly.

Small intestine

The small intestine consists of several feet* of mobile, actively metabolic digestive tubing, highly coiled within the frame-

* A matter of no little disagreement among anatomists, the small intestine has been reported to be as little as 4 ft long and as great as 25 ft. The length is longer in cadavers because the smooth muscle is relaxed. In the living person the length is probably quite variable.

work of the large intestine. The small intestine has three parts: the C-shaped **duodenum,** the **jejunum,** and the **ileum.**

Duodenum. As may be viewed in Fig. 9-9 the duodenum, lying between L1 and L4, begins at the pylorus ("gateway") of the stomach, then curves around the head of the pancreas—both being retroperitoneal—to become the jejunum. Traversing some 19 in. in its travels, the duodenum comes into direct or close contact with a number of viscera in its neighborhood. Now study Fig. 9-9 as you read.

The first part of the duodenum, leaving the pylorus, continues laterally, bumping into the liver and gallbladder as it turns to descend as the second part.

It is convenient that the first part is hinged to the visceral or undersurface of the liver by the hepatoduodenal ligament (lesser omentum), for running in that fold of peritoneum, along with two prominent vessels, is the common bile duct en route to the duodenum from the liver.

The second part of the duodenum becomes retroperitoneal along with the kidney on the right and the pancreas on the left. Functionally, the second part is significant because it receives the secretions from two glandular organs (the liver by way of the common bile duct and the pancreas by way of the pancreatic duct). As you will learn, these secretions (enzymes and bile) profoundly affect the food material entering the duodenum from the stomach as well as significantly lowering the acid level of the internal environment. It is interesting to note that the great majority of duodenal ulcers occur above the duodenal papilla.

The third part heads horizontally anterior to the aorta and inferior vena cava, crossing the midline at about L3, and itself crossed anteriorly by the great mesenteric blood vessels supplying much of the intestines.

The fourth part ascends, peeks out of its parietal peritoneal blanket at about the level of L2, just to the left of the midline, to enter a fold of peritoneum (the mesentery proper), and turn downward to become the jejunum.

The wall of the duodenum, like the rest of the GI tract, consists of mucosa, submucosa, and a muscular tunic, with a partial covering of peritoneum (serosa). Its microscopic appearance, as well as duodenal function, will be discussed shortly. Based on the vertebral levels cited above, can you "map out" the duodenum on yourself?

Jejunum and ileum. Making up the rest of the small intestine, the jejunum (about 8 ft) and ileum (about 11 ft) elaborately coil their way from the upper left quadrant at L2 to the lower right quadrant at L5, all the while secured to the posterior abdominal wall by the quite elastic **mesentery proper** (Fig. 9-5). It is between the peritoneal layers of this mesentery that the mesenteric vessels and lymphatics reach and supply or drain the intestines. Their mobility makes it impossible to place segments of the small bowel precisely within the abdominal cavity. Suffice to say they are generally framed by the colon—from L2 above to the pelvic interior. It seems miraculous that 20 ft of slithering bowel, packed within a space 10 to 14 in. on each side, can survive 70 or more years of digestive processing of foods from cellulose to chili, traumatic shocks and indignities (e.g., infection), erratic body postures, etc., with only an infrequent ulcer or inflammation to show the effects of this prolonged wear and tear.

Microscopic anatomy and function. The microscopic structure of the small intestine, like its fellows the large intestine and stomach, consists of a mucosa, submucosa, muscular tunic, and a partial or complete serous coat (serosa). While the outer three layers remain virtually the same in anatomical character throughout the GI tract, the structure of the inner mucosa progressively reflects the changes in function from esophagus to anal canal. You will recall that the esophageal mucosa is characterized by stratified squamous epithelium, a form of epithelium best suited for the trauma occasionally inflicted by the human diet. The gastric mucosa, incorporating deep pits lined with secretory cells—enzymatic "cauldrons" you could call them—handle the initial breakdown of proteins. The microscopic appearance of small intestinal mucosa, as represented by duodenum, may be seen in Fig. 9-10; study it as you read the following:

The surface of the duodenal wall consists of a carpet of fingerlike projections **(villi)** lining large **circular** or **transverse folds,** looking somewhat like a wrinkled shag rug. These folds and villi, serving to increase the absorptive and secretory area of the intestine, are characteristic of the small intestine, and progressively decrease in number and extent from duodenum to cecum.

Each villus consists of a core of connective tissue (lamina propria) lined with simple columnar epithelium, an offshoot of the muscularis mucosa for mobility, a tiny lymphatic vessel, or lacteal, for draining fat molecules and small proteins, and assorted vessels and nerves. In response to parasympathetic stimulation, the villi collectively contract and bend such that absorption and mechanical digestion are enhanced.

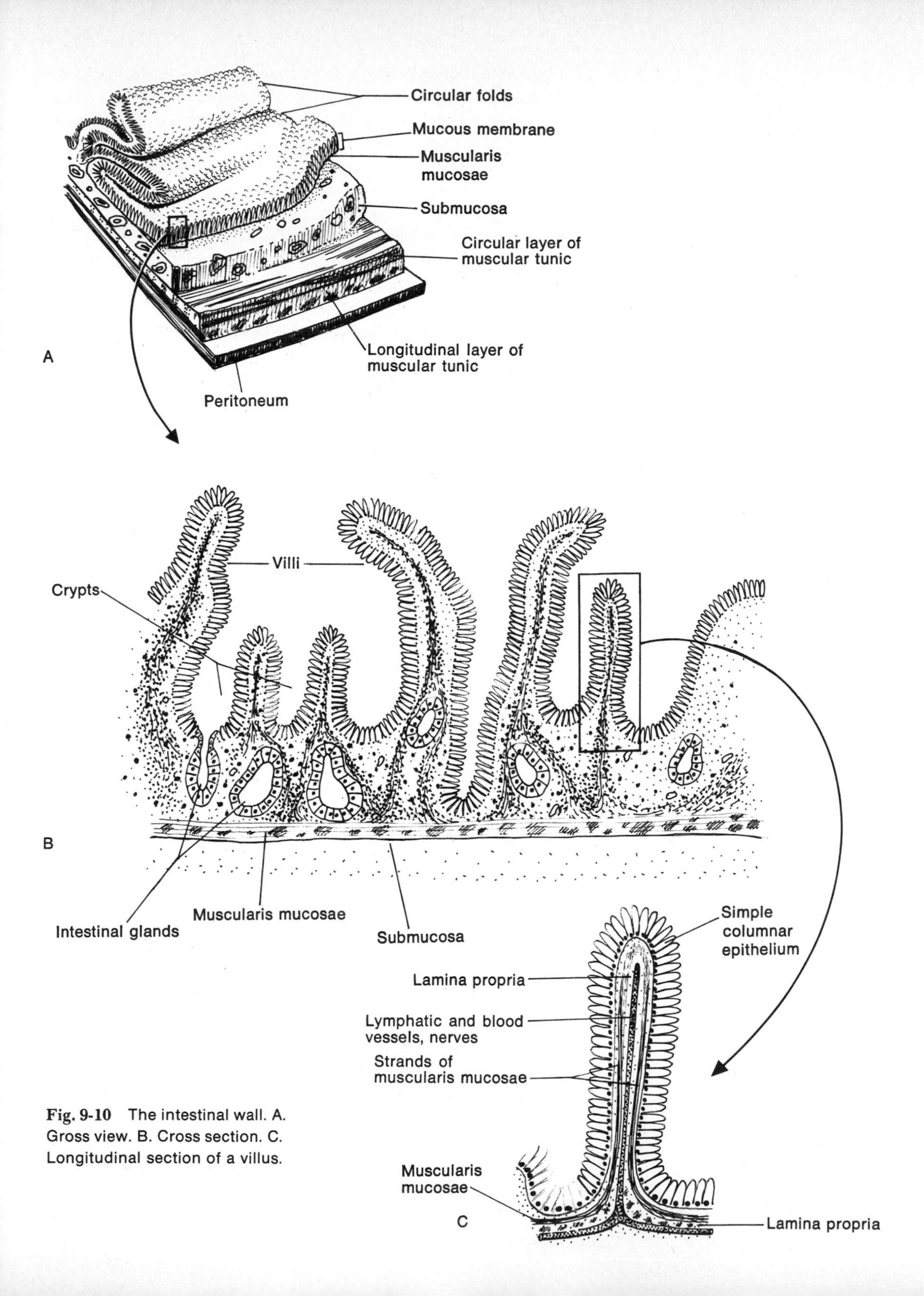

Fig. 9-10 The intestinal wall. A. Gross view. B. Cross section. C. Longitudinal section of a villus.

Intestinal glands open into the spaces between adjacent villi and secrete a variety of digestive enzymes which act on fats (hydrolases), proteins (peptidases), and carbohydrates (saccharidases). Further it is known that secretory cells of the small intestine produce hormones which, in response to the presence of fats and other foodstuff, are released to trigger the pancreas and gallbladder so that they move their secretions to the duodenum via the common bile and pancreatic ducts.

Add to these collective digestive processes of mucus and enzyme secretion the peristaltic contractions of the muscular tunic acting in harmony with the movements of the villi and one has a natural "washing machine"! The result: effective mechanical and chemical digestion of intestinal contents; absorption of the products of this digestive process (fat molecules, varying forms of sugar, amino acids, vitamins, and other complex molecules) by the absorptive columnar cells, which discharge them into the veins and lymphatics of the intestinal wall for transport to the liver courtesy of the hepatic portal circulation; and passage of the "unusables" (cellulose and other nonnutritive matter) on to the large intestine.

The small intestine is also characterized by **nodules of lymphatic tissue** which occur singly or in aggregates in the lamina propria. Principally seen in the ileum and appendix (Fig. 9-11), these masses of lymphocytes resting in reticular connective tissue and drained by small lymphatic vessels, are a part of the body defense system. Proteins and other substances, such as viruses and bacteria, foreign to the body (antigens) but which enter the body through cuts, mouth, anus, etc., must contend with antibodies produced and secreted by lymphatic tissue which can render reasonable concentrations of antigens harmless and nontoxic. In response to bacterial infections of the GI tract, lymphatic nodules and diffuse lymphatic tissue existing throughout the lamina propria of most mucosae enlarge and increase in number to ward off the attack.

Large intestine

The large intestine consists of the **cecum, colon, rectum,** and **anal canal** (Fig. 9-12). Roughly 5 ft in length, it begins in the lower right quadrant of the abdomen where the ileum joins the cecum at about the level of L5. Refer to Fig. 9-12 as you read.

The **cecum** is a saclike affair containing three orifices—one from the ileum, one to the appendix, and one to the ascending colon. The **appendix** is a wormlike (vermiform) blind tube

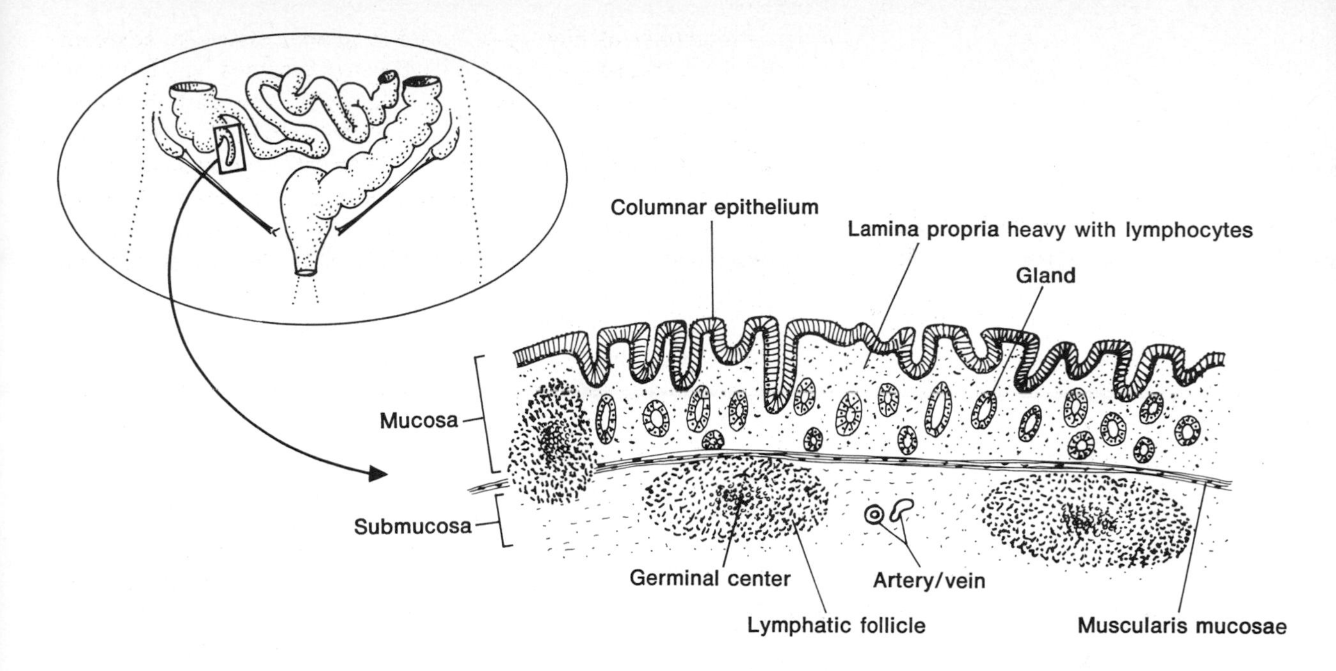

Fig. 9-11 The appendix, longitudinal section. See also Atlas of X-rays, Fig. A-11.

arising from the cecum and has no known digestive function (it may have immunologic significance, a possibility indicated by its large population of lymphatic nodules). To locate the appendix on yourself find the umbilicus and the prominent anterior superior iliac spine. Imagine a line between the two; midway on that line is McBurney's point deep to which often lies the appendix (Fig. 9-1).

The retroperitoneal **ascending colon** passes upward along the right side of the abdominal cavity, bumps into the underside of the liver, turns medially (right colic flexure) to become the **transverse colon** suspended by its own mesentery, the transverse mesocolon (Fig. 9-5). Drooping in the midline, the transverse colon ascends as it approaches the spleen then abruptly turns downward (left colic flexure) to become retroperitoneal as the **descending colon.** At the pelvic brim, the colon makes a sort of S-(sigmoid) turn to drop into the pelvis. Hinged by mesentery (sigmoid mesocolon) the **sigmoid colon** becomes the **rectum** in the pelvis. Passing through and bound by the levator ani muscle, the rectum merges into the **anal canal** with its distal orifice, the **anus.**

The colon is externally characterized by three bands **(taeniae)** of smooth muscle running lengthwise down the in-

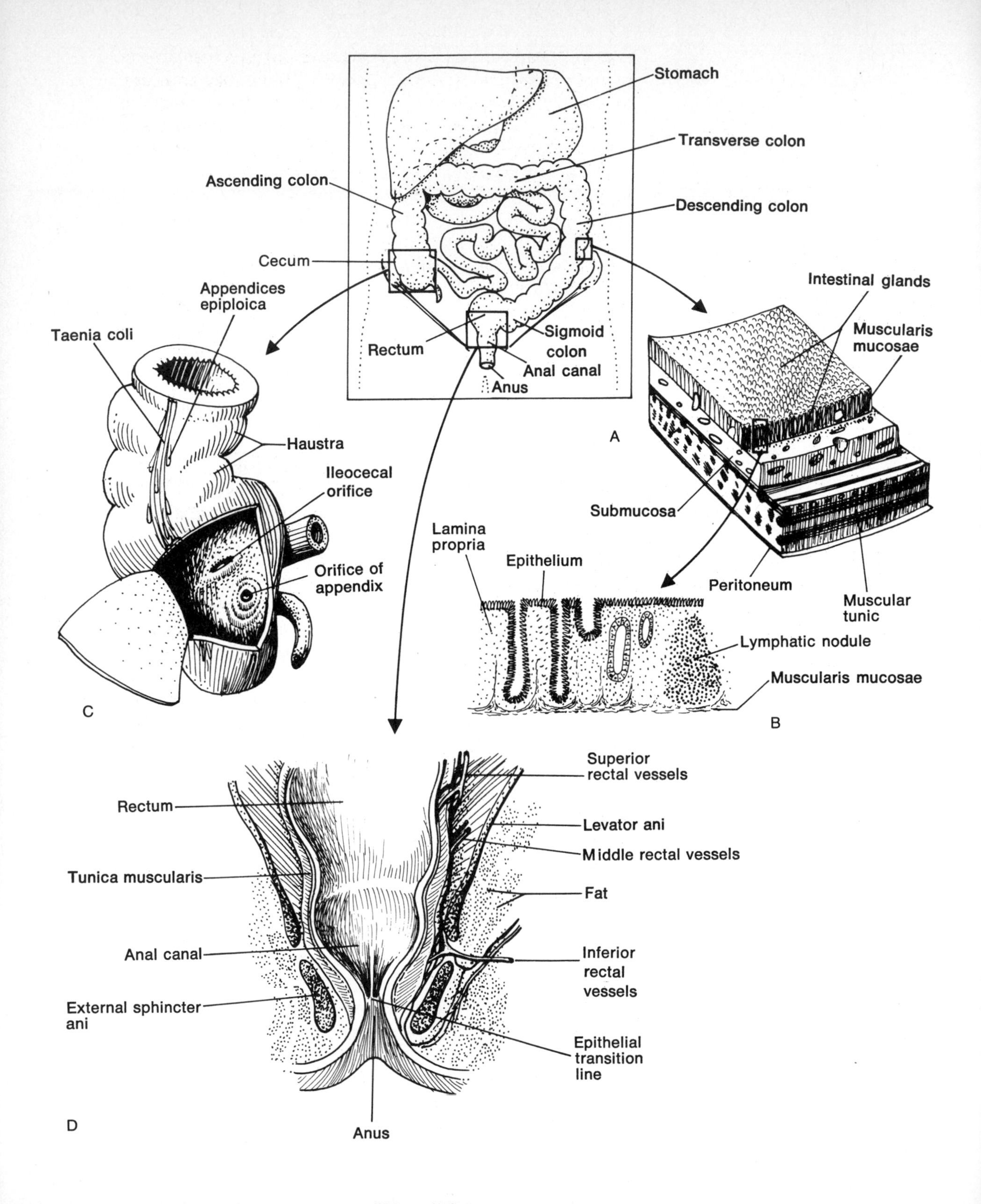
Stomach
Transverse colon
Ascending colon
Descending colon
Cecum
Appendices epiploica
Taenia coli
Rectum
Sigmoid colon
Anal canal
Anus
A
Intestinal glands
Muscularis mucosae
Haustra
Ileocecal orifice
Submucosa
Lamina propria
Epithelium
Orifice of appendix
Peritoneum
Muscular tunic
Lymphatic nodule
Muscularis mucosae
C
B
Superior rectal vessels
Rectum
Levator ani
Middle rectal vessels
Tunica muscularis
Fat
Anal canal
Inferior rectal vessels
External sphincter ani
Epithelial transition line
D
Anus

testine. These bands represent the outer longitudinal layer of the muscular tunic from the appendix (where they arise and to where they may be tracked when seeking the appendix) to the rectum (where the taeniae fan out to cover the entire structure). These taeniae exert tension along the intestine, creating what appears to be a string of pouched sacs (sacculations or **haustra**). Thus there is rarely any problem differentiating normal large bowel from small in the dissecting room.

Projecting from the peritoneal (serosal) covering of the colon are little peritoneum-lined bags of fat (appendices epiploica) not found elsewhere. The significance of these is that in the lower colon of the senior citizens there may be projections of mucosa (diverticula) into these appendices which may become inflamed (diverticulitis). Slight bleeding may follow, presenting slightly bloody stools at defecation.

Microscopic anatomy and function. The mucosa of the large intestine (Fig. 9-12) is lined by masses of intestinal glands supported within lamina propria. This glandular epithelium consists of mucus-secreting cells and absorptive cells. The underlying layers (submucosa and muscular tunic) are similar to its more proximal counterparts. In the anal canal, the intestinal glands decrease in number, and at the anal orifice (anus) there is a smooth transition from simple columnar epithelium-lined mucosa to stratified squamous-lined skin. Surrounding the lower anal canal are sphincter-like masses of skeletal muscle (external sphincter ani) which operate under voluntary control. The interior of the anal canal demonstrates mucosal shelves and vertical columns (Fig. 9-12). In the submucosa of these columns exists an extensive venous plexus which under undue stress can become dilated and inflamed (hemorrhoids).

It is in the large intestine that the developing fecal matter takes shape. This is accomplished by a significant increase in water absorption by the mucosal epithelia. In concert with this action, mucus-secreting cells lay down a film of mucus over the mucosa to protect it against frictional trauma, while at the same time providing easy transport of the feces toward the rectum. At the moment of elimination the sigmoid colon releases its muscular tension and feces fill the previously collapsed rectum and pass on through the anal canal and orifice under the rhythmic contractile efforts of the muscular tunic. Defecation can be stopped in the anal canal by voluntary contraction of external sphincter ani. Willful retention of feces in the intestines is not generally possible.

Fig. 9-12 (opposite) Large intestine. Insert: Large intestine in relation to neighboring viscera. A. Gross view of a section of the large intestinal wall. B. Mucosa of the colon, longitudinal wall. C. Interior of the cecum. D. Frontal section of the rectum and anal canal.

Liver

The liver (L., *hepar*) is the great gland of the abdomen. It weighs 3 lb in health but can weigh 10 lb or more when diseased. It is reddish brown as a result of its extensive vascularity. It takes up most of the upper right and part of the upper left quadrant, and is tucked snugly under the diaphragm and partly fenced in by the lower ribs. The liver developed as an anterior outgrowth of the embryonic digestive tube and ballooned into the ventral mesentery. Consisting of cords of epithelial cells attached to a tree of connective tissue around which there is a network of sinusoids and capillaries, the liver carries on a dozen critical metabolic functions:

1. Absorption of simple sugars; conversion to starch (glycogen) and storage of same; breakdown of glycogen (glycogenolysis) and release of simple sugar into the circulation.
2. Absorption of protein and subsequent detachment and detoxification of nitrogen compounds, e.g., ammonia.
3. Absorption and storage of fat; conversion of fat to glucose (gluconeogenesis*).
4. Detoxification of drugs; conversion of alcohol to carbohydrate (glycogen).
5. Storage of vitamins A and D.
6. Metabolism of products of red blood cell destruction; storage of iron.
7. Production of a chemical soup called **bile,** which passes to the duodenum via the gallbladder to act on fats.
8. Synthesis of urea, a material to be voided from the body by way of the kidneys.
9. Synthesis of fibrinogen, a protein important in clotting.

It would not be surprising to learn, then, that all products of digestion pass out of the GI tract into the liver for processing. They do so by way of a portal circuit of blood to be discussed shortly.

Gross structure. The macroscopic structure of the liver can be studied in Fig. 9-13. The liver consists of four lobes; left, right, caudate, and quadrate. It presents smooth, convex superior and anterolateral surfaces and concave posterior and inferior (visceral) surfaces.

On the visceral surface, just above the gallbladder, lies the **porta** (gateway), the entry for the portal vein and hepatic arteries and the exit for lymphatic vessels and bile ducts. These

* *Gluco,* glucose; *neo,* new; *genesis,* creation of.

structures, you will remember, are shrouded in lesser omentum (gastrohepatic, duodenohepatic ligaments). Being soft and malleable, the liver receives the impressions of its closest neighbors on its visceral surface—the duodenum and kidney on the right, the stomach and esophagus on the left.

The visceral surface slopes downward away from the posterior abdominal wall. The inferior vena cava leaves an impression on the upper posterior surface. Here the hepatic veins discharge into the IVC.

The surface representation of the liver (Fig. 9-1) can be roughly approximated on yourself: on the right from T8 to L2, on the left from T9 sloping obliquely along the right costal margin to L2 on the right. These points vary somewhat during respiration.

Microscopic structure. The basic framework of the liver is simply connective tissue. Lightly but firmly encapsulating the liver exterior, the tissue turns inward at the porta and spreads out like branches of a tree. Supported and demarcated by these branches of connective tissue, sheets of epithelial cells, here called **hepatic cells,** are arranged radially into **lobules.** Refer now to Fig. 9-13 as you read.

The substance of the liver (the structural and functional unit) consists of lobules separated from one another by fibrous septa.

The lobules are radially arranged single sheets (cords) of hepatic (epithelial) cells and adjacent sinusoids. Perforating the center of each lobule is the **central vein,** which drains the sinusoids of the lobule and empties into the hepatic veins. These in turn empty into the IVC.

Branches of the portal vein (conducting blood from the intestines) and hepatic artery (conducting blood from the heart) approach the periphery of the lobule and discharge into the sinusoidal passageways which converge toward the central vein. In this way, hepatic cells receive oxygen as well as molecules of digested food fresh from the GI tract.

Tiny bile canals (canaliculi) arranged delicately on the hepatic cells conduct newly formed bile outward toward the lobule's periphery. The canaliculi then converge from the multitude of lobules to form ducts which ultimately leave the porta as right and left hepatic bile ducts destined for the duodenum.

These three elements—bile duct, artery, vein—and a lymphatic vessel at the periphery of the lobule—constitute a **quadrad.**

The sinusoids are lined by phagocytic cells of Kupffer

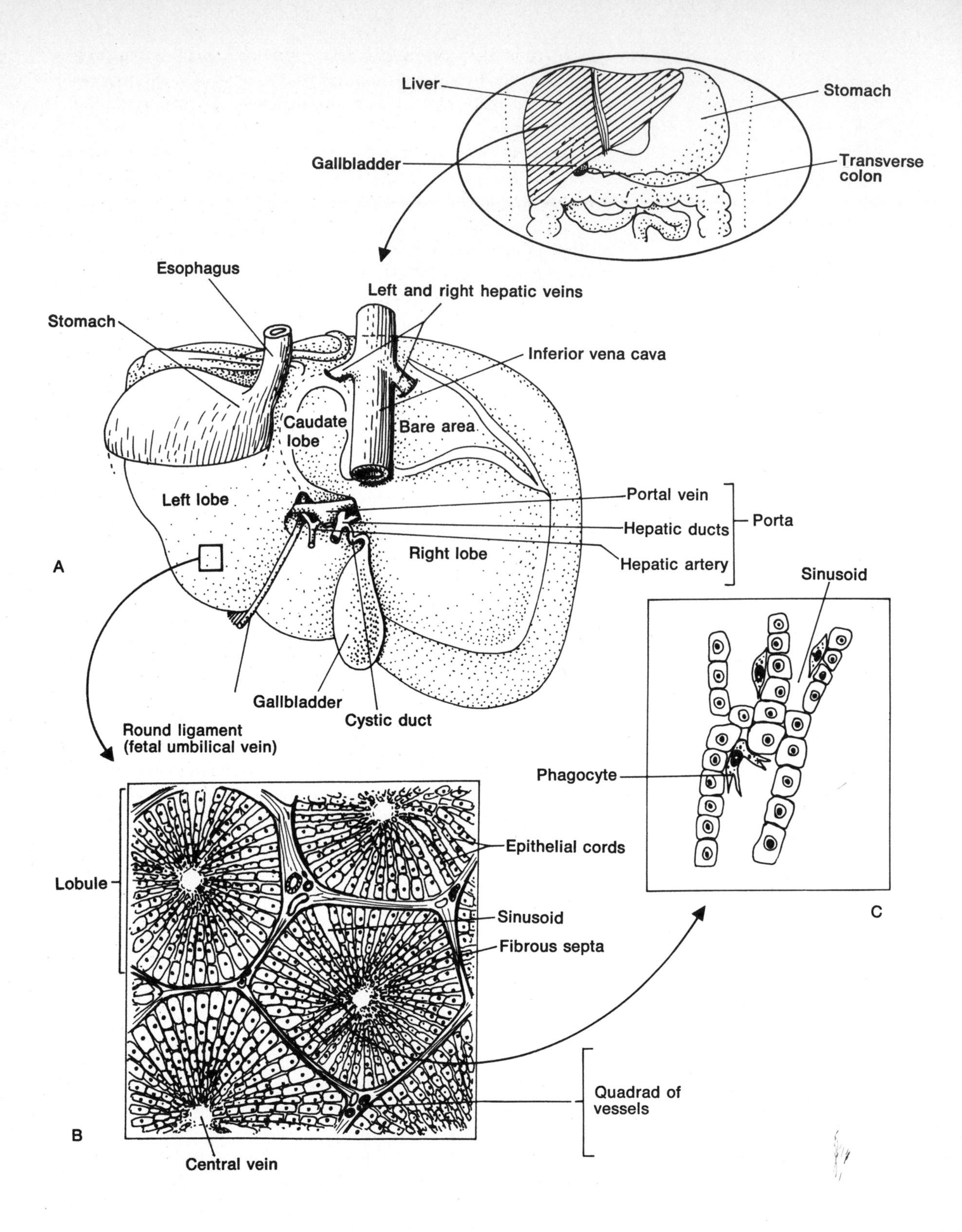
Liver
Stomach
Gallbladder
Transverse colon
Esophagus
Left and right hepatic veins
Stomach
Inferior vena cava
Caudate lobe
Bare area
Left lobe
Portal vein
Hepatic ducts
Hepatic artery
Porta
Right lobe
A
Gallbladder
Cystic duct
Round ligament (fetal umbilical vein)
Sinusoid
Phagocyte
Epithelial cords
Lobule
Sinusoid
Fibrous septa
C
Quadrad of vessels
B
Central vein

which selectively retrieve foreign material and fat passing through. These cells are responsible for phagocytosing large populations of bacteria and viruses entering the vascular system.

The complexity of the liver structure—although easily conceptualized—mirrors its functional complexity. Put simply, metabolic products of intestinal digestion enter the liver by way of the portal vein, pass into the sinusoids and are absorbed by the adjacent hepatic cells. The remaining material continues on to the central vein and to the IVC via the hepatic veins, ultimately reaching the heart.

Gallbladder

The gallbladder is a reservoir of bile collected from the liver. Bound to the visceral surface of the liver by peritoneum, the gallbladder demonstrates three parts (Fig. 9-14): a proximal **neck,** a central **body,** and a distal **fundus.** The last peeks out from under the anterior-inferior border of the liver to touch the greater omentum (Fig. 9-2). The gallbladder communicates with the common bile duct via the **cystic duct** (Fig. 9-14). The gallbladder receives dilute bile from the hepatic ducts, concentrates the bile by a ratio of 10:1, and stores it until a hormone from the duodenum (cholecystokinin) "informs" the gallbladder to contract and discharge bile into the common bile duct for passage to the duodenum. There the bile acts to break down the long-chain fat molecules for future absorption.

The mucosa of the gallbladder is characterized by tall columnar cells with underlying lamina propria. Though glands and a muscularis mucosa are absent, a strong muscular tunic can be seen. The gallbladder is often stained a dark green because of its contents.

Not infrequently, mineralization of bile may create "gallstones." Occlusion of the cystic duct and trauma to the mucosa trigger muscular spasms and a severe pain results. Surgical intervention and removal of the gallbladder (cholecystectomy) may be necessary. There need be no irreparable harm to the patient.

Pancreas

Affectionately termed "sweetbreads" by those who enjoy eating these glands from sheep, the pancreas is a soft, lobulated, overgrown salivary gland secured against the posterior abdominal wall by peritoneum. It is broken down into four parts for descriptive purposes: a **head, neck, body,** and **tail.**

Fig. 9-13 (opposite) The liver. Insert: The liver and its relations. A. Postero-inferior (visceral) surface. B. Cross section showing arrangement of lobules. C. Section of a liver lobule.

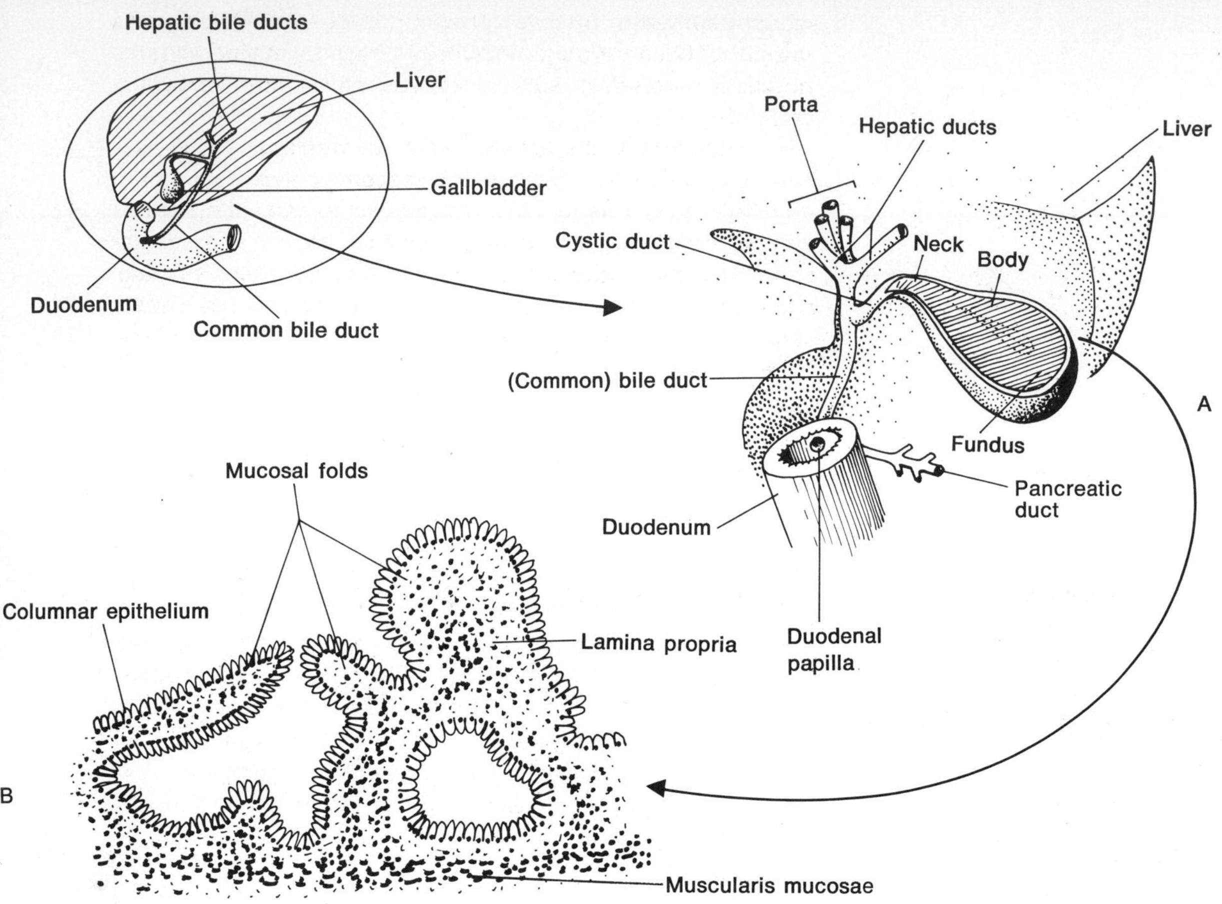

Fig. 9-14 Gallbladder. A. Dissected gallbladder, in situ. B. Mucosal surface of the gallbladder.

You will recall that the head rests in the "C" of the duodenum. The rest of the pancreas proceeds upward and to the left, tapering all the while, to terminate at the hilus of the spleen (Fig. 9-15). The duct of the pancreas, draining the lobules, usually joins the common bile duct before the latter perforates the duodenal wall.

The pancreas, being retroperitoneal, can easily be outlined on the anterior abdominal wall. The head, just an inch or so to the left of the midline, lies at the level of L2. The neck and body ascend obliquely to the left across L1; the tail touches the spleen deep to the 8th rib (T10) just medial to the midclavicular line.

The pancreas is intimately related to several vital organs, giving a clue to the significance of pancreatitis and subsequent spread of infection. These organs include the great vessels (IVC and aorta), mesenteric vessels, splenic vessels, stomach, duodenum, spleen, small intestine, and transverse colon (Fig. 9-2).

Microscopically, the pancreas is an economical two organs in one. The majority of the gland (second in size only to the liver) is taken up by exocrine glands (tubuloalveolar). These glands, supported by loose connective tissue, secrete enzymes into their ducts which are tributaries of the main pancreatic duct. These enzymes (trypsin, amylase, and lipase) break down proteins, carbohydrates, and fats, respectively. They are secreted in response to the hormones secretin and pancreozymin, which are released by the duodenum in response to foodstuffs entering it from the stomach.

The endocrine gland portion of the pancreas consists of a population of "cell islets" distributed apparently randomly

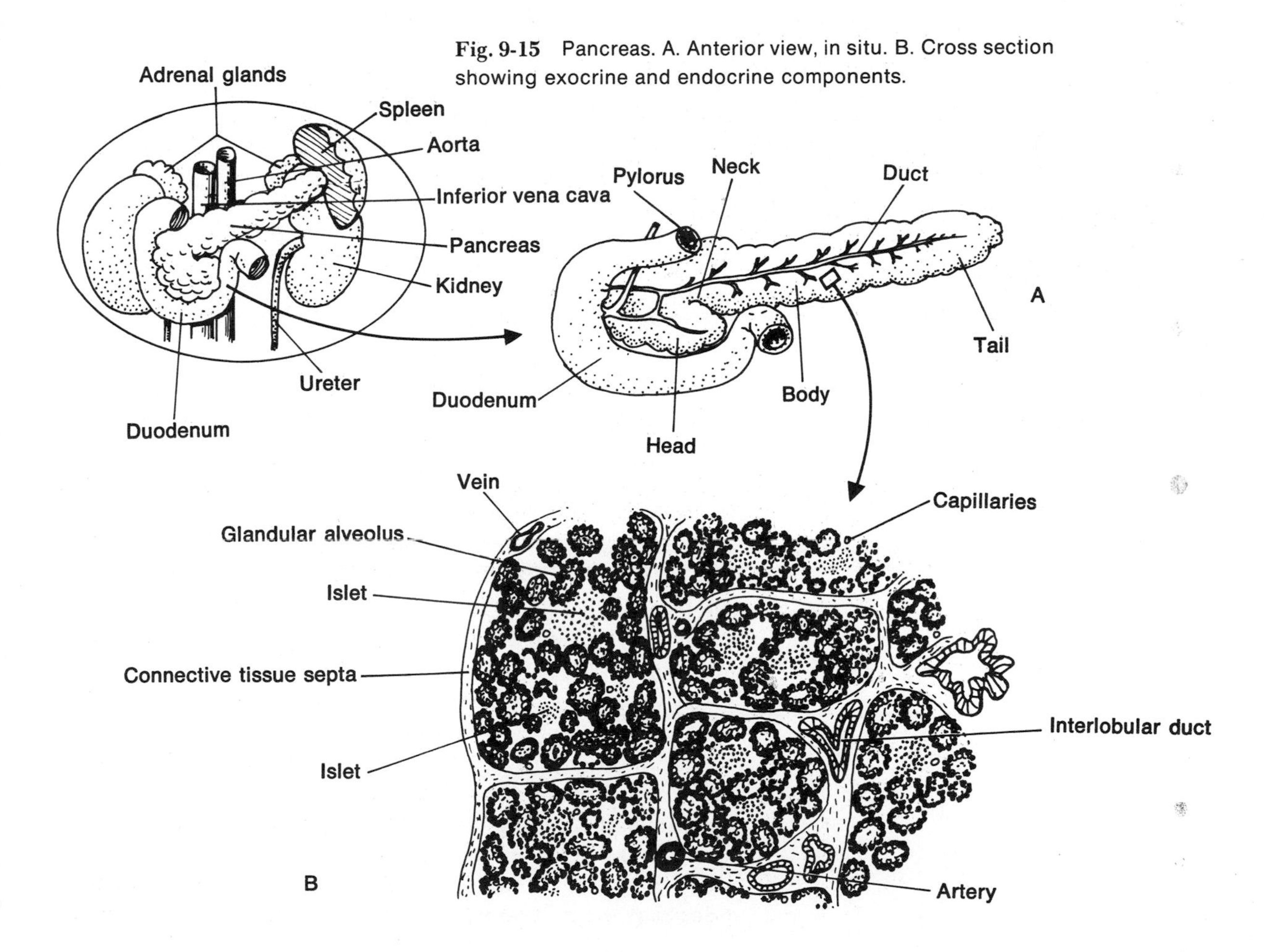

Fig. 9-15 Pancreas. A. Anterior view, in situ. B. Cross section showing exocrine and endocrine components.

throughout the gland. In close association with the epithelial cells of these "islets" are numerous capillaries. These cells produce and secrete two hormones: **insulin** and **glucagon.** Insulin, a polypeptide, influences the transport of sugar molecules desiring entry to the intracellular environment. Exhaustion of this chemical agent results in high concentrations of glucose in the blood (hyperglycemia) and spillage of glucose in the urine (glycosuria). Carbohydrate metabolism is altered abnormally and the resultant condition is termed **diabetes mellitus.**

Glucagon, a polypeptide, acts on the liver cells to increase the breakdown of glycogen and so releases more sugar into the blood.

Spleen

The very delicate, highly vascular spleen, shaped somewhat like a baseball mitt without fingers, occupies a niche in the left upper quadrant flush against the diaphragm under cover of ribs 9 to 11 (vertebral levels T10 to L1). Invested in peritoneum, the spleen is related to the tail of the pancreas, the left kidney, the stomach, and the left colic flexure (Fig. 9-2).

The spleen is an organ of the lymphatic system—in company with the thymus, tonsils, and lymph nodes*. You might suspect that it is concerned, therefore, with an immunologic defense mechanism. Indeed it is, producing lymphocytes and monocytes which are believed to be sources of antibody. Other principal functions of the spleen are related to the complex microscopic anatomy of this organ, which can be visualized by a simple diagram as seen in Fig. 9-16. At first glance the spleen might appear as an overgrown lymph node: it is encapsulated and stuffed with sinuses and lymphatic tissue (reticular tissue and lymphocytes). There the analogy largely ends, for the spleen is a filter of blood, the node a strainer of lymph.

Arteries supplying and veins draining the spleen enter at the **hilus.** The spleen has a thin but tough connective tissue capsule and extensions (septa) of this fibrous capsule divide the spleen into lobules. Within the lobules, two general kinds of tissue are found: masses of lymphatic tissue infiltrated with venous sinuses known as **red pulp,** and nodules of lymphatic tissue, often with active germinal centers perforated by an artery, known as **white pulp.** Blood leaves the arteries, probably percolates through the red pulp, and passes into venous sinuses. The sinuses are lined with phagocytic cells and here

* See Synopsis of Systems, appendix.

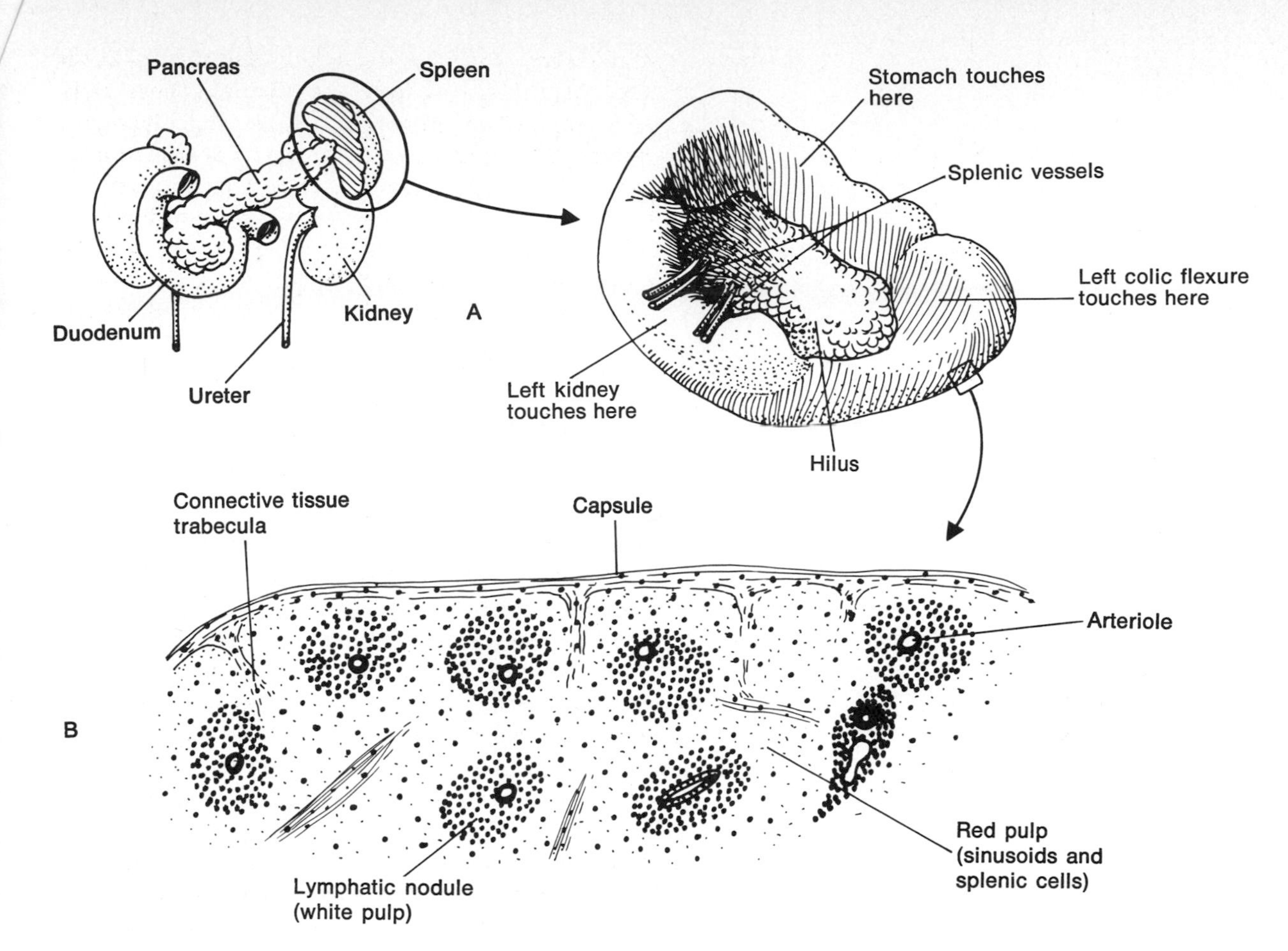

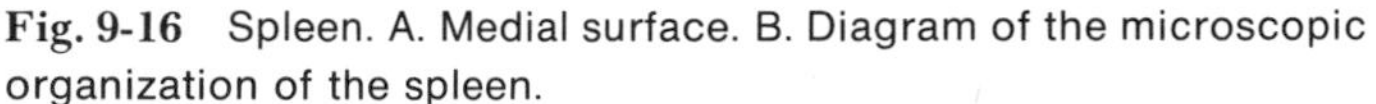

Fig. 9-16 Spleen. A. Medial surface. B. Diagram of the microscopic organization of the spleen.

the blood is strained of old, degenerated blood cells as well as foreign matter. Filtered blood passes out the splenic veins at the hilus.

Old red blood corpuscles are broken up and their iron stored by the splenic cells. Iron is transferred to the liver via the portal system in time of need. The pigment of discarded corpuscles is employed by the liver in the production of bile.

During embryonic life, the spleen is an important source of red and white blood cells. In adult life, it continues to be an important supplier of monocytes and lymphocytes. Normally not essential to life, the spleen is often ruptured in extensive falls, as from high places, and this is then frequently the cause of death. The spleen is considered a reservoir of blood: its artery is relatively large and the volume of the splenic spaces is subject to enlargement, as the capsule and septa of the spleen contain large numbers of elastic fibers. Smooth muscle fibers are also found in these walls. In the dog, and possibly in humans, the spleen may be seen to contract during hemorrhage—suggesting its "dynamic reservoir" role.

NERVE SUPPLY OF THE ALIMENTARY TRACT AND RELATED ORGANS

Happily, motor innervation of the organs of digestion is supplied by the autonomic nervous system. If not, you would have to willfully initiate each stage of the digestive process if you wanted to survive.

You will recall that the parasympathetic ("vegetative") division is generally concerned with glandular secretion, smooth muscle contraction of the digestive viscera (peristalsis), and dilatation of visceral arteries. The sympathetic division induces constriction of the visceral arteries and activates the sphincters at the gastroduodenal and ileocecal orifices. In a number of instances secretomotor activity of the alimentary canal (independent of the ANS) can be initiated merely by the presence of food material bathing the mucosal wall. You should not attempt to understand secretomotor action or peristalsis by creating sharp distinctions between what is sympathetic and what is parasympathetic activity. The two divisions work in concert to carry out the total digestive process, reinforcing the effect of hormones released by the viscera themselves and the ability of the viscera to initiate their own glandular secretions and muscular contractions.

The sensory nerves of the viscera are borrowed from the somatic portion of the peripheral nervous system; the autonomic nervous system is considered a motor system and has no sensory fibers of its own, you will remember (and if you don't, try brushing up on the system in Unit 3).

The innervation plan of the alimentary canal may be viewed in Fig. 9-17. Note that parasympathetic innervation of abdominal viscera is supplied by two different elements: the **vagus nerves** innervate the abdominal viscera to the level of the transverse colon, and the **pelvic splanchnic nerves** from the S2,3,4 levels of the spinal cord innervate all viscera below that level, including pelvic and perineal structures. The vagal fibers ramify within the **aortic plexus** and follow vessels out to the organs they supply. Within the walls of these organs, the pre- and postganglionic neurons synapse. The pelvic splanchnic fibers reach the **pelvic plexus** and follow vessels to the organs they supply.

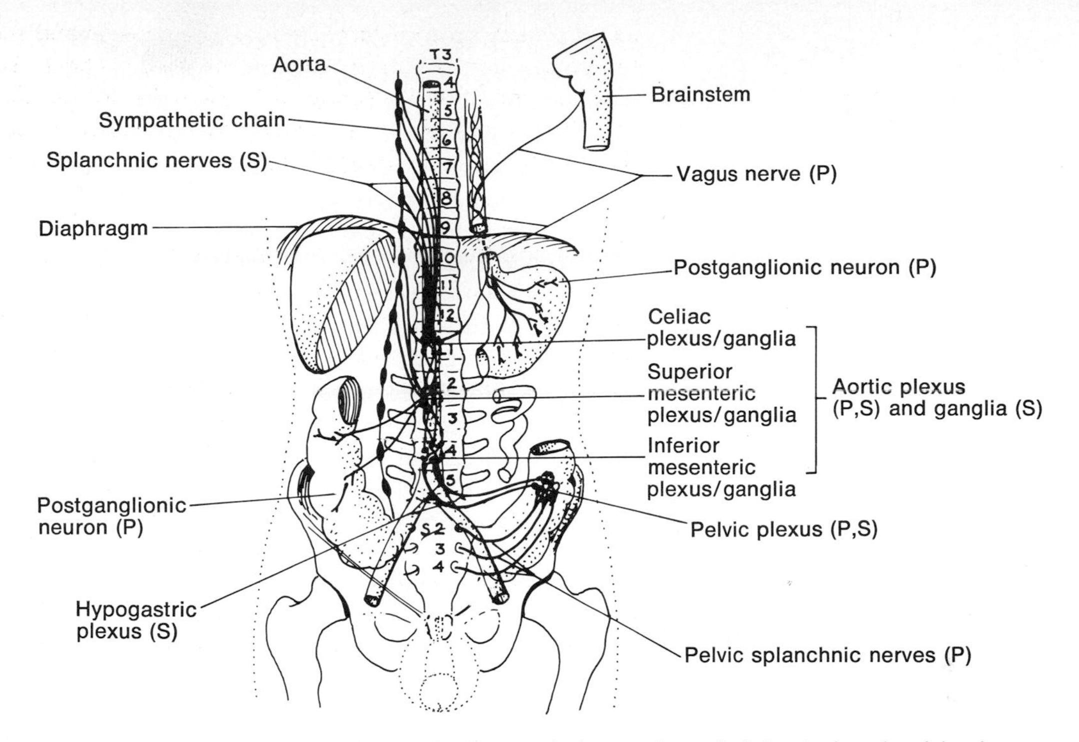

Fig. 9-17 Autonomic innervation of abdominal and pelvic viscera. P, parasympathetic; S, sympathetic. See also Fig. 3-54, page 161.

Sympathetic innervation of the abdominal viscera is derived from the **thoracic splanchnic nerves** (T5 through T12 spinal cord levels). These pass through the diaphragm to enter the aortic plexus on the anterior aspect of the abdominal aorta. Within this plexus, usually associated with the principal visceral branches of the aorta, are **sympathetic ganglia** in which the preganglionic thoracic splanchnic nerves end in synapsis. The postganglionic sympathetic fibers leave the plexus with the vessels and other nerves destined for the organs they supply. Sympathetic fibers to the lower bowel and pelvic and perineal viscera get there via the hypogastric plexus and the pelvic plexus.

In general the sensory fibers (that is, fibers of the peripheral nervous system) ride free with the autonomic fibers, pain fibers passing with the sympathetic division, pressure fibers with the parasympathetic division. Thus, pain from the gallbladder may be mediated by fibers which reach the spinal cord at a level of T5. Since the heart is supplied by sympathetics from as low as the T5 level, pain in the chest due to gallstones could be confused with a heart attack.

VASCULAR SUPPLY OF ABDOMINAL VISCERA

The arterial pattern of the abdominal viscera is made basically simple by the fact that three stems from the abdominal aorta supply the viscera almost entirely. The various branches of these three anastomose rather freely, so a healthy system of collateral circulation exists.

Refer now to Fig. 9-18 as you read.

The **celiac artery** projects off the aorta just above the pancreas at the level of T12 and immediately branches into the **hepatic, splenic,** and **left gastric** arteries. These branches (Fig. 9-18A) supply the stomach, liver, gallbladder, duodenum, pancreas, and spleen.

Deep to the pancreas, the **superior mesenteric artery** springs forward from the aorta at the level of L1, descends, passes over the duodenum between the layers of the mesentery proper, and branches (Fig. 9-18B) to supply the small intestine, appendix, cecum and two-thirds of the colon.

The **inferior mesenteric artery** leaves the aorta to the left of the midline and, passing deep to parietal peritoneum (retroperitoneal), supplies the descending and sigmoid colon as well as the rectum.

Note the number of significant anastomoses between the celiac and superior mesenteric arteries and between the superior and inferior mesenteric arteries.

VENOUS DRAINAGE OF ABDOMINAL VISCERA— THE PORTAL SYSTEM

It has been pointed out that assimilated foodstuffs drawn into the circulation from the alimentary canal go directly to the liver for processing. This implies that venous blood from the GI tract must pass through another organ besides the capillaries before returning to the heart. This extra step is a feature of a **portal system,** as sketched in Unit 3. In this case, the scheme is heart → intestinal capillaries → liver sinusoids → heart. The function of the veins draining the spleen, intestines, stomach, and pancreas is to *transport* the digestive products to the liver; hence the name **hepatic portal system.**

The principal vein of this system is the **portal vein** whose tributaries drain the organs mentioned above. The portal vein enters the porta of the liver and breaks up into interlobular veins, which supply the liver sinusoids (Fig. 9-13). The sinusoids are drained by central veins of the lobules. These are tributaries of the hepatic veins, which empty into the inferior vena cava at the level of the diaphragm.

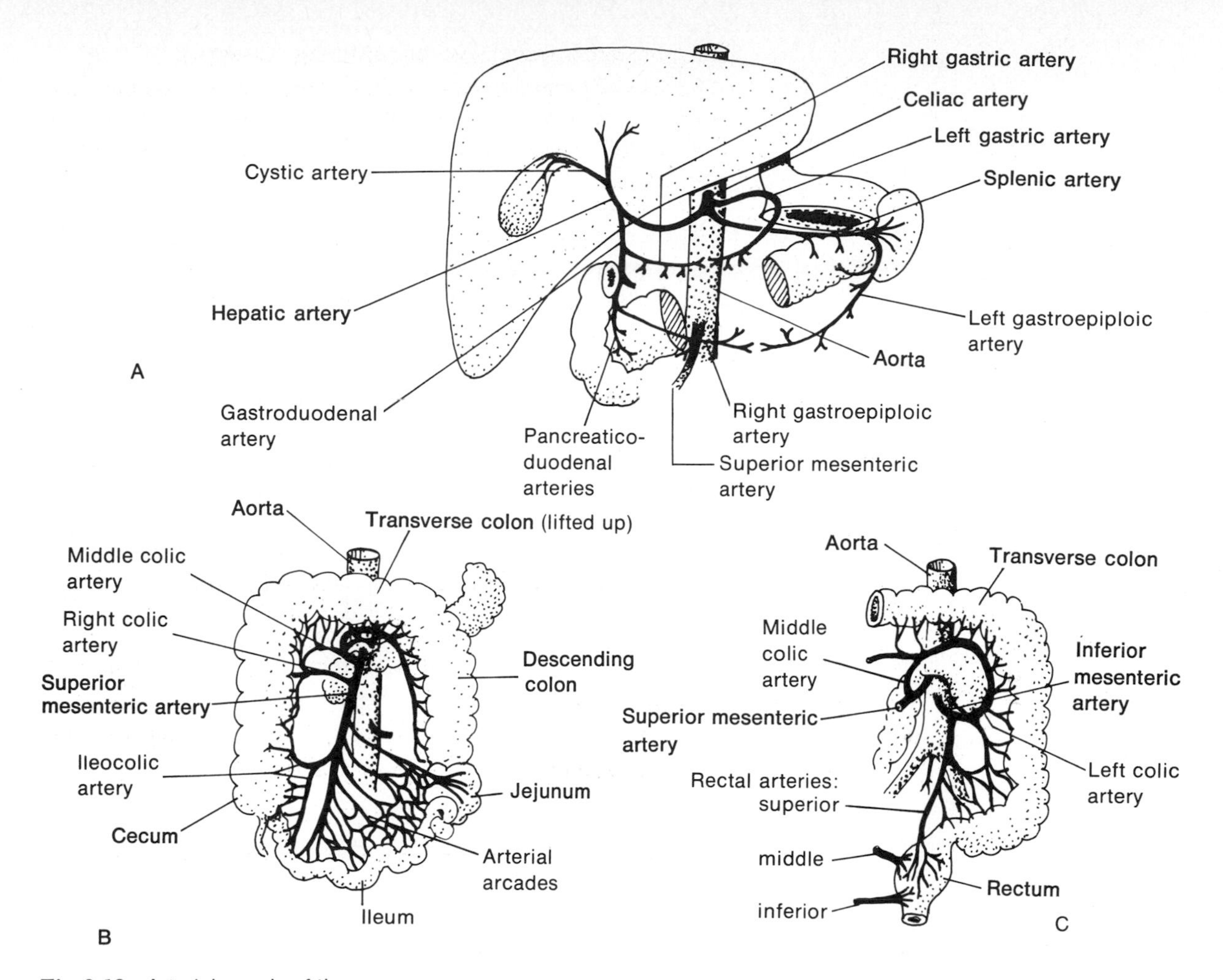

Fig. 9-18 Arterial supply of the alimentary canal and related organs. A. Branches of the celiac artery. B. Distribution of the superior mesenteric artery. C. Distribution of the inferior mesenteric artery. Note the anastomoses among the three principal arteries.

The portal vein and its tributaries, which generally parallel the arteries, may be studied in Fig. 9-19. The portal system of veins communicates at various points with tributaries of the inferior and superior venae cavae. Thus, following occlusion of the portal system, as in advanced cirrhosis of the liver, communicating veins belonging to the IVC, SVC, or azygos systems often dilate—sometimes enormously—to accommodate the increased blood flow.

LYMPHATIC DRAINAGE OF ABDOMINAL VISCERA

In addition to having the normal tasks of lymphatic vessels, abdominal lymphatics absorb and conduct molecules of fat to the venous system. It all begins with tiny lymphatic capillaries in the villi of the small intestine (lacteals). These lacteals drain into the larger vessels of the submucosa and mesentery to enter the sac-like **cisterna chyli** via the intestinal trunks (Fig. 9-20). The cisterna chyli receives lymphatic drainage from the lower limbs and from pelvic and perineal viscera as well as abdominal viscera. Heading superiorly it becomes the **thoracic duct.** Each abdominal organ has its lymphatic vessels and local lymph nodes whose efferent vessels ultimately pass between layers of mesentery or under the parietal peritoneum toward the intestinal trunks. Lymph flow surges through the cisterna chyli and thoracic duct to enter the venous blood at the junction of the left internal jugular and subclavian veins.

QUIZ

1. The interior surface of the stomach is characterized by rugae which serve to increase the secretory and absorptive capacity of the stomach.
2. The digestive target of the stomach is primarily protein.
3. The small intestine consists of three parts: duodenum, ileum, and jejunum.
4. The microscopic structure of the small intestine consists of four layers: submucosa, mucosa,

Fig. 9-19 Plan of venous drainage of the alimentary canal. Hatched areas represent points of anastomoses between portal and caval venous systems. Vessels noted with an asterisk (*) are tributaries of the inferior vena cava. Vessels with a cross (+) are tributaries of the azygos system.

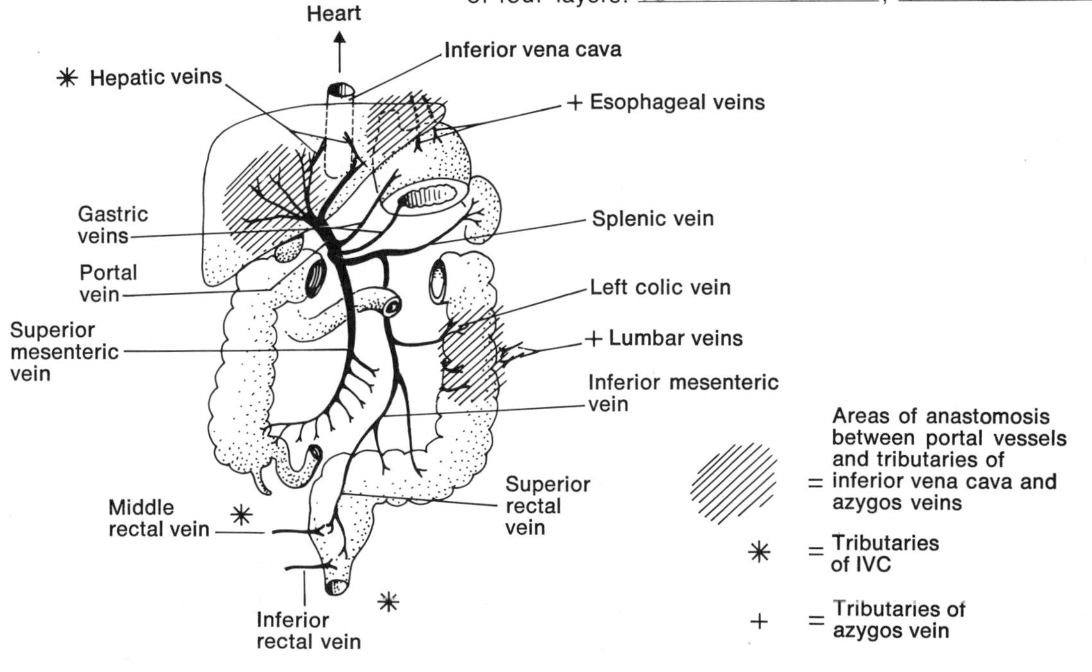

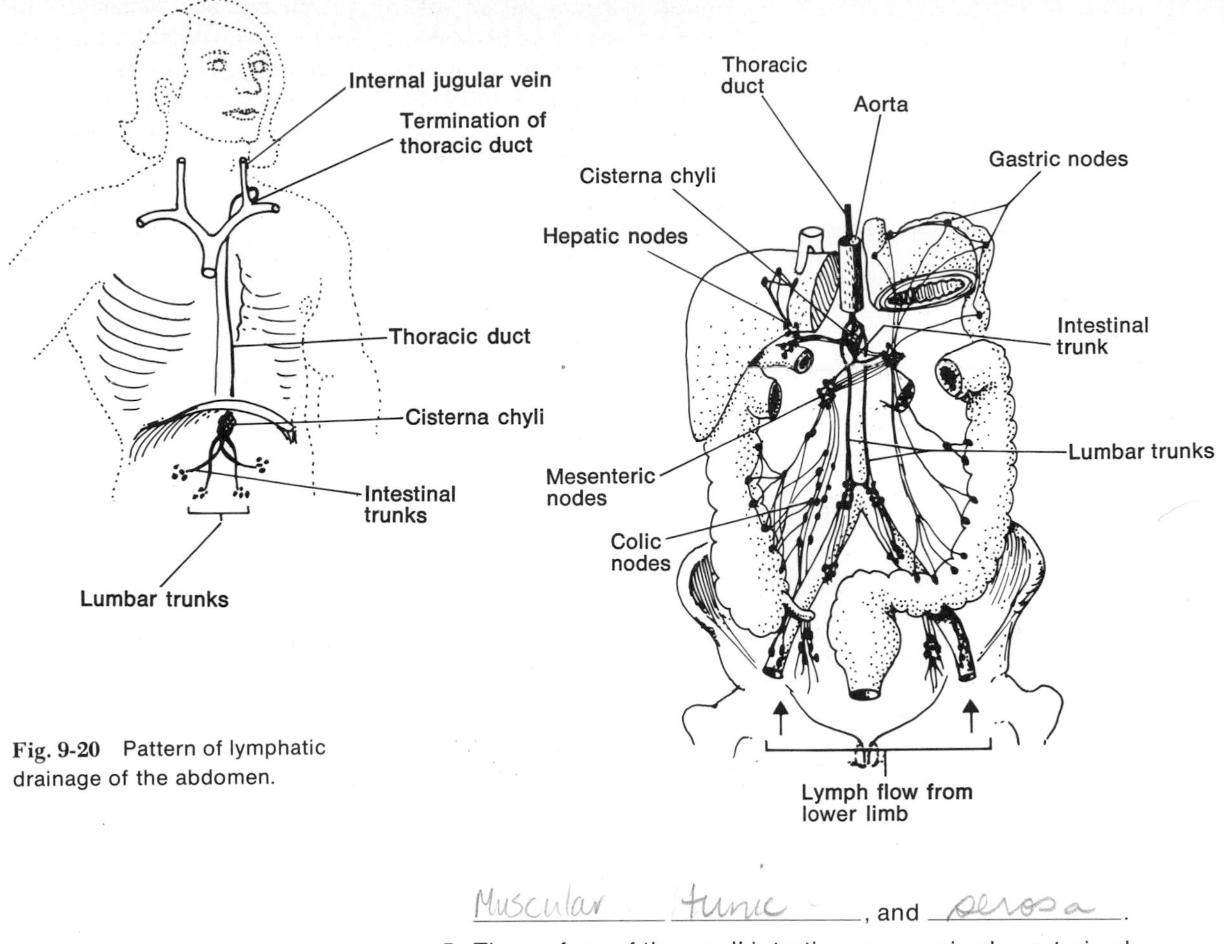

Fig. 9-20 Pattern of lymphatic drainage of the abdomen.

Muscular tunic, and serosa.

5. The surface of the small intestine mucosa is characterized by a carpet of fingerlike projections (villi).
6. The large intestine, or colon, consists of four parts: cecum, colon, rectum and anal canal.
7. It is in the colon that a significant increase in water absorption occurs.
8. The structural and functional unit of the liver is the gland.
9. The pancreas secretes three well-known enzymes: trypsin, amylase, and lipase; and two hormones: secretin and pancreozymin
10. The parasympathetic division of the ANS is represented in the abdominal cavity primarily by the vagus nerve. (and splanchnic nerves)

abdominal viscera of the urinary system

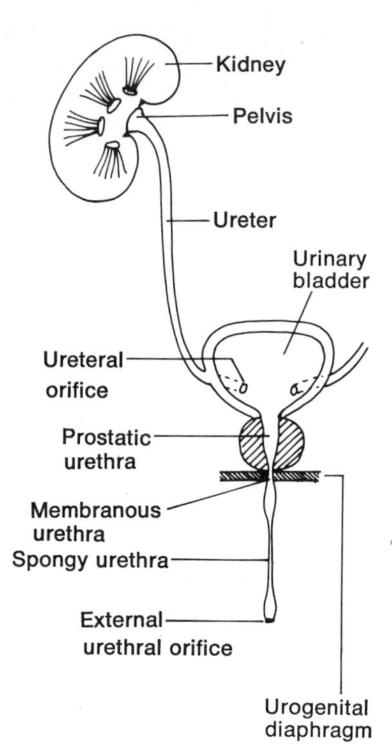

Fig. 9-21 The urinary system.

The urinary system consists of the paired kidneys, ureters, urinary bladder, and urethra (Fig. 9-21). The structures function to maintain and adjust proper fluid and electrolyte balance through reabsorption and secretion as well as excretion of excess water and metabolites. This is primarily the function of the kidney—which is to be considered, along with its ducts, the ureters, in this unit. The bladder and urethra, populating the pelvis, will be considered in Unit 10.

KIDNEYS (L., *renes*)

Like the liver in color and consistency, the brown, bean-shaped, kidneys are fastened in retroperitoneal position by a strong sheet of fascia reinforcing the peritoneum. The right kidney is lower than the left because of the imposing presence of the liver. Each may be outlined on the surface of the back paravertebrally between T12 and L2 (Fig. 9-22). Although partly protected by the costal cage, the kidneys are vulnerable to blows from the back.

A study of the kidney's anterior relations (Fig. 9-23) will give you some idea of the compactness of viscera in the region T12 to L2.

A midline, longitudinal section of the kidney (Fig. 9-24A) nicely demonstrates its four major anatomical subdivisions:

1. An outer rind—the **cortex.** With close examination, the cortex is seen to be punctured by hundreds of thousands of tiny holes—as if made by a pin. The cortex consists of tiny, convoluted tubules, some straight tubules, clusters of renal corpuscles, and a network of capillaries and arterioles.
2. A middle belt of triangular pyramids—the **medulla.** Each pyramid may be seen to consist of longitudinal striations (straight and collecting tubules) raining from the cortex. Columns of cortex dip down between medullary pyramids.
3. An inner set of cuplike structures draining the pyramids—the **minor** and **major calyces.** The smaller

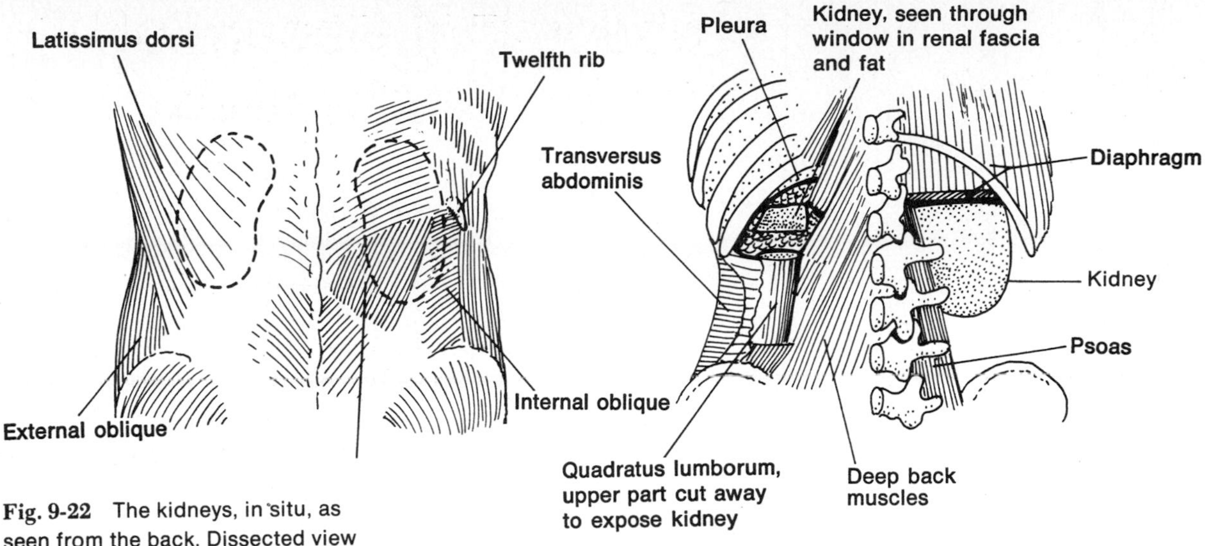

Fig. 9-22 The kidneys, in situ, as seen from the back. Dissected view shown on right. (From R. D. Lockhart, G. F. Hamilton, and F. W. Fyfe, *Anatomy of the Human Body*, 2d ed., J. B. Lippincott, Philadelphia, 1965.)

calyces blend into larger calyces which merge to form the **pelvis** and **ureter** beyond.

4. The concavity of the kidney **(hilus)** disburses the ureter and renal vein and receives the renal artery.

Anatomical basis of kidney function

The microscopic morphology behind the kidney's job of regulating fluids and electrolytes can be followed in Fig. 9-24: The functional unit of the kidney is the **nephron.** Demonstrating a veritable cornucopia of metabolic activity, the nephron consists of:

1. A hollow epithelial ball **(capsule)** into which a tuft of arterioles **(glomerulus)** grew in embryonic development.

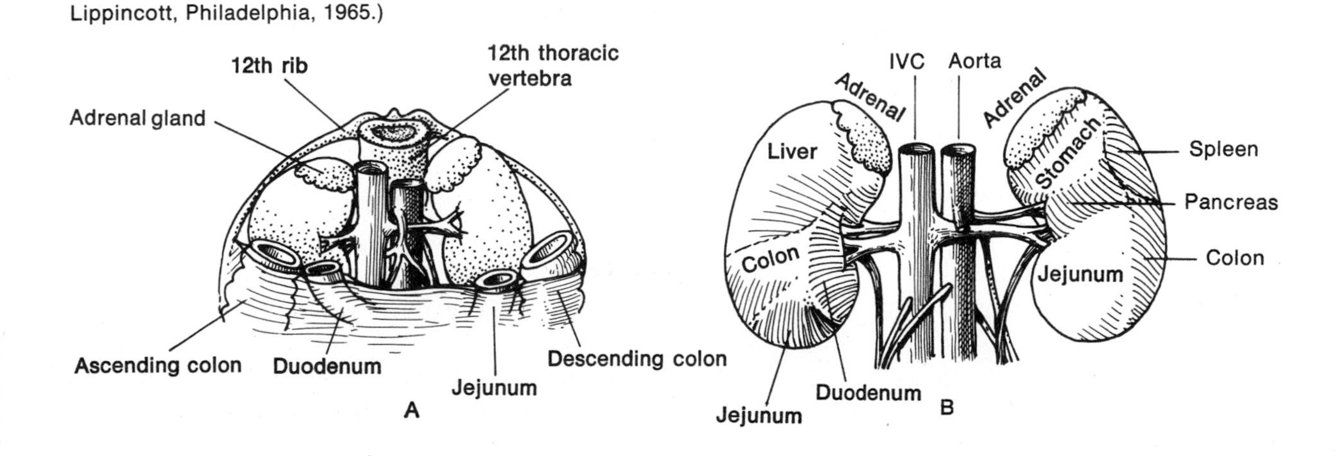

Fig. 9-23 Relations of the kidney. A. Anterior view, in situ. B. Anterior relations. (From R. D. Lockhart, G. F. Hamilton, and F. W. Fyfe, *Anatomy of the Human Body*, 2d ed. J. B. Lippincott, Philadelphia, 1965.)

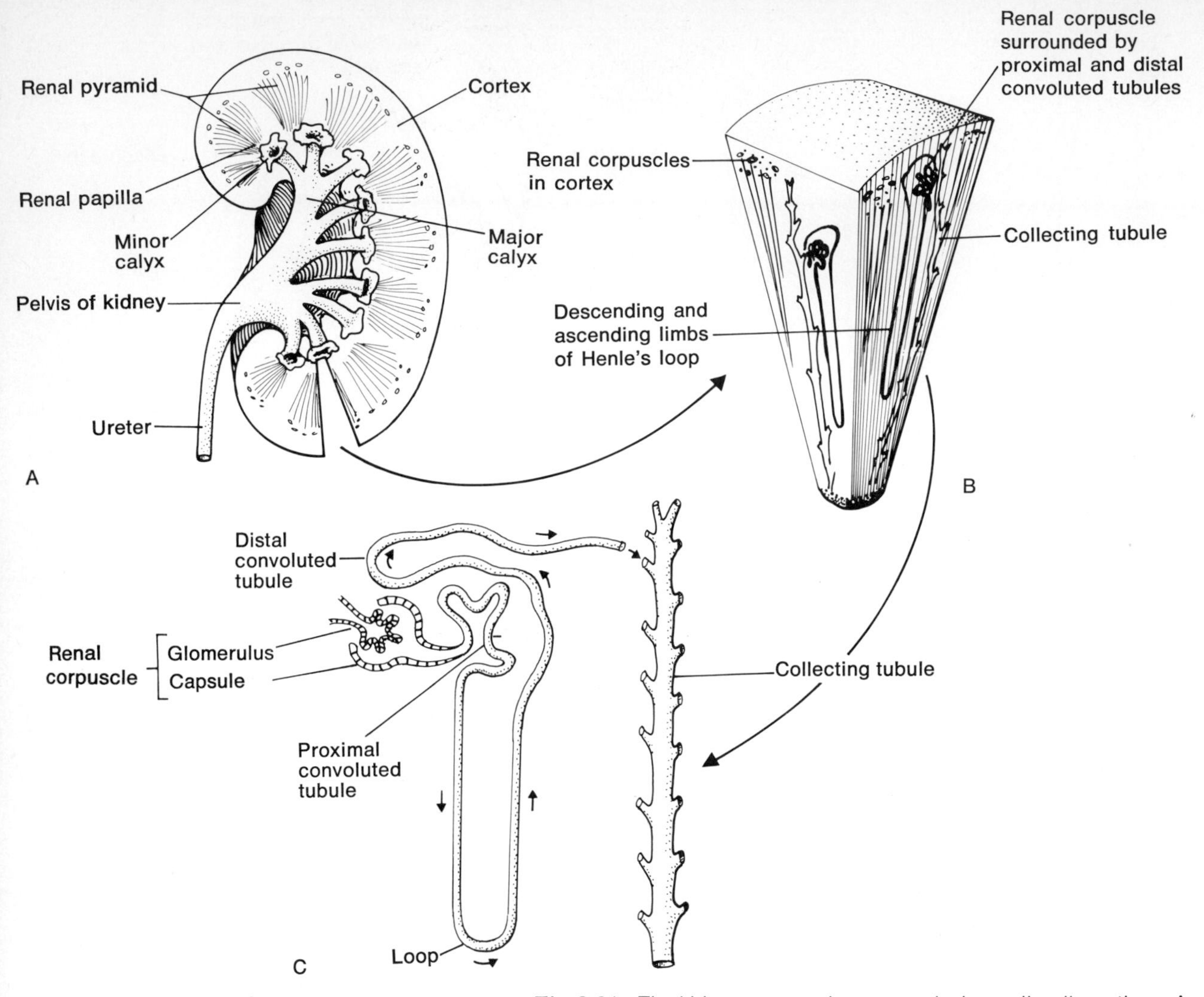

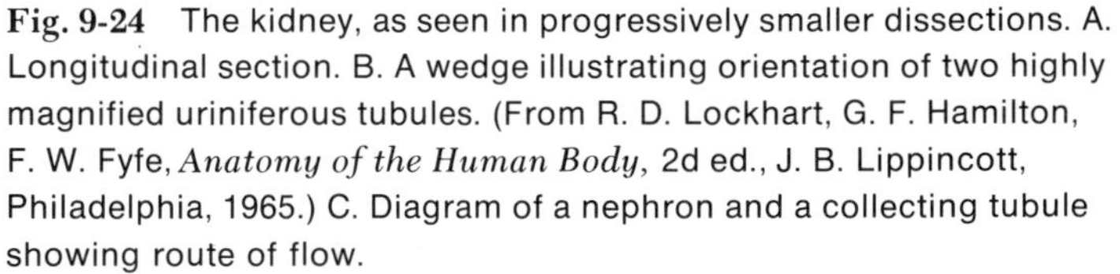

Fig. 9-24 The kidney, as seen in progressively smaller dissections. A. Longitudinal section. B. A wedge illustrating orientation of two highly magnified uriniferous tubules. (From R. D. Lockhart, G. F. Hamilton, F. W. Fyfe, *Anatomy of the Human Body*, 2d ed., J. B. Lippincott, Philadelphia, 1965.) C. Diagram of a nephron and a collecting tubule showing route of flow.

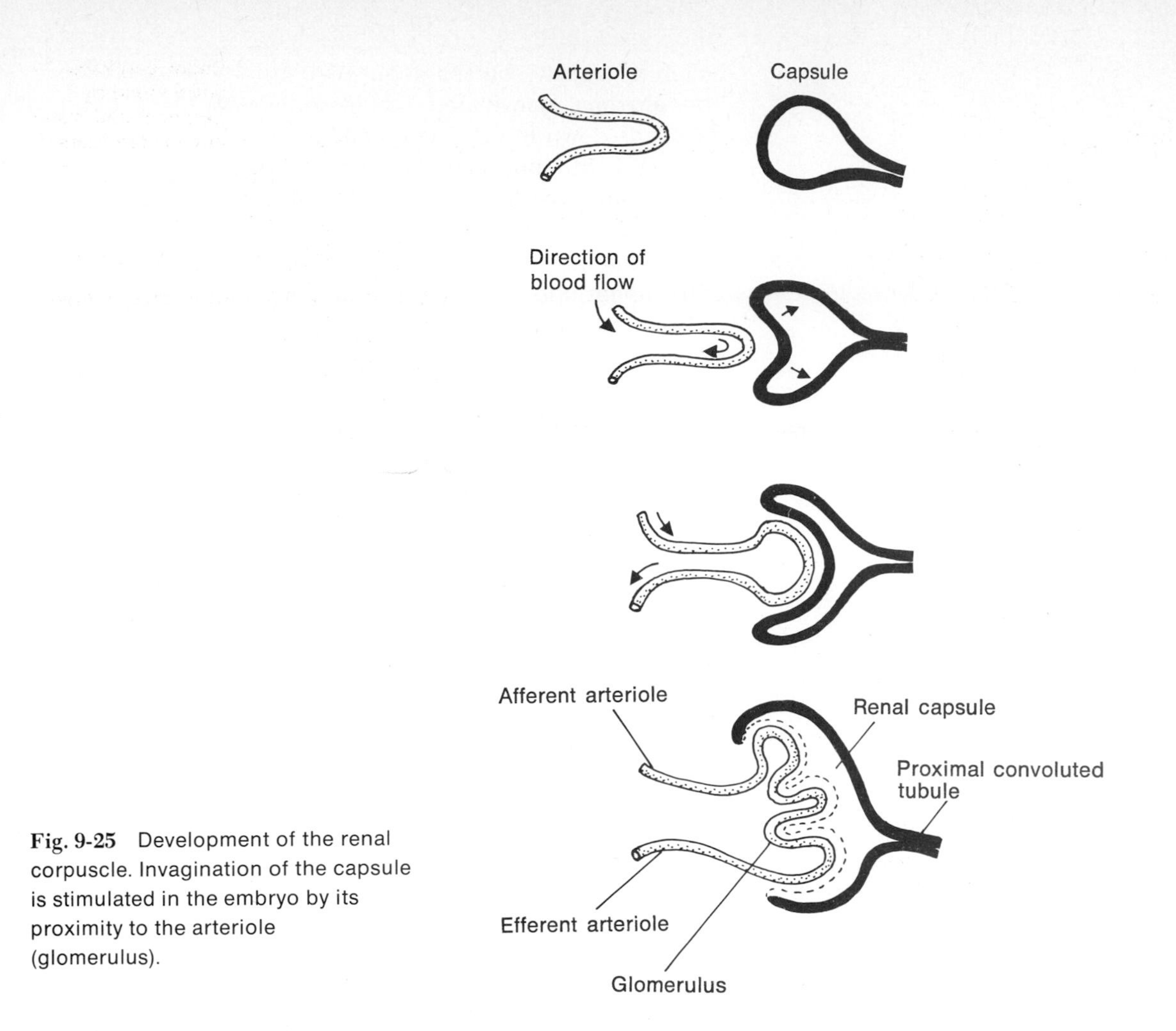

Fig. 9-25 Development of the renal corpuscle. Invagination of the capsule is stimulated in the embryo by its proximity to the arteriole (glomerulus).

2. A length of epithelial-lined tubing, in both convoluted and straight segments, draining the capsule.

A capsule and its glomerulus constitutes a **renal corpuscle.** Numbering roughly one million per kidney, renal corpuscles camp in the cortex, as do most of the convoluted tubules. The embryologic basis for the intimate relations of the glomerulus and capsule can be appreciated in Fig. 9-25. As a consequence of this relationship, the inner simple squamous epithelial (visceral) layer of the capsule acts as a filter for the arteriole. Blood plasma moves into the capsule of the nephron. The blood cells and unfiltered plasma do not. Instead they pass into a peritubular capillary plexus (Fig. 9-26).

The capsule opens into the **proximal convoluted tubule**

which tortuously courses about its capsule; hence its name. The proximal convoluted tubule then straightens out and descends down through the medulla. A million of these and other straight tubules create the striations seen in the medullary pyramid. The descending tubule then loops around to ascend back up toward the capsule of origin. Back at the capsule, the tubule convolutes again **(distal convoluted tubule)** and empties into a collecting tubule or duct.

The renal corpuscle, proximal convoluted tubule, descending limb, loop, ascending limb, and distal convoluted tubule constitute the nephron.

The proximal and distal tubules of the nephron are lined by columnar epithelia with a brushlike border of microvilli on their free surfaces. Such surface adaptations are indicative of highly active secretory or absorptive cells. As the filtrate from the capsule drains down the tubules, certain ions and molecules (glucose, urea, sodium, potassium, chlorides, ammonia, water, etc.) are selectively reabsorbed by the cells lining the tubules and transported to the adjoining extracellular space or capillary. Certain organic molecules are secreted by the tubular cells into the tubular lumen. In the distal convoluted tubules, a great deal of hydrogen and potassium ions are secreted to balance the reabsorption of sodium ions occurring throughout the tubule.

The plasma filtrate—having been added to and subtracted from throughout the length of the nephron—drains from the distal convoluted tubule into the **collecting duct.** Nephrons and collecting ducts together constitute **uriniferous tubules.** Embryologically distinct, collecting ducts are not part of the nephron. Ion exchange is materially reduced in the collecting ducts; water reabsorption from the filtrate by the tubular cells is an activity stimulated by antidiuretic hormone (ADH) secreted by the posterior lobe of the hypophysis.

The thousands of collecting ducts drain into 8 to 12 **minor calyces** which coalesce to form three to five **major calyces.** For every 100 ml of plasma filtrate formed, only 1 ml finds entry to the calyx—more on that momentarily.

The major calyces, **pelvis,** and **ureters** into which they drain are concerned with storage and transport of the newly formed urine. The epithelial lining of these structures reflects this function—a stratified but stretchable urothelium (transitional epithelium). Along these tubes, in which there is an absence of significant metabolic activity, unwanted urine is conducted to that fibromuscular bag of the pelvis, the **bladder,** to await ultimate urination.

The story has not yet been fully told. Reference has been

made to the capillaries associated with the tubules of the medulla and cortex of the kidney. They form the **peritubular capillary plexus,** and if you accept the glomerulus as a "capillary plexus," the peritubular capillary plexus is analogous to the sinusoids of the liver. In effect, we have here a **renal portal system.** The principal features of this system can be studied in Fig. 9-26. Their influence in renal physiology takes on added dimension when one contemplates that this peritubular capillary plexus recovers the 99 percent of plasma filtrate reabsorbed by the tubular epithelial cells.

It is interesting to note that the innervation of the kidney has little direct functional significance; what innervation there is is restricted to vasoconstrictive efforts by sympathetic nerves. The kidneys are largely autoregulated. They are more influenced by hormones than by any other substance or effect. These hormones include **antidiuretic hormone** (acting on the distal convoluted tubules and collecting tubules to in-

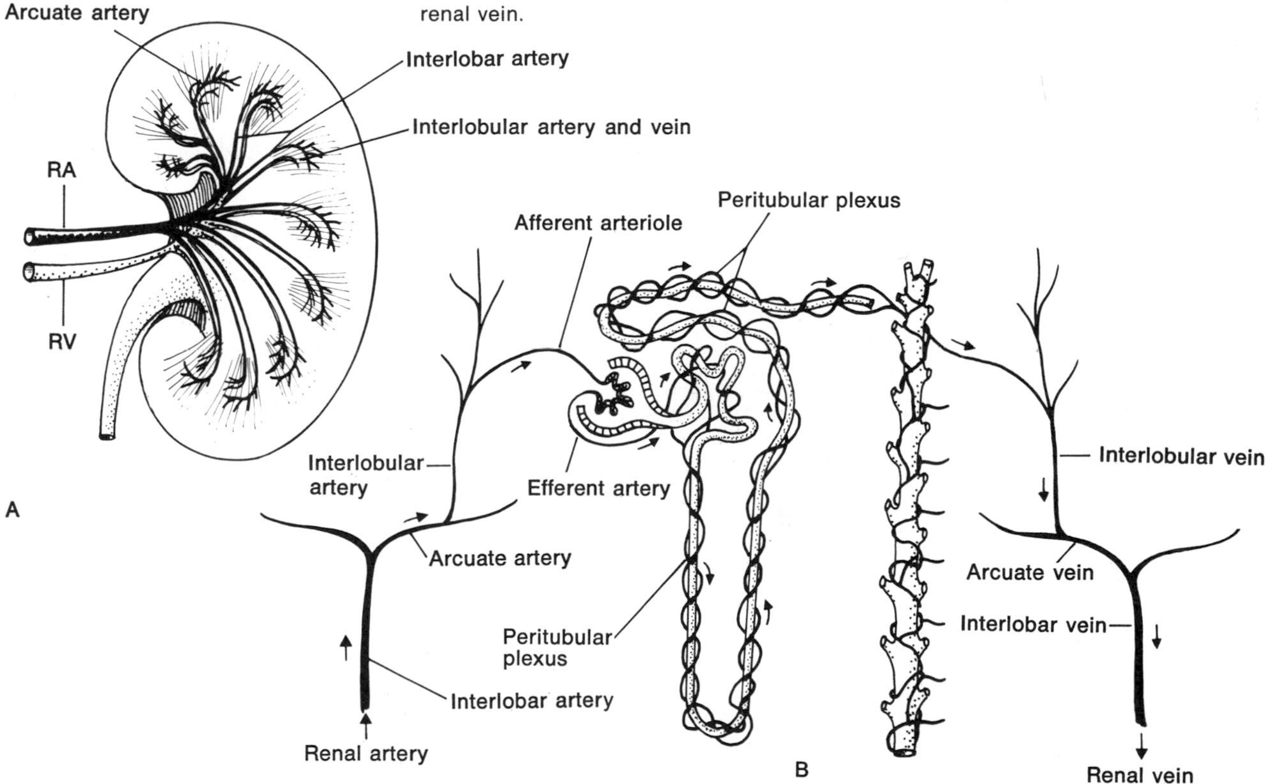

Fig. 9-26 Vascular pattern of the kidney. A. Longitudinal section, showing main branches/tributaries of the renal vessels. B. Renal portal system associated with the renal tubules. RA, renal artery: RV, renal vein.

crease water reabsorption, concentrating urine) and **aldosterone** (a mineralocorticoid from the adrenal cortex which acts to retain sodium ions in the proximal convoluted tubules and increase secretion of hydrogen and potassium ions in the distal convoluted tubules).

A concept of function

The kidney might be considered as a sort of stock exchange (it is assuredly not a sewer!). Blood (the seller) is hustled to the kidney interior where much of its plasma phase (stock certificates) is subjected to high-pressure filtration (initial scrutiny) and maneuvered into a long tubule. Here, the filtrate (stock certificates) is treated to a most intensive screening by the tubular epithelial cells (buyers). These cells can *reabsorb* (buy) elements of the filtrate and retain them or they can *transport* them (sell) to the blood capillaries or extracellular fluid (ECF) adjacent to the tubules. The tubular epithelial cells may also purchase specific elements from the ECF/blood capillaries (seller) and *secrete* them back into the tubular filtrate. Water is reabsorbed (purchased) in wholesale amounts by the tubular cells. Sodium ions ($Na+$) are reabsorbed too—but for a "price"—for electrical equilibrium must be maintained. The price: secretion (unloading) of hydrogen and potassium ions into the filtrate by the tubular cells. This is a very complex affair, for the net effect of all this trading must be equilibrium—a balance of concentrations, electrically and osmotically. Acid-base balance and preservation of water must be achieved. Urea and other undesirable elements must be discharged from the body, it is true, but the constancy of blood and intra- and extracellular fluid composition (fair trade) cannot be sacrificed, for serious sickness (economic depression) will result and death will follow if electrolyte and water equilibrium is not quickly achieved.

Ureter

The ureters are two muscular tubes with a pencil-point passageway lined with uroepithelium. Arising from the renal pelvis, they descend under parietal peritoneum along the posterior abdominal wall (abdominal part), over the iliac crest toward the pelvic brim (iliac part) and into the pelvis to terminate at the single urinary bladder (pelvic part) (Fig. 9-27).

They are of clinical interest for a number of reasons. For one, stones generated in the kidney may pass into the ureters and get stopped there, creating no little pain. As you can see in Fig. 9-27 these stones can be halted in their journey at

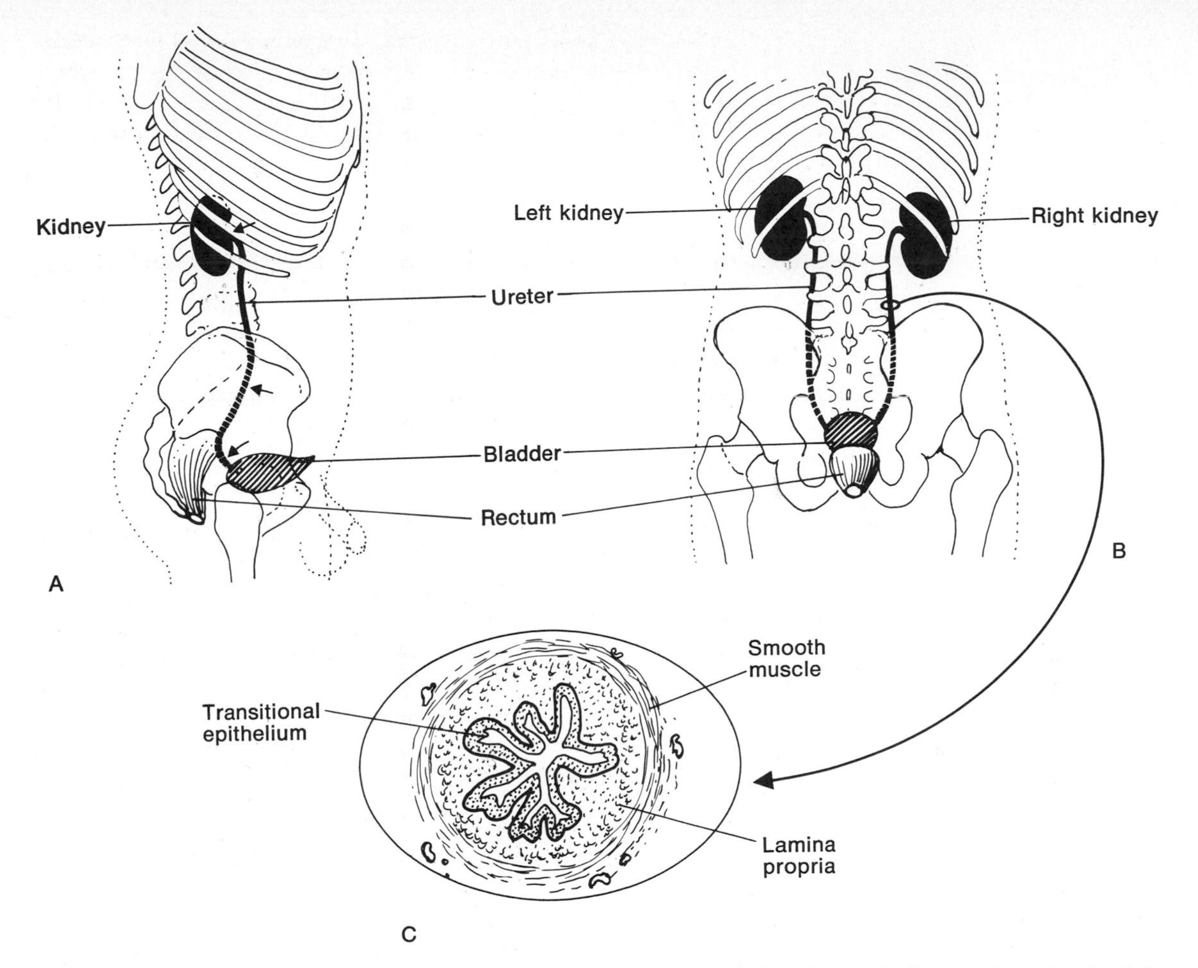

Fig. 9-27 Ureters. A. Lateral view. Arrows indicate points of potential constriction: pelvic-ureteric junction, ureter passing over iliac vessels at pelvic brim, and ureter entering bladder. B. Posterior view. C. Cross section of ureter.

three anatomical sites of constriction. If the uroepithelial lining is torn or destroyed during a stone's trek, connective tissue will quickly move in and take the place of the damaged epithelium (the former multiplying much faster than the latter). Scar tissue and a more restricted (if not occluded) ureteric lumen may follow. Blocked urine may back up into the kidney and seriously complicate what started out to be a minor but painful dilemma.

The ureter is abundantly supplied by arteries and nerves from a variety of sources as it descends to the bladder, e.g., abdominal aorta, autonomic nerves.

abdominal viscera of the endocrine system

The adrenal (suprarenal) glands sit atop the kidneys—enclosed in the same fascial investment—at a vertebral level of about T11. Triangular in shape, the adrenals are supplied by branches of local arteries including the aorta and renal arteries, and drained by tributaries of the inferior vena cava.

When sectioned, the adrenals demonstrate an outer, thick, yellowish **cortex** and a thin, reddish **medulla** (Fig. 9-28). The cortex consists of cords of epithelial cells arranged into three layers, and adjacent capillaries; the medulla, of irregular masses of epithelial cells in close contact with capillaries.

The cortex secretes three groups of hormones, one of which is essential for life: **glucocorticoids, mineralocorticoids,** and **sex hormones.** Glucocorticoids, including cortisol and its metabolites, are related to carbohydrate metabolism; specifically, the breakdown of starch and the formation of glucose (gluconeogenesis). This activity is influenced by adrenocorticotropic hormone (ACTH) from the anterior lobe of the hypophysis. Mineralocorticoids, e.g., aldosterone, control water and electrolyte movement and balance by in-

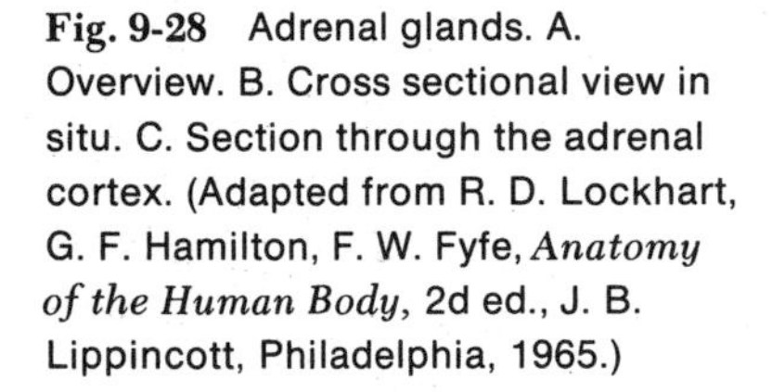

Fig. 9-28 Adrenal glands. A. Overview. B. Cross sectional view in situ. C. Section through the adrenal cortex. (Adapted from R. D. Lockhart, G. F. Hamilton, F. W. Fyfe, *Anatomy of the Human Body*, 2d ed., J. B. Lippincott, Philadelphia, 1965.)

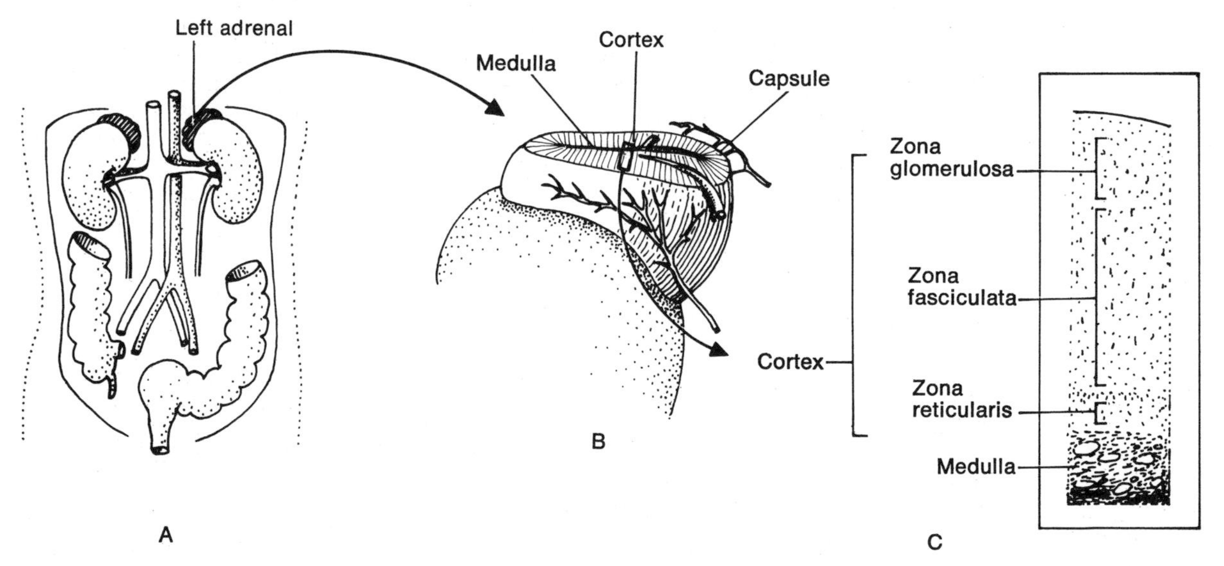

fluencing tubular reabsorption and secretion. This activity is necessary for life. The cortex also secretes male sex hormone (testosterone) which may play a role in sexual development and which, if tumors develop, produces exaggerated masculinity in females. A not very well understood function associated with the adrenal cortex is resistance to stress. Following adrenalectomy, animals cannot withstand the rigors of cold or excess heat; they cannot cope with "flight-or-fight" situations, and in response to these crises they will die.

The medulla secretes **epinephrine** and **norepinephrine.** These hormones activate the sympathetic "flight-or-fight" response as well as affect carbohydrate metabolism. Norepinephrine is also secreted by sympathetic preganglionic neurons at synapses.

UPON REFLECTION

It is a sad commentary that many people carry their bodies around with them for 70 years or so and have not the faintest notion what goes on deep to their navel. Having just tucked away new understanding of the disposition of abdominal viscera, you are *not* among them. You have learned how each segment of the gastrointestinal tract does its job and where it does that job. You have seen how peritoneal membranes support much of the abdominal viscera and how vessels and nerves reach their terminal organs without violating the sanctity of the peritoneal cavity. You have found that the kidney is concerned with the balance of fluids and salts in the body and dismisses only those metabolites and salts not needed at that moment—and in as little precious water as it takes to carry them off. It's amazing what you find when you look inside yourself!

REFERENCES

1. Lockhart, R. D., G. F. Hamilton, and F. W. Fyfe: *Anatomy of the Human Body,* 2d ed., J. B. Lippincott Co., Philadelphia, 1965.
2. Guyton, A. C.: *Textbook of Medical Physiology,* 3d ed., W. B. Saunders Co., Philadelphia, 1970.

clinical considerations

ULCERS OF THE STOMACH AND DUODENUM: PEPTIC ULCERS

The alimentary canal—from esophagus to anus—is lined with epithelium. Any defect or break in that lining constitutes an **ulcer** (L. *ulcus,* a sore). Ulcers can be small or large, but all are characterized by erosion of the epithelial layer and a leakage of fluid (exudate), just as a bag of fluid will leak through a hole in the bag lining. (Remember, we are 60 percent water by weight.) This fluid need not be blood—it can be the extracellular fluid from the subcutaneous tissues.

Histologically an ulcer is often characterized by a crater-shaped depression with the overlying epithelium destroyed (Fig. 9-29). The floor of the ulcer, if it is in the process of healing, is a granular mass of connective tissue, lymphatic tissue, and blood vessels. Collectively this mass of tissues is called granulation tissue. If the ulceration process is not allowed to heal, the ulcer will continue down to the level of blood vessels, and hemorrhage will start (Fig. 9-30B). Once the hemorrhage ceases (by clotting), the floor of the ulcer will

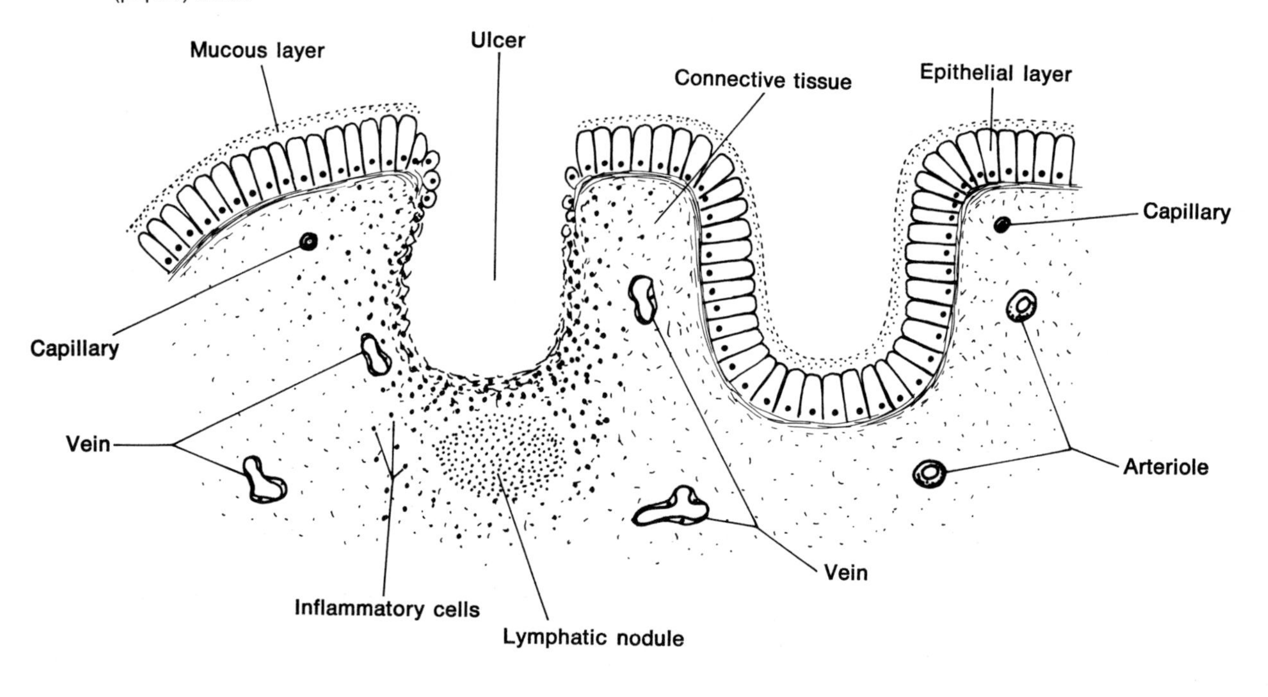

Fig. 9-29 Diagram of a duodenal (peptic) ulcer.

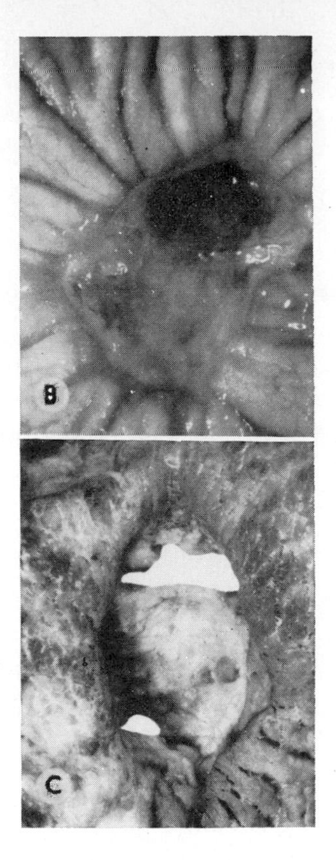

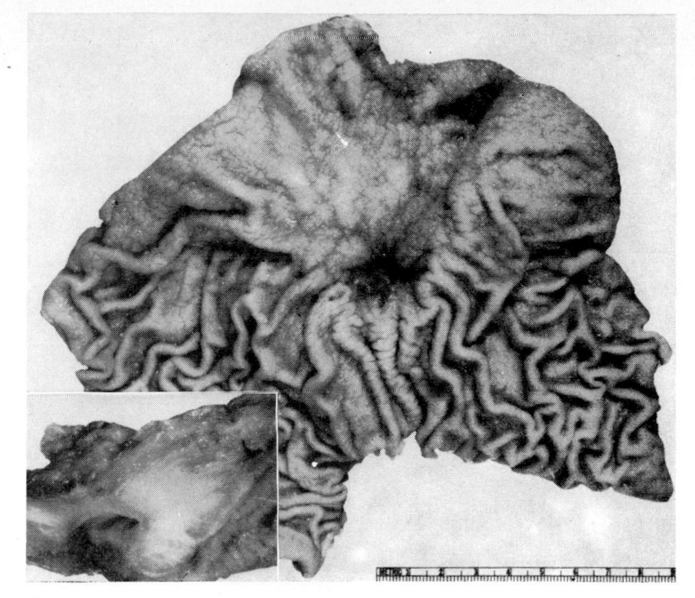

Fig. 9-30 Chronic gastric ulcers. A. Gross view of the interior surface of the stomach. Note the defect in the center of the mucosal surface. The same defect in a section of the surface is shown in the insert. B. An ulcer with a hemorrhage. C. An ulcer which has perforated through the entire gastric wall. (Courtesy of R. P. Morehead, *Human Pathology*, McGraw-Hill, New York, 1965.)

be covered with fluid (exudate) and the granulation process will begin. If the ulcer is not allowed to heal, it may perforate the entire intestinal or gastric wall (Fig. 9-30C), often with disastrous consequences.

Ulcers of the stomach and first part of the duodenum (called **peptic** ulcers) are created by the gastric secretions of the stomach. The ulcers always seem to occur below the source of the gastric juices.

An understanding of the peptic ulcer phenomenon is lacking. It is generally held that increased gastric secretions are initiated by excessive parasympathetic activity. However, it is possible that the mucosa of the stomach or duodenum may be defective and has a lower resistance to the savage attacks of gastric acidity. On the other hand, the degree of acidity may be different among individuals and, in those with a higher level, normal mucosa may be damaged.

The cardinal symptom of the peptic ulcer is a burning or gnawing pain in the mid-upper quadrants of the abdomen which is relieved by food intake, especially milk or seltzers. Apparently such intake gives the acid or enzymes something to "eat on" other than the mucosal lining. In fact, antacids decrease the acid level (increase the pH) and render the enzymes harmless; because of these effects, nonabsorbable antacids represent a most effective means of controlling the ravaging effects of hyperacidity.

unit 10 the pelvis and perineum

the pelvis and its viscera

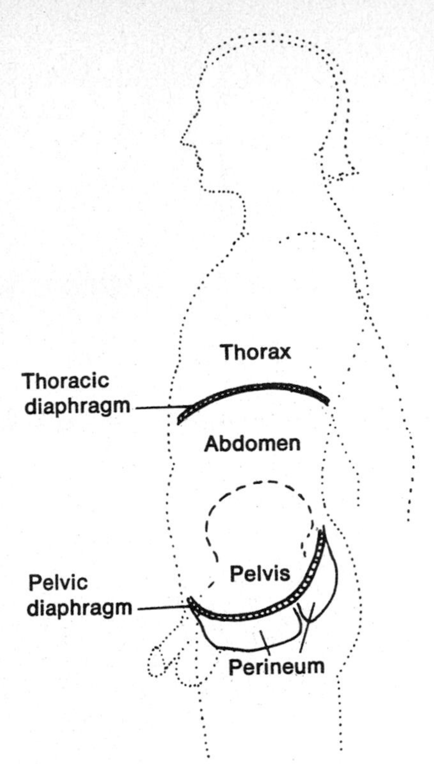

Fig. 10-1 Relationship of the body cavities and the diaphragms that separate them.

Put simply, the pelvis may be likened to a double boiler—the pot above represents the true pelvic cavity and the pot below the perineum. The bottom of the upper pot represents the pelvic diaphragm (Fig. 10-1). Where have you run into a diaphragm before in this course? Well, we're going to study two more before we finish this unit.

The **pelvis,** you remember (and if you don't, review the beginning of Unit 5), is a cavity bounded anteriorly and laterally by the paired hip bones, and posteriorly by the sacrum and coccyx. The pelvis is a direct inferior extension of the abdominal cavity and, therefore, has no roof. It contains:

1. Internal reproductive structures.
2. The urinary bladder.
3. The rectum.
4. Assorted ducts, vessels and nerves.

The floor of the pelvis (the pelvic diaphragm) is muscular and separates the pelvis above from the perineum below.

The perineum is the region within the framework of the bony pelvic outlet. Its floor of skin and fascia—perforated by structures descending from the pelvis—is continuous with the skin and fascia of the anterior abdominal wall, the buttocks, and the thighs.

The perineal spaces are largely fat-filled and encompass:

1. The tube passing distally from the bladder to the exterior (urethra).
2. The tube passing distally from the rectum to the exterior (anal canal and anus).
3. In the female, the distal continuation of the vagina.
4. Miscellaneous glands, muscles, ducts, vessels, and nerves.

PELVIC BOUNDARIES

Refer now to Figs. 10-2 and 10-3 as you read.

The pelvis is divided into a greater space above, the **pelvis major,** and a smaller space below, the **pelvis minor,** by a line circling posteriorly from the upper surface of the pubic bone to the sacrum. This line constitutes the pelvic brim. Forming an incomplete circle around the pelvic interior, the pelvic brim forms the pelvic inlet. The region above the inlet (pelvis

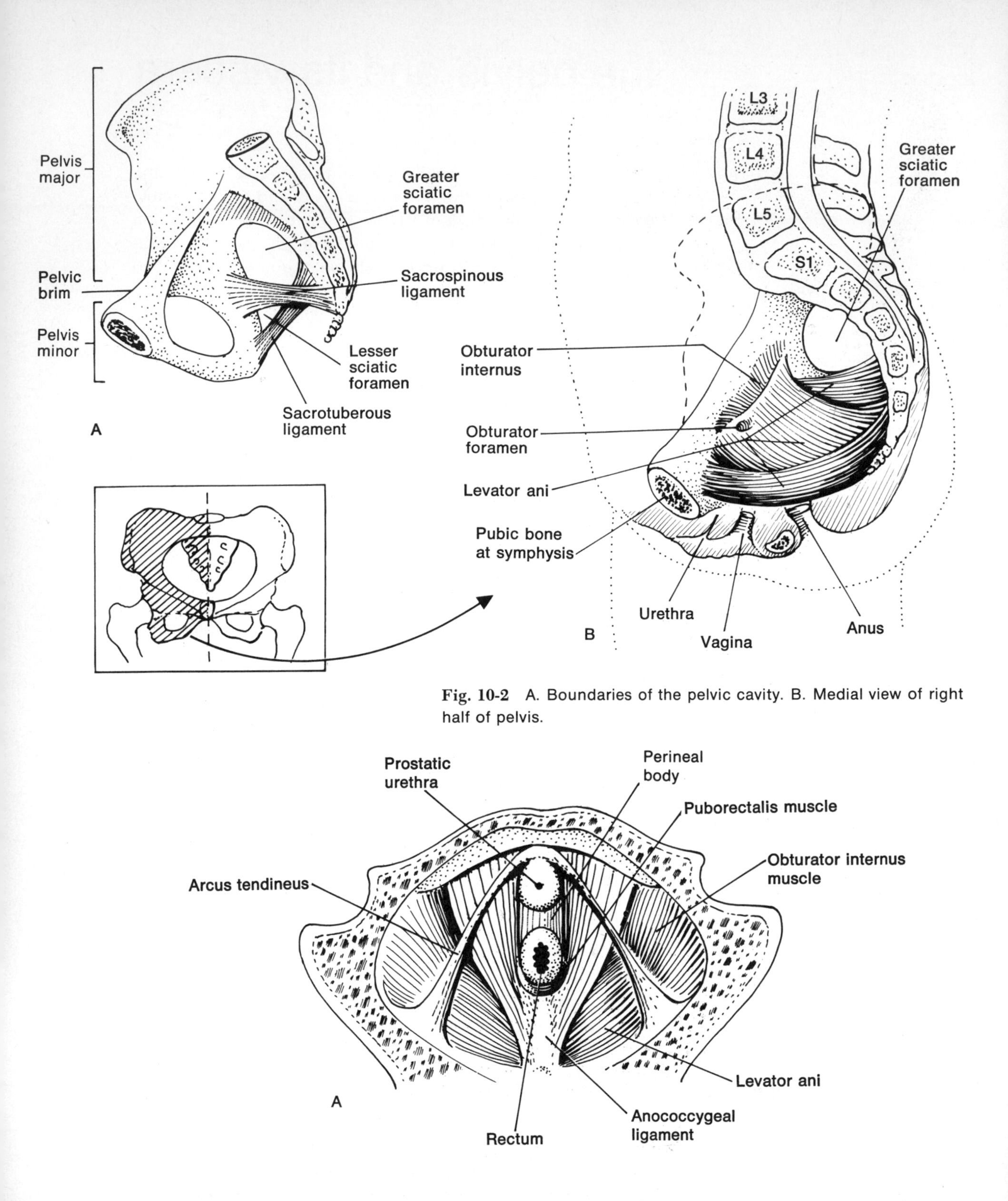

Fig. 10-2 A. Boundaries of the pelvic cavity. B. Medial view of right half of pelvis.

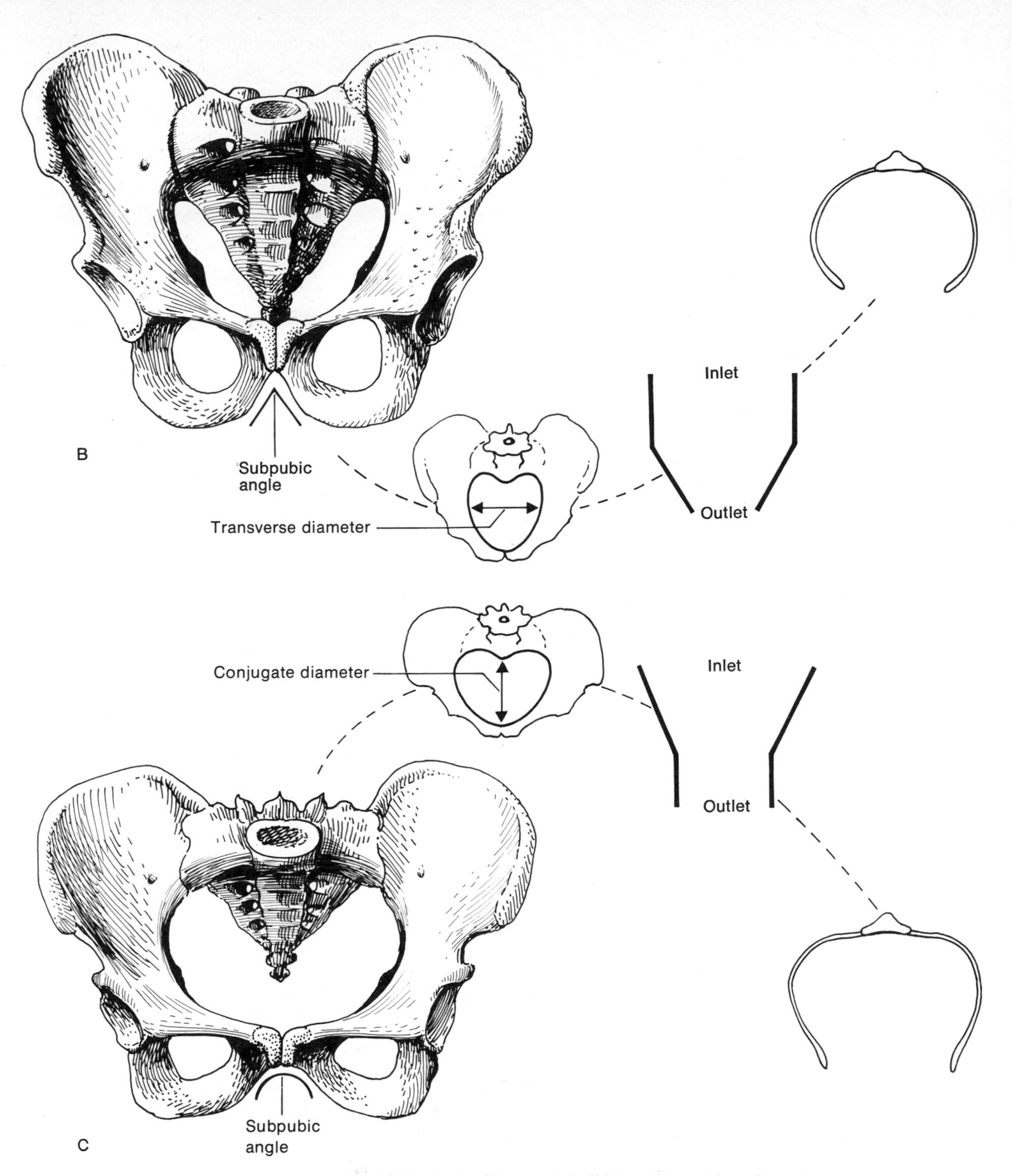

Fig. 10-3 Pelvis. A (opposite). Pelvic walls and floor from above. Anterior view of male (B) and female (C) pelves. Compare the shapes of inlets, passageways, outlets, and diameters. Note the difference in subpubic angles.

major), surrounded in part by the iliac fossae, is actually the lower part of the abdomen. On yourself, feel for the iliac bones and note the relationship to the umbilicus (at the L4 level). The pelvic inlet is inclined upward from anterior to posterior from a level of about the 4th coccygeal vertebra to S1, both well below the level of the iliac crests. The pelvis major is between the level of the iliac crests and the inclined pelvic brim and is, therefore, a part of the abdominal cavity.

The **lateral walls** of the lesser pelvis (minor) consist of:

1. The interior surface of the superior pubic ramus.
2. Part of the iliac bone.
3. Part of the obturator internus muscle covering the obturator canal.
4. Part of the levator ani.
5. Spaces of the sciatic foramina filled with structures exiting and entering laterally.

Referring now to Fig. 10-3, note the **floor** of the pelvis minor: the fascicles of the hammocklike **levator ani.** This complex muscle arises from the ilium and pubis and extends posteromedially to insert on the coccyx via the anococcygeal ligament, forming a sling around the rectum. The levator ani and its fascial envelope constitute the **pelvic diaphragm.** As you can see (Fig. 10-3) it is perforated by the anal canal and the urethra in both sexes and the vagina as well in the female. Levator ani supports the pelvic viscera and, by contracting, resists the rise in intra-abdominal pressure caused by straining during defecation, while lifting heavy objects, or by coughing. It is innervated by the lower sacral nerves. Interestingly, most of levator ani in four-legged beasts functions to wag the tail.

The **posterior wall** of the pelvic interior is largely ligamentous. The pelvic interior has no roof.

So much for the boundaries of the pelvic cavity.

PELVIC STRUCTURES

The **ureters** (Fig. 10-4) extend down behind the peritoneum, first lying on the psoas muscle, then passing over the iliac vessels at the pelvic brim, where they turn forward to enter the bladder. (For internal structure, see Fig. 9-27.)

In the male the ureters pass close to the ductus deferens as they proceed to the bladder. In fact the ductus deferens hooks around each ureter as the former comes in from the abdominal wall.

In the female, the ureters pass along both sides of the

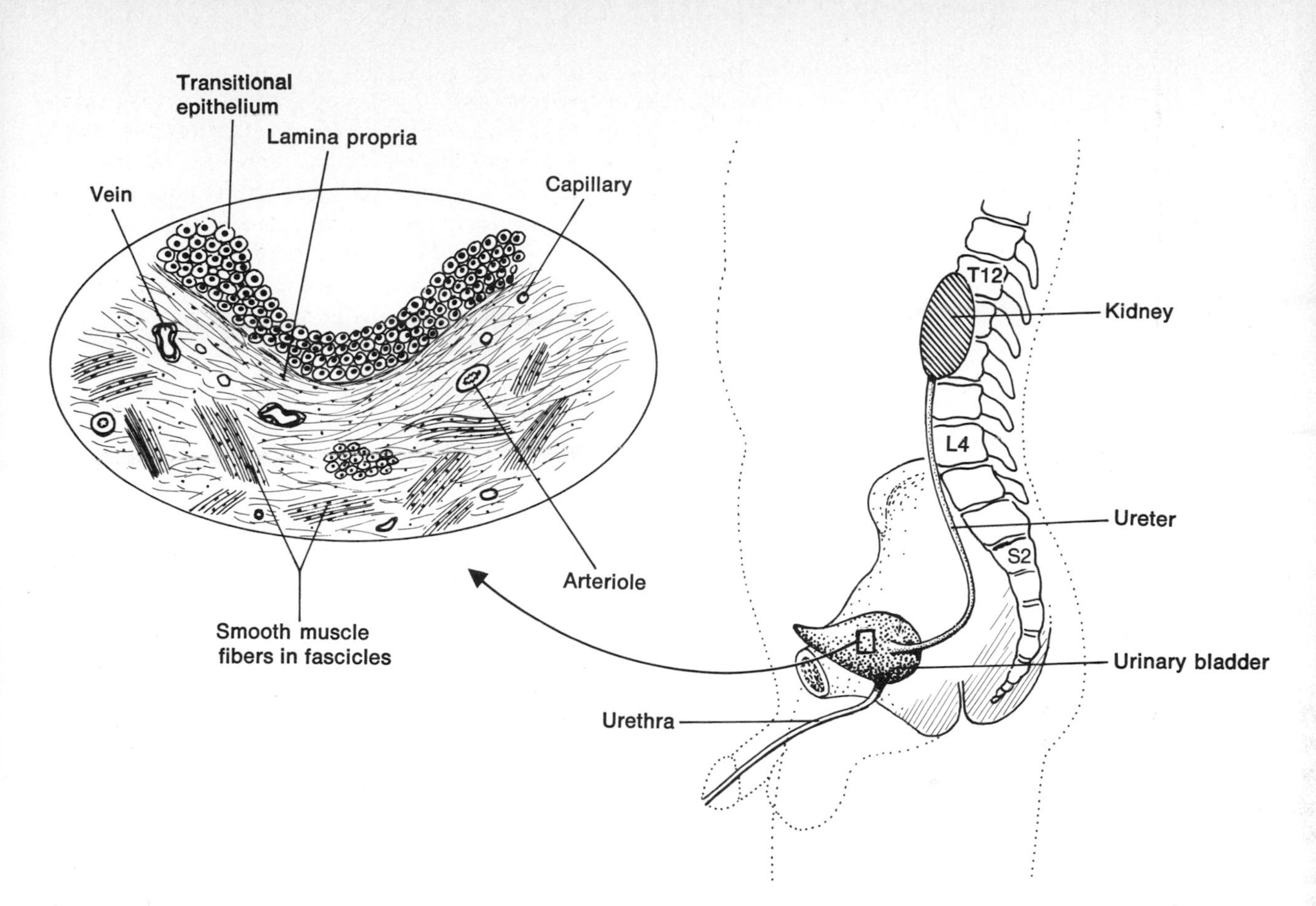

Fig. 10-4 The urinary bladder and related urinary structures.

vagina and the cervix. This has significance surgically: (1) a stone trapped in the lower part of the ureter can be reached through the vagina; (2) when removing the uterus (and cervix) and related connective tissue, the surgeon must take care not to take the ureters as well.

The **bladder** is a fibromuscular bag capable of remarkable distention that stores urine from the kidneys (conducted there via the ureters) and releases it into the urethra following a voluntary signal. The shape of the bladder is variable, depending upon:

1. The volume of urine within.
2. The encroachment of neighboring structures.

By referring to Fig. 10-5, you can see that the bladder may be described as having a base below, an apex anteriorly, two sides inferolaterally and a superior surface. Anterolaterally, the bladder is related to the pubic bones.

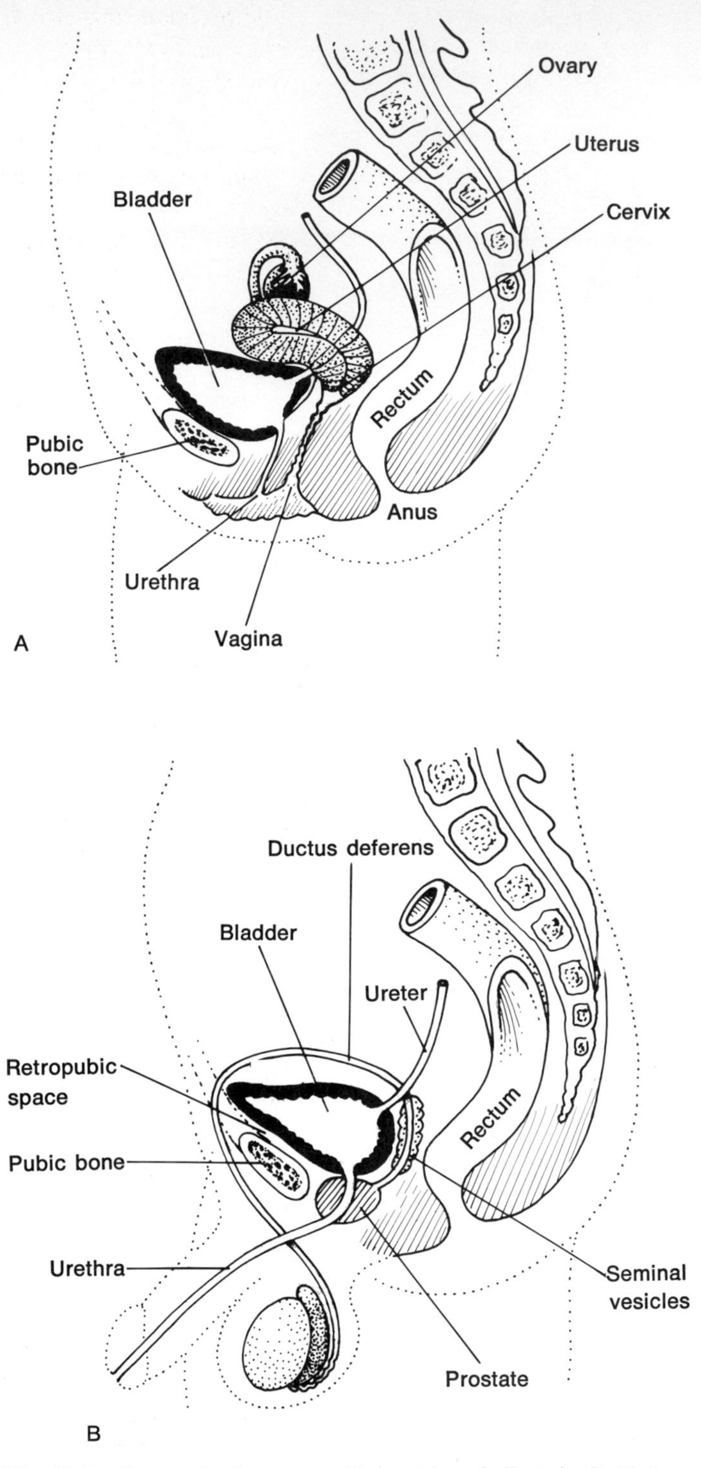

Fig. 10-5 The pelvic viscera, sagittal section. A. Female. B. Male.

In the male, the bladder is anterior to the rectum and separated from it by the ductus deferens, the seminal vesicles, and peritoneal folds.*

In the female, the bladder is nearest anteriorly to the vagina and the cervix of the uterus.

Superiorly, the bladder is related to a peritoneal cover, coils of small intestine, and in the female, the body of the uterus, which is bent forward over the bladder.

Inferiorly, the base of the bladder is in contact directly with the levator ani except in the male, in whom the prostate gland intervenes.

The bladder is anchored to the lateral pelvic walls (by the lateral ligaments), to the anterior abdominal wall (by the median umbilical ligament), and in the male to the prostate gland (in the female, to the pubis). Microscopically (Fig. 10-4) the bladder is lined with a mucous membrane characterized by transitional epithelium (see also Figs. 2-11 and 2-12). Underlying the submucosa, a thick coat of smooth muscle is arranged such that—when contracted—urine is expelled into the urethra.

The movement of urine from the kidneys to the bladder is technically termed urination, in contradistinction to the term of choice describing movement of urine from the bladder to the outside: micturition (a knowledge of such minutiae makes for splendid conversation when trying to impress one's peers!). The bladder is filled irregularly and frequently, so it is probably never empty. The desire to micturate is generated when the volume of urine within the bladder is about 200 cc, although the maximum capacity of the bladder is believed to exceed 500 cc.

A generalized functional analysis of the bladder may be seen in Fig. 10-6:

Figure 10-6A: As urine fills the bladder, the walls are stretched and stretch receptors begin firing. Impulses are conducted to the spinal cord and brain via the pelvic splanchnic nerves.

Figure 10-6B: Reflexly, the bladder contracts in response to the stretching. Simultaneously, one becomes aware of the need because the sensory input has reached the cerebral cortex.

Usually micturition can be initiated by willfully contracting the abdominal musculature. The next time you void, ponder these anatomical and physiological events: Follow-

* The detailed relations of the bladder are complicated by the various folds of peritoneum arising from the peritoneal drapes of the rectum and sides of the pelvis.

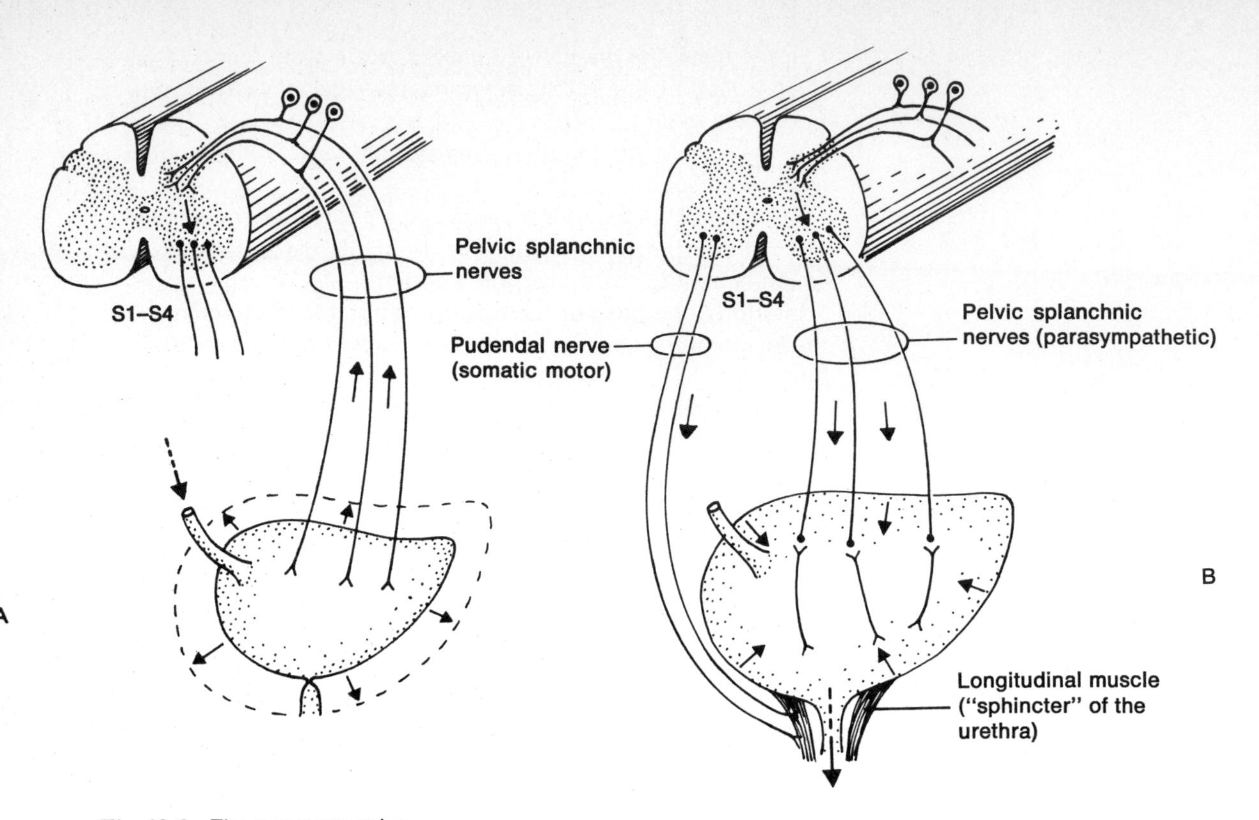

Fig. 10-6 The neuromuscular mechanism of micturition (schematic). A. Afferent phase. B. Efferent phase.

ing contraction of the anterior abdominal wall, the bladder musculature and the longitudinally oriented urethral muscles contract. Simultaneously, part of levator ani supporting the base of the bladder relaxes and the base descends. As the urethral muscle contracts the urethra shortens, its lumen dilates, and micturition occurs. Thus, micturition is a consequence of somatic sensory and motor as well as parasympathetic activity.

The act of micturition can be delayed by a process of inhibition* which is generally learned at about 2 years of age. Anatomically, the brain has not developed sufficiently before this time for voluntary suppression of micturition (or defecation) to occur. Imagine how many babies have been punished because they "wet their bed" when there was no way for them not to have done so?

The **rectum** (L., *rectus,* straight) is the extension of the large intestine, following the more S-shaped (sigmoid) colon of the abdomen. The rectum begins roughly at the midsacral

* Not a modesty kind of inhibition but a process created by a circuit of inhibitory neurons.

level and, following the curve of the sacrum, becomes the anal canal as it pierces the levator ani (pelvic diaphragm) (Fig. 10-7). Draped with peritoneum, the rectum is a structure of significance to the physician, who can palpate numerous neighboring structures by digital examination through it. Some of the significant relations of the rectum can be seen in Fig. 10-8. These structures are, of course, all anterior relations. Posteriorly, the rectum is related to the sacrum. Inferiorly, the rectum is supported by the levator ani (pelvic diaphragm). Laterally, the walls of the pelvis and certain nerves and vessels are adjacent to the rectum.

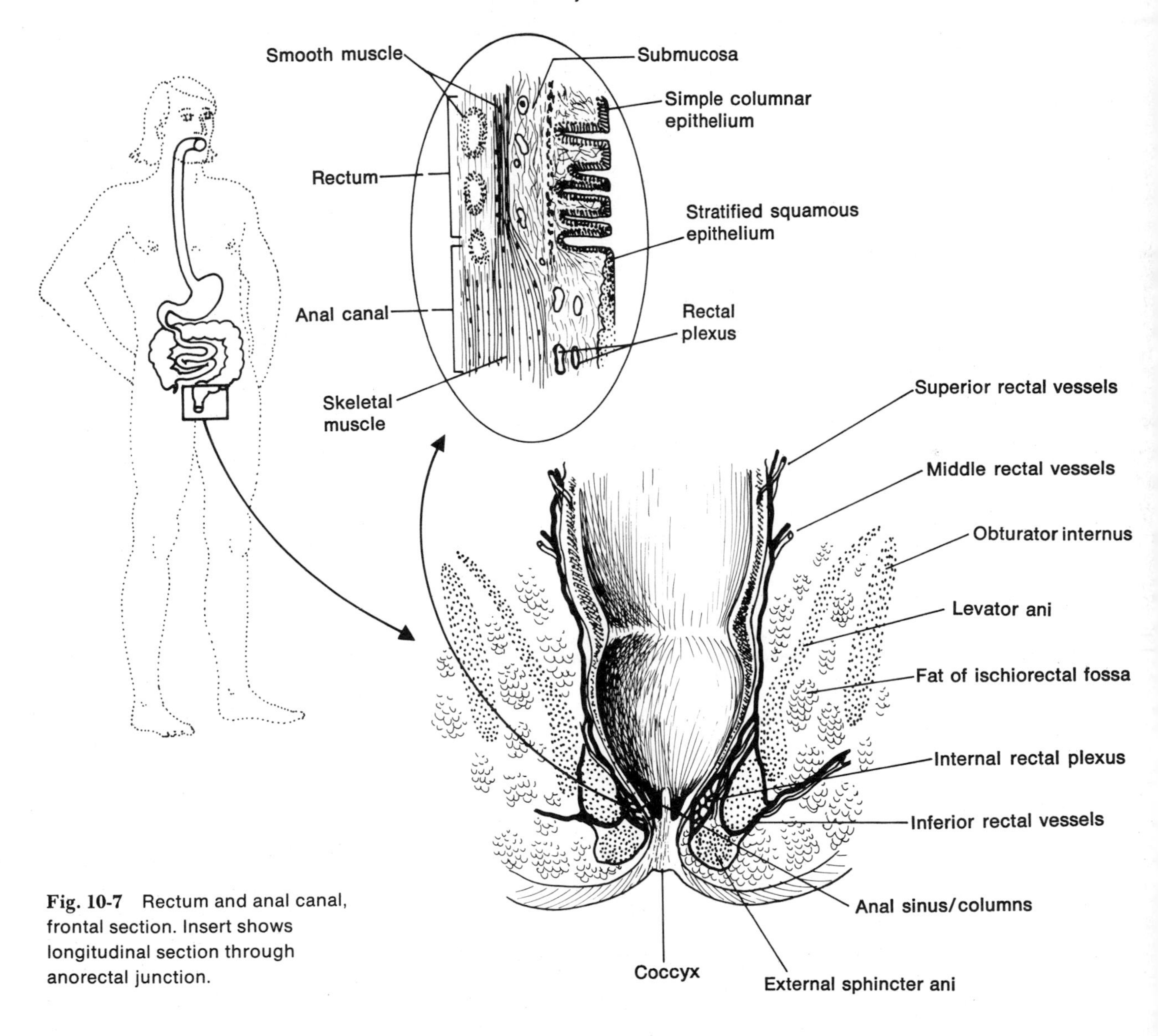

Fig. 10-7 Rectum and anal canal, frontal section. Insert shows longitudinal section through anorectal junction.

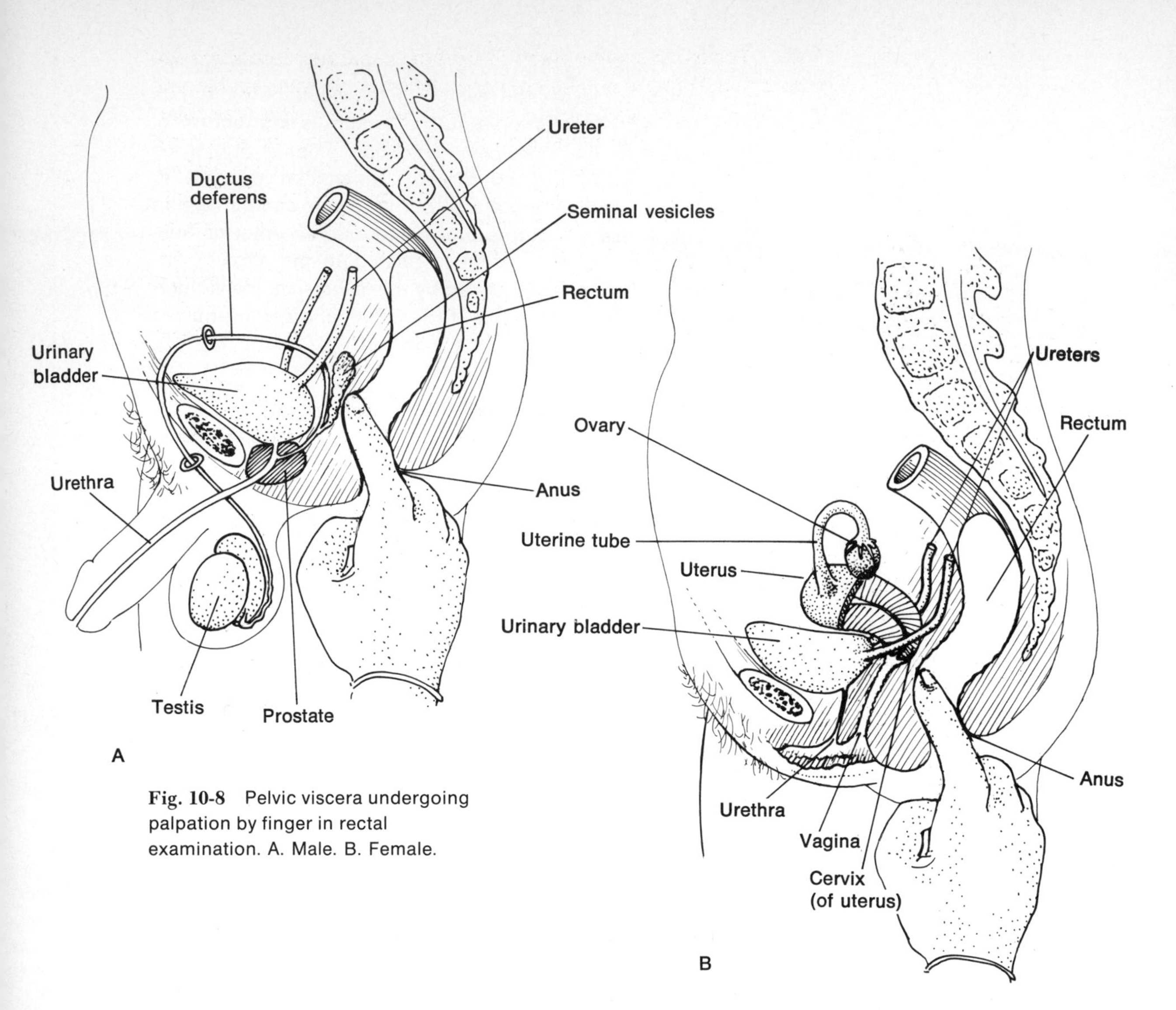

Fig. 10-8 Pelvic viscera undergoing palpation by finger in rectal examination. A. Male. B. Female.

The rectum is bound by fascial ligaments to the sacrum and to the pelvic diaphragm. The latter, in fact, forms a muscular sling around the rectum and attaches to the pubis (puborectalis).

Microscopically the rectum is similar to the colon (Fig. 10-7). At the level of the pelvic diaphragm, the microscopic character of the rectum merges with that of the anal canal.

The **anal canal** represents the last inch and a half of some 20 ft of digestive tubing. The most distal part of the large intestine, it is surrounded by the external sphincter within the ischiorectal fossa of the perineum and opens to the exterior as the **anus.**

The mucous membrane of the anal canal represents a transition from the simple columnar (absorption-type) epithelium to stratified squamous (wear-and-tear type). The muscular coat of the anal canal incorporates two layers—an inner internal sphincter controlling the degree of dilatation and an outer layer merging with the skeletal muscle, the levator ani.

The inner lining of the canal is characterized by columns which contain a venous plexus (Fig. 10-7) (see page 608).

The rectum and anal canal are empty until defecation occurs. Up to that point, fecal matter is retained in the sigmoid colon. During defecation, the following events take place:

1. Intra-abdominal pressure is increased ("straining").
2. The sigmoid colon relaxes and rectal musculature contracts.
3. Sphincters relax and the fecal matter passes.
4. The puborectalis and the two sphincters contract after each passage.

QUIZ

1. The floor of the pelvis is __________________ and separates the pelvis above from the __________________ below.
2. The ________________ _________ muscle and its fascia constitute the pelvic diaphragm. It is innervated by the lower __________________ nerves.
3. The urinary bladder is the reservoir for the ________________ __________________ of the __________________.
4. The ____________________ tunic of the ureter suggests active transport of urine downward.
5. A significant relation of the ureter as it approaches the bladder in the male is the ________________________ __________________.
6. The shape of the bladder is variable, depending partly, upon the __________________ of ____________________ within.
7. Microscopically, the bladder is lined with ________________ epithelium.
8. Micturition is a consequence of ______________________ ________________________ and motor activity as well as __________________ activity.

9. Posteriorly, the rectum is related to the ______________; inferiorly, it is supported by the ______________________ __________.

PELVIC STRUCTURES: MALE INTERNAL ORGANS OF REPRODUCTION

The sum of the reproductive structures suggests an interesting morphological and functional concept—a concept rendering the sequence of structure in both sexes quite reasonable. Let's start with the male:

The principal function of the male reproductive system is to produce germ cells (sperm). Thus the primary sex organ of the male are those paired structures which produce sperm cells: the testes. The secondary but equally important function is to conduct those sperm cells from the site of development (the testes) to the outside of the body. Thus, all the glands and ducts associated with this movement are called accessory structures of reproduction.

Momentarily refer to Fig. 10-9A: one might say that the duct system looks unusually complicated for such a straightforward task. The anatomical basis for this "complexity" is simple: the organs of sperm generation develop within the abdominal cavity. In the latter stages of embryonic development, however, by a process of differential growth, they "descend" into a pouch of the skin that is in fact drawn out from the anterior abdominal wall. This is convenient, since sperm cells would not develop at the temperature of the abdominal cavity. The temperature within the pouch (scrotum) is only about 2°F less than body temperature but is hospitable to development of sperm cells. If you now study the diagram depicting descent of the testes (Fig. 10-10), the evolution of a not really so complicated duct system can be appreciated.

The testes develop within the abdominal cavity at the level of the kidneys. Each testis descends through the anterior abdominal wall via the inguinal ring and canal on its side and enters the scrotum. The canal is occupied by the ductus deferens, the coverings it took along with it during descent, and related vessels and nerves. These structures collectively make up the **spermatic cord.**

Since the testes are derived from tissue well up in the abdominal cavity, you might suspect that the blood and nerve supply to the testes might come from the same level and be dragged down into the scrotum too. In general, this is true.

Next refer to Fig. 10-9 and trace the route a sperm cell would take from testis to the exterior. All of these structures

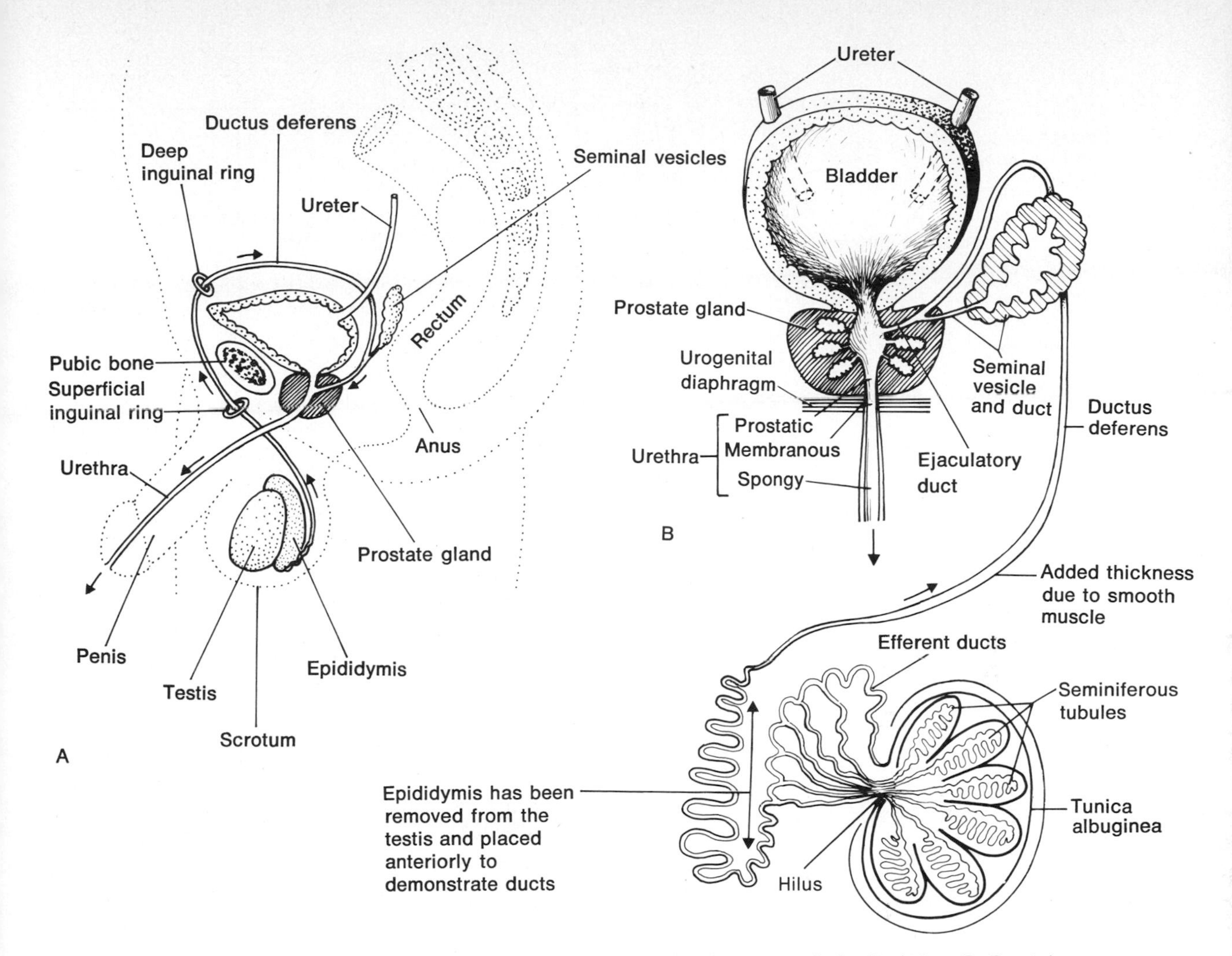

Fig. 10-9 The male reproductive system. A. Sagittal view. B. Frontal view, highly schematic. Arrows trace the pathway of sperm.

and associated glands will now be discussed in some detail with the exception of the urethra. It, the penis, scrotum, and related structures will be presented with the perineum.

Testis

The **testes** are walnut-shaped structures situated in the scrotum. There are normally two of them and the left usually hangs lower than the right. Each testis has a superior and inferior pole (end) and the epididymis sits atop the former.

The outer covering of the testis is a thick fibrous capsule (tunica albuginea) from which septa (walls) of connective

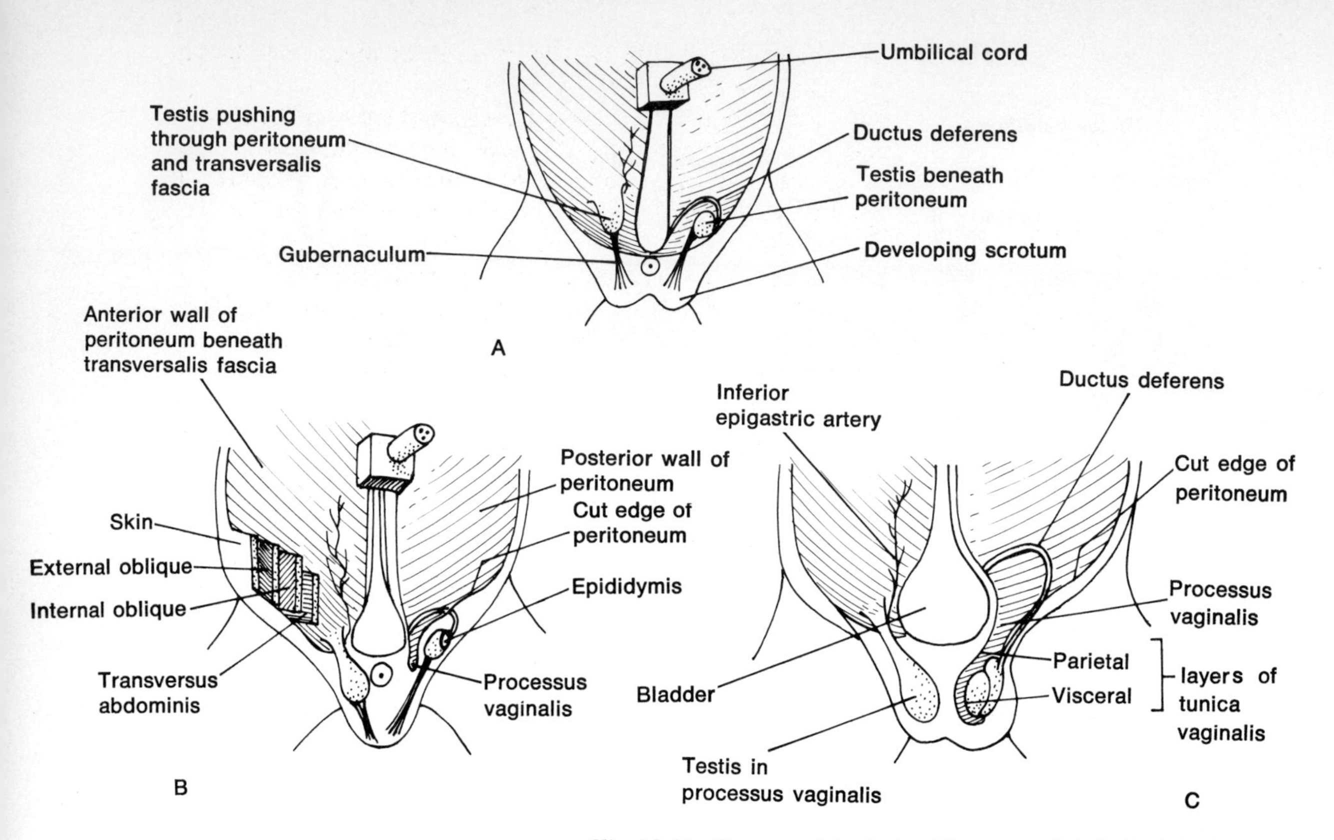

Fig. 10-10 Descent of the testes (diagrammatic). A. Each testis pushes through the abdominal wall, creating a deep inguinal ring and inguinal canal on its side. The gubernaculum leads the way into the scrotum. B. In its descent, the testis takes a portion of each abdominal layer with it into the scrotum. As they descend a canal (processus vaginalis) is formed. C. After birth, it seals itself off. Failure to do so is a congenital defect. A loop of intestine may herniate into the scrotum through the defect. (Adapted from B. M. Patten, *Human Embryology,* 3d ed., McGraw-Hill, New York, 1968.)

tissue project into the lumen of the testis and compartmentalize it. A number of highly coiled **seminiferous tubules** are tucked away into each compartment.

In Fig. 10-9B, it can be observed that the ends of the tubules join in a network at the hilus of the testis. From here efferent ducts conduct the sperm cells to the epididymis.

The testes' seminiferous tubules beget sperm cells (spermatozoa). The manner in which this is accomplished (spermatogenesis) may be visualized in Fig. 10-11.

The primordial germ cells (I) are the **spermatogonia** at the periphery of the tubules.

These spermatogonia develop (throughout) into **primary**

spermatocytes (II). At this point, each of these cells has a full complement of chromosomes—46.

The primary spermatocytes undergo a type of division into **secondary spermatocytes** (III), each of which has one-half the usual chromosome count (23). This process is termed **meiosis** and is a critical stage in sperm development.*

The secondary spermatocytes divide (mitotically) into **spermatids** (IV) which mature into **spermatozoa** (V).

These now-mature sperm, though generally inactive, find their way into the epididymis to await a call to active duty. The supportive cells (of Sertoli) are believed to furnish nourishment for the spermatids.

* The sperm cell (spermatozoon) is programmed to mate with a female germ cell (ovum) to form the first cell of a developing human (zygote). The normal complement of chromosomes in any cell of the body except germ cells is 46. To have more or less of these bearers of inheritance is to have trouble, e.g., mongolism. When the zygote is formed, 23 of the chromosomes are contributed by the male germ cell and 23 by the female. Thus germ cells have one-half the normal complement of 46 chromosomes, i.e., 23.

Fig. 10-11 A. Seminiferous tubules. B. Cross section of one tubule. C. Steps in sperm maturation.

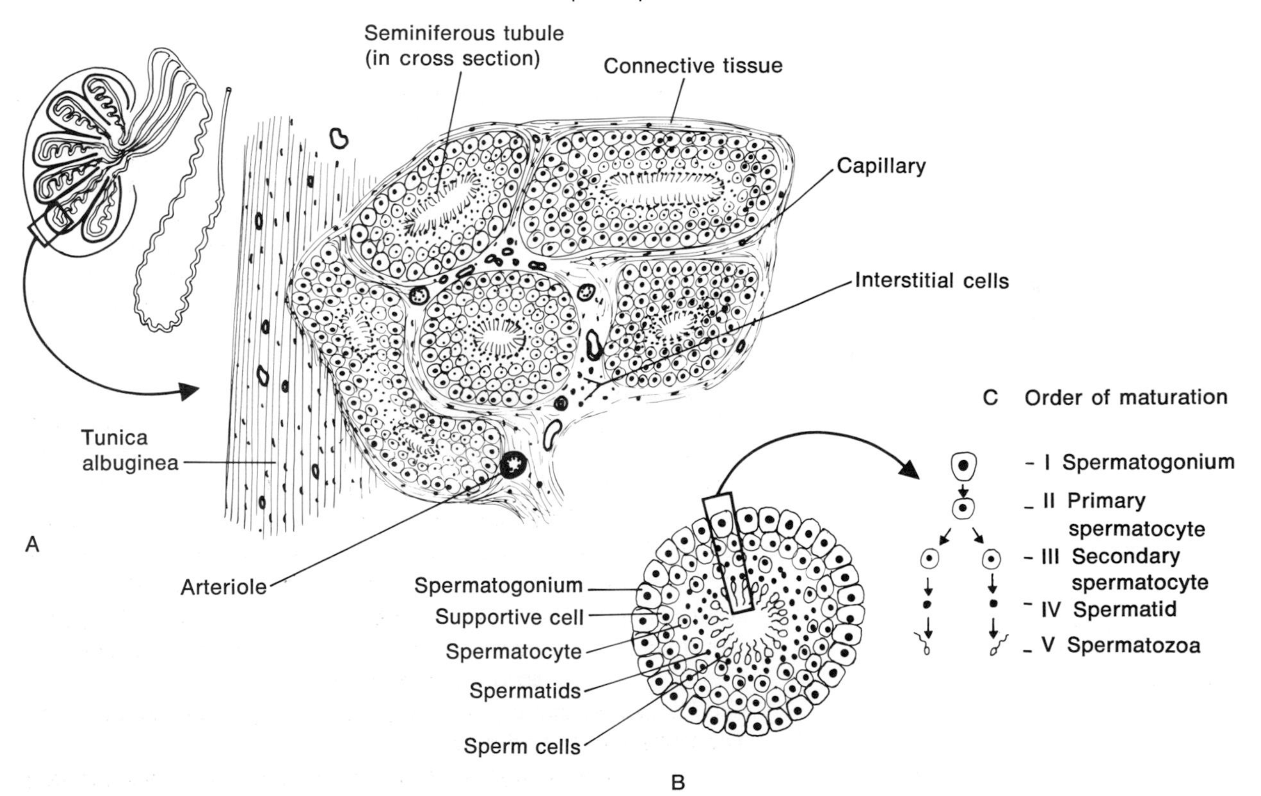

The testes derive their blood supply from the testicular arteries, paired branches of the abdominal aorta. The veins of the testis form a dense plexus along the ductus deferens and ultimately drain into the great inferior vena cava (on the right side) and the renal vein (on the left). The testes—exceptionally sensitive to pain and pressure—derive their rich innervation from the lower thoracic spinal nerves.

Interstitial cells

Interspersed about the sparse connective tissue among the seminiferous tubules is a significant collection of specialized cells and associated capillaries. These **interstitial cells** (Fig. 10-11) are known to secrete the male sex hormone testosterone, which not only stimulates the appearance of secondary sex characteristics at puberty but also has a significant effect on growth in the adolescent. Interestingly, testosterone does *not* have the role of maintaining the sperm-growing epithelium. What does? (FSH.)

What are some of the secondary* sex characteristics brought on by testosterone?

1. Increase in size of the penis.
2. Maturation and increased activity of the reproductive glands, i.e., prostate, seminal vesicles.
3. Enlargement and sharper definition of muscular form.
4. Aggressive mental attitude. Awakening of sexual interest in girls.
5. Deeper voice due to changes in laryngeal structure.
6. Increase in and thickening of sebaceous gland secretions. What manifestations of this phenomenon are commonly seen in young males? (Acne.)
7. Increase in body hair, especially in pubic and axillary regions.

Recent research has shown that differentiation of maleness in the fetus is a function of testosterone acting on the hypothalamus. At birth, the plasma concentration of testosterone decreases and remains at a low level until about 12 years of life, when the pituitary gonadotropins stimulate the production and increased secretion of testosterone. This results in typical changes collectively titled **puberty.**

Epididymis

Perched atop the testis looking like a fore-and-aft cap set at a jaunty angle, the **epididymis** consists of a highly convoluted

* *Primary* sex characteristics are the appropriate organs of reproduction themselves, e.g. testes, glands, ducts, etc.

tubule in which inactive but living sperm are stored. A quick review of Fig. 10-9B shows that the efferent ducts of the testis merge with the tubule of the epididymis. The tubule tortuously descends along the posterior aspect of the testes to the level of the inferior pole, where it turns upward to become the ductus deferens.

Microscopically, the tubule is lined with a pseudostratified epithelium whose cilia are nonmotile (stereocilia). These cilia apparently secrete glycogen droplets for the sustenance of stored spermatozoa. During sexual activity, the sperm are moved out of the epididymis by contractions of smooth muscle fibers oriented about the tubule.

Ductus deferens and ejaculatory ducts

The **ductus deferens** is the principal and largest duct of the male reproductive system. As may be seen in Fig. 10-9:

1. It arises at the tail of the epididymis.
2. Passes out of the scrotum to enter the anterior abdominal wall at the superficial inguinal ring.
3. Traverses the inguinal canal in company with related vessels and nerves as the spermatic cord.
4. Enters the abdominal cavity (via the deep inguinal ring) to arch across the side of the pelvis and pass behind the bladder.
5. Perforates the prostate gland, becomes narrower, receives the duct of the seminal vesicle, and joins the prostatic urethra as the pencil-point ejaculatory duct.

The microscopic character of the ductus deferens includes a thick sheath of smooth muscle enclosing the mucous membrane. What functional significance might this have?

Prostate and seminal vesicles

Collectively these structures are classified as accessory genital glands. The **seminal vesicles** (Fig. 10-12) are actually appendages of the ductus deferens, consisting of an elongated sac lined with secretory epithelium. The **prostate,** on the other hand, is a mass of tubuloalveolar glands (within a connective tissue capsule) whose ducts open into the urethra. In both glands, smooth muscle contributes heavily to their structure.

The seminal vesicles, located behind the bladder, can often be palpated through the rectum. The prostate is at the base of the bladder and encloses the upper (prostatic) urethra. Both glands actively secrete a milky, alkaline material rich in fruc-

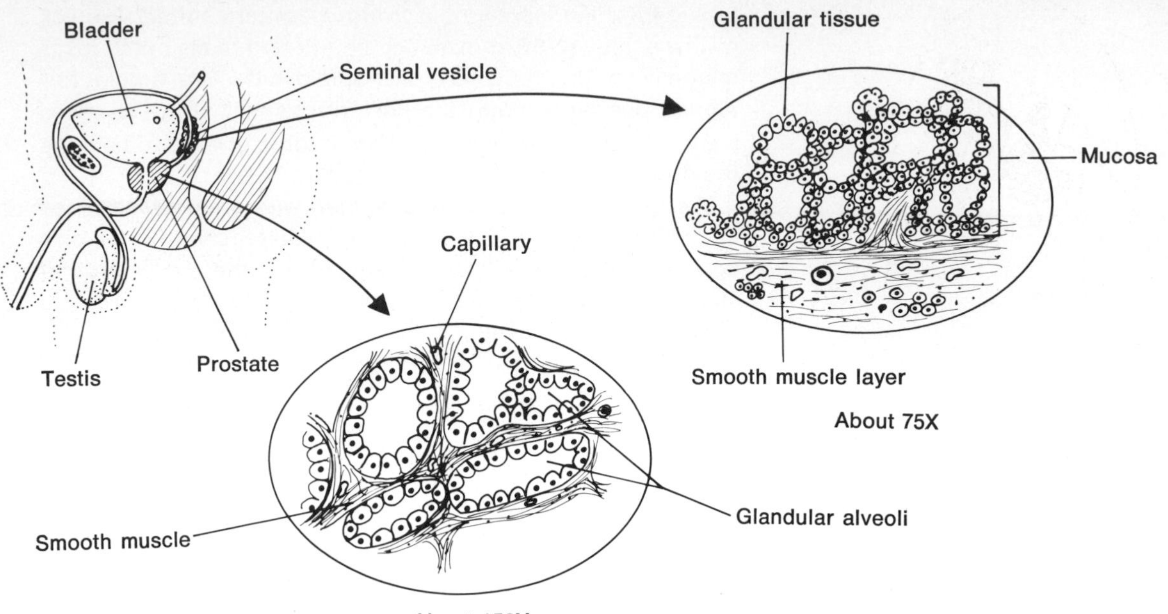

Fig. 10-12 Prostate gland and seminal vesicle.

tose (sugar) which aids sperm motility. Note in Fig. 10-9:

1. The seminal vesicle empties its secretions into the ductus deferens.
2. The ductus then narrows to form the ejaculatory duct, which penetrates the prostate and joins the urethra.
3. The prostate receives the urethra (from the bladder) as well as the ejaculatory ducts.
4. The prostate empties its secretions directly into the urethra.
5. Descending through and exiting anteroinferiorly from the prostate, the urethra penetrates the levator ani to enter the perineum.

The collective function of the structures just discussed will be summarized shortly in coordination with the collective function of female reproductive structures to be discussed.

PELVIC STRUCTURES: FEMALE INTERNAL ORGANS OF REPRODUCTION

Once again we can look at the reproductive system from a conceptual point of view and find that the functional and anatomical order of structures in the female is entirely reasonable. Thus, the anatomy is easily learned. It may also occur to

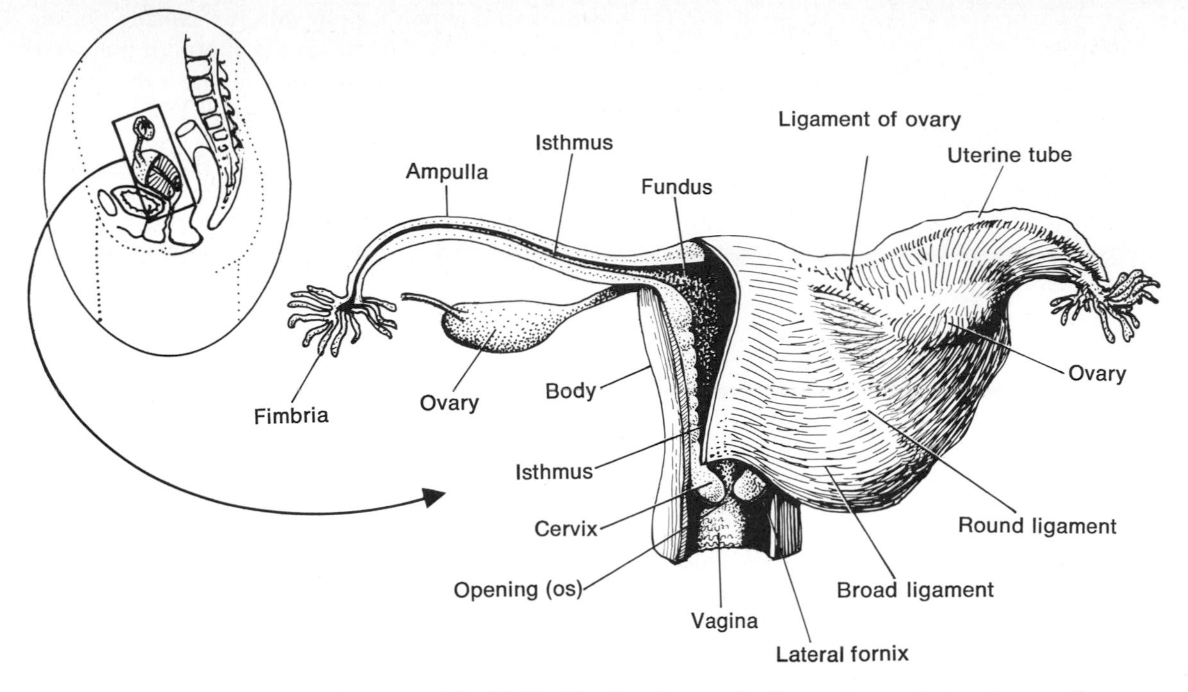

Fig. 10-13 The female reproductive system, anterior view, partly dissected.

you that certain female reproductive structures are homologous to those of the male.

To conceive a human being, a male germ cell must combine with a female germ cell. We have seen in the male a primary sex organ (testis) which generates sperm (germ) cells, and a set of ducts and glands providing the wherewithal for those sperm to be transported out of the body.

The meeting place for the successful union of the sperm with the female germ cell (ovum) is critical. The environment necessary to support the union and subsequent development has not—to date—been artificially duplicated. The place must be warm and moist, and a host of hormones and nutrient media must be on hand. Therefore, an organ carrying on this activity must be (1) located within the body, (2) connected to the outside via a duct so the sperm can reach it, and (3) connected to the organ of female cell generation by a duct.

Figure 10-13 demonstrates that the actual anatomy indeed meets all these criteria—and more:

1. The site of embryonic and fetal development (uterus) is located within the pelvic cavity. It is an internal structure (warm), glandular (moist, nutrient), and richly supplied with blood (carrying hormones).

2. The uterus is in communication with the outside by way of a duct (vagina) which receives the sperm of the male.
3. The uterus is in communication with the primary sex organs (ovaries) via bilateral uterine tubes.

Now to the particulars.

Ovary

The **ovaries** (paired) are located on the lateral walls of the pelvis at the level of the anterior superior iliac spine. Girls, feel for these landmarks and note the relative position of the ovaries. These structures arise from the same tissue in the same place as the male testes,* hence the ovaries and the testes look alike externally, although the former is smaller, flatter, and in the adult, scarred from numerous previous ovulations.

The ovaries are in contact medially with the uterine tubes. Laterally they are suspended, in part, by peritoneum and a number of ligaments. With pregnancy, the ovaries are displaced by the enlarging uterus.

Ovarian function. The ovary really has a two-fold function: to produce female germ cells (ova) and to produce two hormones which act to develop the uterine lining in preparation for implantation of the embryo. The ovary performs these functions cyclically under the direction of hormones from the anterior lobe of the pituitary gland. It takes about 21 days for the ovary to develop a primordial follicle into a mature one with an ovum ready to blast off for the uterine tubes. In the next 10 days, the ovary is concerned with secreting hormones which maintain the uterine lining at a nutritious level for the fertilized ovum. The cycle of the ovary does not correspond precisely to that of the uterus (menstrual cycle), and that is understandable if one remembers that the development of the uterine lining is dependent upon ovarian secretions which necessarily precede it.

Refer now to Fig. 10-14 as you read the following account of the ovarian cycle:

The external surface of the ovary is lined with epithelial cells which are believed to be the primordial ova sufficient in number to last the reproductive lifetime of its owner. Com-

* Both sets of primary sex organs arise from a mass of tissue (gonads) in the abdomen. If the individual has XX sex chromosomes, the developing gonad apparently secretes hormones which stimulate the hypothalamus to promote unisexuality (femaleness). Similarly, an individual with XY chromosomes develops male characteristics.

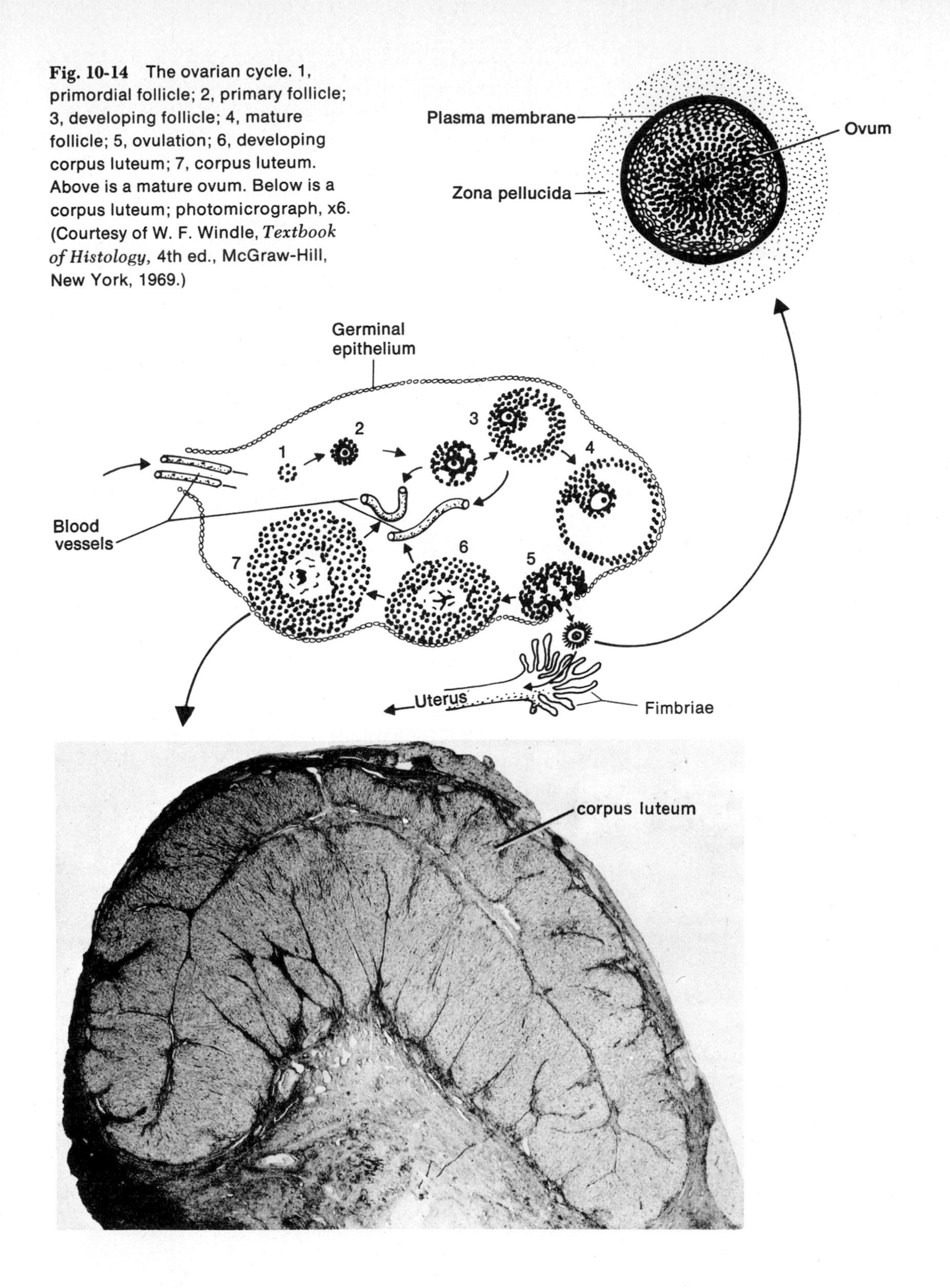

Fig. 10-14 The ovarian cycle. 1, primordial follicle; 2, primary follicle; 3, developing follicle; 4, mature follicle; 5, ovulation; 6, developing corpus luteum; 7, corpus luteum. Above is a mature ovum. Below is a corpus luteum; photomicrograph, x6. (Courtesy of W. F. Windle, *Textbook of Histology*, 4th ed., McGraw-Hill, New York, 1969.)

mencing with the start of a woman's reproductive life, some of these primordial follicles begin to develop, courtesy of follicle-stimulating hormone (FSH) from the pituitary. Many of these will never see the uterine cavity, for only one follicle and its ovum normally achieves full development in each cycle.

As the cells of the developing follicle multiply and mature, the hormone estrogen is secreted into the circulation and the uterus begins to thicken its lining in response. On about the 14th day of the menstrual cycle (about 21 days after the follicle first began to develop), intrafollicular pressure reaches a point where the ovum is literally ejected from the follicle and (because it is near the surface of the ovary) into the peritoneal cavity, to be quickly caught by the fingerlike fimbriae of the uterine tube. This ejection, termed **ovulation,** is believed to be a consequence of the combined action of FSH, estrogen, and another pituitary product, luteinizing hormone.

The remaining follicular cells develop a yellow (luteal), fatty appearance and become known as the **corpus luteum.** The cells of this structure secrete estrogen and progesterone, which help bring the uterine lining to the peak of its development. Glands in the uterine lining begin to secrete—a source of nutrition for the new embryo.

If the ovum has not accepted a sperm cell, the corpus luteum senses that reality and undergoes degeneration—a prelude to disaster for the uterine lining, which will consequently slough, initiating menstruation (see ahead). The degenerating corpus luteum becomes a scar in the ovarian connective tissue (corpus albicans). If conception occurred, the corpus luteum survives to maintain the functional state of the uterus for some 90 days, after which it degenerates while the placenta carries on the hormonal functions.

With the onset of menopause (termination of reproductive activity), the ovary ceases its cyclic production of ova and hormones. The ovary receives its blood from the ovarian arteries, homologues of the testicular arteries, which come directly off the aorta.

The hormones of the ovary have widespread systemic effects, particularly estrogen. It is responsible for such secondary sex characteristics in the female as the development of breasts (fat deposition and duct growth); broadening of the pelvis and the hips and increased fat deposition in the buttocks, thighs, and hips, creating the classical "female figure"; smooth-textured skin; and rapid closure of epiphyseal plates, resulting in shorter stature. Estrogen acts on and matures the organs of reproduction, i.e., uterus, vagina, etc.

Progesterone, secreted by cells of the corpus luteum, has

as its principal effect stimulation of the uterine glands. These glands increase their secretory production in response. Progesterone takes part in stimulating breast development, specifically glandular development. Progesterone also inhibits muscular contractions of the uterus. Women of reproductive age receive indications of reduced progesterone secretion whenever they get uterine cramps just before menstruation.

Uterine tubes

The **uterine tubes** are projections of the uterus and conduct the ejected ova from the ovary to the uterus. The tubes are draped with peritoneum (broad ligament) which extends from the uterus to the lateral pelvic wall (Fig. 10-13).

The mucous membrane of the tubes is characterized by a folded and irregular ciliated columnar epithelium supported by a highly vascular connective tissue, suggesting secretory activity. And indeed, the epithelial cells are secretory, creating a nutrient environment for the ovum that descends toward the uterus. The cilia beat toward the uterine lumen, aiding the ovum in its migration. The mucous membrane is reinforced by a layer of smooth muscle, which probably also aids the ovum's movement.

The sperm cells, having been deposited in the vagina, rapidly (12 to 24 hours) "migrate" to the uterine tubes and there fertilize the ovum.

Uterus

The **uterus** is the sanctuary of the offspring of love, providing a secure residence in which the developing human remains for some 9 months (gestation). Embryologically, the uterus represents the fusion of the two uterine tubes. It is pear-shaped, no bigger than a closed fist in the woman who is not pregnant.

In Fig. 10-13, it can be noted that the uterus has three principal parts:

1. The **fundus** above.
2. The main **body** and lower isthmus.
3. The lower, narrower **cervix.**

The uterus is continuous with the uterine tubes at the fundus and in communication with the vagina at the cervix. The neighbors of the uterus include the rectum posteriorly (with some ileum intervening) and the bladder immediately anterior. Laterally, the uterus is related to the broad ligaments and the structures therein (Fig. 10-13). Ligaments within the broad ligament are responsible for stabilizing the uterus, although the main support probably comes from its attachment to the vagina.

In Fig. 10-8, note the relationship of the uterus to the bladder and vagina. The uterus is generally bent forward (anteflexion) and tilted forward (anteversion) onto the superior surface of the bladder. Thus, the position of the uterus in space is not fixed and varies with increased fullness of the bladder, etc. The probability of conception following sexual intercourse is influenced by the posture of the uterus.

The lumen of the uterus is widest at the body and narrows toward the opening (ostium) into the vagina. The mucous membrane lining the uterus **(endometrium)** is a tremendously variable structure—and changes character regularly on a monthly schedule. These changes, initiated by hormonal stimulation, are related to preparing a place in which the embryo can imbed and develop. Underlying the endometrium is the **myometrium,** composed of some of the longest smooth muscle fibers in the body. During the reproductive cycle, this layer does not change character anything like the endometrium. The myometrium is important at the termination of pregnancy and is responsible for the phenomenon called labor.

The columnar epithelia of the endometrium are largely ciliated. Tubular uterine glands are pronounced throughout the lamina propria, especially in the latter phase of the menstrual cycle. The microscopic character of the endometrium is constantly changing and can best be described chronologically in the menstrual cycle.

Menstrual cycle. This is a cycle of approximately 28 days in which the female reproductive structures prepare for receipt and implantation of a developing embryo. The principal organ of interest in this cycle is the uterus, although the cycle involves the ovaries, the pituitary gland, the hypothalamus, and to a lesser degree, the vagina. The cycle is instigated and maintained by hormones of the pituitary gland following appropriate influences from the hypothalamus. The rhythmic or cyclic nature of this reproductive activity allows the body time to maximally prepare the appropriate organs for the strain of supporting a distinctly separate individual—the embryo. It would be quite a traumatic load on the metabolic activity of the woman who had to maintain her uterus and other related structures in a peak nutritional state month in and month out without a break. Happily that is not the case. In any normal 28-day cycle, the first five days are dedicated to ridding the uterus of the unused endometrium of the previous cycle. The next 14 days are spent rebuilding the uterine endometrium to a peak nutritional and functional level. The following seven days the uterus awaits the arrival and anticipates the implantation of the rapidly dividing fertilized ovum. Of course, if no conception had taken place there will be no fertilized ovum; and the uterine endometrium will then undergo degeneration in the last two days of the cycle. Menstruation is heralded by the appearance of the sloughed endometrium in the vagina, and the uterus begins a new cycle.

The menstrual cycle is peculiar to primates (the apes and women). Other mammalian females (dogs, cats, horses, etc.) have reproductive (estrus) cycles which are not characterized by sloughing of the uterine endometrium (menstruation). The human reproductive cycle commences at about 12 years of age **(menarche)** and ceases at about 45 years of age **(menopause)**. The chronological events of the menstrual cycle may be summarized as follows:

Day 1–5: The uterus has sloughed its endometrium; the debris, some blood, and the "old maid" ovum pass through the cervix and on into the vagina. The amount of menstrual discharge varies from individual to individual, from cycle to cycle. Meanwhile, the ovary has started proliferating a new follicle under the influence of pituitary FSH.

Day 5–14: This is the *proliferative* period of the cycle. The ovary produces a new ovum within its follicle; the uterus begins regenerating its endometrium due to the influence of estrogen from the ovary. As the 14th day approaches, the ovum becomes sensitive to the pressure of its environment and seeks release. On about the 14th day of a classical cycle, the follicle ruptures and the ovum bursts into the peritoneal cavity to be rescued by the fluttering fingers (fimbriae) of the uterine tube. The uterus steadfastly continues its endometrial regeneration.

Day 15–26: Dawn of the *secretory* period of the cycle. The ruptured ovarian follicle, influenced by LH of the pituitary, differentiates into the corpus luteum, secreting progesterone and estrogen. The uterine endometrium, sensing the presence of progesterone, develops glands which soon fill with stored nutrient secretions. If the ovum was fertilized in the distal extremity of the uterine tube (as it normally is) on the 14th day of the cycle, it should enter the uterine cavity and implant about the 21st day. At this time the endometrium is at its peak functional state, having increased in thickness some

400 percent since day 5! If implantation occurred, the corpus luteum will continue its secretions of estrogen and progesterone until the developing placenta can manufacture its own hormones—a period of about 90 days. If implantation did not occur, read on.

Day 27–28: In a manner not clearly defined, the endometrium telegraphs the ovary of its crisis: no implantation. The ovary responds by cutting off the corpus luteum's secretions of progesterone and estrogen—not unlike the way the light company might respond to an unpaid bill—by turning off the lights! With progesterone levels thus reduced, the nutrient material stored in the uterine endometrium starts to be absorbed by the blood since it is no longer being produced. The thick, watery endometrial tissue starts to settle, bending the many coiled arteries to the breaking point. Bleeders spurt momentarily into the endometrium, tearing up the tissue. The muscle layer of the uterus (myometrium) begins to contract in the absence of progesterone. More settling of tissue, more hemorrhage, and finally menstruation is a reality. Actually, the volume of blood loss is low, as the arteries which bleed immediately constrict. Thus the menstruum is not just blood, but is mainly endometrial debris and secretory material. The uterus "weeps" over its loss.

The story of the reproductive cycle does not end on a sad note, however, for as sure as spring follows winter, a new cycle follows the old.

The uterus receives its blood supply by uterine branches of the internal iliac artery. Venous drainage parallels the arterial vessels. These vessels travel in the lateral ligaments of the cervix, less than 1 cm from the ureter on each side.

Vagina

The **vagina** is the accessway to the uterus and its tubes from the external environment. A collapsed fibromuscular passage capable of great distention, it embraces the penis in sexual intercourse, recovers the semen following ejaculation, and is the last place of security and warmth for the newborn, who must pass through it to enter "the real world"—a relatively cold and hostile environment.

Superiorly, the vagina receives the cervix of the uterus; inferiorly, the vagina is open to the outside within the vestibule of the perineum. Here the vagina is often partly closed by a mucous membrane fold (hymen) in the sexually uninitiated woman (Fig. 10-20). At least 75 percent of the 3½-in.-long vagina is in the pelvic cavity, and it is related to the bladder and urethra anteriorly and the rectum and anal canal posteriorly (Fig. 10-8). Digital examination of the vagina often allows palpation of the above structures and of the cervix.

Note that the cervix is snub-nosed and projects down into the vagina some distance (Fig. 10-13). The circular recess created about the cervix within the vagina is called the **fornix.** The posterior aspect of the fornix is deepest and is of significance to the physician, as a speculum in the posterior fornix is merely millimeters away from the peritoneal cavity.

The mucosa of the vagina is lined with stratified squamous epithelium. The character of this epithelial lining undergoes some cyclic changes. The vaginal walls—anterior and posterior—have palpable transverse ridges. The fibromuscular coat of the vagina is continuous with fibers of the levator ani which not only help support the vagina but may have some constrictive ("clasping") function. There are no visible glands in the vagina. Secretions from the vaginal walls have been observed during sexual activity, but these may be diffused fluids from the vasodilated tissues in the vaginal wall.

The vagina is supplied by branches of the uterine artery (upper part) and by direct branches of the internal iliac artery. The upper part of the vagina is served by an autonomic plexus, while the lower part is serviced by the pudendal nerve.

QUIZ

1. The principal function of the male reproductive system is to ______________ ______________ ______________.
2. The testis contains a number of highly coiled ______________ ______________.
3. The interstitial cells of the testis are known to secrete ______________.
4. The epididymis functions to ______________ ______________.
5. The ductus deferens terminates in the prostate gland as the ______________ ______________.
6. The prostate and seminal glands secrete a material which ______________ ______________ ______________.
7. The cyclic changes in internal character of the ovary during each 28 days are a function of ______________ secreted by the ______________ ______________.
8. On about the __________ day of the menstrual cycle the ovum is ejected from the follicle.
9. The sperm cells fertilize the ovum in the ______________ ______________.

perineum

A CONCEPTUAL ANALYSIS

The perineum was defined at the beginning of this unit. Anatomically, the region is complicated. Further, the area is difficult to dissect in the cadaver and, containing more fat than anything else, offers little sense of fulfillment upon completion. But the general organization can be learned without difficulty. Upon reaching that milestone, the more classical considerations of the perineum will be undertaken.

Refer now to Fig. 10-15 and note: the perineum may be visualized as a box rotated 45° to form a three-dimensional diamond-shaped structure. This is the basic shape of the perineum within the framework of the pelvic outlet. Thus the four corners of the perineum are anteriorly, the pubic symphysis; posteriorly, the coccyx; laterally, the ischial tuberosities. The four sides of the perineum may be seen to be the ischiopubic bones anterolaterally, and the sacrotuberous ligaments posterolaterally. The roof of the perineum is the levator ani or pelvic diaphragm (Fig. 10-16). The floor of the perineum is

Fig. 10-15 Boundaries of the perineum and its triangles. See text.

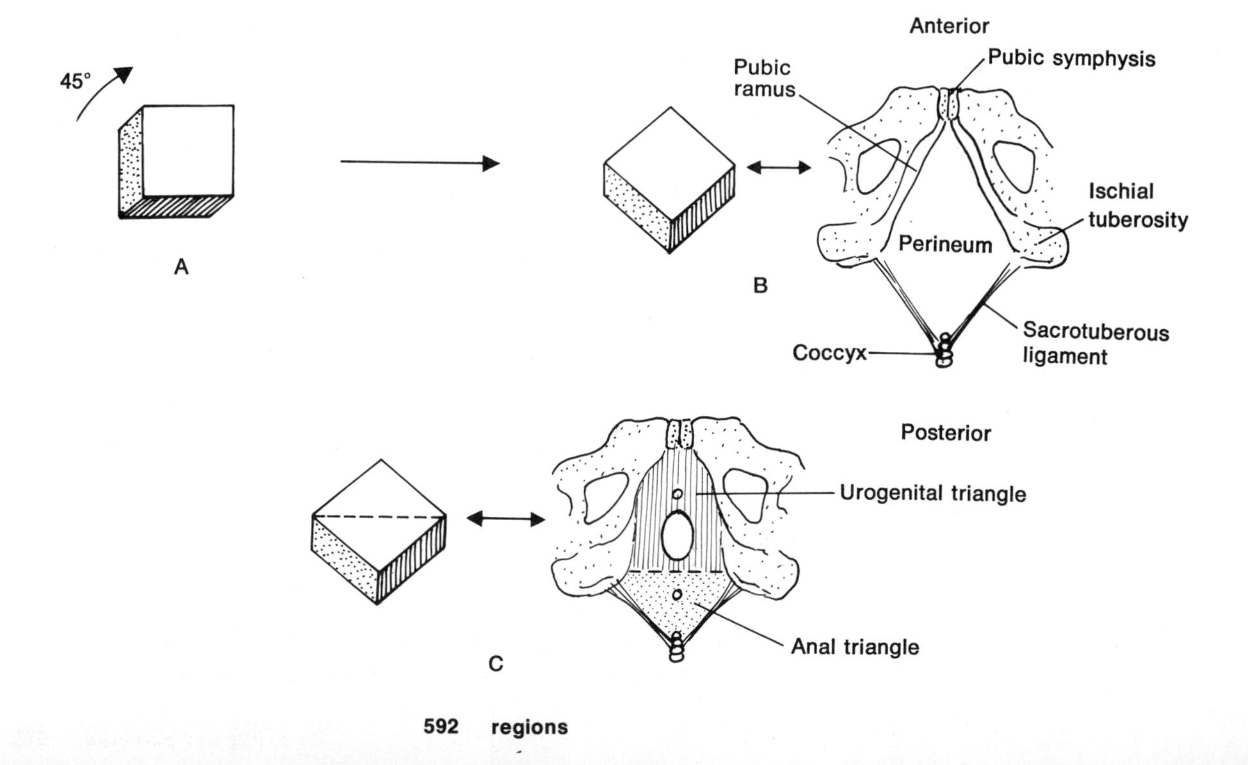

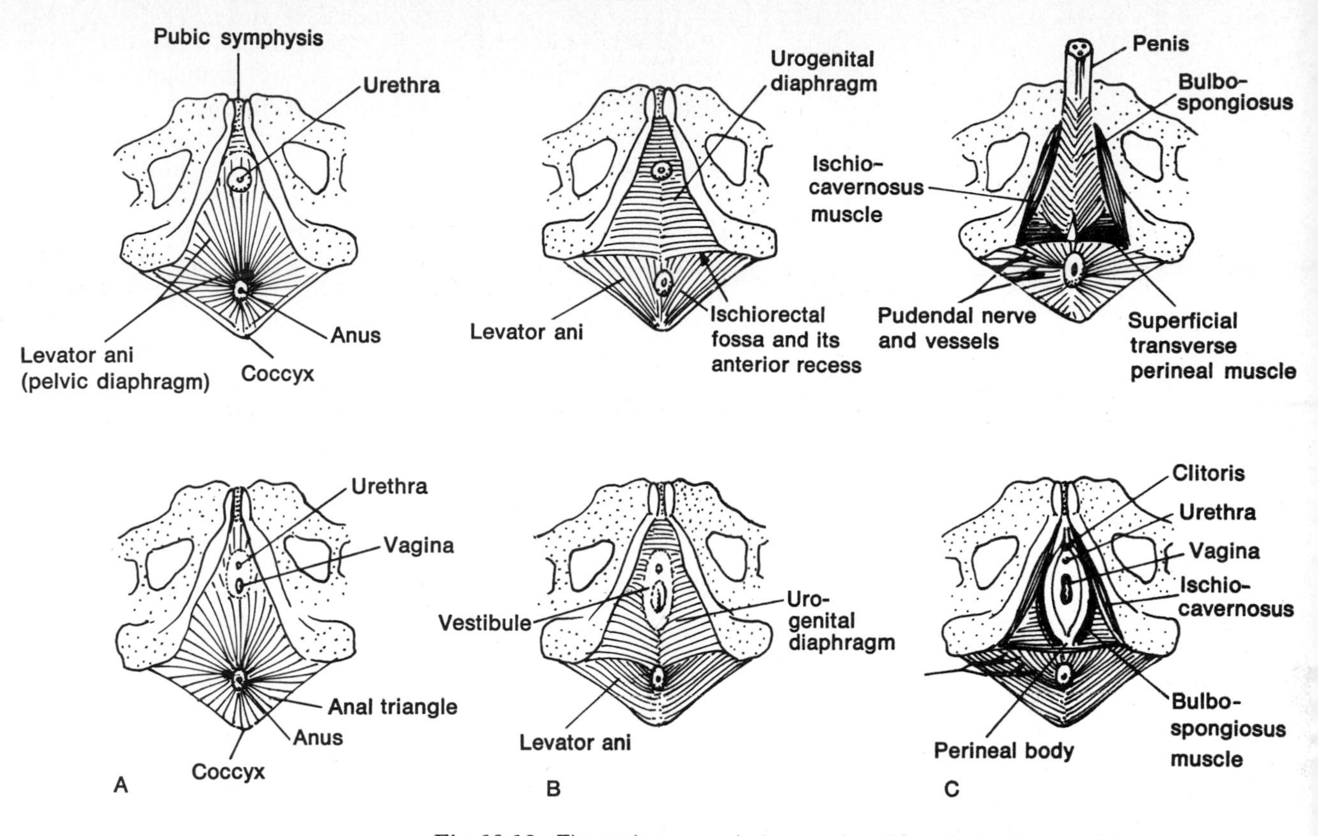

Fig. 10-16 The perineum, male (top row) and female (bottom row), in levels from deep to superficial. A. Roof of the perineum: levator ani. B. The UG diaphragm, or deep perineal space. Between the levator ani and the UG diaphragm is the anterior recess of the ischiorectal fossa. C. Superficial perineal space.

created from skin and superficial fascia and is perforated by passageways from the pelvis (Fig. 10-8).

The four-sided perineum may be bisected into two triangles, anterior and posterior (Fig. 10-15). The anterior triangle contains the ducts of the urinary and genital systems and is, therefore, conveniently titled the **urogenital triangle.** The posterior triangle encompasses the anal canal and so is called the **anal triangle.**

ANAL TRIANGLE

The space of this triangle is called the **ischiorectal fossa,** and its principal occupant is the **anal canal** (Fig. 10-16), previously discussed. The ischiorectal fossa is largely fat-filled. Under pressure of sitting, because it is relatively poorly vascularized, the fossa is an occasional site of abscess. On the lateral walls of the fossa are the fascia-formed pudendal canals conducting nerves and vessels to the perineum from the pelvis via the deep gluteal region (Fig. 10-22).

At the anterior border of the triangle, arising as deep as the levator ani, there is a fibromuscular mass, the **perineal body.** It functions as a site of attachment and support for many of the muscles of the perineum (Fig. 10-16). If it should be damaged during childbirth,* the stability of the passageways of the perineum may be in danger (see Clinical Considerations).

UROGENITAL TRIANGLE

The urogenital (UG) region of the perineum is a space like the ischiorectal fossa with a pertinent exception: a shelf exists which separates the UG region into upper and lower compartments. Such a shelf is properly called a diaphragm, specifically the **urogenital diaphragm** (Fig. 10-16). Refer to Fig. 10-17 and note that the upper compartment is continuous with the more posterior ischiorectal fossa; it is, therefore, referred to as the anterior recess of the fossa. The recess is

* During the time of the baby's passage through the birth canal, the vagina may become so dilated that a shearing pressure can be put on the perineal body. To prevent this possibility, the operator will often make a posterior skin incision from the vagina (episiotomy), creating a larger access way.

Fig. 10-17 The perineal spaces and related fasciae of the male, sagittal section.

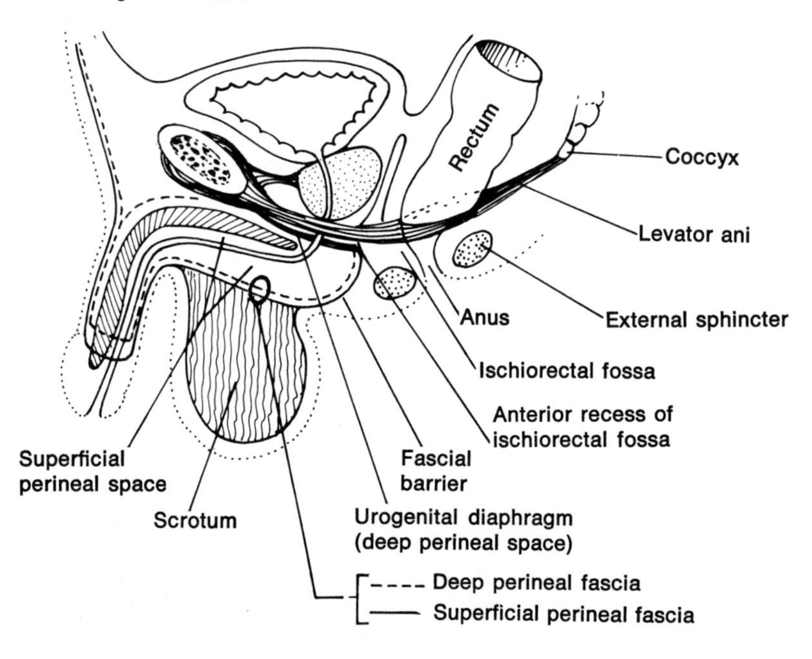

split into two side-by-side recesses by the midline perineal body and the low-slung levator ani, which supports the urethra, the prostate (in the male), and the vagina.

The lower compartment is the **superficial perineal space,** containing muscular and specialized structures related to the external genital organs.

The UG diaphragm itself, therefore, must be, and is, the **deep perineal space** (Fig. 10-17). It consists of skeletal muscle, fascia, and a pair of glands (bulbourethral) in the male. It also encloses the membranous urethra which descended from the bladder. The deep and superficial perineal spaces are sealed off from the ischiorectal fossa by a fascial barrier (Fig. 10-17).

CONTENTS OF THE MALE UROGENITAL REGION

The superficial perineal space of the male includes:

1. Roots of the penis.
2. Muscular envelopes of these roots.
3. Related nerves and vessels.

The deep perineal space consists entirely of the urogenital diaphragm.

The penis will be considered in a moment; here it is enough to mention that it has three fibroelastic roots. Two of these, the **crura,** arise from the underside of the ischial bones. The other, the **bulb,** arises between the crura at the urogenital diaphragm. These erectile bodies, sheathed in skeletal muscle, are properly part of the superficial perineal space.

Study Fig. 10-16 and note that the bulb of the penis, incorporating the urethra, is wrapped in the **bulbospongiosus muscle,** which has its origin in the perineal body. This muscle functions in erection and ejaculation and receives innervation from the pudendal nerve.

Each crus of the penis is wrapped in the **ischiocavernosus muscle,** arising from the ischium. These muscles, which aid in erection, receive innervation from the pudendal nerve.

The slender superficial transverse perineal muscles, which help stabilize the perineal body, are also within the superficial perineal space.

The wall and the contents of the deep perineal space constitute the urogenital diaphragm, which consists of:

1. Deep transverse perineal muscles.
2. Sphincter of the urethra (muscle).
3. Membranous urethra.
4. Bulbourethral glands.

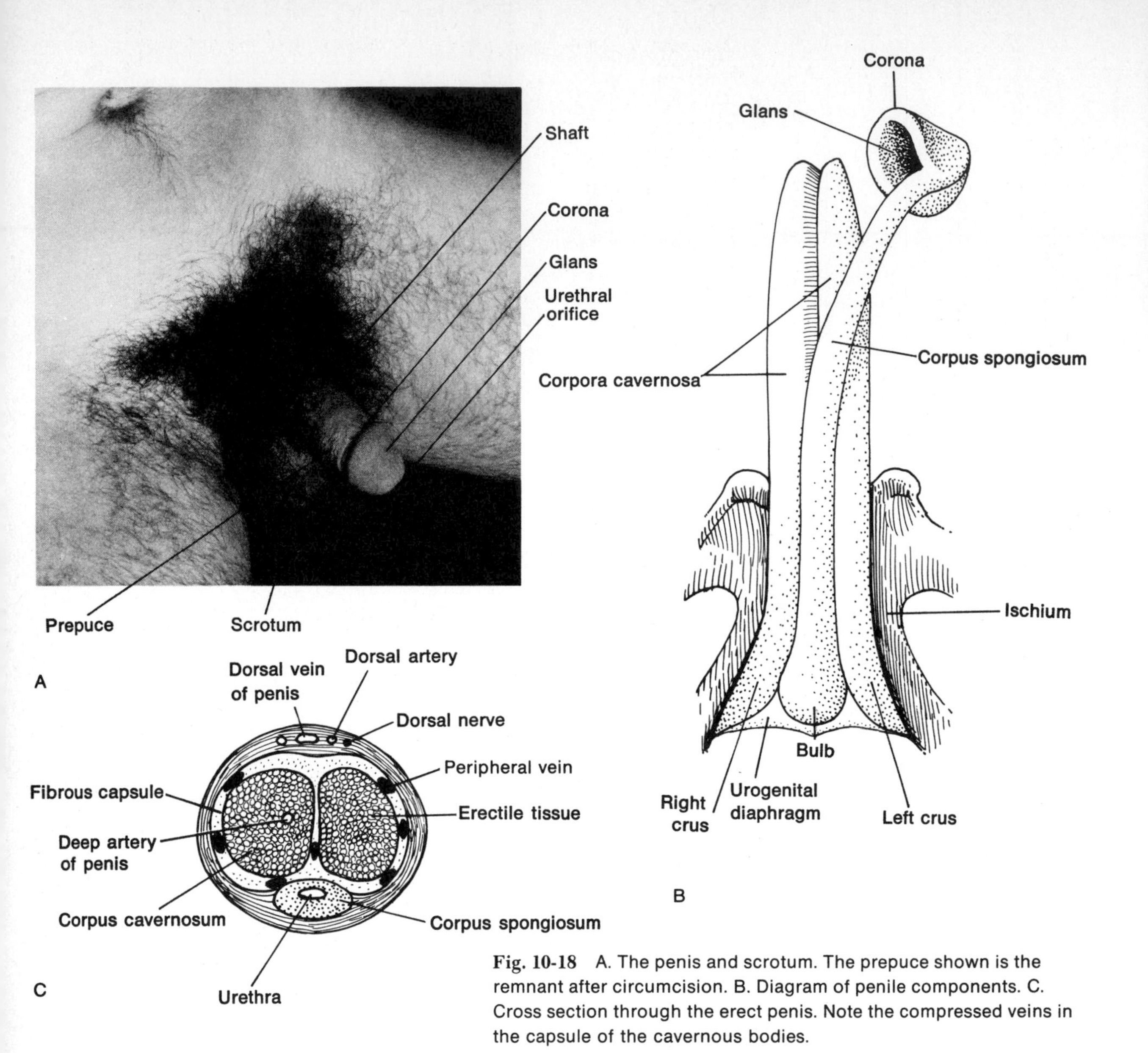

Fig. 10-18 A. The penis and scrotum. The prepuce shown is the remnant after circumcision. B. Diagram of penile components. C. Cross section through the erect penis. Note the compressed veins in the capsule of the cavernous bodies.

MALE EXTERNAL GENITAL ORGANS

The external genital organs are the (1) penis and (2) scrotum.

Penis

The **penis** permits the introduction of sperm into the vagina. Since the vagina is collapsed, the penis must be erect in order to penetrate—and this is what the morphology of the penis is all about.

In Fig. 10-18 it can be seen that the crura are solidly attached to the ischial bones and project downward and medially toward one another within the superficial perineal space.

The bulb of the penis arises between the two crura from the base of the urogenital diaphragm. The membranous urethra leaves the urogenital diaphragm to penetrate the bulb in the superficial perineal space (Fig. 10-17). The bulb is ensheathed by the bulbospongiosus muscle.

The two crura and the bulb come together—the latter under the former—to form the triple-barreled body of the penis. Where the body bends downward the bulb is renamed the **corpus spongiosum,** and the two crura become the **corpora cavernosa.**

The corpus spongiosum terminates distally by flaring out as the mushroom-caplike **glans.** The corpora cavernosa end bluntly against the glans. The penile urethra is conducted by the corpus spongiosum and opens to the outside at the tip of the glans. The two-layered skin variably covering the glans is called the **prepuce** and is usually removed at birth (circumcision).

Note the microscopic nature of the penis as seen in Fig. 10-18. The cavernous bodies and the spongy body consist of large vascular spaces with highly elastic walls of connective tissue and some smooth muscle **(erectile tissue).** Associated with the cavernous spaces are a great number of arteries lined by smooth muscle. At the periphery of the bodies a thick tight band of deep fascia may be seen bandaging the three cavernous bodies together. The veins draining the cavernous spaces are located dorsally just deep to this fascia. This arterial-venous relationship has special functional significance to be appreciated momentarily. A loose layer of superficial fascia and skin surround the tightly bound bodies and move with ease over them and the glans as well.

As the anatomy suggests, the penis is a bloody organ—each erectile body has its own artery, derived from the pudendal artery, a branch of the internal iliac.

The veins draining the spaces are radially arranged and drain into the dorsal vein of the penis.

The nerves of the penis are primarily autonomic to the muscular arteries and sensory to the glans—said by some anatomists to be the most sensitive structure in the body.

Scrotum

The scrotum is a thin-skinned pouch for the testes (Fig. 10-19). It is borrowed from the anterior abdominal wall and is characterized by:

1. Dark pigment.
2. A tunic of smooth muscle, the **dartos** muscle, associated with the nonfatty superficial fascia.
3. A median septum dividing the scrotum into two compartments. A seam, or raphe, visible in the midline of the scrotum can be seen to continue up the underside of the skin of the penis to the glans.*

The scrotal skin has the capacity to change its appearance and physical state in response to the action of the dartos muscle, the aim being to maintain the proper temperature within the scrotum. In cold, the muscle contracts and the scrotum wrinkles, drawing more closely around the testes and so stemming the loss of heat. In warmth, the dartos relaxes, the scrotum loses its tone, and heat is allowed to dissipate.

CONTENTS OF THE FEMALE UROGENITAL REGION

The anatomy of the superficial and deep perineal spaces here is made different from the male by the presence of the vagina and associated external genital organs. Otherwise, the anatomy is similar, if not homologous. See now Fig. 10-16. The urogenital diaphragm is pierced by the urethra anteriorly and the vagina posteriorly. The contents of the urogenital diaphragm are similar to those of the male, including:

1. Deep transverse perineal muscles which help stabilize the important perineal body.
2. The inconsequential "sphincter of the urethra." There is one addition—the vagina—and one subtraction—the bulbourethral glands.

The muscles of the superficial perineal space are homologous with those of the male; except that the bulbospongiosus muscle (overlying the vestibular bulbs) is cleaved as it circumvents the vagina. The superficial trans-

* This scrotal raphe represents a fusion of genital folds which surrounded the embryonic orifices of the urinary system. See the fate of these folds in males and females in Fig. 10-19.

Fig. 10-19 Development of the external genital organs. A–D. Male. A′–D′. Female. Names in parentheses refer to the adult form of the structure indicated.

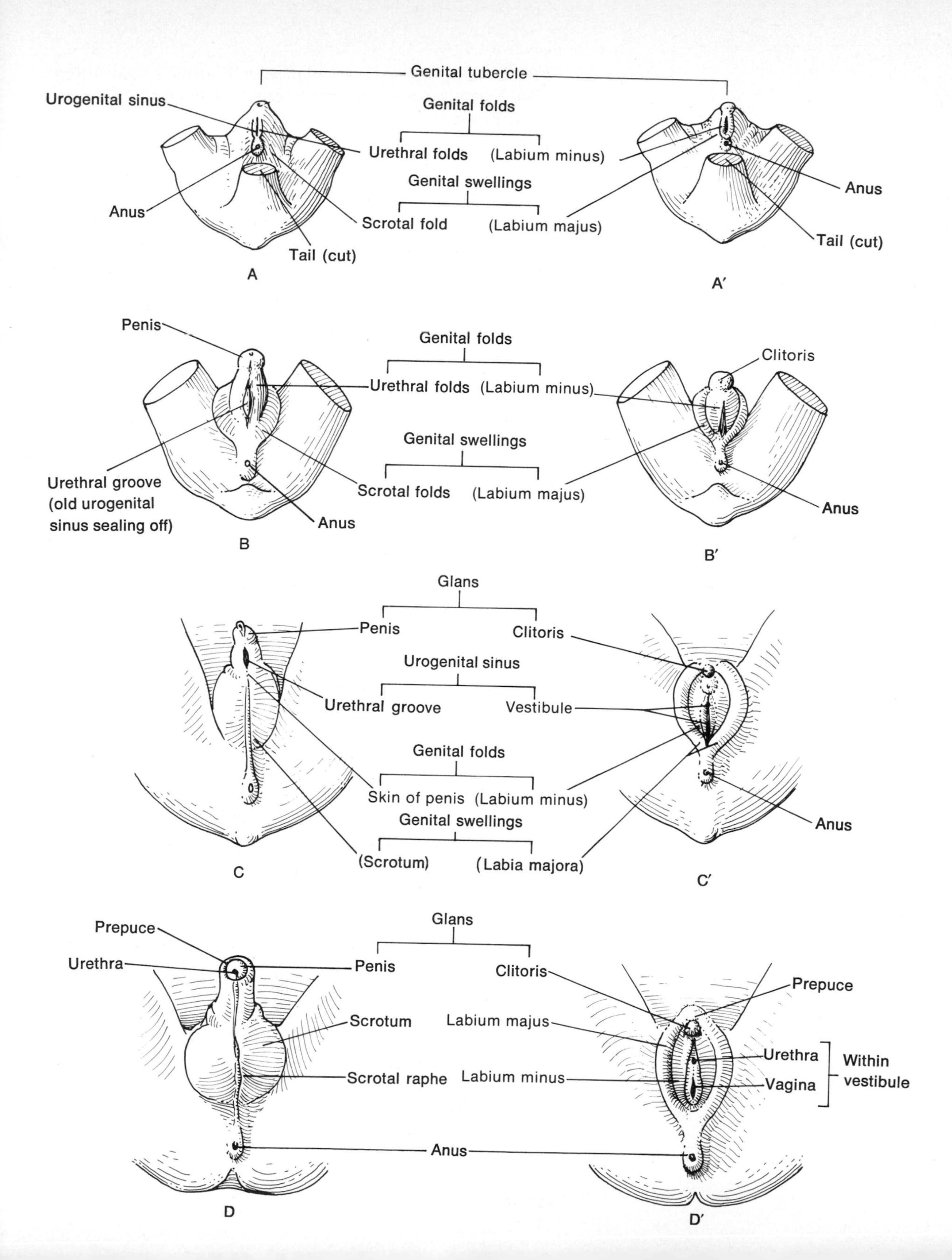

Genital tubercle
Urogenital sinus
Genital folds
Urethral folds
(Labium minus)
Genital swellings
Anus
Scrotal fold
(Labium majus)
Tail (cut)
Anus
Tail (cut)
A
A′
Penis
Genital folds
Clitoris
Urethral folds
(Labium minus)
Genital swellings
Urethral groove
(old urogenital
sinus sealing off)
Scrotal folds
(Labium majus)
Anus
Anus
B
B′
Glans
Penis
Clitoris
Urogenital sinus
Urethral groove
Vestibule
Genital folds
Skin of penis
(Labium minus)
Genital swellings
Anus
(Scrotum)
(Labia majora)
C
C′
Glans
Prepuce
Urethra
Penis
Clitoris
Prepuce
Scrotum
Labium majus
Urethra
Within
vestibule
Scrotal raphe
Labium minus
Vagina
Anus
D
D′

verse muscles appear to stabilize the perineal body. If these muscles were cut in an episiotomy procedure, the stability of the perineal body might be in question.

The pudendal vessels and nerves to the perineum arrive from the pelvis via the pudendal canal of the ischiorectal fossa.

FEMALE EXTERNAL GENITAL ORGANS

The external genital organs (vulva) (Fig. 10-20) are the:

1. **Labia majora** or large lips.
2. **Labia minora** or small lips.
3. **Clitoris.**
4. **Vestibule** or space created by the small lips.
5. **Bulbs of the vestibule.**
6. **Greater vestibular glands** (next to vaginal orifice).

It is interesting to compare the development of the external genital structures of each sex. The parallels in structural development are fascinating and deviations can often be explained by referring back to the embryology.

Refer now to Fig. 10-19 as you read.

Follow the development of the genital tubercle in both sexes from 7 weeks to birth. In the male, it becomes the penis; in the female, the clitoris.

Follow the development of the genital folds. In the male, the folds unite to form the urethral and scrotal raphe. In the female, the cleft remains and the folds enclose it. The folds become the labia minora; the cleft or old urogenital sinus becomes the vestibule into which the urethra and vagina open.

Follow the development of the genital swellings. In the male, these pouch out to accommodate the testes. United by a raphe, they become the scrotum. In the female, the swellings are forced apart by the developing vestibule and labia minora. They become the labia majora.

Now study Fig. 10-20 as you read the following:

The **major labia** consists of two fat-filled folds of skin, covered with hair (beginning at puberty). Anteriorly they merge over the pubic bone to form the pubic mound (mons pubis). The round ligaments of the uterus terminate in these lips just as the spermatic cords end in the homologous scrotum.

The **minor lips** are folds of skin which enclose the vestibule. Posteriorly, they merge with the labia majora. Anteriorly, the lips split to enclose the body of the clitoris. The labia minora

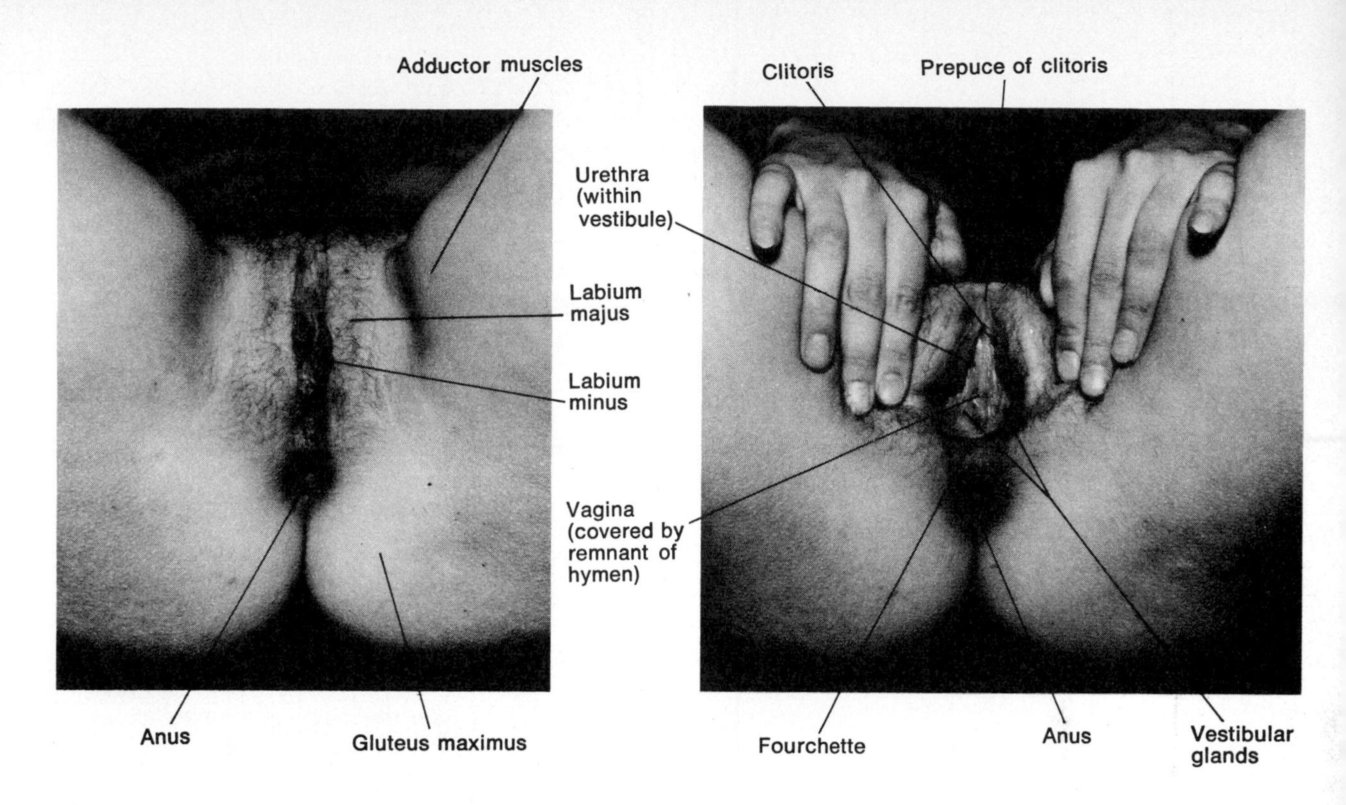

Fig. 10-20 The vulva of a 22-year-old female. A. Labia are closed. B. Labia are separated.

are pressed together by the larger labia and usually hidden except in children.

The **clitoris** – homologue of the penis – consists of two erectile bodies (corpora cavernosa clitoridis) which arise from the ischial rami. Bound in deep fascia, the corpora cavernosa are capped by the glans. The glans is in continuity with the vestibular bulb via a venous plexus. Like the penis, the clitoris is very sensitive to touch and pressure and is capable of erection. The latter function is not immediately visible as the clitoris is hidden and bound down by the frenulum and prepuce from the labia minora.

The **bulbs of the vestibule** consist of a pair of erectile bodies on either side of the vagina. They are homologous with the bulb of the penis. They are covered with the bulbospongiosus muscle, and are continuous anteriorly – in the midline – with the glans of the clitoris. The bulbs are capable of significant distention during sexual stimulation, enhancing vaginal contact with the penis.

The **vestibular glands** are pea-sized structures adjacent to the bulbs. They are homologues of the bulbourethral glands, and empty their secretions at the margin of the vaginal orifice.

the anatomical and functional basis of coitus

Arbitrarily, the sexual act is divided into three phases: precoital activity, coitus, and postcoital activity.

PRECOITAL ACTIVITY

Some precoital activity or foreplay is generally a necessary prerequisite to successful sexual union. This is so simply because the nervous system must prepare the various structures to be involved, e.g., secretion of glands in the cervix, erection of the penis, not to mention stimulation of the psyche. These are, of course, the obvious manifestations of sexual receptivity—some of the not-so-obvious anatomical and functional responses include:

1. Erection of the clitoris.
2. Enlargement of the vestibular bulbs.
3. Swelling of the vaginal walls.
4. Movement of sperm from the testes and epididymis to the urethra (emission).

ERECTION OF PENIS (CLITORIS)

Erection of the penis (clitoris) is a vascular phenomenon. When the male (female) is sexually stimulated, the CNS fires impulses via the parasympathetic autonomics to the smooth muscle of the arteries feeding the three erectile bodies of the penis (clitoris). The muscles lose their tone, and an increased volume of blood is directed into the cavernous spaces, the latter expanding in response. As the bodies enlarge, they compress the peripheral veins against the tight fascial bandage—preventing drainage of blood from the cavernous spaces. Further, the ischiocavernosus and bulbospongiosus muscles encasing the roots of the erectile bodies contract under parasympathetic stimulation and prevent drainage of blood from the cavernous spaces. As a consequence, the penis (clitoris) enlarges, elongates, and becomes hard as cartilage, except for the glans which retains some softness. This latter structure allows a more flexible vaginal penetration and offers some resiliency if and when the glans strikes the cervix during the movements of intercourse.

The clitoris is said to be as sensitive as the penis. Studies indicate that a woman's ability to have an orgasm may often be related to clitoral stimulation.

Simultaneous with erection of the clitoris, the vestibular bulbs (homologues of the bulb of the penis) expand and compress the lateral walls of the vagina. The anterior wall of the vagina swells, protruding somewhat into the vestibule. The cervical endometrial glands secrete profusely, and drain into the upper vagina. There are apparently no glands in the vagina itself, although it has been reported that the vaginal epithelium secretes just before and during coitus. The vestibular glands secrete a mucous material at the orifice of the vagina which facilitates entry of the penis.

EMISSION

With sexual stimulation, the smooth muscle of the male genital ducts and glands rhythmically contracts. Sperm cells—by the millions—are conducted via the following route (Fig. 10-21):

> testis → epididymis → ductus deferens (to the superficial inguinal ring → inguinal canal → deep inguinal ring → across the lateral walls of the abdominopelvic cavity → posterior aspect of the bladder) → ejaculatory ducts (where the secretions of the seminal vesicles join with the sperm) in the substance of the prostate gland → prostatic urethra.

This process of sperm movement to the urethra is called **emission.** It is an involuntary phenomenon and precedes the discharge of semen to the outside (ejaculation). If sexual continence (voluntary restraint of ejaculation) is practiced over a long period of time in spite of occasional stimuli, the muscular walls of the duct system just described may spasm with further stimuli, resulting in pelvic and testicular pain.

COITUS

Study Fig. 10-21 as you read:

Penetration of the vagina by the erect penis can be accomplished in a number of different postures. Postures of intercourse are particularly significant if there is failure to effect conception, or there is the intention to prevent conception, or

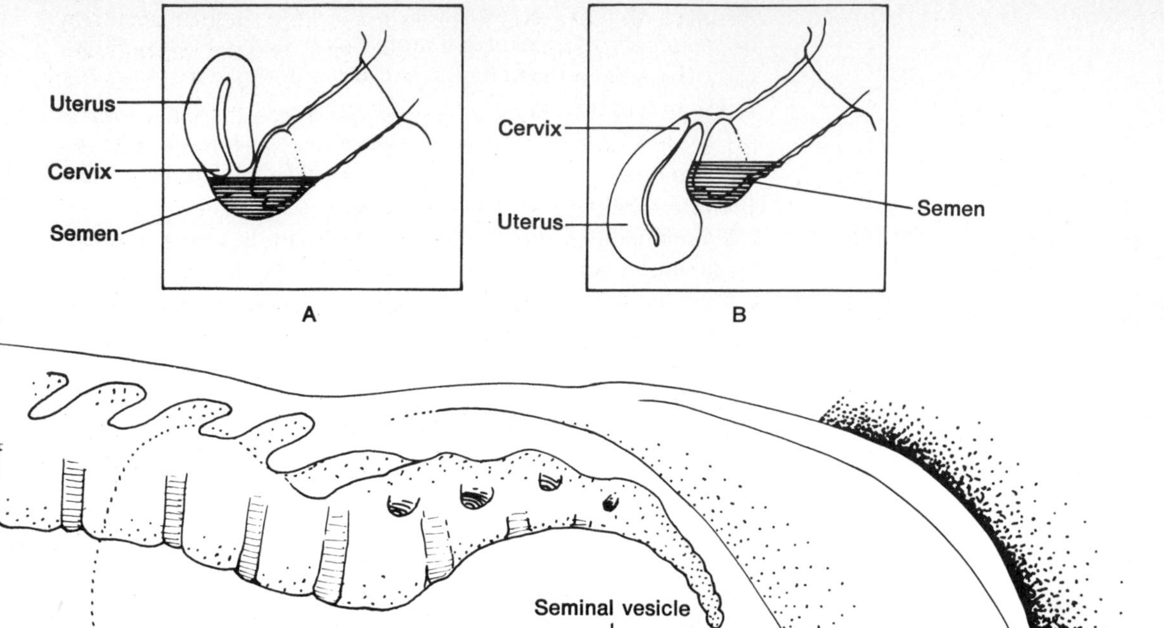
Uterus
Cervix
Semen
A
Cervix
Uterus
Semen
B

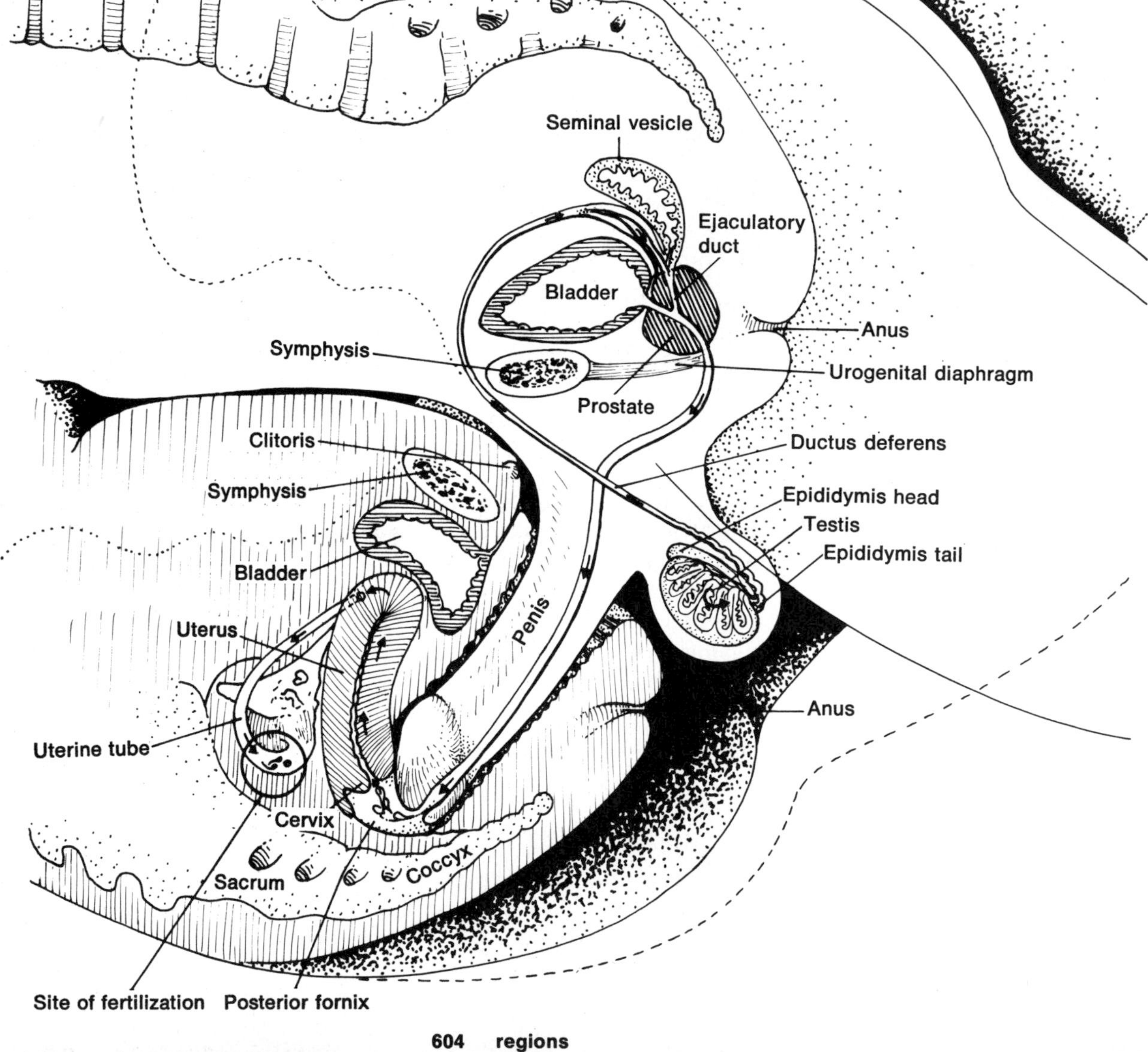
Seminal vesicle
Ejaculatory duct
Bladder
Anus
Symphysis
Urogenital diaphragm
Prostate
Clitoris
Ductus deferens
Symphysis
Epididymis head
Testis
Epididymis tail
Bladder
Uterus
Penis
Anus
Uterine tube
Cervix
Sacrum
Coccyx
Site of fertilization
Posterior fornix

the need to make adjustments for a short vagina, unusually long penis, unusual pelvic or vertebral orientation, retroversion of the cervix, etc.

Within 30 sec or so of vaginal penetration, the penis adds to the vaginal and cervical secretions from contributions of the bulbourethral glands.

The male moves his hips so as to produce pistonlike movements of the penis in the vagina. As a result, sensitive touch receptors at the glans penis, glans clitoridis, and vaginal wall are stimulated and the CNS receives volleys of afferent impulses. Reflexly, glandular secretions pour forth and emotional responses are intensified.

When the penis is in full penetration, the mobile cervix is pushed aside (usually upward and forward) creating a pocket in the quite elastic vagina. This pocket is usually an extension of the posterior fornix. Thus, a 3½-in. vagina can normally accommodate a 6-in. penis.

The sperm cells move into the prostatic urethra and are activated by the presence of nutritious prostatic and seminal vesicle secretions. This milky mixture is called **semen.**

When the CNS is stimulated to a certain point, and when willed by the partners, the climax or **orgasm** occurs. In the male, this is manifested by the ejaculation of semen. Ejaculation is brought about by the rhythmic, spasmodic contraction of the duct and gland musculature and the bulbospongiosus muscle. Once induced, the phenomenon of ejaculation is committed—it cannot be willfully interrupted. The semen—about 2 to 5 cc in volume—charges through the prostatic, membranous, and penile components of the urethra and is ejected in spurts into the upper part of the vagina. Note that in the position of Fig. 10-21 the semen has easy access to the cervix. This is not the case in certain other positions or where the uterus is retroverted. The female climax is similar to the male's emotionally and apparently involves contraction of the

Fig. 10-21 Coitus. The couple is shown with related organs in sagittal section. Her right thigh has been removed to expose her internal structures. Arrows indicate the route that sperm travels to the site of conception. A. The semen pool is near the cervix. This result is achieved in the coital position shown. The uterus is in normal position. B. With a retroverted uterus, semen does not drain easily into the cervix and conception is less likely. (Adapted from R. L. Dickinson, *Human Sex Anatomy*, 2d ed., Williams & Wilkins, Baltimore, 1949.)

perineal musculature. The climax is the height and culmination of the sexual experience.*

POSTCOITAL EVENTS

The penis quickly becomes flaccid following ejaculation. In a sense, it is almost as if the smooth muscle of the reproductive tract has temporarily exhausted itself. In point of fact, the male is usually refractory† to a second erection and ejaculation for a variable period of time.

The sperm—bathed in a nutrient, alkaline medium while in the male duct system—now find themselves in a "hostile," lethal, acid environment. Activated by the alkaline secretions of the prostate and seminal vesicles, the sperm desert the vagina for higher ground—the cervix (Fig. 10-21). If the cervix is not readily accessible to the sperm, the latter will quickly perish.

Moving at a rate of about 2 mm per min, the sperm traverse the 80 mm or so of uterine endometrium and enter the orifices of the uterine tubes. Somewhere in the distal half of the tube, the ovum will come to be surrounded by several hundred thousand sperm cells (Fig. 10-21). One and only one of the sperm will penetrate the acid barrier of the ovum, lose its tail, and offer its 23 chromosomes toward the generation of a new human being.

* The female, of course, cannot "ejaculate"—but she can go through the same emotional and muscular phenomena associated with ejaculation. Orgasm is characterized by a peak in emotional stimulation, rhythmic contraction of perineal musculature, and quickened heart and respiration rates followed by a sharp release/drop-off of emotional tension.

† Refractory state is a transient period of time, e.g., 5 min to an hour, in which the neurons related to a given phenomenon cannot recruit sufficient volleys (impulses) to trigger the act again.

Fig. 10-22 (opposite) Arterial and nerve supply to the pelvis and perineum. A. Posterior view of the deep gluteal region, left side. B. Perineum of a male viewed from below. C. Interior view of the right hip bone.

vessels and nerves of the pelvis and perineum

VESSELS

The principal source of blood to the pelvis and perineum are the **internal iliac arteries,** one of two terminal branches of the common iliac arteries.

Refer now to Fig. 10-22 as you read:

The **internal iliac artery** — on each side — curves medially and

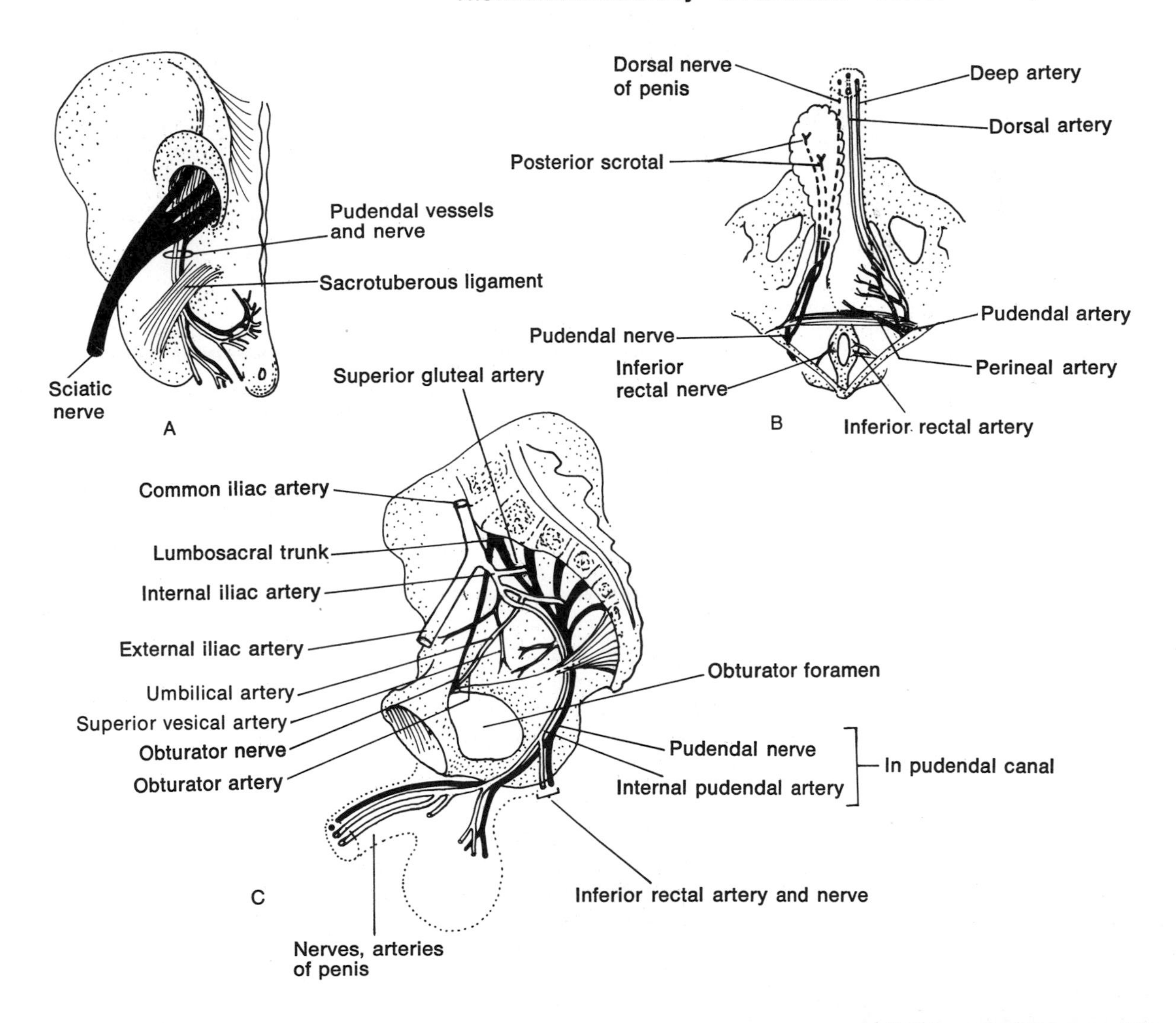

downward after leaving the common iliac, to enter the true pelvis. It divides irregularly into visceral and somatic or parietal branches. The visceral branches include:

1. The umbilical artery (which in the fetus passed up and through the umbilical cord to supply the placenta with unoxygenated blood). It is partly obliterated back from the umbilicus to the bladder and sends off branches to that organ (superior vesical arteries).
2. The inferior vesical artery, which serves the bladder, seminal vesicles, prostate, and ductus deferens.
3. The middle rectal artery to the rectum—one of three arteries to that organ.
4. In the female, the homologue of the male's inferior vesical arteries become the uterine and vaginal arteries.

Parietal branches of the internal iliac artery serve the gluteal region (gluteal arteries), the region of the obturator foramen (obturator artery), the body wall, and the perineum (the internal pudendal artery).

Secondary arteries to the pelvis include the superior rectal artery—this one, you will remember, is a continuation of the inferior mesenteric artery.

You may have noticed that mention of the blood supply to the testes or ovaries is lacking. These organs are supplied by direct branches of the abdominal aorta (testicular or ovarian arteries).

Blood to the perineum arrives through the **internal pudendal artery,** a branch of the internal iliac. Note (Fig. 10-22) that it runs out through the greater sciatic foramen and curves over the ischial spine to enter the ischiorectal fossa via the lesser sciatic foramen. As the internal pudendal passes down and forward, it does so in a fascial tunnel—the pudendal canal—an item of significance to the obstetrician who wishes to administer an anesthetic to ward off the pains of childbirth. The internal pudendal serves the structures of the perineal spaces, the rectum (internal rectal artery), and most of the external genitalia.

The external iliac artery contributes some branches to the scrotum and labia.

The veins, in general, follow the arteries and form dense plexuses within the pelvis. The most significant plexus from a clinical standpoint is that associated with the **rectal veins.** Tributaries of these veins in the anal columns are often subject to trauma by hard fecal stools contributing to inflammation of these veins. As they heal, scar tissue develops. With recurring inflammation the veins become twisted and tortuous (hemorrhoids). The rectum receives and drains blood by three separate routes: the internal iliac, the inferior mesenteric, and the internal pudendal arteries and veins, creating an important site of collateral circulation in event of IVC or portal vein obstruction.

NERVES

As before, visceral structures are supplied by autonomic nerves, musculoskeletal structures by somatic (spinal) nerves. Pelvic viscera receive their motor and sensory innervation via the **pelvic plexus** (see Fig. 9-17). This autonomic plexus receives parasympathetic fibers from S2–4 levels of the spinal cord via pelvic splanchnic nerves. It receives sympathetic fibers from the aortic plexus via the hypogastric plexus (Fig. 9-17). The pelvic splanchnic nerves bring about contraction of the bladder muscles and dilatation of the arteries to the penis/clitoris. Since this latter action causes erection, these nerves are historically referred to as "nervi erigentes." The sympathetic nerves function to initiate emission and preejaculation by causing contraction of the muscular walls of the ductus deferens and the prostatic/seminal gland musculature. Sympathetic nerves also function to constrict arteries in the pelvic and perineal areas.

The **pudendal nerve** from spinal segments S2–4 supplies the skeletal muscles of the perineum. Specifically then, it supplies the bulbospongiosus muscle, whose contraction makes ejaculation possible. Perineal nerves (branches of the pudendal nerve) conduct sensations from the external genital structures to the S2–4 area of the spinal cord. This is the anatomical basis for reflexive (involuntary) ejaculation as may occur in sleep (nocturnal emission).

The testes and ovaries are innervated by autonomic and sensory nerves arising in the aortic plexus (the testis originated on the posterior abdominal wall of the embryo). Therefore, sensations of pain in these structures may be referred to levels as high as T8.

LYMPHATIC DRAINAGE

The lymphatic drainage of the pelvis is of great significance to the clinician, particularly in women. Cancer of the cervix and cancer of the breast make up about 75 percent of the female cancers, and metastases usually occur along lymphatic and venous channels. Older men are subject to carcinoma of the prostate, and lymphatics are a factor in the spread of

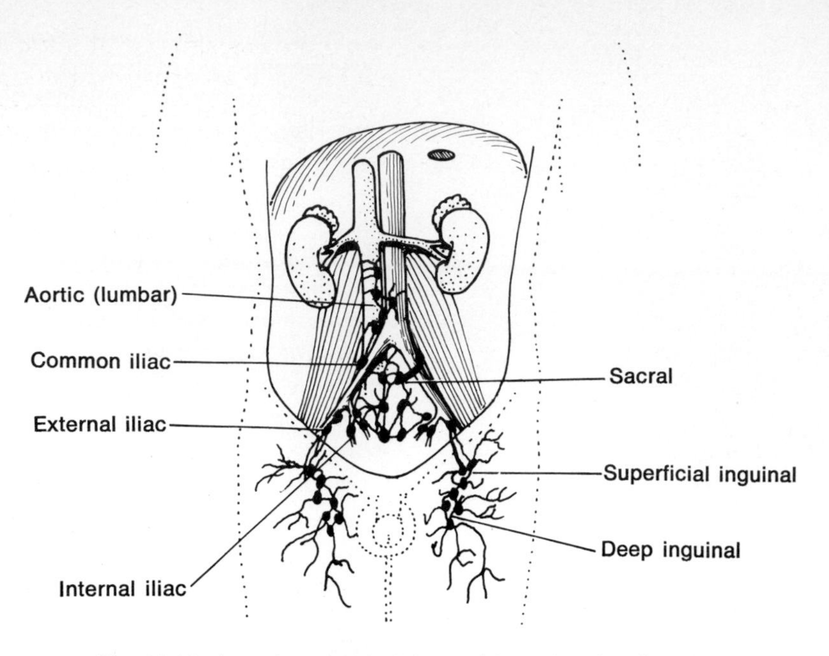

Fig. 10-23 Lymph nodes draining pelvic and perineal viscera.

that disease. Study Fig. 10-23 as you read the following:

Lymph nodes are generally arranged around the major arteries of the pelvis.

The principal collection of nodes within the pelvis is located about the sacral artery and the common, external, and internal iliac arteries. These drain the pelvis and some perineal structures.

The **deep** and **superficial inguinal nodes** drain most of the perineum.

The pelvic and inguinal nodes drain toward the common iliac nodes, which drain into the lumbar trunks. These trunks flow into the **cisterna chyli,** which is the distal extremity of the thoracic duct. This duct and its companion, the right lymphatic duct, empty their contents into the venous system, at the junction of the internal jugular and subclavian veins.

UPON REFLECTION

In our concern for the welfare of our bodies, no region receives more attention than the pelvis and perineum. We use our urinary and lower alimentary apparatus two or more times a day, and today, as always, interest in our organs of generation continues strong. The trick here is to transform casual curiosity into an intellectual one. The mechanisms by which we store and release urine and feces, generate and transport sperm and ova, have sexual relations, and develop babies all involve structures of the pelvis and perineum. And for each organ studied, it is through a knowledge of its anatomy that its physiologic processes can best be appreciated.

REFERENCES

1. Dickinson, R. L.: *Human Sex Anatomy,* 2d ed., Williams & Wilkins Co., Baltimore, 1949.
2. Guyton, A. C.: *Textbook of Medical Physiology,* 3d ed., W. B. Saunders Co., Philadelphia, 1970.
3. Kinsey, A. C. et al.: *Sexual Behavior in the Human Female and Human Male,* 2 vol., W. B. Saunders Co., Philadelphia, 1953.
4. Masters, V. E., and W. H. Johnson: *Human Sexual Response,* Little, Brown, Inc., Boston, 1966.
5. Gardner, E., D. J. Gray, R. O'Rahilly: *Anatomy,* 3d ed., W. B. Saunders Co., Philadelphia, 1969.
6. Goss, C. M.: *Gray's Anatomy of the Human Body,* 28th ed. Lea & Febiger, Philadelphia, 1966.

clinical considerations

PROSTATISM

The prostate gland, you will remember, is located at the apex of the bladder in the male, supported by the pelvic diaphragm. The gland may be seen to have as many as five lobes: two lateral, two posterior, one middle and one anterior (vestigial). The mass of the gland consists of many tubuloalveolar glands wrapped in smooth muscle and elastic connective tissue. The ducts of these glands open onto the surface of the prostatic urethra more or less independently.

With age, the prostate tends to enlarge (hypertrophy), specifically the lateral lobes. Frequently, this hypertrophy encroaches upon the urethra and creates varying degrees of occlusion; and it is this symptom which often brings the person to the physician.

With urethral involvement, urinary retention in the bladder is a consequence, which may lead to an infection (cystitis). As the degree of prostatic hypertrophy increases, secretions of the gland are prevented from draining; the epithelia of the gland become inflamed (prostatitis), causing a great deal of pain.

Prostatitis and prostatic hypertrophy can be diagnosed by

Fig. 10-24 Hyperplasia of the prostate, middle lobe (arrow). The lobe projects into and blocks the urethral lumen. (Courtesy of R. P. Morehead, *Human Pathology*, McGraw-Hill, New York, 1965.)

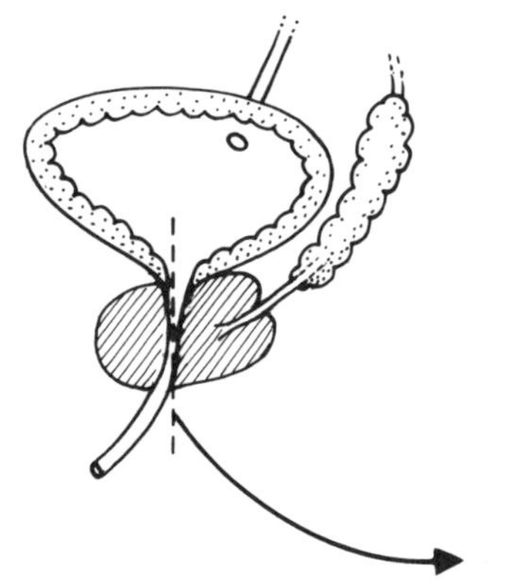

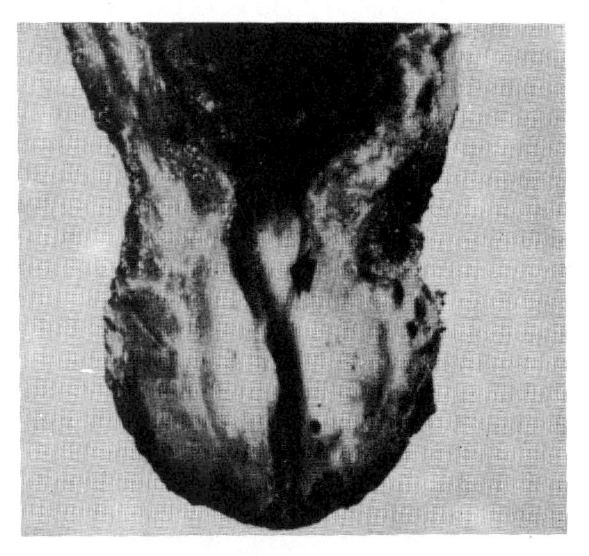

means of a rectal examination, because the enlarged mass is easily palpable. An optical instrument (cystoscope) which allows the physician to visualize the hypertrophy directly can also be passed through the urethra.

Although not necessarily related, cancer of the prostate often occurs in association with prostatic hypertrophy of aged men. The malignancy usually arises in the posterior lobe and is easily palpable through the rectum.

The late onset of prostatic hypertrophy is probably related to a decline in the output of testosterone. Castration of the elderly adult male is accompanied by a decrease in size of an enlarged prostate. The functional activity of the androgens (e.g., testosterone) can often be reduced by estrogen treatment. In fact, estrogens can have a deleterious action on neoplastic cells of the prostate, often preventing or reducing their dispersal.

PROLAPSE OF THE UTERUS

Prolapse is generally defined as a permanent sinking or descent of the uterus from its normal position down into and sometimes through the vagina. You will remember that the uterus is normally flexed and tilted over the bladder (anteflexion, anteversion) and is secured in its place in the pelvis by a number of ligaments, e.g., the broad, lateral cervical, and anterior/posterior ligaments. The pelvic diaphragm (levator ani) and the urogenital diaphragm play a heavy supporting role, as does the perineal body to which the UG diaphragm and other perineal musculature depend for their support. For the uterus "to sink," then, a great deal of ligament and muscle weakening must precede the event. Pregnancy or past multiple pregnancies can contribute to prolapse, for during this time the various ligaments are significantly stretched. Relaxation of the anterior ligaments can induce sufficient retroflexion that the cervix—in alignment with the vagina—can slip down, if other ligaments and muscles are also weakened. Of course, a naturally retroflexed uterus is already predisposed to prolapse. Prolapse or a tendency toward the condition can often be detected during a physical examination in which the patient is asked to strain while supine and the descended (-ing) cervix bulges against the perineal skin, or worse, protrudes through the vaginal orifice. Complete prolapse is often the result of lacerations or the severe weakening of ligaments at childbirth. A prolapsed uterus can be gently reinserted through the vagina; a subsequent surgical operation is then required to mend weakened ligaments and muscle.

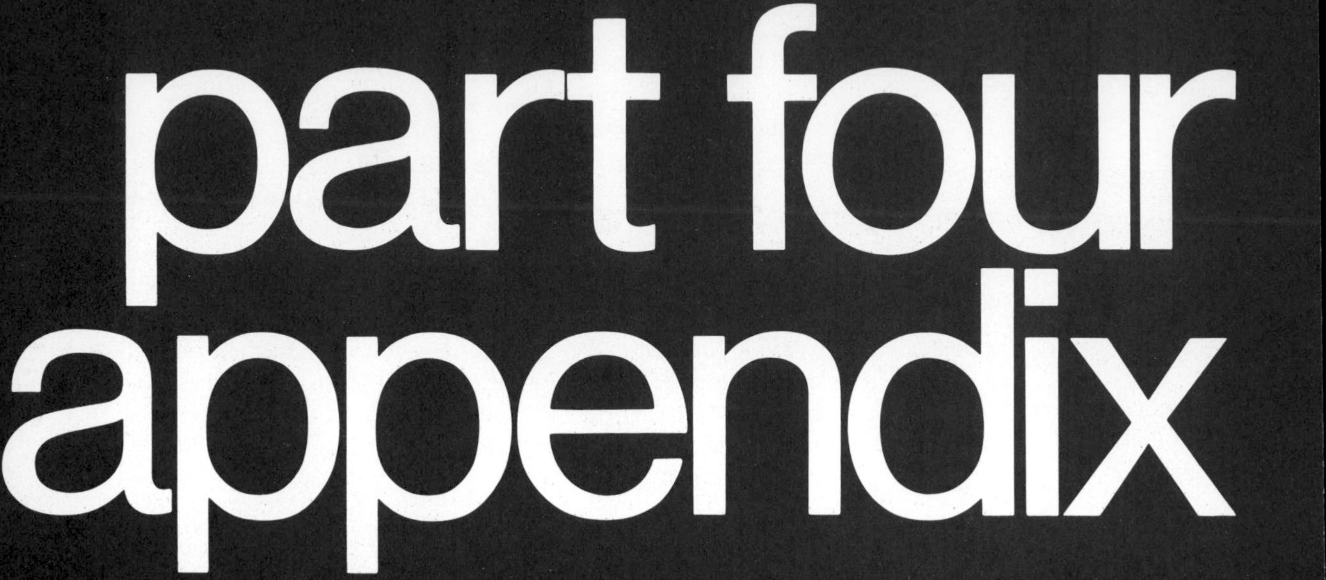
part four
appendix

atlas of x-rays

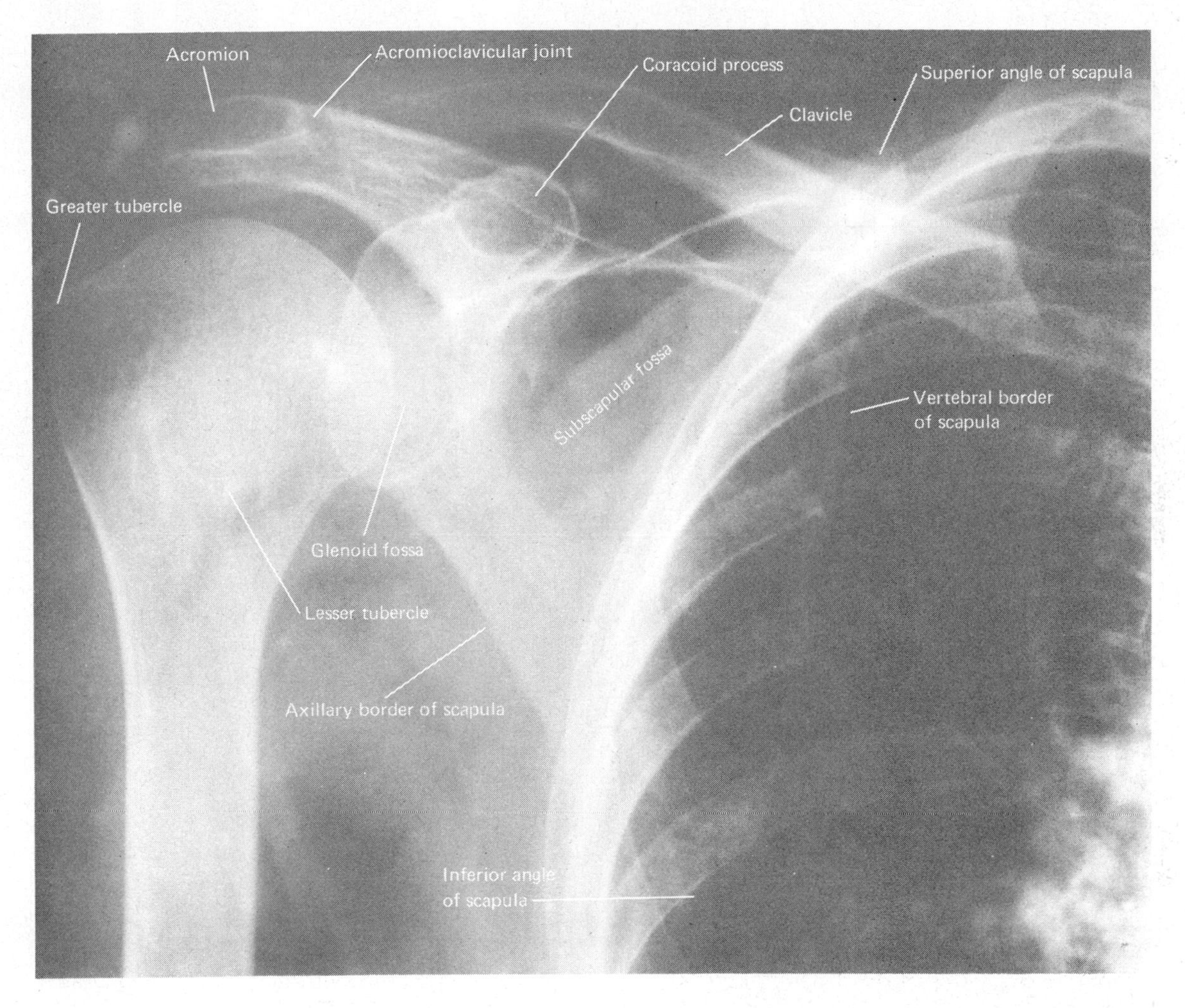

A-1 Anterior view of the scapula, humerus, and shoulder joint.

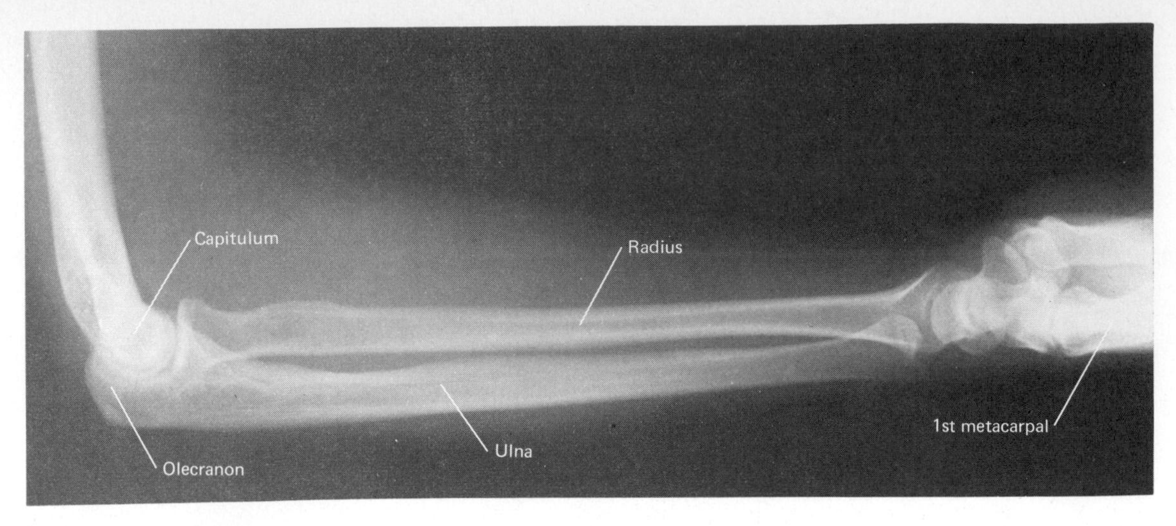

A-2 Lateral view of the radius, ulna, and lower humerus.

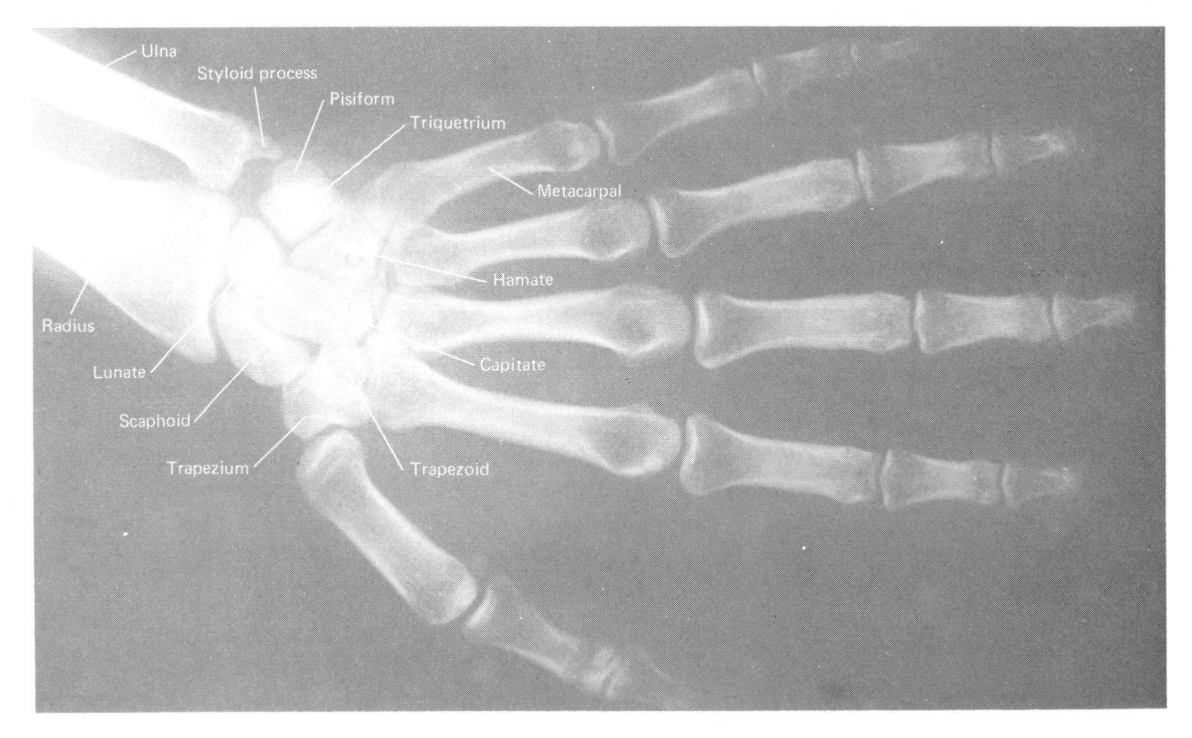

A-3 Palmar aspect of the hand and wrist.

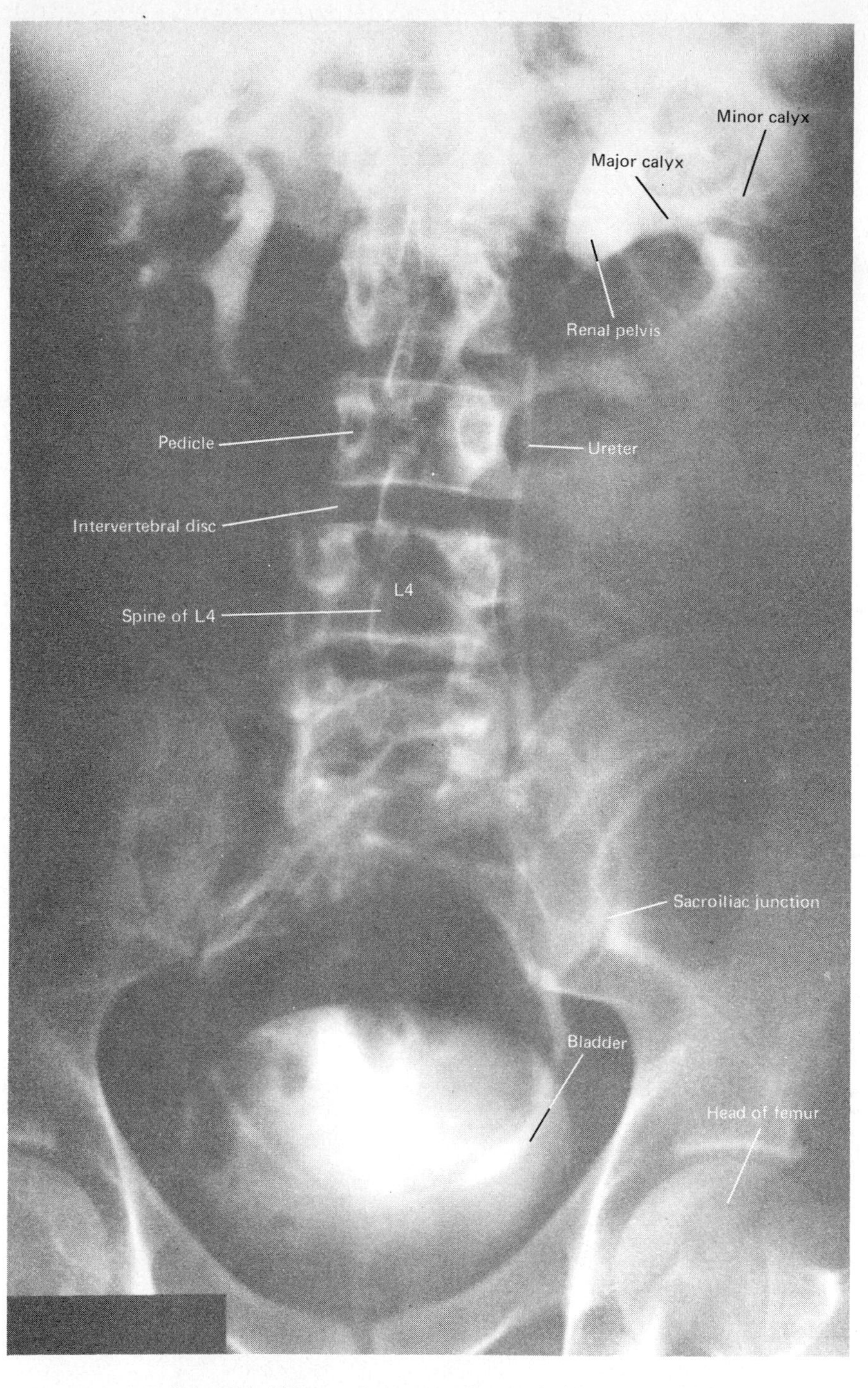

A-4 The kidneys, ureters, urinary bladder (retrograde pyelogram), and hip.

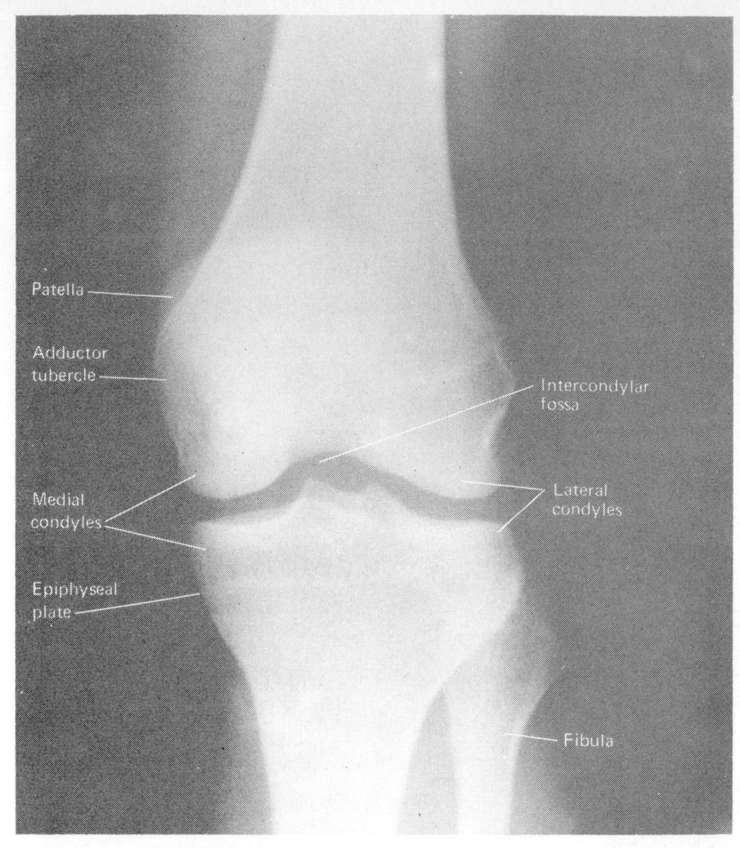

A-5 Anteroposterior radiograph of the knee joint.

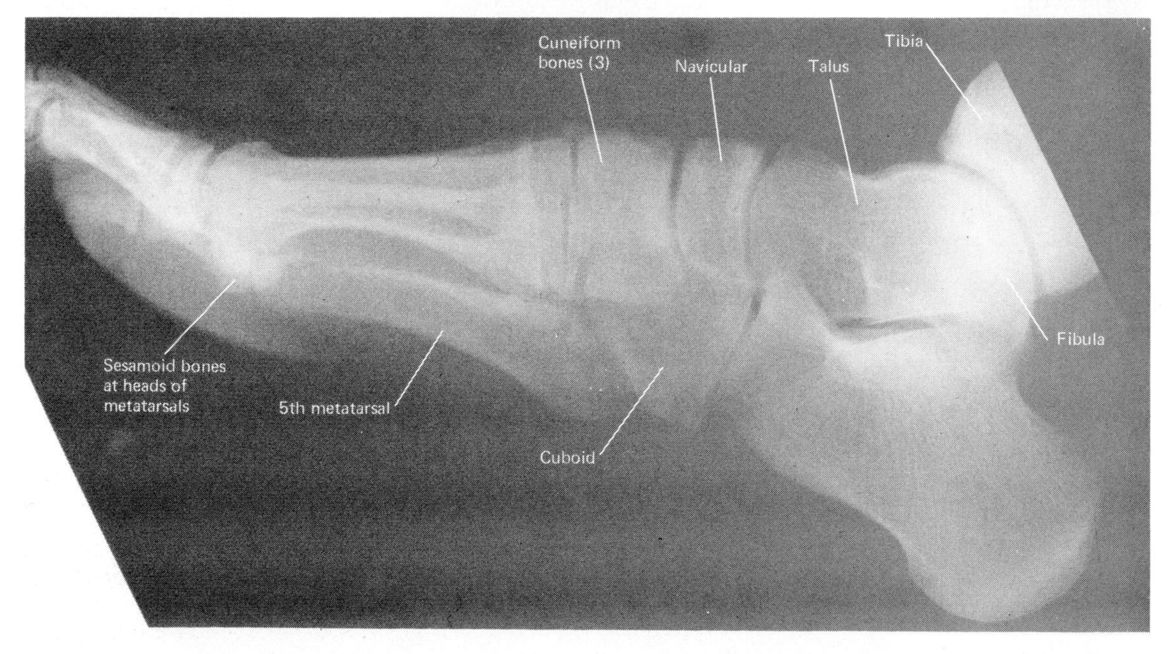

A-6 Medial view of the ankle joint and foot.

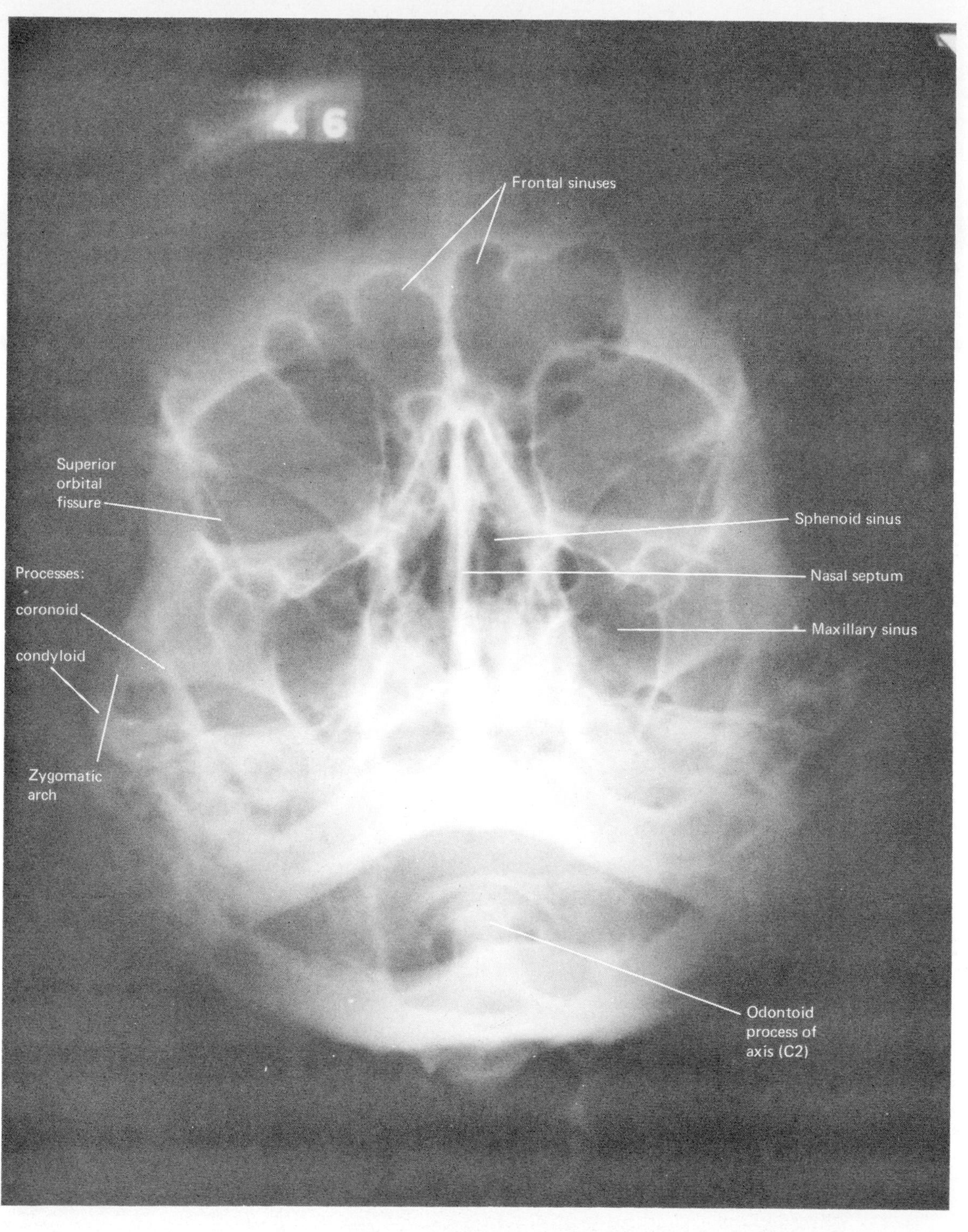

A-7 Frontal view of the head (posteroanterior cranial eccentric position). The patient lacks teeth.

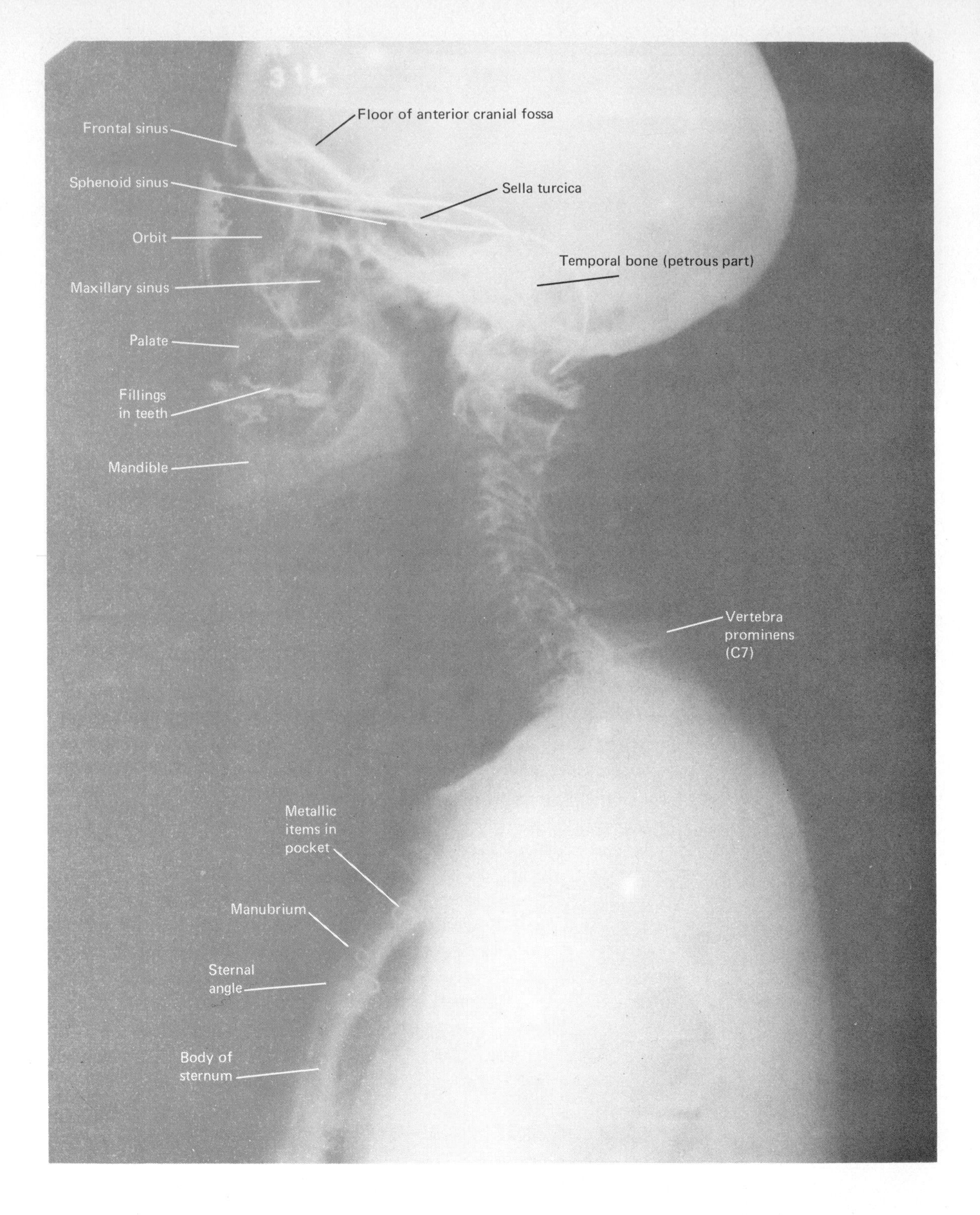

Floor of anterior cranial fossa
Frontal sinus
Sphenoid sinus
Sella turcica
Orbit
Temporal bone (petrous part)
Maxillary sinus
Palate
Fillings in teeth
Mandible
Vertebra prominens (C7)
Metallic items in pocket
Manubrium
Sternal angle
Body of sternum

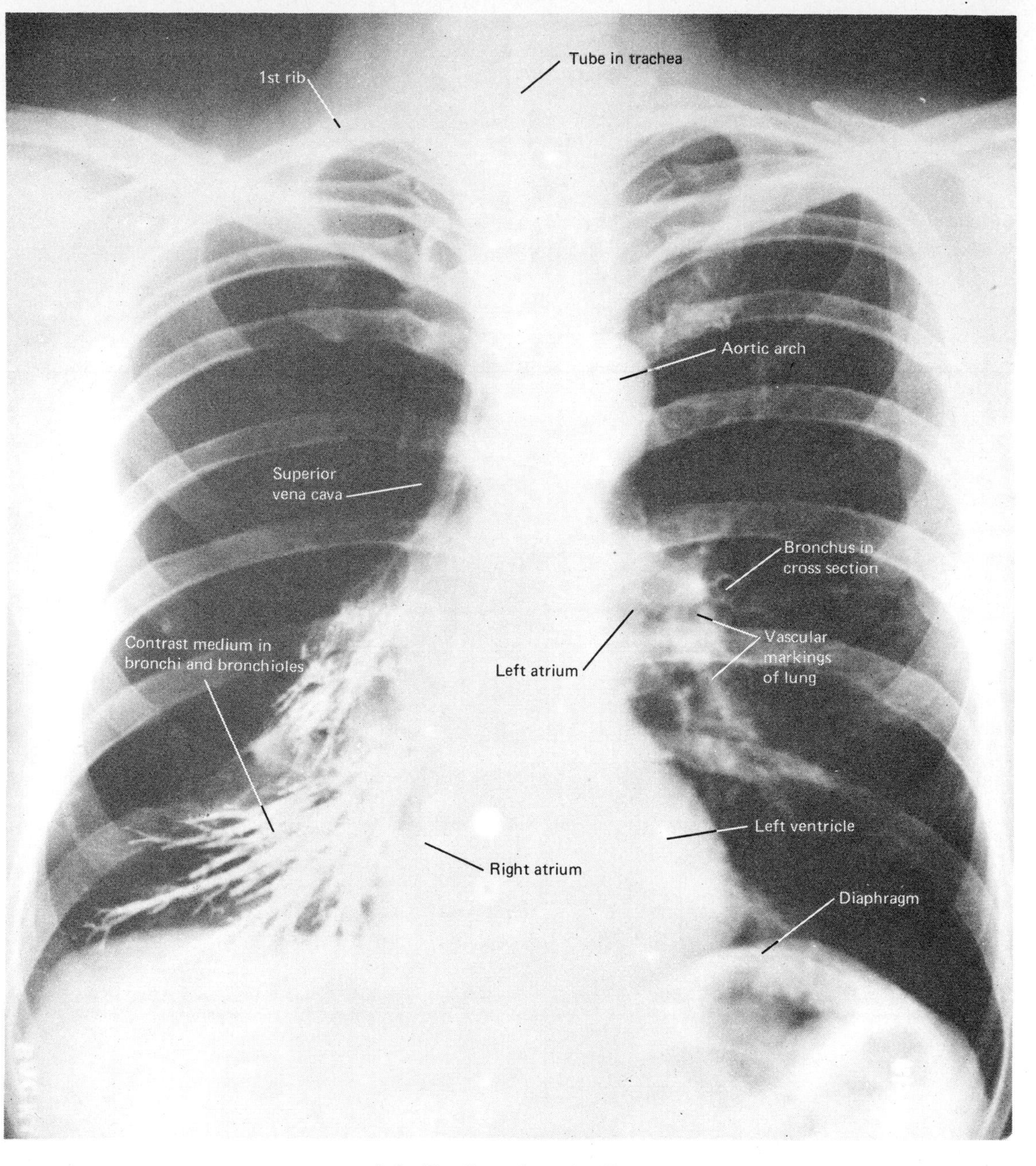

A-9 The thoracic cavity. Right lung is deflated (pneumothorax). Contrast media has been discharged into bronchi and bronchioles by a cathether that can be seen in the trachea.

A-8 (opposite) Lateral view of the head, neck, and upper thorax. Eyeglasses were retained to show relationships not otherwise readily seen.

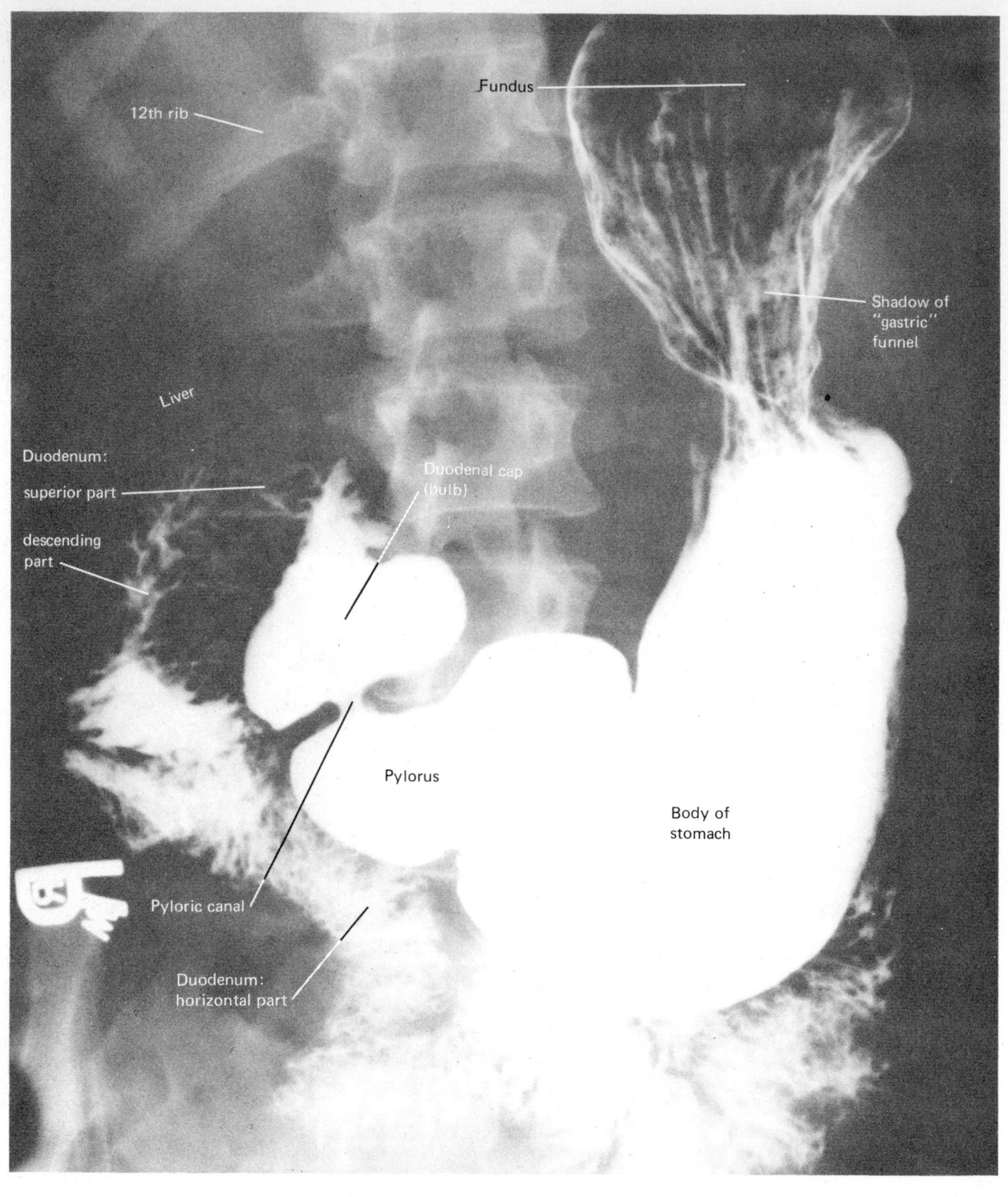

A-10 The stomach and small intestine after ingestion of a barium meal.

A-11 (opposite) The large intestine filled with barium by means of enema.

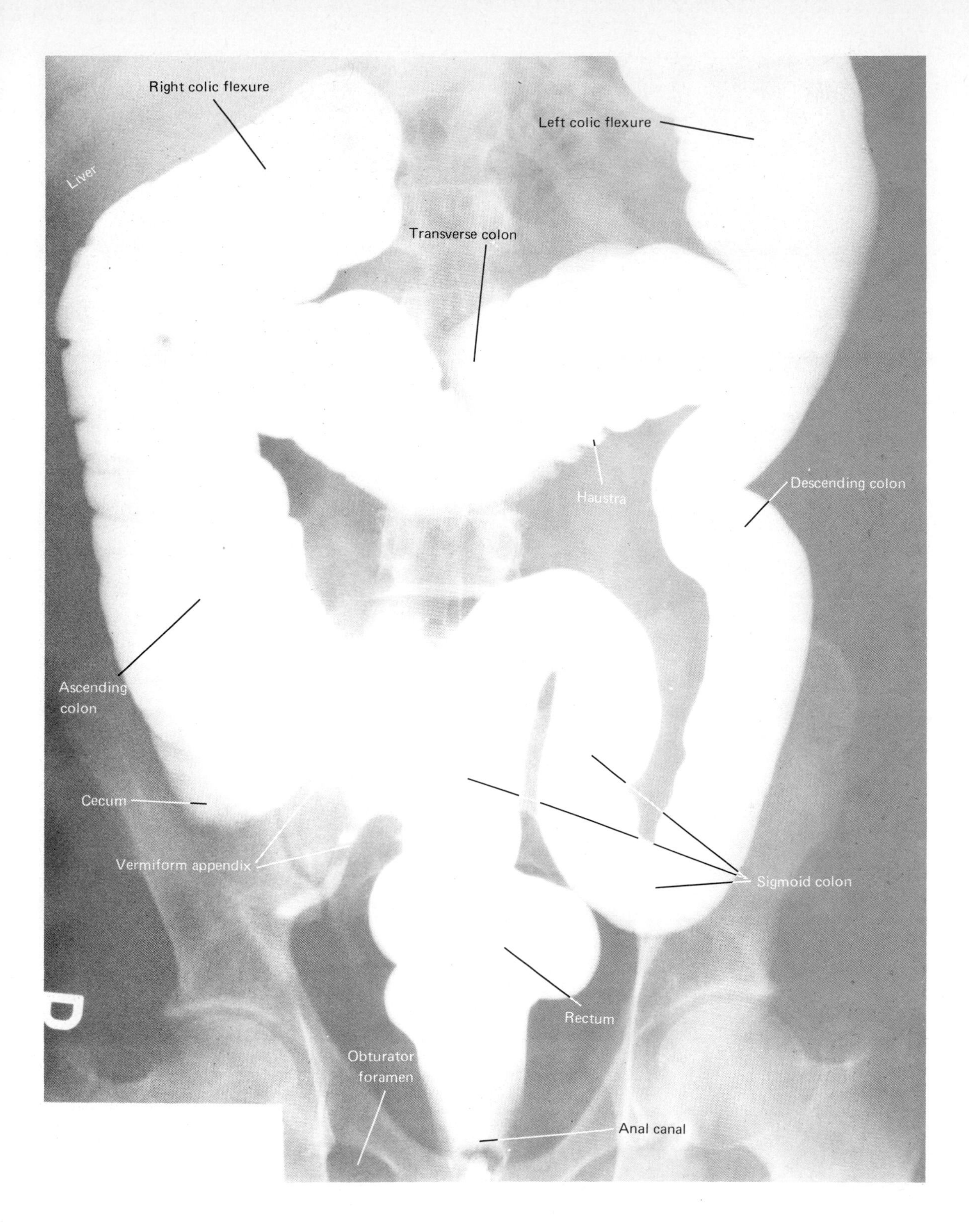
Right colic flexure
Left colic flexure
Liver
Transverse colon
Haustra
Descending colon
Ascending colon
Cecum
Vermiform appendix
Sigmoid colon
Rectum
Obturator foramen
Anal canal
D

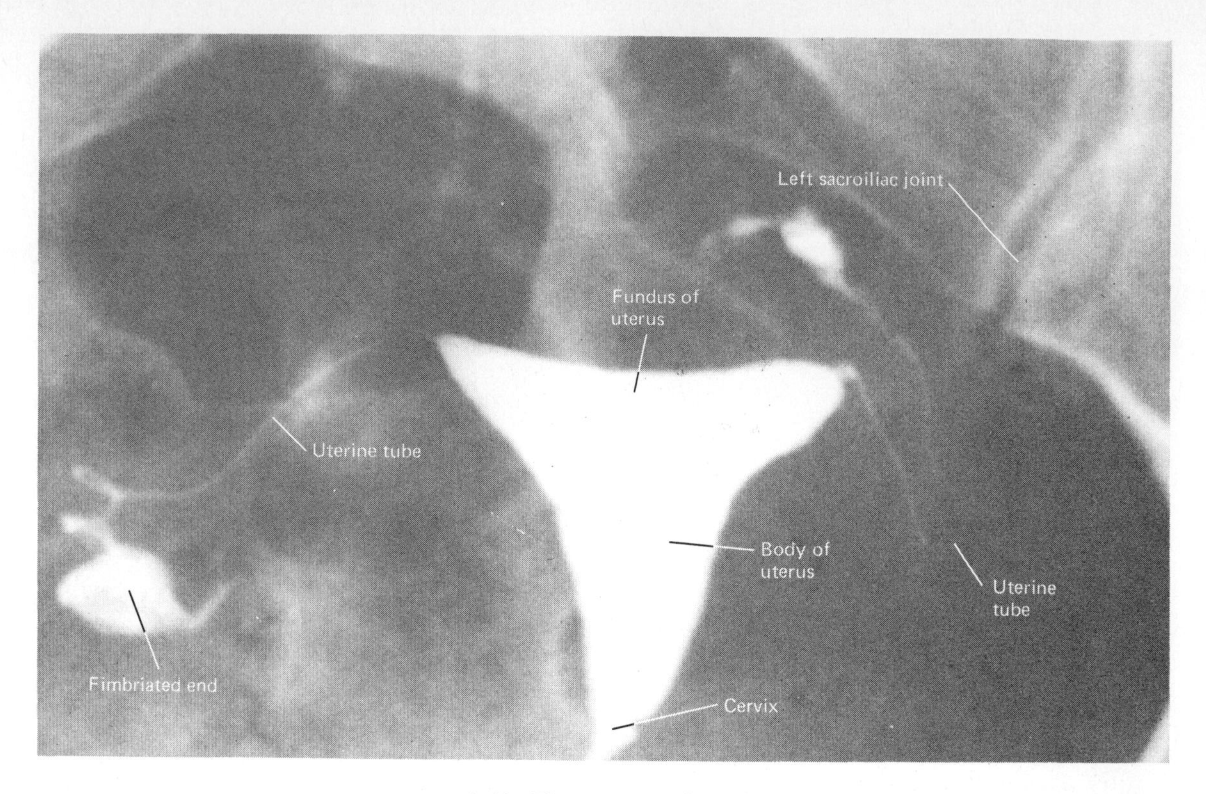

A-12 The uterus and uterine tubes filled with contrast media from the vagina (hysterosalpingogram). Media leaks from the fimbriated ends of the uterine tubes, near the ovaries.

Credits

A-1 Courtesy of Gloria Bowser, Department of Radiology, Baylor College of Medicine.
A-2, A-3, A-5, A-6 Courtesy of Dr. Max Roedel, Methodist Hospital, Houston.
A-4 Courtesy of Dr. Russell Scott, Baylor College of Medicine.
A-7, A-8, A-9 Courtesy of Dr. Heyl G. Tebo, University of Texas Dental Branch, Houston.
A-10, A-11 Courtesy of Dr. Raj Goyal, Baylor College of Medicine.
A-12 Courtesy of Dr. Russell Malinak, Baylor College of Medicine.

synopsis of systems

	Unit
I Osseous system	
A Axial skeleton (80)	3, 6, 7
1 Skull (28)	6
a Cranial bones (8)	6
b Facial bones (14)	6
c Middle-ear ossicles (6)	6
2 Hyoid bone (1)	6
3 Vertebrae (26)	7
4 Sternum (1)	7
5 Ribs (24)	7
B Appendicular skeleton (126)	3, 4, 5
1 Upper limb bones (64)	4
a Clavicle (2)	4
b Scapula (2)	4
c Humerus (2)	4
d Radius and ulna (4)	4
e Carpal bones (16)	4
f Metacarpal bones (10)	4
g Phalanges (28)	4
2 Lower limb bones (62)	5
a Hip bones (2)	5, 10
b Femur (2)	5
c Patella (2)	5
d Tibia and fibula (4)	5
e Tarsal bones (14)	5
f Metatarsal bones (10)	5
g Phalanges (28)	5
II Muscular system	
A Cardiac muscle	2, 8
B Smooth muscle	2, 3, 8 to 10
C Skeletal muscle	2 to 10
1 Of the upper limb	4
a Scapular	4
b Pectoral	4
c Shoulder	4
d Arm	4
e Forearm	4
f Hand	4
2 Of the lower limb	5
a Hip	5
b Thigh	5
c Leg	5
d Foot	5
3 Of the trunk	6, 7, 10
a Deep back	7
b Thoracic wall	7
c Anterolateral abdominal wall	7
d Posterior abdominal wall	7
e Pelvis	10
f Perineum	10
4 Of the head and neck	6
a Anterior cervical	6
b Lateral cervical	6
c Posterior cervical	6
d Mastication	6
e Facial expression	6
III Nervous system	
A Central nervous system	2, 6
1 Telencephalon	6
a Cerebral cortex	6
b White matter	6
c Basal nuclei	6
d Lateral ventricles	6
2 Diencephalon	6
a Thalamus	6
b Hypothalamus	6
c Epithalamus	6
d Third ventricle	6
3 Brainstem	6
a Mesencephalon	6
b Metencephalon	6
c Myelencephalon	6
4 Ventricles	6
5 Spinal cord	6
6 Meninges and CSF	6
B Peripheral nervous system	2 to 10
1 Cranial nerves	3, 6
2 Spinal nerves	3 to 7, 10
a Cervical plexus	6
b Brachial plexus	4
c Intercostal nerves	7
d Lumbosacral plexus	5, 10
3 Autonomic nervous system	3, 6, 8 to 10
a Craniosacral division	3
(1) Cranial nerves III, VII, IX	6
(2) Vagus nerve	6, 8, 9
(3) Sacral nerves	3, 9, 10
b Thoracolumbar division	3
(1) Sympathetic chain	3, 6, 8, 10
(2) Splanchnic nerves	3, 8, 9
c Autonomic plexuses	6, 8 to 10
(1) Pharyngeal	6
(2) Cardiac	8
(3) Pulmonary	8
(4) Esophageal	8
(5) Aortic	9
(6) Pelvic	10
4 Special sensory receptors	6
a Eye	6
b Ear	6
(1) Auditory	6
(2) Vestibular	6
c Olfaction	6
d Taste	6

	Unit
5 General sensory receptors	3
a Exteroceptors	3
b Enteroceptors	3
c Proprioceptors	3
IV Integumentary system	
A Epidermis	3
B Dermis	3
C Appendages	3
D Receptors	3
V Cardiovascular system	
A Heart	8
1 Pericardium	8
2 Chambers	8
3 Coronary system	8
4 Conduction system	8
B Arteries	3 to 10
1 Coronary	8
2 Of the thorax	7, 8
3 Of the abdomen	7, 9
4 Of the pelvis	10
5 Of the perineum	10
6 Of the upper limb	4
7 Of the lower limb	5
8 Of the head and neck	6
C Veins	3 to 10
1 Of the thorax	7, 8
2 Of the abdomen	7, 9
3 Of the pelvis	10
4 Of the perineum	10
5 Of the upper limb	4
6 Of the lower limb	5
7 Of the head and neck	6
8 Sinuses/sinusoids	3
9 Portal system	3, 6, 9
a Hypophyseal	6
b Hepatic	9
c Renal	9
D Capillaries	3
E Blood	2
VI Lymphatic/reticuloendothelial system	
A Lymphatic organs/tissue	3 to 10
1 Thymus gland	8
2 Spleen	9
3 Lymph nodes	3 to 10
4 Tonsils	6
5 Lymphatic nodules	9
B Lymphatic vessels	3 to 10
1 Thoracic duct	8, 9
2 Right lymph duct	8
3 Cisterna chyli	9
4 Lacteals/capillaries	3, 9
C Lymph	3
D Reticuloendothelial system	3
VII Respiratory system	
A Upper respiratory tract	6
1 Nose	6
2 Nasal cavities	6
3 Sinuses	6, 8
4 Pharynx	6
5 Larynx	6
6 Trachea	6, 8
7 Primary bronchi	8

	Unit
B Lower respiratory tract/lungs	8
1 Secondary bronchi	8
2 Tertiary bronchi	8
3 Bronchioles	8
4 Respiratory units	8
a Respiratory bronchioles	8
b Alveolar ducts and sacs	8
c Alveoli	8
5 Bronchopulmonary segments	8
C Pleurae	8
D Diaphragm	7, 8
VIII Digestive system	
A Oral cavity	6
1 Tongue	6
2 Teeth	6
3 Salivary glands	6
B Pharynx	6
C Esophagus	6, 8, 9
D Peritoneum	9
E Stomach	9
F Small intestine	9
G Large intestine	9
H Liver	9
I Gallbladder	9
J Pancreas	9
K Common bile duct	9
IX Urinary system	
A Kidneys	9
B Ureters	9, 10
C Urinary bladder	10
D Urethra	10
X Reproductive system	
A Male organs	10
1 Testes	7, 10
2 Scrotum	7, 10
3 Epididymis	10
4 Ductus deferens	7, 10
5 Seminal vesicles	10
6 Ejaculatory duct	10
7 Prostate gland	10
8 Bulbourethral glands	10
9 Penis	10
B Female organs	10
1 Ovaries	10
2 Uterine tubes	10
3 Uterus	10
4 Vagina	10
5 External genitalia	10
6 Mammary glands	4
XI Endocrine system	
A Hypothalamus	6
B Hypophysis	6
1 Anterior lobe	6, 10
2 Posterior lobe	6
C Thyroid gland	6
D Parathyroid glands	6
E Pancreatic islets	9
F Adrenal glands	9
G Ovary	10
H Testes	10
I Placenta	10
J Gastrointestinal glands	9

index

index

Page numbers in *italic* indicate illustrations.